Lecture Notes in Computer Science 12824

More information about this subseries at http://www.springer.com/series/7412

Josep Lladós · Daniel Lopresti ·
Seiichi Uchida (Eds.)

Document Analysis and Recognition – ICDAR 2021

16th International Conference
Lausanne, Switzerland, September 5–10, 2021
Proceedings, Part IV

 Springer

Editors
Josep Lladós 🆔
Universitat Autònoma de Barcelona
Barcelona, Spain

Daniel Lopresti 🆔
Lehigh University
Bethlehem, PA, USA

Seiichi Uchida 🆔
Kyushu University
Fukuoka-shi, Japan

ISSN 0302-9743   ISSN 1611-3349   (electronic)
Lecture Notes in Computer Science
ISBN 978-3-030-86336-4   ISBN 978-3-030-86337-1   (eBook)
https://doi.org/10.1007/978-3-030-86337-1

LNCS Sublibrary: SL6 – Image Processing, Computer Vision, Pattern Recognition, and Graphics

This Springer imprint is published by the registered company Springer Nature Switzerland AG
The registered company address is: Gewerbestrasse 11, 6330 Cham, Switzerland

Foreword

Our warmest welcome to the proceedings of ICDAR 2021, the 16th IAPR International Conference on Document Analysis and Recognition, which was held in Switzerland for the first time. Organizing an international conference of significant size during the COVID-19 pandemic, with the goal of welcoming at least some of the participants physically, is similar to navigating a rowboat across the ocean during a storm. Fortunately, we were able to work together with partners who have shown a tremendous amount of flexibility and patience including, in particular, our local partners, namely the Beaulieu convention center in Lausanne, EPFL, and Lausanne Tourisme, and also the international ICDAR advisory board and IAPR-TC 10/11 leadership teams who have supported us not only with excellent advice but also financially, encouraging us to setup a hybrid format for the conference.

We were not a hundred percent sure if we would see each other in Lausanne but we remained confident, together with almost half of the attendees who registered for on-site participation. We relied on the hybridization support of a motivated team from the Lule University of Technology during the pre-conference, and professional support from Imavox during the main conference, to ensure a smooth connection between the physical and the virtual world. Indeed, our welcome is extended especially to all our colleagues who were not able to travel to Switzerland this year. We hope you had an exciting virtual conference week, and look forward to seeing you in person again at another event of the active DAR community.

With ICDAR 2021, we stepped into the shoes of a longstanding conference series, which is the premier international event for scientists and practitioners involved in document analysis and recognition, a field of growing importance in the current age of digital transitions. The conference is endorsed by IAPR-TC 10/11 and celebrates its 30th anniversary this year with the 16th edition. The very first ICDAR conference was held in St. Malo, France in 1991, followed by Tsukuba, Japan (1993), Montreal, Canada (1995), Ulm, Germany (1997), Bangalore, India (1999), Seattle, USA (2001), Edinburgh, UK (2003), Seoul, South Korea (2005), Curitiba, Brazil (2007), Barcelona, Spain (2009), Beijing, China (2011), Washington DC, USA (2013), Nancy, France (2015), Kyoto, Japan (2017), and Syndey, Australia (2019).

The attentive reader may have remarked that this list of cities includes several venues for the Olympic Games. This year the conference was be hosted in Lausanne, which is the headquarters of the International Olympic Committee. Not unlike the athletes who were recently competing in Tokyo, Japan, the researchers profited from a healthy spirit of competition, aimed at advancing our knowledge on how a machine can understand written communication. Indeed, following the tradition from previous years, 13 scientific competitions were held in conjunction with ICDAR 2021 including, for the first time, three so-called "long-term" competitions addressing wider challenges that may continue over the next few years.

Other highlights of the conference included the keynote talks given by Masaki Nakagawa, recipient of the IAPR/ICDAR Outstanding Achievements Award, and Mickaël Coustaty, recipient of the IAPR/ICDAR Young Investigator Award, as well as our distinguished keynote speakers Prem Natarajan, vice president at Amazon, who gave a talk on "OCR: A Journey through Advances in the Science, Engineering, and Productization of AI/ML", and Beta Megyesi, professor of computational linguistics at Uppsala University, who elaborated on "Cracking Ciphers with 'AI-in-the-loop': Transcription and Decryption in a Cross-Disciplinary Field".

A total of 340 publications were submitted to the main conference, which was held at the Beaulieu convention center during September 8–10, 2021. Based on the reviews, our Program Committee chairs accepted 40 papers for oral presentation and 142 papers for poster presentation. In addition, nine articles accepted for the ICDAR-IJDAR journal track were presented orally at the conference and a workshop was integrated in a poster session. Furthermore, 12 workshops, 2 tutorials, and the doctoral consortium were held during the pre-conference at EPFL during September 5–7, 2021, focusing on specific aspects of document analysis and recognition, such as graphics recognition, camera-based document analysis, and historical documents.

The conference would not have been possible without hundreds of hours of work done by volunteers in the organizing committee. First of all we would like to express our deepest gratitude to our Program Committee chairs, Joseph Lladós, Dan Lopresti, and Seiichi Uchida, who oversaw a comprehensive reviewing process and designed the intriguing technical program of the main conference. We are also very grateful for all the hours invested by the members of the Program Committee to deliver high-quality peer reviews. Furthermore, we would like to highlight the excellent contribution by our publication chairs, Liangrui Peng, Fouad Slimane, and Oussama Zayene, who negotiated a great online visibility of the conference proceedings with Springer and ensured flawless camera-ready versions of all publications. Many thanks also to our chairs and organizers of the workshops, competitions, tutorials, and the doctoral consortium for setting up such an inspiring environment around the main conference. Finally, we are thankful for the support we have received from the sponsorship chairs, from our valued sponsors, and from our local organization chairs, which enabled us to put in the extra effort required for a hybrid conference setup.

Our main motivation for organizing ICDAR 2021 was to give practitioners in the DAR community a chance to showcase their research, both at this conference and its satellite events. Thank you to all the authors for submitting and presenting your outstanding work. We sincerely hope that you enjoyed the conference and the exchange with your colleagues, be it on-site or online.

September 2021

Andreas Fischer
Rolf Ingold
Marcus Liwicki

Preface

It gives us great pleasure to welcome you to the proceeding Conference on Document Analysis and Recognition (ICD together practitioners and theoreticians, industry research senting a range of disciplines with interests in the latest de document analysis and recognition. The last ICDAR confer Australia, in September 2019. A few months later the CO down the world, and the Document Analysis and Recognitio umbrella of IAPR had to be held in virtual format (DAS 20 ICFHR 2020 in Dortmund, Germany). ICDAR 2021 Switzerland, in a hybrid mode. Thus, it offered the opportu and show that the scientific community in DAR has kept activ

Despite the difficulties of COVID-19, ICDAR 2021 impressive number of submissions. The conference received 3 which 182 were accepted for publication (54%) and, of those presentations (12%) and 142 as posters (42%). Among the ac student as the first author (62%), and 41 were identified a (23%). In addition, a special track was organized in connect of the International Journal on Document Analysis and Re Special Issue received 32 submissions that underwent the revision process. The nine accepted papers were published in were invited to present their work in the special track at ICD

The review model was double blind, i.e. the authors did no reviewers and vice versa. A plagiarism filter was applied to measure of scientific integrity. Each paper received at least three than 1,000 reviews. We recruited 30 Senior Program Committ 200 reviewers. The SPC members were selected based on thei considering that they had served in similar roles in past DAR ev some younger researchers who are rising leaders in the field.

In the final program, authors from 47 different countries China, India, France, the USA, Japan, Germany, and Spain at most popular topics for accepted papers, in order, included tex tion, document image processing, document analysis systems, ha historical document analysis, extracting document semantics, an and recognition. With the aim of establishing ties with other co concept of reading systems at large, we broadened the scope, topics like natural language processing, multimedia docu understanding.

The final program consisted of ten oral sessions, two poster ses one of them given by the recipient of the ICDAR Outstanding A and two panel sessions. We offer our deepest thanks to all who c

and effo
ICDAR
competit
thanks to
Final
impressi
the ICD

Septemb

Organization

Organizing Committee

General Chairs

Andreas Fischer	University of Applied Sciences and Arts Western Switzerland, Switzerland
Rolf Ingold	University of Fribourg, Switzerland
Marcus Liwicki	Luleå University of Technology, Sweden

Program Committee Chairs

Josep Lladós	Computer Vision Center, Spain
Daniel Lopresti	Lehigh University, USA
Seiichi Uchida	Kyushu University, Japan

Workshop Chairs

Elisa H. Barney Smith	Boise State University, USA
Umapada Pal	Indian Statistical Institute, India

Competition Chairs

Harold Mouchère	University of Nantes, France
Foteini Simistira	Luleå University of Technology, Sweden

Tutorial Chairs

Véronique Eglin	Institut National des Sciences Appliquées, France
Alicia Fornés	Computer Vision Center, Spain

Doctoral Consortium Chairs

Jean-Christophe Burie	La Rochelle University, France
Nibal Nayef	MyScript, France

Publication Chairs

Liangrui Peng	Tsinghua University, China
Fouad Slimane	University of Fribourg, Switzerland
Oussama Zayene	University of Applied Sciences and Arts Western Switzerland, Switzerland

Sponsorship Chairs

David Doermann	University at Buffalo, USA
Koichi Kise	Osaka Prefecture University, Japan
Jean-Marc Ogier	University of La Rochelle, France

Local Organization Chairs

Jean Hennebert	University of Applied Sciences and Arts Western Switzerland, Switzerland
Anna Scius-Bertrand	University of Applied Sciences and Arts Western Switzerland, Switzerland
Sabine Süsstrunk	École Polytechnique Fédérale de Lausanne, Switzerland

Industrial Liaison

Aurélie Lemaitre	University of Rennes, France

Social Media Manager

Linda Studer	University of Fribourg, Switzerland

Program Committee

Senior Program Committee Members

Apostolos Antonacopoulos	University of Salford, UK
Xiang Bai	Huazhong University of Science and Technology, China
Michael Blumenstein	University of Technology Sydney, Australia
Jean-Christophe Burie	University of La Rochelle, France
Mickaël Coustaty	University of La Rochelle, France
Bertrand Coüasnon	University of Rennes, France
Andreas Dengel	DFKI, Germany
Gernot Fink	TU Dortmund University, Germany
Basilis Gatos	Demokritos, Greece
Nicholas Howe	Smith College, USA
Masakazu Iwamura	Osaka Prefecture University, Japan
C. V. Javahar	IIIT Hyderabad, India
Lianwen Jin	South China University of Technology, China
Dimosthenis Karatzas	Computer Vision Center, Spain
Laurence Likforman-Sulem	Télécom ParisTech, France
Cheng-Lin Liu	Chinese Academy of Sciences, China
Angelo Marcelli	University of Salerno, Italy
Simone Marinai	University of Florence, Italy
Wataru Ohyama	Saitama Institute of Technology, Japan
Luiz Oliveira	Federal University of Parana, Brazil
Liangrui Peng	Tsinghua University, China
Ashok Popat	Google Research, USA
Partha Pratim Roy	Indian Institute of Technology Roorkee, India
Marçal Rusiñol	Computer Vision Center, Spain
Robert Sablatnig	Vienna University of Technology, Austria
Marc-Peter Schambach	Siemens, Germany

Srirangaraj Setlur	University at Buffalo, USA
Faisal Shafait	National University of Sciences and Technology, India
Nicole Vincent	Paris Descartes University, France
Jerod Weinman	Grinnell College, USA
Richard Zanibbi	Rochester Institute of Technology, USA

Program Committee Members

Sébastien Adam
Irfan Ahmad
Sheraz Ahmed
Younes Akbari
Musab Al-Ghadi
Alireza Alaei
Eric Anquetil
Srikar Appalaraju
Elisa H. Barney Smith
Abdel Belaid
Mohammed Faouzi Benzeghiba
Anurag Bhardwaj
Ujjwal Bhattacharya
Alceu Britto
Jorge Calvo-Zaragoza
Chee Kheng Ch'Ng
Sukalpa Chanda
Bidyut B. Chaudhuri
Jin Chen
Youssouf Chherawala
Hojin Cho
Nam Ik Cho
Vincent Christlein
Christian Clausner
Florence Cloppet
Donatello Conte
Kenny Davila
Claudio De Stefano
Sounak Dey
Moises Diaz
David Doermann
Antoine Doucet
Fadoua Drira
Jun Du
Véronique Eglin
Jihad El-Sana
Jonathan Fabrizio

Nadir Farah
Rafael Ferreira Mello
Miguel Ferrer
Julian Fierrez
Francesco Fontanella
Alicia Fornés
Volkmar Frinken
Yasuhisa Fujii
Akio Fujiyoshi
Liangcai Gao
Utpal Garain
C. Lee Giles
Romain Giot
Lluis Gomez
Petra Gomez-Krämer
Emilio Granell
Mehdi Hamdani
Gaurav Harit
Ehtesham Hassan
Anders Hast
Sheng He
Jean Hennebert
Pierre Héroux
Laurent Heutte
Nina S. T. Hirata
Tin Kam Ho
Kaizhu Huang
Qiang Huo
Donato Impedovo
Reeve Ingle
Brian Kenji Iwana
Motoi Iwata
Antonio Jimeno
Slim Kanoun
Vassilis Katsouros
Ergina Kavallieratou
Klara Kedem

Christopher Kermorvant
Khurram Khurshid
Soo-Hyung Kim
Koichi Kise
Florian Kleber
Pramod Kompalli
Alessandro Lameiras Koerich
Bart Lamiroy
Anh Le Duc
Frank Lebourgeois
Gurpreet Lehal
Byron Leite Dantas Bezerra
Aurélie Lemaitre
Haifeng Li
Zhouhui Lian
Minghui Liao
Rafael Lins
Wenyin Liu
Lu Liu
Georgios Louloudis
Yue Lu
Xiaoqing Lu
Muhammad Muzzamil Luqman
Sriganesh Madhvanath
Muhammad Imran Malik
R. Manmatha
Volker Märgner
Daniel Martín-Albo
Carlos David Martinez Hinarejos
Minesh Mathew
Maroua Mehri
Carlos Mello
Tomo Miyazaki
Momina Moetesum
Harold Mouchère
Masaki Nakagawa
Nibal Nayef
Atul Negi
Clemens Neudecker
Cuong Tuan Nguyen
Hung Tuan Nguyen
Journet Nicholas
Jean-Marc Ogier
Shinichiro Omachi
Umapada Pal
Shivakumara Palaiahnakote

Thierry Paquet
Swapan Kr. Parui
Antonio Parziale
Antonio Pertusa
Giuseppe Pirlo
Réjean Plamondon
Stefan Pletschacher
Utkarsh Porwal
Vincent Poulain D'Andecy
Ioannis Pratikakis
Joan Puigcerver
Siyang Qin
Irina Rabaev
Jean-Yves Ramel
Oriol Ramos Terrades
Romain Raveaux
Frédéric Rayar
Ana Rebelo
Pau Riba
Kaspar Riesen
Christophe Rigaud
Syed Tahseen Raza Rizvi
Leonard Rothacker
Javad Sadri
Rajkumar Saini
Joan Andreu Sanchez
K. C. Santosh
Rosa Senatore
Amina Serir
Mathias Seuret
Badarinath Shantharam
Imran Siddiqi
Nicolas Sidère
Foteini Simistira Liwicki
Steven Simske
Volker Sorge
Nikolaos Stamatopoulos
Bela Stantic
H. Siegfried Stiehl
Daniel Stoekl Ben Ezra
Tonghua Su
Tong Sun
Yipeng Sun
Jun Sun
Suresh Sundaram
Salvatore Tabbone

Kazem Taghva
Ryohei Tanaka
Christopher Tensmeyer
Kengo Terasawa
Ruben Tolosana
Alejandro Toselli
Cao De Tran
Szilard Vajda
Ernest Valveny
Marie Vans
Eduardo Vellasques
Ruben Vera-Rodriguez
Christian Viard-Gaudin
Mauricio Villegas
Qiu-Feng Wang

Da-Han Wang
Curtis Wigington
Liang Wu
Mingkun Yang
Xu-Cheng Yin
Fei Yin
Guangwei Zhang
Heng Zhang
Xu-Yao Zhang
Yuchen Zheng
Guoqiang Zhong
Yu Zhou
Anna Zhu
Majid Ziaratban

Contents – Part IV

Document Classification

Gold-Standard Benchmarks and Data Sets

Historical Document Analysis

Handwriting Recognition

Competition Reports

Scene Text Detection and Recognition

HRRegionNet: Chinese Character Segmentation in Historical Documents with Regional Awareness

Chia-Wei Tang[(⊠)], Chao-Lin Liu, and Po-Sen Chiu

Department of Computer Science, National Chengchi University, Taipei, Taiwan
{106703054,chaolin,107753029}@g.nccu.edu.tw

Abstract. Human beings, as the only species capable of developing high levels of civilization, the transmission of knowledge from historical documents plays an indispensable role in this process. The amount of historical documents accumulated in the last centuries is not to be belittled, and the knowledge they contain is not to be underestimated. However, these historical documents are also facing difficulties in preservation due to various factors. The digitization process was mostly performed manually in the past, but the costs made the process very slow and challenging, so how to automate the digitization process has been the focus of much research previously. The digitization of Chinese historical documents can divide into two main stages: Chinese character segmentation and Chinese character recognition. This study will only focus on Chinese character segmentation in historical documents because only accurate segmentation results can achieve high accuracy in Chinese character recognition. In this research, we further improve the model based on our previously proposed Chinese character detection model, HRCenterNet, by adding a transposed convolution module to restore the output to a higher resolution and use multi-resolution aggregation combine features in different resolutions. In addition, we also propose a new objective function such that the model can more comprehensively consider the features needed to segment Chinese characters during the learning process. In the MTHv2 dataset, our model achieves an IoU score of 0.862 and reaches state-of-the-art. Our source code is available on https://github.com/Tverous/HRRegionNet.

Keywords: Document analytics · Digital humanities · Computer vision · Object detection

1 Introduction

In the natural world where all species coexist, human beings, as the only creature capable of developing a high level of civilization, are different from other creatures in their ability to transmit knowledge. Through the transmission of knowledge, human beings are able to gain the knowledge and experience accumulated by their ancestors and then use these foundations to build more advanced ideas or technologies in the long run. Since the invention of writing, the ability of human beings to transmit knowledge has further

© Springer Nature Switzerland AG 2021
J. Lladós et al. (Eds.): ICDAR 2021, LNCS 12824, pp. 3–17, 2021.
https://doi.org/10.1007/978-3-030-86337-1_1

developed by leaps and bounds, and documents that record this knowledge have become one of the most valuable assets for human history. Therefore, how to preserve these assets through modern technology has become one of the important research topics.

These paper books were not easy to preserve because of their inherent characteristics, which make them easily damaged by various factors, either human-made or natural. With the development of modern technology, the way of retaining knowledge has become different from the past when we could only rely on paper books. Nowadays, we can store such information in electronic media using electronic signals, which significantly reduces the difficulty and cost of storing such information. This process of converting textual content to electronic signals is called digitization. The accumulation of Chinese historical documents and the knowledge contained is not to be underestimated during the past centuries. Due to Chinese historical documents' characteristics, it usually requires domain experts to obtain a high-quality digitalized result, and how to automate this process to reduce the costs is one of the problems that many studies have tried to solve in the past. The digitization process of Chinese historical documents can be divided into two main stages: first, character segmentation, i.e., detecting and segmenting individual Chines characters in the scanned document, and second, character recognition, i.e., identifying the segmented results separately to output a complete and accurate content from the document in electronic format. Since the accuracy of character recognition depends mainly on the accuracy of character segmentation, we will focus only on the Chinese character segmentation in this research.

1.1 A Brief Primer of Chinese Historical Documents

In order to give the reader a better understanding of the difficulties and breakthroughs of this research, in the following, we will provide an overview of the Chinese historical documents.

Various Composition in Chinese Historical Documents

Just as modern books have a variety of printing and layout formats, Chinese history documents also have various formats depending on the content of the record. Although most Chinese books' reading order follows a right-to-left, vertical form of arrangement, in many documents, such a layout format may vary to emphasize part of the content or for aesthetic purpose, etc., resulting in different line spacing, different font sizes, multiple words in a line, characters in a picture, etc., as shown in **Error! Reference source not found.**. All these factors enormously increase the complexity of character segmentation.

Complex Chinese Character

Unlike western language systems, which consists of a small number and easily recognizable alphabets, in the Chinese language system, however, there are tens of thousands of different Chinese characters that make up the language, and these characters can be further split or re-formed into new Chinese characters. For these reasons and the shortage of data, it is challenging to build a robust and general Chinese character detector for historical documents (Fig. 1).

Fig. 1. Although most Chinese historical documents arrange in vertical rows from right to left, there are many variations above this, including scattered character distribution and misalignment between characters, which cause difficulties in processing these documents.

Distortion in Historical Documents

In addition to the above difficulties, another major problem in handling historical documents is that most of them are not well preserved. Unlike modern books, which are printed on solid paper or are maintained in good condition, most of the carriers of Chinese paintings and calligraphy in the past were made of fragile paper and silk, and most of them are too old to have regained their original appearance. Therefore, we have to face noise such as tears, stains, folds, peeling, etc., which significantly increases the uncertainty during the detection.

In this research, we improve our previous proposed model HRCenterNet [1], which uses a parallelization structure to achieve multi-resolution fusion with minimal feature loss and generate the bounding box of individual characters by directly predicting the center keypoint and related attributes of the Chinese characters. We improve the model's architecture by appending a transposed convolution module to the output layer to avoid the loss of features in the prediction due to the large difference between the input and output resolutions; then, we apply multi-resolution aggregation to aggregate the predicted results in different resolution to enable the model to be scalable in predicted results. Meanwhile, we also propose a new objective function, which makes the model better consider the dependency between the predicted center and the related attributes of the Chinese characters during the training process. We achieve an IoU of 86.2 on the MTHv2 dataset during experiments, which makes our model becomes state-of-the-art in the task of Chinese character segmentation in historical documents.

2 Related Works

It is necessary to detect and segment all the individual characters in the document first to automate digitization. However, as mentioned in Sect. 1.1, there are various characteristics of Chinese documents that make previous researches only able to deal with specific forms or limited achievements. In the following, we will review these studies separately.

2.1   Traditional Approach

One of the most common traditional Chinese character segmentation algorithms is the projection-based segmentation method, which first accumulates the pixel values of an axis to obtain a histogram representing the pixel density and then segments it according to a predefined threshold. Repeating this process on different x-y axes, we can obtain the segmentation result of lines or characters by excluding the intervals below the threshold value. Although this algorithm gives good results on partially aligned Chinese text, it is still limited when encountering variable text formats or when using horizontal segmentation of Chinese text plates, and it is a major problem to set a flexible threshold value.

Ptak et al. [2] propose an algorithm that can change the threshold value dynamically to overcome the shortcoming of using a fixed threshold value as the segmentation standard, but the effect is still limited when encountering complex document formats or segmenting individual characters. Xie et al. [3] use the projection-based segmentation method to first segment the lines of characters and then combine them with a Chinese character recognizer for further recognition to segment the individual characters. However, this approach relies on the character detector's accuracy and is difficult to generalize because of the complexity of Chinese characters and the scarcity of data, as mentioned previously.

2.2   Anchor-Based Approach

The problem of detecting characters in the Chinese historical document is similar to detecting objects in complex scenes. Thanks to the success of deep learning in computer vision in recent years, how to generalize the task of object detection to a problem that a machine can learn has been explored in many studies. The most common and best-performing approaches are mostly based on anchor-based models, and the system can be divided into two phases mainly. In the first phase, a model generates numerous anchor boxes through a predefined algorithm or a neural network, and the regions formed by these predefined bounding boxes are called ROIs (Region Of Interests). In the second stage, these ROIs are evaluated by another model, and a probability is given to every anchor box to indicate how likely the anchor containing the desired object. After that, the final bounding boxes can be obtained after filtering the anchor boxes with lower probability. Two-stage detectors such as R-CNN [4] proposes to use a selective search algorithm with region proposals to generate ROIs. Fast R-CNN [5] further proposes to use ROI pooling to generate ROIs by making region proposals on the convolved image to reduce the complexity of the input image. Faster R-CNN [6] offers to implement a region proposal network to transform the problem of generating ROIs into an optimization problem. EfficientDet [7] follows the two-stage detector detection approach and is one of the best-performing detection models with a modified base network. One-stage detectors such as YoLo [8] and SSD [9] generate only a fixed size and number of anchor boxes, compared to two-stage detectors, it reduces the computation time and cost, but the average accuracy is not as good as the two-stage detectors.

Jun et al. [10] and Saha et al. [11] utilize Faster R-CNN [5] for the detection of specific information and graphical images in the document, respectively. Reisswig et al. [12] improve the base network of Faster R-CNN [5] and applies it for character detection in English documents. Baek et al. [13] propose a method that can detect individual characters even when character level annotations are not given by predicting the region score and affinity score.

Although in most cases anchor-based models achieve excellent performance, however, this approach is highly unsuitable for Chinese character detection, mainly because it needs to set many hyperparameters to obtain high accuracy, including the number of anchors to be generated, the size of the anchors, and the aspect ratio of each anchor. These extensive user-defined parameters can easily lead to a model that can only handle documents in a specific format and cannot be generalized. The second problem is that the document usually contains more than two hundred Chinese characters in Chinese historical documents. As a result, the corresponding anchors need to process, resulting in a significant computational resource burden.

2.3 Anchorless Approach

In order to resolve the problem among the anchor-based models, there have been many types of research in recent years to develop various object detection models without the need for anchors or simplify the process. He et al. [14] propose a direct regression-based method on single-stage architecture. It advocates the prediction of bounding box vertices directly from each point within the image. Keserwani et al. [15] propose a centroid-centric vector regression method that combines direct and indirect regression-based approaches by utilizing the geometry of quadrilateral. CornerNet [16] proposes to detect the bounding box by directly predicting the top-left corner and bottom-right corner of the desired object. CenterNet proposed by Duan et al. [17] further proposes to detect the bounding box by predicting three keypoints of the desired object, namely, the center point, top-left corner, and bottom-right corner. Both CornerNet [16] and CenterNet [17] require to apply Associate Embedding [18] to obtain the similarity between different keypoints in order to group different keypoints to form the bounding boxes. However, this approach is not suitable for Chinese character segmentation in historical documents where most of the characters are densely distributed. FoveaBox [19] proposes to form the bounding boxes by predicting the center of the desired objects and the respective distances from the surroundings separately to generate the bounding boxes. CenterNet proposed by Zhou et al. [20] proposes to form a bounding box by directly predicting the center of the desired object and its corresponding attributes such as height and width, etc. Since most Chinese characters are symmetric in width and height to the center, point CenterNet [20] should be more suitable for Chinese character detection than FoveaBox [20]. Moreover, these models' accuracy depends a lot on the precision of predicting the keypoints, so it is also vital to implement a reliable model for Chinese character detection in the historical document.

HRCenterNet [1] combines the approaches of HRNet [21] and CenterNet [20] to segment individual Chinese characters by directly predicting their centroids and their corresponding attributes. Simultaneously, with the parallelized architecture from HRNet [21], HRCenterNet [1] can maintain the consistency between high-level representations and low-level representations and achieve multi-layer feature fusion during the process of feature extraction, without losing features during feature extraction in the last CNN encoder-decoder like architectures. This model achieves state-of-the-art in the task of

Chinese character segmentation in historical documents. However, to save the need for computational resources and reduce the hardness of convergence, HRCenterNet [1] will first compress the input image to one-fourth of its original size, and such a compression process is inappropriate for Chines documents where characters are mostly densely organized. Because when the character distributes densely, the input image will be extremely blurred after preprocessing and compression, seriously affecting the subsequent detection. Another problem is that HRCenterNet [1] treats the centroid's prediction and its related properties as independent relationships in the design of objective functions, but it should be dependent instead of independent since the values of width and height are related to the position of the centroid as well. In this study, we propose a model that aims to solve the above problems and build a more general and accurate system.

3   Our Approach

(See Fig. 2).

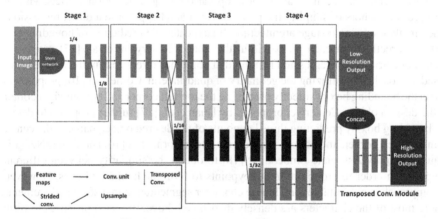

Fig. 2. The architecture of HRRegionNet.

3.1   Center as Characters

Different from the previous anchor-based models that need to use a large number of anchors to generate ROIs and identify them one by one to segment individual Chinese characters. In this study, we transform the problem of Chinese character segmentation into a standard keypoint detection problem, i.e., we use the centroid of each Chinese character to represent individual Chinese characters and ask the model to predict a probability distribution, indicating the likelihood of the character's centroids in per pixel. The corresponding values of other attributes, such as the bounding boxes' height and width, should be predicted from the model in the output as well. This method avoids the problem of hyperparameters tuning and able to achieve fast and accurate detection results.

3.2 Backbone

The backbone of our model is based on our previously proposed model HRCenterNet [1]. The network consists of two components mainly, parallel multi-resolution convolutions and reiterated multi-resolution fusions. With these two components, our model can achieve highly accurate and precise prediction results. Our model's architecture is illustrated in **Error! Reference source not found.**.

Firstly, to lessen the computational resources required by the model, the input image first goes through a stem network that consists of two stride-2 3×3 convolutions to reduce the resolution to 1/4 of the original one, then enters the stage of parallel multi-resolution convolutions. In this stage, the input will pass through a series of building blocks or bottlenecks from ResNet [22] to extract the features, and finally, a stride-2 3×3 convolution will be added to form a new branch, the resolution of this branch is half of the original input resolution, and the stages of parallel multi-resolution convolutions end here. These branches will be added as the network expands and will exist in parallel, preserving high-level and low-level representations simultaneously. After that, the parallel branches will enter the multi-resolution fusions stage, where the feature map between different branches will be upsampled, downsampled, or remain unchanged according to the resolution of the main branches, then combine them together to achieve the effect of exchanging information between different resolutions. The entire network is composed of these two components mainly, which can be repeated several times, depending on the size of the model. In our network, we repeat this process three times, including the preprocessing phase; the whole model can be divided into four stages. The first stage decreases the resolution to 1/4 from the original input image alongside four bottlenecks from ResNet [22], where each block has the number of channels equal to 32, then a new branch is formed via a stride-2 3×3 convolution. Stages 2, 3, and 4 are repeat operations of parallel multi-resolution convolutions combined with multi-resolution fusion, which contain 1, 4, and 3 multi-resolution blocks, respectively. Each multi-resolution block contains four building blocks from ResNet [22], and the number of channels in each block is 2C, 4C, 8C; we set C to 32 in our model.

In the low-resolution output from the model, simply fuse different branches again and take the highest resolution branch after a 1×1 convolution, and a sigmoid function will be the low-resolution prediction result.

3.3 Transposed Convolution Module

In HRCenterNet [1], the input image is compressed to a quarter of its resolution by a stem network composed of two stride-2 3×3 convolutions to accelerate the process of convergence and reduce the computational cost required for training. However, this approach is not the ideal solution for Chinese character segmentation, for the reason that the heatmap representing the probability distribution of Chinese characters is computed by an unnormalized 2D Gaussian function, and when the resolution of the original image differs too much from the output resolution of the model and the characters are overly dense in the image, the resulting probability distribution will be obscure or extremely non-uniform, resulting in a very blurred heatmap, as shown in Fig. 3. And this may cause the model to converge in the wrong direction.

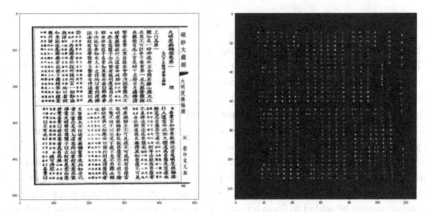

Fig. 3. When the difference in resolution between the input image and the output image is too large, and the characters are densely distributed, the resulting heatmap will be very blurred, and some features of the image will be lost.

To further improve the detection accuracy without increasing the cost of the model as much as possible, instead of directly reducing the standard deviation of the Gaussian function or expanding the model of HRCenterNet [1]. In this study, inspired by HigherHRNet [23], we add a transposed convolution module to the model's output to increase the resolution of the output two times larger than the original one; the transposed convolution module takes the HRCenterNet [1] feature maps and the predicted low-resolution output as input and consists of a stride-1 3 × 3 convolution, a stride-2 3 × 3 convolution, and a stride-1 3 × 3 convolution in order respectively, with a 1 × 1 convolution, and a sigmoid function will be the prediction on the high-resolution output. In order to simplify the model, in addition to the heatmap, which can be represented in probability format, we divide the height and width of the bounding box by the size of the output image to represent them in a probabilistic way. The architecture of our model is shown in **Error! Reference source not found.**.

3.4   Multi-resolution Aggregation

Since our model has both low-resolution output and high-resolution output, in the final prediction, we scale the low-resolution heatmap to the same size as the high-resolution heatmap using the nearest neighbor method. Then we add these two heatmaps and divide them by two to obtain the final heatmap for predicting the centroids of Chinese characters. We expect our model to achieve the scale-aware effect by this method. For the corresponding width, height, and offsets for each individual Chinese character, we directly utilize them from the high-resolution prediction.

3.5 Objective Functions

Heatmap Loss

In our model, we aim to predict the heatmap $\widehat{H}_{xy} \in [0, 1]^{H*W}$ representing the center of each Chinese character with height H and width W, where $\widehat{H}_{xy} = 1$ corresponds to the ground-truth keypoint, while $\widehat{H}_{xy} = 0$ is the background. But it is very difficult to output a heatmap prediction that exactly matches the ground-truth locations. Plus, even if the keypoint is slightly off, if it is very close to the corresponding ground-truth location, it is still probable to produce a prediction close to the original bounding box. As a result, for each ground-truth keypoint, we will give it a radius around the keypoint, and the model will get less penalty for false predictions within this radius than outside it. Given the radius, for each ground-truth keypoint $p \in R^2$, a heatmap corresponding to the ground-truth keypoint can be transformed by an unnormalized 2D Gaussian $H_{xy} = \exp(-\frac{(x-\tilde{p}_x)^2 + (y-\tilde{p}_y)^2}{2\sigma_p^2})$, where σ_p is the value that can be adapted by the radius of the error. An intuitive idea to make the model more accurate in keypoint predictions would be to reduce the value of standard deviation in the Gaussian kernel, but in HigherHRNet [23], it is pointed out that this makes optimization more difficult and may lead to even worse results. Therefore, in our model, the value of σ_p is set to $\frac{1}{10}$ of the desired object's height and width separately, resulting in a probability distribution that is elliptical rather than circular. If the distributions of two objects overlap each other, we take the maximum value between them instead of the average, such that the different Chinese characters can be classified [24].

In order to overcome the problem of imbalanced foreground and background objects, a loss function can be obtained with the focal loss [25]:

$$L_{heatmap} = -\frac{1}{N} \sum_{xy} \begin{cases} \left(1 - \widehat{H}_{xy}\right)^{\alpha} \log\left(\widehat{H}_{xy}\right), & H_{xy} = 1 \\ \left(1 - H_{xy}\right)^{\beta} \left(H_{xy}\right)^{\alpha} \log(1 - \widehat{H}_{xy}), & otherwise \end{cases} \tag{1}$$

where N is the number of desired objects in the image, and α and β are the hyperparameters for focal loss, here we follow [25] setting α to 2 and β to 4 in all our experiments.

Size Loss

In addition to generating a heatmap to represent the center of the Chinese character to form the bounding box, the model also needs to output each desired object's corresponding height and width. Let (h_{xy}, w_{xy}) be the height and width of each desired object in ground-truth keypoint, we use an L1 loss with a mask $M_{xy} \in \{0, 1\}^{H*w}$ indicating whether the pixel has a ground-truth keypoint or not to reduce the computational burden, obtaining the following equation:

$$L_{size} = \frac{1}{N} \sum_{xy} \left| h_{xy} - \hat{h}_{xy} * M_{xy} \right| + \left| w_{xy} - \widehat{w}_{xy} * M_{xy} \right| \tag{2}$$

Offset Loss

To avoid some desired objects are too small, e.g., the desired object's keypoint is between pixels or the inaccurate prediction due to the difference in resolution between input and output. Let (O_{xy}^1, O_{xy}^2) be the offset of the ground-truth keypoint located in (p_{xc}, p_{yc}), where $O_{xy}^1 = p_{xc} \bmod 1$, $O_{xy}^2 = p_{yc} \bmod 1$. Similar to Objective (2), an offset loss function can be formulated as:

$$L_{offset} = \frac{1}{N} \sum_{xy} \left| O_{xy}^1 - \widehat{O}_{xy}^1 * M_{xy} \right| + \left| O_{xy}^2 - \widehat{O}_{xy}^2 * M_{xy} \right| \tag{3}$$

Region Loss

During the prediction of the Chinese character bounding box, the center keypoint of the Chinese character and its height and width should be in a dependent relationship, i.e., the height and the width values should vary with the position of the center keypoint, yet in HRCenterNet [1], the objective functions of both are calculated separately in an independent manner. Although all the objective functions are considered together during backpropagation, there is no guarantee that the model can correctly learn the interdependence between them. Therefore, in this study, we propose a new objective function named regional loss to compare the predicted bounding boxes and the ground-truth values for training. At the same time, in order to speed up the convergence, we only consider the region loss on the ground-truth keypoint, same as the size loss and the offset loss; the whole equation can be formulated as follows:

$$L_{region} = \frac{1}{N} \sum_{xy} \ln \left(\left| M_{xy} - \frac{\widehat{H}_{xy} * \widehat{h}_{xy} * \widehat{w}_{xy}}{H_{xy} * h_{xy} * w_{xy}} \right| + 1 \right) \tag{4}$$

Overall Loss

The overall objective functions can be formulated as follows:

$$L_{all} = \tau_{heatmap} L_{heatmap} + \tau_{size} L_{size} + \tau_{offset} L_{offset} + \tau_{region} L_{region} \tag{5}$$

We set τ_h, τ_s, τ_{offset}, τ_{region} to 1, 5, 10, 0.1 respectively in all our experiments unless specified otherwise.

In this study, our model predicts both low-resolution output and high-resolution output from the model. The overall objective function of the model will be as follows:

$$L_{overall} = L_{all}^{low} + L_{all}^{high} \tag{6}$$

4   Experiments

4.1   Dataset

One of the biggest problems in deep learning applications is the lack of datasets. The MTHv2 dataset provided by [26] consists of about 3500 images with various types of format. Each image has three different types of annotations. The first is line-level annotations, which contain the position of all text lines in the document. The second is character-level annotations, which contain the bounding box coordinates of each Chinese character, and the last is the text content. This research will only use character-level annotations since we will focus on individual Chinese characters' segmentation only.

4.2 Training Details

We implement our model in PyTorch [27]. The network is randomly initialized under the default setting of PyTorch. During training, we first set the input image resolution to 512×512, which leads to an output resolution of 256×267. To reduce overfitting, we adopt random cropping and random grayscale. We use Adam [28] to optimize the Objective (6). We use a batch size of 16 and train the network on 1 T V100 GPU. We train the network for 200 epochs with a learning rate of 1×10^{-6}. We took 1% of the dataset for testing randomly and the rest for training.

4.3 Testing Details

During testing, we take the low-resolution heatmap $\widehat{H}_{low}$ and high-resolution heatmap $\widehat{H}_{high}$ from the model and apply multi-resolution aggregation to obtain a heatmap $\widehat{H}_{low+high} = \frac{(\widehat{H}_{low}+\widehat{H}_{high})}{2}$. Then we can generate the bounding boxes for individual characters after applying the height, width, and offset from the high-resolution output. Then we apply non-maximal suppression (NMS) to these bounding boxes, where we set the confidence score to 0.4 and the IoU threshold to 0.1. Since the resolution of the model output is different from the input image, in order to map the model output back to the input, we multiply the height, width, and offset maps output from the model by $\frac{In_h}{Out_h}$ and $\frac{In_w}{Out_w}$ respectively. Where In_h and In_w are the height and width of the input image, Out_h and Out_w are the height and width of the output image. The final results are shown in Fig. 4.

4.4 Ablation Study

Anchor-Based Approach with Anchorless Approach

To prove that our method and backbone are superior compared to the others, we examine our model alongside with previous state-of-the-art models, as shown in Table 1. All experiments are trained with a batch size of 16 and 200 epochs. For anchor-based models, we follow the default setting shown in the original paper and show that, compared to the anchor-based detection method, the anchorless approach is more general to the task in Chinese character segmentation for historical documents. We also examine other models with different backbones to show that with the parallel architecture, the feature to be preserved can make a significant improvement in the detection results.

Transposed Convolution with Region Loss

To verify that our proposed model and objective function can improve the performance from previous HRCenterNet [1], we compare HRCenterNet [1] and CenterNet [20]

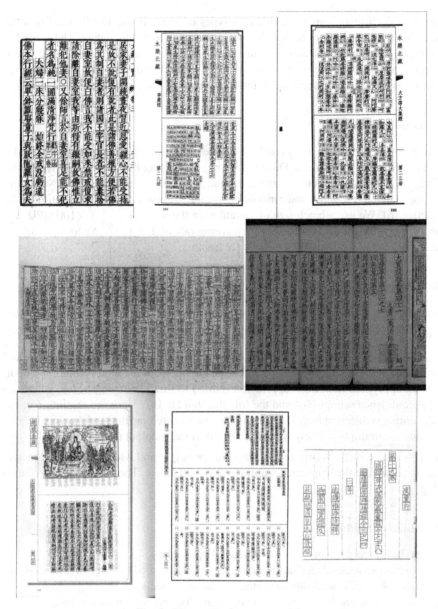

Fig. 4. The results from our model.

with our model. All the experimental settings are the same as in Sect. 4.2 and 4.3. The experimental results are shown in Table 2.

Table 1. Comparisons on MTHv2 dataset. Our approach with a small model, trained from scratch, performs better than previous methods with an acceptable inference time.

Method	Model	#Params	Inference time (ms)	IoU
Anchor-based	Faster R-CNN [6]	27M	54.41	0.516
	EfficientDet-D5 [7]	34.5M	69.56	0.563
	RetinaNet [25]	26.3M	49.82	0.471
	YoLo [8]	12.5M	18.43	0.318
	SSD [9]	11.4M	15.79	0.306
Anchorless	U-Net [29]	32.5M	32.31	0.696
	FPN [30]	26.9M	50.89	0.783
	PSPNet [31]	38.4M	22.74	0.572
	CenterNet [20]	26.2M	23.58	0.732
	HRCenterNet [1]	28.5M	20.56	0.815
	Ours	28.6M	24.37	*0.862*

Table 2. Comparison of the results with and without region loss and transposed convolution module under the same detection method.

Region loss	Model	Output size	IoU
Without region loss	CenterNet [20]	128 × 128	0.732
	HRCenterNet [1]	128 × 128	0.815
	Ours	*256 × 256*	*0.851*
With region loss	CenterNet [20]	128 × 128	0.744
	HRCenterNet [1]	128 × 128	0.823
	Ours	*256 × 256*	*0.862*

5   Conclusion

In this research, we improve the model based on our previously proposed model HRCenterNet [1] by adding a transposed convolution module to enlarge the output resolution to twice the original one, addressing the problem that the difference between the input and output resolutions is too large to cause the generated heatmap to be blurred or some features might be lost in previous. We also propose a new objective function to compare the area difference of the bounding box generated by the predicted keypoint so that the model can learn the dependency relationship between the attributes better and further improve the accuracy. In the MTHv2 testing dataset, our method achieves the best speed and accuracy compared to the others and reaches state-of-the-art.

Acknowledgments. This research has been supported by the contracts MOST-109-2813-C-004-011-E and MOST-107-2200-E-004-009-MY3 from the Ministry of Science and Technology of Taiwan.

References

1. Tang, C.-W., Liu, C.-L., Chiu, P.-S.: HRCenterNet: an anchorless approach to Chinese character segmentation in historical documents. In: 2020 IEEE International Conference on Big Data (Big Data), December 2020, pp. 1924–1930 (2020). https://doi.org/10.1109/BigData50 022.2020.9378051
2. Ptak, R., Żygadło, B., Unold, O.: Projection–based text line segmentation with a variable threshold. Int. J. Appl. Math. Comput. Sci. **27**(1), 195–206 (2017). https://doi.org/10.1515/amcs-2017-0014
3. Xie, Z., et al.: Weakly supervised precise segmentation for historical document images. Neurocomputing **350**, 271–281 (2019). https://doi.org/10.1016/j.neucom.2019.04.001
4. Girshick, R., Donahue, J., Darrell, T., Malik, J.: Rich feature hierarchies for accurate object detection and semantic segmentation. arXiv:1311.2524 [cs], October 2014. Accessed 25 Feb 2021
5. Girshick, R.: Fast R-CNN. arXiv:1504.08083 [cs], September 2015. Accessed 25 Feb 2021
6. Ren, S., He, K., Girshick, R., Sun, J.: Faster R-CNN: Towards Real-Time Object Detection with Region Proposal Networks. arXiv:1506.01497 [cs] (2016)
7. Tan, M., Pang, R., Le, Q.V.: EfficientDet: Scalable and Efficient Object Detection. arXiv: 1911.09070 [cs, eess], July 2020. Accessed 25 Feb 2021
8. Redmon, J., Divvala, S., Girshick, R., Farhadi, A.: You Only Look Once: Unified, Real-Time Object Detection. arXiv:1506.02640 [cs], May 2016. Accessed 25 Feb 2021
9. Liu, W., et al.: SSD: Single Shot MultiBox Detector. arXiv:1512.02325 [cs], vol. 9905, pp. 21–37 (2016). https://doi.org/10.1007/978-3-319-46448-0_2
10. Jun, C., Suhua, Y., Shaofeng, J.: Automatic classification and recognition of complex documents based on Faster RCNN. In: 2019 14th IEEE International Conference on Electronic Measurement Instruments (ICEMI), November 2019, pp. 573–577 (2019). https://doi.org/10.1109/ICEMI46757.2019.9101847
11. Saha, R., Mondal, A., Jawahar, C.V.: Graphical Object Detection in Document Images. arXiv: 2008.10843 [cs], August 2020. Accessed 12 Feb 2021
12. Reisswig, C., Katti, A.R., Spinaci, M., Höhne, J.: Chargrid-OCR: End-to-end Trainable Optical Character Recognition for Printed Documents using Instance Segmentation. arXiv:1909.04469 [cs], February 2020. Accessed 12 Feb 2021
13. Baek, Y., Lee, B., Han, D., Yun, S., Lee, H.: Character region awareness for text detection. In: 2019 IEEE/CVF Conference on Computer Vision and Pattern Recognition (CVPR), Long Beach, CA, USA, June 2019, pp. 9357–9366 (2019). https://doi.org/10.1109/CVPR.2019.00959
14. He, W., Zhang, X.-Y., Yin, F., Liu, C.-L.: Deep direct regression for multi-oriented scene text detection. In: 2017 IEEE International Conference on Computer Vision (ICCV), October 2017, pp. 745–753 (2017). https://doi.org/10.1109/ICCV.2017.87
15. Keserwani, P., Dhankhar, A., Saini, R., Roy, P.P.: Quadbox: quadrilateral bounding box based scene text detection using vector regression. IEEE Access **9**, 36802–36818 (2021). https://doi.org/10.1109/ACCESS.2021.3063030
16. Law, H., Deng, J.: CornerNet: Detecting Objects as Paired Keypoints. arXiv:1808.01244 [cs], March 2019. Accessed 25 Feb 2021

17. Duan, K., Bai, S., Xie, L., Qi, H., Huang, Q., Tian, Q.: CenterNet: Keypoint Triplets for Object Detection. arXiv:1904.08189 [cs], April 2019. Accessed 18 Feb 2021
18. Newell, A., Huang, Z., Deng, J.: Associative Embedding: End-to-End Learning for Joint Detection and Grouping. arXiv:1611.05424 [cs], June 2017. Accessed 25 Feb 2021
19. Kong, T., Sun, F., Liu, H., Jiang, Y., Li, L., Shi, J.: FoveaBox: beyound anchor-based object detection. IEEE Trans. Image Process. **29**, 10 (2020)
20. Zhou, X., Wang, D., Krähenbühl, P.: Objects as Points. arXiv:1904.07850 [cs], April 2019. Accessed 18 Feb 2021
21. Wang, J., et al.: Deep High-Resolution Representation Learning for Visual Recognition. arXiv:1908.07919 [cs], March 2020. Accessed 15 Feb 2021
22. He, K., Zhang, X., Ren, S., Sun, J.: Deep Residual Learning for Image Recognition. arXiv:1512.03385 [cs], December 2015. Accessed 25 Feb 2021
23. Cheng, B., Xiao, B., Wang, J., Shi, H., Huang, T.S., Zhang, L.: HigherHRNet: Scale-Aware Representation Learning for Bottom-Up Human Pose Estimation. arXiv:1908.10357 [cs, eess], March 2020. Accessed 15 Feb 2021
24. Cao, Z., Simon, T., Wei, S.-E., Sheikh, Y.: Realtime Multi-Person 2D Pose Estimation using Part Affinity Fields. arXiv:1611.08050 [cs], April 2017. Accessed 25 Feb 2021
25. Lin, T.-Y., Goyal, P., Girshick, R., He, K., Dollár, P.: Focal Loss for Dense Object Detection. arXiv:1708.02002 [cs], February 2018. Accessed 25 Feb 2021
26. Ma, W., Zhang, H., Jin, L., Wu, S., Wang, J., Wang, Y.: Joint Layout Analysis, Character Detection and Recognition for Historical Document Digitization. arXiv:2007.06890 [cs], July 2020. Accessed 12 Feb 2021
27. Paszke, A., et al.: Automatic differentiation in PyTorch, October 2017. https://openreview.net/forum?id=BJJsrmfCZ. Accessed 03 Mar 2021
28. Kingma, D.P., Ba, J.: Adam: A Method for Stochastic Optimization. arXiv:1412.6980 [cs], January 2017. Accessed 03 Mar 2021
29. Ronneberger, O., Fischer, P., Brox, T.: U-Net: Convolutional Networks for Biomedical Image Segmentation. arXiv:1505.04597 [cs], May 2015. Accessed 03 Mar 2021
30. Lin, T.-Y., Dollár, P., Girshick, R., He, K., Hariharan, B., Belongie, S.: Feature Pyramid Networks for Object Detection. arXiv:1612.03144 [cs], April 2017. Accessed 03 Mar 2021
31. Zhao, H., Shi, J., Qi, X., Wang, X., Jia, J.: Pyramid Scene Parsing Network. arXiv:1612.01105 [cs], April 2017. Accessed 03 Mar 2021

Fast Text vs. Non-text Classification of Images

Jiri Kralicek[✉] and Jiri Matas

Czech Technical University, Prague, Czech Republic
kraliji2@fel.cvut.cz, matas@cmp.felk.cvut.cz

Abstract. We propose a fast method for classifying images as containing text, or with no scene text. The typical application is in processing large image streams, as encountered in social networks, for detection and recognition of scene text. The proposed classifier efficiently removes non-text images from consideration, thus allowing to apply the potentially computationally heavy scene text detection and OCR on only a fraction of the images.

The proposed method, called Fast-Text-Classifier (FTC), utilizes a MobileNetV2 architecture as a feature extractor for fast inference. The text vs. non-text prediction is based on a block-level approach. FTC achieves 94.2% F-measure, 0.97 area under the ROC curve, and 74.8 ms and 8.6 ms inference times for CPU and GPU, respectively. A dataset of 1M images, automatically annotated with masks indicating text presence, is introduced and made public at http://cmp.felk.cvut.cz/data/twitter1M.

Keywords: Text vs. Non-text image classification · Scene text detection · Social network data

1 Introduction

The advent of smart devices in combination with social networks massively increased the volume of data that has been shared in recent years. The data has various forms with image data becoming dominant, mainly on new types of social networks. There is a number of applications where images are processed in order to extract scene text. Since not all images contain text, and high quality, state-of-the-art scene text detectors and recognition methods often employ complex algorithms, direct processing of all images by scene text spotting systems is potentially very time and energy consuming.

This issue can be mitigated by a sequential approach, determining first whether an image contains text and only if it does, the image is passed to a text spotting system that determines the exact location of the text and its transcription. The sequential approach is effective especially when the sorting into "with text" or "without text" image classes is very fast, say an order of magnitude faster then the text detection and recognition itself.

© Springer Nature Switzerland AG 2021
J. Lladós et al. (Eds.): ICDAR 2021, LNCS 12824, pp. 18–32, 2021.
https://doi.org/10.1007/978-3-030-86337-1_2

(a)   (b)   (c)

Fig. 1. The proposed Fast-Text-Classifier (FTC) operates on 3×3 and 5×5 blocks, the grids are shown in (b) and (c). Blocks with color overlays contain text. (Color figure online)

In this paper, we propose a method, called Fast-Text-Classifier, or FTC in short, to efficiently classify image as text/non-text. FTC is fast both on GPUs and CPUs. The FTC classification adopts a block-level approach (Fig. 1). MobileNetV2 [25] is utilized as a backbone for feature extraction. The proposed method is one of the fastest and most accurate methods for the text vs. non-text classification problem; no published method is faster and more accurate at the same time.[1]

To train such a classifier, a large image dataset is highly beneficial, we thus introduce a new dataset, Twitter1M, which we make public. The dataset consists of 1 million images gathered from a social network. Its annotations were automatically created. Manual verification confirmed that the annotation noise level is low. The automatic annotation system (AS) consists of three state-of-the-art scene text detectors, providing three text masks for each image. The AS is in effect used for teacher-student training, where the teacher is very slow, but highly accurate. The FTC student network is very fast, with a slightly lower accuracy.

2   Related Work

2.1   Text Image Classification

Text vs. non-text image classification predicts whether the image contains scene text or not. To the best of our knowledge, only a few papers on text vs. non-text

---

[1] Most methods are both less accurate and slower. The state-of-the-art that outperforms FTC is order of magnitude slower. The single faster published method [35] has lower accuracy than FTC.

image classification have been published. McDonnell and Vladusich [20] proposed an image classification method based on a Fast-learning Shallow CNN, consisting of three hidden layers and a linear output layer; only the final layer weights are learned. Despite focusing primarily on the speed of the learning process, due to its shallow architecture, it also provides fast inference. Alessia et al. [1] proposed an automatic selector of text images based on three handcrafted rules: 1. fast changes in intensity, 2. very frequent luminance changes without a preferred direction, and 3. the presence of a connected zone with an almost constant value of luminance (the background) in a sufficiently wide text area. These methods are limited by the shallow net architecture and handcrafted rules, respectively.

The CNN-Coding method of Zhang et al. [33] combines 330 maximally stable extremal region (MSER), a convolutional neural network (CNN) and a bag of words (BoW) for text image discrimination. The model extracts text candidates by MSER, which are then fed into a CNN model to generate visual features. Last, features are aggregated by BoW to obtain the final representation for natural image.

Bai et al. [4] proposed the MSP-Net and adopted VGG-16 as a backbone for feature extraction, followed by deconvolution in the last three layers to obtain multi-level feature maps. Inspired by Fast R-CNN, the maps are spatially partitioned into blocks of different sizes, followed by the max-pooling layer. The feature vector for each block is fed into the fully-connected layers which make the binary classification for that block. The final classification of the whole image is the logical OR of the individual block classification.

Lyu et al. [21] introduced classification based on CNN and Multi-Dimensional Recurrent Neural Networks (MDRNN). The CNN extracts rich and high-level image representation, while the MDRNN analyzes dependencies along multiple directions and produces block-level predictions.

The classification method of Zhao et al. [35] exploits knowledge distillation. Inspired by MSP-Net, the teacher network uses VGG16 for feature extraction. The last two layers are pooled to fixed-sizes (5×5, 3×3). Region-level feature is linearly mapped in batch into a new space for knowledge distillation. The last layer of the network outputs a prediction for each sub-region in the manner of MSP-Net. The student network contains 6 convolutional layers as the image feature extractor. Multi-scale features are obtained directly from the last two convolution layers. The student network obtains the feature maps with a size of 5×5 or 3×3 after adaptive pooling. Then two-layer fully-connected network process the region level features. The first layer maps the region-level features from the original space to the feature space shared with the teacher network. These region-level features are used for knowledge distillation. The features are used for prediction in the MSP-Net manner. The training process of the student network is divided into two stages. In the first stage, neuron responses of the teacher network are transferred to initialize the student network before classification training. In the second stage, after the network is initialized based on the transfer loss, soft label distillation and adversarial knowledge distillation is used.

Recently, Gupta et al. [7] utilized EAST [36] detector to classify image based on the text detection.

2.2   Scene Text Detectors

The purpose of the scene text detector is to detect and localize text in real-world scene images containing distracting elements. Three scene text detectors, CRAFT [2], TextSnake [18], MaskTextSpotter [13], are used in the annotation system (AS) described in Sect. 3. These detectors were chosen due to their performance, different approaches and code availability. The detectors are described in detail.

Detection of scene text has a unique set of characteristics and challenges such as text diversity, scene complexity, and distortion factors, that require unique methodologies and solutions. Recent deep learning-based methods provide inference in hundreds of millisecond and can be categorized into *bounding-box regression-based, segmentation-based*, and *hybrid* approaches.

Bounding box regression-based methods for text detection consider the text as a common object and aim to predict the candidate bounding boxes directly. Liao et al. [14] proposed TextBoxes, a fully convolutional network adopting VGG-16 [27] as a backbone with outputs aggregated by non-maximum suppression (NMS) [5]. Shi et al. proposed SegLink [26] to handle long lines of Latin and non-Latin text. CTPN [28] adopted Faster R-CNN [23], using vertical anchors to predict the location and text/non-text score of a fixed-width proposals, and then connecting sequential proposals by a recurrent neural network (RNN). Gupta et al. [6] proposed a fully-convolutional regression network inspired by YOLO [22], reducing the false-positive rate by a random-forest classifier. He et al. proposed [11] direct regression by predicting offsets from a given point to determine the vertex coordinates of quadrilateral text boundaries. EAST [36] utilized multi-channel FCN to detect text regions directly defined by rotated boxes or quadrangles, followed by Thresholding and NMS to detect words or text lines. Rotation Region Proposal Networks (RRPN) [19] proposed an anchor strategy based on scale, ration, and angle.

Semantic segmentation aims to classify text regions in images at the pixel level. These methods first extract text blocks from the segmentation map generated by an FCN [17] and then obtain bounding boxes of the text by post-processing. For instance, Zhang et al. [34] utilized FCN to predict the salient map of text regions, and the character centroid information for false text line elimination; text line candidates are extracted from the salient map by MSER. Yao et al. [32] adopted an FCN to produce three kinds of maps: text/non-text regions (words or text lines), individual characters, and character linking orientations.

Baek et al. [2] proposed Character Region Awareness for Text Detection (CRAFT) method that localizes the individual character regions and links the detected characters to a text instance. The features are extracted by the fully-convolutional network, aggregated in manner of the U-Net [24] architecture. The network outputs the region score map for character localization, and the affinity score map to group characters into a single instance. To obtain the final bounding-boxes, three-step post-processing is used: 1. The binary map is created by thresholding region and affinity scores. 2. Connected Component Labeling is performed on the binary map. 3. The rotated rectangles with the minimum area enclosing the connected components are found.

Long et al. [18] proposed TextSnake, which predicts text instances based on geometry attributes, utilizing a fully-convolutional network with feature merging inspired by FPN [15]. The network predicts score maps of *text center line* (TCL) and *text regions* (TR), together with geometry attributes, including r, $cos\theta$ and $sin\theta$. The TCL map is masked by the TR map, and a disjoint set is used to perform instance segmentation. Three-steps striding algorithm [18] is used for the post-processing.

Instance segmentation aims to predict whole text instances instead of parts of the text. For example, inspired by Mask-RCNN, SPCNET [31] uses the mask branch outputs to localize an arbitrary shaped text. To suppress false positives, the method proposed Text-Context Module and Re-Scoring mechanism.

Liao et al. [13] proposed MaskTextSpotter, which provides scene text detection and recognition, however, only the detection output is used in the AS. The method is inspired by Mask R-CNN [9] and consists of four components: a feature pyramid network (FPN) as backbone, a region proposal network (RPN), a Fast RCNN, and a mask branch. The RPN generates text proposals for the Fast R-CNN and the mask branch. RoI Align [9] is adapted to extract the region features of the proposals. The Fast R-CNN includes classification and regression tasks, and outputs bounding boxes. The mask branch consists of three parallel modules: 1. Text Instance Segmentation outputs 1-channel text instance map, 2. Character Segmentation (CS) outputs character maps for each class from classes, where classes represent an alphabet. 3. Spatial Attentional Module (SAM) utilizes Position embedding [30] and Spatial Attention Mechanism [3] with RNN for text prediction. Both SAM and CS modules produce text prediction and its probability.

Hybrid methods combines regression-based and segmentation-based approach to localize text. For example, Single Shot Text Detector (SSTD) [10] proposed an attention mechanism to roughly identify text regions via an attentional map to suppress background inference on the feature level. Liu et al. [16] proposed a Mask R-CNN based framework named Pyramid Mask Text Detector (PMTD) employing a soft text mask for each text instance.

3   Dataset

In this section, a new dataset, Twitter1M, for natural image text classification, is introduced. The section also describes the annotation system (AS). Detectors used in AS were described in Sect. 2.2.

The Twitter1M was collected, in November 2020 in approximately 10 days, by random sampling from millions of images gathered via Twitter API. The dataset consists of 1 million images with resolution higher than 240 pixels, of which approximately 670k do and 330k do not contain text. We split the data into the training and test sets, with sizes 900k and 100k respectively. The training data contains ~602k/298k and the test data ~67k/33k text/non-text images.

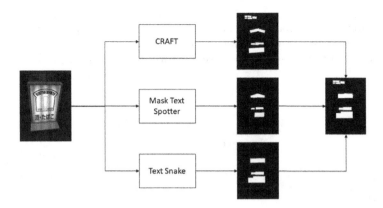

Fig. 2. An overview of the annotation system (AS). The system consists of three state-of-the-art Scene Text Detectors – CRAFT [2], MaskTextSpotter [13], TextSnake [18]. Based on detector prediction, a binary text mask is created. The masks can be concatenated by a function f; in the proposed method, logical OR was used.

The dataset includes printed and handwritten text in various scripts such as Latin, Chinese, Japanese, Korean, Cyrillic, and Indian scripts. The origins of the images are diverse, the dataset includes e.g. documents, phone screenshots, selfies, monuments, product photos, advertisements, etc.

Low resolution and duplicate images were removed before the random sampling. For duplicity removal, we used the following hashing approach: 1. the processed image was converted to grayscale and resized to 8 rows of 9 pixels. 2. The signed difference between horizontally adjacent pixels was compared with 0, resulting in 64 bits. 3. A set of images, with the Hamming distance of the hashes equal to zero, was declared as a set of duplicates, and only one image from the set was used.

The dataset contains three binary masks as an annotation for every single image. The masks describe text areas in the original image, and were created by the annotation system (AS), consisting of three state-of-the-art Scene Text Detectors - CRAFT [2], TextSnake [18] and MaskTextSpotter [13]. The AS converts the output of the text detectors into binary masks of identical size. The process is shown in Fig. 2.

The quality of the AS annotation was assessed by comparison with manual annotations on a random sample of 1000 images. The automatic annotations were incorrect for 4.9% of images, of which 4.5% were false positives and 0.4% false negatives (Table 1). All false-negative annotated images contain borderline cases. The false positives were mainly caused by repetitive patterns resembling text. The false negative vs. false positive trade-off could have been modified by tuning the sensitivity of the detectors, but we did not exploit this and used the default settings. The dataset is available at http://cmp.felk.cvut.cz/data/twitter1M.

Table 1. Confusion matrix of the automatic annotation system labels on a random sample of 1000 images from Twitter1M dataset; 4.9% of annotations are incorrect. Almost all false negative images contained only marginal, very small text areas. False positives were mainly caused by repetitive patterns resembling text.

Predicted	True class	
	Text	Non-text
Text	449	45
Non-text	4	502

4 Model

In this section, we introduce the architecture details of the proposed Fast-Text-Classifier (FTC). The section also describes the main advantage of the Inverted Residual Block [25] used in the proposed model.

The FTC utilizes the well-known MobileNetV2 [25] as a backbone of the network for the feature extraction, and inspired by MSP-Net [4], uses blocks of different sizes for image classification. In the proposed method, 3×3 and 5×5 blocks are used. The model predicts text probability for each block, which are then used for the final image-level prediction – in the proposed method, logical OR was used.

At first, the image is passed to the convolutional layer followed by seven bottleneck layers formed by Inverted Residual Blocks. Some of the bottleneck layers are repeated multiple times, forming 17 layers in total. To obtain richer multi-level information, feature maps from the last two layers are concatenated in depth. In the next step, adaptive max-pooling is applied to obtain 3×3 and 5×5 blocks, which represent sub-regions of the input image. Then, every sub-region is fed into the two-layer fully-connected network (FCN), which acts as a classifier. The FCN outputs a vector of length 34 describing each block, which is then passed to the Softmax function to obtain block-level text probabilities. In the last stage, the probabilities are thresholded and logical OR is applied to obtain the final image-level prediction (Table 2).

We utilized MobileNetV2 as a feature extractor due to the low computational cost and memory footprint provided by Inverted Residual Block. The idea behind Inverted Residual Block is that the "manifold of interest"[2] in neural networks can be embedded in low-dimensional subspaces. Thus, the information carried by the feature maps can be preserved with a lower number of channels, leading to lower computation and memory-requirements. However, deep convolutional neural networks also have non-linear transformations, such as ReLU, and it

---

[2] "Consider a deep neural network consisting of n layers L_i each of which has an activation tensor of dimensions $h_i \times w_i \times d_i$. Lets consider these activation tensors as containers of $h_i \times w_i$ "pixels" with d_i dimensions. Informally, for an input set of real images, we say that the set of layer activations (for any layer L_i) forms a "manifold of interest"." [25].

Table 2. FTC architecture, t – expansion factor, c – output channels, s – stride. The bottleneck is formed by n Inverted Residual Blocks, described in Table 3. The stride of size 2 is applied only in the last Inverted Residual Block. The image is passed to the first convolutional layer, followed by 17 bottlenecks layers for feature extraction. Features from the last two bottleneck layers are concatenated in depth, followed by 3×3 and 5×5 adaptive max-pooling to obtain blocks, which reflects input image sub-regions. Each block is fed into the classifier, which is formed by the two-layer fully-connected network. Softmax provides block-level text probabilities. The final image-level prediction is logical OR function applied to all $3 \times 3 + 5 \times 5 = 34$ block outputs.

Layers	Configurations
Conv2d	c:32, s:2
Bottleneck	$1 \times \{$t:1, c:16, s:1$\}$
Bottleneck	$1 \times \{$t:6, c:24, s:2$\}$
Bottleneck	$2 \times \{$t:6, c:32, s:2$\}$
Bottleneck	$3 \times \{$t:6, c:64, s:2$\}$
Bottleneck	$4 \times \{$t:6, c:96, s:1$\}$
Bottleneck	$3 \times \{$t:6, c:160, s:2$\}$
Bottleneck*	$2 \times \{$t:6, c:160, s:1$\}$
Bottleneck*	$1 \times \{$t:6, c:320, s:1$\}$
Features concatenation in depth from*	
AdaptiveMaxPool2d	3×3, 5×5
Fully-connected	in: 480, out: 1000
Fully-connected	in: 1000, out: 2
Softmax	

Table 3. Inverted Residual Block architecture. The first layer expands features by an expansion factor t and applies ReLU6 non-linear function. Then, depthwise convolution with stride s is applied, followed by ReLU6. At last, pointwise convolution is applied to obtain compressed representation. ReLU6 is a modified rectified linear unit capped to the maximum size of 6.

Input	Operator	Output
$h \times w \times k$	1×1 conv2d, ReLU6	$h \times w \times (tk)$
$h \times w \times tk$	3×3 dwise s, ReLU6	$\frac{h}{s} \times \frac{w}{s} \times (tk)$
$\frac{h}{s} \times \frac{w}{s} \times tk$	linear 1×1 conv2d	$\frac{h}{s} \times \frac{w}{s} \times k'$

has been shown [25] that the information of a channel is inevitably lost when non-linearity transformation collapses the channel. On the other hand, when there is a lot of channels and there is a structure in the activation manifold, the information might be preserved in other channels. This implies that for lower computational cost, it is possible to store the manifold of interest in a lower-dimensional subspace, nevertheless, this low-dimensional subspace must be expanded into higher-dimension space before the non-linear transformation. This is accomplished by an expansion of low-dimensional representation followed by a non-linear function, then a depth-wise convolutional layer and a non-linearity. Finally, higher-dimensional features are compressed back to a lower dimension by point-wise convolution. Details of Inverted Residual Block are described in Table 3.

5 Experiments

This section describes the TextDis[33] benchmark, the training methodology and the evaluation of the inference time and the binary classification performance of FTC. Experiments were conducted on the TextDis and Twitter1M datasets.

TextDis [33]. To the best of our knowledge, TextDis is the only publicly available dataset for natural image text/non-text classification. The splits contain 2000 images for each text and non-text category in the test set, 5302 and 6000 in the training set for text and non-text, respectively. The majority of images contains natural images, however, a small number of scanned documents and digital-born images are present. The dataset contains mostly Chinese text. The annotation of text images is in the form of clockwise-ordered box coordinates.

Training. The model was trained on the 913 002 images of TextDis and Twitter1M training sets; no augmentation was used. The images were resized to 360px resolution. The training was performed at the block-level, with block grid sizes 3×3 and 5×5, i.e. on 34 blocks in total. Block-level annotation y_i was obtained as:

$$y_i = \begin{cases} 1, & \text{if } x \geq \tau \\ 0, & \text{otherwise} \end{cases} \tag{1}$$

where x denotes text area coverage in the block, and τ is the minimum text area coverage. We set 1% as a minimum to label block as text, thus $\tau = 0.01$.

The Adam optimizer [12] was used, the learning rate was set to 10^{-3}. To obtain the text block probability, the network outputs were passed to the softmax function. The cross-entropy objective function was used:

$$\mathcal{L} = -\sum_{i=1}^{n}(y_i log(p_i) + (1 - y_i)log(1 - p_i)) \tag{2}$$

where n denotes the number of blocks (in our experiment, $n = 34$), y_i and p_i denote the ground truth and predicted probability respectively.

Evaluation. We conducted several experiments on both TextDis and Twitter1M datasets. FTC was compared with CNN Coding [33], MDRNN [21], MSP-Net [4], Zhao et al. [35], Gupta et al. [7], i.e. with all text/non-text we were aware of.

The performance was measured by precision, recall, F-measure and the Receiver Operating Characteristic (ROC). To obtain block-level ground truth, we followed the approach described in Eq. 1 with $\tau = 0.01$. The image-level annotation and prediction were obtained by logical OR function across all 34 blocks for each image. The experiments were conducted using Intel Xeon E5-2620 (CPU) and NVIDIA RTX 2080Ti (GPU). The model was implemented in PyTorch 1.4.0 with CUDA version 10.2.

Table 4 shows the qualitative results on TextDis (Fig. 3). The Fast-Text-Classifier outperforms the method proposed by Zhao et al. [35] by 1.2% F-measure, providing higher recall with a slightly increased time cost. For CPU timing, the same CPU as published in [35] was used. Unlike [35], the proposed method is simple, straightforward and trainable directly without the requirement of multiple steps. In comparison with the method of Gupta et al. [7], which is derived directly from EAST [36] detector, our method provides 10 times faster inference time.

Table 5 shows that the performance impact of the introduced Twitter1M dataset is significant. Table 6 depicts model results on the Twitter1M dataset, where we note that the test set was not annotated manually and it thus contains noisy labels. Figure 4 shows the Receiver Operating Characteristic curve of FTC with a 0.97 area under the curve, which demonstrates the quality of the proposed method.

Table 4. Performance on TexDis the test set: recall, precision, F-measure and speed on CPU (where available) and GPU. The proposed Fast-Text-Classifier, abbreviated FTC, was trained on the TextDis + Twitter1M datasets. Note: *results published in [21], **results published in [4].

Method	Recall (%)	Precision (%)	F-measure (%)	Time (CPU) ms	Time (GPU) ms
LLC* [29]	77.4	83.9	80.5	–	300.00
SPP-Net** [8]	83.9	84.1	84.0	–	160.00
CNN Coding [33]	90.3	89.8	90.1	–	460.00
MDRNN [21]	93.3	90.4	91.8	–	90.00
MSP-Net [4]	95.4	93.7	94.6	9530.00	130.00
Gupta et al. [7]	97.8	94.9	96.3	–	100.00
				slow ↑	fast ↓
Zhao et al. [35]	94.0	92.0	93.0	54.03	1.40
FTC	94.6	93.9	94.2	74.80	8.60

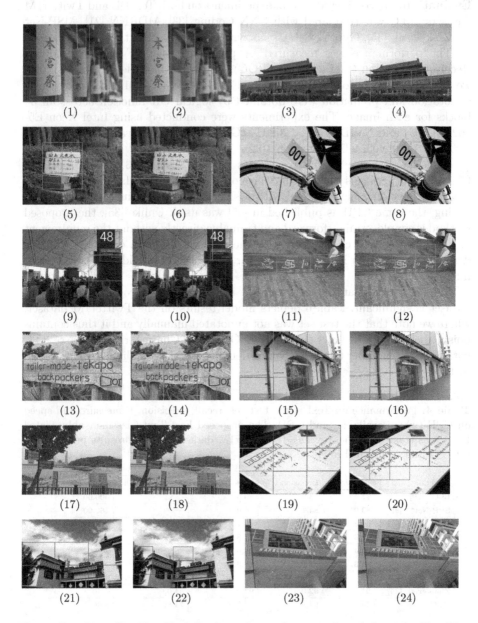

Fig. 3. Fast-Text-Classifier block-level results on images selected from the TextDis dataset, at resolutions 3×3 (odd columns) and 5×5 (even columns). The green and red boxes mark true and false positive predictions, respectively, according to the ground truth TextDis labels. Note that some of the false positives are in fact mislabeled true positives, e.g. in images 2, 9, 10, 18, 19, 20. (Color figure online)

Table 5. Performance of FTC on the TextDis dataset: recall, precision, and F-measure, in % for blocks (rows 3 × 3 and 5 × 5) and at the image image-level – "combined". No augmentation was used in training.

Trained on:	TextDis			TextDis + Twitter1M		
Blocks	Recall	Precision	F-measure	Recall	Precision	F-measure
3 × 3	58.7	60.1	59.4	88.3	75.8	81.5
5 × 5	54.9	52.0	53.4	86.7	71.8	78.6
Combined	84.0	77.4	80.6	94.6	93.9	94.2

Table 6. Performance of FTC on the Twitter1M dataset: recall, precision, and F-measure, in % for blocks (rows 3 × 3 and 5 × 5) and at the image image-level – "combined". No augmentation was used in training.

Trained on:	TextDis			TextDis + Twitter1M		
Blocks	Recall	Precision	F-measure	Recall	Precision	F-measure
3 × 3	61.6	79.0	69.2	92.4	93.1	92.7
5 × 5	58.2	72.3	64.5	92.2	91.2	91.7
Combined	80.3	84.7	82.5	92.7	95.3	94.0

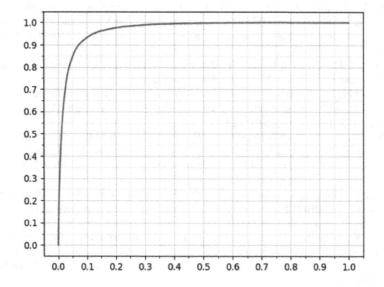

Fig. 4. ROC curve of FTC on the TextDis test set. The area under the curve is 0.97.

6 Conclusion

We have proposed Fast-Text-Classifier (FTC), a method for natural image text/non-text classification. Unlike MSP-Net [4], the proposed method uses

Inverted Residual Blocks for fast and efficient feature extraction; multi-level feature maps are obtained from the last two layers. Only 3×3 and 5×5 blocks are used, block-level features are passed directly to the FCN classifier

The proposed method outperforms the fastest competing method by 1.4% in F-measure and provides $10\times$ faster inference than the most accurate published method. FTC is one of the fastest and most accurate text/non-text classifiers; no method is faster and more accurate at the same time. As a second contribution, we have introduced and made public the Twitter1M dataset for natural image text/non-text classification.

Acknowledgement. This research was supported by Research Center for Informatics (project CZ.02.1.01/-0.0/0.0/16_019/0000765) and CTU student grant SGS20/171/OHK3/-3T/13.

References

1. Alessi, N.G., Battiato, S., Gallo, G., Mancuso, M., Stanco, F.: Automatic discrimination of text images. In: Sampat, N., Motta, R.J., Blouke, M.M., Sampat, N., Motta, R.J. (eds.) Sensors and Camera Systems for Scientific, Industrial, and Digital Photography Applications IV, vol. 5017, pp. 351–359. International Society for Optics and Photonics, SPIE (2003). https://doi.org/10.1117/12.476747
2. Baek, Y., Lee, B., Han, D., Yun, S., Lee, H.: Character region awareness for text detection. In: 2019 IEEE/CVF Conference on Computer Vision and Pattern Recognition (CVPR), pp. 9357–9366, June 2019. https://doi.org/10.1109/CVPR.2019.00959
3. Bahdanau, D., Cho, K., Bengio, Y.: Neural machine translation by jointly learning to align and translate. arXiv 1409, September 2014
4. Bai, X., Shi, B., Chengquan, Z., Cai, X., Qi, L.: Text/non-text image classification in the wild with convolutional neural networks. Pattern Recogn. (2016). https://doi.org/10.1016/j.patcog.2016.12.005
5. Bodla, N., Singh, B., Chellappa, R., Davis, L.S.: Soft-NMS - improving object detection with one line of code. In: 2017 IEEE International Conference on Computer Vision (ICCV), pp. 5562–5570 (2017). https://doi.org/10.1109/ICCV.2017.593
6. Gupta, A., Vedaldi, A., Zisserman, A.: Synthetic data for text localisation in natural images. In: 2016 IEEE Conference on Computer Vision and Pattern Recognition (CVPR), pp. 2315–2324 (2016)
7. Gupta, N., Jalal, A.S.: Text or non-text image classification using fully convolution network (FCN). In: 2020 International Conference on Contemporary Computing and Applications (IC3A), pp. 150–153 (2020). https://doi.org/10.1109/IC3A48958.2020.233287
8. He, K., Zhang, X., Ren, S., Sun, J.: Spatial pyramid pooling in deep convolutional networks for visual recognition. IEEE Trans. Pattern Anal. Mach. Intell. **37**(9), 1904–1916 (2015). https://doi.org/10.1109/TPAMI.2015.2389824
9. He, K., Gkioxari, G., Dollár, P., Girshick, R.B.: Mask R-CNN. In: 2017 IEEE International Conference on Computer Vision (ICCV), pp. 2980–2988 (2017)
10. He, P., Huang, W., He, T., Zhu, Q., Qiao, Y., Li, X.: Single shot text detector with regional attention. In: 2017 IEEE International Conference on Computer Vision (ICCV), pp. 3066–3074 (2017)

11. He, W., Zhang, X., Yin, F., Liu, C.: Deep direct regression for multi-oriented scene text detection. In: 2017 IEEE International Conference on Computer Vision (ICCV), pp. 745–753 (2017). https://doi.org/10.1109/ICCV.2017.87
12. Kingma, D., Ba, J.: Adam: a method for stochastic optimization. In: International Conference on Learning Representations, December 2014
13. Liao, M., Lyu, P., He, M., Yao, C., Wu, W., Bai, X.: Mask textspotter: an end-to-end trainable neural network for spotting text with arbitrary shapes. IEEE Trans. Pattern Anal. Mach. Intell. **PP**, 1 (2019). https://doi.org/10.1109/TPAMI.2019.2937086
14. Liao, M., Shi, B., Bai, X., Wang, X., Liu, W.: Textboxes: a fast text detector with a single deep neural network. In: AAAI (2017)
15. Lin, T., Dollár, P., Girshick, R., He, K., Hariharan, B., Belongie, S.: Feature pyramid networks for object detection. In: 2017 IEEE Conference on Computer Vision and Pattern Recognition (CVPR), pp. 936–944 (2017). https://doi.org/10.1109/CVPR.2017.106
16. Liu, J., Liu, X., Sheng, J., Liang, D., Li, X., Liu, Q.: Pyramid mask text detector. arXiv abs/1903.11800 (2019)
17. Long, J., Shelhamer, E., Darrell, T.: Fully convolutional networks for semantic segmentation. In: 2015 IEEE Conference on Computer Vision and Pattern Recognition (CVPR), pp. 3431–3440 (2015)
18. Long, S., Ruan, J., Zhang, W., He, X., Wu, W., Yao, C.: TextSnake: a flexible representation for detecting text of arbitrary shapes. In: ECCV (2018)
19. Ma, J., et al.: Arbitrary-oriented scene text detection via rotation proposals. IEEE Trans. Multimed. **20**(11), 3111–3122 (2018)
20. McDonnell, M.D., Vladusich, T.: Enhanced image classification with a fast-learning shallow convolutional neural network. In: 2015 International Joint Conference on Neural Networks (IJCNN), pp. 1–7 (2015)
21. Lyu, P., Shi, B., Zhang, C., Bai, X.: Distinguishing text/non-text natural images with multi-dimensional recurrent neural networks. In: 2016 23rd International Conference on Pattern Recognition (ICPR), pp. 3981–3986 (2016)
22. Redmon, J., Divvala, S., Girshick, R., Farhadi, A.: You only look once: unified, real-time object detection. In: 2016 IEEE Conference on Computer Vision and Pattern Recognition (CVPR), pp. 779–788 (2016). https://doi.org/10.1109/CVPR.2016.91
23. Ren, S., He, K., Girshick, R., Sun, J.: Faster R-CNN: towards real-time object detection with region proposal networks. IEEE Trans. Pattern Anal. Mach. Intell. **39**(6), 1137–1149 (2017)
24. Ronneberger, O., Fischer, P., Brox, T.: U-Net: convolutional networks for biomedical image segmentation. In: Navab, N., Hornegger, J., Wells, W.M., Frangi, A.F. (eds.) MICCAI 2015. LNCS, vol. 9351, pp. 234–241. Springer, Cham (2015). https://doi.org/10.1007/978-3-319-24574-4_28
25. Sandler, M., Howard, A., Zhu, M., Zhmoginov, A., Chen, L.: MobileNetV2: inverted residuals and linear bottlenecks. In: 2018 IEEE/CVF Conference on Computer Vision and Pattern Recognition, pp. 4510–4520 (2018)
26. Shi, B., Bai, X., Belongie, S.: Detecting oriented text in natural images by linking segments. In: 2017 IEEE Conference on Computer Vision and Pattern Recognition (CVPR), pp. 3482–3490 (2017). https://doi.org/10.1109/CVPR.2017.371
27. Simonyan, K., Zisserman, A.: Very deep convolutional networks for large-scale image recognition. arXiv:1409.1556, September 2014

28. Tian, Z., Huang, W., He, T., He, P., Qiao, Yu.: Detecting text in natural image with connectionist text proposal network. In: Leibe, B., Matas, J., Sebe, N., Welling, M. (eds.) ECCV 2016. LNCS, vol. 9912, pp. 56–72. Springer, Cham (2016). https://doi.org/10.1007/978-3-319-46484-8_4

29. Wang, J., Yang, J., Yu, K., Lv, F., Huang, T., Gong, Y.: Locality-constrained linear coding for image classification. In: 2010 IEEE Computer Society Conference on Computer Vision and Pattern Recognition, pp. 3360–3367 (2010). https://doi.org/10.1109/CVPR.2010.5540018

30. Wojna, Z., et al.: Attention-based extraction of structured information from street view imagery. In: 2017 14th IAPR International Conference on Document Analysis and Recognition (ICDAR), vol. 01, pp. 844–850 (2017). https://doi.org/10.1109/ICDAR.2017.143

31. Xie, E., Zang, Y., Shao, S., Yu, G., Yao, C., Li, G.: Scene text detection with supervised pyramid context network. In: Proceedings of the AAAI Conference on Artificial Intelligence, vol. 33, pp. 9038–9045, July 2019

32. Yao, C., Bai, X., Sang, N., Zhou, X., Zhou, S., Cao, Z.: Scene text detection via holistic, multi-channel prediction. arXiv, June 2016

33. Zhang, C., Yao, C., Shi, B., Bai, X.: Automatic discrimination of text and non-text natural images. In: 2015 13th International Conference on Document Analysis and Recognition (ICDAR), pp. 886–890 (2015). https://doi.org/10.1109/ICDAR.2015.7333889

34. Zhang, Z., Chengquan, Z., Shen, W., Yao, C., Liu, W., Bai, X.: Multi-oriented text detection with fully convolutional networks. In: 2016 IEEE Conference on Computer Vision and Pattern Recognition (CVPR), pp. 4159–4167, June 2016. https://doi.org/10.1109/CVPR.2016.451

35. Zhao, M., Wang, R., Yin, F., Zhang, X., Huang, L., Ogier, J.: Fast text/non-text image classification with knowledge distillation. In: 2019 International Conference on Document Analysis and Recognition (ICDAR), pp. 1458–1463 (2019)

36. Zhou, X., et al.: East: an efficient and accurate scene text detector. In: 2017 IEEE Conference on Computer Vision and Pattern Recognition (CVPR), pp. 2642–2651 (2017). https://doi.org/10.1109/CVPR.2017.283

Mask Scene Text Recognizer

Haodong Shi⬛, Liangrui Peng$^{(\boxtimes)}$⬛, Ruijie Yan⬛, Gang Yao⬛,
Shuman Han⬛, and Shengjin Wang⬛

Beijing National Research Center for Information Science and Technology
Department of Electronic Engineering, Tsinghua University, Beijing, China
{shd20,yrj17,yg19}@mails.tsinghua.edu.cn,
{penglr,hanshum,wgsgj}@tsinghua.edu.cn

Abstract. Scene text recognition is a challenging sequence modeling problem. In this paper, a novel mask scene text recognizer (MSTR) is proposed to incorporate a supervised learning task of predicting text image mask into a CNN (convolutional neural network)-Transformer framework for scene text recognition. The incorporated mask predicting branch is connected in parallel with the CNN backbone, and the predicted mask is used as attention weights for the feature maps output by the CNN. We investigate three variants of the incorporated mask predicting branches, i.e. a) mask branch which predicts text foreground image mask; b) boundary branch which predicts boundaries of characters in the input images; c) fused mask and boundary branches with different fusion schemes. Experimental results on seven commonly used scene text recognition datasets show that our method with fused mask and boundary branches has outperformed previous state-of-the-art methods.

Keywords: Scene text recognition · Text image mask · Character image boundary · Feature map · Attention weight

1 Introduction

Scene text recognition is one of the key research topics in optical character recognition (OCR) research field. With the emergence of deep learning technologies, the research community has witnessed significant advancements in scene text recognition tasks [30,31,45], especially for recognizing irregular text with complex backgrounds in natural scene images [19,22,31,42].

In deep learning based scene text recognition frameworks, convolutional neural network (CNN) is usually used to convert scene text images into feature maps. However, the commonly used CNN does not distinguish the foreground and the background of scene text image, thus cannot fully utilize the characteristics of the text foreground regions.

Research on biological mechanism of human early visual system has shown that segmentation of scene images into foreground and background is an initial step of visual content identification [24]. In both traditional machine learning [34] and deep learning [3,9] perspectives, foreground detection is important.

© Springer Nature Switzerland AG 2021
J. Lladós et al. (Eds.): ICDAR 2021, LNCS 12824, pp. 33–48, 2021.
https://doi.org/10.1007/978-3-030-86337-1_3

Especially, Mask R-CNN [9] which introduces an additional mask branch for predicting foreground regions in images has achieved tremendous success in object detection and related tasks. Boundary-preserving Mask R-CNN [3] can further improve the performance of object detection. As far as we know, although Mask R-CNN has been adopted in scene text detection [40], no research work has been reported to incorporate mask branches in deep learning based frameworks for scene text recognition.

Inspired by Mask R-CNN, we propose to design a multi-task neural network framework for scene text recognition, including a basic sequence modeling task and an additional text image mask prediction task. We call this method as mask scene text recognizer (MSTR). A CNN-Transformer framework is adopted for the sequence modeling task. A mask predicting branch is introduced to convert the feature maps output by the CNN into the foreground mask of the input scene text instance. The predicted mask is used as attention weights for the feature maps output by the CNN.

We investigate three variants of the mask predicting branches, i.e. a) mask branch, which predicts text foreground image mask; b) boundary branch, which predicts boundaries of characters in the input images; c) fused mask and boundary branches, which predict the text foreground image mask and the boundaries of characters in the input images simultaneously. For the fusion scheme of the mask and boundary branches, different numbers of shortcut connections to add mask features to boundary features are explored.

In order to compare our method with other state-of-the-art methods, we follow the evaluation protocol to train our model on two public synthetic datasets: the MJSynth [11] dataset and the SynthText [8] dataset, then test the model on seven public English scene text datasets (IIIT5k, SVT, IC03, IC13, IC15, SVTP, and CUTE).

One problem to implement the supervised learning for mask prediction is that there is no groundtruth for text image mask and character image boundary in publicly available scene text datasets. To solve this problem, we use the text transcriptions in the training set and true type fonts to generate the pseudo-groundtruth of text image masks.

The experimental results on seven public scene text datasets show that MSTR with feature maps weighted by the fused predicted text image masks and character image boundaries has achieved better performance compared with previous state-of-the-art methods.

To summarize, the main contributions of this paper are as follows:

- A novel scene text recognizer with multi-task-learning strategy by adding additional mask branch to predict text image mask is proposed.
- The predicted text image mask is used as attention weights for feature maps, which help the model to focus on the features in foreground area.
- Different schemes for adding mask predicting branches are further explored, including mask branch, boundary branch, and the fusion of mask and boundary branches.

- Experiments on commonly used scene text recognition datasets show that our method has outperformed previous state-of-the-art methods.

2   Related Work

2.1   Deep Learning Based Scene Text Recognition

Deep learning based scene text recognition approaches can be classified into two categories. The first category of approaches is segmentation-based schemes [12, 36,37,43,44], which needs a character detector to localize each character, then uses a classifier for character recognition. Additional character-level annotations are needed for training the character detector. The second category of approaches is sequence modeling based schemes, including recurrent neural network (RNN) with connectionist temporal classification (CTC) decoding [7,18,30] and the attention-based encoder-decoder framework [1,2,4,15,30,46]. For attention-based encoder-decoder framework, Transformer model [35] with self-attention mechanism has shown promising performance, which is adopted as the baseline model in our paper.

2.2   Mask R-CNN for Object Detection

Mask R-CNN [9] is a successful object detection and instance segmentation framework. In addition to the classification and bounding box regression branches, Mask R-CNN adds a mask branch to predict the foreground mask of each detected instance, which brings a significant performance improvement. Mask R-CNN has already been used in scene text detection [40]. To our knowledge, it is the first time for our paper to propose a scene text recognizer with mask branch.

Boundary-preserving Mask R-CNN [3] is proposed to leverage boundary information to achieve better instance segmentation performance. It offers an upgrade of mask branch, which has two interfused parts to learn boundaries and masks simultaneously. As learning boundary-preserving mask is effective for object detection tasks, we borrow this idea to our scene text recognizer.

3   Methodology

3.1   System Framework

The system framework of MSTR is shown in Fig. 1, which includes a feature extraction module with mask predicting branches and an encoder-decoder model.

For feature extraction module, a U-Shaped CNN with additional mask and boundary branches is shown in Fig. 1. The mask and boundary branches are introduced to predict the foreground mask image and character boundaries for an input image. The outputs of the mask and boundary branches are further used as spatial-based attention weights for feature maps output by the CNN.

For the encoder-decoder model, we use a Transformer [35] for sequence modeling of scene text recognition task.

In the training stage, we use cross entropy loss for the task of sequence modeling and binary cross entropy loss for the tasks of predicting the foreground mask image and character boundaries.

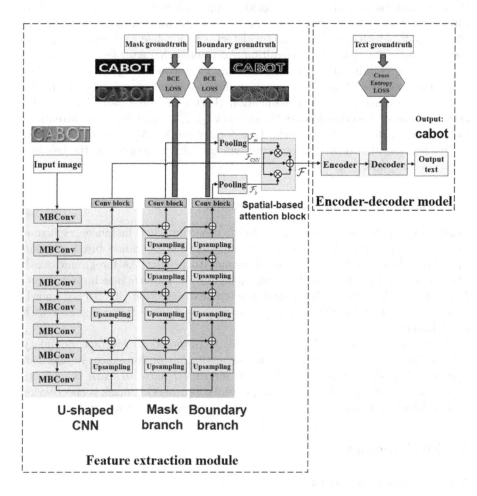

Fig. 1. System framework

3.2   Feature Extraction

U-Shaped CNN. The feature extractor in the proposed system framework is based on EffficientNet [33] which contains seven mobile inverted bottleneck convolutional blocks (MBConv blocks) [29,32]. The MBConv block is an Inverted Residual Block with depthwise separable convolutions originally used in MobileNetV2 [29].

We modify the EfficientNet-B3 [33] into a U-shaped CNN structure by adding an upsampling branch, as is shown in Fig. 1.

Mask Predicting Branches. The mask predicting branches are parallel with the upsampling branch in the U-shaped CNN. We propose three variants of mask predicting branches, i.e. a) mask branch which predicts text foreground image mask; b) boundary branch which predicts the boundary of character images; c) fused mask and boundary branches, which predict the text foreground image mask and the boundaries of characters in the input images simultaneously.

The mask or boundary branch has the similar network structure with the upsampling branch in the U-shaped CNN. Compared with the upsampling branch in U-shaped CNN, two additional upsampling steps are added in the mask branch and boundary branch respectively. The Conv blocks used after the upsampling steps in Fig. 1 are all three-layer CNNs.

The final weighted feature maps $\mathcal{F}$ can be calculated by using the output of the U-shaped CNN $\mathcal{F}_{CNN}$, the predicted text image mask $\mathcal{F}_m$, and the predicted character image boundary $\mathcal{F}_b$, such as

$$\mathcal{F} = \mathcal{F}_{CNN} \times (1 + \lambda_m \mathcal{F}_m + \lambda_b \mathcal{F}_b) \tag{1}$$

where λ_m or λ_b is a hyperparameter, which is set to 1 if corresponding branch is used, or set to 0 if corresponding branch is not used.

Fusion Schemes for Mask Predicting Branches. When mask branch and boundary branch are used simultaneously, we can further explore different schemes of fusion shortcut connections to add mask features to boundary features. The different configurations of the fusion schemes are shown in Fig. 2.

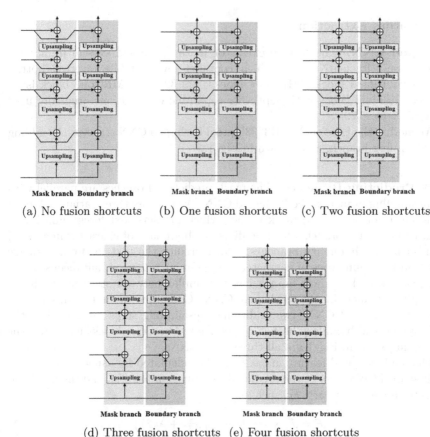

Fig. 2. Comparison with different number of fusion shortcut connections for adding mask features to boundary features.

Loss Functions for Mask Predicting Branches. The loss functions of mask branch $\mathcal{L}_{mask}$ and boundary branch $\mathcal{L}_{boundary}$ are pixel-level binary cross-entropy losses for each mask and boundary instance. The calculation of $\mathcal{L}_{mask}$ is shown in Eq. (2).

$$\mathcal{L}_{mask} = -\frac{1}{MN}\sum_{i=1}^{M}\sum_{j=1}^{N}(q_{ij}log\ p_{ij} + (1-q_{ij})log\ (1-p_{ij})) \qquad (2)$$

where M and N are the height and width of the predicted text image mask respectively. p_{ij} is the predicted pixel value at the i-th row and j-th column. q_{ij} is the groundtruth of the corresponding pixel in the text image mask. The calculation of $\mathcal{L}_{boundary}$ is similar to $\mathcal{L}_{mask}$.

If the scheme of fused mask and boundary branches is used, both $\mathcal{L}_{mask}$ and $\mathcal{L}_{boundary}$ are used for training.

In order to use the same evaluation protocol to compare our model with other state-of-the art methods, our model is trained on two public synthetic datasets: the MJSynth (MJ) [11] and SynthText (ST) [8] datasets, which contain more than 16 million synthetic scene text word images in total. As no groundtruth for text image masks is available in these two datasets, we produce the pseudo-groundtruth of binary mask for each image according to the transcripts of the datasets and selected true type fonts. We use the binarized images of the self-generated samples as the groundtruth for text image masks. We further use Canny operator to get the boundary of character images for each samples.

3.3   Encoder-Decoder Model

Before the feature maps are input into the encoder-decoder model, the feature maps are converted into feature vectors after adding positional encoding [35].

The encoder-decoder Transformer [35] framework with self-attention mechanism is used. The encoder converts feature vectors into hidden representations. To utilize the temporal dependencies of output texts from both forward and backward directions, a bi-directional decoder is adopted. The cross-entropy loss function is defined in Eq. (3):

$$\mathcal{L}_{sequence} = -\frac{1}{2T} \sum_{r=1}^{2} \sum_{t=1}^{T} log \ p(y_t^{(r)}|I) \tag{3}$$

where r denotes the forward and backward decoder direction, T is the sequence length, I is an input image, and y_t $(t = 1, 2......T)$ is the corresponding groundtruth text.

3.4   Multi-task Learning

Multi-task learning methods have been successfully adopted in many studies such as [3,6,16,17,39]. In our MSTR, we introduce additional mask loss and boundary loss for multi-task learning. The loss function is defined in Eq. (4). The weight μ_{mask} or $\mu_{boundary}$ is set to 1 if the corresponding mask or boundary branch is used, or set to 0 if the corresponding mask or boundary branch is not used.

$$\mathcal{L} = \mathcal{L}_{sequence} + \mu_{mask}\mathcal{L}_{mask} + \mu_{boundary}\mathcal{L}_{boundary} \tag{4}$$

where $\mathcal{L}_{sequence}$ is defined in Eq. (3), mask branch loss $\mathcal{L}_{mask}$ is defined in Eq. (2), and the calculation of boundary branch loss $\mathcal{L}_{boundary}$ is similar to $\mathcal{L}_{mask}$.

4 Experiments

In our experiments, we first compare the recognition accuracy of our MSTR with other state-of-the-art methods. To verify the effectiveness of each component in MSTR, we further conduct several ablation studies. Finally, error analysis is given for our method.

4.1 Experiment Setup

Images are normalized to 64 × 256 pixels, and batch size is set to 96 in the training stage. The number of final output channels is 384 in CNN module. Two-layer transformer network structures are used in the encoder-decoder model. ADADELTA [47] is adopted for optimization. Our model is implemented by using the PyTorch [25] framework and trained on an Nvidia Tesla V100 graphics card with 32GB memory.

The baseline model is a network without mask and boundary branches. To shorten the training time, we first train our baseline model from scratch and load pre-trained baseline model to train mask branch and boundary branch for one additional epoch. Data augmentation is also used in our training process such as rotating, stretching and blurring the images.

4.2 Datasets

(a) Original images

(b) Our generated pseudo text image masks

(c) Our generated pseudo character image boundaries

Fig. 3. Examples of scene text images in MJSynth and SynthText datasets with our generated pseudo text image masks and character image boundaries.

We train our model on two public synthetic datasets: the MJSynth [11] dataset and the SynthText [8] dataset, which include more than 16 million English scene text images in total. The corresponding generated pseudo-groundtruths of text image masks and character image boundaries are shown in Fig. 3. Note that the pseudo-groundtruth does not strictly match the original image. We test our model on seven public English scene text datasets which contains four regular text datasets IIIT5k-Words (IIIT5k) [23], Street View Text (SVT) [36],

ICDAR 2003 (IC03) [20], and ICDAR 2013 (IC13) [14], and three irregular text datasets ICDAR2015 (IC15) [13], SVT-Perspective (SVTP) [26], and
CUTE80(CUTE) [28].

4.3 Comparison with State-of-the-Art Methods

We compare the performance of our MSTR with previous state-of-art methods.
Word accuracy is adopted as the performance evaluation index. Table 1 shows
that our model has outperformed previous state-of-the-art methods on both
regular and irregular scene text recognition datasets.

Table 1. Word recognition accuracy (%) across different methods and datasets. MJ
and ST represent the MJSynth [11] dataset and the SynthText [8] dataset. The results
of the best performance are marked in **Bold**.

Model	Training data	IIIT5k	SVT	IC03	IC13	IC15	SVTP	CUTE
CRNN (Shi et al.) [30]	MJ+ST	82.9	81.6	93.1	89.2	64.2	70.0	65.5
AON (Cheng et al.) [5]	MJ+ST	87.0	82.8	91.5	–	68.2	73.0	76.8
MORAN (Luo et al.) [21]	MJ+ST	91.2	88.3	95.0	92.4	68.8	76.1	77.4
DAN (Wang et al.) [38]	MJ+ST	94.3	89.2	95.0	93.9	74.5	80.0	84.4
ESIR (Zhan et al.) [48]	MJ+ST	93.3	90.2	–	91.3	76.9	79.6	83.3
ASTER (Shi et al.) [31]	MJ+ST	93.4	93.6	94.5	91.8	76.1	78.5	79.5
SE-ASTER (Qiao et al.) [27]	MJ+ST	93.8	89.6	–	92.8	80.0	81.4	83.6
MEAN(Yan et al.) [41]	MJ+ST	95.9	94.3	95.9	95.1	79.7	86.8	87.2
AutoSTR (Zhang et al.) [49]	MJ+ST	94.7	90.9	93.3	94.2	81.8	81.7	–
RobustScanner (Yue et al.) [46]	MJ+ST	95.3	88.1	–	94.8	77.1	79.5	90.3
SRN (Yu et al.) [45]	MJ+ST	94.8	91.5	–	95.5	82.7	85.1	87.8
Baseline	MJ+ST	95.6	92.9	95.4	95.9	84.9	85.4	89.6
MSTR	MJ+ST	**96.1**	**94.4**	**96.0**	**96.5**	**85.8**	**88.4**	**91.7**

4.4 Ablation Studies

For ablation studies, we use a smaller dataset by randomly selecting 260k images
from the 16 million images in MJSynth and SynthText. We first compare the
performance of the mask branch and boundary branch. Then we explore the
detailed fusion schemes for the fused mask and boundary branches. Finally, we
compare the performance of our MSTR with different backbones.

Using Mask Branch or Boundary Branch. In this ablation study, only one
mask predicting branch is added to the baseline model. We study the following
three factors:

- Different branches: mask or boundary.
- Whether to use the predicted text image masks or character image boundaries
  as attention weights for feature maps.

– Different down-sampling methods for the predicted text image masks or character image boundaries before using as attention weights: max-pooling or average-pooling.

As shown in Table 2, using the boundary branch can achieve slightly higher accuracy than using the mask branch. Using the predicted text image masks or character image boundaries as the attention weights for feature maps can bring performance improvements. For the down-sampling methods, max-pooling is relatively better than average-pooling.

Table 2. Word recognition accuracy (%) of different variants of MSTR. The baseline model is a network without any additional branch. "att" refers to using masks or boundaries to weight feature maps. Different pooling methods (Avgpooling or Maxpooling) are adopted to reduce the branch output size.

Method	Regular text datasets				Irregular text datasets		
	IIIT5	SVT	IC03	IC13	IC15	SVTP	CUTE
Baseline	77.7	74.8	84.3	85.1	60.3	60.3	54.2
Mask	80.1	77.4	86.3	85.3	65.2	65.3	59.0
Mask-att(Avgpool)	76.8	74.2	82.8	83.1	61.6	61.6	56.6
Mask-att(Maxpool)	79.3	77.7	87.4	86.1	64.6	65.7	**63.2**
Boundary	80.3	76.8	**88.1**	86.5	67.0	66.7	59.0
Boundary-att(Avgpool)	**80.5**	78.1	87.7	86.2	66.3	**67.0**	60.8
Boundary-att(Maxpool)	80.2	**79.4**	87.7	**87.4**	**67.3**	66.2	61.1

Fused Mask and Boundary Branches. We further add mask branch and boundary branch simultaneously, then explore the methods with different numbers of fusion shortcut connections to add mask features to boundary features. The comparison results of the performance with different number of fusion shortcut connections are shown in Table 3. The best result is achieved with two fusion shortcut connections.

Table 3. Word recognition accuracy (%) of different fusion schemes for adding mask features to boundary features.

Number of fusion shortcuts	Regular text datasets				Irregular text datasets		
	IIIT5	SVT	IC03	IC13	IC15	SVTP	CUTE
0	80.4	78.2	87.2	85.0	66.8	67.3	60.4
1	80.2	77.3	87.5	86.1	66.6	67.0	61.1
2	**81.8**	**78.9**	88.2	**87.4**	**68.0**	**69.0**	62.8
3	81.1	77.9	**88.5**	86.3	66.2	67.6	**63.2**
4	81.1	78.2	86.5	85.4	66.0	67.1	62.5

Different Backbone. We further replace the EfficientNet [33] with ResNet-50 [10] to investigate the robustness of our method with different backbone. The results are shown in Table 4. We can see that our method also brings improvements with ResNet-50 [10] backbone in most datasets. Comparing the results in Table 1 and Table 4, using EfficientNet [33] backbone is better than using ResNet-50 [10] backbone.

Table 4. Word recognition accuracy (%) of our MSTR using ResNet-50 [10] as backbone. MJ and ST represent the MJSynth [11] dataset and the SynthText [8] dataset.

Model	Training data	IIIT5k	SVT	IC03	IC13	IC15	SVTP	CUTE
Baseline	MJ+ST	92.3	89.8	94.7	93.7	**82.4**	**83.1**	87.2
MSTR	MJ+ST	**93.1**	**90.6**	**95.2**	**94.3**	82.2	82.3	**87.8**

4.5 Visualization

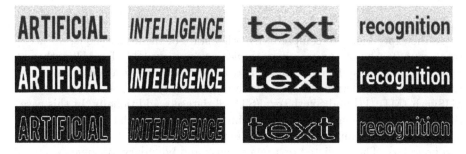

Fig. 4. Examples of self-generated samples and the corresponding groundtruth of text image masks and character image boundaries.

To improve the performance of the mask and boundary prediction, we utilize a self-generated small scale synthetic dataset with mask and boundary groundtruth, as is shown in Fig. 4. In this case, the samples used to train our MSTR have accurate groundtruth of text image mask and boundary. The predicted text image mask and boundary for some examples in the CUTE dataset are shown in Fig. 5 (b) and (c) respectively.

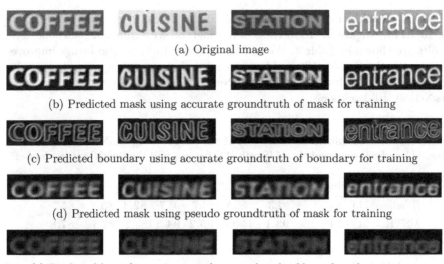

(a) Original image

(b) Predicted mask using accurate groundtruth of mask for training

(c) Predicted boundary using accurate groundtruth of boundary for training

(d) Predicted mask using pseudo groundtruth of mask for training

(e) Predicted boundary using pseudo groundtruth of boundary for training

Fig. 5. Examples of images in the test set (CUTE) and corresponding predicted text image masks and character image boundaries.

In the experiments to compare with other state-of-the-art methods, the accurate groundtruths of the training samples are not available for MJSynth and SynthText datasets. To overcome this difficulty, we use the transcripts of the two datasets to generate the groudtruth of text image mask and boundary for MJSynth (MJ) [11] and SynthText (ST) [8] datasets additionally. In this case, the groundtruth of mask and boundary image does not match the original image accurately, which can be regarded as pseudo-groundtruth of text image mask and boundary. The predicted text image mask and boundary for some examples in the CUTE dataset are show in Fig. 5 (d) and (e) respectively. In practice, it is feasible to use a self-generated synthetic dataset with accurate mask and boundary groundtruth to pretrain the feature extraction module.

4.6   Error Analysis

The recognition errors are roughly categorized into four types: wrongly recognized similar characters, missing characters, character segmentation errors and text instances in complex backgrounds. Some failure cases are shown in Fig. 6. Wrongly recognized similar characters are the most common mistakes. Missing characters often occur at the beginning and end of a word. There is room to improve our MSTR in recognizing characters with touched strokes or text instances in complex background.

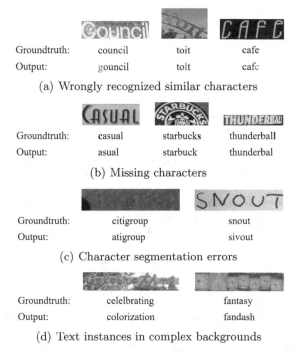

Groundtruth: council   toit   cafe
Output: gouncil   tolt   cafc

(a) Wrongly recognized similar characters

Groundtruth: casual   starbucks   thunderball
Output: asual   starbuck   thunderbal

(b) Missing characters

Groundtruth: citigroup   snout
Output: atigroup   sivout

(c) Character segmentation errors

Groundtruth: celelbrating   fantasy
Output: colorization   fandash

(d) Text instances in complex backgrounds

Fig. 6. Failure cases of MSTR.

5   Conclusion

In this paper, we propose a novel MSTR, which utilizes text image mask and character image boundary information to facilitate text recognition in natural scene images. Experiments on seven public English scene text datasets show that our model has outperformed previous state-of-the-art methods on both regular and irregular scene text images. In our future work, we will explore more possible schemes of multi-task learning for recognizing low quality scene text images.

Acknowledgements. The authors thank Jie Li, Yuqi Zhang and Ruixue Zhang (Shanghai Pudong Development Bank) for useful discussions. The authors also thank the anonymous reviewers for their valuable comments to improve the quality of the paper. This work was supported by National Key R&D Program of China and a grant from the Institute for Guo Qiang, Tsinghua University.

References

1. Bahdanau, D., Cho, K., Bengio, Y.: Neural machine translation by jointly learning to align and translate. arXiv preprint arXiv:1409.0473 (2014)
2. Bai, F., Cheng, Z., Niu, Y., et al.: Edit probability for scene text recognition. In: CVPR, pp. 1508–1516 (2018)

3. Cheng, T., Wang, X., Huang, L., Liu, W.: Boundary-preserving mask R-CNN. In: Vedaldi, A., Bischof, H., Brox, T., Frahm, J.-M. (eds.) ECCV 2020. LNCS, vol. 12359, pp. 660–676. Springer, Cham (2020). https://doi.org/10.1007/978-3-030-58568-6_39

4. Cheng, Z., Bai, F., Xu, Y., et al.: Focusing attention: towards accurate text recognition in natural images. In: ICCV, pp. 5076–5084 (2017)

5. Cheng, Z., Xu, Y., Bai, F., et al.: AON: towards arbitrarily-oriented text recognition. In: CVPR, pp. 5571–5579 (2018)

6. Dong, D., Wu, H., He, W., Yu, D., Wang, H.: Multi-task learning for multiple language translation. In: IJCNLP, pp. 1723–1732 (2015)

7. Graves, A., Fernández, S., Gomez, F., et al.: Connectionist temporal classification: labelling unsegmented sequence data with recurrent neural networks. In: ICML, pp. 369–376 (2006)

8. Gupta, A., Vedaldi, A., Zisserman, A.: Synthetic data for text localisation in natural images. In: CVPR, pp. 2315–2324 (2016)

9. He, K., Gkioxari, G., Dollár, P., Girshick, R.: Mask R-CNN. In: ICCV, pp. 2961–2969 (2017)

10. He, K., Zhang, X., Ren, S., Sun, J.: Deep residual learning for image recognition. In: CVPR, pp. 770–778 (2016)

11. Jaderberg, M., Simonyan, K., Vedaldi, A., et al.: Synthetic data and artificial neural networks for natural scene text recognition. arXiv preprint arXiv:1406.2227 (2014)

12. Jaderberg, M., Vedaldi, A., Zisserman, A.: Deep features for text spotting. In: Fleet, D., Pajdla, T., Schiele, B., Tuytelaars, T. (eds.) ECCV 2014. LNCS, vol. 8692, pp. 512–528. Springer, Cham (2014). https://doi.org/10.1007/978-3-319-10593-2_34

13. Karatzas, D., Gomez-Bigorda, L., Nicolaou, A., et al.: ICDAR 2015 competition on robust reading. In: ICDAR, pp. 1156–1160 (2015)

14. Karatzas, D., Shafait, F., Uchida, S., et al.: ICDAR 2013 robust reading competition. In: ICDAR, pp. 1484–1493 (2013)

15. Lee, C.Y., Osindero, S.: Recursive recurrent nets with attention modeling for OCR in the wild. In: CVPR, pp. 2231–2239 (2016)

16. Liu, P., Qiu, X., Huang, X.: Adversarial multi-task learning for text classification. arXiv preprint arXiv:1704.05742 (2017)

17. Liu, S., Johns, E., Davison, A.J.: End-to-end multi-task learning with attention. In: CVPR, pp. 1871–1880 (2019)

18. Liu, W., Chen, C., Wong, K.Y.K., et al.: STAR-Net: a spatial attention residue network for scene text recognition. In: BMVC, p. 7 (2016)

19. Liu, Y., Chen, H., Shen, C., He, T., Jin, L., Wang, L.: ABCNet: real-time scene text spotting with adaptive Bezier-curve network. In: CVPR, pp. 9809–9818 (2020)

20. Lucas, S.M., Panaretos, A., Sosa, L., et al.: ICDAR 2003 robust reading competitions: entries, results, and future directions. IJDAR 7(2–3), 105–122 (2005)

21. Luo, C., Jin, L., Sun, Z.: MORAN: a multi-object rectified attention network for scene text recognition. Pattern Recogn. 90(2–3), 109–118 (2019)

22. Lyu, P., Yang, Z., Leng, X., et al.: 2D attentional irregular scene text recognizer. arXiv preprint arXiv:1906.05708 (2019)

23. Mishra, A., Alahari, K., Jawahar, C.: Scene text recognition using higher order language priors. In: BMVC, pp. 1–11 (2012)

24. Papale, P., et al.: Foreground enhancement and background suppression in human early visual system during passive perception of natural images. bioRxiv: 109496 (2017)

25. Paszke, A., Gross, S., Massa, F., et al.: PyTorch: an imperative style, high-performance deep learning library. In: NeurIPS, pp. 8024–8035 (2019)
26. Phan, T.Q., Shivakumara, P., Tian, S., et al.: Recognizing text with perspective distortion in natural scenes. In: ICCV, pp. 569–576 (2013)
27. Qiao, Z., Zhou, Y., Yang, D., Zhou, Y., Wang, W.: Seed: semantics enhanced encoder-decoder framework for scene text recognition. In: CVPR, pp. 13528–13537 (2020)
28. Risnumawan, A., Shivakumara, P., Chan, C.S., et al.: A robust arbitrary text detection system for natural scene images. Expert Syst. Appl. **41**(18), 8027–8048 (2014)
29. Sandler, M., Howard, A., Zhu, M., Zhmoginov, A., Chen, L.C.: MobileNetV2: inverted residuals and linear bottlenecks. arXiv:1801.04381 (2018)
30. Shi, B., Bai, X., Yao, C.: An end-to-end trainable neural network for image-based sequence recognition and its application to scene text recognition. IEEE Trans. Pattern Anal. Mach. Intell. **39**(11), 2298–2304 (2017)
31. Shi, B., Yang, M., Wang, X., et al.: ASTER: an attentional scene text recognizer with flexible rectification. IEEE Trans. Pattern Anal. Mach. Intell. **41**(9), 2035–2048 (2019)
32. Tan, M., Chen, B., Pang, R., et al.: MNASNET: platform-aware neural architecture search for mobile. In: CVPR, pp. 2820–2828 (2019)
33. Tan, M., Le, Q.V.: EfficientNet: rethinking model scaling for convolutional neural networks. arXiv preprint arXiv:1905.11946 (2019)
34. Thakore, D.G., Trivedi, A.: Prominent boundaries and foreground detection based technique for human face extraction from color images containing complex background. In: ICVGIP, pp. 15–20 (2011)
35. Vaswani, A., Shazeer, N., Parmar, N., et al.: Attention is all you need. In: NIPS, pp. 5998–6008 (2017)
36. Wang, K., Babenko, B., Belongie, S.: End-to-end scene text recognition. In: ICCV, pp. 1457–1464 (2011)
37. Wang, T., Wu, D.J., Coates, A., Ng, A.Y.: End-to-end text recognition with convolutional neural networks. In: ICPR, pp. 3304–3308 (2012)
38. Wang, T., Zhu, Y., Jin, L., et al.: Decoupled attention network for text recognition. arXiv preprint arXiv:1912.10205 (2019)
39. Wu, R., Feng, M., Guan, W., Wang, D., Lu, H., Ding, E.: A mutual learning method for salient object detection with intertwined multi-supervision. In: CVPR, pp. 8150–8159 (2019)
40. Xiao, S., Peng, L., Yan, R., An, K., Yao, G., Min, J.: Sequential deformation for accurate scene text detection. In: Vedaldi, A., Bischof, H., Brox, T., Frahm, J.-M. (eds.) ECCV 2020. LNCS, vol. 12374, pp. 108–124. Springer, Cham (2020). https://doi.org/10.1007/978-3-030-58526-6_7
41. Yan, R., Peng, L., Xiao, S., Yao, G., Min, J.: MEAN: multi-element attention network for scene text recognition. In: ICPR, pp. 6850–6857 (2020)
42. Yang, X., He, D., Zhou, Z., Kifer, D., Giles, C.L.: Learning to read irregular text with attention mechanisms. In: IJCAI, pp. 3280–3286 (2017)
43. Yao, C., Bai, X., Liu, W.: A unified framework for multioriented text detection and recognition. IEEE Trans. Image Process. **23**(11), 4737–4749 (2014)
44. Yao, C., Bai, X., Shi, B., Liu, W.: Strokelets: a learned multi-scale representation for scene text recognition. In: CVPR, pp. 4042–4049 (2014)
45. Yu, D., Li, X., Zhang, C., et al.: Towards accurate scene text recognition with semantic reasoning networks. In: CVPR, pp. 12113–12122 (2020)

46. Yue, X., Kuang, Z., Lin, C., Sun, H., Zhang, W.: RobustScanner: dynamically enhancing positional clues for robust text recognition. In: Vedaldi, A., Bischof, H., Brox, T., Frahm, J.-M. (eds.) ECCV 2020. LNCS, vol. 12364, pp. 135–151. Springer, Cham (2020). https://doi.org/10.1007/978-3-030-58529-7_9
47. Zeiler, M.D.: ADADELTA: an adaptive learning rate method. arXiv preprint arXiv:1212.5701 (2012)
48. Zhan, F., Lu, S.: ESIR: end-to-end scene text recognition via iterative image rectification. In: CVPR, pp. 2059–2068 (2019)
49. Zhang, H., Yao, Q., Yang, M., Xu, Y., Bai, X.: AutoSTR: efficient backbone search for scene text recognition. arXiv:2003.06567 (2020)

Rotated Box Is Back: An Accurate Box Proposal Network for Scene Text Detection

Jusung Lee[ID], Jaemyung Lee[ID], Cheoljong Yang[ID], Younghyun Lee[ID], and Joonsoo Lee[✉][ID]

Vision AI Lab, AI Center, NCSOFT, Gyeonggi-do, Republic of Korea
{jusunglee,jaemyunglee,cjyang,younghyunlee,jslee509}@ncsoft.com

Abstract. Scene text detection is a challenging task because it must be able to handle text in various fonts and from various perspective views. This makes it difficult to use rectangular bounding boxes to detect text locations accurately. To detect multi-oriented text, rotated bounding box-based methods have been explored as an alternative. However, they are not as accurate for scene text detection as rectangular bounding box-based methods. In this paper, we propose a novel region-proposal network to suggest rotated bounding boxes and an iterative region refinement network for final scene text detection. The proposed region-proposal network predicts rotated box candidates from pixels and anchors, which increases recall by creating more candidates around the text. The proposed refinement network improves the accuracy of scene text detection by correcting the differences in the locations between the ground truth and the prediction. In addition, we reduce the backpropagation time by using a new pooling method called rotated box crop and resize pooling. The proposed method achieved state-of-the-art performance on ICDAR 2017, that is, an f-score of 75.0% and competitive results with f-scores of 86.9% and 92.4% on ICDAR 2015 and ICDAR 2013, respectively. Furthermore, our approach achieves a significant increase in performance over previous methods based on rotated bounding boxes.

Keywords: Scene text detection · Multi-oriented text detection · Multi-lingual text detection

1 Introduction

Scene text detection is in high demand in industrial applications, such as automation, product search, and document analysis. Owing to the development of deep learning-based approaches, scene text detection has undergone significant improvements, and the algorithms have come closer to commercialization. However, the use of scene text detection in the real world involves challenges. Scene image characteristics, such as text size, text shape, camera perspective, and multi-lingual text, make handling scene text detection difficult.

© Springer Nature Switzerland AG 2021
J. Lladós et al. (Eds.): ICDAR 2021, LNCS 12824, pp. 49–63, 2021.
https://doi.org/10.1007/978-3-030-86337-1_4

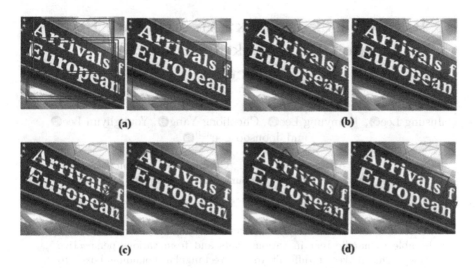

Fig. 1. Examples of the limitations of text detection methods with different bounding box shapes. For each method, the left image shows the candidate bounding boxes and the right image shows the final detection results. (a) Rectangular-based, (b) rbox-based, (c) qbox-based, and (d) proposed methods.

To address these problems, box-based text detection methods using various shapes such as a rectangular box, rotated box (rbox), or quad box (qbox) have been proposed. Methods that employ a rectangular bounding box with instance segmentation [1–3] are commonly used and achieve state-of-the-art performance on several datasets. However, as shown in Fig. 1(a), this approach has fundamental problems because of the limits of the non-maximum suppression of candidate rectangular bounding boxes. Hence, text detection methods based on rboxes have been proposed to overcome these limitations [4–6]. [4] introduced a rotated anchor box with a skew intersection over union (IoU). A rotated anchor box with a skew IoU can effectively deal with text at multiple orientations. However, it fails to predict the rotation parameter, as shown in Fig. 1 (b), because this task is difficult. Methods based on qboxes [5,6], which are similar to rboxes, have also been proposed. Because a qbox requires more parameters than an rbox, parameter estimation is more difficult, and this approach can produce inaccurate results, as shown in Fig. 1 (c). As these examples demonstrate, although the consideration of a large number of parameters facilitates greater shape representation, achieving high performance is difficult owing to the high complexity of learning.

In this paper, we propose a novel rbox-based text detection method that both handles multi-oriented text and detects it accurately. To achieve high accuracy in text detection while using the rbox, we propose the multi-proposal network (MPN) and rbox refinement network (RRN). The MPN increases the variety of candidate text bounding boxes, and the RRN refines inaccurate bounding boxes and removes false positives. The MPN predicts candidate rboxes using an anchor- and pixel-based methods. The anchor-based method generally has a higher precision value because it can detect the text area accurately through predefined anchors. In contrast, by detecting the text area based on the center-pixel, the pixel-based method is robust in detecting overlapped texts and generally has a higher recall value. To improve the performance by utilizing both methods, MPN combines these two methods. The RRN predicts the offset of the candidate rboxes from the ground truth and classifies whether the boxes enclose text or non-text regions. The iterative rbox refinement identifies bounding boxes that are more accurate and closer to the ground truth. We also propose rbox crop and resize (RCR) pooling, which rotates cropped and resized feature maps. RCR pooling is faster during training than rotational region-of-interest (RRoI) pooling but has an equivalent f-measure score.

The contributions of this work are summarized as follows:

– We propose a novel rbox-based text detection framework that is optimal for text detection in scenes that have text at multiple orientations.
– Thanks to MPN and RRN, our method suppresses inaccurate text bounding boxes and increases recall, which is generally low in existing rbox-based methods.
– The proposed RCR pooling is simple to implement, fast to train, and can replace RRoI pooling. The performance of RCR pooling was almost identical to that of RRoI pooling in our experiments.
– When used with MPN and RRN, the rbox-based text detection method is highly effective on multi-lingual datasets (ICDAR 2017) [7]. Moreover, it performs better than previous rbox-based text detection methods on horizontal (ICDAR 2013) [8], multi-oriented (ICDAR 2015) [9], and multi-lingual (ICDAR 2017) scene text datasets.

2   Related Work

The development of scene text detection has followed three different approaches. Each approach has its advantages and disadvantages. However, progress has been made to address these issues.

Semantic Segmentation-Based Approach. The semantic segmentation-based approach is a general approach for scene text detection. Because per-pixel prediction is more suitable for arbitrary text detection, several methods have been developed based on this approach. [3] successfully identified adjacent text instances by gradually expanding the boundaries of instances based on multiple segmentation results. Their network, the Progressive Scale Expansion Network, can detect arbitrary shape text instances by accurately separating adjacent instances. Despite the many research studies that have attempted to differentiate text instances, the semantic segmentation-based method still struggles to separate text instances from a segmentation map.

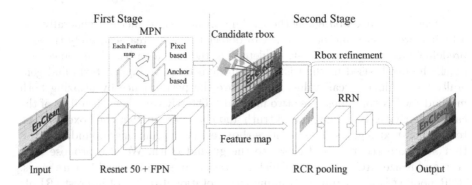

Fig. 2. Overview of the proposed architecture, which consists of a two-stage network. The first stage predicts candidate rboxes using pixel- and anchor-based methods. The second stage classifies whether the rbox is text or non-text and refines the geometry parameters of the candidate rbox from the first-stage network.

Instance Segmentation-Based Approach. One emerging approach for scene text detection is based on instance segmentation. The seminal method is Mask R-CNN [10]. Although Mask R-CNN was originally designed for general object detection, [1] enriched features using global segmentation maps. Additional, more representative features were inserted in the Mask R-CNN pipeline, and their re-scoring strategy improved the method's accuracy. Although Mask R-CNN achieves a promising performance, the non-maximum suppression used to remove overlapped boxes can remove true positives, especially for arbitrary-oriented text.

Regression-Based Approach. The regression-based approach considers text detection as a bounding-box regression problem. [4] laid the foundation for a rotated representation of anchor-based methods. The skew IoU can calculate the exact IoU between two rboxes. RRoI pooling is an appropriate pooling method for proposing rbox candidates. With these methods, [4] presented effective results for arbitrary-oriented text detection. [11] approached text detection using a fusion of a pixel-based rbox and an anchor-based rectangular bounding box to increase diversity; however, as they mentioned, anchor-based rectangles have the "anchor matching" dilemma. [12] proposed iteratively refining the qbox, which enhances the performance of regression-based methods. Because of the refinement of the quad point, a proper corresponding-point matching algorithm is required to match the ground truth and the qbox predictions.

3   Method

3.1   Overview

Our proposed architecture, shown in Fig. 2, is divided into two stages. The first stage is the rbox proposal network MPN. It uses a feature pyramid network

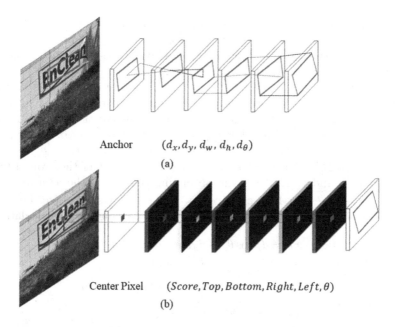

Anchor $(d_x, d_y, d_w, d_h, d_\theta)$

(a)

Center Pixel $(Score, Top, Bottom, Right, Left, \theta)$

(b)

Fig. 3. Proposed MPN. (a) The anchor-based method uses predefined anchors and predicts the deltas between the ground truth and anchors to (x, y, w, h, θ) (b) The pixel-based method predicts the direct location of the ground truth for $(t_p, b_p, r_p, l_p, \theta_p)$ in the center area of the score map.

(FPN) to handle various text scales. The architecture predicts candidate rboxes using two methods: anchor- and pixel-based methods. The second stage classifies the rbox as text or nontext and refines geometry parameters for the rbox candidate by the MPN. The input of the second stage consists of cropped feature maps generated by RCR pooling, which creates fixed-size feature maps when the candidate rboxes have been provided. The second stage outputs iteratively refine geometry parameters that specify the final rbox.

3.2   MPN

The MPN, shown in Fig. 3, predicts rotated bounding-box candidates through two methods: an anchor-based method and a pixel-based method. The anchor-based method predicts the candidate rbox parameters by using predefined anchors and offset parameters, calculated as follows:

$$
\begin{aligned}
x_A &= x_a + w_a d_x, \quad y_A = y_a + h_a d_y, \\
w_A &= w_a e^{d_w}, \quad h_A = h_a e^{d_h}, \quad \theta_A = \theta_a + d_\theta
\end{aligned}
\tag{1}
$$

where $(x_a, y_a, w_a, h_a, \theta_a)$ is a predefined rotated anchor parameter, and $(d_x, d_y, d_w, d_h, d_\theta)$ is an offset parameter predicted by the anchor-based network. x and y denote x- and y-coordinate of the center point respectively; w and h denote the width and height of the rbox, respectively; and θ denotes the angle between the x-axis and the bottom-line of the rbox. To generate predefined anchor, we follow the FPN strategy [13]. Each stage has five three-tuple anchors (scale, ratio, theta), namely, $(1, 1, 0)$, $(1, \frac{1}{2}, 0)$, $(1, 2, 0)$, $(1, \frac{1}{2}, \frac{\pi}{4})$, and $(1, \frac{1}{2}, -\frac{\pi}{4})$. The method predicts the score pair $S = (S_0, S_1)$, where S_0 indicates that the candidate region is a nontext region and S_1 indicates that it is a text region.

In contrast, the pixel-based method predicts the parameters of candidate rbox directly inside a text area. Let $\boldsymbol{p_{cp}} = \{\boldsymbol{x_{cp}, y_{cp}}\}$ be a set of center pixels and $\boldsymbol{x_{cp}} = \{x_{cp} : C_x - d_c \leq x_{cp} \leq C_x + d_c\}$, $\boldsymbol{y_{cp}} = \{y_{cp} : C_y - d_c \leq y_{cp} \leq C_y + d_c\}$, where $\{C_x, C_y\}$ denotes the average coordinates of the ground truth of the text area and d_c is a constant specifying the size of the area in pixels. We set $d_c = \frac{\sqrt{Area_{text}}}{16} + 4$, where $Area_{text}$ is the text instance size. From the set of center pixels, the pixel-based candidate rbox parameters are calculated as follows:

$$x_P = d_x \cos \theta_p - d_y \sin \theta_p + x_{cp}$$
$$y_P = d_y \sin \theta_p + d_y \cos \theta_p + y_{cp}$$
$$w_P = l_p + r_p, \quad h_P = t_p + b_p, \quad \theta_P = \theta_p \tag{2}$$
$$s.t. \quad d_x = \frac{r_p - l_p}{2}, \quad d_y = \frac{b_p - t_p}{2}$$

where $(t_p, b_p, r_p, l_p, \theta_p)$ denote the geometry parameters predicted by the pixel-based network. The score pair of candidate is identical to that of the anchor-based method. the pixel score map has five levels and text instances are assigned by text size in the ranges of $(0, 64^2)$, $(32^2, 128^2)$, $(64^2, 256^2)$, $(128^2, 512^2)$, and $(256^2, 2000^2)$.

3.3   RRN

As in Mask R-CNN, the pooled feature maps based on the first-stage candidate rboxes are input to the RRN. In our case, we pool feature maps to an 8×8 size, and we apply simple convolution to extract the refined parameters and score. The refined rbox parameters are calculated as follows:

$$x_{rf} = x + wr_x, \quad y_{rf} = y + hr_y,$$
$$w_{rf} = we^{r_w}, \quad h_{rf} = he^{r_h}, \quad \theta_{rf} = \theta + r_\theta \tag{3}$$

where (x, y, w, h, θ) denote the candidate rbox parameters that are generated in the first stage, and $(r_x, r_y, r_w, r_h, r_\theta)$ denote the refined parameters predicted by the RRN. Like the MPN in the first stage, RRN also predicts the text score pair $S = (S_0, S_1)$.

3.4   RCR Pooling

RCR pooling (Fig. 4) crops the minimum rectangular area from the rbox, resizes it to a fixed size, and rotates the pooled feature map. Unlike the conventional pooling method, which requires a longer backpropagation time to calculate the sampling point, it quickly calculates the gradients because it only uses scaling and rotation transforms.

Fig. 4. RCR pooling in the image domain.

3.5 Loss Function

The anchor loss L_{anchor} is calculated as follows:

$$L_{anchor} = L_{delta_a} + L_{score_a} \qquad (4)$$

The anchor-offset loss L_{delta_a} is represented as

$$L_{delta_a} = \sum_{i \in (x,y,w',h')} \|d_i^* - d_i\| + \min_{j \in (0,1,2,3)} \|\theta^* - \theta_j\| \qquad (5)$$

where d is the offset parameter in Eq. 1 and d^* is the ground truth of each offset parameter. θ_j is the angle between the x-axis and the j-th side of the rbox. w' and h' denote the width and height of the rbox, respectively, as determined by the parameter j that minimize $\|\theta^* - \theta_j\|$. The score loss L_{score_a} is the same as that used in [4], represented as:

$$L_{score_a} = -\log(p_a) \qquad (6)$$

where $p_a = (p_0, p_1)$ is the probability over (S_0, S_1) computed by the softmax function.

For the losses associated with the pixel-based network loss L_{pixel}, the geometry loss consists of L_{AABB}, L_θ is the same as that of [5], and the score loss L_{score_p} is similar to the anchor-based score loss L_{score_a}, represented as:

$$L_{AABB} = -\log IoU(\hat{R}, R^*), \quad L_\theta = 1 - \cos(\hat{\theta} - \theta^*)$$
$$L_{score_p} = -\log(p_p) \qquad (7)$$

where $\hat{R}$ and R^* denote the predicted axis-aligned rectangular box described in Eq. 2 and its corresponding ground truth respectively. $\hat{\theta}$ and θ^* denote the predicted angle and its ground truth respectively. $p_p = (p_0, p_1)$ is the probability over (S_0, S_1) computed by the pixel-based network. To generate the ground truth for this score, we set the pixels inside the area enclosing the center pixel to 1 and the pixels outside this area to 0. The pixel-based network loss is represented as:

$$L_{pixel} = L_{AABB} + L_\theta + L_{score_p} \qquad (8)$$

When calculating the RRN loss L_{refine}, we use the skew IoU as the matching metric. A value above 0.5 is considered positive, whereas a value below 0.5 is considered negative. Negative samples are excluded from the calculation of delta

Fig. 5. Visualization of the results of each stage. (Top row) Candidate bounding-box results from the MPN. The cyan boxes were proposed by the anchor-based method and the yellow boxes were proposed by the pixel-based method. (Bottom row) Results refined by the RRN. The red boxes indicate the first refinement results, and the green boxes represent the second refinement results. (Color figure online)

loss L_{delta_r}. The score loss L_{score_r} and delta loss L_{delta_r} are calculated via same methods as those of L_{score_a} and L_{delta_a}. For the L_{delta_a} calculation, the anchor offset parameter of d in Eq. 5 can be replaced by the refinement offset parameter of r in Eq. 3. This is represented as follows:

$$L_{refine} = L_{delta_r} + L_{score_r} \qquad (9)$$

Overall, our entire loss function L is defined as follows:

$$L = L_{anchor} + L_{pixel} + L_{refine} \qquad (10)$$

The final loss L is the summation of the anchor-based network losses, pixel-based network losses, and RRN loss. If rbox refinement iterations are performed during training, then L_{refine_i} can be added to the loss.

4   Experiments

4.1   Datasets

ICDAR 2017 Multi-lingual Scene Text. ICDAR 2017 is one of the largest datasets for text detection [7]. It contains multi-lingual, multi-oriented text instances in natural scene images. The dataset has 7,200 training images, 1,800 validation images, and 9,000 test images. The annotation of each text instance consists of the four vertices of the bounding-box quadrilateral, language of the text, and its transcription. ICDAR 2017 is the most challenging dataset in scene text detection, and many researchers consider the results obtained on this dataset to be an important indicator for comparing methods. In this study, we used the 9,000 training and 1,800 validation images for training.

ICDAR 2015 Incident Scene Text. ICDAR 2015 is a multi-oriented text dataset for scene text detection [9]. Because the images were incidentally captured by Google Glass, the scenes are diverse and contain oriented blurry text instances. The dataset is composed of 1,000 training images and 500 test images. Annotations provide the four vertices of the bounding-box quadrilaterals, similar to ICDAR 2017.

ICDAR 2013 Focused Scene Text. ICDAR 2013 is a natural scene text dataset of horizontal text [8]. It consists of 229 images for training and 223 images for testing. Unlike the first two datasets, the annotation of each text instance is in the form of rectangular bounding-box coordinates composed of the top-left and bottom-right vertices.

4.2 Implementation Details

We used ResNet-50 as a backbone to compare our method with other methods. For ICDAR 2017, we pre-trained SynthText [14] for 200k iterations with warm-up weights taken from the ImageNet pre-trained model. The batch size was two, the optimizer was Adam, and the learning rate was 10^{-4} with a weight decay of 0.94 every 50k iterations. After training SynthText, we trained on the ICDAR 2017 dataset for 350k iterations. To train the iterative refinement network, we added losses L_{refine_1} and L_{refine_2} and trained for an additional 400k iterations. For ICDAR 2015, we fine-tuned our ICDAR 2017 model for 500k iterations with the same learning rate and decay settings. For ICDAR 2013, we used the ICDAR 2017 model, similar to the approach taken in [1].

Cropping, rotating, scaling, flipping, and blur augmentations were used for training the model. We randomly cropped the images to more than half their size and rotated them in the range $-15°$ to $15°$. Multi-scale training was conducted using images with sizes of 448, 512, 768, and 1024. The images were blurred using Gaussian filtering, and horizontal and vertical flipping were also applied to the images.

The inference resolutions for ICDAR 2017, ICDAR 2015, and ICDAR 2013 are (1,600, 2,000), (1,100, 2,000), and (700, 900), respectively, where the resolution is (minimum of the shortest side, maximum of the largest side). The inference batch size is 1, and the GPU was a NVIDIA GTX 2080TI. The inference speed was 6.25 frames per second at a resolution of (1,280, 704).

Table 1. Detection results for the anchor-only, pixel-only, and MPN models with two rbox refinement iterations on ICDAR 2017. F: f-measure, P: precision, R: recall, Anchor: anchor-based method, Pixel: pixel-based method.

Anchor	Pixel	F	P	R
✓		72.8	85.0	63.7
	✓	73.6	80.7	67.7
✓	✓	75.0	83.4	68.2

Table 2. Detection results for different numbers of rbox refinement iterations on ICDAR 2017. RI: number of refinement iterations.

Method	RI	F	P	R
MP	0	68.3	79.8	59.6
MP + RR	1	70.4	82.9	61.3
MP + RR	2	75.0	83.4	68.2

4.3   Ablation Study

Effectiveness of the MPN. We performed three experiments to evaluate the effectiveness of the MPN. First, we assessed the performance when predicting rbox using the pixel- and anchor-based methods, respectively. Thereafter, we compared the achieved performance to that of the proposed MPN-based method. The results are shown in Table 1. The MPN achieves an f-measure that is 2.2% and 1.4% higher than those of the anchor- and pixel-based models, respectively.

Effectiveness of the RRN. The RRN is crucial when datasets with various text sizes and languages are used for evaluation. The rbox refinement both refines a rotated text bounding box and scores the refined text region, considerably increasing the f-measure on ICDAR 2017. In Table 2, two rbox refinement iterations improve the f-measure by 6.7% with respect to no iterations and 4.6% with respect to one iteration.

Evaluation of RRoI Pooling. When θ is large, the pooled features obtained by RCR pooling include more of the extraneous area outside of the text RoI than

Table 3. Effect of the pooling algorithm on the ICDAR 2017 results. TI: training time per iteration when the batch size is two and multi-scale training is applied.

Method	F	P	R	TI(s)
RCR pooling	75.0	83.4	68.2	0.6
RRoI pooling	75.0	82.9	68.5	0.8

Fig. 6. Qualitative results for our method. (a) Detection results from ICDAR 2017. (b) Detection results from ICDAR 2015. (c) Detection results from ICDAR 2013.

those obtained by RRoI pooling, which could degrade performance. To prove that our RCR algorithm is equivalent to RRoI pooling, we re-implemented the RRoI pooling algorithm and used it in our model. The results in Table 3 reveal no difference in the results when RRoI or RCR pooling are used, even though the training speed of RCR pooling is faster.

4.4   Comparisons with Other Methods

We compared our method with previously proposed state-of-the-art rotated bounding box-based text detection methods to show that our method is a substantial improvement over existing approaches. The benchmark datasets ICDAR 2017, ICDAR 2015, and ICDAR 2013 are widely used for measuring text detection performance. Our method achieves state-of-the-art performance on ICDAR 2017 and very competitive results on ICDAR 2015, and ICDAR 2013 when compared with other state-of-the-art methods.

Table 4. Results for rotated bounding box-based text detection methods on ICDAR 2017, ICDAR 2015 and ICDAR 2013.

Method	ICDAR 2017			ICDAR 2015			ICDAR 2013		
	F	P	R	F	P	R	F	P	R
EAST [5]	–	–	–	80.7	83.2	78.3	–	–	–
EAST++ [15]	72.8	80.4	66.6	–	–	–	–	–	–
RRPN [4]	62.3	71.1	55.5	80	84	77	91	95	88
TextBoxes++ [6]	–	–	–	81	87	76	80	86	74
Ours	**75.0**	**83.4**	**68.2**	**86.9**	**89.7**	**84.2**	**92.4**	**95.9**	**89.1**

Performance on ICDAR 2017. In this dataset, our method has an f-measure score that is 12.7% higher than that of RRPN, which is a rotated anchor-based method. Moreover, our method's f-measure score is 2.2% higher than that of EAST [5], which is a pixel-based rotated bounding-box method. The results (Table 4) show that our method outperforms previous methods. Examples of each stage of the detection results are shown in Fig. 5, and Table 5 compares the results of all the state of-the-art methods. Most results were obtained using a ResNet-50 backbone, and single-scale results are reported to fairly compare the methods. Our method obtains an f-measure score of 75%, which is state of the art.

Performance on ICDAR 2015. ICDAR 2015 is a multi-oriented text dataset. It has blurry text instances and several types of noise in the images. Our method has an f-measure score that is 6.9% higher than that of RRPN and 6.2% higher than that of EAST. Our method also achieves a 5.9% improvement over Textboxes++ (Table 4). In the results in Table 5, the performance of our method is 2.8% lower than that of CCNet, which uses a character-level annotation for Synthtext. Because ICDAR 2015 has several blurry text instances, attaching a word or character recognition branch to the main network, as FOTS, CCNet, and Mask TextSpotter do, improves performance with respect to detection-only models. Our method achieves an f-measure score that is slightly lower (by 0.7%) than that of Pixel-Anchor, which is a state-of-the-art detection-only method.

Performance on ICDAR 2013. On this dataset, our method has a slightly better f-measure than RRPN (1.4% higher) and a huge performance increase (12.4%) with respect to TextBoxes++ (Table 4). The results in Table 5 reveal that our method obtains slightly higher f-measure scores than the method based on rectangle bounding boxes with instance segmentation. These results demonstrate that a rotated bounding box is still effective on horizontal text and can perform slightly better than the rectangle bounding box-based methods. Qualitative results on each dataset are shown in Fig. 6.

Table 5. Detection results on ICDAR 2017, ICDAR 2015, and ICDAR 2013. *: a different backbone was used. **: character annotations in the synthetic data were used in training. The IC13 metric was used for the ICDAR 2013 evaluation.

Type	Method	ICDAR 2017			ICDAR 2015			ICDAR 2013		
		F	P	R	F	P	R	F	P	R
Reg.	AF-RPN [16]	70	75	66	86	89	83	92	94	90
Reg.	FOTS [17]	67.2	80.9	57.5	87.9	91.0	85.1	88.2	–	–
Reg.	Pixel-Anchor [11]	68.1	79.5	59.5	87.6	88.3	87.0	–	–	–
Reg.	LOMO [12]	68.5	78.8	60.6	87.2	91.3	83.5	–	–	–
Reg.	RRD [18]	–	–	–	82.2	85.6	79.0	81	88	75
Reg.	ContourNet [19]	–	–	–	86.9	87.6	86.1	–	–	–
Reg.	FEN* [20]	–	–	–	–	–	–	91.3	93.6	89.1
Seg.	CRAFT** [21]	73.9	80.6	68.2	86.9	89.8	84.3	–	–	–
Seg.	PSENet [3]	70.8	73.7	68.2	85.7	86.9	84.5	–	–	–
Seg.	PSENet* [3]	72.2	75.3	69.2	–	–	–	–	–	–
Seg.	GNNets [22]	–	–	–	85.1	88.0	82.3	–	–	–
Seg.	GNNets* [22]	74.5	79.6	70.0	88.5	90.4	86.7	–	–	–
Seg.	DB-ResNet-50 [23]	74.7	83.1	67.9	87.3	91.8	83.2	–	–	–
Seg.	TextField [24]	–	–	–	82.4	84.3	80.5	–	–	–
Seg.	PixelLink* [25]	–	–	–	83.7	85.5	82.0	84.5	86.4	83.6
Seg.	SAE [26]	–	–	–	86.6	88.3	85.0	–	–	–
Seg.	Xue et al. [27]	–	–	–	–	–	–	87.4	87.8	86.9
Reg. & Seg.	CCNet** [28]	73.4	77.0	**70.1**	**89.7**	91.1	**88.3**	–	–	–
Inst Seg.	SPCNet [1]	70.0	73.4	66.9	87.2	88.7	85.8	92.1	93.8	**90.5**
Inst Seg.	SAST [29]	68.7	70.0	67.5	86.9	86.7	87.0	–	–	–
Inst Seg.	FAN [2]	73.3	78.7	68.6	84.9	88.8	81.8	–	–	–
Inst Seg.	Mask TextSpotter** [30]	–	–	–	86.0	**91.6**	81.0	–	–	–
Inst Seg.	IncepText [31]	–	–	–	85.3	90.5	80.6	–	–	–
	Ours	**75.0**	**83.4**	68.2	86.9	89.7	84.2	**92.4**	**95.9**	89.1

5   Conclusion

In this paper, we presented a novel rbox-based scene text detection framework that suppresses inaccurately rotated bounding boxes when given a variety of bounding box candidates. We demonstrated that it achieves substantial increases in performance with respect to previously proposed rotated bounding box-based text detection methods on benchmark datasets. These results show that rotation remains important in text detection.

Acknowledgements. This work was supported by Institute of Information & communications Technology Planning & Evaluation (IITP) grant funded by the Korea government (MSIT) (No. 1711125972, Audio-Visual Perception for Autonomous Rescue Drones).

References

1. Xie, E., Zang, Y., Shao, S., Yu, G., Yao, C., Li, G.: Scene text detection with supervised pyramid context network. In: Proceedings of the AAAI Conference on Artificial Intelligence, vol. 33, pp. 9038–9045 (2019)
2. Huang, Z., Zhong, Z., Sun, L., Huo, Q.: Mask R-CNN with pyramid attention network for scene text detection. In: 2019 IEEE Winter Conference on Applications of Computer Vision (WACV), pp. 764–772. IEEE (2019)
3. Wang, W., et al.: Shape robust text detection with progressive scale expansion network. In: Proceedings of the IEEE Conference on Computer Vision and Pattern Recognition, pp. 9336–9345 (2019)
4. Ma, J., et al.: Arbitrary-oriented scene text detection via rotation proposals. IEEE Trans. Multimed. **20**(11), 3111–3122 (2018)
5. Zhou, X., et al.: East: an efficient and accurate scene text detector. In: Proceedings of the IEEE Conference on Computer Vision and Pattern Recognition, pp. 5551–5560 (2017)
6. Liao, M., Shi, B., Bai, X.: Textboxes++: a single-shot oriented scene text detector. IEEE Trans. Image Process. **27**(8), 3676–3690 (2018)
7. Nayef, N., et al.: ICDAR 2017 robust reading challenge on multi-lingual scene text detection and script identification-RRC-MLT. In: 2017 14th IAPR International Conference on Document Analysis and Recognition (ICDAR), vol. 1, pp. 1454–1459. IEEE (2017)
8. Karatzas, D., et al.: ICDAR 2013 robust reading competition. In: 2013 12th International Conference on Document Analysis and Recognition, pp. 1484–1493. IEEE (2013)
9. Karatzas, D., et al.: ICDAR 2015 competition on robust reading. In: 2015 13th International Conference on Document Analysis and Recognition (ICDAR), pp. 1156–1160. IEEE (2015)
10. He, K., Gkioxari, G., Dollár, P., Girshick, R.: Mask R-CNN. In: Proceedings of the IEEE International Conference on Computer Vision, pp. 2961–2969 (2017)
11. Li, Y., et al.: Pixel-anchor: a fast oriented scene text detector with combined networks. arXiv preprint arXiv:1811.07432 (2018)
12. Zhang, C., et al.: Look more than once: an accurate detector for text of arbitrary shapes. In: Proceedings of the IEEE Conference on Computer Vision and Pattern Recognition, pp. 10552–10561 (2019)
13. Lin, T.Y., Dollár, P., Girshick, R., He, K., Hariharan, B., Belongie, S.: Feature pyramid networks for object detection. In: Proceedings of the IEEE Conference on Computer Vision and Pattern Recognition, pp. 2117–2125 (2017)
14. Gupta, A., Vedaldi, A., Zisserman, A.: Synthetic data for text localisation in natural images. In: Proceedings of the IEEE Conference on Computer Vision and Pattern Recognition, pp. 2315–2324 (2016)
15. ICDAR 2017 competition on multi-lingual scene text detection and script identification. https://rrc.cvc.uab.es/?ch=8&com=evaluation&task=1
16. Zhong, Z., Sun, L., Huo, Q.: An anchor-free region proposal network for faster R-CNN-based text detection approaches. Int. J. Document Anal. Recogn. (IJDAR) **22**(3), 315–327 (2019)
17. Liu, X., Liang, D., Yan, S., Chen, D., Qiao, Y., Yan, J.: FOTS: fast oriented text spotting with a unified network. In: Proceedings of the IEEE Conference on Computer Vision and Pattern Recognition, pp. 5676–5685 (2018)

18. Liao, M., Zhu, Z., Shi, B., Xia, G.S., Bai, X.: Rotation-sensitive regression for oriented scene text detection. In: Proceedings of the IEEE Conference on Computer Vision and Pattern Recognition, pp. 5909–5918 (2018)

19. Wang, Y., Xie, H., Zha, Z.J., Xing, M., Fu, Z., Zhang, Y.: ContourNet: taking a further step toward accurate arbitrary-shaped scene text detection. In: Proceedings of the IEEE/CVF Conference on Computer Vision and Pattern Recognition, pp. 11753–11762 (2020)

20. Zhang, S., Liu, Y., Jin, L., Luo, C.: Feature enhancement network: a refined scene text detector. arXiv preprint arXiv:1711.04249 (2017)

21. Baek, Y., Lee, B., Han, D., Yun, S., Lee, H.: Character region awareness for text detection. In: Proceedings of the IEEE Conference on Computer Vision and Pattern Recognition, pp. 9365–9374 (2019)

22. Xu, Y., et al.: Geometry normalization networks for accurate scene text detection. In: Proceedings of the IEEE International Conference on Computer Vision, pp. 9137–9146 (2019)

23. Liao, M., Wan, Z., Yao, C., Chen, K., Bai, X.: Real-time scene text detection with differentiable binarization. In: AAAI, pp. 11474–11481 (2020)

24. Xu, Y., Wang, Y., Zhou, W., Wang, Y., Yang, Z., Bai, X.: TextField: learning a deep direction field for irregular scene text detection. IEEE Trans. Image Process. **28**(11), 5566–5579 (2019)

25. Deng, D., Liu, H., Li, X., Cai, D.: PixelLink: detecting scene text via instance segmentation. arXiv preprint arXiv:1801.01315 (2018)

26. Tian, Z., et al.: Learning shape-aware embedding for scene text detection. In: Proceedings of the IEEE Conference on Computer Vision and Pattern Recognition, pp. 4234–4243 (2019)

27. Xue, C., Lu, S., Zhan, F.: Accurate scene text detection through border semantics awareness and bootstrapping. In: Proceedings of the European Conference on Computer Vision (ECCV), pp. 355–372 (2018)

28. Xing, L., Tian, Z., Huang, W., Scott, M.R.: Convolutional character networks. In: Proceedings of the IEEE International Conference on Computer Vision, pp. 9126–9136 (2019)

29. Wang, P., et al.: A single-shot arbitrarily-shaped text detector based on context attended multi-task learning. In: Proceedings of the 27th ACM International Conference on Multimedia, pp. 1277–1285 (2019)

30. Lyu, P., Liao, M., Yao, C., Wu, W., Bai, X.: Mask TextSpotter: an end-to-end trainable neural network for spotting text with arbitrary shapes. In: Proceedings of the European Conference on Computer Vision (ECCV), pp. 67–83 (2018)

31. Yang, Q., et al.: IncepText: a new inception-text module with deformable PSROI pooling for multi-oriented scene text detection. arXiv preprint arXiv:1805.01167 (2018)

Heterogeneous Network Based Semi-supervised Learning for Scene Text Recognition

Qianyi Jiang$^{(\boxtimes)}$ (ID), Qi Song (ID), Nan Li (ID), Rui Zhang (ID), and Xiaolin Wei (ID)

Meituan, Beijing, China
{jiangqianyi02,songqi03,linan21,zhangrui36,weixiaolin02}@meituan.com

Abstract. Scene text recognition is an important research topic in both academia and industry. Remarkable progress has been achieved by recent works, which are based on abundant labeled data for model training. Obtaining text images is a relatively easy process, but labeling them is quite expensive. To alleviate the dependence on labeled data, semi-supervised learning which combines labeled and unlabeled data seems to be a reasonable solution, and is proved to be effective for image classification. However, due to the apparent differences between scene text recognition and image classification, there are very few related works which incorporate semi-supervised learning with scene text recognition. To the best of our knowledge, this paper is the first time to propose a semi-supervised learning method for scene text recognition. In our framework, given a text image, two heterogeneous networks are adopted to produce independent predictions from different perspectives, namely, context information and visual representation. Then, a discriminator is introduced for pseudo-labeling and instance selection using the predictions. With these two modules, the generalization performance and the precision of pseudo-labeling are improved. Comprehensive experiments show that our framework leverages the unlabeled data effectively, and brings consistent improvements with different unlabeled data scales on both English and Chinese datasets.

Keywords: Semi-supervised learning · Scene text recognition · Pseudo-labeling

1 Introduction

Scene text recognition has drawn considerable research interest in recent years due to its numerous applications, *e.g.*, shop sign reading, intelligent inspection, and image searching. Benefiting from the development of deep learning, many scene text recognition methods have achieved notable success [20, 22, 26, 32]. It is noteworthy that most state-of-art approaches need a large amount of labeled data for training. The manual annotation of such large-scale data is costly, and there are abundant unlabeled text images on the Internet that are not fully

© Springer Nature Switzerland AG 2021
J. Lladós et al. (Eds.): ICDAR 2021, LNCS 12824, pp. 64–78, 2021.
https://doi.org/10.1007/978-3-030-86337-1_5

utilized. Semi-supervised learning (SSL), which can make good use of the unlabeled data, becomes an appealing paradigm and has been successfully applied in image classification. However, due to the apparent differences between scene text recognition and image classification, the representative SSL methods [19,27,28] have some practical difficulties for being implemented.

Firstly, some SSL approaches for image classification use self-labeling, which regard predictions of unknown data as pseudo-labels and select reliable instances by confidence scores [27,28]. Nevertheless, the pseudo-labels for scene text recognition are much complicated since the sequence-to-sequence recognition is involved. Besides, the confidence measurement of text recognition is much difficult and sometimes unreliable. Secondly, recent SSL methods use consistency regularization under the assumption that the predictions remain invariant to input noise [2,8]. However, scene text recognition is sensitive to disturbances. Some aggressive data augmentations, such as random cropping, shifting, rotation, and geometry transformation cannot be transferred to text recognition as they render the image unreadable.

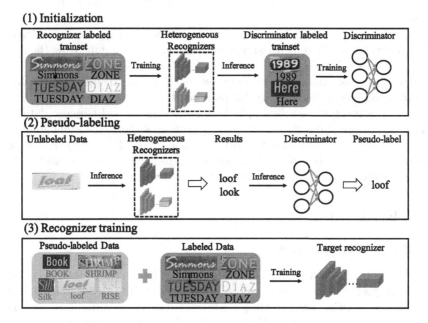

Fig. 1. Illustration of our proposed framework, which consists of three phases. (1) Initialization: we train the heterogeneous networks on part of the labeled data and infer them on the rest for predictions. The discriminator is trained on the prediction results. (2) Pseudo-labeling: The unlabeled data is fed into the trained heterogeneous networks and the discriminator sequentially for pseudo-labeling and instance selection. A working example is given for generating the pseudo-label of "loof". (3) Recognizer training: the target recognizer is trained on the combination of labeled and pseudo-labeled data.

Considering the difficulties mentioned above, in our work, we introduce a novel SSL framework for scene text recognition as shown in Fig. 1. Two heterogeneous networks are adopted to produce independent predictions from different perspectives. Then, a discriminator is employed for pseudo-labeling and instance selection based on the predictions from heterogeneous networks. In detail, the heterogeneous networks include a target recognizer and a surrogate recognizer integrated with different architectures. The target recognizer adopts the attentional sequence recognition to capture context information between characters, while the surrogate recognizer uses CTC decoder [6] and focuses on the visual representations. Instead of directly using the identical results from two recognizers as pseudo-labels, we propose a discriminator that uses two prediction results and the complementary representations to produce confidence scores. Then, by adopting a threshold, valid pseudo-labeled instances are selected as the training samples. In summary, the contributions of this study are as follows:

- We propose a novel SSL framework for scene text recognition based on two key modules, namely, the heterogeneous networks and the discriminator. As far as we know, it is the first time to propose an SSL framework in scene text recognition.
- The heterogeneous networks provide different perspectives and complementary information, thereby avoiding confirmation bias and improving the accuracy of pseudo-labeling.
- We design a discriminator to make full use of the geometric information, probability, and recognition content for pseudo-labeling and instance selection.
- Our SSL framework can be applied on both English and Chinese datasets. The experimental results on different benchmarks show consistent improvements over the models only trained on labeled data. Furthermore, We perform exhaustive experiments that provide empirical data on the relationship between the unlabeled data scale and SSL performance.

2   Related Work

2.1   Scene Text Recognition

Early works of scene text recognition mainly rely on hand-crafted features [18, 29, 31] and the performance of these methods is limited. With the rise of deep learning, recent approaches tend to treat text recognition as a sequence-to-sequence recognition problem. Most of them are based on an asymmetrical encoder-decoder framework and can be categorized into two mainstreams, *i.e.*, Connectionist Temporal Classification (CTC) [6] based and attention-based methods. CTC is originally proposed by Graves *et al.* for speech recognition which removes the need for pre-segmented training data and post-processed outputs. Inspired by this approach, Baoguang *et al.* [21] introduce CTC into scene text recognition by integrating with both convolutional neural network (CNN) and recurrent neural network (RNN). Furthermore, Liu *et al.* [14] present the STAR-Net for achieving better performance on distorted and irregular scene text images. The

STAR-Net adopts a spatial transformer to rectify input text images and utilizes a deeper CNN with residual structures. The attention-based method for sequence recognition is first proposed in [13] which introduces a novel model called R2AM. Afterward, Cheng et al. [5] present the Focusing Attention Network (FAN) to alleviate attention shifts. It should be noted that all the methods mentioned above are fully supervised. The scene text recognition itself is a weakly supervised learning problem in which the character positions in a text line are unknown. Thus, rare text recognition works introduce SSL because of the difficulty of combining both weakly supervised learning and SSL.

2.2 Semi-supervised Learning (SSL)

Recent approaches for SSL mainly rely on two components, pseudo-labeling and consistency regularization. Pseudo-labeling utilizes the model itself to obtain artificial labels from unlabeled data. The generalization performance of the model can be improved when trained by labeled and pseudo-labeled data. Consistency regularization uses unlabeled data with perturbations. The assumption is that the model should output similar predictions for the same input image with different disturbances. Dong et al. [3] propose a network called Tri-net which adopts three modules to exploit unlabeled data. Pseudo-labels are generated from any two of the models, then they are fed to the rest model for refinement. In [28], an SSL pipeline based on a "Teacher-student" paradigm is presented for pseudo-labeling by leveraging an extensive collection of unlabeled images. Qiao et al. [19] introduces the Deep Co-Training framework, which imposes the view difference constraint by training each network to be resistant to the adversarial examples. Berthelot et al. [2] produce an algorithm called MixMatch, which utilizes many tricks for consistency regularization, such as data augmentation and mixing labeled and unlabeled data using MixUp [33]. A method named FixMatch [24] is presented for image classification by combining both pseudo-labeling and consistency regularization.

In scene text recognition, for the aforementioned reason, there are no methods that incorporate SSL. Speech recognition can also be regarded as a sequence-to-sequence recognition problem. Recently some exploratory SSL works [1,4,10] have emerged but the researches results are limited.

3 Our Method

Algorithm 1 delineates our overall SSL framework. We depart from the literature in three phases: initialization, pseudo-labeling, and recognizer training. In detail, first, all the modules are initialized using labeled datasets. Then the unlabeled data is pseudo-labeled and filtered. Finally, the scene text recognizer is trained on the combination of labeled and pseudo-labeled data. We will introduce the details of every component next.

Algorithm 1. Heterogeneous network based semi-supervised framework

---
Input:

    Unlabeled dataset $\mathcal{U} = \{(x_u)\}$;

    Labeled dataset $\mathcal{L} = \{(x_l, y_l)\}$ dividing into recognizer trainset $\mathcal{L}_r$ and discriminator trainset $\mathcal{L}_d$;

    Confidence threshold η;

Output:

    The updated target recognizer $\hat{\Theta}_t$

1: *Recognizer initialization*: training initial target recognizer Θ_t and surrogate recognizer Θ_s on $\mathcal{L}_r$;

2: Inferring Θ_t, Θ_s on $\mathcal{L}_d$. Θ_t outputs recognition results α and conditional probability $\mathbf{p}$. Θ_s outputs α' and $\mathbf{p}'$;

3: *Discriminator initialization*: training discriminator Λ on $(\mathcal{L}_d, \alpha, \mathbf{p}, \alpha', \mathbf{p}')$;

4: *Pseudo-labeling and instance selection*: inferring Θ_t, Θ_s and Λ on $\mathcal{U}$ to generate pseudo-labels. Threshold η is applied to obtain pseudo-labeled dataset $\hat{\mathcal{U}} = \{(\hat{x}_u, \hat{y}_u)\}$;

5: *Recognizer training*: training target recognizer $\hat{\Theta}_t$ on $\hat{\mathcal{U}} \cup \mathcal{L}_r$;

6: **return** $\hat{\Theta}_t$;

3.1   Initialization

Data Partition. We have a labeled dataset $\mathcal{L} = \{(x_l, y_l)\}$ with x_l the text image and y_l the text annotation which is a sequence. We also have a large-scale unlabeled dataset $\mathcal{U} = \{(x_u)\}$. We divide $\mathcal{L}$ into a discriminator trainset $\mathcal{L}_d$ of relative small number of images and a recognizer trainset $\mathcal{L}_r$ with the rest of the images.

Heterogeneous Recognizers. The previous SSL works train models from different initial conditions or different data sources to obtain different views and complementary information. However, in scene text recognition, instead of predicting a single tag, inferring a sequence of characters is more sophisticated. Meanwhile, scene text recognition is a one-view data task, therefore the conventional SSL based on two views data is infeasible. Considering the problems mentioned above and to avoid the neural networks from collapsing into each other, we adopt two heterogeneous networks in our framework to give predictions from different perspectives. In detail, the target recognizer is an attention-based network, which aims to be improved by our framework. And a CTC-based network is used as a surrogate recognizer, which infers recognition results with a different decoding mechanism. As illustrated in Fig. 2, Both the target and surrogate recognizers adopt the encoder-decoder structure.

    In the target recognizer, A CNN is used to extract visual features from input images in the encoder. Then a stacked Bi-LSTMs is utilized for modeling contextual dependencies from the visual features. As for the decoder, the target recognizer applies an attentional RNN like [13] to obtain predictions with more context information. In the inference stage, the decoder of target recognizer proceeds iteratively until it encounters the End Of Sequence (EOS) token.

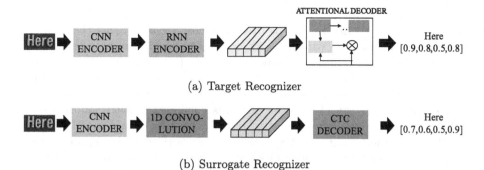

(a) Target Recognizer

(b) Surrogate Recognizer

Fig. 2. The illustration of the target and surrogate recognizers.

The surrogate recognizer is different from the target one, there are no RNNs in the encoder and decoder. Visual features are extracted in the encoder by CNNs and directly fed into the CTC decoder. In this style, the recognition results are inferred from visual information, and no context information is introduced. Thus the predictions of the target and surrogate recognizers can be regarded as complementary. The target and surrogate recognizers are both trained on $\mathcal{L}_r$ independently for initialization.

Discriminator. Many early SSL works regard the consistent results of different models as pseudo-labels. In scene text recognition, this naive approach is limited by the number of valid pseudo-labels, as the two recognition sequences are rare to be identical. To improve the efficiency by using the complementary information, we design a discriminator for pseudo-labeling and instance selection. We consider four aspects for distinguishing the correct result from two predictions, namely, probability features, text features, geometry features, and comparative features.

Given a sample $(x_d, y_d) \in \mathcal{L}_d$ with x_d size of $W \times H$, the heterogeneous recognizers output the prediction sequence results α and α' along with the conditional probability sequences $\mathbf{p}$ and $\mathbf{p}'$.

$$\Theta_t(x_d) = (\alpha, \mathbf{p}), \alpha = (\alpha_1, \alpha_2...\alpha_D), \mathbf{p} = (p_1, p_2, ...p_D) \tag{1}$$

$$\Theta_s(x_d) = (\alpha', \mathbf{p}'), \alpha' = (\alpha'_1, \alpha'_2...\alpha'_K), \mathbf{p}' = (p'_1, p'_2, ...p'_K) \tag{2}$$

D and K represent the sequence length.

The input vector $\mathbf{V}$ of the discriminator is a concatenation of the following features:

(1) $\min(p_1, p_2, ...p_D)$ (2) $\min(p'_1, p'_2, ...p'_K)$ (3) mean $(p_1, p_2, ...p_D)$
(4) mean $(p'_1, p'_2, ...p'_K)$ (5) $D' = \frac{D}{\text{Constant}}$ (6) $K' = \frac{K}{\text{Constant}}$
(7) Char width $= \frac{W}{D}$ (8) Char width$' = \frac{W}{K}$ (9) Normalized Edit Distance

(1)–(4) are probability features: *mean* illustrates the overall confidence and *min* reflects the most likely misrecognized character. (5) and (6) are the normalized number of characters (divide by a *Constant* to make them between 0 to 1) in recognition results which are related to the text features. (7) and (8) are the character widths calculated according to the recognition results and geometry features. (9) is the normalized edit distance between two results that illustrates the comparative feature. The above elements combine geometric information, probability, and recognition content to provide different measures of how reliable the recognition results are. For example, when both recognizers produce the same recognition result with high probabilities, the result tends to be correct.

We apply a multi-layer perceptron as the discriminator which is presented in Eq. 3 and 4. $\mathbf{W}$ is the learnable parameter matrix in the discriminator. The classification result I has four possibilities: both recognitions are right (BR), both are wrong (BW), only the target recognition is right (TR) and only the surrogate recognition is right (SR). It is worth noting that being "right" means the recognition result is utterly identical to the label y_d. We use softmax as the activation function and the largest value of $\mathbf{S}$ is taken as the confidence score s.

$$\mathbf{S} = softmax(\mathbf{WV}) \tag{3}$$

$$I = \arg\max(\mathbf{S}), s = \max(\mathbf{S}) \tag{4}$$

During the initialization phase, we first infer $\mathcal{L}_d$ on the two recognizers obtained in the previous subsection. The ground truth and features are computed according to the prediction results and are fed into the discriminator for training.

3.2 Pseudo-labeling and Instance Selection

The pseudo-labeling is processed on the unlabeled dataset $\mathcal{U}$. We first infer our trained target and surrogate recognizers on $\mathcal{U}$. Afterward, the feature vectors are extracted, and they are fed into the trained discriminator. The pseudo-labels are selected from α and α' according to the classification result.

$$\text{pseudo-label}\ (x_u) = \begin{cases} \alpha & \text{if } (I = BR \text{ or } TR) \text{ and } s > \eta \\ \alpha' & \text{if } I = SR \text{ and } s > \eta \\ \text{abandon} & \text{otherwise} \end{cases} \tag{5}$$

The instance selection is processed according to the confidence score s. There are two cases when the sample should be abandoned. Firstly, when the two recognition results are classified as BW. Secondly, the sample whose classification confidence score is below the threshold η.

3.3 Recognizer Training

After the procedures mentioned above, the pseudo-labeled data is obtained. To train a text recognizer with better performance, we mix the pseudo-labeled data with the labeled trainset $\mathcal{L}_r$ in this phase. The new target recognizer is trained on the mixed dataset from scratch. It should be noted that we can do these three steps (initialization, pseudo-labeling, and recognizer training) more than once to improve recognition performance iteratively.

4 Experiments

4.1 Datasets

We conduct experiments on English and Chinese datasets. For English, two synthetic datasets (**MJSynth** [9] and **SynthText** [7]) are thoroughly mixed for training. MJSynth consists of about 9 million images covering 90k English words. Random transformations and other effects are applied to every word image. SynthText is intended for scene text detection and we crop about 7 million word-level instances from the trainset. There are five testsets for English:

ICDAR 2003 (IC03) [16] contains 1,156 images for training and 1,110 images for evaluation. Following Wang *et al.* [25], we discard images that contain non-alphanumeric characters or have less than three characters. The number of images after filtering is 867.

Street View Text (SVT) [25] consists of 249 images collected from Google Street View. The testset is cropped from these samples and contained 647 images. Many images are severely corrupted by noise and blur or have very low resolutions.

ICDAR 2013 (IC13) [12] inherits most images from IC03. The trainset and testset have 848 and 1,095 images separately. The filtered testset contains 1,015 images.

ICDAR 2015 (IC15) [11] includes 2,077 images for the test. The original testset is referred to as IC15-2077. There is a wildly used subset that has 1,811 images (refer to as IC15-1811). Images with non-alphanumeric characters, extreme transformation, and curved text are filtered.

IIIT5K-Words (IIIT) [17] is gathered from the Internet. It contains 3,000 cropped word images for testing.

For Chinese, we collect three public datasets which are RCTW [23], ReCTS [15] and MSRA-TD500 [30]. Our model is trained on the training data of RCTW and ReCTS and tested on all the three datasets.

RCTW [23] contains 12,263 annotated samples. We crop over 120,000 text images according to the detection ground truth, and the recognition ground truth is regarded as the annotation.

ReCTS [15] is a competition for Chinese text reading on the signboards. The trainset consists of 20,000 images and the testset includes 5,000 images.

MSRA-TD500 [30] is a dataset contains both Chinese and English. There are 200 images in the testset with only text line level annotations. We crop, filter and annotate the test images for a 556 text lines testset for recognition.

4.2 Implementation Details

The structures of the target and surrogate recognizers are inspired by [21,22]. While some simplifications are adopted to accelerate the training procedure. A ResNet34 is adopted as the CNN for the target and surrogate recognizers. For Chinese, the pooling strides in the last two blocks are changed to 2×1. For English, the pooling strides in the last three blocks are changed to 2×1. For the target network, the RNN encoder is a two-layer Bi-LSTMs where the number of each hidden unit is 512. The attentional Bi-LSTMs used in the decoder has 1024 hidden units. In the surrogate network, a convolutional layer with 1×3 kernel and $1,024$ channels is adopted after the CNN. Greedy searching is employed for both Chinese and English experiments. The discriminator is a three-layer fully connected network and the hidden unit number in each layer is 10. The final prediction is calculated from a softmax function, and the default value of confidence threshold η is 0.8. The English charset has 62 classes, including digits, uppercase, and lowercase letters. When evaluating the trained model on benchmarks, the predictions are normalized to case insensitive, and punctuation marks are discarded. The Chinese charset includes 5,108 characters and symbols collected from ReCTS and RCTW.

The numbers of labeled data in recognizer initialization are 1 million and 78,000 for English and Chinese respectively. The corresponding discriminators are trained on 10,000 English and 5,000 Chinese samples. Then the rest data are treated as unlabeled data for pseudo-labeling. The ADADELTA is applied in the training procedure for both target and surrogate recognizers with a batch size of 64. For the target recognizer training, the initial learning rate is set to 1.0 for the first 3 epochs and decreased by a factor of 10 for the 4th epoch. The training of the surrogate recognizer is similar to the target recognizer's, except the initial learning rate is set to 0.01. For English and Chinese, the input images are resized to 32×100 and 32×320 respectively. Random rotation ($[-3°, +3°]$), elastic deformation, random hue, brightness and contrast are applied for data augmentation[1].

4.3 Evaluation Results

Accuracy Improvement. As shown in Table 1, the accuracies of initial target recognizers are regarded as the baselines. The best results obtained from multiple training experiments with different unlabeled data sizes are noted. Our approach provides a significant improvement (+1.93% to +4.68%) overall baselines that only trained on limited labeled data. Moreover, our method achieves comparable results with the fully-supervised model (all the training data are labeled). The accuracy gap is less than 1.64%. Chinese recognizer benefits more from our SSL framework. The initial Chinese recognizer is trained on much fewer samples than the English recognizer, and the increase in performance is especially pronounced when only a few labeled samples are available for training. Besides, the

---

[1] https://github.com/mdbloice/Augmentor.

Table 1. Comparison of models trained by our SSL framework (the 3rd row) and other fully-supervised models trained on the complete (the 1st row) or partial labeled data (the 2nd row). The gain is the absolute accuracy improvement over the initial recognizers. The unlabeled data scales are given in the last row.

Testsets	ReCTS	MSRA	IC03	IC13	IC15-1811	IC15-2077	SVT	IIIT
Fully-supervised	77.40	64.56	92.24	89.92	69.19	64.94	85.51	82.04
Initial recognizer	71.28	58.45	88.80	86.27	66.17	61.76	80.51	79.00
Best recognizer	75.77	63.13	91.26	88.36	68.69	63.69	84.23	81.15
Gains	**+4.49**	**+4.68**	**+2.46**	**+2.09**	**+2.52**	**+1.93**	**+3.72**	**+2.15**
Unlabeled data (10^4)	13.5	10.8	100.0	100.0	100.0	100.0	80.0	80.0

Chinese dataset consists of real images, which represent a more diverse visual distribution.

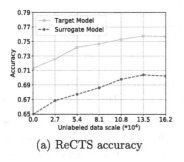
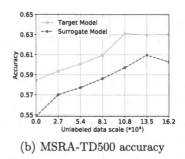

(a) ReCTS accuracy   (b) MSRA-TD500 accuracy

Fig. 3. The accuracies of target and surrogate recognizers on Chinese testsets (ReCTS and MSRA-TD500) vs. unlabeled data scale.

Impact of Pseudo-labeled Data Scale. We want to further investigate the relationship between the unlabeled data scale and the recognizer accuracy. Figure 3 and Fig. 4 show the accuracies of Chinese and English recognizers as a function of the unlabeled dataset size.

For all the testsets, the consistent accuracy improvements on the target and surrogate recognizers are observed every time the dataset size increases until reaching a certain dataset scale. This result implies two facts. First, the improvement demonstrates that leveraging unlabeled data by our SSL framework is effective for both English and Chinese recognition. Second, the accuracy improvement becomes less as the unlabeled dataset size grows further, which indicates that there is a limit for leveraging unlabeled data. For the Chinese testset ReCTS, the target model is possibly reaching its saturation point when 135,000 unlabeled data (70,000 pseudo-labeled data after filtering) are added. The plots on English testsets show that the saturation point is around 0.8 to 1 million unlabeled data (0.74 to 0.93 million pseudo-labeled data after filtering). Specifically, we observe that for both Chinese and English recognizers, given limited labeled data whose

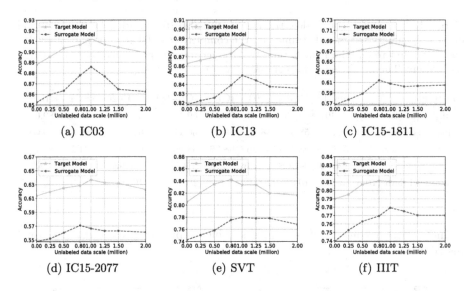

Fig. 4. Target and surrogate recognizers accuracies on English testsets (IC03, IC13, IC15, SVT and IIIT) vs. the unlabeled data scale.

size is between 100,000 and 1 million, the best performance is achieved when the amount of unlabeled data is almost the same as that of the labeled data.

Effect of Each Component. We now study the effect of the heterogeneous networks and the discriminator. Considering that our SSL framework generates pseudo-labels for unlabeled samples, we introduce two indexes to evaluate the pseudo-labeling performance. The pseudo-labeling precision (PLP) and the pseudo-labeling utilization (PLU) are defined as follows:

$$\text{PLP} = \frac{\text{Correct pseudo-labels number}}{\text{Total pseudo-labels number}} \qquad (6)$$

$$\text{PLU} = \frac{\text{Total pseudo-labels number}}{\text{Total unlabeled data number}} \qquad (7)$$

PLP reflects the error rate of the pseudo-labeling, which is correlated to the confirmation bias and thus affects the recognition accuracy. PLU tells the proportion of unlabeled data converted to valid pseudo-labeled data. The more valid images involved in training, the better the generalization performance of the recognizer.

As shown in Table 2, considering the number and structure of the recognizers and whether to use a discriminator in pseudo-labeling, there are four variants. For those variants without a discriminator, the pseudo-labels are generated when two results are consistent and we use the average probability as a threshold for instance selection. We compare PLPs by controlling the threshold to make PLUs at the same level.

We can observe that for both English and Chinese datasets, our framework achieves the best PLP given the same PLU. The "Naive-hetero" method for

Table 2. PLPs and PLUs of pseudo-labeling on English (left) and Chinese (right) datasets with different frameworks. **R** and **Λ** represent recognizer and discriminator separately. Using two models with the same or different structures (2-h) and generates the pseudo-labels without a discriminator is referred to as "Naive-homo" and "Naive-hetero". The integration of the discriminator for "Naive-homo" is referred to as "Homo" method. "Teacher-student": using a single model for pseudo-labeling and training.

Method	**R**	**Λ**	PLP	PLU		Method	**R**	**Λ**	PLP	PLU
Ours	2-h	✓	91.63	92.46		Ours	2-h	✓	83.70	52.49
Naive-hetero	2-h	✗	97.59	75.78		Naive-hetero	2-h	✗	81.95	52.44
Homo	2	✓	89.74	92.46		Homo	2	✓	80.49	52.44
Naive-homo	2	✗	89.84	92.43		Naive-homo	2	✗	79.27	52.49
Teacher-student	1	✗	88.94	92.48		Teacher-student	1	✗	78.66	52.46

English is an exception since its PLU is already lower than the given value before instance selection. Our experimental results prove that both the heterogeneous networks and the discriminator play important roles in improving the precision of pseudo-labeling while ensuring the data utilization. Especially the heterogeneous networks, which show significant improvement of PLP when compared with the "Homo" method. In our framework, even though the target recognizer is designed to be more powerful than the surrogate one (as reflected in the plots of Fig. 3 and Fig. 4), it has some limitations. Due to the implementation of LSTMs which learn the character dependencies, the target recognizer is sometimes misled by semantic information. Besides, the attentional sequence recognition applied in the decoder has the problem of misalignment due to the attention drifts. The surrogate recognizer, which focuses more on the visual information, is a complement to the target recognizer. Therefore, by aggregating two predictions from different perspectives, the pseudo-labeling process can thus be enhanced. For English and Chinese experiment, there are 20% and 33% pseudo-labels which are generated by the surrogate recognizers when the two predictions are different.

Furthermore, to validate the relation between our proposed indexes (PLP/PLU) and accuracy, we compare our method with the classic self-training framework "Teacher-student" [28] on two Chinese testsets. For a fair comparison, the structure of the teacher and student model is the same as our target recognizer. As shown in Table 3, our method outperforms their pipeline in terms of PLP and recognizer accuracy. This also reflects the importance of PLP and PLU during pseudo-labeling.

Effect of Hyper-parameter. We now study the only one hyper-parameter in our framework, the confidence threshold η, which is relevant to the error rate of pseudo-labeling and model performance. In this part, we conduct some experiments to discuss the parameter setting.

Firstly, we study the impact of η on the PLP and PLU and different values of η are adopted in pseudo-labeling. As illustrated in Table 4, the variation of η

Table 3. The comparison between our SSL framework and the Teacher-student on Chinese testsets.

	Ours	Teacher-student
PLP	**83.70%**	78.66%
PLU	**52.49%**	52.46%
ReCTS	**75.69%**	74.38%
MSRA	**63.05%**	60.25%

Table 4. PLPs and PLUs on English and Chinese trainsets with different η.

η	Chinese		English	
	PLP (%)	PLU (%)	PLP (%)	PLU (%)
0.0	80.22	61.25	91.45	92.82
0.5	80.23	61.23	91.57	92.72
0.8	83.70	52.49	91.63	92.46
0.9	83.75	32.42	91.68	92.30

has little effect on the English unlabeled data. The value of η moves from 0.0 to 0.9, while the variation of PLP and PLU are only $+0.23\%$ and -0.52%. The possible reason is that the trainset for English consists of all synthetic images, which show little diversity. While the Chinese training data is all captured in the real-world, the PLP and PLU vary drastically with different η.

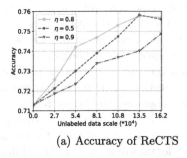

(a) Accuracy of ReCTS

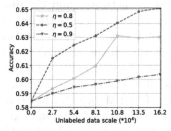

(b) Accuracy of MSRA-TD500

Fig. 5. Surrogate recognizers accuracies obtained with different η on ReCTS and MSRA-TD500 testsets vs. unlabeled data scale.

Secondly, to further explore the relation between η, unlabeled data scale and accuracy, we conduct corresponding experiments on Chinese datasets. The experiment settings are the same as the first ones. As shown in Fig. 5, the accuracies are plotted as a function of unlabeled data scale with $\eta \in \{0.5, 0.8, 0.9\}$. The performance improvement of our framework is consistent under different settings. However, the value of η has an ineligible impact on the accuracy. On ReCTS, $\eta = 0.8$ achieves the best performance, while on MSRA-TD500 the best value is 0.5. According to our analysis, setting a higher η can reduce the errors in pseudo-labels (namely increase the PLP), but the PLU drops drastically. This will harm the model's generalization and affect the accuracy. On the contrary, if we set η to a lower value, the PLU can be increased and the PLP will decline. More noisy labels will be brought into the pseudo-labeled data and the accuracy can also be influenced. In conclusion, the choice of η is a trade-off between PLP and PLU.

5   Conclusion

This paper introduces a novel framework for leveraging unlabeled images via SSL to improve scene text recognition performance. First, two heterogeneous networks are utilized to exploit complementary information from visual representation and character dependencies respectively. Second, a discriminator is proposed to improve the pseudo-labeling and instance selection based on the predictions from two heterogeneous networks. In this way, the quality and quantity of the pseudo-labeled data are enhanced, which results in the generalization and performance improvement of the scene text recognizer. Comprehensive experiments demonstrate that our SSL framework can effectively improve both Chinese and English recognition accuracy. Moreover, extensive studies about the effectiveness of each component in our framework and performance vs. unlabeled data scale are also conducted to explore the law of SSL in scene text recognition. Due to the limitation of time and resources, we only leverage 100 thousand to millions of unlabeled data. As for future directions, a larger scale unlabeled data for scene text recognition can be investigated and new insights from SOTA SSL studies will be applied.

References

1. Baskarφ, M.K., Watanabe, S., Astudilloπ, R., HoriY, T.: Semi-supervised sequence-to-sequence ASR using unpaired speech and text. arXiv preprint arXiv:1905.01152 (2019)
2. Berthelot, D., Carlini, N., Goodfellow, I., Papernot, N., Oliver, A., Raffel, C.A.: Mixmatch: a holistic approach to semi-supervised learning. In: NIPS (2019)
3. Chen, D.D., Wang, W., Gao, W., ZhiHua, Z.: Tri-net for semi-supervised deep learning. In: IJCAI (2018)
4. Chen, Y., Wang, W., Wang, C.: Semi-supervised ASR by end-to-end self-training. arXiv preprint arXiv:2001.09128 (2020)
5. Cheng, Z., Bai, F., Xu, Y., Zheng, G., Pu, S., Zhou, S.: Focusing attention: towards accurate text recognition in natural images. In: ICCV (2017)
6. Graves, A., Fernández, S., Gomez, F., Schmidhuber, J.: Connectionist temporal classification: labelling unsegmented sequence data with recurrent neural networks. In: ICML (2006)
7. Gupta, A., Vedaldi, A., Zisserman, A.: Synthetic data for text localisation in natural images. In: CVPR (2016)
8. Hady, M.F.A., Schwenker, F.: Co-training by committee: a new semi-supervised learning framework. In: 2008 IEEE International Conference on Data Mining Workshops (2008)
9. Jaderberg, M., Simonyan, K., Vedaldi, A., Zisserman, A.: Synthetic data and artificial neural networks for natural scene text recognition. CoRR (2014)
10. Kahn, J., Lee, A., Hannun, A.: Self-training for end-to-end speech recognition. In: ICASSP (2020)
11. Karatzas, D., et al.: ICDAR 2015 competition on robust reading. In: ICDAR (2015)
12. Karatzas, D., et al.: ICDAR 2013 robust reading competition. In: ICDAR (2013)
13. Lee, C.Y., Osindero, S.: Recursive recurrent nets with attention modeling for OCR in the wild. In: CVPR (2016)

14. Liu, W., Chen, C., Wong, K.Y.K., Su, Z., Han, J.: Star-Net: a spatial attention residue network for scene text recognition. In: BMVC (2016)
15. Liu, X., et al.: ICDAR 2019 robust reading challenge on reading Chinese text on signboard. arXiv preprint arXiv:1912.09641 (2019)
16. Lucas, S., et al.: ICDAR 2003 Robust Reading Competitions: Entries, Results and Future Directions. ICDAR (2005)
17. Mishra, A., Alahari, K., Jawahar, C.: Scene text recognition using higher order language priors. In: BMVC (2012)
18. Neumann, L., Matas, J.: Real-time scene text localization and recognition. In: CVPR (2012)
19. Qiao, S., Shen, W., Zhang, Z., Wang, B., Yuille, A.: Deep co-training for semi-supervised image recognition. In: ECCV (2018)
20. Qiao, Z., Zhou, Y., Yang, D., Zhou, Y., Wang, W.: Seed: semantics enhanced encoder-decoder framework for scene text recognition. In: Proceedings of the IEEE/CVF Conference on Computer Vision and Pattern Recognition, pp. 13528–13537 (2020)
21. Shi, B., Bai, X., Yao, C.: An end-to-end trainable neural network for image-based sequence recognition and its application to scene text recognition. TPAMI (2017)
22. Shi, B., Yang, M., Wang, X., Lyu, P., Yao, C., Bai, X.: Aster: an attentional scene text recognizer with flexible rectification. TPAMI (2018)
23. Shi, B., et al.: ICDAR 2017 competition on reading Chinese text in the wild (RCTW-17). In: ICDAR (2017)
24. Sohn, K., et al.: Fixmatch: simplifying semi-supervised learning with consistency and confidence (2020)
25. Wang, K., Babenko, B., Belongie, S.: End-to-end scene text recognition. In: ICCV (2011)
26. Wang, T., et al.: Decoupled attention network for text recognition. In: AAAI (2020)
27. Xie, Q., Luong, M.T., Hovy, E., Le, Q.V.: Self-training with noisy student improves ImageNet classification. In: CVPR (2020)
28. Yalniz, I.Z., Jégou, H., Chen, K., Paluri, M., Mahajan, D.: Billion-scale semi-supervised learning for image classification. arXiv preprint arXiv:1905.00546 (2019)
29. Yao, C., Bai, X., Liu, W.: A unified framework for multioriented text detection and recognition. TIP (2014)
30. Yao, C., Bai, X., Liu, W., Ma, Y., Tu, Z.: Detecting texts of arbitrary orientations in natural images. In: CVPR (2012)
31. Yao, C., Bai, X., Shi, B., Liu, W.: Strokelets: a learned multi-scale representation for scene text recognition. In: CVPR (2014)
32. Yu, D., et al.: Towards accurate scene text recognition with semantic reasoning networks. In: CVPR (2020)
33. Zhang, H., Cisse, M., Dauphin, Y.N., Lopez-Paz, D.: Mixup: beyond empirical risk minimization. In: ICLR (2018)

Scene Text Detection with Scribble Line

Wenqing Zhang[1], Yang Qiu[1], Minghui Liao[1], Rui Zhang[2],
Xiaolin Wei[2], and Xiang Bai[1(✉)]

[1] Huazhong University of Science and Technology, Wuhan, China
{wenqingzhang,yqiu,mhliao,xbai}@hust.edu.cn
[2] Meituan, Beijing, China
{zhangrui36,weixiaolin02}@meituan.com

Abstract. Scene text detection, which is one of the most popular topics in both academia and industry, can achieve remarkable performance with sufficient training data. However, the annotation costs of scene text detection are huge with traditional labeling methods due to the various shapes of texts. Thus, it is practical and insightful to study simpler labeling methods without harming the detection performance. In this paper, we propose to annotate the texts with scribble lines instead of polygons for text detection. It is a general labeling method for texts with various shapes and requires low labeling costs. Furthermore, a weakly-supervised scene text detection framework is proposed to use the scribble lines for text detection. The experiments on several benchmarks show that the proposed method bridges the performance gap between the weakly labeling method and the original polygon-based labeling methods, with even better performance. We will release the weak annotations of the benchmarks in our experiments and hope it will benefit the field of scene text detection to achieve better performance with simpler annotations.

Keywords: Scene text detection · Weak annotation · Scribble line

1 Introduction

Scene text detection, which is a fundamental step for text reading, aims to locate the text instances in scene images. It has become one of the most popular research topics in academia and industry for a long time. In practice, detecting text in natural scene images is a basic task for various real-world applications, such as autonomous navigation, image/video understanding, and photo transcription.

Due to the variety of scene text shapes (*e.g.* horizontal texts, multi-oriented texts, and curved texts), there is a high requirement for text location in scene text detection. In the early methods, they only focus on horizontal texts and adopt the axis-aligned bounding boxes to locate texts [8], which is very similar to general object detection methods. In the following years, MSRA-TD500 dataset [30] and ICDAR 2015 dataset [7] adopt multi-oriented quadrilaterals to better locate the multi-oriented texts. Recently, in the Total-Text dataset [2] and CTW1500

© Springer Nature Switzerland AG 2021
J. Lladós et al. (Eds.): ICDAR 2021, LNCS 12824, pp. 79–94, 2021.
https://doi.org/10.1007/978-3-030-86337-1_6

Fig. 1. Different annotations of the (a) CTW1500, (b) ICDAR 2015, and (c) Total-Text datasets. Green: original annotations; Red: our scribble line annotations. (Color figure online)

dataset [31], polygon annotations are used to outline texts with various shapes, and the Total-Text dataset even provides pixel-level text segmentation annotations. Hence, for the general scene text detection task, accurate labeling is time-consuming and laborious.

In this paper, we research into the difficulty of scene text detection that how to use a simpler labeling method to achieve the same performance as the current state-of-the-art methods. Similar to Wu *et al.* [27], we adopt scribble lines to annotate texts in a coarse manner. As shown in Fig. 1, different from polygon annotations, these scribble lines are simply annotated with points, and roughly depict the text instances. Compared with the previous labeling methods, the proposed labeling method has the following advantages: (1) It is a simpler labeling method with fewer point coordinates and faster labeling speed, as shown in Fig. 2. (2) It is a general labeling method for scene text detection because it can be applied to texts with various shapes, including horizontal texts, multi-oriented texts, and curved texts. (3) Based on these scribble line annotations, our experiment results demonstrate that we can achieve the performance of the state-of-the-art methods.

To compensate for the loss of edge information from our weak annotations, we propose a new scene text detection framework, which can be trained in a weakly supervised manner. Especially, an online proposal sampling method and a boundary reconstruction module are further introduced for character prediction and post-processing, respectively. We conduct extensive experiments on several benchmarks to demonstrate the effectiveness of our scene text detection method, which can bridge the performance gap between our weakly labeling method and the strongly labeling methods, with even better performance.

The contributions of this paper are three-fold.

- We propose a weakly labeling method for scene text detection with much lower labeling costs, which is a general method for texts with various shapes.

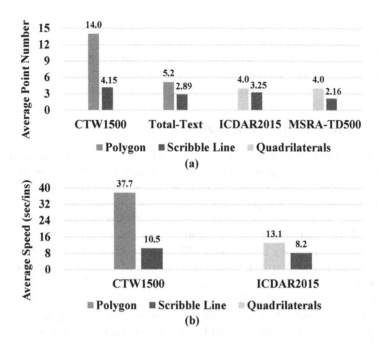

Fig. 2. Annotation cost comparison by (a) the average labeled points for each text instance and (b) the average labeling speed for each text instance, which is obtained from 10 annotators on average.

- A weakly-supervised scene text detection framework is proposed for the weak annotations. It outperforms the state-of-the-art methods that use original annotations on the Total-Text dataset and the CTW1500 dataset.
- We will release our annotations and hope that it will benefit the research community in the field of weakly supervised scene text detection to achieve better performance with lower annotation costs.

2 Related Work

2.1 Annotations for Scene Text Detection

The traditional annotations for scene text detection can be roughly classified into three types. The first one is the axis-aligned bounding box that is mostly used to label the horizontal text datasets, such as the ICDAR 2013 dataset [8]. The second type is the quadrilateral or oriented rectangular bounding box for multi-oriented text datasets, such as the ICDAR 2015 dataset [7] and the MSRA-TD500 dataset [30]. The third type is the polygonal bounding boxes for arbitrary-shaped text datasets, such as the Total-text dataset [2] and the CTW1500 dataset [31]. The above-mentioned annotations may not include detailed shape information (e.g. the first two types) for text instances of irregular shapes or require heavy labeling costs (e.g. the third type).

Recently, Wu *et al.* [27] propose to draw continuous lines for text detection and use two coarse masks for text instances and background, respectively. However, continuous lines need more storage space than points and they are hard to draw without touch devices such as touch screens or touch pens. Thus, we propose a new line-level labeling method to overcome these shortcomings.

2.2   Scene Text Detection Algorithms

Fully Supervised Text Detection. The methods for scene text detection can be roughly divided into regression-based and segmentation-based methods. The regression-based methods are mainly inspired by general object detection methods [14, 19]. TextBoxes [10] proposes a scene text detector based on SSD [14]. TextBoxes++ [9] proposes to predict the multi-oriented texts by quadrilateral regression. EAST [34] and DDR [6] propose to directly regress the offsets for text instances in the pixel level. RRD [12] proposes to apply rotation-sensitive and rotation-invariant features for the regression and classification respectively. The segmentation-based methods are mainly based on FCN [16] to predict text instances in the pixel level. PSE-Net [23] proposes to segment text instances by progressively expanding kernels at different scales. DB [11] propose a segmentation network with a differentiable binarization module, and improve on both accuracy and speed.

Weakly Supervised Text Detection. Due to the high annotation requirement for texts with various shapes, it is a difficulty that how to reduce the annotation costs without harming the detection performance. WeText [22] proposes a weakly supervised learning technique for text detection with a small fully labeled dataset and a large unlabeled dataset. CRAFT [1] proposes a text detection method by using characters and the affinity between characters, and it uses word-level annotations of real data to generate pseudo labels for fine-tuning. Wu *et al.* [27] propose a segmentation-based network to use the proposed coarse masks for training, but there is a large gap between their performance and those using full masks.

3   Methodology

In this section, we first introduce our proposed weakly labeling method and then describe the proposed weakly supervised scene text detection method which performs text detection with the weak annotations.

3.1   Weakly Labeling Method

Our proposed labeling method is annotating each text instance with a few points, as shown in Fig. 3. The labeling rule of a text instance is: (1) For horizontal or oriented text instances, two points near the centers of the first and the last

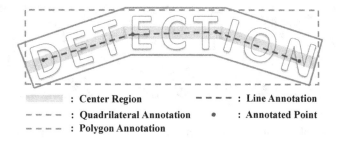

: Center Region   ---- : Line Annotation
---- : Quadrilateral Annotation   • : Annotated Point
---- : Polygon Annotation

Fig. 3. The illustration of our proposed weakly labeling method. We annotate each text instance with a scribble line composed of several points.

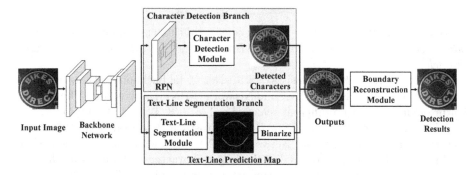

Fig. 4. The pipeline of our proposed method. The backbone network is a ResNet with a Feature Pyramid Network (FPN) structure. The character detection branch and text-line segmentation branch are used to predict character boxes and text-lines, respectively. The final detection results are generated by the boundary reconstruction module.

characters are annotated; (2) For irregular-shaped text instances, we annotate several points near the vertices of the center line. The annotated points of a text instance are saved as a set of coordinates and can be connected into a scribble line. We keep the extreme blurry regions with the original annotation as they occupy only a small proportion and are not used in the training.

3.2 Weakly Supervised Scene Text Detection

Compared with multi-oriented quadrilaterals and polygons, the proposed weak annotations lose the edge information, which is indispensable to previous scene text detection methods. Hence, we propose a weakly supervised scene text detection framework, as shown in Fig. 4. In the following parts of this section, we introduce the **Network Architecture**, **Pseudo Label Generation**, **Training Strategy**, **Optimization**, and **Inference**, respectively.

Fig. 5. An example of the pseudo label generation: (a) Predict character boxes by a pre-trained model. (b) Remove boxes (red) with scores lower than T_{pseudo} and the "Unknown" class. (c) Remove boxes (yellow) without overlap with any scribble lines. (d) The final pseudo labels for characters. (Color figure online)

Network Architecture. The network consists of three parts, including a *Backbone Network*, a *Character Detection Branch*, and a *Text-Line Segmentation Branch*.

The *Backbone Network* consists of a ResNet-50 [5] backbone with deformable convolution [3] and a Feature Pyramid Network (FPN) [13] structure. The output feature maps are with 5 different scales (*i.e.* 1/4, 1/8, 1/16, 1/32, 1/64).

The *Character Detection Branch* is a two-stage detection network, which consists of a Region Proposal Network (RPN) [19] and a character detection module. The RPN is used to predict the coarse character proposals. The character detection module classifies the character proposals into different classes and regresses the offsets of the coarse proposals for refinement.

The *Text-Line Segmentation Branch* predicts a probability map for the text-line location, on which the score of each pixel indicates the probability that it belongs to the center line of the text. The probability map is supervised by the scribble line. The feature maps in different scales are fed into a fully convolutional module with up-sampling. All feature maps are up-sampled by Nearest Neighbor interpolation to quarter size of the input image and concatenated in the channel-wise. Then, the concatenated feature map is fed into a convolution layer and two transposed convolution layers, to predict a probability map for the text-line location.

Pseudo Label Generation. We propose to generate pseudo labels for the character detection branch from the model pre-trained with synthetic data [4], since the real data is only annotated with our weakly labeling method while the character-level annotations can be easily obtained from the synthetic data. As shown in Fig. 5, the pseudo label generation consists of four steps: (1) Use the pre-trained model to predict character boxes, scores, and classes of the real training data. (2) Remove the character boxes with scores less than a threshold T_{pseudo} or with the "Unknown" class. (3) Remove the character boxes without overlap with any scribble lines. (4) The remained boxes are saved as the pseudo labels for training.

Training Strategy. The model is first pre-trained with synthetic data and then fine-tuned with both synthetic data and real data.

For the character detection branch, the Region Proposal Network (RPN) [19] is only fine-tuned with synthetic data to avoid introducing the noise of wrong pseudo labels in the first stage of character detection. The second stage is fine-tuned with synthetic and real data with our online proposal sampling. For real data, the positive samples are selected from the proposals matched with pseudo labels, and the negative samples are selected by the following two rules: (1) Those do not match any labels; (2) Those have no overlap with any scribble lines or difficult text instances; Then, the rest of the proposals are ignored in the training. Due to the domain gap between synthetic data and real data, the pre-trained model can not generate pseudo labels for all characters as shown in Fig. 5. The proposals, which can not match any labels but have overlap with scribble lines, are the potential positive ones. Hence, we should not regard these proposals as the negative, or they will have a negative influence on the detection performance.

The text-line segmentation branch is supervised with the scribble line ground-truths of synthetic data and real data.

Optimization. The objective function consists of three parts, which is defined as follows.

$$Loss = L_{rpn} + L_{char} + L_{line}, \tag{1}$$

where the L_{rpn} includes binary cross entropy loss for character classification and smooth L1 loss for character box regression, which is similar to it in [19], so we do not describe it in detail.

We apply a smooth L1 loss for the axis-aligned character box regression and a cross-entropy loss for character classification, which is defined as follows.

$$L_{char} = L_{regress} + L_{classify} = L1_{smooth}\{\widehat{B}, B\} + L_{CE}, \tag{2}$$

where B refers to the predicted offsets from positive character proposals to associated ground-truths and $\widehat{B}$ is the target offsets, which are selected by our online proposal sampling.

The L_{line} is the loss of the text-line segmentation module, and we adopt a binary cross-entropy loss for the text-line prediction map. Due to the imbalance between the number of text pixels and non-text pixels, we apply the online hard negative mining strategy proposed in [21] to avoid the bias, and set the ratio of positive samples and negative samples as 1:3. It should be noted that the pixels in the difficult text instances will not be sampled.

We define L_{line} as follows,

$$L_{line} = \frac{1}{N}\sum_{i \in P}(1 - y_i)\log(1 - x_i) + y_i \log x_i, \tag{3}$$

where P refers to the sampled set, and N is the number of samples.

Algorithm 1. Boundary Reconstruction

Input: Detected Character boxes $\mathbf{B} = \{B_i\}$ and scores $\mathbf{S} = \{S_i\}$; Detected Text-Lines
    $\mathbf{L} = \{L_j\}$;
Output: Text detection results $\mathbf{T}$
 1: **for** each pair of (B_i, S_i) in $\mathbf{B}$ and $\mathbf{S}$ **do**
 2:    If $S_i <$ a threshold T_{infer}: remove B_i from $\mathbf{B}$
 3:    If B_i has the maximal overlap with L_j in $\mathbf{L}$: put B_i into group G_j
 4: **end for**
 5: **for** All boxes $\mathbf{B}_j$ in each group G_j **do**
 6:    $D = \frac{1}{N} \sum_{k=0}^{N-1} \sqrt{h_k * w_k}$, where h_k and w_k are the height and width of the box
    in $\mathbf{B}_j$.
 7:    Expand the contour of the associated text-line L_i to reconstruct the text boundary T_i by a distance D
 8: **end for**
 9: Output detection results $\mathbf{T} = \{T_i\}$

The text-line segmentation ground-truth is generated by connecting the annotated points of each text instance into a scribble line and draw it with *thickness* $= 5$ on a map.

Inference. In the inference period, we apply a boundary reconstruction algorithm to generate accurate text boundaries from the predicted text scribble lines and the character boxes, as shown in Algorithm 1.

In the previous character-based text detection algorithms [1,20], they propose to use the "link" or "affinity" to connect detected characters into the final detection results. However, noisy or missing characters probably lead to wrong or split detection results. As shown in Algorithm 1, the final detection results are generated by the geometry information of character boxes on average. It can suppress the impact of the noise, and will not suffer from the missing characters. Hence, our method is more robust compared with the previous weakly supervised methods and can achieve better performance.

4 Experiment

4.1 Datasets

The synthetic dataset is used for pre-training our model, and we fine-tune on each real dataset respectively. We use our weak annotations for training and the original annotations for evaluation.

SynthText [4] is a synthetic dataset including 800k images. We use the no-cost character bounding boxes with classes and the text-lines composed of characters' center points.

Total-Text [2] is a word-level based English text dataset. It consists of 1255 training images and 300 testing images, which contain horizontal texts, multi-oriented texts, and curved texts.

Fig. 6. The visualization of the network outputs and the final detection results in different cases, including horizontal texts, multi-oriented texts, curved texts, vertical texts, long texts, and multi-lingual texts. The images in the first row show the detected characters and text-lines, and those in the second row show the text detection results after post-processing.

CTW1500 [31] is a curved English text dataset that consists of 1000 training images and 500 testing images. All the text instances are annotated with 14 vertices.

ICDAR2015 [7] contains images that are captured by Google Glasses casually, and most of them are severely distorted or blurred. There are 1000 training images and 500 testing images, which are annotated with quadrilaterals.

MSRA-TD500 [30] is a multi-lingual long text dataset for Chinese and English. It includes 300 training images and 200 testing images with arbitrary orientations. Following previous works [11,17], we include HUST-TR400 [29] as training data in the fine-tuning.

4.2 Implementation Details

The training procedure can be roughly divided into 2 steps: (1) We pre-train our model on SynthText for 300k iterations. (2) For each benchmark dataset, we use the pre-trained model to generate pseudo labels for characters, and fine-tune the model with pseudo labels and weak annotations for 300k. For each mini-batch, we sample the training images with a fixed ratio 1:1 for SynthText and the benchmark data.

The models are trained on two Tesla V100 GPUs with a batch size of 2. We optimize our model by SGD with a weight decay of 0.0001 and a momentum of 0.9. The initial learning rate is set to 0.0025, and it is decayed by a factor 0.1 at iteration 100k and 200k. For training, the shorter sides of training images are randomly resized to different scales (*i.e.* 600, 800, 1000, 1200). For inference, the

Table 1. Detection results on the Total-Text dataset and CTW1500 dataset. "P", "R" and "F" refer to precision, recall and f-measure. *using line-level weak annotations.

Method	Total-Text			CTW1500		
	P	R	F	P	R	F
TLOC [31]	–	–	–	77.4	69.8	73.4
TextSnake [17]	82.7	74.5	78.4	67.9	**85.3**	75.6
TextField [28]	81.2	79.9	80.6	83.0	79.8	81.4
PSE-Net [23]	84.0	78.0	80.9	84.8	79.7	82.2
LOMO [32]	88.6	75.7	81.6	**89.2**	69.6	78.4
ATRR [25]	80.9	76.2	78.5	80.1	80.2	80.1
CRAFT [1]	87.6	79.9	83.6	86.0	81.1	83.5
PAN [24]	89.3	81.0	85.0	86.4	81.2	83.7
DB [11]	87.1	82.5	84.7	86.9	80.2	83.4
ContourNet [26]	86.9	83.9	85.4	83.7	84.1	83.9
Zhang *et al.* [33]	86.5	**84.9**	85.7	85.9	83.0	84.5
*Texts as Lines [27]	78.5	76.7	77.6	83.8	80.8	82.3
*****Proposed Method**	**89.7**	83.5	**86.5**	87.2	85.0	**86.1**

shorter sides of the input images are usually resized to 1200. In both training and inference, the upper limit of the longer sides is 2333. Moreover, we set the thresholds in our experiments as follows: (1) T_{pseudo} for pseudo label generation is set to 0.9; (2) T_{infer} for inference is usually set to 0.5;

4.3 Comparisons with Previous Algorithms

To compare with previous fully supervised methods, we evaluate our scene text detection method on four standard benchmarks. As shown in Fig. 6, we provide some qualitative results of texts in different cases, including horizontal texts, multi-oriented texts, curved texts, vertical texts, long texts, and multi-lingual texts.

Curved Text Detection. As shown in Table 1, we demonstrate the effectiveness of our method. On CTW1500 and Total-Text, our detector trained with our weak annotations outperforms the state-of-the-art fully supervised method by 1.6% and 0.8% on accuracy, respectively. The curved texts usually need higher annotation costs, but our weakly supervised method can achieve better performance with simpler annotations. Especially, compared with [27], our method improves a lot and bridges the performance gap between the weakly supervised methods and the fully supervised methods.

Multi-lingual Text Detection. To evaluate our scene text detection method in other languages, we conduct an experiment on the MSRA-TD500 dataset. Due to the lack of Chinese texts in SynthText, we use a public code to generate

Table 2. Detection results on the MSRA-TD500 dataset and ICDAR2015 dataset. *using line-level weak annotations.

Method	MSRA-TD500			ICDAR2015		
	P	R	F	P	R	F
DDR [6]	–	–	–	82.0	80.0	81.0
EAST [34]	87.3	67.4	76.1	83.6	73.5	78.2
TLOC [15]	84.5	77.1	80.6	–	–	–
Corner [18]	87.6	76.2	81.5	**94.1**	70.7	80.7
TextSnake [17]	83.2	73.9	78.3	84.9	80.4	82.6
TextField [28]	87.4	75.9	81.3	84.3	83.9	84.1
PSE-Net [23]	–	–	–	86.9	84.5	85.7
LOMO [32]	–	–	–	91.3	83.5	87.2
ATRR [25]	85.2	82.1	83.6	89.2	86.0	87.6
CRAFT [1]	88.2	78.2	82.9	89.8	84.3	86.9
PAN [24]	84.4	**83.8**	84.1	84.0	81.9	82.9
DB [11]	**91.5**	79.2	84.9	91.8	83.2	**87.3**
ContourNet [26]	–	–	–	87.6	**86.1**	86.9
Zhang et al. [33]	88.1	82.3	**85.1**	88.5	84.7	86.6
*Texts as Lines [27]	80.6	74.1	77.2	81.7	77.1	79.4
*Proposed Method	87.7	80.8	84.1	83.1	85.7	84.4

Table 3. Detection results on the Total-Text dataset with different annotations by different annotators.

Annotation	P	R	F
Annotation-1	89.8	82.6	86.0
Annotation-2	88.2	83.7	85.9
Annotation-3	89.7	83.5	86.5

ten thousand synthetic images with Chinese texts. After pre-training on both SynthText and our synthetic data, we fine-tune our model on the MSRA-TD500 dataset. As shown in Table 2, our method achieves a very close performance to the state-of-the-art methods.

Multi-oriented Text Detection. Most of the texts in ICDAR2015 are small, and some of them are severely distorted or blurred, which limits the performance of our character detection branch. Especially, our character labels of real data are generated by a model pre-trained on synthetic data, so it is much more difficult to generate accurate pseudo labels with the weak annotations. However, as shown in Table 2, our method still achieves a close performance compared

Fig. 7. The illustration of the annotations by different annotators (in different colors).

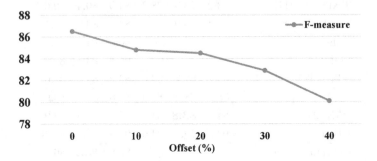

Fig. 8. Detection results on the Total-Text dataset with different manual noise, achieved by applying random offsets to the annotated coordinates.

with fully supervised methods, and outperforms the previous line-based weakly supervised method a lot.

4.4 Annotation Deviation

Our proposed labeling method is annotating each text instance with a scribble line, and different annotators tend to annotate different point coordinates by their intuitions. Hence, we arrange different annotators to label the training data individually, and it is easy to find the differences among different sets of weak annotations as shown in Fig. 7. To show the robustness of our scene text detection method to these different annotations, we conduct experiments on Total-Text. As shown in Table 3, the models trained with different annotations consistently achieve the state-of-the-art performance on the Total-Text dataset.

Furthermore, we add random noise to the Annotation-3 to study the influence of annotation deviation on the detection performance. For each point coordinate (x, y), we randomly modify it into $(x \pm R_x * H, y \pm R_y * H)$, where R_x and R_y are randomly selected in $[-\frac{1}{2}\text{Offset}, +\frac{1}{2}\text{Offset}]$ and H is the height of each text instance. Comparing Fig. 8 and Table 3, we find that the deviation from different annotators has less negative influence than the random noise of 10% offset, and the detection performance is still over 80% f-measure when the annotations are seriously disturbed by 40% offset.

Table 4. Detection performance of pre-trained model and fine-tuned models. The evaluation metric is "F-measure".

Model	ICDAR-15	MSRA-TD500	Total-Text	CTW1500
Pre-trained	62.1	56.2	66.3	57.0
Fine-tuned	84.4	84.1	86.5	86.1
Gain	**+22.3**	**+27.9**	**+20.2**	**+29.1**

Table 5. Detection results on the Total-Text dataset and the CTW1500 dataset with different multi-class classification (short for "MC") manners. "BF": The character classes include "Background" and "Foreground"; "All": The character classes for both synthetic and real data include "Background", 10 numbers, 26 letters, and "Unknown"; "All+BF": The character classes for synthetic data include "Background", 10 numbers, 26 lowercase letters, and "Unknown", and those for real data only include "Background" and "Foreground".

Dataset	MC	P	R	F
Total-Text	BF	88.4	79.5	83.7
	All	89.3	81.5	85.2
	All+BF	89.7	83.5	**86.5**
CTW1500	BF	87.6	80.8	84.1
	All	86.9	83.8	85.3
	All+BF	87.2	85.0	**86.1**

4.5 The Effectiveness of Weak Annotation

As shown in Table 4, we evaluate our pre-trained model on several benchmarks, and find that it can not perform well on real-world data, which is much worse than the models fine-tuned with weak annotation. Hence, it is hard for the pre-trained model to generate reliable pseudo labels without introducing more information, and our proposed weak annotation is effective and helpful to scene text detection.

4.6 Discussion

Multi-class Character Detection. As shown in Table 5, we study the character classification by three different settings and find that "All+BF" achieves the best performance. Since the pseudo labels of character classes generated by the pre-trained model are not as accurate as the real labels in synthetic data, we only use "Background" and "Foreground" classification for real data instead of "All" classes classification, which makes the model easier to optimize. Moreover, it is easy to change our method into an end-to-end method in the second manner. We have directly obtained the recognition results by sorting the characters from left to right of each text instance, but we can only achieve 67.0 f-measure on Total-Text, which is far from the state-of-the-art end-to-end method. In the future work, it needs the recognition annotations and better post-processing to achieve better performance for the text spotting task.

Comparison with Line-Based Text Detection Methods. Compared with Texts as Lines [27], our method has improved in the following aspects:

(1) Our weakly labeling method can save on both the annotation costs and the storage space. Their weak annotations contain two images of coarse masks for texts and background, while our annotations only consist of point coordinates for each scribble lines.

(2) Our network can use character information without involving any extra cost. Their text detection method only depends on segmentation, but ignores the character information. Although our proposed annotations lose the edge information of text instances, our character detection branch and the boundary reconstruction module can compensate for the weakness with the geometric information of detected characters. Demonstrated by our experiments, our method improves a lot on accuracy, and even outperforms state-of-the-art methods on the Total-Text dataset and the CTW1500 dataset.

Limitation. For the character detection branch, it is difficult to detect the distorted or small characters, which is the reason for the unremarkable performance on the ICDAR 2015 dataset. However, due to our proposed post-processing, there is no need to detect all the characters in a text instance, which reduces the negative influence caused by the disadvantages of the detection branch to some extent.

5 Conclusion

In this paper, we research into the difficulty that how to use a simpler labeling method for scene text detection without harming the detection performance. We propose to annotate each text instance with a scribble line in order to reduce the annotation costs, while it is a general labeling method for texts with various shapes. Moreover, we present a weakly supervised framework for scene text detection, which can be trained with our weak annotations of real data and no-cost annotations of synthetic data. By our experiments on several benchmarks, we demonstrate that it is possible to achieve state-of-the-art performance based on the weak annotations, even better. For research purposes, we will release the annotations of different benchmarks in our experiments, and we hope it will benefit the field of scene text detection to achieve better performance with simpler annotations.

Acknowledgement. This work was supported by Meituan, NSFC 61733007, the National Program for Support of Top-notch Young Professionals and the Program for HUST Academic Frontier Youth Team 2017QYTD08.

References

1. Baek, Y., Lee, B., Han, D., Yun, S., Lee, H.: Character region awareness for text detection. In: Proceedings of CVPR, pp. 9365–9374 (2019)
2. Ch'ng, C.K., Chan, C.S.: Total-text: a comprehensive dataset for scene text detection and recognition. In: Proceedings of ICDAR, vol. 1, pp. 935–942 (2017)
3. Dai, J., et al.: Deformable convolutional networks. In: Proceedings of ICCV, pp. 764–773 (2017)
4. Gupta, A., Vedaldi, A., Zisserman, A.: Synthetic data for text localisation in natural images. In: Proceedings of CVPR, pp. 2315–2324 (2016)
5. He, K., Zhang, X., Ren, S., Sun, J.: Deep residual learning for image recognition. In: Proceedings of CVPR, pp. 770–778 (2016)
6. He, W., Zhang, X.Y., Yin, F., Liu, C.L.: Deep direct regression for multi-oriented scene text detection. In: Proceedings of ICCV, pp. 745–753 (2017)
7. Karatzas, D., et al.: ICDAR 2015 competition on robust reading. In: Proceedings of ICDAR, pp. 1156–1160 (2015)
8. Karatzas, D., et al.: ICDAR 2013 robust reading competition. In: Proceedings of ICDAR, pp. 1484–1493 (2013)
9. Liao, M., Shi, B., Bai, X.: Textboxes++: a single-shot oriented scene text detector. TIP 27(8), 3676–3690 (2018)
10. Liao, M., Shi, B., Bai, X., Wang, X., Liu, W.: Textboxes: a fast text detector with a single deep neural network. In: Proceedings of AAAI, pp. 4161–4167 (2017)
11. Liao, M., Wan, Z., Yao, C., Chen, K., Bai, X.: Real-time scene text detection with differentiable binarization. In: Proceedings of AAAI, pp. 11474–11481 (2020)
12. Liao, M., Zhu, Z., Shi, B., Xia, G., Bai, X.: Rotation-sensitive regression for oriented scene text detection. In: Proceedings of CVPR, pp. 5909–5918 (2018)
13. Lin, T.Y., Dollár, P., Girshick, R., He, K., Hariharan, B., Belongie, S.: Feature pyramid networks for object detection. In: Proceedings of CVPR, pp. 2117–2125 (2017)
14. Liu, W., et al.: SSD: single shot multibox detector. In: Leibe, B., Matas, J., Sebe, N., Welling, M. (eds.) ECCV 2016. LNCS, vol. 9905, pp. 21–37. Springer, Cham (2016). https://doi.org/10.1007/978-3-319-46448-0_2
15. Liu, Y., Jin, L., Zhang, S., Luo, C., Zhang, S.: Curved scene text detection via transverse and longitudinal sequence connection. PR 90, 337–345 (2019)
16. Long, J., Shelhamer, E., Darrell, T.: Fully convolutional networks for semantic segmentation. In: Proceedings of CVPR, pp. 3431–3440 (2015)
17. Long, S., Ruan, J., Zhang, W., He, X., Wu, W., Yao, C.: TextSnake: a flexible representation for detecting text of arbitrary shapes. In: Ferrari, V., Hebert, M., Sminchisescu, C., Weiss, Y. (eds.) ECCV 2018. LNCS, vol. 11206, pp. 19–35. Springer, Cham (2018). https://doi.org/10.1007/978-3-030-01216-8_2
18. Lyu, P., Yao, C., Wu, W., Yan, S., Bai, X.: Multi-oriented scene text detection via corner localization and region segmentation. In: Proceedings of CVPR, pp. 7553–7563 (2018)
19. Ren, S., He, K., Girshick, R., Sun, J.: Faster R-CNN: towards real-time object detection with region proposal networks. IEEE Trans. Pattern Anal. Mach. Intell. 39(6), 1137–1149 (2016)
20. Shi, B., Bai, X., Belongie, S.: Detecting oriented text in natural images by linking segments. In: Proceedings of CVPR, pp. 2550–2558 (2017)
21. Shrivastava, A., Gupta, A., Girshick, R.: Training region-based object detectors with online hard example mining. In: Proceedings of CVPR, pp. 761–769 (2016)

22. Tian, S., Lu, S., Li, C.: WeText: scene text detection under weak supervision. In: Proceedings of ICCV, pp. 1492–1500 (2017)
23. Wang, W., et al.: Shape robust text detection with progressive scale expansion network. In: Proceedings of CVPR, pp. 9336–9345 (2019)
24. Wang, W., et al.: Efficient and accurate arbitrary-shaped text detection with pixel aggregation network. In: Proceedings of ICCV, pp. 8440–8449 (2019)
25. Wang, X., Jiang, Y., Luo, Z., Liu, C.L., Choi, H., Kim, S.: Arbitrary shape scene text detection with adaptive text region representation. In: Proceedings of CVPR, pp. 6449–6458 (2019)
26. Wang, Y., Xie, H., Zha, Z.J., Xing, M., Fu, Z., Zhang, Y.: ContourNet: taking a further step toward accurate arbitrary-shaped scene text detection. In: Proceedings of CVPR, pp. 11753–11762 (2020)
27. Wu, W., Xing, J., Yang, C., Wang, Y., Zhou, H.: Texts as lines: text detection with weak supervision. Math. Prob. Eng. **2020** (2020)
28. Xu, Y., Wang, Y., Zhou, W., Wang, Y., Yang, Z., Bai, X.: Textfield: learning a deep direction field for irregular scene text detection. TIP **28**(11), 5566–5579 (2019)
29. Yao, C., Bai, X., Liu, W.: A unified framework for multioriented text detection and recognition. TIP **23**(11), 4737–4749 (2014)
30. Yao, C., Bai, X., Liu, W., Ma, Y., Tu, Z.: Detecting texts of arbitrary orientations in natural images. In: Proceedings of CVPR, pp. 1083–1090 (2012)
31. Yuliang, L., Lianwen, J., Shuaitao, Z., Sheng, Z.: Detecting curve text in the wild: new dataset and new solution. arXiv preprint arXiv:1712.02170 (2017)
32. Zhang, C., et al.: Look more than once: an accurate detector for text of arbitrary shapes. In: Proceedings of CVPR, pp. 10552–10561 (2019)
33. Zhang, S.X., et al.: Deep relational reasoning graph network for arbitrary shape text detection. In: Proceedings of CVPR, pp. 9699–9708 (2020)
34. Zhou, X., et al.: East: an efficient and accurate scene text detector. In: Proceedings of CVPR, pp. 5551–5560 (2017)

EEM: An End-to-end Evaluation Metric for Scene Text Detection and Recognition

Jiedong Hao(✉), Yafei Wen, Jie Deng, Jun Gan, Shuai Ren, Hui Tan,
and Xiaoxin Chen

VIVO AI Lab, Shenzhen, China

Abstract. An objective and fair evaluation metric is fundamental to
scene text detection and recognition research. Existing metrics cannot
handle properly one-to-many and many-to-one matchings that arise naturally from the bounding box granularity inconsistency issue. They also
use thresholds to match the ground truth and detection boxes, which
leads to unstable matching result. In this paper, we propose a novel End-
to-end Evaluation Metric (EEM) to tackle these problems. EEM handles
one-to-many and many-to-one matching cases more reasonably and is
threshold-free. We design a simple yet effective method to find matching groups from the ground truth and detection boxes in an image. We
further employ a label merging method and use normalized scores to evaluate the performance of end-to-end text recognition methods more fairly.
We conduct extensive experiments on the ICDAR2015, RCTW dataset,
and a new general OCR dataset covering 17 categories of real-life scenes.
Experimental results demonstrate the effectiveness and fairness of the
proposed evaluation metric.

Keywords: Evaluation metric · Text detection and recognition ·
End-to-end

1 Introduction

End-to-end text recognition is drawing more and more attention [7,10–12,17,
20] in the research community recently. To evaluate the performance of end-
to-end text recognition systems, we need to first obtain reasonable matching
result between detection (DT) and ground truth (GT) text boxes. Based on the
matching result, edit distance or F-score can be further computed to indicate
the overall performance of the systems.

Compared with evaluation of general object recognition task [3,5,15], evaluation of end-to-end text recognition is much more challenging. In general object
recognition, object boxes are rectangular, and the label for a box is an object
class. In text recognition, the shape of text boxes may be rectangular, quadrilateral, or even curved, with arbitrary orientation. Label length of text boxes also
vary drastically: a text box may contain only one character or a long text line

J. Lladós et al. (Eds.): ICDAR 2021, LNCS 12824, pp. 95–108, 2021.
https://doi.org/10.1007/978-3-030-86337-1_7

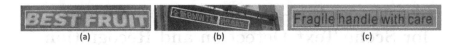

Fig. 1. (a) One-to-one (b) one-to-many (c) many-to-one matching cases. GT and DT boxes are colored in red and yellow respectively. We will use this convention throughout the paper. (Color figure online)

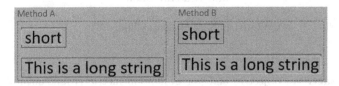

Fig. 2. Two different end-to-end methods with different detection and recognition results will have the same evaluation score if NED is used.

with more than 50 characters. The highly diversified shapes and label lengths of boxes introduce three challenging issues for evaluation metrics.

The first issue is bounding box granularity inconsistency. When annotating GT boxes, it is common and acceptable that a text line is either annotated with a line-level box or several word-level boxes. In detection results, text instances may have a single DT box or several smaller DT boxes. As a result, when matching GT with DT boxes, we frequently encounter one-to-many (OM, one GT matches several DT boxes) and many-to-one (MO, several GT match one DT box) or even many-to-many (MM, multiple GT and DT matches) matches, besides the normal one-to-one (OO, one GT matches one DT box) matches. This issue is especially common in non-Latin languages, e.g., Chinese. See Fig. 1 for some matching cases. Currently, in text box matching stage, existing end-to-end evaluation metrics [4,6,8,18] only consider OO matchings, thus producing unreasonable matching result for OM and MO matchings. We show a case in Fig. 1(c). If existing metrics were used, GT boxes would be matched against the DT box one by one, and neither of the four GT boxes matches the DT box. Thus the DT box will be a false detection, and all GT boxes will not be recalled, which is unreasonable, since the GT and DT boxes can be matched together as a MO matching.

The second issue is the unreasonable results caused by threshold. In Fig. 1(a), we show an OO matching. Since IoU between the GT and DT box is below threshold (0.5), they cannot match each other. The GT box is not recalled, and DT box is considered a false detection. Considering that one word is recalled and can be correctly recognized, this matching result is apparently not appropriate. Moreover, using threshold makes the matching result unstable, an IoU of 0.501 and 0.499 will result in completely different matchings.

The third issue is fairness of the evaluation score. F-score [4,8] requires that GT and DT label match exactly. Long DT label that has only one character error compared to GT label will be treated as false recognition, which is unreasonable.

The normalized edit distance (NED) score [18], which is based on edit distance, also has its own problems. Suppose for the same image, method A and B has detection results shown in Fig. 2, and the detected texts are perfectly recognized. Since method B recalls long texts that are perfectly recognized, its evaluation score should be higher than that of method A. However, the NED scores for method A and B are the same, which is unfair.

To address the weaknesses of existing evaluation metrics, we propose an **End-to-end Evaluation Metric (EEM)**. We tackle the OM, MO and MM matching issues by employing a simple yet effective text box matching method. We first find the best matching DT for each GT box, and vice versa. To find possible OM and MO matchings, we merge the matching box pairs into larger matching groups. Our method involves no threshold and can handle the OM, MO and MM matchings more elegantly than existing metrics. To the best of our knowledge, this is the first time that OM and MO matchings are considered in an end-to-end metric.

In order to compute the evaluation score for OM and MO matchings found in the box matching step, we further propose a new label merging scheme to merge GT and DT labels in a matching group into a single GT and DT label respectively. We also propose a new normalized score (NS) to evaluate the overall performance of end-to-end recognition systems more reliably.

The major contributions of EEM metric are summarized as follows:

- We propose a novel text box matching method that can better handle OO, OM, MO and MM matchings. Our method needs no threshold, and does not require additional annotations.
- We design a new label merging method to merge the GT and DT labels in a matching group. We propose a new normalized score (NS) to better quantify the performance of end-to-end text recognition systems.
- We conduct extensive experiments on ICDAR 2015, RCTW and a new general OCR dataset. Experimental results verify the effectiveness and fairness of EEM. We will also make the general OCR dataset, code for EEM metric as well as the recognition model publicly available.

2   Related Work

2.1   Text Box Matching

The simplest matching scheme is used by the IoU (intersection over union) metric [8,14] in which only OO matchings are considered. IoU score is used to decide whether two boxes match:

$$IoU(r_1, r_2) = \frac{Area(r_1 \cap r_2)}{Area(r_1 \cup r_2)}, \tag{1}$$

in which r_1 and r_2 are the rectangles representing the GT and DT boxes. A DT box matches a GT box if their IoU score is above 0.5. Otherwise, it is considered as false detection. The IoU metric fails to deal with complex matching cases such

Fig. 3. MO and OM matching cases that Deteval and TIoU cannot handle well.

as MO, OM and MM matches, which appear frequently in real text detection results.

The Deteval metric [9,21] aims at addressing the OM and MO matching cases in detection results. Its matching strategy is built on the concept of *area precision* and *area recall*, which are defined as follows:

$$P_{AR}(G_i, D_j) = \frac{Area(G_i \cap D_j)}{Area(D_j)}, \tag{2}$$

$$R_{AR}(G_i, D_j) = \frac{Area(G_i \cap D_j)}{Area(G_i)}, \tag{3}$$

in which G_i and D_j are GT and DT boxes, respectively.

Two thresholds t_p (0.4) and t_r (0.8) are used to impose quality constraints on the precision and recall of the detected boxes, respectively. Specifically, the constraints for different matching cases are as follows:

OO Match: The area precision and recall between GT and DT boxes must satisfy the threshold constraints.

OM Match: Two conditions must be satisfied: (1) Area precision for each DT box must be above t_p. (2) The sum of area recalls for all the DT boxes must be above t_r.

MO Match: Two conditions must be satisfied: (1) Area recall for each GT box must be above t_r. (2) The sum of area precisions for all the GT boxes must be above t_p.

Deteval fails to handle MO and OM matches if the GT and DT boxes do not meet the threshold constraints. In Fig. 3 (a) and (b), we give two such examples. In these two examples, since the sum of area recalls of all the DT boxes with the GT box is below t_r, the DT boxes fail to match the corresponding GT boxes.

TIoU [13] try to solve the OM and MO issues by providing both word and textline level annotations. The DT boxes are matched against both annotations in two steps. A threshold of 0.5 is used in the word and textline level matching step, which fails to match some DT and GT boxes due to the threshold constraints. We show a textline level and a word level GT annotation with the same detection result in Fig. 3 (c) and (d). In the first step, the DT boxes are matched to the textline-level annotation. Both the two DT boxes fail to match the GT box due to low IoU ($IoU < 0.5$). In the second word-level matching step, only the left DT box can match the leftmost GT box ($IoU > 0.5$). After the two steps, the right DT box will be regarded as false detection, and all word level GT boxes except the leftmost one are counted as missed GTs. It is unreasonable since the DT boxes and GT boxes can match quite well even if they do no satisfy the threshold constraints. Besides, since TIoU requires additional annotations, it is also labor intensive thus infeasible for large datasets.

2.2   Evaluation Score

After text box matching step, two variants of evaluation score are often used. The first is F-score [1,4,8,20], in which DT is considered a match if it satisfies the IoU metric and its label matches the GT label (ignoring the case). The other score is based on edit distance. In practice, total edit distance [6], average edit distance (AED) or normalized edit distance (NED) [18] are used. NED is often preferred since it can give an evaluation score in the $[0,1]$ range. NED can be calculated using the following formula:

$$NED = 1 - \frac{1}{N} \sum_{i=1}^{N} \frac{edit\_dist(gt_i, dt_i)}{max(len(gt_i), len(dt_i))} \tag{4}$$

in which gt_i and dt_i are the GT label and DT label for a matching pair and N is the number of matching pairs in the dataset.

2.3   Evaluation Metric for End-to-end Text Recognition

Currently, there are mainly two kinds of evaluation metrics for end-to-end text recognition. The first metric is the ICDAR metric [4,8]. The second metric is the RCTW metric [6,18]. Both the two metrics only consider OO matchings and ignores OM and MO matchings, which is the same as the matching method used by IoU metric. They differ in the evaluation score used: the ICADR metric uses F-score, while RCTW uses edit distance based scores.

3   Proposed Method

3.1   Text Box Matching

Our text box matching method consists of two steps: pair matching and set merging.

3.1.1   Pair Matching For an image, the GT and DT text boxes are represented as set G and D,

$$G = \{G_i\}, \ i = 1, 2, \cdots, |G|, \tag{5}$$
$$D = \{D_i\}, \ i = 1, 2, \cdots, |D|, \tag{6}$$

in which $|G|$ and $|D|$ represents the number of boxes in set G and D respectively. In this step, we find the best matching DT box for each GT box and vice versa. *Best matching* DT for a GT box is chosen by finding the DT box that has

---

Algorithm 1: Pair matching

Input: $\{G_i\}_{i=1}^{|G|}$, $\{D_i\}_{i=1}^{|D|}$
Output: N matching set, where $N = |G| + |D|$

1 **for** $i \leftarrow 1$ **to** $|G|$ **do**
2 $\quad maxArea \leftarrow 0$;
3 $\quad$ Initialize a new set $\{G_i\}$;
4 $\quad$ **for** $j \leftarrow 1$ **to** $|D|$ **do**
5 $\quad\quad$ **if** $Area(G_i \cap D_j) > maxArea$ **then**
6 $\quad\quad\quad maxArea = Area(G_i \cap D_j)$;
7 $\quad\quad\quad$ Update the set to $\{G_i, D_j\}$;
8 $\quad\quad$ **end if**
9 $\quad$ **end for**
10 **end for**
11 **for** $j \leftarrow 1$ **to** $|D|$ **do**
12 $\quad maxArea \leftarrow 0$;
13 $\quad$ Initialize a new set $\{D_j\}$;
14 $\quad$ **for** $i \leftarrow 1$ **to** $|G|$ **do**
15 $\quad\quad$ **if** $Area(G_i \cap D_j) > maxArea$ **then**
16 $\quad\quad\quad maxArea = Area(G_i \cap D_j)$;
17 $\quad\quad\quad$ Update the set to $\{D_j, G_i\}$;
18 $\quad\quad$ **end if**
19 $\quad$ **end for**
20 **end for**

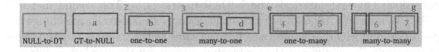

Fig. 4. An illustration of all the matching cases we can handle with our evaluation metric. DT and GT boxes are indexed in yellow digits and red alphabet characters, respectively. (Color figure online)

maximum area intersection with this GT box and vice versa. The procedure is summarized in Algorithm 1.

The idea of finding the best matching box is intuitive and consistent with the original idea of IoU. The difference is that we do not use a threshold, which will make sure that every DT and GT box can find a best matching box, regardless of actual IoU. By discarding the threshold, we can get rid of the problems introduced by threshold as discussed in Sect. 1.

After this step, we get N matching set, where $N = |G| + |D|$. The number of elements in a matching set may be one (a particular GT or DT box matches no other box) or two (where the second element is the best match for the first element).

To better illustrate the matching algorithm, we show an example of all the matching cases successfully handled by EEM in Fig. 4. We assume that all these

Table 1. Best matching DT for GT (a) and best matching GT for DT (b).

GT	a		b	c	d	e	f	g
DT	NULL		2	3	3	5	6	6

(a)

DT	1		2	3	4	5	6	7
GT	NULL		b	c	e	e	g	g

(b)

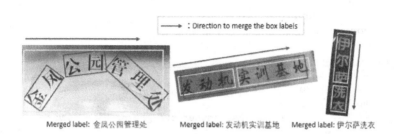

Fig. 5. Label merging order for horizontal and vertical texts and merged labels.

cases appear in one image. The matching result for GT and DT boxes are shown in Table 1a and b, respectively. After the matching step, the matching sets are:

$$\{a\}, \{b, 2\}, \{c, 3\}, \{d, 3\}, \{e, 5\}, \{f, 6\}, \{g, 6\},$$
$$\{1\}, \{2, b\}, \{3, c\}, \{4, e\}, \{5, e\}, \{6, g\}, \{7, g\}.$$

3.1.2 Set Merging As shown in Tables 1a and b, a DT box can be the best match for multiple GT boxes and vice versa. We aim at merging matching sets sharing common elements into larger groups in this step. A matching group is formed by merging recursively smaller groups until different groups have no common elements. In Algorithm 2, we give a formal presentation of this procedure.

After step 2, we get a few matching groups. The matching sets in step 1 now become

$$\{1\}, \{a\}, \{b, 2\}, \{c, d, 3\}, \{e, 4, 5\}, \{f, g, 6, 7\},$$

which correspond to the different matching cases shown in Fig. 4. For example, matching group $\{e, 4, 5\}$ and $\{c, d, 3\}$ correspond to OM and MO matching respectively.

By employing the set merging method, different cases (OO, OM, MO, MM, NULL-to-DT and GT-to-NULL) are all well handled. We will show some qualitative examples in Sect. 4.

3.2 Label Merging and Evaluation Score

3.2.1 Label Merging Since we consider both OM, MO and MM matching cases in EEM, we need to properly calculate the edit distance for a matching

---

Algorithm 2: Set merging

Input: A list of matching set: S_i, $i = 1, 2, \cdots, N$
Output: Several matching groups

1 **for** $i \leftarrow 1$ **to** N **do**
2     $marked[i] \leftarrow False$
3 **end for**
4 /* After the loop, sets which have not been marked are the final matching groups.   */
5 **for** $i \leftarrow 1$ **to** N **do**
6   **if** $marked[i]$ **then**
7   continue;
8   **end if**
9   **for** $j \leftarrow 1$ **to** N **do**
10   **if** $marked[j]$ or $i == j$ **then**
11   continue
12   **end if**
13   /* $hasCommon()$ checks whether two groups have common elements   */
14   **if** hasCommon (S_i, S_j) **then**
15   merge S_j into S_i;
16           $marked[j] = True$;
17   **end if**
18   **end for**
19 **end for**

---

group. We use a label merging method to concatenate labels of GT and DT boxes in the same matching group into one GT and DT label[1]. The edit distance of a matching group is then calculated between the new GT and DT label.

In the label merging step, it is important to decide the order in which to concatenate the labels of GT or DT boxes. We first calculate the rotation angle θ of the box with the largest area in a matching group. If $\theta \leq 45°$, We assume that characters are written horizontally from left to right. We then sort boxes of the same type, i.e., GT boxes or DT boxes, by the x coordinates of their upper left vertices and concatenate their labels respectively. If $\theta > 45°$, we assume the characters are written vertically from top to bottom. We sorted the boxes by the y coordinate of their upper left vertices and merge their labels. In Fig. 5, we show two examples of how the label merging method works.

3.2.2 Evaluation Score The performance of end-to-end text recognition systems is often compared in terms of F-score or edit distance. F-score is not a fair measure for long texts since it requires exact match. We choose edit distance in our EEM metric. After merging the labels of GT and DT boxes from the same matching group, we calculate the edit distance between the new GT label

---

[1] If there is no GT or DT box in a matching group, the GT or DT label is an empty string.

Table 2. Statistics of matching and the number of penalty. *Other* Means NULL-to-DT and GT-to-NULL cases.

	ICDAR15	CTW-12k	General OCR
OO	643	752	2547
OM	22	79	430
MO	96	6	34
Other	82	343	491
#penalty	5	22	57

and DT label for each matching group. Total edit distance on the dataset can be calculated by summing up the edit distance of all matching groups in the dataset.

In order to give an evaluation score in the range of $[0, 1]$, we introduce the Normalized Score (NS):

$$NS = 1 - \frac{\sum_{i=1}^{N} \text{edit\_dist}(gt_i, dt_i)}{\sum_{i=1}^{N} \max(\text{len}(gt_i), \text{len}(dt_i))}, \tag{7}$$

in which gt_i and dt_i are the merged GT and DT labels of a matching group, and N is the number of matching groups on the dataset.

NED pays more attention to line level, while NS cares more about character level. NS is more reliable for most cases, especially for the case mentioned in Fig. 2. The NED scores for method A and B are both 0.5 ($1 - \frac{0+1}{2} = 0.5$ and $1 - \frac{1+0}{2} = 0.5$), while NS scores are $1 - \frac{0+21}{5+21} = \frac{5}{26}$ and $1 - \frac{5+0}{5+21} = \frac{21}{26}$ respectively. The NS score for method B is higher than method A, which is more reasonable since method B recognizes more words than method A.

There are indeed a small proportion of matching cases that may give over-segmented or under-segmented results, as shown by examples in Fig. 5. For these cases, it may be appropriate to add some penalties so that the scores for these cases should be lower. The main difficulty is that it is hard to know which matching cases need penalty and to what extent. To assess the impact of this issue on fairness of EEM, we have randomly selected 200 images from the three datasets, and list the relevant statistics in Table 2. The proportion of OM&MO cases that need penalty account for less than 2% in all the matching cases. So their impact on the EEM score is insignificant.

4   Experimental Results and Analysis

4.1   Datasets and Evaluation Metrics Compared

We use three datasets in our experiments, ICDAR 2015 [8], CTW-12k [18] and a general OCR dataset collected by ourselves.

Fig. 6. Example images from the general OCR dataset. The GT annotations are shown in red quadrilateral boxes. (Color figure online)

ICDAR 2015 is an incidental scene text dataset. It contains only English words and has only one image category. The CTW-12k dataset is created specifically for Chinese text detection and recognition. It has limited image category, and the images are most scene image or born-digital image.

To increase the diversity of images in the dataset and test performance of different end-to-end recognition systems reliably, we have collected a general OCR dataset that contains images of 17 categories, such as Power Point slide, URL, receipt etc. It consists of 2760 images from different real life scenes. The texts in this dataset are mostly Chinese and English characters. We show some example images from the general OCR dataset in Fig. 6.

We compared EEM with two evaluation metrics: RCTW metric [18] and ICDAR metric [4,8]. For fair comparison with EEM, we replace the NED score of RCTW with NS score based on Eq. (7).

4.2 Experimental Details

In order to form an end-to-end text recognition system, we need to combine text detection algorithms with recognition algorithms or use end-to-end text recognition methods. We choose three detection algorithms: CTPN [19], EAST [23] and PixelLink [2], and two text recognition algorithms: LSTM-based CRNN algorithm [16] and the attention-based algorithm ASTER [17]. The combination of different detection and recognition algorithms gives us six different end-to-end recognition systems. We also compared the performance of the two end-to-end text recognition algorithms: MaskTextSpotter [10] and CharNet [22].

For the detection algorithms (PixelLink, EAST and CTPN), the detection results for ICDAR 2015 are from the authors' published models. The detection models for CTW-12k and general OCR dataset are trained by ourselves. For CTW-12k, we only use the train-val set since there is no GT annotations for test set. We select 1000 random images from the original train set for testing and use the rest for training. For general OCR dataset, the detection models are trained using our internal data.

For the recognition algorithms, we trained the CRNN and attention-based models using more than 10M images to improve the generalization ability of the models. The training images are from public datasets and a synthetic dataset that supports both the Chinese and English languages.

For the end-to-end text recognition algorithms (MaskTextSpotter and CharNet), the detection as well as the recognition results are provided by the authors.

Table 3. Experimental results on ICDAR 2015 dataset.

Methods	ICDAR			RCTW	EEM
	P	R	F	NS	NS
PixelLink + CRNN	0.447	0.430	0.438	0.586	0.703
EAST + CRNN	0.454	0.414	0.433	0.579	0.707
CTPN + CRNN	0.367	0.255	0.301	0.394	0.568
MaskTextSpotter	0.600	0.610	0.605	0.668	0.787
CharNet	0.794	0.612	0.691	0.686	0.751

Table 4. Experimental results on the general OCR dataset.

Methods	ICDAR			RCTW	EEM
	P	R	F	NS	NS
PixelLink + CRNN	0.525	0.619	0.568	0.774	0.854
EAST + CRNN	0.520	0.497	0.508	0.763	0.809
CTPN + CRNN	0.447	0.419	0.432	0.666	0.801
PixelLink + Attention	0.435	0.639	0.517	0.740	0.823
EAST + Attention	0.545	0.521	0.532	0.761	0.800
CTPN + Attention	0.477	0.447	0.461	0.660	0.793

4.3   Results and Analysis

We compare eight different end-to-end recognition systems on the three datasets, and get 14 data points for each of the three evaluation metrics. The results are shown in Tables 3, 4 and 5. We can observe that the EEM scores for different recognition systems are consistently higher than the corresponding RCTW scores. This is due to the better handling of OM and MO matchings, as well as the OO matchings. Under EEM, more GT and DT boxes are correctly matches instead of being ignored. As a result, the total edit distance is reduced, and the NS score boosted.

The performance of different end-to-end systems can also be evaluated more fairly by employing the EEM metric. Denote the first three end-to-end systems in Table 4 as S1, S2 and S3. According to the RCTW metric, the performance gain of S1 compared to S2 is insignificant (1.44%), and the performance gain of S2 compared to S3 is huge (14.6%). However, by the EEM metric, the performance gain of S1 compared to S2 is big (5.4%), while the performance gain of S2 compared to S3 is negligible (0.998%). Similarly, in Table 3, according the ICDAR metric, CharNet is better than MaskTextSpotter (+14.2% in F-score). By the EEM metric, the performance of MaskTextSpotter is actually better than CharNet (+4.79% in F-score).

Table 5. Experimental results on CTW-12k dataset.

Methods	ICDAR			RCTW	EEM
	P	R	F	NS	NS
PixelLink + CRNN	0.325	0.247	0.281	0.540	0.623
EAST + CRNN	0.322	0.173	0.225	0.464	0.542
CTPN + CRNN	0.220	0.141	0.172	0.310	0.475

Fig. 7. OO matching cases where GT and DT are not matched by other evaluation metrics. The GT, DT labels, and the evaluation results using different metrics are listed below each image.

To gain a visual understanding of the effectiveness of EEM, we also present some qualitative matching examples randomly picked from many similar matching cases in Figs. 7, 8 and 9. From these matching cases, we can see that:

OO Matches are Handled More Properly. Both the matching method used by ICDAR and RCTW metric impose the same threshold constraints for OO matches. For OO matches that do not satisfy the constraints, GT and DT boxes are not matched. Considering that some words can be recognized from the DT box even if its IoU with GT box is below the threshold, it is unreasonable to ignore this match. Take Fig. 7 (a) for an example, the IoU between the DT and GT box is 0.326. Both GT and DT box will match to None. So the DT box is considered a false detection and the GT box is considered a miss. For RCTW metric, the number of characters in both the GT and DT boxes will be added to edit distance, leading to unreasonably large edit distance (37 in this case). The ICDAR metric is also far from reasonable, since its F-score is zero.

OM and MO Matches are Better Handled. We show some OM and MO matching cases in Figs. 8 and 9 that are handled better by EEM. For the OM cases, most of the DT boxes in each example will be matched to NULL based on the matching rule of ICDAR and RCTW metric. For example, in Fig. 8 (a), the IoU between the two DT boxes and the GT box is 0.409 and 0.376. By the RCTW and ICDAR metric, the two DT boxes are not matched to the GT box ($IoU < 0.5$). The number of characters in the two DT boxes and the GT box will be all added to the edit distance. The ICDAR precision and recall are both zero, leading to zero F-score.

<table>
<tr><td>IoU: 0.409, 0.376</td><td>IoU: .0.074, 0.082, 0.072, 0.067, 0.085, 0.067</td><td>IoU: 0.373, 0.514</td></tr>
<tr><td>GT: lingbi stone</td><td>GT: 把健康带回家</td><td>GT: 翰林</td></tr>
<tr><td>DT: lingbi, stone</td><td>DT: 把 健 康 带 回 家</td><td>DT: 輸、 林</td></tr>
<tr><td>Merged DT: lingbistone</td><td>merged DT: 把健康带回家</td><td>merged DT: 翰林</td></tr>
<tr><td>ICDAR F-score: 0</td><td>ICDAR F-score: 0</td><td>ICDAR F-score: 0</td></tr>
<tr><td>RCTW edit distance: 24</td><td>RCTW edit distance: 12</td><td>RCTW edit distance: 2</td></tr>
<tr><td>Ours edit distance: 1</td><td>Ours edit distance: 0</td><td>Ours edit distance: 0</td></tr>
<tr><td>(a)</td><td>(b)</td><td>(c)</td></tr>
</table>

Fig. 8. OM matching cases successfully handled by our evaluation metrics. The IoU values between DT boxes and GT box are listed from left to right.

<table>
<tr><td>IoU: 0.403, 0.412</td><td>IoU: 0.538, 0.351</td><td>IoU: 0.274, 0.251, 0.245</td></tr>
<tr><td>GT:多种平台. 各种费率</td><td>GT: 新系统、 IC卡</td><td>GT: 鲁巷, 渔具, 总汇</td></tr>
<tr><td>DT:多种平台 各种费率</td><td>DT: 新系统 IC卡</td><td>DT: 鲁巷渔具总汇</td></tr>
<tr><td>Merged GT: 多种平台各种费率</td><td>Merged GT: 新系统IC卡</td><td>merged GT: 鲁巷渔具总汇</td></tr>
<tr><td>ICDAR F-score: 0</td><td>ICDAR F-score: 0</td><td>ICDAR F-score: 0</td></tr>
<tr><td>RCTW edit distance: 16</td><td>RCTW edit distance: 6</td><td>RCTW edit distance: 12</td></tr>
<tr><td>Ours edit distance: 1</td><td>Ours edit distance: 1</td><td>Ours edit distance: 0</td></tr>
<tr><td>(a)</td><td>(b)</td><td>(c)</td></tr>
</table>

Fig. 9. MO matching cases successfully handled by our evaluation metric. The IoU values between GT boxes and DT box are listed from left to right.

EEM metric works equally well for the MO matching cases while the ICDAR and RCTW metric work poorly. For example, in Fig. 9(a), since the IoU values between the DT and the two GT boxes are both below threshold, the DT box matches neither of the two GT boxes. The DT and GT boxes match to None by the ICDAR and RCTW metric, which leads to zero F-score for ICDAR and unreasonably large edit distances for RCTW metric.

5   Conclusion and Future Work

We presented an end-to-end evaluation metric (EEM) for evaluating the performance of end-to-end text recognition systems. We propose for the first time a novel yet simple method to find the matching groups in an image, which can handle various complicated OM and MO matchings and is threshold-free. We employ a label merging algorithm to merge the GT and DT labels for matching groups. We propose a normalized score to better quantify the performance of different end-to-end recognition systems. Quantitative and qualitative results on ICDAR 2015, CTW-12k and a new general OCR dataset show that EEM is reliable and fair compared to other metrics.

If detection alone evaluation metric is needed, our pair matching methods can be used to help solve the matching problems, like OM/MO matching and threshold issue described in introduction. We also suggest using e2e way to evaluate the performance of detection algorithms by introducing the same recognition model in EEM.

There are some limited OM&MO cases that violate our assumption during label merging stage (Sect. 3.2.1). One possible direction is to invent a new method that can rectify wrong merging orders. As for the penalty issue (Sect. 3.2.2), we will try to find proper ways to add penalties dynamically if necessary.

References

1. https://rrc.cvc.uab.es/?ch=4&com=tasks#end-to-end

2. Deng, D., Liu, H., Li, X., Cai, D.: PixelLink: detecting scene text via instance segmentation. In: AAAI (2018)
3. Girshick, R.: Fast R-CNN. In: ICCV, pp. 1440–1448 (2015)
4. Gomez, R., et al.: ICDAR2017 robust reading challenge on COCO-text. In: ICDAR, vol. 01, pp. 1435–1443 (2017)
5. He, K., Gkioxari, G., Dollár, P., Girshick, R.: Mask R-CNN. In: ICCV, pp. 2961–2969 (2017)
6. He, M., et al.: ICPR2018 contest on robust reading for multi-type web images. In: ICPR, pp. 7–12 (2018)
7. He, T., Tian, Z., Huang, W., Shen, C., Qiao, Y., Sun, C.: An end-to-end textspotter with explicit alignment and attention. In: CVPR, pp. 5020–5029 (2018)
8. Karatzas, D., et al.: ICDAR 2015 competition on robust reading. In: ICDAR, pp. 1156–1160 (2015)
9. Karatzas, D., et al.: ICDAR 2013 robust reading competition. In: ICDAR. pp. 1484–1493 (2013)
10. Liao, M., Lyu, P., He, M., Yao, C., Wu, W., Bai, X.: Mask textspotter: an end-to-end trainable neural network for spotting text with arbitrary shapes. IEEE Trans. Pattern Anal. Mach. Intell. (2019)
11. Liao, M., Shi, B., Bai, X.: Textboxes++: a single-shot oriented scene text detector. IEEE Trans. Image Process. 27(8), 3676–3690 (2018)
12. Liu, X., Liang, D., Yan, S., Chen, D., Qiao, Y., Yan, J.: FOTS: Fast oriented text spotting with a unified network. In: CVPR, pp. 5676–5685 (2018)
13. Liu, Y., Jin, L., Xie, Z., Luo, C., Zhang, S., Xie, L.: Tightness-aware evaluation protocol for scene text detection. In: CVPR, pp. 9612–9620 (2019)
14. Nayef, N., et al.: ICDAR2017 robust reading challenge on multi-lingual scene text detection and script identification - RRC-MLT. In: ICDAR, vol. 01, pp. 1454–1459 (2017)
15. Ren, S., He, K., Girshick, R., Sun, J.: Faster R-CNN: towards real-time object detection with region proposal networks. In: NeurIPS, pp. 91–99 (2015)
16. Shi, B., Bai, X., Yao, C.: An end-to-end trainable neural network for image-based sequence recognition and its application to scene text recognition. IEEE Trans. Pattern Anal. Mach. Intell. 39(11), 2298–2304 (2017)
17. Shi, B., Yang, M., Wang, X., Lyu, P., Yao, C., Bai, X.: Aster: an attentional scene text recognizer with flexible rectification. IEEE Trans. Pattern Anal, Mach. Intell. (2018)
18. Shi, B., et al.: ICDAR2017 competition on reading Chinese text in the wild (RCTW-17). In: ICDAR, vol. 1, pp. 1429–1434 (2017)
19. Tian, Z., Huang, W., He, T., He, P., Qiao, Yu.: Detecting text in natural image with connectionist text proposal network. In: Leibe, B., Matas, J., Sebe, N., Welling, M. (eds.) ECCV 2016. LNCS, vol. 9912, pp. 56–72. Springer, Cham (2016). https://doi.org/10.1007/978-3-319-46484-8_4
20. Wang, K., Babenko, B., Belongie, S.J.: End-to-end scene text recognition. In: ICCV, pp. 1457–1464 (2011)
21. Wolf, C., Jolion, J.M.: Object count/area graphs for the evaluation of object detection and segmentation algorithms. In: IJDAR, vol. 8, pp. 280–296 (2006)
22. Xing, L., Tian, Z., Huang, W., Scott, M.R.: Convolutional character networks. In: ICCV (2019)
23. Zhou, X., et al.: EAST: an efficient and accurate scene text detector. In: CVPR, pp. 2642–2651 (2017)

SynthTIGER: Synthetic Text Image GEneratoR Towards Better Text Recognition Models

Moonbin Yim[1], Yoonsik Kim[1], Han-Cheol Cho[1], and Sungrae Park[2]

[1] CLOVA AI Research, NAVER Corporation, Seongnam-si, South Korea
{moonbin.yim,yoonsik.kim90,han-cheol.cho}@navercorp.com
[2] Upstage AI Research, Yongin-si, South Korea
sungrae.park@upstage.ai

Abstract. For successful scene text recognition (STR) models, synthetic text image generators have alleviated the lack of annotated text images from the real world. Specifically, they generate multiple text images with diverse backgrounds, font styles, and text shapes and enable STR models to learn visual patterns that might not be accessible from manually annotated data. In this paper, we introduce a new synthetic text image generator, SynthTIGER, by analyzing techniques used for text image synthesis and integrating effective ones under a single algorithm. Moreover, we propose two techniques that alleviate the long-tail problem in length and character distributions of training data. In our experiments, SynthTIGER achieves better STR performance than the combination of synthetic datasets, MJSynth (MJ) and SynthText (ST). Our ablation study demonstrates the benefits of using sub-components of SynthTIGER and the guideline on generating synthetic text images for STR models. Our implementation is publicly available at https://github.com/clovaai/synthtiger.

Keywords: Optical character recognition · Synthetic text image generator · Scene text synthesis · Scene text recognition · Synthetic dataset

1 Introduction

Optical character recognition (OCR) is a technology extracting machine-encoded texts from text images. It is a fundamental function for visual understanding and has been used in diverse real-world applications such as automatic number plate recognition [16], business document recognition [3,4,15] and passport recognition [9]. In the deep learning era [10,11], OCR performance has been dramatically improved by learning from large-scale data consisting of image and text pairs. In general, OCR uses large-scale data consisting of synthetic text images because it is virtually impossible to manually gather and annotate real text images that cover the exponential combinations of diverse characteristics such as text length, fonts, and backgrounds.

© Springer Nature Switzerland AG 2021
J. Lladós et al. (Eds.): ICDAR 2021, LNCS 12824, pp. 109–124, 2021.
https://doi.org/10.1007/978-3-030-86337-1_8

Fig. 1. Word box images generated by synthesis engines. MJ provides diverse text styles but there is no noise from other texts. The examples of ST are cropped from a scene text image including multiple text boxes and they includes some part of other texts. Although our synthesis engine generates word box images as like MJ, its examples includes text noises observed in examples of ST.

OCR in the wild consists of two sub-tasks, scene text detection (STD) and scene text recognition (STR). They require similar but different training data. Since STD has to localize text areas from backgrounds, its training example is a raw scene or document snapshot containing multiple texts. In contrast, STR identifies a character sequence from a word box image patch that contains a single word or a line of words. It requires a number of synthetic examples to cover the diversity of styles and texts that might exist in the real world. This paper focuses on synthetic data generation for STR to address the diversity of textual appearance in a word box image.

There are two popular synthesis engines, MJ [5] and ST [2]. MJ is a text image synthesis engine that generates word box images by processing multiple rendering modules such as font rendering, border/shadow rendering, base coloring, projective distortion, natural data blending, and noise injection. By focusing on generating word box images rather than scene text images, MJ can control all text styles, such as font color and size, used in its rendering modules but the generated word box images cannot fully represent text regions cropped from a real scene image. In contrast, ST [2] generates scene text images that includes multiple word boxes on a single scene image. ST identifies text regions and writes texts upon the regions by processing font rendering, border/shadow rendering, base coloring, and poisson image editing. Since word boxes are cropped from a scene text image, the identified word box images can include text noises from other word boxes as like real STR examples. However, there are some constraints on choosing text styles because background regions identified for text rendering may not be compatible with some text rending functions (e.g., too small to use big font size).

To take advantage of both approaches, recent STR research [1,19] simply integrates datasets generated by both MJ and ST. However, the simple data integration not only increases the total number of training data but also causes a bias on co-covered data distributions of both synthesis engines. Although the integration provides better STR performance than the individuals, there is still room for improvement by considering a better method combining the benefits of MJ and ST.

In this paper, we introduces a new synthesis engine, referred to as **Synth**etic **T**ext **I**mage **GE**nerato**R** (SynthTIGER), for better STR models. As like MJ, SynthTIGER generates word box images without the style constraints of ST. However, as like ST, it adopts additional noise from other text regions, which can occur during cropping a text region from a scene image. Figure 1 shows synthesized examples with MJ, ST, and SynthTIGER. The examples of SynthTIGER may contain parts of other texts as like those of ST. In our apples-to-apples comparison, using SynthTIGER shows better performance than the cases of using MJ and ST respectively. Moreover, SynthTIGER provides comparable performance to the integrated dataset of MJ and ST even though only one synthesis engine is utilized.

Furthermore, we propose two methods to alleviate skewed data distribution for infrequent characters and short/lengthy words. Previous synthesis engines generate text images by randomly sampling target texts from a pre-defined lexicon. Due to the low sampling chance of infrequent characters and very short/long words, trained models often poorly perform on these kinds of words. SynthTIGER uses length augmentation and infrequent character augmentation methods to address this problem. As shown in experiment results, these methods improve STR performance over rare and short/long words.

Finally, this paper provides an open-source synthesis engine and a new synthetic dataset that shows better STR performance than the combined dataset of MJ and ST. Our experiments under fair comparisons to baseline engines prove the superiority of SynthTIGER. Furthermore, ablative studies on rendering functions describe how rendering processes in SynthTIGER contribute to improving STR performance. The experiments on data distributions of lengths and characters show the importance of synthetic text balancing. The official implementation of SynthTIGER is open-source and the synthesized dataset is publicly available.

2   Related Work

In STR, it has become a standard practice to use synthetic datasets. We introduce previous studies providing synthetic data generation algorithms for STR [5] and other tasks that can be exploited for STR [2, 8, 12].

MJ [5] is one of the most popular data generation algorithms (and the dataset generated with that approach) for STR. It produces an image patch containing a single word. In detail, the algorithm consists of six stages. *Font rendering* stage randomly selects a font and font properties such as size, weight and underline. Then, it samples a word from a pre-defined vocabulary and renders it on the foreground image layer following a horizontal line or a random curve. *Border/shadow rendering* step optionally adds inset border, outset border, or shadow image layers with a random width. The following *base coloring* stage fills these image layers with different colors. *Projective distortion* stage applies a random, full-projective transformation to simulate the 3D world. *Natural data blending* step blends these image layers with a randomly sampled crop of an image from the ICDAR 2003 and SVT datasets. Finally, *Noise* stage injects various noise such as blur and JPEG compression artifacts into the image. While MJ is known to generate text images useful enough to train STR models, it is not clear how much each stage contributes to its success.

Synthetic datasets for scene text detection (STD) can be used for STR by cropping text regions from synthesized images. The main difference of STD data generation algorithms from STR is that STD must consider the geometry of backgrounds to create realistic images. ST [2] is the most successful STD data generation algorithm. In detail, it first samples a text and a background image. Then, it segments the image based on color and texture to obtain well-defined regions (e.g., surfaces of objects). Next, it selects the region for the text and fills the text (and optionally outline) with a color based on the region's color. Finally, the text is rendered using a randomly selected font and transformation according to the region's orientation. However, the use of off-the-shelf segmentation techniques for text-background alignment can produce erroneous predictions and result in unrealistic text images. Recent studies like SynthText3D [8] and UnrealText [12] address this problem by synthesizing images with 3D graphic engines. Experiment results show that text detection performance can be notably improved by using synthesized text images without text alignment error. However, it is not clear whether these datasets generated from the virtual 3D world can benefit text recognition task.

3   SynthTIGER

SynthTIGER consists of two major components: text selection and text rendering modules. The text selection module is used to sample a target text, t, from a pre-defined lexicon, L. Then, the text rendering module generates a text image by using multiple fonts F, backgrounds (textures) B, and a color map C. In this section, we first describe the text rendering process and then the target text selection process.

3.1   Text Rendering Process

Synthesized text images should reflect realness of texts in both a "micro" perspective of a word box image and a "macro" perspective of a scene-level text image. The rendering process of SynthTIGER generates text-focused images for the micro-level perspective, but it additionally adapts noises of the macro-level perspective. Specifically, SynthTIGER renders a target text and a noisy text and combines them to reflect the realness of the text regions (in a wild, a part of a word appearance can be included in a region for another word). Figure 2 overviews the modules of SynthTIGER engine. It consists of five procedures: (a) text shape selection, (b) text style selection, (c) transformation, (d) blending, and (e) post-processing. The first three processes, (a), (b), and (c), are separately applied to the foreground layer for a target text and the mid-ground layer for a noise text. In the (d) step, the two layers are combined with a background to represent a single synthesized image. Finally, the (e) adds realistic noises. The followings introduce each module in detail.

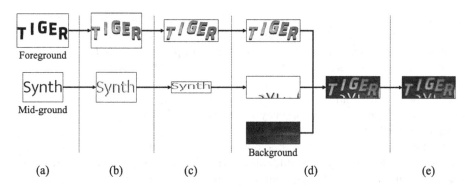

Fig. 2. Overview of SynthTIGER rendering processes consisting of (a) text shape selection, (b) text style selection, (c) transformation, (d) blending, and (e) post-processing. Foreground text presents a ground-truth text of a generated image and mid-ground text is used as a high frequency visual noises with lines and dots from visual appearance of characters. Foreground and mid-ground texts are rendered independently in (a), (b), and (c) and the background and the texts are combined in (d). Finally, vision noises are added upon the combined image in (e).

(a) Text Shape Selection. Text shape selection decides a 2-dimensional shape of a 1-dimensional character sequence. This process first identify individual character shapes of a target text **t** and then renders them upon a certain line on 2D space in the left-to-right order.

To reveal visual appearances of the characters, a font is randomly selected from a pool of font styles **F** and each character is rendered upon individual boards with randomly chosen font size and thickness. To add diversity of font styles, elastic distortion [20] is applied to the rendered characters.

Defining a spatial order of characters is essential to map characters upon 2-dimensional space. For straight texts, SynthTIGER basically aligns character boards in the left-to-right order with a certain margin between the boards. For curved texts, SynthTIGER places the character boards on a parabolic curve. The curvature of the curve is identified by the maximum height-directional gaps between the centers of the boards. The maximum gap is randomly chosen and the middle points of the target text are allocated on the centroid of the parabolic curve. The character boards upon the curve are rotated with a slope of the curve under a certain probability.

(b) Text Style Selection. This part chooses colors and textures of a text, and injects additional text effects such as bordering, shadowing and extruding texts.

A color map **C** is an estimation of a real distribution over colors of text images. It can be identified by clustering colors of real text images. It usually consists of 2, or 3 clusters with the mean gray-scale colors and their standard deviation (s.t.d). MJ and ST also utilize this color map identified from ICDAR03 dataset [13] and IIIT dataset [14], respectively. In our work, we adapt to the color map used in ST. The color selection from the color map is conducted sequentially in an order of a cluster and a color based on the mean and the s.t.d. Once a color is selected, SynthTIGER changes the color of the character appearances.

The colors of texts in the real world is not simply represented with a single color. SynthTIGER uses multiple texture sources, **B**, to reflect the realness of text colors. Specifically, it picks up a random texture from **B**, performs a random crop of the texture, and use it as a texture of the text appearance of the synthetic image. In this process, transparency of the texture is also randomly chosen to diversify the effect of textures.

In the real world, the characters' boundary exhibits diverse patterns depending on text styles, text background, and environmental conditions. We can simulate the boundary styles by applying text border, shadow, and extruding effects. SynthTIGER randomly chooses one of these effects and applies it to the text. All required parameters such as effect size and color will be sampled randomly from a pre-defined range.

(c) Transformation. The visual appearance of the same scene text image can be significantly different depending on the view angle. Moreover, these text images detected by different OCR engines or labeled by multiple human annotators will have different patterns. SynthTIGER generates synthesized images reflecting these characteristics by utilizing multiple transformation functions.

In detail, SynthTIGER provides stretch, trapezoidate, skew and rotate transformations. Their functions are explained below.

- *Stretch* adjusts the width or height of the text images.
- *Trapezoidate* choose an edge of the text image and then adjust its length.
- *Skew* tilts the text image to one of the four directions such as the right, left, top and bottom.
- *Rotate* turns the text image clockwise or anticlockwise.

SynthTIGER applies one of these transformations to the text image with necessary parameter values randomly sampled.

Finally, SynthTIGER adds random margins to simulate the diverse results of text detectors. The margins are independently applied to the top, bottom, left, and right of the image.

(d) Blending. The blending process first creates a background image by randomly sampling color and texture from the color map, **C**, and the texture database, **B**. It randomly changes the transparency of the background texture to diversify the impact of the background. Secondly, it creates two text images, foreground and mid-ground, with the same rendering processes but different random parameters. The first one contains a target text and the second one carries a noise text. The next step is to combine the mid-ground and background images. The blending process first crops the background image to match the text image size. Then, it randomly shifts the noise text in the mid-ground and makes the non-textual area transparent. Finally, it merges two images by using one of multiple blending methods: normal, multiply, screen, overlay, hard-light, soft-light, dodge, divide, addition, difference, darken-only, and lighten-only. The last step is to overlay the foreground text image on the merged background. The target text area with a little margin is kept non-transparent to distinguish between the target text and the noise text. During this process, it also uses one of the blending methods aforementioned.

A synthesized image created through these steps from (a) to (d) might not be a good text-focused image for several reasons. For example, its text and background color happen to be indistinguishable because they are chosen independently. To address this problem, we adopted Flood-Fill algorithm[1]. We apply this algorithm starting from a pixel inside the target text, count the number of text boundary pixels visited, and calculate the ratio of the visited text boundary pixels to the number of all boundary pixels. This process is repeated until all target text pixels are used. If this ratio exceeds a certain threshold, we conclude that the target text and background are indistinguishable and discard the generated image.

(e) Post-processing. Post-processing is conducted to finalize the synthetic data generation. In this process, SynthTIGER injects general visual noises such as gaussian noise, gaussian blur, resize, median blur and JPEG compression.

3.2   Text Selection Strategy

The previous methods, MJ and ST, randomly sample target texts from a user-provided lexicon. In contrast, SynthTIGER provides two additional strategies to control the text length distribution and character distribution of a synthesized dataset. It alleviates the long-tail problem inherited from the use of a lexicon.

Text Length Distribution Control. The length distribution of texts randomly sampled from a lexicon does not represent the true distribution of a real world text data. To alleviate this problem, SynthTIGER performs text length distribution augmentation with the probability $p_{l.d.}$ where $l.d.$ stands for length distribution. The augmentation process first randomly chooses the target text length between 1 and the pre-defined maximum value. Then, it randomly samples a word from the lexicon. If the word matches the target length, SynthTIGER uses it as a target text. For longer words, it simply cuts off extra rightmost characters. For shorter words, it samples a new word and attaches it to the right of the previous one until the concatenated word matches or exceeds the target length. The rightmost extra characters will be cut off. Text length augmentation, however, should be used with caution because the generated texts are mostly nonsensical. In the experiment section, we show that text length augmentation with $p_{l.d.} = 0.5$ increases STR accuracy more than 2%, while making little difference with $p_{l.d.} = 1.0$.

---

[1] https://en.wikipedia.org/wiki/Flood_fill.

Character Distribution Control. Languages such as Chinese and Japanese use a large number of characters. A synthesized dataset for such a language often lacks enough amount of samples for rare characters. To deal with this problem, SynthTIGER conducts character distribution augmentation with the probability $p_{c.d}$ where *c.d.* stands for character distribution. When the augmentation is triggered, it randomly chooses a character from vocabulary and samples a word having that character. In the experiments, we show that character distribution augmentation with $p_{c.d.}$ between 0.25 and 0.5 improves STR performance for both scene and document domains.

4 Experimental Results

This section consists of (S4.1) our experimental settings for both synthetic data generation and STR model development, (S4.2) comparison to popular synthetic datasets, (S4.3) apples-to-apples comparison of synthesis engines with the common resources, (S4.4) ablative studies on rendering functions, (S4.5) experiments on controlling text distributions.

4.1 Experimental Settings

To compare synthetic data generation engines, synthetic datasets are built with them and then STR performances are evaluated from the models trained with the generated datasets. Here, we describe the resources used in synthetic data generation engines and the training and evaluating settings of STR models.

Resources for Synthetic Data Generation. To build synthetic datasets, multiple resources, **L**, **F**, **B**, and **C**, are required. Table 1 describes the resources used in MJ and ST as well as in our experiments. As can be seen, MJ and ST are built with their own resources. MJ utilizes a lexicon combining Hunspell corpus[2] and ground-truths of real STR examples from ICDAR(IC), SVT, and IIIT datasets. MJ also uses textures and its color map of IC03 and SVT. In contrast, ST does not use the ground-truth information except for the color map from IIIT. SynthTIGER utilizes a lexicon consisting of texts of MJ and ST dataset and uses the same textures and color map with ST. They have different number of fonts that are available from google fonts[3]. Common* in the table uses an another lexicon from Wikipedia to evaluate all synthesis engines without ground-truth information of real STR test examples and test sets except for the color map. For our Japanese STR tasks, we utilize a Japanese lexicon (84M) from Wikipedia and Twitter, 382 fonts, $\mathbf{B}_{ST}$, and $\mathbf{C}_{ST}$.

---

[2] http://hunspell.github.io/.
[3] https://fonts.google.com/.

Table 1. Resources used to build synthetic datasets of Latin texts. Common* indicates a setting for apples-to-apples comparison between synthesis engines. "×3" indicates text augmentation for capitalized, upper-cased, and lower-cased words.

	Lexicon (**L**)	Font (**F**)	Texture (**B**)	Color map (**C**)
MJ [5]	Hunspell + test-sets of IC, SVT, IIIT(90K × 3)	1,400 fonts	IC03, SVT train-set (358)	IC03 train-set
ST [2]	Newsgroup20 (366K)	1,200 fonts	Crawling (8,010)	IIIT word dataset
SynthTIGER	MJ + ST (197K×3)	3,568 fonts	$\mathbf{B}_{ST}$	$\mathbf{C}_{ST}$
Common*	Wikipedia (19M×3)	$\mathbf{F}_{SynthTIGER}$	$\mathbf{B}_{ST}$	$\mathbf{C}_{ST}$

Experimental Settings for Training and Evaluating STR Models. In this paper, we evaluate synthetic dataset by training a STR model with them and evaluating the trained model on real STR examples. We choose BEST [1] as our base model since it is generally used as well as its implementation is publicly available. All synthetic datasets built for our experiments consists of 10M word box images. The public datasets, MJ and ST, contains 8.9M and 7M word box images respectively and they are also evaluated with the same process.

The BEST model are trained only with synthetic datasets. The training and evaluation is conducted with the STR test-bed[4]. Most of experimental settings follows the training protocol of Baek et al. [1] except for the input image size of 32 by 256.

The evaluation protocol is also the same with [1]. Specifically, we test two STR scenario depending on languages: one is Latin and the other is Japanese. For the Latin case, character vocabulary consists of 94 including both alphanumeric and special characters. STR models are evaluated on test-sets of STR benchmarks; 3,000 images of IIIT5k [14], 647 images of SVT [21], 1,110 images of IC03 [13], 1,095 images of IC13 [7], 2,077 images of IC15 [6], 645 images of SVTP [17], and 288 images of CUTE80 [18]. We also test performances on business documents with our in-house 38,493 images. We only evaluate on alphabets and digits due to in-consistent labels of the benchmark datasets. For the Japanese case, the vocabulary consists of 6,723 characters including alphanumeric, special, hiragana/katakana, and some Chinese characters. The evaluation is conducted on our in-house datasets; 40,938 images of scenes and 38,059 images of Japanese business documents.

4.2   Comparison on Synthetic Text Data

Table 2 compares STR performances of BEST [1] models trained with synthetic text images from MJ, ST, and SynthTIGER. As reported in previous works, the combination of MJ and ST shows better performances than their single usages. SynthTIGER always provides a better STR performance than the single usages

---

[4] https://github.com/clovaai/deep-text-recognition-benchmark.

Table 2. Benchmark performances of BEST [1] trained from synthetic text images. The amount of MJ, ST, MJ+ST, our data are 8.9M, 7M, 15.9M, and 10M, respectively.

Dataset	Regular				Irregular			Total
	IIIT5k	SVT	IC03	IC13	IC15	SVTP	CUTE80	
MJ [5]	83.4	84.5	85.6	83.5	66.0	73.0	64.6	78.3
ST [2]	86.1	82.5	90.7	89.8	64.5	69.1	60.1	79.7
MJ [5] + ST [2]	90.9	87.2	**92.1**	91.2	**72.9**	**77.8**	73.6	85.1
SynthTIGER (Ours)	**93.2**	**87.3**	90.5	**92.9**	72.1	77.7	**80.6**	**85.9**

Table 3. Benchmark performances of BEST [1] trained from synthetic generators with *the same resources*. The total amount of MJ*, ST*, MJ*+ST*, and Ours* are identical to 10M.

Dataset	Regular				Irregular			Total
	IIIT5k	SVT	IC03	IC13	IC15	SVTP	CUTE80	
MJ*	87.1	81.9	83.6	86.1	62.9	69.8	57.3	78.3
ST*	79.0	80.4	76.8	79.5	59.6	66.2	59.4	72.8
MJ* + ST*	89.5	83.9	**87.0**	**89.3**	68.1	**74.6**	72.9	**82.1**
SynthTIGER* (Ours*)	**89.8**	**84.5**	84.2	87.9	**69.5**	73.8	**74.0**	**82.1**

of MJ and ST. Interestingly, ours achieves comparable or better performance than combined data (MJ+ST). It should be noted that the amount of combined training data is 1.5 times larger than ours.

4.3   Comparison on Synthetic Text Image Generators with Same Resources

Since the outputs of the engines depend on the resources such as fonts, textures, color maps, and a lexicon, we provide fair comparisons, referred as to '*', by setting the same resources in Table 3. To present a fair comparison, we set the total amount of comparison data as 10M. For example, the total amount of MJ*+ST* is 10M and other comparisons such as MJ*, ST*, and Ours* are also 10M. In Table 3, ours shows clear improvement from single usages of MJ* and ST*. Also, ours have comparable performance with combined datasets.

We collect some examples from the test benchmarks where ours provides correct predictions. In Fig. 3, ours has robust predictions when the text-images contain high-frequency noises such as lines, complex backgrounds, and parts of other characters. Although complexly combined functions could contribute to the correct predictions, we believe employing the proposed mid-ground also result in robust performance. We will provide more descriptions about effects of mid-ground in (S4.4) with the ablative study.

Table 4. Latin and Japanese recognition performance on scene and document images.

Dataset	Latin		Japanese	
	Scene	Document	Scene	Document
MJ* + ST*	**82.1**	82.5	60.0	83.8
Ours*	**82.1**	**85.6**	**60.1**	**86.8**

Fig. 3. Correctly predicted cases of ours. The predictions are positioned on the right side of the images. The texts of the first row are the predictions of MJ*+ST*; The texts of the second row are the predictions of Ours* (-) Mid-ground text that is described in ablative study; The texts of the third row are the predictions of Ours*.

We extend the experiments on Japanese to compare the language generalization performance between ours and previous generators. Moreover, we demonstrate the performance of document images to confirm the extensibility of another domain. In these experiments, all engines share the same resources because MJ and ST have not provided Japanese. Table 4 shows that ours achieves comparable or better performance than combined datasets. Specifically, ours achieves much better performance on document image with 3.0 pp improvements. Since document images usually contain high-frequency noises such as scan noise and part of characters that are included in other lines or paragraphs, we believe the use of mid-ground could cope with these noises.

4.4 Ablative Studies on Rendering Functions

We investigate the effects on the STR performance of rendering function by excluding each function. Although, we do not propose and optimize all the rendering functions, we believe these ablative studies are significant to STR fields. This is because detailed investigations of rendering functions have not been reported and also can be a guide to subsequent data generation researches. To help intuitive understanding of each function, we present some visual examples of rendering function in Fig. 4. As presented in Table 5, the texture blending, transformation, margin, and post-processing critically impact the performance. Moreover, the proposed mid-ground text enhances both regular and irregular benchmarks. Other rendering functions also contribute to the STR performances.

To show the effects of the proposed mid-ground, we present some examples in Fig. 3 where baseline (Ours*) can correctly predict the results, but "(-)Mid-ground text" cannot. These figures show that the proposed mid-ground can

Curved text	Text effect	Transformation	Mid-ground	Post-processing

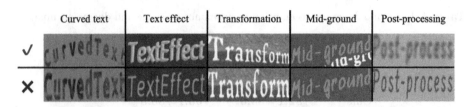

Fig. 4. Rendering effect visualization of SynthTIGER modules. The images on the top show the cases when the effects are applied. The bottom images represent the cases when the effects are off. All functions are essential to reveal the realness of the synthetic text images.

Table 5. The performance of rendering functions. (-) indicates exclusion from baseline. The color exclusion indicates using random color selection and the blending exclusion means using "normal" blending

	Regular	Irregular	Total
Baseline	87.8	70.9	82.1
(-) Curved text	87.2 (−0.6)	70.2 (−0.7)	81.4 (−0.7)
(-) Elastic distortion	87.7 (−0.1)	68.4 (−2.5)	81.1 (−1.0)
(-) Color map	87.6 (−0.2)	70.7 (−0.2)	81.9 (−0.2)
(-) Texture blending	84.5 (−3.3)	66.0 (−4.9)	78.2 (−3.9)
(-) Text effect	87.3 (−0.5)	67.7 (−3.2)	80.6 (−1.5)
(-) Transformation	87.3 (−0.5)	64.4 (−6.5)	79.5 (−2.6)
(-) Margin	87.1 (−0.7)	66.7 (−4.2)	80.2 (−1.9)
(-) Mid-ground text	87.4 (−0.4)	69.6 (−1.3)	81.4 (−0.7)
(-) Blending modes	88.0 (+0.2)	70.2 (−0.7)	81.9 (−0.2)
(-) Visibility check	87.1 (−0.7)	71.3 (+0.4)	81.8 (−0.3)
(-) Post-processing	86.1 (−1.7)	59.0 (−11.9)	76.9 (−5.2)

help to handle more diverse real-world scenes that could degrade the recognition performances.

4.5 Experiments on Text Selection

Experiments on Text Length Distribution Control. As presented in Fig. 5(a), we found that short and long length texts in Latin training data are insufficient to cover real-world texts and length distribution between training and evaluation data are quite different. To alleviate these problems, we apply text length distribution augmentation to Latin, and thus, the augmented distribution can cover a wide range of length texts as a red graph. We also present the STR performance according to the change of $p_{l.d.}$ in Table 6. We find that the best performance is achieved when $p_{l.d.}$ sets 50% and it is comparable with "Optimized" that is regarded as upper bound. Specifically, "Optimized" makes

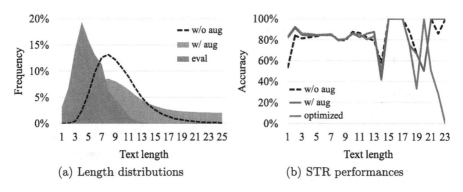

(a) Length distributions   (b) STR performances

Fig. 5. (a) Text length distribution of training data without augmentation (dashed), with augmentation (red) and evaluation data (blue). (b) Accuracy by the length of the model without length augmentation (dashed), model with 50% applied (red) and model with length augmentation optimized to length distribution of evaluation data (blue). The blue line drops sharply for long texts because texts longer than 15 characters in the evaluation dataset rarely exist. (Color figure online)

Table 6. STR accuracy according to the change of probability of length distribution augmentation. "Optimized" indicates the length augmentation is optimized to length distribution of evaluation data.

Probability	0%	25%	50%	75%	100%	Optimized
Accuracy	82.1	83.9	**84.2**	82.7	82.0	84.9

the distribution of training data have a similar distribution of evaluation data by controlling target length. Figure 5(b) shows that the proposed augmentation prevents critical performance degradation for very short and long length texts.

Experiments on Character Distribution Control. As presented in Fig. 6(a), Japanese characters, which is composed of thousands of characters, have an unbalanced long-tail problem. However, these low-presented characters in training data are frequently exhibited in evaluation data. For example, ¥, which is a currency sign with red-circled in the figure, contained words are rarely presented in training data, but, they are frequently exhibited in evaluation data. To relieve this problem, we apply character augmentation method guaranteeing the minimum numbers of examples including rare characters as red histogram in Fig. 6(a). We also present text recognition performances according to the change of $p_{c.d.}$ in Table 7. It shows the character distribution augmentation greatly improves scene and document performances when the $p_{c.d.}$ sets 50%. It can be seen in Fig. 6(b) that the proposed augmentation works for recognizing rare character included words.

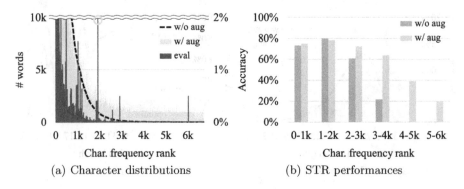

(a) Character distributions   (b) STR performances

Fig. 6. (a) The black dashed line shows the relationship between a specific Japanese character (X-axis) and the number of words involving that character (Y-axis) in the training data. Note that characters at the X-axis are sorted in descending order by the number of their occurrence. The red and blue histograms show the same relationship with the character distribution augmented training data and the evaluation data. (b) The red and gray histograms indicate the STR accuracy (Y-axis) for a subset of the text vocabulary (X-axis). Note that the values at X-axis (e.g., 0-1k) stands for a set of any words involving a character that occurs *min-max* times in the training data. (Color figure online)

Table 7. Recognition accuracy according to the change of probability of character distribution augmentation.

Probability	0%	25%	50%	75%	100%
Scene	60.1	61.9	**62.6**	59.4	54.5
Document	86.8	**88.1**	87.8	87.2	83.2

5   Conclusion

Synthesizing text images is essential to learn a general STR model by simulating diversity of texts in the real-world. However, there has been no guideline for the synthesis process. This paper addresses the issues by introducing a new synthesis engine, SynthTIGER, for STR. SynthTIGER solely shows better or comparable performance when compared to existing synthetic datasets and its rendering functions are evaluated under a fair comparison. SynthTIGER also addresses biases on text distribution of synthetic datasets by providing two text selection methods over lengths and characters. Our experiments on rendering methods and text distributions show that controlling text styles and text distributions of synthetic dataset affects to learn more generalizable STR models. Finally, this paper contributes to OCR community by providing an open-sourced synthesis engine and a new synthetic dataset.

References

1. Baek, J., et al.: What is wrong with scene text recognition model comparisons? Dataset and model analysis. In: International Conference on Computer Vision (ICCV) (2019)
2. Gupta, A., Vedaldi, A., Zisserman, A.: Synthetic data for text localisation in natural images. In: Proceedings of the IEEE Conference on Computer Vision and Pattern Recognition, pp. 2315–2324 (2016)
3. Hwang, W., et al.: Post-OCR parsing: building simple and robust parser via bio tagging. In: Workshop on Document Intelligence at NeurIPS 2019 (2019)
4. Hwang, W., Yim, J., Park, S., Yang, S., Seo, M.: Spatial dependency parsing for 2D document understanding. arXiv preprint arXiv:2005.00642 (2020)
5. Jaderberg, M., Simonyan, K., Vedaldi, A., Zisserman, A.: Synthetic data and artificial neural networks for natural scene text recognition. In: Workshop on Deep Learning, NIPS (2014)
6. Karatzas, D., Gomez-Bigorda, L., et al.: ICDAR 2015 competition on robust reading. In: ICDAR, pp. 1156–1160 (2015)
7. Karatzas, D., et al.: ICDAR 2013 robust reading competition. In: ICDAR, pp. 1484–1493 (2013)
8. Liao, M., Song, B., Long, S., He, M., Yao, C., Bai, X.: SynthText3D: synthesizing scene text images from 3D virtual worlds. Science China Inf. Sci. **63**(2), 1–14 (2020). https://doi.org/10.1007/s11432-019-2737-0
9. Limonova, E., Bezmaternykh, P., Nikolaev, D., Arlazarov, V.: Slant rectification in Russian passport OCR system using fast Hough transform. In: Ninth International Conference on Machine Vision (ICMV 2016), vol. 10341, p. 103410P. International Society for Optics and Photonics (2017)
10. Liu, X., Meng, G., Pan, C.: Scene text detection and recognition with advances in deep learning: a survey. Int. J. Doc. Anal. Recognit. (IJDAR) **22**(2), 143–162 (2019). https://doi.org/10.1007/s10032-019-00320-5
11. Long, S., He, X., Yao, C.: Scene text detection and recognition: the deep learning era. Int. J. Comput. Vision **129**(1), 161–184 (2021). https://doi.org/10.1007/s11263-020-01369-0
12. Long, S., Yao, C.: UnrealText: synthesizing realistic scene text images from the unreal world. arXiv preprint arXiv:2003.10608 (2020)
13. Lucas, S.M., Panaretos, A., Sosa, L., Tang, A., Wong, S., Young, R.: ICDAR 2003 robust reading competitions. In: ICDAR, pp. 682–687 (2003)
14. Mishra, A., Alahari, K., Jawahar, C.: Scene text recognition using higher order language priors. In: BMVC (2012)
15. Motahari, H., Duffy, N., Bennett, P., Bedrax-Weiss, T.: A report on the first workshop on document intelligence (DI) at NeurIPS 2019. ACM SIGKDD Explor. Newsl. **22**(2), 8–11 (2021)
16. Patel, C., Shah, D., Patel, A.: Automatic number plate recognition system (ANPR): a survey. Int. J. Comput. Appl. **69**(9) (2013)
17. Phan, T.Q., Shivakumara, P., Tian, S., Tan, C.L.: Recognizing text with perspective distortion in natural scenes. In: ICCV, pp. 569–576 (2013)
18. Risnumawan, A., Shivakumara, P., Chan, C.S., Tan, C.L.: A robust arbitrary text detection system for natural scene images. ESWA **41**, 8027–8048 (2014)
19. Shi, B., Yang, M., Wang, X., Lyu, P., Yao, C., Bai, X.: Aster: an attentional scene text recognizer with flexible rectification. IEEE Trans. Pattern Anal. Mach. Intell. **41**(9), 2035–2048 (2018)

20. Simard, P.Y., Steinkraus, D., Platt, J.C., et al.: Best practices for convolutional neural networks applied to visual document analysis. In: ICDAR, vol. 3. Citeseer (2003)
21. Wang, K., Babenko, B., Belongie, S.: End-to-end scene text recognition. In: ICCV, pp. 1457–1464 (2011)

Fast Recognition for Multidirectional and Multi-type License Plates with 2D Spatial Attention

Qi Liu[1], Song-Lu Chen[1], Zhen-Jia Li[1], Chun Yang[1], Feng Chen[2], and Xu-Cheng Yin[1]([✉])

[1] University of Science and Technology Beijing, Beijing, China
{qiliu7,songluchen,zhenjiali}@xs.ustb.edu.cn,
{chunyang,xuchengyin}@ustb.edu.cn
[2] EEasy Technology Company Ltd., Zhuhai, China

Abstract. The multi-type license plate can be roughly classified into two categories, i.e., one-line and two-line. Many previous methods are proposed for horizontal one-line license plate recognition and consider license plate recognition as a one-dimensional sequence recognition problem. However, for multidirectional and two-line license plates, the features of adjacent characters may mix together when directly transforming a license plate image into a one-dimensional feature sequence. To solve this problem, we propose a two-dimensional spatial attention module to recognize license plates from a two-dimensional perspective. Specifically, we devise a lightweight and effective network for multidirectional and multi-type license plate recognition in the wild. The proposed network can work in parallel with a fast running speed because it does not contain any time-consuming recurrent structures. Extensive experiments on both public and private datasets verify that the proposed method outperforms state-of-the-art methods and achieves a real-time speed of 278 FPS. Our codes are available at https://github.com/qiLiu77/SALPR.

Keywords: Multidirectional license plate recognition · Multi-type license plate recognition · 2D spatial attention · 278 FPS

1 Introduction

License plate recognition (LPR) can be applied to many real applications, such as traffic control, vehicle search, and toll station management [15,19,23]. There are many types of license plates (LPs) in real scenarios. As shown in Fig. 1, multi-type LPs could be roughly classified into two categories, i.e., one-line and two-line. Besides, license plates in the wild are multidirectional due to different shooting angles. However, many existing approaches are only aim at horizontal one-line license plate recognition. Therefore, it is still challenging to recognize multidirectional and multi-type license plates.

J. Lladós et al. (Eds.): ICDAR 2021, LNCS 12824, pp. 125–139, 2021.
https://doi.org/10.1007/978-3-030-86337-1_9

(a)

(b)

Fig. 1. Examples of multi-type LPs with multiple directions: (a) one-line LPs, and (b) two-line LPs.

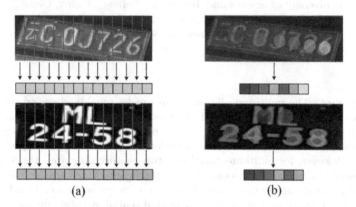

(a)   (b)

Fig. 2. Illustration of extracting visual features in the 1D or 2D space. (a) The images are transformed into 1D feature sequences, and the features of some adjacent characters are mixed. (b) The features of each LP character are extracted in the 2D space by our proposed method. Different colors represent different character orders.

Many previous LPR methods [5,12,26,31] consider license plate recognition as a one-dimensional sequence recognition problem. They transform input images into one-dimentional (1D) sequences by Convolutional Neural Networks (CNNs) for visual features extraction and utilize Connectionist Temporal Classification (CTC) loss [7] or sequential attention mechanism [2] to predict characters. However, these methods are not suitable for multidirectional and two-line license plate recognition. As shown in Fig. 2(a), the features of adjacent characters may mix together when directly transforming LP images into 1D features. Besides, many methods [16,20,31] propose to rectify LP images or features to the horizontal direction by a rectification module for multidirectional license plate recognition. However, the rectification process has a large amount of calculation and is time-consuming. In addition, only a few methods are proposed for multi-type especially two-line license plate recognition. However, these methods [6,10] are based on character segmentation, requiring lots of character location annotations and complicated post-processing.

We choose to use a two-dimensional (2D) attention mechanism to develop a fast and effective network for multidirectional and multi-type license plate recognition. Different from existing sequential attention algorithms on 2D visual feature [8,13,18], we consider LPR as an image classification problem that the prediction of each character in the LP is independent and parallel. Specifically, we propose a 2D spatial attention module without any recurrent structures to extract the most distinctive features for each character in the 2D space (see Fig. 2(b)). Based on this, we devise a fast and effective network for multidirectional and multi-type license plate recognition. In detail, the proposed network is composed of three components, i.e., a CNN-based encoder, a 2D spatial attention module, and a fully connected (FC) layer based decoder. Given an input image, the network first extracts visual features using the encoder, then aligns the whole LP features with each character by the 2D spatial attention module by reweighting features, and finally decodes attentioned features to predict characters.

To summarize, our main contributions are as follows:

- We propose a 2D spatial attention module to align the whole LP features with each character in a 2D space. The 2D spatial attention module could extract the most distinctive features for each character independently and in parallel.
- We devise a lightweight and effective network for license plate recognition. The network only contains 1.2M parameters and achieves a speed of 278 FPS on an Nvidia GTX 1080Ti GPU.
- The proposed network is well applied to multidirectional and multi-type license plate recognition and achieves state-of-the-art performance on both public and private LP datasets.

2   Related Work

2.1   Multidirectional License Plate Recognition

License plates in real scenes are multidirectional due to various shooting angles. To recognize multidirectional LPs, Zherzdev et al. [31] propose to introduce STN [9] before recognizer to rectify input images to horizontal LPs. Wang et al. [26] also introduce STN as preprocessing but use bi-directional recurrent neural networks (BRNNs) and CTC loss for license plate recognition. Lu et al. [16] propose to predict the tilted angle of an LP firstly, and rotate LP features by the predicted angle. Qin et al. [20] propose to use projective transformation to rectify multidirectional LPs to horizontal. To sum up, all of these approaches need an extra rectified module to rectify input images or features. Instead, our proposed method does not need any rectified module before the recognizer. Besides, Zhang et al. [30] propose a sequential attention module on 2D visual feature for multidirectional license plate recognition, but it needs time-consuming RNNs and only achieves a speed at 40 FPS. Our proposed method is based on 2D spatial attention and is much faster than sequential attention-based approaches.

2.2   Multi-type License Plate Recognition

Multi-type license plates can be classified into two categories: one-line and two-line. Spanhel et al. [25] devises eight classifiers, each of which is for one character of fixed location in LP. For various one-line license plate, they insert blank

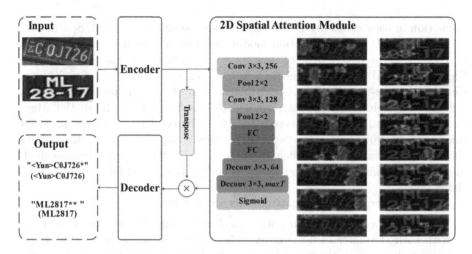

Fig. 3. Overall architecture of our proposed network. "Conv" and "Deconv" stand for convolutional layer and deconvolution layer with kenerl size and channel. "Pool", "FC", and "Transpose" refer to max pooling layer with kenerl size, FC layer, and matrix transpose, respectively. "maxT" is the max length of LP texts during training, and the operator of "⊗" means matrix multiplication. "*" in the output results represents a blank character and it will be removed in the testing phase.

characters to the ground truth of license plate text according to the layout of different license plates. With the popularity of CTC loss in text recognition [3,4,24], many LPR approaches based on CTC [12,20] are proposed to recognize one-line license plates with variable length. However, the above methods are not suitable for two-line license plate recognition. There are some methods [6,10] for multi-type license plate recognition, including both one-line and two-line. However, these methods are based on character segmentation and needs plenty of character-level annotations for training. Compared to the previous multi-type LPR methods, we propose a unified net-work for both one-line and two-line license plate recognition and do not need any character-level labels.

Kessentini et al. [10] propose a two-stage network for both one-line and two-line license plate recognition. They use YOLOv2 [21] to detect and recognize each character of the license plate and needs plenty of character-level annotations for training, which is labor-intensive. Compared to the previous multi-type LPR methods, we propose a unified network for both one-line and two-line license plate recognition and do not need any character-level labels.

2.3   Fast License Plate Recognition

Fast license plate recognition methods focus on both the inference speed and the performance. Spanhel et al. [25] abandon RNN and adopt eight branches of fully connected layers to predict characters. Similar to [25], Wang et al. [28] propose a lightweight backbone and devises three classifiers for license plate recognition.

Zherzdev et al. [31] also abandon RNN and propose a lightweight LPR network with CTC loss. However, these methods are only suitable for one-line license plate recognition. Instead, our method not only achieves a comparable speed but also can recognize multi-type license plates, including both one-line and two-line license plates.

3   Method

Our proposed method aims at recognizing multidirectional and multi-type license plates with a fast speed. Given an input image, the proposed network first extracts visual features of the whole LP and then calculates the attention weights of each character in the 2D space. According to the calculated attention weights, the network calculates the attentioned features which represent the most distinctive features for each character. Finally, the network decodes the attentioned features to characters in parallel. As shown in Fig. 3, our proposed network contains three components, i.e., the encoder, the 2D spatial attention module, and the decoder. The following sections will illustrate the details of these three components.

3.1   Encoder

We adopt a lightweight backbone from Holistic CNN (HC) [25] to extract visual features. Specifically, we abandon all the FC layers used as character classifiers and change the downsampling rate of features by alternating the number and the kernel size of pooling layers. Precisely, we empirically choose a downsampling rate r of 4, and the detailed structure of the encoder is shown in Table 1. The input is set to be gray images with the fixed size (W, H) of (96, 32).

Table 1. Description of the encoder. The stride and the padding of each convolutional layer are 1. A batchnorm layer and a RELU layer are added to each convolutional layer.

Name	Layer Type	Channel	Size	Input	Output
conv1-1	convolution	32	3×3	$96 \times 32 \times 1$	$96 \times 32 \times 32$
conv1-2	convolution	32	3×3	$96 \times 32 \times 32$	$96 \times 32 \times 32$
conv1-3	convolution	32	3×3	$96 \times 32 \times 32$	$96 \times 32 \times 32$
conv2-1	convolution	64	3×3	$96 \times 32 \times 32$	$96 \times 32 \times 64$
conv2-2	convolution	64	3×3	$96 \times 32 \times 64$	$96 \times 32 \times 64$
conv2-3	convolution	64	3×3	$96 \times 32 \times 64$	$96 \times 32 \times 64$
pool1	max pooling	64	2×2	$96 \times 32 \times 64$	$48 \times 16 \times 64$
conv3-1	convolution	128	3×3	$48 \times 16 \times 64$	$48 \times 16 \times 128$
conv3-2	convolution	128	3×3	$48 \times 16 \times 128$	$48 \times 16 \times 128$
conv3-3	convolution	128	3×3	$48 \times 16 \times 128$	$48 \times 16 \times 128$
pool2	max pooling	128	2×2	$48 \times 16 \times 128$	$24 \times 8 \times 128$

3.2   2D Spatial Attention Module

The features extracted by the encoder contain the information of the total characters in an LP image. We choose to utilize attention mechanism to obtain the most distinctive features for each character, which will learn the proper alignment of the whole LP features with each character.

The Sequential Attention Mechanism. is proposed for sequence recognition problems. The most commonly used attention algorithms for LP and text recognition is based on [2]. The attention weights α_{tj} of t time is calculated as Eq. (1), where $f_h()$, $f_a()$, and $f_{softmax}()$ represent an RNN cell, scoring function and softmax function, respectively. y_{t-1} is the output of the last time, and h_j represents the j-th hidden state of the input.

$$\begin{cases} s_t = f_h\left(s_{t-1}, y_{t-1}, \sum_{j=1}^{T_x} \alpha_{tj} h_j\right) \\ \alpha_{tj} = f_{softmax}\left(f_a\left(s_{t-1}, h_j\right)\right) \end{cases} \tag{1}$$

It can be found that attention weights are coupled with previous outputs. Therefore, when the previous character is wrongly predicted, the attention weights of the current time might be wrong. In addition, since there is little semantics between LP characters, the recurrent structure for modeling the relationship of adjacent characters is unnecessary. In view of the above two aspects, we consider LPR as an image classification task and propose a novel spatial attention module to predict each character in the LP independently and in parallel.

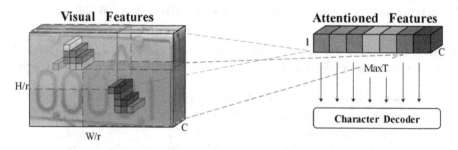

Fig. 4. The function sketch map of our proposed spatial attention algorithm. Attentioned features for each character are the weighted sum of the visual features according to the calculated attention weights. The different colors on attentioned features represent different character orders and the different colors on visual features represent different attention weights. "MaxT" refers to the max length of LP texts in the training phase. (Color figure online)

The proposed 2D spatial attention module aims to extract the most distinctive features for every character in the 2D space. As illustrated in Fig. 4, for every character in an LP image, its corresponding attentioned features are the weighted sum of the visual features according to the calculated attention weights. Then, all the attentioned features are sent to the decoder in parallel by concatenating features together along the width dimension.

To ensure both global and local information could be covered, we adopt two kinds of operators, i.e., convolution (Conv) and FC. Specifically, the 2D spatial attention module takes the visual features $F \in \mathbb{R}^{C \times H/r \times W/r}$ extracted from encoder as input and calculates attention weights $A \in \mathbb{R}^{maxT \times H/r \times W/r}$ as Eq. (2),

$$A = f_{Sig}\left(f_{Deconv}\left(f_{FC}\left(f_{Conv-Pool}\left(F\right)\right)\right)\right) \tag{2}$$

where C is the channel of visual features, $maxT$ is the max length of LP texts in the dataset, $f_{Sig}()$, $f_{Deconv}()$, $f_{FC}()$ and $f_{Conv-Pool}()$ refers to sigmoid operator, deconvolution operator, fully connected operator, and convolution with max pooing operator. Detailed configuration is described in Fig. 3. In this way, the attention weights of different characters are calculated independently and in parallel. For the k-th character of the LP, its attention weight $A_k \in \mathbb{R}^{H/r \times W/r}$ is the k-th slice of A in channel dimension, and its attentioned feature $F_{A(k)} \in \mathbb{R}^{C \times 1 \times 1}$ is computed as,

$$F_{A(k,j)} = \sum_{m,n=0}^{m=H/r,n=W/r} A_{k,m,n} \times F_{j,m,n} \ , j \in [0, C-1] \tag{3}$$

To realize parallel computation of attentioned feature, we first reshape A to $A' \in \mathbb{R}^{maxT \times (H/r \times W/r)}$, and F to $F' \in \mathbb{R}^{C \times (H/r \times W/r)}$, then utilize matrix multiplication to calculate attentioned features for all the characters simultaneously (see Eq. (4)), where F^T represents the transpose matrix of F.

$$F_A = A'\left(F'\right)^T \tag{4}$$

3.3 Decoder

The decoder of the proposed method is a character classifier. Considering that there is little semantic relation between adjacent characters, we adopt one FC layer as the decoder for convenience. The cross-entropy loss is employed for training. Because the length of the LP text changes with the LP type, we increase a blank character to deal with variable lengths of LPs. For the LP with fewer characters than $maxT$, blank characters are added to the end of ground truth until the length reaching $maxT$ in the training phase.

4 Experiment

4.1 Datasets

The existing public datasets are basically composed of one-line LPs. To demonstrate the effectiveness of our proposed method on multi-type LPs, including

(a)   (b)

(c)

Fig. 5. Examples of private LP datasets: (a) synthetic LPs, (b) examples of MBLP, captured by mobile phones, and (c) examples of DRLP, collected by driving records.

one-line and two-line LPs, we evaluate our method on both public datasets and private datasets which cover several types of LP.

SynthLP is generated according to the layouts of Chinese license plates in the way of using Python Image Library(PIL). Besides, we add random affine transformation, Gaussian noise, and RGB channel addition noise to the generated images. SynthLP covers two types of LP and 31 provinces in China (see Fig. 5(a)). Each type of LPs contains 80000 images.

MBLP is collected by mobile phones in different cities of China. After careful annotation, a total of 26426 LP images are obtained. According to different layouts of LP, MBLP is categorized into three subsets (see Fig. 5(b)), i.e., one-line plates with a green background (l1-g), one-line plates with a yellow background (l1-y), and two-line plates with a yellow background (l2-y). The number of LPs of each subset is 8412, 5514, and 12500 sequentially. The ratio of training set to testing set is 9:1.

DRLP is collected by drive recorders in different cities of China. After careful annotation, a total of 50536 LP images are obtained. According to different layouts of LP, DRLP is classified into five subsets (see Fig. 5(c)), i.e., one-line plates with a blue background (l1-b), two-line plates with a yellow background (l2-y), two-line plates with a black background (l2-bl), one-line plates with a black background containing 6 characters (l1-bl6), and one-line plates with a black background containing 7 characters (l1-bl7). The number of LPs of each subset is 33980, 2946, 736, 5791, and 7083 sequentially. The ratio of training set to testing set is 9:1.

CCPD [29] is currently the largest public LP dataset. It totally contains 8 subsets divided by the complexity and photographing condition of LP images, such as tilt, weather and so on. Among them, 10k images are used for training, and 15k images are used for testing.

CDLP [30] is collected from the internet and only for testing. It contains 1200 images covering different photographing conditions, vehicle types, and region codes of Chinese provinces.

4.2   Implementation Details

For testing on MBLP and DRLP, due to the different layout of LPs and the unbalanced number of various types of LPs, it is difficult to converge by training only on these two sets. To avoid using character location labels, we use large amounts of synthetic data to pretrain the network to ensure convergence. Concretely, the network was first pretrained on SynthLP for 8 epochs and was finetuned on the combined training datasets of MBLP and DRLP for 100 epochs.

For testing on CCPD, we use the same evaluation criteria as in [29]. We trained a YOLOv3 [22] detector on the training set of CCPD to extract the bounding boxes of LPs and achieve a performance of precision=0.998, recall=0.999. Our proposed network is trained on LP images on CCPD training datasets cropped by annotations for 50 epochs.

For testing on CDLP, we follow [30] to use the model trained on CCPD to evaluate CDLP.

In our experiments, $maxT$ is set to 8. The network is trained with ADAM optimizer [11]. The initial learning rate is set to be 1e-3 and multiplies 0.5 every 20000 iterations. We adopt a batch size of 128 and a fixed size of (96, 32) for the input images. We do not adopt any data augment strategy for training. And all the experiments are conducted on an NVIDIA GTX 1080Ti GPU with 11GB memory.

4.3   Experiments on Benchmarks

Table 2. License plate recognition accuracy on MBLP and DRLP. The subsets represent different types of LPs. ‡ means that it is implemented by ourselves.

Datasets			MBLP			DRLP				
Methods (images)	Dim	All	l1-g	l1-y	l2-y	l1-b	l2-y	l2-b	l1-b6	l1-b7
		7700	842	552	1250	3398	295	74	580	709
CTC [1]	1D	83.54	91.69	86.77	84.56	78.25	75.25	63.51	90.00	95.35
Atten1D [1]	1D	83.92	91.81	87.14	85.76	78.46	75.59	56.76	91.03	95.49
Atten2D [1]	2D	89.71	**96.31**	**94.93**	91.12	85.13	79.32	83.78	**95.17**	97.74
HC [25] ‡	2D	87.94	94.18	90.04	90.08	84.14	**80.68**	78.54	91.21	95.63
Ours	2D	**90.23**	96.19	**94.93**	**92.72**	**85.63**	80.33	**85.13**	95.00	**97.88**

Performances on MBLP and DRLP: With reference to [1], we implemented the three most commonly used text recognition algorithms for LPR, i.e., CTC-based, 1D attention-based, and 2D attention-based. Except for the difference of some convolutional kernel size, the encoder networks of these implementations are the same as ours. The results are shown in Table 2. Comparing the results of 1D-based methods with 2D-based methods, it shows that the 2D-based method

Table 3. License plate recognition accuracy on CCPD. † means that LP images used for testing are cropped by the ground truths. "Rot.", "Wea.", and "Cha." represent "Rotate", "Weather", and "Challenge", respectively.

Methods (images)	All	Base 100k	DB 20k	FN 20k	Rot. 10k	Tilt 10k	Wea. 10k	Cha. 10k	Time (ms)
SSD [14]+HC [25]	95.2	98.3	96.6	95.9	88.4	91.5	87.3	83.8	25.6
TE2E [12]	94.4	97.8	94.8	94.5	87.9	92.1	86.8	81.2	310
PRnet [29]	95.5	98.5	96.9	94.3	90.8	92.5	87.9	85.1	11.7
MORAN [17]	98.3	99.5	98.1	98.6	98.1	98.6	97.6	86.5	18.2
DAN [27]	96.6	98.9	96.1	96.4	91.9	93.7	95.4	83.1	19.3
Zhang et al. [30]	98.5	99.6	98.8	98.8	96.4	97.6	98.5	88.9	24.9
Ours	**98.74**	**99.73**	**99.05**	99.23	97.62	**98.4**	**98.89**	88.51	**3.6**
Ours †	98.73	99.67	99.04	**99.24**	**97.66**	98.26	98.83	**89.17**	

has significant superiority over the 1D-based methods, which reveals that decoding on 2D features is more suitable for LPR. Furthermore, among the three 2D-based methods, our proposed method performs better in recognizing multi-type LPs, especially for two-line LPs.

Performances on CCPD: As illustrated in Table 3, our approach achieves the best performance, especially on non-horizontal subsets, such as "Rotate", "Tilt". Meanwhile, the proposed method achieves a fast inference speed of 3.6 ms per image, i.e., 278 FPS. Due to the difference between annotated and detected bounding boxes, the LP images cropped by these two kinds of bounding boxes is different. The performance on the two kinds of LP images fluctuates slightly, which indicates our proposed method is robust to the variation of LP boundaries.

Table 4. License plate recognition accuracy on CDLP. RC refers to region code, i.e., the Chinese characters representing the Chinese provinces.

Methods	CDLP	
Criterion	ACC	ACC w/o RC
PRnet [29]	66.5	78.9
Zhang et al. [30]	70.8	86.1
Ours	**80.3**	**91.7**

Performances on CDLP: Table 4 shows our method significantly outperforms the other methods. It indicates that our approach could extract the most distinctive feature for each character and has good generalization for different photographing conditions. Besides, there is almost one kind of region code of LPs in CCPD, whereas CDLP covers many region codes. Therefore, the accuracy with region code is relatively low.

4.4   Ablation Study

To further demonstrate the effectiveness of our proposed 2D spatial attention algorithm on multidirectional and multi-type LPs, we conduct ablation experiments on MBLP, DRLP, and CCPD. "Rotate" and "Tilt" in CCPD are combined as the non-horizontal subset. Moreover, we divide the combination of MBLP and DRLP into the one-line subset and the two-line subset according to the number of LP text lines.

Table 5. Ablation study on the 2D spatial attention modules with local or global operators. "Atten." represents attention module.

Methods	Local	Global	CCPD		MBLP&DRLP		
			All	Non-horizontal	All	One-line	Two-line
Ours (w/o. Atten.)			98.53	94.33	87.96	88.29	86.72
Ours (w/o. FC)	√		98.63	97.61	88.66	88.73	88.38
Ours (w/o. Conv)		√	98.66	97.74	89.19	89.01	89.87
Ours	√	√	**98.73**	**97.96**	**90.23**	**92.26**	**90.12**

The proposed attention module contains two different types of operators, namely local and global operators. We first learn the influence of different types of operators. Besides, we remove the attention module and add several convolutional layers to validate the effectiveness of the attention module. In the way of adjusting the hyperparameters of the operators, such as kernel size, we ensure that different models have a comparable quantity of parameters. As shown in Table 5 and Fig. 6, the proposed attention module could boost accuracy on both non-horizontal and multi-type LP datasets. Besides, the attention module using both global and local operators is better than using only one of them.

Table 6. Ablation study on the 2D spatial attention modules with different numbers of operators. "Configuration" represents the structure of attention modules, where "×" means the number of operators. "Non-hor." is noted for non-horizontal.

Methods	Configuration	CCPD		MBLP&DRLP		
		All	Non-hor.	All	one-line	two-line
Ours-1	Conv × 1-Fc × 1-Deconv × 1	98.04	96.48	87.05	87.46	85.48
Ours-2	Conv × 2-Fc × 2-Deconv × 2	**98.73**	**97.96**	**90.23**	**92.26**	90.12
Ours-3	Conv × 3-Fc × 3-Deconv × 3	98.65	97.72	90.15	90.10	**90.36**

Furthermore, the suitable number of operators for the 2D spatial attention module is learned. We implement three spatial attention modules with different numbers of Conv, FC, and Deconv layers. Table 6 indicates that the performance grows with the increase of operators and tends to be saturated when the number of each operator reaching 2.

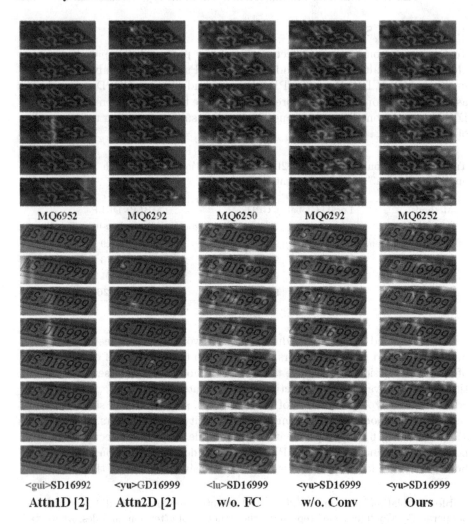

Fig. 6. Examples of attention visualization. Top to bottom: 1st to 6th or 8th character. Each column represents a different attention algorithm. Red: wrong predictions. (Color figure online)

4.5 Attention Visualization

The attention weights obtained by the spatial attention module are visualized. We also visualize the attention weights of the 1D and 2D attention text recognition algorithms for comparison. As illustrated in Fig. 6, the 1D attention weights are prone to be inaccurate, while the attention weights in 2D are more precise. Moreover, different from sequential attention algorithms, the 2D spatial attention module pays attention to not only the character areas but also the background. Therefore, we infer that moderate background information may be helpful for character recognition.

<hei> AC3B00   <ning> A89786   MT4287   <meng> A3600M

Fig. 7. Examples of wrong predictions. Red: wrong predictions.

4.6 Failure Analysis

There are some failure cases in Fig. 7. Our proposed method is not capable of processing LPs with large tilted angles (greater than 45 degrees). Besides, due to complex lighting conditions, similar characters in LPs are easy to be misclassified, such as "8" and "0", "0" and "D", and "1" and "7".

5 Conclusion

In this paper, we propose a 2D spatial attention module for multidirectional and multi-type license plate recognition. Furthermore, we develop a lightweight and effective LPR network, which contains no RNNs and achieves 278 FPS. Extensive experiments demonstrate the superiority of our approach. For future research, we will add some regular terms to constrain the attention weights to be more precise.

Acknowledgement. The research is supported by National Natural Science Foundation of China (62006018).

References

1. Baek, J., et al.: What is wrong with scene text recognition model comparisons? Dataset and model analysis. In: ICCV 2019, pp. 4714–4722. IEEE (2019). https://doi.org/10.1109/ICCV.2019.00481
2. Bahdanau, D., Cho, K., Bengio, Y.: Neural machine translation by jointly learning to align and translate. In: Bengio, Y., LeCun, Y. (eds.) ICLR 2015 (2015)
3. Breuel, T.M.: High performance text recognition using a hybrid convolutional-LSTM implementation. In: ICDAR 2017, pp. 11–16. IEEE (2017). https://doi.org/10.1109/ICDAR.2017.12
4. Chen, K., et al.: A compact CNN-DBLSTM based character model for online handwritten Chinese text recognition. In: ICDAR 2017, pp. 1068–1073. IEEE (2017). https://doi.org/10.1109/ICDAR.2017.177
5. Duan, S., Hu, W., Li, R., Li, W., Sun, S.: Attention enhanced ConvNet-RNN for Chinese vehicle license plate recognition. In: Lai, J.-H., et al. (eds.) PRCV 2018. LNCS, vol. 11257, pp. 417–428. Springer, Cham (2018). https://doi.org/10.1007/978-3-030-03335-4_36
6. Gou, C., Wang, K., Yao, Y., Li, Z.: Vehicle license plate recognition based on extremal regions and restricted Boltzmann machines. IEEE Trans. Intell. Transp. Syst. **17**(4), 1096–1107 (2016)

7. Graves, A., Fernández, S., Gomez, F.J., Schmidhuber, J.: Connectionist temporal classification: labelling unsegmented sequence data with recurrent neural networks. In: ICML 2006, vol. 148, pp. 369–376. ACM (2006)

8. Huang, Y., Luo, C., Jin, L., Lin, Q., Zhou, W.: Attention after attention: Reading text in the wild with cross attention. In: ICDAR 2019, pp. 274–280. IEEE (2019). https://doi.org/10.1109/ICDAR.2019.00052

9. Jaderberg, M., Simonyan, K., Zisserman, A., Kavukcuoglu, K.: Spatial transformer networks. In: Advances in Neural Information Processing Systems 28: Annual Conference on Neural Information Processing Systems 2015, pp. 2017–2025 (2015)

10. Kessentini, Y., Besbes, M.D., Ammar, S., Chabbouh, A.: A two-stage deep neural network for multi-norm license plate detection and recognition. Expert Syst. Appl. **136**, 159–170 (2019)

11. Kingma, D.P., Ba, J.: Adam: a method for stochastic optimization. In: Bengio, Y., LeCun, Y. (eds.) ICLR 2015 (2015)

12. Li, H., Wang, P., Shen, C.: Toward end-to-end car license plate detection and recognition with deep neural networks. IEEE Trans. Intell. Transp. Syst. **20**(3), 1126–1136 (2019)

13. Li, H., Wang, P., Shen, C., Zhang, G.: Show, attend and read: a simple and strong baseline for irregular text recognition. In: AAAI 2019, pp. 8610–8617. AAAI Press (2019). https://doi.org/10.1609/aaai.v33i01.33018610

14. Liu, W., et al.: SSD: single shot multibox detector. In: Leibe, B., Matas, J., Sebe, N., Welling, M. (eds.) ECCV 2016. LNCS, vol. 9905, pp. 21–37. Springer, Cham (2016). https://doi.org/10.1007/978-3-319-46448-0_2

15. Liu, X., Ma, H., Li, S.: PVSS: a progressive vehicle search system for video surveillance networks. J. Comput. Sci. Technol. **34**(3), 634–644 (2019). https://doi.org/10.1007/s11390-019-1932-x

16. Lu, N., Yang, W., Meng, A., Xu, Z., Huang, H., Huang, L.: Automatic recognition for arbitrarily tilted license plate. In: ICVIP 2018, pp. 23–28. ACM (2018). https://doi.org/10.1145/3301506.3301547

17. Luo, C., Jin, L., Sun, Z.: MORAN: a multi-object rectified attention network for scene text recognition. Pattern Recognit. **90**, 109–118 (2019)

18. Ly, N.T., Nguyen, C.T., Nakagawa, M.: An attention-based end-to-end model for multiple text lines recognition in Japanese historical documents. In: ICDAR 2019, pp. 629–634. IEEE (2019). https://doi.org/10.1109/ICDAR.2019.00106

19. Martínez-Carballido, J., Alfonso-López, R., Ramírez-Cortés, J.M.: License plate digit recognition using 7×5 binary templates at an outdoor parking lot entrance. In: CONIELECOMP 2011, pp. 18–21. IEEE (2011)

20. Qin, S., Liu, S.: Towards end-to-end car license plate location and recognition in unconstrained scenarios. CoRR **abs/2008.10916** (2020)

21. Redmon, J., Farhadi, A.: YOLO9000: better, faster, stronger. In: CVPR 2017, pp. 6517–6525. IEEE Computer Society (2017). https://doi.org/10.1109/CVPR.2017.690

22. Redmon, J., Farhadi, A.: Yolov3: an incremental improvement. CoRR abs/1804.02767 (2018). http://arxiv.org/abs/1804.02767

23. Shao, W., Chen, L.: License plate recognition data-based traffic volume estimation using collaborative tensor decomposition. IEEE Trans. Intell. Transp. Syst. **19**(11), 3439–3448 (2018)

24. Shi, B., Bai, X., Yao, C.: An end-to-end trainable neural network for image-based sequence recognition and its application to scene text recognition. IEEE Trans. Pattern Anal. Mach. Intell. **39**(11), 2298–2304 (2017)

25. Spanhel, J., Sochor, J., Juránek, R., Herout, A., Marsik, L., Zemcík, P.: Holistic recognition of low quality license plates by CNN using track annotated data. In: AVSS 2017, pp. 1–6. IEEE Computer Society (2017). https://doi.org/10.1109/AVSS.2017.8078501

26. Wang, J., Huang, H., Qian, X., Cao, J., Dai, Y.: Sequence recognition of Chinese license plates. Neurocomputing **317**, 149–158 (2018)

27. Wang, T., et al.: Decoupled attention network for text recognition. In: AAAI 2020, pp. 12216–12224. AAAI Press (2020)

28. Wang, Y., Bian, Z., Zhou, Y., Chau, L.: Rethinking and designing a high-performing automatic license plate recognition approach. CoRR abs/2011.14936 (2020). https://arxiv.org/abs/2011.14936

29. Xu, Z., et al.: Towards end-to-end license plate detection and recognition: a large dataset and baseline. In: Ferrari, V., Hebert, M., Sminchisescu, C., Weiss, Y. (eds.) ECCV 2018. LNCS, vol. 11217, pp. 261–277. Springer, Cham (2018). https://doi.org/10.1007/978-3-030-01261-8_16

30. Zhang, L., Wang, P., Li, H., Li, Z., Shen, C., Zhang, Y.: A robust attentional framework for license plate recognition in the wild. IEEE Trans. Intell. Transp. Syst. **PP**(99), 1–10 (2020)

31. Zherzdev, S., Gruzdev, A.: LPRNet: license plate recognition via deep neural networks. CoRR abs/1806.10447 (2018). http://arxiv.org/abs/1806.10447

A Multi-level Progressive Rectification Mechanism for Irregular Scene Text Recognition

Qianying Liao[1], Qingxiang Lin[3], Lianwen Jin[1,2(✉)], Canjie Luo[1], Jiaxin Zhang[1], Dezhi Peng[1], and Tianwei Wang[1]

[1] School of Electronic and Information Engineering, South China University of Technology, Guangzhou 510641, China
{eelqy,eelwjin,eedzpeng}@mail.scut.edu.cn
[2] Guangdong Artificial Intelligence and Digital Economy Laboratory (Pazhou Lab), Guangzhou, China
[3] Tencent Technology (Shenzhen) Co., Ltd., Shenzhen, China

Abstract. Rectifying irregular texts into regular ones is a promising approach for improving scene text recognition systems. However, most existing methods only perform rectification at the image level once. This may be insufficient for complicated deformations. To this end, we propose a multi-level progressive rectification mechanism, which consists of global and local rectification modules at the image level and a refinement rectification module at the feature level. First, the global rectification module roughly rectifies the entire text. Then, the local rectification module focuses on local deformation to achieve a more fine-grained rectification. Finally, the refinement rectification module rectifies the feature maps to achieve supplementary rectification. In this way, the text distortion and interference from the background are gradually alleviated, thus benefiting subsequent recognition. The entire rectification stage is trained in an end-to-end weakly supervised manner, requiring only images and their corresponding text labels. Extensive experiments demonstrate that the proposed rectification mechanism is capable of rectifying irregular scene texts flexibly and accurately. The proposed method achieves state-of-the-art performance for three testing datasets including IIIT5K, IC13 and SVTP.

Keywords: Optical character recognition (OCR) · Deep learning · Irregular scene text recognition

1 Introduction

Scene text recognition plays an essential role in a wide range of practical applications, such as autonomous driving, instant translation, and image searching.

This research is supported in part by NSFC (Grant No.: 61936003, 61771199), GD-NSF (no. 2017A030312006).

J. Lladós et al. (Eds.): ICDAR 2021, LNCS 12824, pp. 140–155, 2021.
https://doi.org/10.1007/978-3-030-86337-1_10

(a) Pipeline of the existing rectification methods

(b) Pipeline of the proposed rectification method

Fig. 1. Difference between other rectification methods and ours.

Thanks to advancements in deep learning, significant progress has been made in scene text recognition over the past years. However, recognizing irregular scene text is still regarded as a very challenging task owing to the varied shapes and distorted patterns. Rectifying an image containing irregular text is a promising solution to address this issue, which converts the text into a more regular shape and improves the recognition performance.

In recent years, many researchers have proposed various rectification methods. Most methods perform rectification only once at the image level. Liu et al. [22] employed affine transformation while Shi et al. [32] employed Thin-Plate Spline (TPS) [1] transformation to rectify the entire text image. These two methods focus on the global deformation of the text rather than the local deformation of details. To deal with local deformation, Luo et al. [24] divided the image into multiple patches and predicted patch-wise offsets. Ideally, this method can handle arbitrary irregular texts. However, it is still difficult to rectify the text image locally without any global geometric constraints, which limits the effectiveness of rectification. Rather than methods that preform single rectification, Zhan et al. [43] designed an iterative rectification pipeline, demonstrating the comparative superiority of multiple rectification. However, this method employs only TPS-based transformation for iterative rectification, which ignores local details. Regardless of single or multiple rectification, these methods perform rectification only at the image level and employ only one rectification mode (see Fig. 1(a)), which cannot fully exploit the advantages of global and local transformation.

Therefore, we propose a novel multi-level progressive rectification mechanism. We adopt three different rectification modules to achieve text transformation progressively, performing at both the image and feature levels (see Fig. 1(b)). First, inspired by the Spatial Transformation Network (STN) [12], we adopt affine transformation in the global rectification module to capture the global shape of the text and rectify the shape into a more horizontal one, which achieves rough rectification. Then, inspired by Luo et al. [24], the local rectification module predicts offsets for each pixel of the image to rectify curved text, which achieves a more fine-grained rectification. Finally, inspired by Huang et al. [10], we also adopt a rectification module for feature-level rectification. Specifically, the refinement rectification module predicts attention maps for the feature maps to perceive more informative features, which achieves supplementary rectification at the feature level. Notably, the final rectification module performs rectification implicitly, which does not result in a visual change in shape. Through the multi-level progressive rectification mechanism, the text distortion and background interference are gradually alleviated, and thus the subsequent recognition network can achieve better performance.

The contributions of this paper are summarized as follows:

1) We propose a novel multi-level progressive rectification mechanism. It consists of three different rectification modules, including global and local rectification modules at the image level and a refinement rectification module at the feature level. The three modules work progressively, yielding accurate rectification results.
2) The proposed rectification mechanism can be integrated into mainstream recognizers to improve recognition performance, indicating the generalization of the mechanism.
3) The proposed rectification modules are optimized in an end-to-end weakly supervised manner, requiring only word-level annotations. Extensive experiments demonstrate the superiority of the proposed rectification mechanism on public benchmarks.

2   Related Work

Scene text recognition is one of the most important problems in the field of computer vision. Inspired by speech recognition and machine translation, numerous methods based on encoder-decoder frameworks have been proposed to recognize text in natural scene images. Shi et al. [30] combined convolutional neural networks (CNNs) with recurrent neural networks (RNNs) to construct an end-to-end trainable network under the supervision of Connectionist Temporal Classification (CTC) [6] loss. Lee et al. [15] introduced an attention-based decoder to automatically align feature areas with targets. Cheng et al. [2] proposed a focus mechanism to avoid attention drift, but it requires extra character-level bounding box annotations.

The methods mentioned above mainly focus on regular scene text recognition. They are inadequate for recognizing irregular scene text suffering from various types of distortions and complex backgrounds. In recent years, irregular scene text recognition has become a popular research topic. Existing methods can be divided into the following categories. One is to recognize irregular text from a two-dimensional (2D) perspective. Such methods [16,18,34] localize and classify characters on 2D feature maps. Another is to improve the representative ability of features, namely noise suppression. Liu et al. [23] and Luo et al. [25] guided the features of the original image, based on generative adversarial networks, toward those of a clean image without background noise. Huang et al. [10] employed vertical and horizontal attention modules to enable the model to focus on informative regions rather than background noise. The last category

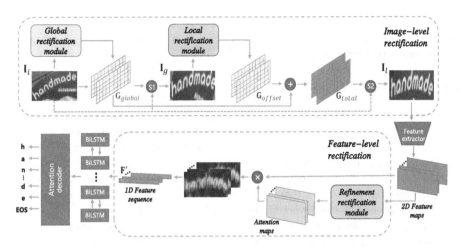

Fig. 2. Overall architecture of our method. Here, $\mathbf{I}_i$ represents the input image. $\mathbf{I}_g$, $\mathbf{I}_l$ and $\mathbf{F}'$ represent the results of the global, local and refinement rectification, respectively. $\mathbf{G}_{global}$, $\mathbf{G}_{offset}$ and $\mathbf{G}_{total}$ represent the global sampling grid, offset grid, and total sampling grid, respectively. $S1$ and $S2$ represent the samplers in the global rectification module and local rectification module respectively. The blue dotted lines represent the sampling processing. For image-level rectification, both rectified images, namely $\mathbf{I}_g$ and $\mathbf{I}_l$, are sampled from the input image $\mathbf{I}_i$.

involves adopting a rectification mechanism to rectify irregular text. Such methods [19,22,24,32,39,43] aim to eliminate distortion and remove the background to aid the subsequent recognizer. Liu et al. [22] employed affine transformation to rectify irregular text in a natural image. Because affine transformation only includes rotation, scaling, and translation operations, its rectification capability is limited. To handle more complex distortions, Shi et al. [32] regressed a series of fiducial points around the text and then adopted the TPS [1] transformation to rectify the text. Owing to the limitation of regression precision of the fiducial points, the rectification method is unstable. Later, Zhan et al. [43] proposed a polynomial function to represent the text shape and iteratively rectified the image. This method suggests that multiple rectification is better than single rectification. Yang et al. [39] generated more precise fiducial points by adding symmetrical geometric constraints. However, extra character-level annotations are required. To further develop a flexible rectification approach, Luo et al. [24] introduced a multi-object rectification network to rectify irregular text by directly predicting the patch-wise offsets of a scene image. However, the training process is very unstable, which increases the regression difficulty of the rectification module. Therefore, irregular text recognition remains challenging.

3   Methodology

The proposed rectification mechanism is shown in Fig. 2. In particular, the global and local rectification modules are cascaded to perform rectification at the image level, whereas the refinement rectification module is applied after the feature extractor of the recognition network to implicitly rectify the feature maps. The proposed rectification mechanism is integrated into an attention-based recognition network to recognize irregular text in scene images.

3.1 Global Rectification Module

We employ affine transformation to handle global distortion and rectify the text into a more horizontal one. This module takes the input image $\mathbf{I}_i \in \mathbf{R}^{C_i \times H_i \times W_i}$ with height H_i, width W_i and C_i channels and outputs image $\mathbf{I}_g$. Specifically, the localization network takes input image $\mathbf{I}_i$ and predicts T_θ, namely a transformation matrix containing six parameters. The localization network consists of six convolutional layers and two fully-connected layers, as presented in Table 1. Notably, all convolutional layers are followed by a batch normalization layer and a ReLU layer. We set the input source image size to (64, 200) to maintain a high resolution before rectification, but the image is downsampled to a low resolution (16, 50) before the convolutional layers to reduce calculation.

Table 1. Architecture of the global rectification module.

Type	Configurations	Size
Input	–	$1 \times 64 \times 200$
Resize	–	$1 \times 16 \times 50$
Conv	k: 3×3 s: 1×1 p: 1×1 maps: 32	$32 \times 16 \times 50$
MaxPooling	k: 2×2 s: 2×2	$32 \times 8 \times 25$
Conv	k: 3×3 s: 1×1 p: 1×1 maps: 64	$64 \times 8 \times 25$
Conv	k: 3×3 s: 1×1 p: 1×1 maps: 128	$128 \times 8 \times 25$
MaxPooling	k: 2×2 s: 2×2	$128 \times 4 \times 12$
Conv	k: 3×3 s: 1×1 p: 1×1 maps: 256	$256 \times 4 \times 12$
Conv	k: 3×3 s: 1×1 p: 1×1 maps: 256	$256 \times 4 \times 12$
MaxPooling	k: 2×2 s: 2×2	$256 \times 2 \times 6$
Conv	k: 3×3 s: 1×1 p: 1×1 maps: 256	$256 \times 2 \times 6$
FC	c_i: $256 \times 2 \times 6$, c_o: 256	256
FC	c_i: 256, c_o: 6	6
Grid	–	$2 \times 32 \times 100$

Here k, s, p, maps, c_i and c_o are the kernel, stride, padding size, number of convolution kernels, number of input units, and number of output units, respectively.

Subsequently, the pointwise transformation between input image $\mathbf{I}_i$ and output image $\mathbf{I}_g$ is

$$\begin{bmatrix} x_i \\ y_i \end{bmatrix} = T_\theta \begin{bmatrix} x_g \\ y_g \\ 1 \end{bmatrix}, \tag{1}$$

where (x_g, y_g) is the target coordinate of the output image $\mathbf{I}_g$ and (x_i, y_i) is the source coordinate of the input image $\mathbf{I}_i$. The global sampling grid $\mathbf{G}_{global}$ is generated based on the above equation, which contains two channels representing the x-coordinate and y-coordinate, respectively.

Based on sampling grid $\mathbf{G}_{global}$, the rectified image $\mathbf{I}_g$ is sampled from the input image $\mathbf{I}_i$ using bilinear interpolation,

$$I_{g(x_g,y_g)} = \sum_{n}^{H_i} \sum_{m}^{W_i} I_{i(n,m)} \max\left(0, 1 - |x_i - n|\right) \max\left(0, 1 - |y_i - m|\right), \quad (2)$$

where $I_{g(x_g,y_g)}$ is the pixel value at location (x_g, y_g) of the rectified image $\mathbf{I}_g$ and $I_{i(n,m)}$ is the pixel value at location (n, m) of the input image $\mathbf{I}_i$. Next, the output image $\mathbf{G}_{global}$ is fed into the local rectification module for a more fine-grained rectification.

3.2   Local Rectification Module

Through the global rectification module, the irregular text becomes more horizontal. However, deformation, such as the curvature of the text caused by the irregular placement of characters, still exists. Hence, we adopt a local rectification module to perform a more fine-grained rectification. This module takes the image $\mathbf{I}_g \in \mathbf{R}^{C_g \times H_g \times W_g}$ with height H_g, width W_g and C_g channels as input and outputs image $\mathbf{I}_l$.

The module predicts offset for each pixel of the input image $\mathbf{I}_g$ and obtains the offset grid $\mathbf{G}_{offset}$. Similar to the global sampling grid $\mathbf{G}_{global}$, the offset grid $\mathbf{G}_{offset}$ also contains two channels which represent the x-coordinate and y-coordinate, respectively. We adopt five convolutional layers to regress the offset value of each pixel, as presented in Table 2. Notably, all convolutional layers are followed by a batch normalization layer and a ReLU layer.

Table 2. Architecture of the local rectification module.

Type	Configurations	Size
Input	–	$1 \times 32 \times 100$
MaxPooling	k: 2×2 s: 2×2	$1 \times 16 \times 50$
Conv	k: 3×3 s: 1×1 p: 1×1 maps: 64	$64 \times 16 \times 50$
MaxPooling	k: 2×2 s: 2×2	$64 \times 8 \times 25$
Conv	k: 3×3 s: 1×1 p: 1×1 maps: 128	$128 \times 8 \times 25$
MaxPooling	k: 2×2 s: 2×2	$128 \times 4 \times 12$
Conv	k: 3×3 s: 1×1 p: 1×1 maps: 64	$64 \times 4 \times 12$
Conv	k: 3×3 s: 1×1 p: 1×1 maps: 16	$16 \times 4 \times 12$
Conv	k: 3×3 s: 1×1 p: 1×1 maps: 2	$2 \times 4 \times 12$
MaxPooling	k: 2×2 s: 1×1	$2 \times 3 \times 11$
Tanh	–	$2 \times 3 \times 11$
Resize	–	$2 \times 32 \times 100$

Here k, s, p and maps are the kernel, stride, padding size, and number of convolution kernels, respectively.

The global sampling grid $\mathbf{G}_{global}$ and offset grid $\mathbf{G}_{offset}$ are summed as follows:

$$\mathbf{G}_{total(c,i,j)} = \mathbf{G}_{offset(c,i,j)} + \mathbf{G}_{global(c,i,j)}, c = 1, 2 \quad (3)$$

where (i, j) is the position of the i-th row and j-th column and $\mathbf{G}_{total}$ is the total sampling grid. Similar to the sampling process in the global rectification module, we generate the rectified image $\mathbf{I}_l$. Notably, the rectified image $\mathbf{I}_l$ is sampled directly from the input source image $\mathbf{I}_i$, instead of the rectified image $\mathbf{I}_g$ from the global rectification module. This avoids the low image resolution caused by multiple samplings.

3.3 Refinement Rectification Module

Through the above rectifications at the image level, most background interference is effectively removed. However, background that is visually similar to the text may remain. To further reduce the recognition difficulty, we perform rectification on feature maps. The feature-level rectification performs rectification implicitly, which does not result in a visual change.

Specifically, the feature extractor generates 2D feature maps $\mathbf{F} \in \mathbf{R}^{C_f \times H_f \times W_f}$ with height H_f, width W_f and C_f channels. Taking the 2D feature maps $\mathbf{F}$ as an input, the module conducts an attention mechanism to select features corresponding to text for each column and outputs a feature sequence $\mathbf{F}' \in \mathbf{R}^{C_f \times 1 \times W_f}$. We denote each column of the feature map $\mathbf{F}$ as $f_i \in \mathbf{R}^{C_f \times H_f}$. Then, the highlighted feature f_i' in the feature sequence $\mathbf{F}'$ is generated as:

$$f_i' = \sum_{k=1}^{H_f} \beta_{i,k} f_{i,k},$$ (4)

where $f_{i,k}$ represents the k-th vector of f_i, and $\beta_i = \{\beta_{i,1}, \beta_{i,2}, ..., \beta_{i,H_f}\}$ represents the attention mask of f_i. Each attention value $\beta_{i,j}$ in β_i is calculated as:

$$\beta_{i,j} = \frac{\exp(v_{i,j})}{\sum_{k=1}^{H_f} \exp(v_{i,k})},$$ (5)

where $v_i = \{v_{i,1}, v_{i,2}, ..., v_{i,H_f}\}$ represents the evaluation score of f_i. We adopt three convolutional layers to generate an evaluation map $v = \{v_1, v_2, ..., v_{W_f}\}$, as shown in Table 3. Note that all the convolutional layers are followed by a batch normalization layer and a ReLU layer.

Table 3. Architecture of the refinement rectification module.

Type	Configurations	Size
Input	–	$512 \times 8 \times 25$
Conv	k: 3×3 s: 1×1 p: 1×1 maps: 512	$512 \times 8 \times 25$
Conv	k: 3×3 s: 1×1 p: 1×1 maps: 512	$512 \times 8 \times 25$
Conv	k: 3×3 s: 1×1 p: 1×1 maps: 1	$1 \times 8 \times 25$
Tanh	–	$1 \times 8 \times 25$
Expand	–	$512 \times 8 \times 25$

Here, k, s, p and maps are the kernel, stride, padding size, and number of convolution kernels, respectively. 'Expand' means that the evaluation maps with single channel are extended to multiple channels by a replication.

The module highlights features containing text content and suppresses noise, thus greatly reducing confusion in the subsequent decoding process. Finally, the one-dimensional (1D) feature sequence $\mathbf{F}' = \{f_1', f_2', ..., f_{W_f}'\}$ is generated by the refinement rectification module for further decoding.

3.4   Encoder-Decoder Recognition Framework

We adopt the popular encoder-encoder framework [2,15] in the field of scene text recognition. In the encoder, we employ a modified 45-layer ResNet [32] to extract 2D visual features, followed by the refinement module to generate 1D highlighted visual features. Subsequently, two layers of bidirectional LSTM [5] are employed to generate the sequence features $H = (h_1, h_2, ..., h_L)$ with context relationship, where L is the length of the sequence features.

An attention-based decoder is used to accurately align the target and the label. It generates the target sequence $(y_1, y_2, ..., y_N)$, where N is the length of the target sequence. The decoder stops processing when it predicts an end-of-sequence token "EOS". At time step t, the decoder outputs the probability distribution y_t:

$$y_t = Softmax(Ws_t + b),$$ (6)

where s_t is the hidden state at time step t, computed by

$$s_t = RNN(y_{t-1}, g_t, s_{t-1}),$$ (7)

where RNN represents a recurrent network (e.g. GRU [4]), and g_t represents the weighted sum of sequential feature vectors,

$$g_t = \sum_{i=1}^{L} (\alpha_{t,i} h_i),$$ (8)

where $\alpha_{t,i}$ denotes a vector of attention weights,

$$e_{t,i} = W_e \tanh\left(W_s s_{t-1} + W_h h_i + W_p p_{t-1} + W_c c_i\right),$$ (9)

$$\alpha_{t,i} = \frac{\exp(e_{t,i})}{\sum_{j=1}^{L}(\exp(e_{t,j}))},$$ (10)

where W_s, W_h, W_p and W_c are trainable parameters. Here, we borrow the idea of position embedding [17,45] to improve the localization accuracy of the attention decoder. Specifically, we introduce two additional variables, namely p_{t-1} and c_i, for learning localization. p_{t-1} is the position vector with the greatest attention value at time step $t-1$,

$$p_{t-1} = E_p(\underset{0 \leq m < L}{\arg\max}\, \alpha_{t-1,m}),$$ (11)

where $E_p \in \mathbf{R}^{L \times M}$ is the position embedding matrix and M represents the length of the position vector. c_i is the coverage attention based on the sum of all past attention probabilities,

$$c_i = \sum_{\tau=0}^{t-1} \alpha_{\tau,i}. \tag{12}$$

Inspired by [32], we introduce a bidirectional decoder. The loss function is the average of losses in both directions:

$$L_{total} = -\frac{1}{2} \sum_t \left(\log P_{l2r} \left(y_t \mid \mathbf{I}_i \right) + \log P_{r2l} \left(y_t \mid \mathbf{I}_i \right) \right), \tag{13}$$

where $\mathbf{I}_i$ denotes the input source image, y_t denotes the ground truth of the t-th character, and P_{l2r} and P_{r2l} are the predicted distributions of the left-to-right decoder and the right-to-left decoder, respectively.

4 Experiments

4.1 Datasets

The training set consists of two synthesized natural scene text datasets, namely Synth90k [11] and SynthText [7], which contain 8 million and 6 million text line samples, respectively. Six published testing datasets were used to verify the effectiveness of the method. The performances of all the methods are measured by word accuracy.

IIIT5K-Words (IIIT5K) [26] contains 3,000 cropped word images for testing. These images were collected on the Internet.

Street View Text (SVT) [35], consisting of 647 word images, was collected from Google Street View. Many images are severely corrupted by noise and blur, or have very low resolutions.

ICDAR2013 (IC13) [14] contains 1,015 cropped text images from 288 scene images with annotations.

ICDAR 2015 (IC15) [13] contains 2,077 cropped images collected by Google Glasses.

SVT-Perpective (SVTP) [28] contains 645 cropped images for testing. These images were collected from side-view angle snapshots from Google Street View. Therefore, most of the images are perspective distorted.

CUTE [29] contains 80 high-resolution images captured from natural scenes. It was specifically collected for evaluating the performance of the curve text recognition. It contains 288 cropped natural images for testing.

4.2 Implementation Details

Network: The details of the three rectification modules are listed in Table 1, Table 2 and Table 3, respectively. For the global rectification module, the weight of the last fully-connected layer was set to zero and the bias was set to [1, 0, 0, 0, 1, 0], ensuring that the first rectification module is identity transformation at the initial stage of training. For the recognition framework, we adopt an encoder borrowed from ASTER [32]. To obtain the 2D feature maps of size (8, 25), we set the stride to 1×1 in the third to fifth residual blocks of the 45-layer ResNet instead of 2×1. Other configurations in the encoder are the same as those in [32].

Implementation: With the ADADELTA [42] optimization method, we initially set the learning rate to 1.0 and decrease it to 0.1 and 0.01 at steps 600,000 and 800,000, respectively. To stablize the training, the learning rate of the global and local rectification modules is 0.1 times the base learning rate. The batch size was set to 64. We pre-trained the recognizer for 20,000 iterations for stable training and fast convergence. Subsequently, the entire network was trained for 1,000,000 iterations.

Table 4. Validation of each rectification module. Here, ARN represents the attention-based recognition network, and GR, LR, and RR represent the global rectification module, the local rectification module and the refinement rectification module, respectively.

Method				IIIT5k	SVT	IC13	IC15	SVTP	CUTE
ARN	GR	LR	RR						
① ✓				92.5	87.3	92.1	73.1	79.1	78.8
② ✓	✓			93.0	88.7	93.6	76.5	83.3	81.6
③ ✓		✓		93.0	88.7	92.0	73.7	79.4	80.6
④ ✓			✓	93.5	90.1	93.7	75.9	83.1	81.9
⑤ ✓	✓	✓		93.5	89.8	94.2	76.7	83.7	83.0
⑥ ✓	✓	✓	✓	**94.2**	**90.4**	**94.2**	**78.4**	**85.1**	**84.7**

Rectification Mechanism. We conducted ablation studies to verify the effectiveness of the three rectification modules. For the recognizer, We employed the ResNet-18 [8] as the feature extractor and the vanilla attention-based decoder with only a single decoding direction. As shown in Table 4, the recognition accuracy increases with the progressive combination of the three modules. Specifically, the accuracy increases by 5.3% for IC15, 6.0% for SVTP, and 5.9% for CUTE.

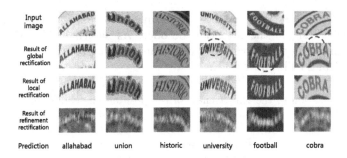

Fig. 3. Visualization of rectification results. The first row represents the input images, and the second to fourth rows represent the results of the global, local and refinement rectification, respectively. The red dotted circle indicates information loss caused by global rectification, but the informative regions are recovered after local rectification. (Color figure online)

4.3   Ablation Study

Given the visualization results (see Fig. 3 and Fig. 4), we find that the three rectification modules work in a complementary manner. First, the global rectification
and the local rectification are complementary. The former is the basement of the
latter (see the third and fifth experiments in Table 4), whereas the latter is the
supplement of the former (see the second and fifth experiments in Table 4). As
shown in Fig. 4, the global rectification module rectifies the entire text roughly,
which benefits the subsequent local rectification. As shown in Fig. 3, the local
rectification module performs rectification based on global rectification. The curvature of the text is significantly reduced and the text occupies a larger area in
the image after local rectification. In particular, we find that information loss
may be caused by global rectification, but the informative regions are recovered
after local rectification (see the last three columns in Fig. 3). Second, feature-
level rectification is the supplement for the image-level rectification, referring to
the fifth and sixth experiments in Table 4. As shown in the last row on Fig. 3,
the feature-level rectification module highlights the text information and suppresses noise to further reduce the disturbance from the background noise, which
significantly reduces the recognition difficulty.

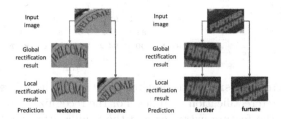

Fig. 4. As a basis, global rectification improves the effect of local rectification. The
blue branch on the left indicates that the local rectification is performed after global
rectification, and the red branch on the right indicates that the local rectification is
directly performed on the input image. (Color figure online)

Attention Decoder. We conducted experiments to verify the effectiveness of
position embedding in the attention decoder. As shown in Table 5, position
embedding improves the recognition performance on four testing datasets including SVT, IC13, SVTP and CUTE.

Table 5. Verification of the effectiveness of the position embedding in the attention
decoder.

Variants	Regular text			Irregular text		
	IIIT5K	SVT	IC13	IC15	SVTP	CUTE
Without pos.	95.2	91.5	94.8	79.6	85.9	85.4
With pos.	95.2	**91.8**	**96.0**	79.6	**86.2**	**86.8**

4.4   Comparisons with Other Rectification Methods

To verify the superiority of our rectification mechanism, we quantitatively compare four existing rectification methods with ours. For a fair comparison, STAR-Net, ASTER, MORAN, STAN, and our method employ the same recognition network. Specifically, STAR-Net and ASTER rectify the entire text as a whole, focusing on the global deformations. Nevertheless, MORAN and STAN attempt to divide the text into several patches, and only estimate local deformations. Our method can handle both global and local deformations, and rectify the feature maps implicitly. The recognition performance of our method is significantly better than that of the other methods on three irregular datasets (see Table 6).

Table 6. Comparison of the rectification mechanism of STAR-Net, ASTER, MORAN, STAN and our method. Here, the rectification modules are integrated into the same recognition network proposed in ASTER.

Method	IC15	SVTP	CUTE
STAR-Net [22]	71.7	76.7	76.4
ASTER [32]	76.1	78.5	79.5
MORAN [24]	73.9	79.7	81.9
STAN [19]	76.1	82.0	82.3
Ours	**79.6**	**85.9**	**85.4**

Table 7. Comparison with state-of-the-art methods. No lexicon is used. '*' indicates the method that uses extra data besides Synth90k [11] and SynthText [7] to train the model or extra annotations besides word-level text to supervise the training of the network. The **bold** represents the best recognition results. '†' denotes the best recognition performance of using extra training datasets or extra annotations.

Method	Regular text			Irregular text		
	IIIT5K	SVT	IC13	IC15	SVTP	CUTE
Shi et al. [30]: CRNN	81.2	82.7	89.6	–	–	–
Liu et al. [22]: STAR-Net	83.3	83.6	89.1	–	73.5	–
Shi et al. [31]: RARE	81.9	81.9	88.6	–	71.8	59.2
*Cheng et al. [2]: FAN	87.4	85.9	93.3	85.3	–	–
Cheng et al. [3]: AON	87.0	82.8	–	68.2	73.0	76.8
*Liu et al. [21]: Char-Net	92.0	85.5	91.1	74.2	78.9	–
Shi et al. [32]: ASTER	93.4	89.5	91.8	76.1	78.5	79.5
Luo et al. [24]: MORAN	91.2	88.3	92.4	68.8	76.1	77.4
*Li et al. [16]: SAR	91.5	84.5	94.0	78.8	86.4	89.6
*Liao et al. [18]	91.9	86.4	91.5	–	–	79.9
Zhan et al. [43]: ESIR	93.3	90.2	91.3	76.9	79.6	83.3
Xie et al. [37]	–	–	–	68.9	70.1	82.6
*Yang et al. [39]: ScRN	94.4	88.9	93.9	78.7	80.8	87.5
*Wan et al. [33]: TextScanner	†**95.7**	92.7	94.9	†**83.5**	84.8	†**91.6**
Wang et al. [36]: DAN	94.3	89.2	93.9	74.5	80.0	84.4
*Hu et al. [9]: GTC	95.5	†**92.9**	94.3	82.5	86.2	92.3
Lin et al. [19]: STAN	94.1	90.6	92.8	76.7	82.2	83.3
Luo et al. [25]	94.1	90.6	94.2	78.5	82.2	87.8
*Yu et al. [40]	94.8	91.5	95.5	82.7	85.1	87.8
*Litman et al. [20]: SCATTER	93.7	92.7	93.9	82.2	†**86.9**	87.5
*Yue et al. [41]: RobustScanner	95.4	89.3	94.1	79.2	82.9	92.4
*Qiao et al. [27]: SEED	93.8	89.6	92.8	80.0	81.4	83.6
Yan et al. [38]: PlugNet	94.4	**92.3**	95.0	**82.2**	84.3	85.0
Zhang et al. [44]: AutoSTR	94.7	90.9	94.2	81.8	81.7	–
Ours	**95.2**	91.8	**96.0**	79.6	**86.2**	86.8

4.5   Integration with Mainstream Recognizers

Because the three rectification modules are pluggable, our rectification mechanism can be flexibly integrated into mainstream recognizers. First, we integrated the proposed rectification mechanism into the attention-based recognition network mentioned in Sect. 3.4 and compare its performance with state-of-the-art methods. Our method only requires text images and their corresponding text labels for training, rather than any extra supervision (e.g., character position annotations or pixel-level annotations). Under fair comparison, our method achieves state-of-the-art performance on three datasets including IIIT5K, IC13 and SVTP. In particular, our method outperforms the state-of-the-art methods with extra annotations or training datasets on IC13. On IIIT5K and SVTP, our method achieves performance comparable to that of the state-of-the-art methods with extra annotations or training datasets (Table 7).

To verify the generalization of our rectification mechanism, we integrated it into another mainstream recognition method, namely the CTC-based recognition method. We chose the CRNN [30] as the baseline. Table 8 demonstrates that the performance of the CRNN [30] improves significantly for all datasets after adding our rectification mechanism. The proposed rectification mechanism is flexible and can be integrated with most general recognition methods, and we believe that better results can be achieved if we employ a stronger recognizer.

Table 8. Verification of generalization performance in the CTC-based recognition network. Here, CRNN presents the results reported in [30], CRNN* presents the reproduced results, and '+Ours' presents the results of the CRNN with our rectification mechanism.

Method	Regular text			Irregular text		
	IIIT5K	SVT	IC13	IC15	SVTP	CUTE
CRNN [30]	81.2	82.7	89.6	–	–	–
CRNN*	90.7	77.7	90.0	68.1	68.2	68.1
+Ours	92.3	79.3	91.1	70.9	70.7	71.9

5   Conclusions

In this paper, we propose a multi-level progressive rectification mechanism to rectify irregular texts. Our mechanism consists of three rectification modules performing at the image and feature levels. Through the progressive rectification, the deformation of the text is gradually alleviated and the interference of the background is inhibited. Therefore, the recognition network can focus on the informative area with less disturbance from background noise. Extensive experiments and visualization results demonstrate the effectiveness of the proposed mechanism. Moreover, our rectification mechanism can be integrated with state-of-the-art recognizers to achieve higher accuracy, indicating the generalization of the mechanism.

References

1. Bookstein, F.L.: Principal warps: thin-plate splines and the decomposition of deformations. TPAMI **11**(6), 567–585 (1989)
2. Cheng, Z., Bai, F., Xu, Y., Zheng, G., Pu, S., Zhou, S.: Focusing attention: towards accurate text recognition in natural images. In: ICCV, pp. 5086–5094 (2017)
3. Cheng, Z., Xu, Y., Bai, F., Niu, Y., Pu, S., Zhou, S.: AON: towards arbitrarily-oriented text recognition. In: CVPR, pp. 5571–5579 (2018)
4. Cho, K., et al.: Learning phrase representations using RNN encoder-decoder for statistical machine translation. In: Proceedings of EMNLPS, pp. 1724–1734 (2014)
5. Graves, A., Liwicki, M., Fernández, S., Bertolami, R., Bunke, H., Schmidhuber, J.: A novel connectionist system for unconstrained handwriting recognition. TPAMI **31**(5), 855–868 (2009)
6. Graves, A., Fernández, S., Gomez, F., Schmidhuber, J.: Connectionist temporal classification: labelling unsegmented sequence data with recurrent neural networks. In: ICML, pp. 369–376 (2006)
7. Gupta, A., Vedaldi, A., Zisserman, A.: Synthetic data for text localisation in natural images. In: CVPR, pp. 2315–2324 (2016)
8. He, K., Zhang, X., Ren, S., Sun, J.: Deep residual learning for image recognition. In: CVPR, pp. 770–778 (2016)
9. Hu, W., Cai, X., Hou, J., Yi, S., Lin, Z.: GTC: guided training of CTC towards efficient and accurate scene text recognition. In: AAAI, pp. 11005–11012 (2020)
10. Huang, Y., Luo, C., Jin, L., Lin, Q., Zhou, W.: Attention after attention: reading text in the wild with cross attention. In: ICDAR, pp. 274–280 (2019)
11. Jaderberg, M., Simonyan, K., Vedaldi, A., Zisserman, A.: Synthetic data and artificial neural networks for natural scene text recognition. arXiv preprint arXiv:1406.2227 (2014)
12. Jaderberg, M., Simonyan, K., Zisserman, A., et al.: Spatial transformer networks. In: NeurIPS, vol. 28, pp. 2017–2025 (2015)
13. Karatzas, D., et al.: ICDAR 2015 competition on robust reading. In: ICDAR, pp. 1156–1160 (2015)
14. Karatzas, D., et al.: ICDAR 2013 robust reading competition. In: ICDAR, pp. 1484–1493 (2013)
15. Lee, C., Osindero, S.: Recursive recurrent nets with attention modeling for OCR in the wild. In: CVPR, pp. 2231–2239 (2016)
16. Li, H., Wang, P., Shen, C., Zhang, G.: Show, attend and read: A simple and strong baseline for irregular text recognition. In: AAAI, vol. 33, 8610–8617 (2019)
17. Liao, M., Lyu, P., He, M., Yao, C., Wu, W., Bai, X.: Mask TextSpotter: an end-to-end trainable neural network for spotting text with arbitrary shapes. IEEE Trans. Pattern Anal. Mach. Intell. **43**(2), 532–548 (2021). https://doi.org/10.1109/TPAMI.2019.2937086
18. Liao, M., et al.: Scene text recognition from two-dimensional perspective. In: AAAI, vol. 33, pp. 8714–8721 (2019)
19. Lin, Q., Luo, C., Jin, L., Lai, S.: STAN: A sequential transformation attention-based network for scene text recognition. Pattern Recogn. **111**, 107692 (2021)
20. Litman, R., Anschel, O., Tsiper, S., Litman, R., Mazor, S., Manmatha, R.: SCATTER: selective context attentional scene text recognizer. In: CVPR, pp. 11962–11972 (2020)
21. Liu, W., Chen, C., Wong, K.Y.K.: Char-Net: a character-aware neural network for distorted scene text recognition. In: AAAI, vol. 1, p. 4 (2018)

22. Liu, W., Chen, C., Wong, K.Y.K., Su, Z., Han, J.: STAR-Net: a spatial attention residue network for scene text recognition. In: BMVC, vol. 2, p. 7 (2016)
23. Liu, Y., Wang, Z., Jin, H., Wassell, I.: Synthetically supervised feature learning for scene text recognition. In: ECCV, pp. 435–451 (2018)
24. Luo, C., Jin, L., Sun, Z.: MORAN: a multi-object rectified attention network for scene text recognition. Pattern Recogn. **90**, 109–118 (2019)
25. Luo, C., Lin, Q., Liu, Y., Jin, L., Shen, C.: Separating content from style using adversarial learning for recognizing text in the wild. Int. J. Comput. Vis. **129**, 960–976 (2021). https://doi.org/10.1007/s11263-020-01411-1
26. Mishra, A., Srivastava, V.: Cognition based selection and categorization of maintenance engineer (agent) using artificial neural net and data mining methods. In: 2012 CSI Sixth International Conference on Software Engineering, pp. 1–11 (2012)
27. Qiao, Z., Zhou, Y., Yang, D., Zhou, Y., Wang, W.: SEED: semantics enhanced encoder-decoder framework for scene text recognition. In: CVPR, pp. 13528–13537 (2020)
28. Quy Phan, T., Shivakumara, P., Tian, S., Lim Tan, C.: Recognizing text with perspective distortion in natural scenes. In: ICCV, pp. 569–576 (2013)
29. Risnumawan, A., Shivakumara, P., Chan, C.S., Tan, C.L.: A robust arbitrary text detection system for natural scene images. Expert Syst. Appl. **41**(18), 8027–8048 (2014)
30. Shi, B., Bai, X., Yao, C.: An end-to-end trainable neural network for image-based sequence recognition and its application to scene text recognition. TPAMI **39**(11), 2298–2304 (2017)
31. Shi, B., Wang, X., Lyu, P., Yao, C., Bai, X.: Robust scene text recognition with automatic rectification. In: CVPR, pp. 4168–4176 (2016)
32. Shi, B., Yang, M., Wang, X., Lyu, P., Yao, C., Bai, X.: ASTER: an attentional scene text recognizer with flexible rectification. TPAMI **41**(9), 2035–2048 (2018)
33. Wan, Z., He, M., Chen, H., Bai, X., Yao, C.: TextScanner: reading characters in order for robust scene text recognition. In: AAAI, vol. 34, pp. 12120–12127 (2020)
34. Wan, Z., Xie, F., Liu, Y., Bai, X., Yao, C.: 2D-CTC for scene text recognition. arXiv preprint arXiv:1907.09705 (2019)
35. Wang, K., Babenko, B., Belongie, S.: End-to-end scene text recognition. In: ICCV, pp. 1457–1464 (2011)
36. Wang, T., et al.: Decoupled attention network for text recognition. In: AAAI, pp. 12216–12224 (2020)
37. Xie, Z., Huang, Y., Zhu, Y., Jin, L., Liu, Y., Xie, L.: Aggregation cross-entropy for sequence recognition. In: CVPR, pp. 6538–6547 (2019)
38. Mou, Y., et al.: PlugNet: degradation aware scene text recognition supervised by a pluggable super-resolution unit. In: Vedaldi, A., Bischof, H., Brox, T., Frahm, J.-M. (eds.) ECCV 2020. LNCS, vol. 12360, pp. 158–174. Springer, Cham (2020). https://doi.org/10.1007/978-3-030-58555-6_10
39. Yang, M., et al.: Symmetry-constrained rectification network for scene text recognition. In: ICCV, pp. 9147–9156 (2019)
40. Yu, D., et al.: Towards accurate scene text recognition with semantic reasoning networks. In: CVPR, pp. 12113–12122 (2020)
41. Yue, X., Kuang, Z., Lin, C., Sun, H., Zhang, W.: RobustScanner: dynamically enhancing positional clues for robust text recognition. In: Vedaldi, A., Bischof, H., Brox, T., Frahm, J.-M. (eds.) ECCV 2020. LNCS, vol. 12364, pp. 135–151. Springer, Cham (2020). https://doi.org/10.1007/978-3-030-58529-7_9
42. Zeiler, M.D.: ADADELTA: an adaptive learning rate method. arXiv preprint arXiv:1212.5701 (2012)

43. Zhan, F., Lu, S.: ESIR: end-to-end scene text recognition via iterative image rectification. In: CVPR, pp. 2059–2068 (2019)
44. Zhang, H., Yao, Q., Yang, M., Xu, Y., Bai, X.: Efficient backbone search for scene text recognition. In: ECCV (2020)
45. Zhang, J., et al.: Watch, attend and parse: an end-to-end neural network based approach to handwritten mathematical expression recognition. Pattern Recogn. **71**, 196–206 (2017)

Representation and Correlation Enhanced Encoder-Decoder Framework for Scene Text Recognition

Mengmeng Cui[1(✉)], Wei Wang[1], Jinjin Zhang[1], and Liang Wang[1,2]

[1] Institute of Automation, Chinese Academy of Sciences (CASIA), Beijing, China
{mengmeng.cui,jinjin.zhang}@cripac.ia.ac.cn,
{wangwei,wangliang}@nlpr.ia.ac.cn
[2] School of Artificial Intelligence, University of Chinese Academy of Sciences (UCAS), Beijing, China

Abstract. Attention-based encoder-decoder framework is widely used in the scene text recognition task. However, for the current state-of-the-art (SOTA) methods, there is room for improvement in terms of the efficient usage of local visual and global context information of the input text image, as well as the robust correlation between the scene processing module (encoder) and the text processing module (decoder). In this paper, we propose a Representation and Correlation Enhanced Encoder-Decoder Framework (RCEED) to address these deficiencies and break performance bottleneck. In the encoder module, local visual feature, global context feature, and position information are aligned and fused to generate a small-size comprehensive feature map. In the decoder module, two methods are utilized to enhance the correlation between scene and text feature space. 1) The decoder initialization is guided by the holistic feature and global glimpse vector exported from the encoder. 2) The feature enriched glimpse vector produced by the Multi-Head General Attention is used to assist the RNN iteration and the character prediction at each time step. Meanwhile, we also design a Layernorm-Dropout LSTM cell to improve model's generalization towards changeable texts. Extensive experiments on the benchmarks demonstrate the advantageous performance of RCEED in scene text recognition tasks, especially the irregular ones. The source code will be available https://github.com/Mona9955/RCEED-ICDAR2021.

Keywords: STR · Sequence-to-sequence · Multi-head attention · Layernorm & dropouts

1 Introduction

Scene Text Recognition (STR) refers to the text recognition of natural scene images captured by camera. Compared with traditional Optical Character Recognition (OCR) systems dedicated to high-quality document images, STR techniques are developed for the outdoor images and applied in a wider range of

© Springer Nature Switzerland AG 2021
J. Lladós et al. (Eds.): ICDAR 2021, LNCS 12824, pp. 156–170, 2021.
https://doi.org/10.1007/978-3-030-86337-1_11

fields, such as street view positioning, image advertisement filtering, bill recognition, et al. Due to the randomness in the process of capturing text images in natural scenes, STR has many challenges in practical applications, including uneven lighting and focusing caused image quality degradation, complex image background, occluded and incomplete characters. Moreover, the characters themselves also have diverse font types, font sizes and colors; the text distribution is irregular, many of which are perspective, distorted or oriented. Therefore, as a complex problem, STR has been extensively studied in industry and academia.

STR is divided into regular and irregular text recognition tasks according to the text distribution. Modern technical solutions mainly include Connectionist Temporal Classification (CTC) based methods [1], attention-based encoder-decoder framework [2,3] and the combination of both [4,5]. These methods only need word-level annotations and robust to complicated scene text images. CTC-based methods solve the problem of misalignment between the input image and the target outputs but can not leverage the contextual dependency between characters, thus it is mainly used for horizontal text recognition. Comparatively, attention mechanism is a good way to strengthen the relevance between visual and semantic features and improve the interpretability of the model, making it a suitable choice for irregular STR scenarios. The major architectures of the attention-based encoder-decoder framework include the sequence-to-sequence models which adopt 1D attention mechanism [2] or 2D attention mechanism [6] in the decoder, and the transformer-based models [3,7,8].

However, there are still shortcomings for the existing attention-based encoder-decoder framework. 1D attention methods generally use RNN layer(s) to model contextual dependencies but lose apparent information of the text [2]. Although 2D attention is able to handle the irregular spatial distribution of the text, its performance is greatly restricted by the size of the encoded feature map [6]. Since the inter-character dependence in STR is weaker than the inter-word dependence in machine translation, the self-attention design in transformer-based models which targets at building long-range dependencies may not achieve the expected performance [3], but increases the parameters due to its multiple fully-connected layers and the multi-layer stacking structure of decoder. The common reasons for these problems lie in the information loss of the global and local feature, as well as the weak relevance between the encoder working on the visual space and the decoder working on the language space.

Therefore, we propose the **R**epresentation and **C**orrelation **E**nhanced **E**ncoder-**D**ecoder Framework (RCEED). The encoder generates a comprehensive representation of local visual feature and global context feature. The decoder utilizes the Multi-Head General Attention mechanism to capture an enriched glimpse of the encoded feature. The initialization manner and the efficient workflow of the decoder increase the correlation between the visual feature and the decoded characters. Our main contributions are summarized as follows:

1. In the encoder module, a representation enhanced feature map is obtained by combining the visual, context and position information. The encoded feature map has a small size corresponding to the spatial distribution of characters.

2. In the decoder module, a holistic feature and the a global glimpse vector are introduced from the encoder to guide the initialization of the decoder. The intuitive workflow enables the glimpse vector to participate in the update of the decoded hidden state and the character prediction at the same time. These integrated designs make the model achieve SOTA performance in public benchmarks, especially the irregular ones.
3. We devise a Mulit-Head General Attention mechanism to capture the main information and the supplementary information of the encoded feature with fewer operations and parameters.
4. We specially design the LD-LSTM cell as basic block to form the RNN layers of the encoder and the decoder. The LD-LSTM can balance independence and relevance between characters and improve model's generalization for irregular texts, which is very important for the STR applications.

2   Related Work

Early text image recognition is oriented to the document recognition scenario which has a clear picture and a fixed pattern. People use binarization method [9] and sliding window method [10] for individual character detection, and then integrate the characters into words by dynamic programming. These methods are vulnerable to the background noise, and unable to use the global context information. Later works tend to treat the text image recognition as a sequence recognition problem. These methods are more capable of the complex STR tasks and mainly divided into two categories, the CTC-based methods and the attention based methods. CRNN [1] utilizes the CNN and RNN to generate feature sequence from the visual information, and CTC to align the characters predicted by the RNN decoder. Attention based encoder-decoder models like RARE [11] and R^2AM [12] are developed to introduce the attention-mechanism from machine translation [13] to solve the image-based sequence recognition problem. Focusing attention network [14] is raised to fix the attention drifting caused by complex scenes or low image quality. With similar targets, Bai et al. introduce the edit probability [15] method to alleviate character missing or superfluous in text recognition. In addition, model as a combination of CTC and attention based methods [5] also performs well on regular scene text datasets.

In recent years, many approaches have been proposed regarding the more challenging irregular scene text recognition task. The first type is the rectification based methods. ASTER [2] combines the Thin-Plate Spline (TPS) method [16] and the Spatial Transformer Network (STN) [17] to form the rectification network. The line-fitting transformation method is proposed in ESIR [18] which employs iterative rectification to improve the performance. MORAN [19] proposes a pixel transform method to make a smooth conversion to the text images. The other type is the character level methods. Models like Char-Net [20] and Mask TextSpotter [21] detect and rectify the individual characters, which requires additional character-level supervision. The last type is the attention based encoder-decoder frameworks. 2D feature map is used for both the sequence-to-sequence models [6,22] and the transformer-based models [3,7]. The expanded focus range contributes to the recognition of characters with arbitrary shape and position. RobustScanner [22] introduces a positional enhancement branch to the 2D attentional encoder-decoder structure proposed in SAR [6], leveraging the positional information to make prediction during decoding process. MASTER [7] employs the non-local network as the encoder of the transformer-based structure to capture longer contextual dependencies.

3   Model Architecture

As presented in Fig. 1, there are three important components of RCEED: the Rectification Network which redistributes the characters, the Representation Enhanced Encoder which combines the local visual feature and global context feature, the Multi-Head General Attention Decoder which increases the correlation between the visual space and the language space. Two basic composition methods are utilized in the encoder and decoder, including the LD-LSTM cell which improves model's generalization and the Multi-Head General Attention mechanism which makes effective use of main and supplementary information.

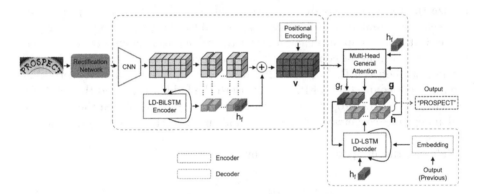

Fig. 1. Architecture of the RCEED. The input image is transformed by the Rectification Network before fed into the encoder. The encoder combines the visual feature from CNN, the context feature from the LD-BiLSTM layer and the position information to generate the comprehensive feature map ($\mathbf{v}$). The holistic feature (h_f) and the global glimpse vector (g_f) from encoder are used to guide decoder initialization. At each decoding step t, the glimpse vector g_t ($\in \mathbf{g}$) is calculated by MHGAT based on the current hidden state h_t ($\in \mathbf{h}$), and exported to the next LD-LSTM iteration to generate h_{t+1}. Both g_t and h_t are used to make the current prediction.

3.1   Rectification Network

Thin-Plat Spline[16] is a 2D interpolation method which makes conversion with minimal bending energy based on a set of corresponding control points of two pictures. Similar to the STN in [2], our rectification network utilizes a lightweight

CNN to generate the control points of the source image, associating them with the pre-defined control points of the target image through TPS to accomplish the rectification process. We also find that the rectification network is able to adjust the width and spacing of characters adaptively, which reduces the divergence of input text images and improves the alignment between receptive fields and characters.

3.2 Basic Composition Methods for Encoder-Decoder Framework

In order to facilitate understanding, before the illustration of the Encoder and Decoder, we introduce two methods which are important components of them.

Layernorm-Dropout LSTM Cell (LD-LSTM). The Long Short-Term Memory (LSTM) [23] is widely used in the machine translation models [13]. Considering the difference between the text recognition task and the language processing task which has strong semantic dependencies between tokens, we specially design the Layernorm-Dropout LSTM Cell.

Firstly, we add layernorm [24] to the current input and the previous hidden state of the LSTM cell to speed up the convergence of training process and regularize the network. Secondly, on account of the relatively weak dependencies between characters in the text image, we introduce the dropout function [25] to reduce feature co-adaptation and improve the presentation of important information. Different from the conventional way of applying dropout in the feed-forward connections between RNN layers [26], we design a per-step dropout method in the recurrent connections of RNN cells. Both the hidden state and the cell state are sampled by the dropout masks with probability p to balance the relevance and independence between characters, and improve the generalization at the same time.

The layernorm operation is given by:

$$LN(\mathbf{x}; \alpha, \beta) = \frac{(\mathbf{x} - \mu)}{\sigma} \odot \alpha + \beta \tag{1}$$

$$\mu = \frac{1}{D} \sum_{i=1}^{D} x_i, \quad \sigma = \sqrt{\frac{1}{D} \sum_{i=1}^{D} (x_i - \mu)^2} \tag{2}$$

Where x_i is the i_{th} element of $\mathbf{x}$ with length D. α and β are defined as gain and bias parameters. In this paper, we set β as zero. Then the LD-LSTM cell is defined as:

$$\begin{pmatrix} f_t \\ i_t \\ o_t \\ \hat{c}_t \end{pmatrix} = LN(W_x x_t; \alpha_1, \beta_1) + LN(W_h h_{t-1}; \alpha_2, \beta_2) \tag{3}$$

$$c_t = Dropout\left(sigm(f_t) \odot c_{t-1} + sigm(i_t) \odot tanh(\hat{c}_t), \, p\right) \tag{4}$$

$$h_t = Dropout\,(sigm(o_t) \odot tanh(c_t),\ p) \tag{5}$$

Where W_x and W_h are weight matrixes of the input x_t and the hidden state h_{t-1}. $\odot$ denotes the element-wise product operation. p is the probability of each element of the vector being zero.

Multi-Head General Attention (MHGAT). The commonly used attention functions include general attention, additive attention, dot product attention et al. Unlike most sequence-to-sequence text recognition models which utilize additive attention mechanism to build connection between encoder and decoder [2,4,6], we use general function [27] to reduce the computational complexity. Compared with additive attention [13] which needs two-step operations to obtain the attention weights(add first and then multiply by the transform matrix), general attention only needs one-step matrix multiplication operation (Eq. (6)). Given $\mathbf{v}' = [v_1', v_2', ..., v_N']$ as a splitted part of the flattened comprehensive feature map $\mathbf{v}$ with length N, the formulation of the general attention mechanism for $\mathbf{v}'$ can be expressed as follows:

$$score(h_t', \mathbf{v}') = \frac{\mathbf{v}' W_a' h_t'}{d_{v'}^2} \tag{6}$$

$$\mathbf{a_t'} = softmax(score(h_t', \mathbf{v}')) \in \mathbb{R}^N \tag{7}$$

$$GeneralAttention(h_t', \mathbf{v}') = \sum_{i=1}^{N} a_{t,i}' v_i' \tag{8}$$

Where $a_{t,i}'$ is the i_{th} element of the attention weights $\mathbf{a_t'}$. $W_a' \in \mathbb{R}^{d_v \times d_{v'}}$ is a parameter matrix with d_v and $d_{v'}$ representing the dimensions of $\mathbf{v}$ and $\mathbf{v}'$. In order to build a harder attention module to suppress the background noise, we set the a relatively large scale factor, which is $d_{v'}^2$ under 8 parallel heads.

Due to the diversity of character size and distribution, there is misalignment between the feature map obtained after CNN downsampling and the character visual information. Therefore, we adopt the multi-head attention (shown in Fig. 4(a)) to increase the attention flexibility and reduce information loss. Given the hidden state h_t at time step t as a query, the Multi-Head General Attention for $\mathbf{v}$ is generated by:

$$g_t = MultiHead(h_t, \mathbf{v}) = Concat(head_1, ..., head_m) \tag{9}$$

$$where\ head_j = GeneralAttention(h_t W_{h,j}, \mathbf{v_j}),\ \mathbf{v_j} \in Split(\mathbf{v}, m)$$

The parameters are $W_{h,j} \in \mathbb{R}^{d \times \frac{d_v}{m}}$, where d denotes the dimension of h_t, m is the number of parallel heads. $\mathbf{v_j}$ is obtained by splitting m times along d_v. g_t as a concatenation of the attention heads refers to the "glimpse" of the encoded feature where the current step attends to. Figure 2 visualizes the 8-heads attention maps at each decoding step. Head 8 and head 4 are respectively in charge of the main and the supplementary attention tasks, and accurately aligned with the target characters.

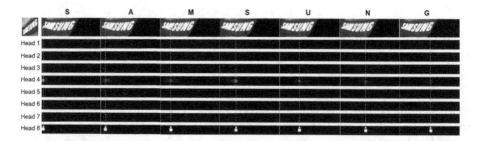

Fig. 2. Visualization of the 8-heads attention map at each decoding step. Head 8 is in charge of the main information and completely aligned with the decoded characters. Head 4 pays attention to the supplementary information around the target area.

3.3   Representation Enhanced Encoder

The encoder of RCEED consists of two parts: the Resnet [18] based CNN backbone (presented in Table 1) which extracts local visual feature, and the single-layer LD-BiLSTM which outputs global context feature. Limited by the receptive field, the CNN encoder can not well represent the contextual information. Therefore, BiLSTM [23] layer with the LD-LSTM Cell as block(abbreviated as LD-BiLSTM) is utilized to generate the context feature sequence, which is based on the intermediate feature sequence produced by average pooling on the visual feature column vectors. The final hidden state is taken as the holistic feature.

However, due to the abstractness of the LD-BiLSTM operation, the complex shape and spatial information of the text is lost. Thus, as shown in Fig. 3, we add the context feature sequence and the corresponding visual feature column vectors along the row axis, getting the comprehensive feature representation.

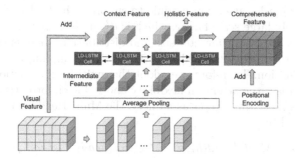

Fig. 3. Structure of the representation enhanced encoder. The intermediate feature sequence is obtained by applying average pooling on the visual feature column vectors. After the column-aligned addition between the visual feature and the context feature, as well as the positional encoding operation, a comprehensive feature map is generated.

Since the character position distribution along the vertical direction is mainly divided into three parts: upper, middle, and lower. Meanwhile, for the sake of reducing computational consumption, we set the feature map size as 3×20,

Table 1. The configuration of the ResNet based CNN feature extractor. The stride and padding for the convolutional layers are all set to 1. The 'k' and 's' in the max-pooling layers refer to kernel size and stride. Height and width of the output of each layer are presented in the last column.

Layer	Configuration	Output
Conv	$3 \times 3,\ 64$	48×160
Conv	$3 \times 3,\ 128$	48×160
Max-pooling	$k : 2 \times 2,\ s : 2 \times 2$	24×80
Residual block	$\begin{bmatrix} 3 \times 3,\ 256 \\ 3 \times 3,\ 256 \end{bmatrix} \times 1$	24×80
Conv	$3 \times 3,\ 256$	24×80
Max-pooling	$k : 2 \times 2,\ s : 2 \times 2$	12×40
Residual block	$\begin{bmatrix} 3 \times 3,\ 256 \\ 3 \times 3,\ 256 \end{bmatrix} \times 2$	12×40
Conv	$3 \times 3,\ 256$	12×40
Max-pooling	$k : 2 \times 2,\ s : 2 \times 2$	6×20
Residual block	$\begin{bmatrix} 3 \times 3,\ 512 \\ 3 \times 3,\ 512 \end{bmatrix} \times 5$	6×20
Conv	$3 \times 3,\ 512$	6×20
Max-pooling	$k : 2 \times 2,\ s : [2, 1] \times [2, 1]$	3×20
Residual block	$\begin{bmatrix} 3 \times 3,\ 512 \\ 3 \times 3,\ 512 \end{bmatrix} \times 3$	3×20
Conv	$3 \times 3,\ 512$	3×20

where 3 corresponds to the 3 types vertical position. Compared with [6,7,22], our feature map size is only a quarter of theirs. Similar to [28], we consider the influence of position information and use sinusoid function to encode absolute as well as relative position information to the comprehensive feature map.

3.4   Recurrent Multi-Head General Attention Decoder

As shown in Fig. 4(b), the decoder is built on a sequence-to-sequence architecture with the LD-LSTM cell and the Multi-Head General Attention as basic blocks. Before the decoding starts, given the flattened comprehensive feature map $\mathbf{v}$ and the holistic feature h_f from the encoder as inputs, a global glimpse vector g_f related to h_f is derived from MHGAT. h_f and g_f represent the global information of the encoded feature map and serve as the guidance for the initialization of decoder, which helps to focus on the explicit part at the first decoding step and improve the accuracy. Then, the LD-LSTM operation is performed fed with g_f, h_f, and a start token y_s. Note that the LD-LSTM is slightly changed in the decoder, the word embedding of the start token and sequential outputs $\mathbf{y} = [y_s, y_1, ..., y_{T-1}]$ is added to Eq. (3).

$$(h_t, c_t) = \begin{cases} LD - LSTM(g_f, h_f, y_s) & t = 1 \\ LD - LSTM(g_{t-1}, h_{t-1}, y_{t-1}) & t > 1 \end{cases} \qquad (10)$$

Next, the Multi-Head General Attention mechanism is performed to generate the glimpse vector related to the current hidden state h_t.

$$g_t = MultiHead(h_t, \mathbf{v}) \qquad (11)$$

Finally, the predicted character is calculated by:

$$y_t = \phi(h_t, g_t) = argmax(softmax(W_o[h_t; g_t] + b_o)) \qquad (12)$$

The hidden state h_t and current glimpse vector g_t are concatenated and passed through the linear transformation to make the final prediction to 63 classes, including 10 digits, 52 case sensitive letters and the 'EOS' token. Then the hidden state, cell state and glimpse vector are recovered and passed to the next step. The decoder works iteratively until the 'EOS' token is predicted or the maximum time step T is arrived.

Attention based sequence-to-sequence models [2,4,21] generally use the decoding workflow based on [13], which is $h_{t-1} \rightarrow \mathbf{a_t} \rightarrow g_t \rightarrow h_t$, where the glimpse vector g_t is related to the last time step and cannot be directly used for the current prediction. On the other hand, the workflow of RCEED is more intuitive, which is $(h_{t-1}, g_{t-1}) \rightarrow h_t \rightarrow \mathbf{a_t} \rightarrow g_t$. The decoding workflow of SAR [6] is comparable to us. Nevertheless, the first round RNN iteration of SAR does not make predictions, and its glimpse vector is not utilized in the RNN calculation. In our design, $\mathbf{g}$ as representation of a certain part of the encoded feature directly participates in the recurrent iteration of $\mathbf{h}$ and the character prediction. Therefore, a content enriched expression is obtained and the correlation between the visual feature and the character sequence is enhanced.

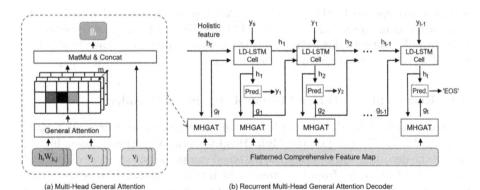

(a) Multi-Head General Attention   (b) Recurrent Multi-Head General Attention Decoder

Fig. 4. (a) Structure of the multi-head general attention. (b) Structure of the recurrent multi-head general attention decoder. When decoding starts, the holistic feature h_f and global glimpse vector g_f from encoder are input to the LD-LSTM cell together with the start token y_s to get the hidden state h_1. h_1 and its related glimpse vector g_1 are both used to make the first prediction. Then decoding iteration continues based on the previous outputs until the 'EOS' token is predicted.

4   Experiments

4.1   Datasets

We train our RCEED on part of the two synthetic datasets: MJSynth and SynthText, and the training sets of the four scene text datasets, i.e., IIIT5K, ICDAR2013, SVT, ICDAR2015. Evaluation results are based on the six standard benchmarks, including three regular datasets and three irregular datasets.

MJSynth (MJ) [29] is also named Synth90k. MJ has 9 million synthetic images of english words, each of them is annotated with a word-level ground truth. We use 4.8 million images from MJ for training.

SynthText (ST) [30] is another synthetic dataset which is widely used in text detection and recognition. We crop the word patches from the background images and random select 4.8 million out of 7 million word images for training.

IIIT5K [31] is a natural scene text dataset cropped from the Google search images. It is divided into 2000 training and 3000 testing images.

SVT [10] consists of 257 training and 647 testing data cropped from the Google Street View images. SVT is challenging for the blur or noise interference.

ICDAR2013 [32] contains 848 images for training and 1095 images for evaluation. We filter out the images that only contain the non-alphanumeric characters, resulting in a test set with 1078 images.

ICDAR2015 [33] has 4468 word patches for training and 2077 for testing. IC15 is a benchmark dataset for irregular scene text recognition. The images are captured from the Google Glasses under arbitrary angles. Therefore, most words are irregular and have a changeable perspective.

SVT-Perspective (SVTP) [34] has 645 cropped scent text images for testing. It is also from the Google Street View but has more perspective words than SVT.

CUTE80 (CUTE) [35] consists of 288 word patches. CUTE is widely used for STR model evaluation for various curved text images.

4.2   Implementation Details

We build our model on the Tensorflow framework. All experiments are conducted on 2 NVIIDA Titan X GPUs with 12 GB memory. The training dataset consists of 4.8 million from MJ, 4.8 million from ST, and 50k from the mixed training sets of IIIT5K, IC13, IC15 and SVT. Data augmentation methods such as perspective distortion and color transformation are applied to the mixed real datasets. The batch size is set to 42, including 20 from the MJ, 20 from the ST and 2 from the mixed dataset. The training process only needs 2.2 epochs in total with word-level annotation. Adam is utilized for optimization, with the addition of cross-entropy loss and l2 regularization loss(with the coefficient of 1e−4) as the objective function. Learning rate is set to 1e−4 per GPU, and reduced to 1e−5 in the last 0.2 epoch.

The input size of our model is 48×160. Height is fixed, if the image width is shorter than target width after resize with the original scale, we apply the padding operation, else we change the scale and simply reshape the image to the target size. The LD-BiLSTM in the encoder and LD-LSTM in the decoder are both single-layer, with the dropout rate of 0.1 and 0.5 respectively. The dimensions of the visual feature/context feature/decoded hidden state/word embedding matrix are set to 512/512/512/256. The maximum time step T is 27.

At the inference stage, no lexicon is used. For the sake of efficiency, we simply rotate 90 degree clockwise for the images of which the height is larger than 2 times of the width, instead of applying two directions rotation as previous studies [2,4,6,7,22]. Besides, unlike [2,6], we do not use the beam search method to improve the performance. The decoder works in the forward direction, rather than the bidirectional strategy which is adopted in [2,3].

Table 2. Recognition accuracy of our method and other sate-of-the-art methods on the six public datasets. All the results listed are lexicon free. The approaches which need character-level annotations are marked *. In each column, the best performance is shown in **bold** font, and the second best result is underlined. Our model achieves the best accuracy in regular dataset SVT and the most challenging irregular dataset IC15, the second highest accuracy in regular dataset IC13 and curve dataset CUTE.

Method	Regular text			Irregular text		
	IIIT5K	SVT	IC13	IC15	SVTP	CUTE
CRNN (2015) [1]	81.2	82.7	89.6	–	–	–
R^2AM (2016) [12]	78.4	80.7	90.0	–	–	–
RARE (2016) [11]	81.9	81.9	88.6	–	–	–
FAN (2017) [14]*	87.4	85.9	93.3	70.6	–	–
NRTR (2017) [8]	86.5	88.3	94.7	–	–	–
EP (2018) [15]*	88.3	87.5	94.4	73.9	–	–
ESIR (2018) [18]	93.3	90.2	91.3	76.9	79.6	83.3
Char–Net (2018) [20]*	92.0	85.5	91.1	74.2	78.9	–
Mask TextSpotter (2018) [21]*	95.3	**91.8**	**95.3**	78.2	83.6	88.5
ASTER (2018) [2]	93.4	89.5	91.8	76.1	78.5	79.5
ZOU, L et al. (2019) [5]	85.4	84.5	91.0	–	–	–
MORAN (2019) [19]	91.2	88.3	92.4	68.8	76.1	77.4
MASTER (2019) [7]	95.0	90.6	**95.3**	79.4	84.5	87.5
SCATTER (2020) [4]	92.9	89.2	93.8	81.8	84.5	85.1
Yang, L et al. (2020) [3]	94.7	88.9	93.2	74.0	80.9	85.4
SAR (2019) [6]	95.0	91.2	94.0	78.8	**86.4**	89.6
SAR 3 × 20	94.0	90.6	93.1	76.2	83.7	87.5
RobustScanner (2020) [22]	**95.4**	89.3	94.1	79.2	82.9	**92.4**
RCEED (Ours)	94.9	91.8	94.7	82.2	83.6	91.7

4.3   Experimental Results

In this section, we test the model performance on regular and irregular STR public datasets, and make comparison with the SOTA methods. As shown in Table 2, RCEED achieves the best performance in regular dataset SVT and irregular dataset IC15, and the second highest accuracy in regular dataset IC13 and curve dataset CUTE. Compared with competitors SAR [6] and RobustScanner [22] which are baseline and up-to-date methods targeting at irregular scene text recognition and trained with synth&real datasets, RCEED performs accuracy increases of 3pp to 3.4pp in the most challenging irregular dataset IC15 and outperforms them by 4 out of 6 benchmarks, while the size of the encoded feature map is only 1/4 of theirs. Compared with SAR modeled with the same-size encoded feature map(SAR 3 × 20), our model outperforms it in 5 benchmarks with 6pp ahead in maximum. Experimental results demonstrate that RCEED has got rid of the performance bottleneck caused by small-size feature map, and is competitive in both the regular and the irregular datasets.

Figure 5 shows the success and failure cases of RCEED. Success cases in the first row demonstrate that the proposed model has a robust ability to deal with blur, variation, occlusion, distortion, uneven light and other difficult situations. However, when the problems become serious, failure cases such as over-prediction and less-prediction appear. It can be seen that image quality still has a significant impact on the recognition results.

Fig. 5. Success and failure cases of RCEED. GT refers to the ground truth, Pred refers to the predicted results.

4.4   Ablation Studies

In this section, we conduct a series of comparison experiments to analyse the effectiveness of the key contributions in RCEED. Evaluation results of 8 conditions are listed in Table 3.

Impact of Comprehensive Feature. We evaluate the performance when visual feature and context feature are used alone as the input of the decoder. As shown in Row 2 and 3, single visual and single context features both have lower accuracy of up to −1.6pp on the regular dataset and −2.8pp on the irregular dataset compared with the comprehensive feature, indicating that the feature fusion operation plays an important role in improving model performance.

Impact of Layernorm-Dropout LSTM Cell. Row 1 reveals the model performance when the LD-LSTM design in the encoder and decoder are both replaced with normal LSTM cells. Compared with Row 8, the model implemented with normal LSTM cells shows an accuracy drop of -1.3pp on the largest irregular dataset IC15. While in the largest regular dataset IIIT5K, the accuracy is 0.6pp higher and equals to the best performance [22]. The dropout design in the LD-LSTM improves model's ability in dealing with changeable characters, but is not compatible with regular texts with stable features. Overall, LD-LSTM cell still outperforms the normal one in most benchmarks.

Impact of Guided Initialization of Decoder. Row 4 lists the results when the decoder is initialized with zero vectors instead of the holistic feature and global glimpse vector from the encoder. Accuracy shows a drop up to -1.6pp in the five out of six public datasets, which demonstrates that the global information from the encoder is helpful for improving the decoding accuracy.

Impact of Glimpse Vector for Prediction. We build a comparison model which does not use glimpse vector for prediction. As shown in Row 5, the performance degradation is more serious in irregular datasets(up to -1.8pp), proving that visual information has a greater impact on recognizing complex texts.

Impact of Heads Number. We compare the evaluation results when attention heads number m is set to 1, 4 and 8 in Row 6, 7, 8. The 1 head condition is equivalent to not using the multi-head design. We set $m = 8$ in our model for the superior performance on irregular datasets.

Table 3. Ablation studies by changing model structures and hyper-parameters. LD refers to models w/o the Layernorm-Dropout method in the RNN layers. VF/CF denote whether the Visual Feature and Context Feature are involved in the encoded feature map. GI/GP analyze the impact of the Guided Initialization of decoder and the Glimpse vector for Prediction. Heads column presents the conditions with different number of heads. All the comparison models are trained from scratch with word-level annotations.

Cond	LD	VF	CF	GI	GP	Heads	IIIT5K	SVT	IC13	IC15	SVTP	CUTE
1	×	✓	✓	✓	✓	8	95.4	91.5	94.7	80.9	84.9	89.9
2	✓	×	✓	✓	✓	8	94.7	90.9	93.1	80.0	82.6	88.9
3	✓	✓	×	✓	✓	8	94.9	91.6	93.7	80.4	82.8	90.3
4	✓	✓	✓	×	✓	8	95.3	90.9	94.6	80.6	83.3	90.6
5	✓	✓	✓	✓	×	8	94.9	92.6	94.3	80.9	82.9	89.9
6	✓	✓	✓	✓	✓	1	94.3	91.6	94.3	80.8	83.6	91.7
7	✓	✓	✓	✓	✓	4	95.0	92.4	94.1	81.1	83.7	89.5
8	✓	✓	✓	✓	✓	8	94.9	91.8	94.7	82.2	83.6	91.7

5   Conclusion

In this work, we propose a representation and correlation enhanced encoder-decoder framework for scene text recognition. The encoder enhances model's representation ability through the aligning and fusing operation between the local visual feature and the global context feature. The decoder strengthens the correlation between the encoded comprehensive feature and the decoded character sequence through the guided initialization and the efficient workflow. Essential components including the Multi-Head General Attention mechanism and the LD-LSTM cell are designed to reduce feature deficiency and improve the generalization towards changeable texts. The model breaks the constraint of feature map size and has a superior performance on public benchmarks. In future research, we will develop an end-to-end integrated detection and recognition model for the text spotting task and develop advanced applications of the visual-semantic interaction.

Acknowledgments. This work is supported by National Natural Science Foundation of China (61976214, 61721004, 61633021), and Science and Technology Project of SGCC Research on feature recognition and prediction of typical ice and wind disaster for transmission lines based on small sample machine learning method.

References

1. Shi, B., Bai, X., Yao, C.: An end-to-end trainable neural network for image-based sequence recognition and its application to scene text recognition. PAMI **39**(11), 2298–2304 (2016)
2. Shi, B., Yang, M., Wang, X., Lyu, P., Yao, C., Bai, X.: ASTER: an attentional scene text recognizer with flexible rectification. PAMI **41**(9), 2035–2048 (2018)
3. Yang, L., Wang, P., Li, H., Li, Z., Zhang, Y.: A holistic representation guided attention network for scene text recognition. Neurocomputing **414**, 67–75 (2020)
4. Litman, R., Anschel, O., Tsiper, S., Litman, R., Mazor, S., Manmatha, R.: SCAT-TER: selective context attentional scene text recognizer. In: CVPR, pp. 11962–11972 (2020)
5. Zuo, L.Q., Sun, H.M., Mao, Q.C., Qi, R., Jia, R.S.: Natural scene text recognition based on encoder-decoder framework. IEEE Access **7**, 62616–62623 (2019)
6. Li, H., Wang, P., Shen, C., Zhang, G.: Show, attend and read: A simple and strong baseline for irregular text recognition. In: AAAI, vol. 33, pp. 8610–8617 (2019)
7. Lu, N., Yu, W., Qi, X., Chen, Y., Gong, P., Xiao, R.: MASTER: multi-aspect non-local network for scene text recognition. arXiv preprint arXiv:1910.02562 (2019)
8. Sheng, F., Chen, Z., Xu, B.: NRTR: a no-recurrence sequence-to-sequence model for scene text recognition. In: ICDAR, pp. 781–786. IEEE (2019)
9. Casey, R.G., Lecolinet, E.: A survey of methods and strategies in character segmentation. PAMI **18**(7), 690–706 (1996)
10. Wang, K., Babenko, B., Belongie, S.: End-to-end scene text recognition. In: ICCV, pp. 1457–1464. IEEE (2011)
11. Shi, B., Wang, X., Lyu, P., Yao, C., Bai, X.: Robust scene text recognition with automatic rectification. In: CVPR, pp. 4168–4176 (2016)
12. Lee, C.Y., Osindero, S.: Recursive recurrent nets with attention modeling for OCR in the wild. In: CVPR, pp. 2231–2239 (2016)
13. Bahdanau, D., Cho, K., Bengio, Y.: Neural machine translation by jointly learning to align and translate. arXiv preprint arXiv:1409.0473 (2014)

14. Cheng, Z., Bai, F., Xu, Y., Zheng, G., Pu, S., Zhou, S.: Focusing attention: towards accurate text recognition in natural images. In: ICCV, pp. 5076–5084 (2017)
15. Bai, F., Cheng, Z., Niu, Y., Pu, S., Zhou, S.: Edit probability for scene text recognition. In: CVPR, pp. 1508–1516 (2018)
16. Bookstein, F.L., Green, W.D.K.: A thin-plate spline and the decomposition of deformations. Math. Methods Med. Imaging **2**, 14–28 (1993)
17. Jaderberg, M., Simonyan, K., Zisserman, A., Kavukcuoglu, K.: Spatial transformer networks. arXiv preprint arXiv:1506.02025 (2015)
18. Zhan, F., Lu, S.: ESIR: end-to-end scene text recognition via iterative image rectification. In: CVPR, pp. 2059–2068 (2019)
19. Luo, C., Jin, L., Sun, Z.: MORAN: a multi-object rectified attention network for scene text recognition. Pattern Recogn. **90**, 109–118 (2019)
20. Liu, W., Chen, C., Wong, K.Y.: Char-net: A character-aware neural network for distorted scene text recognition. In: AAAI, vol. 32 (2018)
21. Liao, M., Lyu, P., He, M., Yao, C., Wu, W., Bai, X.: Mask TextSpotter: an end-to-end trainable neural network for spotting text with arbitrary shapes. PAMI **43**(2), 532–548 (2021)
22. Yue, X., Kuang, Z., Lin, C., Sun, H., Zhang, W.: RobustScanner: dynamically enhancing positional clues for robust text recognition. In: Vedaldi, A., Bischof, H., Brox, T., Frahm, J.-M. (eds.) ECCV 2020. LNCS, vol. 12364, pp. 135–151. Springer, Cham (2020). https://doi.org/10.1007/978-3-030-58529-7_9
23. Hochreiter, S., Schmidhuber, J.: Long short-term memory. Neural Comput. **9**(8), 1735–1780 (1997)
24. Ba, J.L., Kiros, J.R., Hinton, G.E.: Layer normalization. arXiv preprint arXiv:1607.06450 (2016)
25. Hinton, G.E., Srivastava, N., Krizhevsky, A., Sutskever, I., Salakhutdinov, R.R.: Improving neural networks by preventing co-adaptation of feature detectors. arXiv preprint arXiv:1207.0580 (2012)
26. Zaremba, W., Sutskever, I., Vinyals, O.: Recurrent neural network regularization. arXiv preprint arXiv:1409.2329 (2014)
27. Luong, M.T., Pham, H., Manning, C.D.: Effective approaches to attention-based neural machine translation. arXiv preprint arXiv:1508.04025 (2015)
28. Vaswani, A., et al.: Attention is all you need. arXiv preprint arXiv:1706.03762 (2017)
29. Jaderberg, M., Simonyan, K., Vedaldi, A., Zisserman, A.: Synthetic data and artificial neural networks for natural scene text recognition. arXiv preprint arXiv:1406.2227 (2014)
30. Gupta, A., Vedaldi, A., Zisserman, A.: Synthetic data for text localisation in natural images. In: CVPR, pp. 2315–2324 (2016)
31. Mishra, A., Alahari, K., Jawahar, C.V.: Scene text recognition using higher order language priors. In: BMVC-British Machine Vision Conference. BMVA (2012)
32. Karatzas, D., et al.: ICDAR 2013 robust reading competition. In: ICDAR, pp. 1484–1493. IEEE (2013)
33. Karatzas, D., et al.: ICDAR 2015 competition on robust reading. In: ICDAR, pp. 1156–1160. IEEE (2015)
34. Phan, T.Q., Shivakumara, P., Tian, S., Tan, C.L.: Recognizing text with perspective distortion in natural scenes. In: ICCV, pp. 569–576 (2013)
35. Risnumawan, A., Shivakumara, P., Chan, C.S., Tan, C.L.: A robust arbitrary text detection system for natural scene images. Expert Syst. Appl. **41**(18), 8027–8048 (2014)

FEDS - Filtered Edit Distance Surrogate

Yash Patel$^{(\boxtimes)}$ [iD] and Jiří Matas [iD]

Visual Recognition Group, Czech Technical University in Prague,
Prague, Czech Republic
{patelyas,matas}@fel.cvut.cz

Abstract. This paper proposes a procedure to train a scene text recognition model using a robust learned surrogate of edit distance. The proposed method borrows from self-paced learning and filters out the training examples that are hard for the surrogate. The filtering is performed by judging the quality of the approximation, using a ramp function, enabling end-to-end training. Following the literature, the experiments are conducted in a post-tuning setup, where a trained scene text recognition model is tuned using the learned surrogate of edit distance. The efficacy is demonstrated by improvements on various challenging scene text datasets such as IIIT-5K, SVT, ICDAR, SVTP, and CUTE. The proposed method provides an average improvement of 11.2% on total edit distance and an error reduction of 9.5% on accuracy.

1 Introduction

Supervised deep learning has benefited many tasks in computer vision, to the point that these methods are now commercially used. It involves training a neural network on a task-specific dataset annotated by humans. The training is performed with a loss function that compares the model output with the expected output. The loss function choice is driven by multiple factors such as the application-defined objective, generalization to out-of-distribution samples, and constrained by the training algorithm.

Deep neural networks are trained by back-propagating gradients [37], which requires the loss function to be differentiable. However, the task-specific objective is often defined via an evaluation metric, which may not be differentiable. The evaluation metric's design is to fulfill the application requirements, and for the cases where the evaluation metric is differentiable, it is directly used as a loss function. For scene text recognition (STR), accuracy and edit distance are popular evaluation metric choices. Accuracy rewards the method if the prediction exactly matches the ground truth. Whereas edit distance (ED) is defined by counting addition, subtraction, and substitution operations, required to transform one string into another. As shown in Fig. 1, accuracy does not account for partial correctness. Note that the low *ED* errors from M2 can be easily corrected by a dictionary search in a word-spotting setup [31]. Therefore, edit distance is a better metric, especially when the state-of-the-art is saturated on the benchmark datasets [17,18,30].

© Springer Nature Switzerland AG 2021
J. Lladós et al. (Eds.): ICDAR 2021, LNCS 12824, pp. 171–186, 2021.
https://doi.org/10.1007/978-3-030-86337-1_12

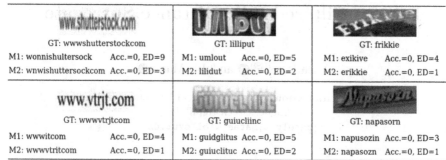

GT: wwwshutterstockcom	GT: lilliput	GT: frikkie
M1: wonnishultersock Acc.=0, ED=9	M1: umlout Acc.=0, ED=5	M1: exikive Acc.=0, ED=4
M2: wnwishuttersockcom Acc.=0, ED=3	M2: lilidut Acc.=0, ED=2	M2: erikkie Acc.=0, ED=1
GT: wwwvtrjtcom	GT: guiucliinc	GT: napasorn
M1: wwwitcom Acc.=0, ED=4	M1: guidglitus Acc.=0, ED=5	M1: napusozin Acc.=0, ED=3
M2: wwwvtritcom Acc.=0, ED=1	M2: guiuclituc Acc.=0, ED=2	M2: napasozn Acc.=0, ED=1

M1: Total Accuracy = 0, Total ED = 30   M2: Total Accuracy = 0, Total ED = 10

Fig. 1. Accuracy and edit distance comparison for different predictions of scene text recognition (STR). For the scene text images, green shows the ground truth, red shows the prediction from a STR model M1 and blue shows the predictions from another STR model M2. For these examples accuracy ranks both the models equally, however, it can be clearly seen that for the predictions in blue vocabulary search or Google search will succeed. (Color figure online)

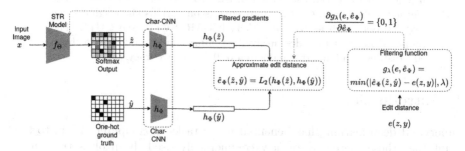

Fig. 2. Overview of the proposed post-tuning procedure. x is the input to the STR model $f_\Theta(x)$ with output $\hat{z}$. y is the ground truth, $\hat{y}$ is the ground truth expressed as one-hot, $e(z, y)$ is the evaluation metric, $\hat{e}_\Phi(\hat{z}, \hat{y})$ is the learned surrogate and $g_\lambda(e, \hat{e}_\Phi)$ is the filtering function. The approximations from the learned surrogate are checked against the edit distance by the filtering function. The STR model is not trained on the samples where the surrogate is incorrect.

When the evaluation metric is non-differentiable, a proxy loss is employed, which may not align well with the evaluation metric. Edit distance is computed via dynamic programming and is non-differentiable. Therefore, it can not be used as a loss function for training deep neural networks. The proxy loss used for training STR models is per-character cross-entropy or Connectionist Temporal Classification (CTC) [7]. The models trained with cross-entropy or CTC may have a sub-optimal performance on edit distance as they optimize a different objective.

The aforementioned issue can be addressed by learning a surrogate, e.g. [32], where a model trained with the proxy loss is post-tuned on a learned surrogate of the evaluation metric. In [32], post-tuning has shown significant improvement in performance on the evaluation metric. While Patel *et al.* [32] have paid attention to learning the surrogate, none was given to robustly train the neural network with the surrogate. In the training procedure, [32] assumes that the learned surrogate robustly estimates the edit distance for all samples. Since the surrogate is learned via supervised training, it is prone to overfitting on the training distribution and may fail on out-of-the-distribution samples. In hope for better generalizability of the surrogate, [32] makes use of a data generator to train the surrogate, which requires extra engineering effort. This paper shows that the learned edit distance surrogate often fails, leading to noisy training.

As an improvement, this paper proposes **F**iltered **E**dit **D**istance **S**urrogate. In FEDS, the STR model is trained only on the samples where the surrogate approximates the edit distance within a small error bound. This is achieved by computing the edit distance for a training sample and comparing it with the approximation from the surrogate. The comparison is realized by a ramp-function, which is piece-wise differentiable, allowing for end-to-end training. Figure 2 provides an overview of the proposed method. The proposed training method simplifies the training and eliminates the need for a data generator to learn a surrogate.

The rest of the paper is structured as follows. Related work is reviewed in Sect. 2, the technique for robustly training with the learned surrogate of ED is presented in Sect. 3, experiments are shown in Sect. 4 and the paper is concluded in Sect. 5.

2   Related Work

Scene text recognition (STR) is the task of recognizing text from images against complex backgrounds and layouts. STR is an active research area; comprehensive surveys can be found in [1, 26, 49]. Before deep learning, STR methods focused on recognizing characters via sliding window, and hand-crafted features [43, 44, 48]. Deep learning based STR methods have made a significant stride in improving model architectures that can handle both regular (axis-aligned text) and irregular text (complex layout, such as perspective and curved text). Selected relevant methods are discussed subsequently.

Convolutional Models for STR. Among the first deep learning STR methods was the work of Jaderberg *et al.* [15], where a character-centric CNN [20] predicts a text/no-text score, a character, and a bi-gram class. Later this work was extended to word-level recognition [13] where the CNN takes a fixed dimension input of the cropped word and outputs a word from a fixed dictionary. Bušta *et al.* [3,4] proposed a fully-convolutional STR model, which operates on variable-sized inputs using bi-linear sampling [14]. The model is trained jointly with a detector in a multi-task learning setup using CTC [7] loss. Gomez*et al.* [5] trains an embedding model for word-spotting, such that, the euclidean distance between the representations of two images corresponds to the edit-distance between their text strings. This embedding model differs from FEDS as it operates on images instead of STR model's predictions and is not used to train a STR model.

Recurrent Models for STR. Shi *et al.* [38] and He *et al.* [11] were among the first to propose end-to-end trainable, sequence-to-sequence models [42] for STR. An image of a cropped word is seen as a sequence of varying length, where convolutional layers are used to extract features and recurrent layers to predict a label distribution. Shi *et al.* [39] later combined the CNN-RNN hybrid with spatial transformer network [14] for better generalizability on irregular text. In [40], Shi *et al.* adapted Thin-Plate-Spline [2] for STR, leading to an improved performance on both regular and irregular text (compared to [39]). While [39, 40] rectify the entire text image, Liu *et al.* [25] detects and rectifies each character. This is achieved via a recurrent RoIWarp layer, which sequentially attends to a region of the feature map that corresponds to a character. Li *et al.* [21] passed the visual features through an attention module before decoding via an LSTM. MaskTextSpotter [22] solves detection and recognition jointly; the STR module consists of two branches while the first uses local visual features, the second utilizes contextual information in the form of attention. Litman *et al.* [24] utilizes a stacked block architecture with intermediate supervision during training, which improves the encoding of contextual dependencies, thus improving the performance on the irregular text.

Training Data. Annotating scene text data in real images is complex and expensive. As an alternative, STR methods often use synthetically generated data for training. Jaderberg *et al.* [15] generated 8.9 million images by rendering fonts, coloring the image layers, applying random perspective distortion, and blending it to a background. Gupta *et al.* [9] placed rendered text on natural scene images; this is achieved by identifying plausible text regions using depth and segmentation information. Patel *et al.* [33] further extended this to multi-lingual text. The dataset of [9] was proposed for training scene text detection; however, it is also useful for improving STR models [1]. Long *et al.* [27] used a 3D graphics engine to generate scene text data. The 3D synthetic engine allows for better text region proposals as scene information such as normal and objects meshes are available. Their analysis shows that compared to [9], more realistic looking diverse images (contains shadow, illumination variations, etc.) are more useful for STR models. As an alternative to synthetically generate data, Janouskova *et al.* [16] leverages weakly annotated images to generate pseudo scene text labels. The approach uses an end-to-end scene text model to generate initial labels, followed by a heuristic neighborhood search to match imprecise transcriptions with weak annotations.

As discussed, significant work has been done towards improving the model architectures [1, 3, 4, 13, 14, 23, 24, 34, 38–40, 45, 47, 50, 51, 53] and obtaining data for training [6, 9, 13, 16, 27].

Limited attention has been paid to the loss function. Most deep learning based STR methods rely on per-character cross-entropy or CTC loss functions [1, 7]. While in theory and under an assumption of infinite training data, these loss functions align with accuracy [19], there is no concrete evidence of their alignment with edit-distance. In comparison to the related work, this paper makes an orthogonal contribution, building upon learning surrogates [32], this paper proposes a robust training procedure for better optimization of STR models on edit distance.

3   FEDS: Filtered Edit Distance Surrogate

3.1   Background

The samples for training the scene text recognition (STR) model are drawn from a distribution $(x, y) \sim U_D$. Here, x is the image of a cropped word, and y is the corresponding transcription. An end-to-end trainable deep model for STR, denoted by $f_\Theta(x)$ predicts a soft-max output $\hat{z} = f_\Theta(x)$, $f_\Theta : \mathbb{R}^{W \times H \times 1} \to \mathbb{R}^{|A| \times L}$. Here W and H are the dimensions of the input image, A is the set of characters, and L is the maximum possible length of the word.

For training, the ground truth y is converted to one-hot representation $\hat{y}^{|A| \times L}$. Cross entropy (CE) is a popular choice of the loss function [1], which provides the loss for each character:

$$CE(\hat{z}, \hat{y}) = -\frac{1}{L|A|} \sum_{i=1}^{L} \sum_{j=1}^{|A|} \hat{y}_{i,j} log(\hat{z}_{i,j}) \qquad (1)$$

Patel _et al._ [32] learns the surrogate of edit distance via a learned deep embedding h_Φ, where the Euclidean distance between the prediction and the ground truth corresponds to the value of the edit distance, which provides the edit distance surrogate, denoted by $\hat{e}_\Phi$:

$$\hat{e}_\Phi(\hat{z}, \hat{y}) = \|h_\Phi(\hat{z}) - h_\Phi(\hat{y})\|_2 \qquad (2)$$

where h_Φ is the Char-CNN [32,54] with parameters Φ. Note that the edit distance surrogate is defined on the one-hot representation of the ground truth and the soft-max prediction from the STR model.

3.2   Learning Edit Distance Surrogate

Objective. To fairly demonstrate the improvements using the proposed FEDS, the loss for learning the surrogate is the same as LS-ED [32]:

1. The learned edit distance surrogate should correspond to the value of the edit distance:

$$\hat{e}_\Phi(\hat{z}, \hat{y}) \approx e(z, y) \qquad (3)$$

   where $e(z, y)$ is the edit distance defined on the string representation of the prediction and the ground truth.
2. The first order derivative of the learned edit distance surrogate with respect to the STR model prediction $\hat{z}$ is close to 1:

$$\left\| \frac{\partial \hat{e}_\Phi(\hat{z}, \hat{y})}{\partial \hat{z}} \right\|_2 \approx 1 \qquad (4)$$

Bounding the gradients (Eq. 4) has shown to enhance the training stability for Generative Adversarial Networks [8] and has shown to be useful for learning the surrogate [32].

Both objectives are realized and linearly combined in the training loss:

$$loss(\hat{z}, \hat{y}) = w_1 \left\| \left(\hat{e}_\Phi(\hat{z}, \hat{y}) - e(z, y) \right\|_2^2 + w_2 \left(\left\| \frac{\partial \hat{e}_\Phi(\hat{z}, \hat{y})}{\partial \hat{z}} \right\|_2 - 1 \right)^2 \qquad (5)$$

Training Data. Patel *et al.* [32] uses two sources of data for learning the surrogate - the pre-trained STR model and a random generator. The random generator provides a pair of words and their edit distance and ensures uniform sampling in the range of the edit distance. The random generator helps the surrogate to generalize better, leading to an improvement in the final performance of the STR model.

The proposed FEDS does not make use of a random generator, reducing the effort and the computational cost. FEDS learns the edit distance surrogate only on the samples obtained from the STR model:

$$(\hat{z}, \hat{y}) \sim f_\Theta(x) \mid (x, y) \sim U_D \qquad (6)$$

3.3   Robust Training

The filtering function g_λ is defined on the surrogate and the edit distance, parameterized by a scalar λ that acts as a threshold to determine the quality of the approximation from the surrogate. The filtering function is defined as:

$$g_\lambda(e(z, y), \hat{e}_\Phi(\hat{z}, \hat{y})) = \min(|\hat{e}_\Phi(\hat{z}, \hat{y}) - e(z, y)|, \lambda) \mid \lambda > 0 \qquad (7)$$

The filtering function is piece-wise differentiable, as can be seen in Fig. 3. For the samples where the quality of approximation from the surrogate is low, the gradients are zero, and the STR model is not trained on those samples. Whereas for samples where the quality of the approximation is within the bound of λ, the STR model is trained to minimize the edit distance surrogate.

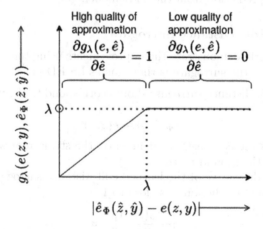

Fig. 3. The filtering function enforces zero gradients for the samples that are hard for the surrogate (low quality of approximation). STR model is trained only on the samples where the quality of the approximation from the edit distance surrogate is high.

Learning of the ED surrogate $\hat{e}_\Phi$ and post-tuning of the STR model $f_\Theta(x)$ are conducted alternatively. The surrogate is learned first for I_a number of iterations while the STR model is fixed. Subsequently, the STR model is trained using the

---

Algorithm 1. Post-tuning with FEDS

Inputs: Supervised data D, evaluation metric e.

Hyper-parameters: Number of update steps I_a and I_b, learning rates η_a and η_b, number of epochs E.

Objective: Robustly post-tune the STR model, *i.e.*, $f_\Theta(x)$ and learn the edit distance surrogate, *i.e.*, $\hat{e}_\Phi$.

1: *Initialize* $\Theta \leftarrow$ pre-trained weights, $\Phi \leftarrow$ random weights.
2: **for** epoch $= 1,...,E$ **do**
3:     **for** i $= 1,...,I_a$ **do**
4:         sample, $(x,y) \sim U_D$
5:         inference, $\hat{z} = f_{\Theta^{epoch-1}}(x)$
6:         compute loss, $l_{\hat{e}} = loss(\hat{z},\hat{y})$ (Eq. 5)
7:         update ED surrogate, $\Phi^i \leftarrow \Phi^{i-1} - \eta_a \frac{\partial l_{\hat{e}}}{\partial \Phi^{i-1}}$
8:     **end for**
9:     $\Phi \leftarrow \Phi^{I_a}$
10:     **for** i $= 1,...,I_b$ **do**
11:         sample, $(x,y) \sim U_D$
12:         inference, $\hat{z} = f_{\Theta^{i-1}}(x)$
13:         compute ED from the surrogate, $\hat{e} = \hat{e}_{\Phi^{epoch}}(\hat{z},\hat{y})$ (Eq. 2)
14:         compute ED, $e = e(z,y)$
15:         computer loss, $l_f = g_\lambda(e,\hat{e})$ (Eq. 7)
16:         update STR model, $\Theta^i \leftarrow \Theta^{i-1} - \eta_b \frac{\partial(l_f)}{\partial \Theta^{i-1}}$
17:     **end for**
18:     $\Theta \leftarrow \Theta^{I_b}$
19: **end for**

---

surrogate and the filtering function, while the ED surrogate parameters are kept fixed. Algorithm 1 and Fig. 2 demonstrate the overall training procedure with FEDS.

4   Experiments

4.1   FEDS Model

The model for learning the deep embedding, *i.e.*, h_Φ is kept same as [32]. A Char-CNN architecture [54] is used with five $1D$ convolution layers, LeakyReLU [46] and two FC layers. The embedding model, h_Φ, maps the input to a 1024 dimensions, $h_\Phi : \mathbb{R}^{|A| \times L} \rightarrow \mathbb{R}^{1024}$. Feed forward (Eq. 2), generates embeddings for the ground-truth $\hat{y}$ (one-hot) and model prediction $\hat{z}$ (soft-max) and an approximation of edit distance is computed by L_2 distance between the two embedding.

4.2   Scene Text Recognition Model

Following the survey on STR, [1], the state-of-the-art model ASTER is used [40], which contains four modules: (a) transformation, (b) feature extraction,

(c) sequence modeling, and (d) prediction. Baek *et al.* [1] provides a detailed analysis of STR models and the impact of different modules on the performance.

Transformation. Operates on the input image and rectifies the curved or tilted text, easing the recognition for the subsequent modules. The two popular variants include Spatial Transformer [14] and Thin Plain Spline (TPS) [40]. TPS employs a smooth spline interpolation between a set of fiducial points, which are fixed in number. Following the analysis of Shi *et al.* [1,40], the STR model used employs TPS.

Feature Extraction. Involves a Convolutional Neural Network [20], that extracts the features from the image transformed by TPS. Popular choices include VGG-16 [41] and ResNet [10]. Follwoing [1], the STR model used employs ResNet for the ease of optimization and good performance.

Sequence Modeling. Captures the contextual information within a sequence of characters; this module operates on the features extracted from a ResNet. The STR model used employs BiLSTM [12].

Prediction. The predictions are made based on the identified features of the image. The prediction module depends on the loss function used for training. CTC loss requires the prediction to by sigmoid, whereas cross-entropy requires the prediction to be a soft-max distribution over the set of characters. The design of FEDS architecture (Sect. 4.1) requires a soft-max distribution.

FEDS and LS-ED [32] are investigated with the state-of-the-art performing configuration of the STR model, which is *TPS-ResNet-BiLSTM-Attn*.

4.3   Training and Testing Data

The STR models are trained on synthetic and pseudo labeled data and are evaluated on real-world benchmarks. Note that the STR models are not fine-tuned on evaluation datasets (same as [1]).

Training Data. The experiments make use of the following synthetic and pseudo labeled data for training:

- **MJSynth** [15] (synthetic). 8.9 million synthetically generated images, obtained by rendering fonts, coloring the image layers, applying random perspective distortion, and blending it to a background.
- **SynthText** [9] (synthetic). 5.5 million text instance by placing rendered text on natural scene images. This is achieved by identifying plausible text regions using depth and segmentation information.
- **Uber-Text** [16] (pseudo labels). 138K real images from Uber-Text [55] with pseudo labels obtained using [16].
- **Amazon book covers** [16] (pseudo labels). 1.5 million real images from amazon book covers with pseudo labels obtained using [16].

Testing Data. The models trained purely on the synthetic and pseudo labelled datasets are tested on a collection of real datasets. This includes regular scene text - IIIT-5K [29], SVT [43], ICDAR'03 [28] and ICDAR'13 [18], and irregular scene text ICDAR'15 [17], SVTP [35] and CUTE [36].

4.4   Implementation Details

The analysis of the proposed FEDS and LS-ED [32] is conducted for two setups of training data. First, similar to [1], the STR models are trained on the union of the synthetic data obtained from MJSynth [15], and SynthText [9] resulting in a total of 14.4 million training examples. Second, additional pseudo labeled data [16] is used to obtain a stronger baseline.

The STR models are first trained with the proxy loss, *i.e.*, cross-entropy for $300K$ iterations with a mini-batch size of 192. The models are optimized using ADADELTA [52]. Once the training is complete, these models are tuned with FEDS (Algorithm 1) on the same training set for another $20K$ iterations. For learning the edit distance surrogate the weights in the loss (Eq. 5) are set as $w_1 = 1, w_2 = 0.1$. Note that the edit distance value is a non-negative integer, therefore, optimal range for λ is $(0, 0.5)$. Small value of λ filters out substantial number of samples, slowing down the training, whereas, large values of λ allows a noisy training. Therefore, the threshold for the filtering function (Eq. 7) is set as $\lambda = 0.25$, *i.e.*, in the middle of the optimal range.

4.5   Quality of the Edit Distance Surrogate

Figure 4 shows a comparison between the edit distance and the approximation from the surrogate. As the training progresses, the approximation improves,

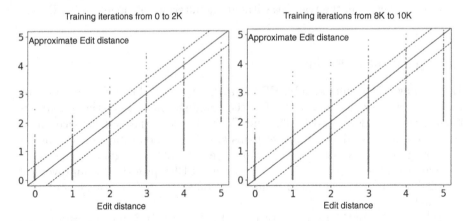

Fig. 4. A comparison between the true edit distance and the approximated edit distance is shown. Each point represents a training sample for the STR model. The solid line represents an accurate approximation of the edit distance. The dotted lines represent the filtering in FEDS. **Left:** Plot for the first 2K iterations of the STR model training. **Right:** Plot for iterations from 8K to 10K of the STR model training. (Color figure online)

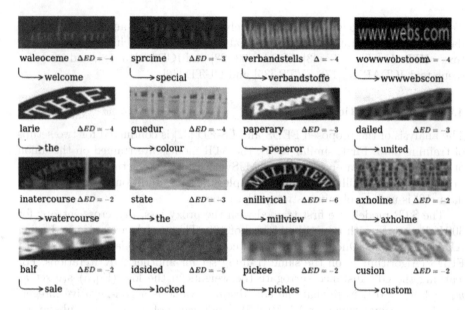

Fig. 5. Randomly chosen examples from the test set where FEDS improves the STR model trained with cross-entropy. Red shows the incorrect predictions from the baseline model, blue shows the correct prediction after post-tuning with FEDS and the arrow indicates post-tuning with FEDS. (Color figure online)

i.e., more samples are closer to the solid line. The dotted lines represent the filtering in FEDS, *i.e.*, only the samples between the dotted lines contribute to the training of the STR model. Note that the surrogate fails for a large fraction of samples; therefore, the training without the filtering (as done in LS-ED [32]) is noisy.

4.6 Quantitative Results

Table 1 shows the results with LS-ED [32] and the proposed FEDS in compression with the standard baseline [1,40]. For the training, only the synthetic datasets [9,15] are used. Both LS-ED [32] and FEDS improve the performance on all evaluation metrics. Most significant gains are observed on total edit distance as the surrogate approximates it. In comparison with LS-ED, significant gains are observed with the proposed FEDS. On average, FEDS provides an improvement of 11.2% on the total edit distance and 0.98% on accuracy (an equivalent of 9.5% error reduction).

Table 2 presents the results with LS-ED [32] and FEDS in compression with a stronger baseline [16]. For the training, a combination of synthetic [9,15] and pseudo labelled [16] data is used. LS-ED [32] provides a limited improvement of 2.91% on total edit distance whereas FEDS provides a significant improvement of 7.90% and an improvement of 1.01% on accuracy (equivalently 7.9% error

Table 1. STR model trained with MJSynth [15] and SynthText [9]. Evaluation on IIIT-5K [29], SVT [43], IC'03 [28], IC'13 [18], IC'15 [17], SVTP [35] and CUTE [36]. The results are reported using accuracy **Acc.** (higher is better), normalized edit distance **NED** (higher is better) and total edit distance **TED** (lower is better). Relative gains are shown in blue and relative declines in red.

Test data	Loss function	↑ Acc.	↑ NED	↓ TED
IIIT-5K (3000)	Cross-Entropy [1]	87.1	0.959	772
	LS-ED [32]	88.0 +1.03%	0.962 +0.31%	680 +11.9%
	FEDS	88.8 +1.95%	0.966 +0.72%	591 +23.44%
SVT (647)	Cross-Entropy [1]	87.2	0.953	175
	LS-ED [32]	87.3 +0.11%	0.954 +0.10%	161 +8.00%
	FEDS	88.7 +1.72%	0.957 +0.41%	147 +16.0%
IC'03 (860)	Cross-Entropy [1]	95.1	0.981	105
	LS-ED [32]	95.3 +0.21%	0.982 +0.10%	89 +15.2%
	FEDS	95.4 +0.31%	0.983 +0.20%	87 +17.1%
IC'03 (867)	Cross-Entropy [1]	95.1	0.982	102
	LS-ED [32]	95.2 +0.10%	0.983 +0.10%	90 +11.7%
	FEDS	95.0 −0.10%	0.981 −0.10%	81 +20.5%
IC'13 (857)	Cross-Entropy [1]	92.9	0.979	110
	LS-ED [32]	93.9 +1.07%	0.981 +0.20%	97 +11.8%
	FEDS	93.8 +0.96%	0.985 +0.61%	99 +10.0%
IC'13 (1015)	Cross-Entropy [1]	92.2	0.966	140
	LS-ED [32]	93.1 +0.97%	0.969 +0.31%	123 +12.1%
	FEDS	92.6 +0.43%	0.969 +0.31%	118 +15.7%
IC'15 (1811)	Cross-Entropy [1]	77.9	0.915	880
	LS-ED [32]	78.2 +0.38%	0.915 -	851 +3.29%
	FEDS	78.5 +0.77%	0.919 +0.43%	820 +6.81%
IC'15 (2077)	Cross-Entropy [1]	75.0	0.884	1234
	LS-ED [32]	75.3 +0.39%	0.883 −0.11%	1210 +1.94%
	FEDS	75.7 +0.93%	0.888 +0.45%	1176 +4.70%
SVTP (645)	Cross-Entropy [1]	79.2	0.912	340
	LS-ED [32]	80.0 +1.01%	0.915 +0.32%	327 +3.82%
	FEDS	80.9 +2.14%	0.919 +0.76%	307 +9.70%
CUTE (288)	Cross-Entropy [1]	74.9	0.881	221
	LS-ED [32]	75.6 +0.93%	0.885 +0.45%	204 +7.69%
	FEDS	75.3 +0.53%	0.891 +1.13%	197 +10.8%
TOTAL	Cross-Entropy [1]	85.6	0.941	4079
	LS-ED [32]	86.1 +0.61%	0.942 +0.18%	3832 +6.05%
	FEDS	86.5 +0.98%	0.946 +0.48%	3623 +11.2%

Table 2. STR model trained with MJSynth [15], SynthText [9] and pseudo labelled [16] data. Evaluation on IIIT-5K [29], SVT [43], IC'03 [28], IC'13 [18], IC'15 [17], SVTP [35] and CUTE [36]. The results are reported using accuracy **Acc.** (higher is better), normalized edit distance **NED** (higher is better) and total edit distance **TED** (lower is better). Relative gains are shown in blue and relative declines in red.

Test data	Loss function	↑ Acc.	↑ NED	↓ TED
IIIT-5K (3000)	Cross-Entropy [1]	91.7	0.973	550
	LS-ED [32]	91.8 +0.14%	0.973 –	539 +2.00%
	FEDS	92.2 +0.54%	0.975 +0.20%	479 +12.9%
SVT (647)	Cross-Entropy [1]	91.8	0.970	107
	LS-ED [32]	91.8 –	0.971 +0.10%	100 +6.54%
	FEDS	92.1 +0.32%	0.971 +0.10%	102 blue+4.67%
IC'03 (860)	Cross-Entropy [1]	95.6	0.984	85
	LS-ED [32]	95.5 −0.01%	0.983 −0.10%	91 −7.05%
	FEDS	96.2 +0.62%	0.986 +0.20%	73 +14.1%
IC'03 (867)	Cross-Entropy [1]	95.7	0.984	89
	LS-ED [32]	95.7 +0.03%	0.984 –	91 −2.24%
	FEDS	96.4 +0.73%	0.987 +0.30%	77 +13.4%
IC'13 (857)	Cross-Entropy [1]	95.4	0.988	65
	LS-ED [32]	96.3 +1.03%	0.989 +0.10%	55 +15.3%
	FEDS	96.5 +1.15%	0.989 +0.10%	57 +12.3%
IC'13 (1015)	Cross-Entropy [1]	94.1	0.975	97
	LS-ED [32]	94.8 +0.82%	0.975 –	87 +10.3%
	FEDS	95.3 +1.27%	0.975 –	90 +7.21%
IC'15 (1811)	Cross-Entropy [1]	82.8	0.939	614
	LS-ED [32]	83.2 +0.56%	0.939 –	599 +2.44%
	FEDS	83.8 +1.20%	0.942 +0.31%	578 +5.86%
IC'15 (2077)	Cross-Entropy [1]	80.0	0.908	961
	LS-ED [32]	80.4 +0.54%	0.908 –	944 +1.76%
	FEDS	80.9 +1.12%	0.91 +0.22%	929 +3.32%
SVTP (645)	Cross-Entropy [1]	82.4	0.930	271
	LS-ED [32]	83.4 +1.22%	0.933 +0.32%	258 +4.79%
	FEDS	84.0 +1.94%	0.935 +0.53%	248 +8.48%
CUTE (288)	Cross-Entropy [1]	77.3	0.883	211
	LS-ED [32]	77.3 +0.06%	0.885 +0.22%	197 +6.63%
	FEDS	79.0 +2.19%	0.898 +1.69%	176 +16.5%
TOTAL	Cross-Entropy [1]	88.7	0.953	3050
	LS-ED [32]	89.0 +0.41%	0.954 +0.62%	2961 +2.91%
	FEDS	89.6 +1.01%	0.956 +0.35%	2809 +7.90%

reduction). Furthermore, LS-ED [32] declines the performance on ICDAR'03 [28] dataset.

4.7 Qualitative Results

Figure 5 shows randomly picked qualitative examples where FEDS leads to an improvement in the edit distance. Notice that the predictions from the baseline model are incorrect in all the examples. After post-tuning with FEDS, the predictions are correct, *i.e.*, perfectly match with the ground truth.

Figure 6 shows hand-picked examples where FEDS leads to a maximum increase in the edit distance (ED increases). Notice that in these examples, the predictions from the baseline model are also incorrect. Furthermore, the input images are nearly illegible for a human.

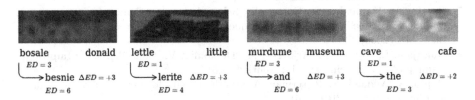

bosale   donald   lettle   little   murdume   museum   cave   cafe
| ED = 3 | ED = 1 | ED = 3 | ED = 1
→ besnie ΔED = +3   → lerite ΔED = +3   → and ΔED = +3   → the ΔED = +2
ED = 6   ED = 4   ED = 6   ED = 3

Fig. 6. Four worst examples out of 12K samples in the test set where FEDS leads to an increase in the edit distance. Red shows the incorrect predictions, green shows the ground truth and the arrow indicates post-tuning with FEDS. (Color figure online)

5   Conclusions

This paper makes an orthogonal contribution to the trend of scene text recognition progress. It proposes a method to robustly post-tune a STR model using a learned surrogate of edit distance. The empirical results demonstrate an average improvement of 11.2% on total edit distance and an error reduction of 9.5% on accuracy on a standard baseline [1]. Improvements of 7.9% on total edit distance and an error reduction of 10.3% on accuracy are shown on a stronger baseline [16].

Project link: yash0307.github.io/FEDS_ICDAR2021.

Acknowledgement. This research was supported by Research Center for Informatics (project CZ.02.1.01/0.0/0.0/16_019/0000765 funded by OP VVV), CTU student grant (SGS20/171/OHK3/3T/13), Project StratDL in the realm of COMET K1 center Software Competence Center Hagenberg and Amazon Research Award.

References

1. Baek, J., et al.: What is wrong with scene text recognition model comparisons? Dataset and model analysis. In: ICCV (2019)
2. Bookstein, F.L.: Principal warps: thin-plate splines and the decomposition of deformations. TPAMI **11**, 567–585 (1989)
3. Busta, M., Neumann, L., Matas, J.: Deep textspotter: an end-to-end trainable scene text localization and recognition framework. In: ICCV (2017)
4. Bušta, M., Patel, Y., Matas, J.: E2E-MLT - an unconstrained end-to-end method for multi-language scene text. In: Carneiro, G., You, S. (eds.) ACCV 2018. LNCS, vol. 11367, pp. 127–143. Springer, Cham (2019). https://doi.org/10.1007/978-3-030-21074-8_11
5. Gómez, L., Rusinol, M., Karatzas, D.: LSDE: levenshtein space deep embedding for query-by-string word spotting. In: ICDAR (2017)
6. Gomez, R., Biten, A.F., Gomez, L., Gibert, J., Karatzas, D., Rusiñol, M.: Selective style transfer for text. In: ICDAR (2019)
7. Graves, A., Fernández, S., Gomez, F., Schmidhuber, J.: Connectionist temporal classification: labelling unsegmented sequence data with recurrent neural networks. In: ICML (2006)
8. Gulrajani, I., Ahmed, F., Arjovsky, M., Dumoulin, V., Courville, A.C.: Improved training of wasserstein gans. In: NeurIPS (2017)
9. Gupta, A., Vedaldi, A., Zisserman, A.: Synthetic data for text localisation in natural images. In: CVPR (2016)
10. He, K., Zhang, X., Ren, S., Sun, J.: Deep residual learning for image recognition. In: CVPR (2016)
11. He, P., Huang, W., Qiao, Y., Loy, C., Tang, X.: Reading scene text in deep convolutional sequences. In: AAAI (2016)
12. Hochreiter, S., Schmidhuber, J.: Long short-term memory. Neural Comput. **9**, 1735–1780 (1997)
13. Jaderberg, M., Simonyan, K., Vedaldi, A., Zisserman, A.: Reading text in the wild with convolutional neural networks. IJCV **116**, 1–20 (2016)
14. Jaderberg, M., Simonyan, K., Zisserman, A., et al.: Spatial transformer networks. In: NeurIPS (2015)
15. Jaderberg, M., Vedaldi, A., Zisserman, A.: Deep features for text spotting. In: Fleet, D., Pajdla, T., Schiele, B., Tuytelaars, T. (eds.) ECCV 2014. LNCS, vol. 8692, pp. 512–528. Springer, Cham (2014). https://doi.org/10.1007/978-3-319-10593-2_34
16. Janoušková, K., Matas, J., Gomez, L., Karatzas, D.: Text recognition - real world data and where to find them. In: ICPR (2021)
17. Karatzas, D., et al.: ICDAR 2015 competition on robust reading. In: ICDAR (2015)
18. Karatzas, D., et al.: ICDAR 2013 robust reading competition. In: ICDAR (2013)
19. Lapin, M., Hein, M., Schiele, B.: Loss functions for top-k error: analysis and insights. In: CVPR (2016)
20. LeCun, Y., Bottou, L., Bengio, Y., Haffner, P.: Gradient-based learning applied to document recognition. Proc. IEEE **86**, 2278–2324 (1998)
21. Li, H., Wang, P., Shen, C., Zhang, G.: Show, attend and read: a simple and strong baseline for irregular text recognition. In: AAAI (2019)
22. Liao, M., Lyu, P., He, M., Yao, C., Wu, W., Bai, X.: Mask textspotter: an end-to-end trainable neural network for spotting text with arbitrary shapes. TPAMI (2019)

23. Liao, M., et al.: Scene text recognition from two-dimensional perspective. In: AAAI (2019)
24. Litman, R., Anschel, O., Tsiper, S., Litman, R., Mazor, S., Manmatha, R.: Scatter: selective context attentional scene text recognizer. In: CVPR (2020)
25. Liu, W., Chen, C., Wong, K.Y.K.: Char-net: a character-aware neural network for distorted scene text recognition. In: AAAI (2018)
26. Long, S., He, X., Yao, C.: Scene text detection and recognition: the deep learning era. IJCV **129**, 161–184 (2020)
27. Long, S., Yao, C.: UnrealText: synthesizing realistic scene text images from the unreal world. In: CVPR (2020)
28. Lucas, S.M., Panaretos, A., Sosa, L., Tang, A., Wong, S., Young, R.: ICDAR 2003 robust reading competitions. In: ICDAR (2003)
29. Mishra, A., Alahari, K., Jawahar, C.: Scene text recognition using higher order language priors. In: BMVC (2012)
30. Nayef, N., et al.: ICDAR 2019 robust reading challenge on multi-lingual scene text detection and recognition-RRC-MLT-2019. In: ICDAR (2019)
31. Patel, Y., Gomez, L., Rusiñol, M., Karatzas, D.: Dynamic lexicon generation for natural scene images. In: Hua, G., Jégou, H. (eds.) ECCV 2016. LNCS, vol. 9913, pp. 395–410. Springer, Cham (2016). https://doi.org/10.1007/978-3-319-46604-0_29
32. Patel, Y., Hodaň, T., Matas, J.: Learning surrogates via deep embedding. In: Vedaldi, A., Bischof, H., Brox, T., Frahm, J.-M. (eds.) ECCV 2020. LNCS, vol. 12375, pp. 205–221. Springer, Cham (2020). https://doi.org/10.1007/978-3-030-58577-8_13
33. Patel, Y., Bušta, M., Matas, J.: E2E-MLT-an unconstrained end-to-end method for multi-language scene text (2018)
34. Qiao, Z., Zhou, Y., Yang, D., Zhou, Y., Wang, W.: SEED: semantics enhanced encoder-decoder framework for scene text recognition. In: CVPR (2020)
35. Quy Phan, T., Shivakumara, P., Tian, S., Lim Tan, C.: Recognizing text with perspective distortion in natural scenes. In: ICCV (2013)
36. Risnumawan, A., Shivakumara, P., Chan, C.S., Tan, C.L.: A robust arbitrary text detection system for natural scene images. Expert Syst. Appl. **41**, 8027–8048 (2014)
37. Rumelhart, D.E., Hinton, G.E., Williams, R.J.: Learning representations by back-propagating errors. Nature **323**, 533–536 (1986)
38. Shi, B., Bai, X., Yao, C.: An end-to-end trainable neural network for image-based sequence recognition and its application to scene text recognition. TPAMI **39**, 2298–2304 (2016)
39. Shi, B., Wang, X., Lyu, P., Yao, C., Bai, X.: Robust scene text recognition with automatic rectification. In: CVPR (2016)
40. Shi, B., Yang, M., Wang, X., Lyu, P., Yao, C., Bai, X.: ASTER: an attentional scene text recognizer with flexible rectification. PAMI **41**, 2035–2048 (2018)
41. Simonyan, K., Zisserman, A.: Very deep convolutional networks for large-scale image recognition. arXiv preprint arXiv:1409.1556 (2014)
42. Sutskever, I., Vinyals, O., Le, Q.V.: Sequence to sequence learning with neural networks. In: NeurIPS (2014)
43. Wang, K., Babenko, B., Belongie, S.: End-to-end scene text recognition. In: ICCV (2011)
44. Wang, K., Belongie, S.: Word spotting in the wild. In: Daniilidis, K., Maragos, P., Paragios, N. (eds.) ECCV 2010. LNCS, vol. 6311, pp. 591–604. Springer, Heidelberg (2010). https://doi.org/10.1007/978-3-642-15549-9_43

45. Wang, T., et al.: Decoupled attention network for text recognition. In: AAAI (2020)
46. Xu, B., Wang, N., Chen, T., Li, M.: Empirical evaluation of rectified activations in convolutional network. In: CoRR (2015)
47. Yang, M., et al.: Symmetry-constrained rectification network for scene text recognition. In: ICCV (2019)
48. Yao, C., Bai, X., Shi, B., Liu, W.: Strokelets: a learned multi-scale representation for scene text recognition. In: CVPR (2014)
49. Ye, Q., Doermann, D.: Text detection and recognition in imagery: a survey. TPAMI **37**, 1480–1500 (2014)
50. Yu, D., Li, X., Zhang, C., Liu, T., Han, J., Liu, J., Ding, E.: Towards accurate scene text recognition with semantic reasoning networks. In: CVPR (2020)
51. Yue, X., Kuang, Z., Lin, C., Sun, H., Zhang, W.: RobustScanner: dynamically enhancing positional clues for robust text recognition. In: Vedaldi, A., Bischof, H., Brox, T., Frahm, J.-M. (eds.) ECCV 2020. LNCS, vol. 12364, pp. 135–151. Springer, Cham (2020). https://doi.org/10.1007/978-3-030-58529-7_9
52. Zeiler, M.D.: ADADELTA: an adaptive learning rate method. In: CoRR (2012)
53. Zhan, F., Lu, S.: ESIR: end-to-end scene text recognition via iterative image rectification. In: CVPR (2019)
54. Zhang, X., Zhao, J., LeCun, Y.: Character-level convolutional networks for text classification. In: NeurIPS (2015)
55. Zhang, Y., Gueguen, L., Zharkov, I., Zhang, P., Seifert, K., Kadlec, B.: Uber-text: a large-scale dataset for optical character recognition from street-level imagery. In: CVPR Workshop (2017)

Bidirectional Regression
for Arbitrary-Shaped Text Detection

Tao Sheng⬤ and Zhouhui Lian$^{(\boxtimes)}$⬤

Wangxuan Institute of Computer Technology, Peking University, Beijing, China
{shengtao,lianzhouhui}@pku.edu.cn

Abstract. Arbitrary-shaped text detection has recently attracted increasing interests and witnessed rapid development with the popularity of deep learning algorithms. Nevertheless, existing approaches often obtain inaccurate detection results, mainly due to the relatively weak ability to utilize context information and the inappropriate choice of offset references. This paper presents a novel text instance expression which integrates both foreground and background information into the pipeline, and naturally uses the pixels near text boundaries as the offset starts. Besides, a corresponding post-processing algorithm is also designed to sequentially combine the four prediction results and reconstruct the text instance accurately. We evaluate our method on several challenging scene text benchmarks, including both curved and multi-oriented text datasets. Experimental results demonstrate that the proposed approach obtains superior or competitive performance compared to other state-of-the-art methods, e.g., 83.4% F-score for Total-Text, 82.4% F-score for MSRA-TD500, etc.

Keywords: Arbitrary-shaped text detection · Text instance expression · Instance segmentation

1 Introduction

In many text-related applications, robustly detecting scene texts with high accuracy, namely localizing the bounding box or region of each text instance with high Intersection over Union (IoU) to the ground truth, is fundamental and crucial to the quality of service. For example, in vision-based translation applications, the process of generating clear and coherent translations is highly dependent on the text detection accuracy. However, due to the variety of text scales, shapes and orientations, and the complex backgrounds in natural images, scene text detection is still a tough and challenging task.

With the rapid development of deep convolutional neural networks (DCNN), a number of effective methods [34–36,38,41,44] have been proposed for detecting texts in scene images, achieving promising performance. Among all these DCNN-based approaches, the majority of them can be roughly classified into two categories: regression-based methods with anchors and segmentation-based

© Springer Nature Switzerland AG 2021
J. Lladós et al. (Eds.): ICDAR 2021, LNCS 12824, pp. 187–201, 2021.
https://doi.org/10.1007/978-3-030-86337-1_13

methods. Regression-based methods are typically motivated by generic object detection [2,11,21,30], and treat text instances as a specific kind of object. However, it is difficult to manually design appropriate anchors for irregular texts. On the contrary, segmentation-based methods prefer to regard scene text detection as a segmentation task [4,8,20,23] and need extra predictions besides segmentation results to rebuild text instances. Specifically, as shown in Fig. 1, MSR [38] predicts central text regions and distance maps according to the distance between each predicted text pixel and its nearest text boundary, which brings the state-of-the-art scene text detection accuracy. Nevertheless, the performance of MSR is still restricted by its relatively weak capability to utilize the context information around the text boundaries. That is to say, MSR only uses the pixels inside the central areas for detection, while ignores the pixels around the boundaries which include necessary context information. Moreover, the regression from central text pixels makes the positions of predicted boundary points ambiguous because there is a huge gap between them, requiring the network to have a large receptive field.

Fig. 1. The comparison of detection results of MSR [38] and our method. In the first row, MSR predicts central text regions (top right) and distance maps according to the distance between each predicted text pixel and its nearest text boundary (top left). In the second row, we choose the pixels around the text boundary for further regression. The pixels for regression are marked as light blue. (Color figure online)

To solve these problems, we propose an arbitrary-shaped text detector, which makes full use of context information around the text boundaries and predicts the boundary points in a natural way. Our method contains two steps: 1) input images are feed to the convolutional network to generate text regions, text kernels, pixel offsets, and pixel orientations. 2) at the stage of post-processing, the previous predictions are fused one by one to rebuild the final results. Unlike MSR, we choose the pixels around the text boundaries for further regression (Fig. 1 bottom). Due to the regression computed in two directions, either from the external or the internal text region to the text boundary, we call our method Bidirectional Regression. For a fair comparison, we choose a simple and commonly-used network structure as the feature extractor, ResNet50 [12] for backbone and FPN [19] for detection neck. Then, the segmentation branch predicts text regions and text kernels, while the regression branch predicts pixel offsets and pixel orientations which help to separate adjacent text lines. Also, we propose a novel post-processing algorithm to reconstruct final text instances accurately. We conduct extensive experiments on four challenging benchmarks including Total-Text [5], CTW1500 [41], ICDAR2015 [14], and MSRA-TD500 [40], which demonstrate that our method obtains superior or comparable performance compared to the state of the art.

Major contributions of our work can be summarized as follows:

1) To overcome the drawbacks of MSR, we propose a new kind of text instance expression which makes full use of context information and implements the regression in a natural manner.
2) To get complete text instances robustly, we develop a novel post-processing algorithm that sequentially combines the predictions to get accurate text instances.
3) The proposed method achieves superior or comparable performance compared to other existing approaches on two curved text benchmarks and two oriented text benchmarks.

2   Related Work

Convolutional neural network approaches have recently been very successful in the area of scene text detection. CNN-based scene text detectors can be roughly classified into two categories: regression-based methods and segmentation-based methods.

Regression-based detectors usually inherit from generic object detectors, such as Faster R-CNN [30] and SSD [21], directly regressing the bounding boxes of text instances. TextBoxes [16] adjusts the aspect ratios of anchors and the scales of convolutional kernels of SSD to deal with the significant variation of scene texts. TextBoxes++ [15] and EAST [44] further regress quadrangles of multi-oriented texts in pixel level with and without anchors, respectively. For better detection of long texts, RRD [18] generates rotation-invariant features for classification and rotation-sensitive features for regression. RRPN [27] modifies Faster R-CNN by adding rotation to proposals for titled text detection. The methods mentioned above achieve excellent results in several benchmarks. Nevertheless, most of them suffer from the complex anchor settings or the inadequate description of irregular texts.

Segmentation-based detectors prefer to treat scene text detection as a semantic segmentation problem and apply a matching post-processing algorithm to get the final polygons. Zhang et al. [42] utilized FCN [23] to estimate text regions and further distinguish characters with MSER [29]. In PixelLink [6], text/non-text and links predictions in pixel level are carried out to separate adjacent text instances. Chen et al. [3] proposed the concept of attention-guided text border for better training. SPCNet [36] and Mask TextSpotter [25] adopt the architecture of Mask R-CNN [11] in the instance segmentation task to detect the texts with arbitrary shapes. In TextSnake [24], text instances are represented with text center lines and ordered disks. MSR [38] predicts central text regions and distance maps according to the distance between each predicted text pixel and its nearest text boundary. PSENet [35] proposes progressive scale expansion algorithm, learning text kernels with multiple scales. However, these methods all lack the rational utilization of context information, and thus often result in inaccurate text detection.

Some previous methods [3,25,37] also try to strengthen the utilization of context information according to the inference of the text border map or the perception of the whole text polygon. Regrettably, they only focus on pixels inside the text polygon and miss pixels outside the label. Different from existing methods, a unique text expression is proposed in this paper to force the network to extract foreground and background information simultaneously and produce a more representative feature for precise localization.

3   Methodology

In this section, we first compare the text instance expressions of common object detection models, a curved text detector MSR [38] and ours. Then, we describe the whole network architecture of our proposed method. Afterwards, we elaborate on the post-processing algorithm and the generation procedure of arbitrary-shaped text instances. Finally, the details of the loss function in the training phase are given.

3.1   Text Instance Expression

Bounding Box. A robust scene text detector must have a well-defined expression for text instances. In generic object detection methods, the text instances are always represented as bounding boxes, namely rotated rectangles or quadrangles, whose shapes heavily rely on vertices or geometric centers. However, it is difficult to decide vertices or geometric center of a curved text instance, especially in natural images. Besides, as shown in Fig. 2(a), the bounding box (orange solid line) can not fit the boundary (green solid line) of the curved text well and introduces a large number of background noises, which could be problematic for detecting scene texts. Moreover, at the stage of scene text recognition, the features extracted in this way may confuse the model to obtain incorrect recognition results.

MSR does not have this expression problem because as shown in Fig. 2(b), it predicts central text regions (violet area) and offset maps according to the distance (orange line with arrow) between each predicted text pixel and its nearest text boundary. Note that the central text regions are derived from the original text boundary (green solid line), which help to separate adjacent words or text lines. Obviously, a central text region S_{kernel} can be discretized into a set of points $\{p_1, p_2, \ldots, p_T\}$, and T is the number of points. Naturally, the offset

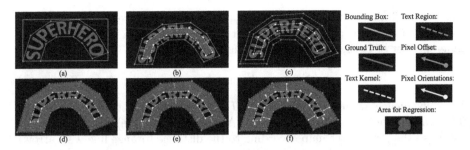

Fig. 2. The comparison of three text instance expressions: (a) bounding boxes, (b) MSR's expression [38], and (c–f) our Bidirectional regression. (Color figure online)

between p_i and its nearest boundary can be represented as $\{\Delta p_i | i = 1, 2, \ldots, T\}$. Further, the contour of the text instance can be represented with the point set,

$$R = \{p_i + \Delta p_i | i = 1, 2, \ldots, T\}. \tag{1}$$

We observe that MSR focuses on the location of center points and the relationship between center and boundary points, while ignores the crucial context information around the text boundaries and chooses inappropriate references as the starts of offsets. More specifically, 1) the ignored features around the boundary are the most discriminative part of the text instance's features extracted from the whole image, because foreground information inside the boundary and background information outside the boundary are almost completely different and easy to be distinguished. 2) compared to the points around the boundary, the center ones are farther away from the boundary, which can not give enough evidences to decide the position of the scene text instance. In other words, if we can utilize the context information well and choose better references, we will possibly be able to address MSR's limitations and further improve the detection performance.

Bidirectional Regression (Ours). We now define our text instance expression so that the generated feature maps can have more useful contextual information for detection. We use four predictions to represent a text instance, including the text region, text kernel, pixel offset, and pixel orientation. As shown in Fig. 2(c), inspired by PSENet [35], the text kernel is generated by shrinking the annotated polygon P (green solid line) to the yellow dotted line using the Vatti clipping algorithm [33]. The offset d of shrinking is computed based on the perimeter and area of the original polygon P:

$$d = \frac{\text{Area}(P) \times (1 - \alpha^2)}{\text{Perimeter}(P)}, \tag{2}$$

where α is the shrink ratio, set to 0.6 empirically. To contain the background information, we enlarge the annotated polygon P to the blue dotted line in the same way and produce the text region. The expansion ratio β is set to 1.2 in this work. Similarly, the text kernel S_{kernel} and the text region S_{text} can be treated as point sets, and further the text border (violet area in Fig. 2(d)) is the difference set of both, which can be formulated as:

$$S_{border} = S_{text} - S_{kernel}, \tag{3}$$

For convenience, we use $\{p_i | i = 1, 2, \ldots, T'\}$ to represent the text border, where T' is the number of points in this area. As shown in Fig. 2(e) and Fig. 2(f), we establish the pixel offset map according to the distance (orange line with arrow) between each text border point and its nearest text boundary, and the pixel orientation map according to the orientation (white line with arrow) from each text border point to its nearest text kernel point. Note that if two instances overlap, the smaller one has higher priority. Like MSR, we use sets $\{\Delta p_i | i = 1, 2, \ldots, T'\}$ and $\{\vec{\theta_i} | i = 1, 2, \ldots, T'\}$ to represent the pixel offset and the pixel orientation respectively, where $\vec{\theta_i}$ is a unit vector. Similar to Eq. 1, the final predicted contour can be formulated as follows:

$$R = \{p_i + \Delta p_i | \exists p_k \in S_{kernel}, D(p_i, p_k) < \gamma \text{ and}$$
$$\frac{\overrightarrow{p_i p_k}}{|p_i p_k|} \cdot \vec{\theta_i} > \epsilon, i = 1, 2, \ldots, T'\}, \tag{4}$$

where $D(\cdot)$ means the Euclidean distance between two points. γ and ϵ are constants and we choose $\gamma = 3.0, \epsilon = \cos(25°) \approx 0.9063$ empirically. That is to say, if there exists a text border point that is close enough to the text kernel point, and the vector composed of these two points has a similar direction with the predicted pixel orientation, then the text border point will be shifted according to the amount of the predicted pixel offset, and be treated as the boundary point of the final text instance.

Comparing the areas for regression in MSR and our approach (Fig. 2(b) (e)), we can see that our proposed method chooses the points much closer to the boundary so that we do not require the network to have large receptive fields when detecting large text instances. Meanwhile, our model learns the foreground and background features macroscopically, and mixes them into a more discriminating one to localize the exact position of scene texts. To sum up, we design a powerful expression for arbitrary-shaped text detection.

3.2   Network Architecture

From the network's perspective, our model is surprisingly simple, and Fig. 3 illustrates the whole architecture. For a fair comparison with MSR, we employ ResNet-50 [12] as the backbone network to extract initial multi-scale features from input images. A total of 4 feature maps are generated from the Res2, Res3, Res4, and Res5 layers of the backbone, and they have strides of 4, 8, 16, 32 pixels with respect to the input image respectively. To reduce the computational cost and the network complexity, we use 1×1 convolutions to reduce the channel

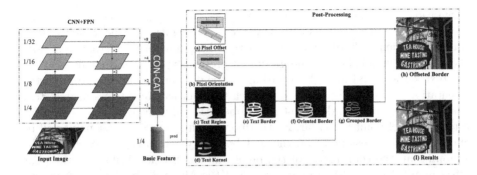

Fig. 3. An overview of our proposed model. Our method contains two components, the CNN-based network and the post-processing algorithm. (a) We use ResNet-50 and FPN to extract the feature pyramid, and concatenate them into a basic feature for further prediction. (b) The post-processing algorithm takes two forecasts as inputs and produces a new one every step, reconstructing the scene texts with arbitrary shapes finally. (Color figure online)

number of each feature map to 256. Inspired by FPN [19], the feature pyramid is enhanced by gradually merging the features of adjacent scales in a top-down manner. Then, the deep but thin feature maps are fused by bilinear interpolation and concatenation into a basic feature, whose stride is 4 pixels and the channel number is 1024. The basic feature is used to predict text regions, text kernels, pixel offsets, and pixel orientations simultaneously. Finally, we apply a sequential and efficient post-processing algorithm to obtain the final text instances.

3.3 Post-Processing

As described in the above subsection and illustrated in Fig. 3(a–d), four predictions are generated from the basic feature, followed by the post-processing algorithm. The text region can describe the text instance coarsely but can not separate two adjacent instances (Fig. 3(c)). In contrast, the text kernel can separate them but can not describe them (see Fig. 3(d)). Therefore, we use text kernels to determine the coarse position, then use pixel orientations to classify the ungrouped text region points, and use pixel offsets to slightly modify the contours.

We first find the connected components in the text kernel map, and each connected component represents the kernel of a single text instance. For better visualization, different kernels are painted with different colors (see Fig. 3(d)). Moreover, the pixel offset map and the pixel orientation map are difficult to be visualized, so we replace the real ones with the diagrammatic sketches in a fictitious scene (see Fig. 3(a) (b)). Then, we combine the text region and the text kernel, obtaining the text border (see Fig. 3(e)), which is the difference set of two predictions and meanwhile the aggregation of the ungrouped text region points. Combined with the pixel orientation, each text border point has its own orientation. As shown in Fig. 3(f), four colors (yellow, green, violet, and blue) represent the four directions (up, down, left, and right) correspondingly. After- wards, the oriented border and the text kernel are combined together to classify the text border points into the groups of previously connected components in the text kernel (see Fig. 3(g)) according to the difference between the predicted orientation and the orientation from each text border point to its nearest text kernel point. A text border point should be deserted if the distance to its near- est text kernel point is too far. Furthermore, each grouped border point will be shifted to its nearest point on the text boundary, which can be calculated by summing up the coordinates of the point and the predicted offset in the pixel offset map (see Fig. 3(h)). Finally, we adopt the Alpha-Shape Algorithm to pro- duce concave polygons enclosing the shifted points of each text kernel group (see Fig. 3(I)), precisely reconstructing the shapes of the text instances in scene images. Through this effective post-processing algorithm, we can detect scene texts with arbitrary shapes fast and accurately, which is experimentally proved in Sect. 4.

3.4 Loss Function

Our loss function can be formulated as:

$$\mathcal{L} = \mathcal{L}_{text} + \lambda_1 \mathcal{L}_{kernel} + \lambda_2 (\mathcal{L}_{offset} + \mathcal{L}_{orientation}), \tag{5}$$

where $\mathcal{L}_{text}$ and $\mathcal{L}_{kernel}$ denote the binary segmentation loss of text regions and text kernels respectively, $\mathcal{L}_{offset}$ denotes the regression loss of pixel offsets, and $\mathcal{L}_{orientation}$ denotes the orientation loss of pixel orientations. λ_1 and λ_2 are normalization constants to balance the weights of the segmentation and regression loss. We set them to 0.5 and 0.1 in all experiments.

The prediction of text regions and text kernels is basically a pixel-wise binary classification problem. We follow MSR and adopt the dice loss [28] for this part. Considering the imbalance of text and non-text pixels in the text regions, Online Hard Example Mining (OHEM) [32] is also adopted to select the hard non-text pixels when calculating $\mathcal{L}_{text}$.

The prediction of the distance from each text border point to its nearest text boundary is a regression problem. Following the regression for bounding boxes in generic object detection, we use the Smooth L1 loss [9] for supervision, which is defined as:

$$\mathcal{L}_{offset} = \frac{1}{|S_{border}|} \sum_i \text{Smooth}_{L1}(\Delta p_i - \Delta p_i^*), \tag{6}$$

where Δp_i and Δp_i^* denote the predicted offset and the corresponding ground truth, respectively, $\text{Smooth}_{L1}(\cdot)$ denotes the standard Smooth L1 loss, and $|S_{border}|$ denotes the number of text border points. Moreover, the prediction of the orientation from each text border point to its nearest text kernel point can also be treated as a regression problem. For simplicity, we adopt the cosine loss defined as follows:

$$\mathcal{L}_{orientation} = \frac{1}{|S_{border}|} \sum_i \left(1 - \overrightarrow{\theta_i} \cdot \overrightarrow{\theta_i^*}\right), \tag{7}$$

where $\overrightarrow{\theta_i} \cdot \overrightarrow{\theta_i^*}$ denote the dot product of the predicted direction vector and its ground truth, which is equal to the cosine value of the angle between these two orientations. Note that we only take the points in the text border into consideration when calculating $\mathcal{L}_{offset}$ and $\mathcal{L}_{orientation}$.

4 Experiments

4.1 Datasets

SynthText [10] is a synthetical dataset containing more than 800,000 synthetic scene text images, most of which are annotated at word level with multi-oriented rectangles. We pre-train our model on this dataset.

Total-Text [5] is a curved text dataset which contains 1,255 training images and 300 testing images. The texts are all in English and contain a large number of horizontal, multi-oriented, and curved text instances, each of which is annotated at word level with a polygon.

CTW1500 [41] is another curved text dataset that has 1,000 images for training and 500 images for testing. The dataset focuses on curved texts, which are largely in English and Chinese, and annotated at text-line level with 14-polygons.

ICDAR2015 [14] is a commonly-used dataset for scene text detection, which contains 1,000 training images and 500 testing images. The dataset is captured by Google Glasses, where text instances are annotated at word level with quadri-laterals.

MSRA-TD500 [40] is a small dataset that contains a total of 500 images, 300 for training and the remaining for testing. All captured text instances are in English and Chinese, which are annotated at text-line level with best-aligned rectangles. Due to the rather small scale of the dataset, we follow the previous works [24,44] to add the 400 training images from HUST-TR400 [39] into the training set.

4.2 Implementation Details

The following settings are used throughout the experiments. The proposed method is implemented with the deep learning framework, Pytorch, on a regular GPU workstation with 4 Nvidia Geforce GTX 1080 Ti. For the network architecture, we use the ResNet-50 [12] pre-trained on ImageNet [7] as our backbone. For learning, we train our model with the batch size of 16 on 4 GPUs for 36K iterations. Adam optimizer with a starting learning rate of 10^{-3} is used for optimization. We use the "poly" learning rate strategy [43], where the initial rate is multiplied by $(1 - \frac{iter}{max\_iter})^{power}$, and the "power" is set to 0.9 in all experiments. For data augmentation, we apply random scale, random horizontal flip, random rotation, and random crop on training images. We ignore the blurred texts labeled as DO NOT CARE in all datasets. For others, Online hard example mining (OHEM) is used to balance the positive and negative samples, and the negative-positive ratio is set to 3. We first pre-train our model on SynthText, and then fine-tune it on other datasets. The training settings of the two stages are the same.

4.3   Ablation Study

To prove the effectiveness of our proposed expression for text instances, we carry out ablation studies on the curved text dataset Total-Text. Note that, all the models in this subsection are pre-trained on SynthText first. The quantitative results of the same network architecture with different expressions (with corresponding post-processing algorithms) are shown in Table 1.

Table 1. The results of models with different expressions over the curved text dataset Total-Text. "β" means the expansion ratio of the text region.

Expression	β	Precision	Recall	F-score
MSR	–	84.7	77.3	80.8
Ours	1.0	85.8	79.1	82.3
Ours	1.2	**87.0**	**80.1**	**83.4**

Table 2. The results of models with different expressions over the varying IoU thresholds from 0.5 to 0.9. "F_x" means the IoU threshold is set to "x" when evaluating. The measure of "F" follows the Total-Text dataset, setting TR to 0.7 and TP to 0.6 threshold for a fairer evaluation.

Expression	F	$F_{0.5}$	$F_{0.6}$	$F_{0.7}$	$F_{0.8}$	$F_{0.9}$
MSR	80.8	**88.2**	85.5	76.4	48.2	5.7
Ours	**83.4**	87.9	**85.6**	**78.4**	**56.9**	**15.2**

To better analyze the capability of the proposed expression, we replace our text instance expression in the proposed text detector with MSR's. The F-score of the model with the MSR's expression (the first row in Table 1) drops 2.6% compared to our method (the third row in Table 1), which indicates the effectiveness of our text instance expression clearly. To prove the necessity of introducing background pixels around the text boundary, we adjust the expansion ratio β from 1.2 to 1.0 (the second row in Table 1). We can see that the F-score value increases by 1.1% when the model extracts the foreground and background features macroscopically. Furthermore, to judge whether the text border pixels are the better references for the pixel offsets or not, we compare the model with MSR's expression and ours without enlarging text instances, and notice that ours makes about 1.5% improvement on F-score. We further analyze the detection accuracy between the model with MSR's expression and ours by varying the evaluation IoU threshold from 0.5 to 0.9. Table 2 shows that our method defeats the competitor for most IoU settings, especially in high IoU levels, indicating that our predicted polygons fit text instances better.

4.4   Comparisons with State-of-the-Art Methods

Curved Text Detection. We first evaluate our method over the datasets Total-Text and CTW1500 which contain many curved text instances. In the testing phase, we set the short side of images to 640 and keep their original aspect ratio. We show the experimental results in Table 3. On Total-Text, our method achieves the F-score of 83.4%, which surpasses all other state-of-the-art methods by at least 0.5%. Especially, we outperform our counterpart, MSR, in F-score by over 4%. Analogous results can be found on CTW1500. Our method obtains 81.8% in F-score, the second-best one of all methods, which is only lower than PSENet [35] but surpasses MSR by 0.3%. To sum up, our experiments conducted on these two datasets demonstrate the advantages of our method when detecting text instances with arbitrary shapes in complex natural scenes. We visualize our detection results in Fig. 4(a) (b) for further inspection.

Table 3. Experimental results on the curved-text-line datasets Total-Text and CTW1500.

Method	Total-text			CTW1500		
	P	R	F	P	R	F
SegLink [31]	30.3	23.8	26.7	42.3	40.0	40.8
EAST [44]	50.0	36.2	42.0	78.7	49.1	60.4
Mask TextSpotter [25]	69.0	55.0	61.3	–	–	–
TextSnake [24]	82.7	74.5	78.4	67.9	**85.3**	75.6
CSE [22]	81.4	79.1	80.2	81.1	76.0	78.4
TextField [37]	81.2	79.9	80.6	83.0	79.8	81.4
PSENet-1s [35]	84.0	78.0	80.9	84.8	79.7	**82.2**
SPCNet [36]	83.0	**82.8**	82.9	–	–	–
TextRay [34]	83.5	77.9	80.6	82.8	80.4	81.6
MSR (Baseline) [38]	83.8	74.8	79.0	85.0	78.3	81.5
Ours	**87.0**	80.1	**83.4**	**85.7**	78.2	81.8

Oriented Text Detection. Then we evaluate the proposed method over the multi-oriented text dataset ICDAR2015. In the testing phase, we set the short

Table 4. Experimental results on the oriented-text-line dataset ICDAR2015 and long-straight-text-line dataset MSRA-TD500.

Method	ICDAR2015				MSRA-TD500		
	P	R	F	FPS	P	R	F
SegLink [31]	73.1	76.8	75.0	–	86.6	70.0	77.0
RRPN [27]	82.0	73.0	77.0	–	82.0	68.0	74.0
EAST [44]	83.6	73.5	78.2	**13.2**	87.3	67.4	76.1
Lyu et al. [26]	**94.1**	70.7	80.7	3.6	87.6	76.2	81.5
DeepReg [13]	82.0	80.0	81.0	–	77.0	70.0	74.0
RRD [18]	85.6	79.0	82.2	6.5	87.0	73.0	79.0
PixelLink [6]	82.9	81.7	82.3	7.3	83.0	73.2	77.8
TextSnake [24]	84.9	80.4	82.6	1.1	83.2	73.9	78.3
Mask TextSpotter [25]	85.8	81.2	83.4	4.8	–	–	–
PSENet-1s [35]	86.9	**84.5**	85.7	1.6	–	–	–
CRAFT [1]	89.8	84.3	86.9	8.6	88.2	78.2	**82.9**
DB [17]	91.8	83.2	**87.3**	12.0	**90.4**	76.3	82.8
MSR (Baseline) [38]	86.6	78.4	82.3	4.3	87.4	76.7	81.7
Ours	82.6	81.9	82.2	**13.2**	83.7	**81.1**	82.4

side of images to 736 for better detection. To fit its evaluation protocol, we use a minimum area rectangle to replace each output polygon. The performance on ICDAR2015 is shown in Table 4. Our method achieves the F-score of 82.2%, which is on par with MSR while the FPS of ours is 3 times of MSR. Indeed, MSR adopts the multi-scale multi-stage detection network, a particularly time-consuming architecture, so it is no surprise that its speed is lower than Ours. Compared with state-of-the-art methods, our method is not as well as some competitors (e.g. PSENet [35], CRAFT [1], DB [17]), but our method has the fastest inference speed (13.2 fps) and keeps a good balance between accuracy and latency. The qualitative illustrations in Fig. 4(c) show that the proposed method can detect multi-oriented texts well.

Long Straight Text Detection. We also evaluate the robustness of our proposed method on the long straight text dataset MSRA-TD500. During inference, the short side of images is set to 736 for a fair comparison. As shown in Table 4, our method achieves 82.4% in F-score, which is 0.7% better than MSR and comparable to the best-performing detectors DB and CRAFT. Therefore, our method is robust for detecting texts with extreme aspect ratios in complex scenarios (see Fig. 4(d)).

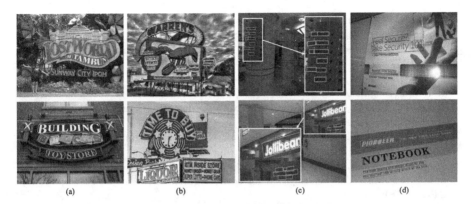

Fig. 4. The qualitative results of the proposed method. Images in columns (a)–(d) are sampled from the datasets Total-Text, CTW1500, ICDAR2015, and MSRA-TD500 respectively. The green polygons are the detection results predicted by our method, while the blue ones are ground-truth annotations. (Color figure online)

5   Conclusion

In this paper, we analyzed the limitations of existing segmentation-based scene text detectors and proposed a novel text instance expression to address these limitations. Moreover, considering their limited ability to utilize context information, our method extracts both foreground and background features for robust detection. The pixels around text boundaries are chosen as references of the predicted offsets for accurate localization. Besides, a corresponding post-processing algorithm is introduced to generate the final text instances. Extensive experiments demonstrated that our method achieves the performance superior or comparable to other state-of-the-art approaches on several publicly available benchmarks.

Acknowledgements. This work was supported by Beijing Nova Program of Science and Technology (Grant No.: Z191100001119077), Center For Chinese Font Design and Research, and Key Laboratory of Science, Technology and Standard in Press Industry (Key Laboratory of Intelligent Press Media Technology).

References

1. Baek, Y., Lee, B., Han, D., Yun, S., Lee, H.: Character region awareness for text detection. In: CVPR, pp. 9365–9374 (2019)
2. Cai, Z., Vasconcelos, N.: Cascade R-CNN: delving into high quality object detection. In: CVPR, pp. 6154–6162 (2018)
3. Chen, J., Lian, Z., Wang, Y., Tang, Y., Xiao, J.: Irregular scene text detection via attention guided border labeling. SCIS **62**(12), 220103 (2019)

4. Chen, L.C., Papandreou, G., Kokkinos, I., Murphy, K., Yuille, A.L.: DeepLab: semantic image segmentation with deep convolutional nets, atrous convolution, and fully connected CRFs. IEEE Trans. Pattern Anal. Mach. Intell. **40**(4), 834–848 (2017)

5. Ch'ng, C.K., Chan, C.S.: Total-text: A comprehensive dataset for scene text detection and recognition. In: ICDAR, vol. 1, pp. 935–942 (2017)

6. Deng, D., Liu, H., Li, X., Cai, D.: PixelLink: detecting scene text via instance segmentation. In: AAAI (2018)

7. Deng, J., Dong, W., Socher, R., Li, L.J., Li, K., Fei-Fei, L.: ImageNet: a large-scale hierarchical image database. In: CVPR, pp. 248–255 (2009)

8. Ding, H., Jiang, X., Shuai, B., Qun Liu, A., Wang, G.: Context contrasted feature and gated multi-scale aggregation for scene segmentation. In: CVPR, pp. 2393–2402 (2018)

9. Girshick, R.: Fast R-CNN. In: ICCV, pp. 1440–1448 (2015)

10. Gupta, A., Vedaldi, A., Zisserman, A.: Synthetic data for text localisation in natural images. In: CVPR, pp. 2315–2324 (2016)

11. He, K., Gkioxari, G., Dollár, P., Girshick, R.: Mask R-CNN. In: ICCV, pp. 2961–2969 (2017)

12. He, K., Zhang, X., Ren, S., Sun, J.: Deep residual learning for image recognition. In: CVPR, pp. 770–778 (2016)

13. He, W., Zhang, X.Y., Yin, F., Liu, C.L.: Deep direct regression for multi-oriented scene text detection. In: ICCV, pp. 745–753 (2017)

14. Karatzas, D., et al.: ICDAR 2015 competition on robust reading. In: ICDAR, pp. 1156–1160 (2015)

15. Liao, M., Shi, B., Bai, X.: TextBoxes++: a single-shot oriented scene text detector. IEEE Trans. Image Process. **27**(8), 3676–3690 (2018)

16. Liao, M., Shi, B., Bai, X., Wang, X., Liu, W.: TextBoxes: a fast text detector with a single deep neural network. In: AAAI, pp. 4161–4167 (2017)

17. Liao, M., Wan, Z., Yao, C., Chen, K., Bai, X.: Real-time scene text detection with differentiable binarization. In: AAAI, vol. 34, pp. 11474–11481 (2020)

18. Liao, M., Zhu, Z., Shi, B., Xia, G.S., Bai, X.: Rotation-sensitive regression for oriented scene text detection. In: CVPR, pp. 5909–5918 (2018)

19. Lin, T.Y., Dollár, P., Girshick, R., He, K., Hariharan, B., Belongie, S.: Feature pyramid networks for object detection. In: CVPR, pp. 2117–2125 (2017)

20. Liu, S., Qi, L., Qin, H., Shi, J., Jia, J.: Path aggregation network for instance segmentation. In: CVPR, pp. 8759–8768 (2018)

21. Liu, W., et al.: SSD: single shot multibox detector. In: Leibe, B., Matas, J., Sebe, N., Welling, M. (eds.) ECCV 2016. LNCS, vol. 9905, pp. 21–37. Springer, Cham (2016). https://doi.org/10.1007/978-3-319-46448-0_2

22. Liu, Z., Lin, G., Yang, S., Liu, F., Lin, W., Goh, W.L.: Towards robust curve text detection with conditional spatial expansion. In: CVPR, pp. 7269–7278 (2019)

23. Long, J., Shelhamer, E., Darrell, T.: Fully convolutional networks for semantic segmentation. In: CVPR, pp. 3431–3440 (2015)

24. Long, S., Ruan, J., Zhang, W., He, X., Wu, W., Yao, C.: TextSnake: a flexible representation for detecting text of arbitrary shapes. In: ECCV, pp. 20–36 (2018)

25. Lyu, P., Liao, M., Yao, C., Wu, W., Bai, X.: Mask textspotter: an end-to-end trainable neural network for spotting text with arbitrary shapes. In: ECCV, pp. 67–83 (2018)

26. Lyu, P., Yao, C., Wu, W., Yan, S., Bai, X.: Multi-oriented scene text detection via corner localization and region segmentation. In: CVPR, pp. 7553–7563 (2018)

27. Ma, J., et al.: Arbitrary-oriented scene text detection via rotation proposals. IEEE Trans. Image Process. **20**(11), 3111–3122 (2018)
28. Milletari, F., Navab, N., Ahmadi, S.A.: V-net: Fully convolutional neural networks for volumetric medical image segmentation. In: 3DV, pp. 565–571 (2016)
29. Neumann, L., Matas, J.: A method for text localization and recognition in real-world images. In: ACCV, pp. 770–783 (2010)
30. Ren, S., He, K., Girshick, R., Sun, J.: Faster R-CNN: towards real-time object detection with region proposal networks. In: NIPS, pp. 91–99 (2015)
31. Shi, B., Bai, X., Belongie, S.: Detecting oriented text in natural images by linking segments. In: CVPR, pp. 2550–2558 (2017)
32. Shrivastava, A., Gupta, A., Girshick, R.: Training region-based object detectors with online hard example mining. In: CVPR, pp. 761–769 (2016)
33. Vatti, B.R.: A generic solution to polygon clipping. CACM **35**(7), 56–63 (1992)
34. Wang, F., Chen, Y., Wu, F., Li, X.: TextRay: contour-based geometric modeling for arbitrary-shaped scene text detection. In: ACM-MM, pp. 111–119 (2020)
35. Wang, W., et al.: Shape robust text detection with progressive scale expansion network. In: CVPR, pp. 9336–9345 (2019)
36. Xie, E., Zang, Y., Shao, S., Yu, G., Yao, C., Li, G.: Scene text detection with supervised pyramid context network. In: AAAI, vol. 33, pp. 9038–9045 (2019)
37. Xu, Y., Wang, Y., Zhou, W., Wang, Y., Yang, Z., Bai, X.: TextField: learning a deep direction field for irregular scene text detection. IEEE Trans. Image Process. **28**(11), 5566–5579 (2019)
38. Xue, C., Lu, S., Zhang, W.: MSR: multi-scale shape regression for scene text detection. In: IJCAI, pp. 989–995 (2019)
39. Yao, C., Bai, X., Liu, W.: A unified framework for multioriented text detection and recognition. IEEE Trans. Image Process. **23**(11), 4737–4749 (2014)
40. Yao, C., Bai, X., Liu, W., Ma, Y., Tu, Z.: Detecting texts of arbitrary orientations in natural images. In: CVPR, pp. 1083–1090 (2012)
41. Yuliang, L., Lianwen, J., Shuaitao, Z., Sheng, Z.: Detecting curve text in the wild: new dataset and new solution. arXiv preprint arXiv:1712.02170 (2017)
42. Zhang, Z., Zhang, C., Shen, W., Yao, C., Liu, W., Bai, X.: Multi-oriented text detection with fully convolutional networks. In: CVPR, pp. 4159–4167 (2016)
43. Zhao, H., Shi, J., Qi, X., Wang, X., Jia, J.: Pyramid scene parsing network. In: CVPR, pp. 2881–2890 (2017)
44. Zhou, X., et al.: EAST: an efficient and accurate scene text detector. In: CVPR, pp. 5551–5560 (2017)

Document Classification

VML-HP: Hebrew Paleography Dataset

Ahmad Droby[1(✉)], Berat Kurar Barakat[1], Daria Vasyutinsky Shapira[1],
Irina Rabaev[2], and Jihad El-Sana[1]

[1] Ben-Gurion University of the Negev, Beer-Sheva, Israel
{drobya,berat,dariavas}@post.bgu.ac.il, el-sana@cs.bgu.ac.il
[2] Shamoon College of Engineering, Beer-Sheva, Israel
irinar@ac.sce.ac.il

Abstract. This paper presents a public dataset, VML-HP, for Hebrew
paleography analysis. The VML-HP dataset consists of 537 document
page images with labels of 15 script sub-types. Ground truth is manually created by a Hebrew paleographer at a page level. In addition, we
propose a patch generation tool for extracting patches that contain an
approximately equal number of text lines no matter the variety of font
sizes. The VML-HP dataset contains a train set and two test sets. The
first is a typical test set, and the second is a blind test set for evaluating algorithms in a more challenging setting. We have evaluated several
deep learning classifiers on both of the test sets. The results show that
convolutional networks can classify Hebrew script sub-types on a typical
test set with accuracy much higher than the accuracy on the blind test.

Keywords: Paleography · Handwritten style analysis · Hebrew
medieval manuscripts · Script type classification · Learning-based
classification · Convolutional neural network

1 Introduction

Robust and accurate algorithms in document image analysis can be developed
and compared by the public availability of labeled datasets. A vital document
image analysis task is to provide solutions for the study of ancient and medieval
handwriting.

Paleography (from Greek "palaios" - "old" and "graphein" - "to write") is
a study of handwriting. Throughout history, different script types were used for
different types of manuscripts; these script types appeared, developed, and dis-
appeared as time went by. The classification of script types started in the middle
ages, and the contemporary paleography research of Latin, Greek, and Hebrew
scripts emerged in the mid-20th century. An experienced librarian who works
with medieval or ancient manuscripts knows to recognize and read the scripts
of a given collection. A researcher specially trained to recognize and compare
all medieval and ancient script types and sub-types is called a paleographer.
The paleographic analysis is used to determine the place and date of writing

© Springer Nature Switzerland AG 2021
J. Lladós et al. (Eds.): ICDAR 2021, LNCS 12824, pp. 205–220, 2021.
https://doi.org/10.1007/978-3-030-86337-1_14

manuscripts that have no date, fragmentary and damaged manuscripts, etc. In some cases, it is possible, by comparison, to identify the scribe, to check the manuscripts' authenticity, or derive other essential data.

Contemporary Hebrew paleography emerged in the mid-1950s, concurrently with modern Latin paleography. The theoretical basis of Hebrew paleography is formulated in the works of Malachi Beit-Arié, Norman Golb, Benjamin Richler, Colette Sirat [2,3,19,20,23,27]. Hebrew manuscripts have a stereotyped nature of handwriting, which is a product of the cultural and pedagogical convention. A scribe was required to emulate the writing of his master until the forms of their writing become indistinguishable. This stereotypical script of a specific region is called a script type; script sub-types reflect the time of writing the manuscript or its type, whereas the separation of the hands within a single script sub-type is referred to as a unique handwriting style.

The digital era has enabled Hebrew manuscript images to be accessible publicly. This paper exploits these manuscript images to introduce the first publicly available Hebrew paleography dataset called VML-HP (Visual Media Lab - Hebrew Paleography). We believe that Hebrew paleography dataset is an important resource for developing a large-scale paleographic analysis of Hebrew manuscripts as well as for evaluating and benchmarking the analysis of algorithms for script classification. A trained paleographer can only describe a limited number of manuscripts, the number of such paleographers is very small, and there are still manuscript collections without even a good basic catalogue. We believe that a proper algorithm will become an essential tool in manuscripts' research.

Contemporary Hebrew paleography identifies six main-types of scripts: Byzantine, Oriental, Yemenite, Ashkenazi, Italian, Sephardic. Each main script type may contain up to three sub-types of scripts: square, semi-square, cursive (Fig. 1). In total, there are 15 script sub-types, which are included in the VML-HP dataset. Figure 2 shows example document patches for each Hebrew script sub-type from the dataset. The VML-HP dataset contains a train set and two test sets. The first test set is called the typical test set and is composed of unseen pages from the manuscripts used in the train set. The second test set is called the blind test set and is composed of unseen pages from the manuscripts not used in the train set. Currently, the VML-HP dataset can be downloaded from http://www.cs.bgu.ac.il/~berat/.

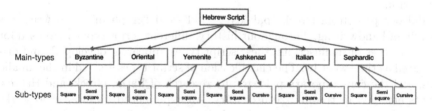

Fig. 1. Medieval Hebrew script has six main-types, and each main-type has up to three sub-types. In total, there are 15 sub-types of Hebrew script.

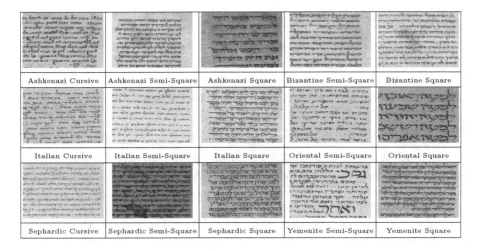

Fig. 2. Example document image patches of the 15 Hebrew script types.

In this paper, we investigate two problems: (1) the classification of the 15 script types available in the VML-HP dataset, and (2) the distinction between square and cursive scripts, regardless of the script main-type. Distinguishing between square and cursive scripts can help to identify important sections of a manuscript, which are often written in a different script sub-type. Such sections include colophon (an inscription at the end, less often at other places, of the manuscript that tells when, where, and by whom it was written), the owner's notes and more. Since colophon and other text notes are the primary sources of information about the manuscript, knowing exactly where they are located on the page would be a great help for a researcher.

We provide baseline results for the two classification problems using several convolutional neural networks. The networks are trained to classify patches that are extracted from the pages in the VML-HP dataset. Each network is evaluated on the blind and typical test sets at patch level and page level, where a page is classified based on the majority vote of its patches. Unsurprisingly, all of the networks achieved significantly higher accuracy on the typical test set compared to the blind test set, with ResNet50 giving the best performance. In addition, we explore preprocessing the input patches and applying different data augmentation strategies to improve the results.

The rest of the paper is organized as follows: Sect. 2 is a short survey of the related literature. Section 3 describes the theoretic foundation, collection and properties of the VML-HP dataset. Section 4 explains two ground truth formats provided with the dataset. In Sect. 5 we evaluate several deep learning classifiers on VML-HP.

2   Related Work

During the last decade, various computer vision techniques were employed for paleography analysis. Earlier methods used manually defined features, mostly based on textural, grapheme-based descriptors, and their combination [12–14].

Over the recent decade, deep learning methods have achieved new standards in many research frontiers. The early work [4,10] trained CNNs to classify the writing styles and used the penultimate layer activations as features. Such supervised methods require a lot of labeled data. Christlein *et al.* [5] and Hosoe *et al.* [15] showed that deep activations learned in an unsupervised manner can perform better. In a writing style classification dataset, the training classes are different from the test ones. To handle this difference, Keglevic *et al.* [17] propose to use a triplet CNN that measures the similarity of two image patches instead of training a classification network on surrogate classes. The work of Abdalhaleem *et al.* [1] examines the in-writer variations in a manuscript. Their model is based on Siamese convolutional neural networks, where the model is trained to learn the tiny changes in a person's writing style.

Deep learning methods won first place in the competitions on the classification of medieval handwritings in Latin script [7,8] held in 2016 and 2017. The objective of the competition was to classify medieval Latin scripts into 12 classes according to their writing style. Also, another level of classification - the date classification - was added in [7]. The results show that deep learning models can classify Latin script types with acceptable accuracy, more than 80% on homogeneous document collections (TIFF format only) and about 60% on heterogeneous documents (JPEG and TIFF images). Studer *et al.* [24] studied the effect of ImageNet pre-training for different historical document analysis tasks, including style classification of Latin manuscripts. They investigated a number of well-known architectures: VGG19 [22], Inception V3 [25], ResNet152 [11] and DenseNET12 [16]. The models were trained from scratch and obtained 39% to 46% accuracy rate on the script classification task, while the pre-trained models obtained a 49%−55% accuracy rate.

Early research on the paleographic classification of Hebrew documents is described in [26], where authors apply computerized tools on the documents from the Rabbanite Cairo Genizah collection. They construct a dictionary based on k-means clustering and represent each document using a set of descriptors based on the constructed prototypes. Dot product is applied between the descriptors of two documents to measure their similarity. The results depend crucially on the methods used to construct the dictionary. Dhali *et al.* [9] apply textural and grapheme-based features together with support vector regression to estimate a date of the ancient manuscripts from Dead Sea Scrolls collection.

This paper utilizes deep learning models to classify medieval Hebrew manuscripts according to their script type. Our dataset mainly includes manuscripts from SfarData collection (see Sect. 3). In some rare cases, we added manuscripts from other collections to balance the number of representative documents in each class.

3   Construction of VML-HP

VML-HP dataset is built upon a theoretic foundation of contemporary Hebrew paleography. It contains 537 accurately labeled, high-resolution manuscript page images that are hierarchically organized into six main-types and 15 sub-types of Hebrew script (Fig. 1). A Hebrew paleographer collected and labeled these manuscript page images manually.

3.1   SfarData

SfarData[1], an ongoing database project based on the Hebrew Paleography project, in cooperation with the Israeli Academy of Sciences and Humanities and the National French Center for Scientific Research (CNRS), was initiated by Malachi Beit-Arié in the 1970s. Paleograpic and codicological criteria of our project are derived from this site. Sfardata aims to locate, classify, and identify all existing dated Hebrew manuscripts written before 1540. Today it includes almost 5000 manuscripts, which makes about 95% of the known dated medieval Hebrew manuscripts. This database is currently hosted at the site of the National Library of Israel. It includes the codicological and paleographical features of the manuscripts obtained *in situ*, i.e., in the libraries in which they are kept. The project intends to study and classify these features to expose the historical typology of Hebrew manuscripts.

The most important collection of digitized and microfilmed Hebrew manuscripts belongs to the Institute for Microfilmed Hebrew manuscripts at the National Library of Israel. The Institute has been collecting microfilms (now digital photos) of Jewish manuscripts for decades, and its goal is to obtain digital copies of all Hebrew manuscripts worldwide. Today, the Institute hosts more than 70,000 microfilms and thousands of digital images, which makes more than 90% of the known Hebrew manuscripts in the world.

3.2   Collecting Manuscript Pages

Initially, the primary criterion for selecting manuscripts for our project was the fact that they were described in the SfarData. In cases when this turned out to be impossible, we chose manuscripts based on the SfarData criteria. Two sub-types proved to be particularly problematic: the Oriental square and Ashkenazi cursive. The Oriental square is the oldest Hebrew script sub-type. Most manuscripts written in this script type are not complete (collection of fragments). The better-preserved ones are kept in the National Library of Russia (Firkowicz manuscripts' collections), whose collections have not yet been entirely digitized. The Ashkenazi cursive, on the other hand, is a very common script with lots of manuscripts. However, it developed and began to be actively used only shortly before 1540, and thus there are not enough examples in the SfarData.

---

[1] http://sfardata.nli.org.il/.

Pages in the VML-HP dataset were extracted from high-quality digitized manuscripts. Among the manuscripts described in SfarData, we gave first preference to those kept in the National Library of Israel. When this was impossible, we used manuscripts from other libraries, first and foremost the British Library and the Bibliothèque Nationale de France, which have vast collections of digitized manuscripts available for download. Whenever good quality digital photos of manuscripts were not available or were of insufficient quantity for some Hebrew script sub-types, we turned to microfilms from the collection of the Institute for Microfilmed Hebrew manuscripts at the National Library of Israel.

The initial dataset that is described in this paper is relatively small. The reason is that all the manuscripts were manually picked up by our team's paleographer. Each chosen manuscript, except that it met the requirements described in the previous paragraph, had to be written in a typical (and not deviated) script sub-type that it stood for, had to have one script per page (and not multiple scripts on one page), etc. This meant weeks and even months of work. Also, the number of manuscripts of the required quality and available for download turned out to be very limited.

3.3   Properties of VML-HP

VML-HP aims to provide complete coverage of the Hebrew paleography study. It contains accurately labeled page images for each of the 15 sub-type scripts. The VML-HP dataset contains a train set and two test sets. The first test set is called the typical test set and is composed of unseen pages from manuscripts used in the train set. The second test set is called the blind test set and is composed of unseen pages from manuscripts that are not used in the train set. The blind test set is more challenging as it comes from another distribution than the train set's distribution; however, it is necessary for evaluating and benchmarking algorithms in a real-world scenario. Table 1 shows the distributions of the number of pages per main-type and sub-type scripts in the train set and two test sets. VML-HP is constructed with the goal that all the discriminator features of a script type are included in the page images with all possible variable appearances. To our knowledge, this is the first accurately labeled Hebrew paleography dataset available to the document image analysis research community.

4   Ground Truth of VML-HP

VML-HP dataset images are accurately labeled by a Hebrew paleographer at page level. However, page level labels are not always fully useful for a computer algorithm because historical document images suffer from several issues, such as physical degradation, ink bleed through, ink degradation, and image noise. In addition, hand painted motifs commonly appear in mediaval manuscripts. Therefore, we provide a clean patch generation tool that generates image patches with approximately five text lines. And we include additional ground truth format which stores the coordinates of bounding polygons around the text regions into

Table 1. The distributions of the number of pages per main-type and sub-type scripts in the train set and two test sets.

Main-Type	Sub-Type	Train	Typical Test	Blind Test	Total
Ashkenazi	Square	16	4	10	30
	Semi-Square	16	3	10	29
	Cursive	16	4	10	30
	Total	48	11	30	89
Byzantine	Square	16	4	10	30
	Semi-Square	16	4	10	30
	Total	32	8	20	60
Italian	Square	16	4	10	30
	Semi-Square	16	4	10	30
	Cursive	16	4	10	30
	Total	48	12	30	90
Oriental	Square	64	14	10	98
	Semi-Square	16	4	10	30
	Total	80	18	20	118
Sephardic	Square	16	4	10	30
	Semi-Square	24	6	10	40
	Cursive	16	4	10	30
	Total	56	14	30	100
Yemenite	Square	24	6	10	40
	Semi-Square	24	6	10	40
	Total	48	12	20	80
Total		312	75	150	537

PAGE-XML [6,18] files. The PAGE-XML files and the clean patch generation tool are available together with the dataset at http://www.cs.bgu.ac.il/~berat/.

4.1   Clean Patch Generation Algorithm

Often occurring non-text elements cause a naive patch generation algorithm (that only considers the foreground area) to generate patches with irrelevant or limited features (Fig. 3). Moreover, varying font-sizes across manuscripts may lead to low level cues like the number of text lines or sizes of characters in a patch.

To ensure that the classifier algorithm extracts the desired features, script shape features in our case, we propose a clean patch generation algorithm that can generate patches containing pure text regions and an approximately equal number of text lines. A document image patch needs to include the maximum possible amount of script features while still fitting the memory requirements.

Fig. 3. Example patches generated by a naive algorithm. Some patches contain irrelevant features, some patches contain only a few characters, and others contain no text at all.

According to paleographers, a patch would be sufficient to figure out the script type if it contains approximately five text lines. We first calculate the patch size $s \times s$ that includes approximately five text lines for each page. Then, we randomly generate patches of the size $s \times s$ from this page and resize them to the appropriate size based on the network and memory requirement. Finally, we validate the generated patches according to the additional criteria described below.

Extracting Patches with n Text Lines. To calculate the patch size s that includes n text lines, we extract k random patches of the size equal to one-tenth of the page height. A patch of this size usually includes several text lines. The desired patch size is given by $s = h/10 \times n/m$, where h is the height of the page, n is the average targeted number of lines, and m is the actual average number of lines in the k extracted patches. The number of lines in a given patch is computed by counting the peaks of the y profile using Savitzky-Golay filter [21]. We used $n = 5$ and $k = 20$.

Patch Validation. Each extracted patch is validated according to the following conditions:

- The foreground area should be at least 20% of the total patch area and not exceed 70% of the total patch area. This condition eliminates almost empty patches and patches with large spots, stains, or decorations.
- The patch should contain at least 30 connected components. This condition eliminates patches with few foreground elements.
- The variance of the x and y profiles denoted by σ_x and σ_y, respectively, should satisfy the conditions $\sigma_x \leq T_x$, $\sigma_y \geq T_y$. Assuming horizontal text lines, the variance of the x profile should be relatively low. During our experiments we set $T_x = 1500$ and $T_y = 500$
- The following inequality should be satisfied:

$$0.5 \leq \frac{\sum_{i=0}^{\frac{v}{2}} P_x(i)}{\sum_{i=\frac{v}{2}}^{v} P_x(i)} \leq 1.5 \tag{1}$$

Where v is the number of values in the x profile and $P_x(i)$ is the i-th value. This condition eliminates the patches with text lines that occupy only a fraction of a patch.

4.2 Clean Patch Generation Results

Figure 4 shows some example output patches from the clean patch generation algorithm, and Fig. 5 illustrates that the generated patches are sampled uniformly over the text regions.

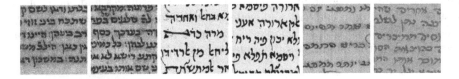

Fig. 4. Example output patches from the clean patch generation algorithm.

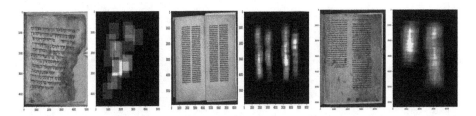

Fig. 5. Heat maps illustrate the location distribution of the generated patches covers the text regions.

5 Evaluation of Deep Learning Classifiers on VML-HP

In this section, we report the results of several deep learning classifiers. These experiments provide baseline results for potential benchmarking and underline that real-world problems are significantly challenging. First, we demonstrate the necessity of having two test sets, the typical test set and the blind test set. Then, we introduce the setting used in all the experiments, followed by evaluating different types of convolutional networks. Finally, we investigate the influence of preprocessing the input patches and the influence of different data augmentation strategies.

5.1 Real World Challenge

A dataset is usually split randomly into training, validation, and test sets. In such a scenario, pages belonging to the same manuscript may appear in the training and test sets. While this is a standard scheme, such a split can lead to misleading results. The model can learn to identify features specific to a manuscript, such as a background texture, ink color, or handwriting style. These features, as a whole, can mistakenly be used for classification. To assess this, we created a blind test set containing pages from manuscripts that are not present in the train set.

The necessity of a blind test set is visualized by training a pretrained ResNet50 and embedding the extracted feature vectors onto 2D space using t-SNE. Figure 6 shows the train, typical test, and blind test set clusters before the training. Figure 7 shows the train, typical test, and blind test set clusters after the training. The training embeds the train set samples and the typical test set samples onto compact and well-separated clusters relative to the fuzzy clusters of the blind test set samples. This shows the hardness of discovering a pattern among the blind test set samples.

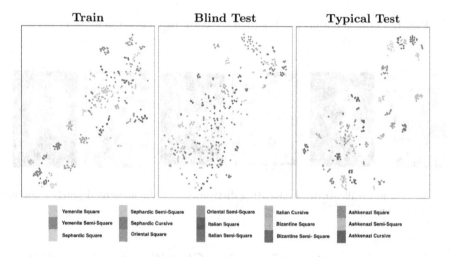

Fig. 6. Initial distribution embedding of the train, typical test and blind test sets before training.

5.2   Experimental Setting

We experiment with several convolutional network architectures. In all the experiments, we train the network on the training set and test it on both test sets, typical and blind. First, we generate 150K train patches, 10K typical test patches, and 10K blind test patches of size 350 × 350 using the clean patch generation algorithm proposed in Sect. 4.1. Input patches are normalized in terms of their pixel values. The objective training function is cross-entropy loss and is minimized using the Adam optimizer algorithm. We continue training until there is no improvement in validation loss with five epochs' patience and save the model with the least validation loss for testing.

Train	Blind Test	Typical Test

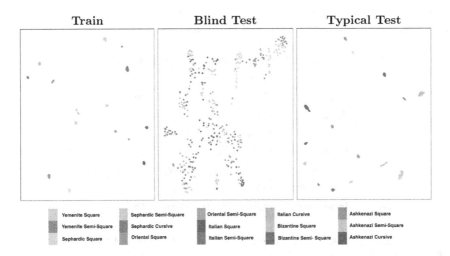

Fig. 7. Final distribution embedding of the train, typical test and blind test sets after training.

Classification results are evaluated by patch and page levels accuracy. For the page level accuracy, the label of a page is computed by taking the majority vote of the predictions of 15 patches from the page.

5.3 Effect of Network Type

Table 2 shows the accuracy results for classifying 15 script sub-types using different convolutional networks, comparing the typical test set and the blind test set at patch and page levels. The results indicate that the typical test set patches and pages are easier to classify. At nearly all levels and sets, the performance of the ResNet50 classifier is consistently higher but does not surpass 40% accuracy on the blind test set. The difference between the typical test set accuracy and the blind test set accuracy indicates overfitting on irrelevant features. However, the random guess accuracy of 15 classes is 7.6%, indicating that the network extracts script type-dependent features and improves the random classification accuracy. We argue that script type classification is an expressible function, but the network needs more data to learn this function.

Table 3 shows the accuracy results for classifying square and cursive script sub-types using different convolutional networks, comparing the typical test set and the blind test set at patch and page levels. The typical test set accuracy is fully saturated, whereas there is little room for improvement at blind test set accuracy. This result strengthens the above argument because decreasing the number of classes from 15 to two increases the number of samples per class, leading to higher accuracy.

Table 2. Patch and page level accuracies on typical test set and blind test set using different network architectures for classifying 15 script sub-types.

	Patch level		Page level	
	Typical	Blind	Typical	Blind
DenseNet	97.97	32.95	98.63	38.36
AlexNet	91.99	27.03	93.15	28.28
VGG11	99.16	35.55	100	35.63
SqueezeNet	98.03	30.38	98.63	29.45
ResNet18	97.07	30.95	98.63	34.25
ResNet50	99.55	36.15	98.63	39.73
Inception v3	94.94	26.41	95.89	26.71

Table 3. Patch and page level accuracies on typical test set and blind test set using different network architectures for classifying square and cursive script sub-types.

	Patch level		Page level	
	Typical	Blind	Typical	Blind
DenseNet	99.85	87.06	100	83.72
AlexNet	99.48	88.01	100	91.86
VGG11	99.93	86.45	100	88.37
ResNet18	99.65	86.85	100	87.21
ResNet50	99.99	90.58	100	94.19
SqueezeNet	98.03	82.45	100	86.05
Inception v3	99.16	82.06	100	80.23

5.4 Effect of Preprocessing

As the irrelevant background features might lead to poor training, we preprocess the patches by applying a bilateral filter and a bandpass filter, reducing the amount of information passed to the network through the background pixels. Hence, improving the overfitting of a convolutional network on spurious background frequencies. Table 4 shows the effect of preprocessing using a ResNet50 network on 15 script sub-types. As validated in the results, preprocessing further boosts the performance to 42.1% and 49.3% accuracy at patch and page levels, respectively.

Table 4. Effect of preprocessing at patch and page level accuracies using a ResNet50 network on blind test set for classifying 15 script sub-types.

Preprocessing	Patch level	Page level
x	36.15%	39.73%
v	42.10%	49.30%

5.5   Effect of Augmentation

It is known that augmenting the training data increases the model's ability to overcome overfitting. We experimented using various combinations of augmentation methods, such as random scaling by a factor between 1.0 and 1.2, random rotation by a degree between $-30°$ and $30°$, and random horizontal flipping.

Table 5 shows the effect of augmentation at patch and page levels accuracy results on the blind test set for classifying 15 script sub-types using a ResNet50 network. Random rotation and horizontal flipping boost the accuracy at both the patch and page levels. However, random scaling only improves the patch level accuracy. In addition, we can conclude that applying all augmentation strategies at once is counterproductive. Perhaps it biases the network through the train set distribution, which is very dissimilar to the data distribution that the model is tested on.

Furthermore, we experimented with combining preprocessing and augmentation. However, doing so did not improve the results, and in some cases, it worsened it. We hypothesize that in some cases, combining preprocessing and augmentation results in the loss of text features, which reduced the classification accuracy.

Table 5. Effect of augmentation by patch and page levels accuracy results on blind test set for classifying 15 script sub-types using a ResNet50 network.

Augmentations			Patch level	Page level
Scaling	Rotation	H. Flip		
✗	✗	✗	36.15%	39.73%
✓	✗	✗	38.76%	32.88%
✗	✓	✗	37.11%	42.47%
✗	✗	✓	38.30%	43.84%
✓	✓	✓	32.06%	30.82%

5.6   Comparing Deep Neural Networks Against a Paleographer Expert

To compare the deep learning networks against a paleographer expert, we performed a classification experiment. In this experiment, we ask a paleographer to classify 75 document patches according to the 15 script sub-types. The document patches were randomly chosen from the set of patches used in our experiments, five from each sub-type. The accuracy rate of the paleographer expert is 70%. We can draw two conclusions from this experiment. First, the problem was challenging for the human expert due to the unusual format: paleographers work with manuscripts and pages, not patches. Second, there is a large room for improvement for automatic classification. As we previously pointed, we expect that training the networks on a larger dataset will improve the classification rate on the blind test set. A potential limitation of this experiment is that it was performed only with one paleographer expert. Unfortunately, the number of Hebrew paleographers is extremely small, and we did not want to involve the paleographers who created the SfarData. However, we do not expect a larger experiment to change the results significantly.

6 Conclusion

Automatic paleographic analysis of historical documents is a challenging task, and benchmark datasets lie at the heart of the development, assessment, and comparison of the algorithms. This paper introduces a medieval Hebrew manuscripts dataset, VML-HP dataset, which includes 537 pages labeled with 15 script sub-types. The VML-HP dataset contains a train set, typical and blind test sets. The VML-HP is the first publicly available Hebrew paleographic dataset.

We report baseline results of several established deep learning classification networks. Results show that there is a big room for improvement on the blind test set, whereas the typical test set is an easier mission. In addition, we showed that preprocessing the input patches by applying a bilateral filter and a bandpass filter boosts the model's performance. Furthermore, we explored different data augmentation strategies.

Acknowledgment. This research was partially supported by The Frankel Center for Computer Science at Ben-Gurion University. The participation of Dr. Vasyutinsky Shapira in this project is funded by Israeli Ministery of Science, Technology and Space, Yuval Ne'eman scholarship n. 3-16784.

References

1. Abdalhaleem, A., Barakat, B.K., El-Sana, J.: Case study: fine writing style classification using Siamese neural network. In: 2018 IEEE 2nd International Workshop on Arabic and Derived Script Analysis and Recognition (ASAR), pp. 62–66. IEEE (2018)
2. Beit-Arié, M.: Hebrew codicology. Tentative Typology of Technical Practices Employed in Hebrew Dated Medieval Manuscripts, Jerusalem (1981)
3. Beit-Arié, M., Engel, E.: Specimens of mediaeval Hebrew scripts, vol. 3. Israel Academy of Sciences and Humanities (1987, 2002, 2017)
4. Christlein, V., Bernecker, D., Maier, A., Angelopoulou, E.: Offline writer identification using convolutional neural network activation features. In: Gall, J., Gehler, P., Leibe, B. (eds.) GCPR 2015. LNCS, vol. 9358, pp. 540–552. Springer, Cham (2015). https://doi.org/10.1007/978-3-319-24947-6_45
5. Christlein, V., Gropp, M., Fiel, S., Maier, A.: Unsupervised feature learning for writer identification and writer retrieval. In: 2017 14th IAPR International Conference on Document Analysis and Recognition (ICDAR), vol. 1, pp. 991–997. IEEE (2017)
6. Clausner, C., Pletschacher, S., Antonacopoulos, A.: Aletheia-an advanced document layout and text ground-truthing system for production environments. In: ICDAR, pp. 48–52. IEEE (2011)

7. Cloppet, F., Eglin, V., Helias-Baron, M., Kieu, C., Vincent, N., Stutzmann, D.: ICDAR 2017 competition on the classification of medieval handwritings in Latin script. In: 2017 14th IAPR International Conference on Document Analysis and Recognition (ICDAR), vol. 1, pp. 1371–1376. IEEE (2017)

8. Cloppet, F., Eglin, V., Stutzmann, D., Vincent, N., et al.: ICFHR 2016 competition on the classification of medieval handwritings in Latin script. In: 2016 15th International Conference on Frontiers in Handwriting Recognition (ICFHR), pp. 590–595. IEEE (2016)

9. Dhali, M.A., Jansen, C.N., de Wit, J.W., Schomaker, L.: Feature-extraction methods for historical manuscript dating based on writing style development. Pattern Recogn. Lett. **131**, 413–420 (2020)

10. Fiel, S., Sablatnig, R.: Writer identification and writer retrieval using the fisher vector on visual vocabularies. In: 12th International Conference on Document Analysis and Recognition, pp. 545–549. IEEE (2013)

11. He, K., Zhang, X., Ren, S., Sun, J.: Deep residual learning for image recognition. In: Proceedings of the IEEE Conference on Computer Vision and Pattern Recognition, pp. 770–778 (2016)

12. He, S., Samara, P., Burgers, J., Schomaker, L.: Discovering visual element evolutions for historical document dating. In: 2016 15th International Conference on Frontiers in Handwriting Recognition (ICFHR), pp. 7–12. IEEE (2016)

13. He, S., Samara, P., Burgers, J., Schomaker, L.: Historical manuscript dating based on temporal pattern codebook. Comput. Vis. Image Underst. **152**, 167–175 (2016)

14. He, S., Sammara, P., Burgers, J., Schomaker, L.: Towards style-based dating of historical documents. In: 2014 14th International Conference on Frontiers in Handwriting Recognition, pp. 265–270. IEEE (2014)

15. Hosoe, M., Yamada, T., Kato, K., Yamamoto, K.: Offline text-independent writer identification based on writer-independent model using conditional autoencoder. In: 2018 16th International Conference on Frontiers in Handwriting Recognition (ICFHR), pp. 441–446. IEEE (2018)

16. Huang, G., Liu, Z., Van Der Maaten, L., Weinberger, K.Q.: Densely connected convolutional networks. In: Proceedings of the IEEE Conference on Computer Vision and Pattern Recognition, pp. 4700–4708 (2017)

17. Keglevic, M., Fiel, S., Sablatnig, R.: Learning features for writer retrieval and identification using triplet CNNs. In: 2018 16th International Conference on Frontiers in Handwriting Recognition (ICFHR), pp. 211–216. IEEE (2018)

18. Pletschacher, S., Antonacopoulos, A.: The page (page analysis and ground-truth elements) format framework. In: ICPR, pp. 257–260. IEEE (2010)

19. Richler, B.: Hebrew manuscripts in the Vatican library: catalogue. Hebrew manuscripts in the Vatican Library, pp. 1–790 (2008)

20. Richler, B., Beit-Arié, M.: Hebrew manuscripts in the biblioteca palatina in parma: catalogue; palaeographical and codicological descriptions (2011)

21. Savitzky, A., Golay, M.J.: Smoothing and differentiation of data by simplified least squares procedures. Anal. Chem. **36**(8), 1627–1639 (1964)

22. Simonyan, K., Zisserman, A.: Very deep convolutional networks for large-scale image recognition. arXiv preprint arXiv:1409.1556 (2014)

23. Sirat, C.: Hebrew Manuscripts of the Middle Ages. Cambridge University Press, Cambridge (2002)

24. Studer, L., et al.: A comprehensive study of ImageNet pre-training for historical document image analysis. In: 2019 International Conference on Document Analysis and Recognition (ICDAR), pp. 720–725. IEEE (2019)

25. Szegedy, C., et al.: Going deeper with convolutions. In: Proceedings of the IEEE Conference on Computer Vision and Pattern Recognition, pp. 1–9 (2015)
26. Wolf, L., Potikha, L., Dershowitz, N., Shweka, R., Choueka, Y.: Computerized paleography: tools for historical manuscripts. In: 2011 18th IEEE International Conference on Image Processing, pp. 3545–3548. IEEE (2011)
27. Yardeni, A., et al.: The Book of Hebrew Script: History, Palaeography, Script Styles, Calligraphy & Design. Carta Jerusalem, Jerusalem (1997)

Open Set Authorship Attribution Toward Demystifying Victorian Periodicals

Sarkhan Badirli[1]([📧])(iD), Mary Borgo Ton[2](iD), Abdulmecit Gungor[3], and Murat Dundar[4](iD)

[1] Computer Science Department, Purdue University, West Lafayette, IN, USA
[2] Illinois University Library, University of Illinois Urbana-Champaign, Champaign, IL, USA
[3] Intel, Hillsboro, OR, USA
[4] Computer and Information Science Department, IUPUI, Indianapolis, IN, USA

Abstract. Existing research in computational authorship attribution (AA) has primarily focused on attribution tasks with a limited number of authors in a closed-set configuration. This restricted set-up is far from being realistic in dealing with highly entangled real-world AA tasks that involve a large number of candidate authors for attribution during test time. In this paper, we study AA in historical texts using a new data set compiled from the Victorian literature. We investigate the predictive capacity of most common English words in distinguishing writings of most prominent Victorian novelists. We challenged the closed-set classification assumption and discussed the limitations of standard machine learning techniques in dealing with the open set AA task. Our experiments suggest that a linear classifier can achieve near perfect attribution accuracy under closed set assumption yet, the need for more robust approaches becomes evident once a large candidate pool has to be considered in the open-set classification setting.

Keywords: Author attribution · Open-set classification · Victorian literature

1 Introduction

Deriving its name from Queen Victoria (1837–1901) of Great Britain, Victorian literature encompasses some of the most widely-read English writers, including Charles Dickens, the Brontë sisters, Arthur Conan Doyle and many other eminent novelists. The most prolific Victorian authors shared their freshest interpretations of literature, religion, politics, social science and political economy in monthly periodicals (W. F. Poole, 1882). To avoid damaging their reputations as novelists, almost ninety percent of these articles were either written anonymously or under pseudonyms. As the pseudonym gave authors greater license to express their personal, political, and artistic views, anonymous essays were often honest and outspoken, particularly on controversial issues. However, this publication

© Springer Nature Switzerland AG 2021
J. Lladós et al. (Eds.): ICDAR 2021, LNCS 12824, pp. 221–235, 2021.
https://doi.org/10.1007/978-3-030-86337-1_15

strategy presents literary critics and historians with a significant challenge. As Houghton notes, knowing the author's identity can radically reshape interpretations of anonymously-authored articles, particularly those that include political critique. Furthermore, the intended audience of an anonymous essay cannot be accurately identified without knowing the true identity of the contributor (Walter Houghton, 1965).

To address this problem, Walter Houghton worked in collaboration with staff members, a board of editors, librarians and scholars from all over the world to pioneer the traditional approach to authorship attribution. In 1965, they created a 5-volume journal of *Wellesley Index to Victorian Periodicals*, named after Houghton's during his time in Wellesley College. Since then additions and corrections to the Wellesley Index [3] are recorded in the Curran Index [12]. Together, the Wellesley and the Curran Indices have become the primary resources for author indexing in Victorian periodicals.

In his introduction to the Wellesley Index [3], the editor in chief Dr. Houghton describes the main sources of evidence used for author indexing. 'In making the identifications, we have not relied on stylistic characteristics, it was external evidence: passages in published biographies and letters, collections of essays which are reprints of anonymous articles, marked files of the periodicals, publishers' lists and account books, and the correspondence of editors and leading contributors in British archives.' Such an approach draws from the disciplinary strengths of historically-oriented literary criticism. However, this method of author identification cannot be used on a larger scale, for this approach requires a massive amount of human labor to track all correspondence related to published essays, the vast majority of which appear only in manuscript form or in edited volumes in print. Furthermore, the degree of ambiguity in these sources, combined with the incomplete nature of correspondence archives, makes a definitive interpretation difficult to achieve.

Our study focuses on the stylistic characteristics previously overlooked in these indices. Even though author attribution using stylistic characteristics may become impractical when the potential number of contributors is on the order of thousands, we demonstrate that attribution with a high degree of accuracy is still possible based on usage and frequency of the most common words as long as one is interested in the works of a select few contributors with an adequate number of known work available for training. To support this, we focused our study on 36 of the most eminent and prolific writers of the Victorian era who have 4 or more published books that are accessible through Gutenberg Project [2]. However, the fact that any given work can belong to one of the few select contributors as well as any one of the thousands of potential contributors still poses a significant technical challenge for traditional machine learning, especially for Victorian periodical literature. Less than half of the articles in these periodicals were written by well-known writers; the rest was contributed by thousands of less known individuals many of those first occupation is not literature. The technical challenges posed by this diverse pool of authors can only be addressed with an open-set classification framework.

Unlike its traditional counterpart, computational Authorship Attribution (AA) identifies the author of a given text out of possible candidates using statistical/machine learning methods. Thanks to the seminal work [30] of Mostellar and Wallace in the second half of the 20th century, AA has become one of the most prevalent application areas of modern Natural Language Processing (NLP) research. With the advances in internet and smartphone technology, the accumulation of text data has accelerated at an exponential rate in the last two decades. Abundant text data available in the form of emails, tweets, blogs and electronic version of books [1,2] has opened new venues for authorship attribution and led to a surge of interest in AA among academic communities as well as corporate establishments and government agencies.

Most of the large body of existing work in AA utilizes methods that expose subtle stylometric features. These methods include n-grams, word/sentence length and vocabulary richness. When paired with word distributions, these approaches have achieved promising results performing various AA tasks. What this paper explores deviates from this trend by asking a simple yet pressing question: Can we identify authors of texts written by the world's most prominent writers based on the distributions of the words most commonly used in daily life? As a corollary, which factors play the central role in shaping predictive accuracy? Toward achieving this end, we have investigated writings of most renowned Victorian era novelists by quantifying their writing patterns in terms of their usage frequency of the most common English words. We investigated the effect of different variables on the performance of the classifier model to explore the strengths and limitations of our approach. The experiments are run in an open-set classification setup to reflect real world characteristics of our multi-authored corpus. Unattributed articles in this corpus of Victorian texts are identified as belonging to one of the 46 known novelists or classified as an unknown author. Results are evaluated using Wellesley index as a reference standard.

2   Related Work

Computational AA has offered compelling analyses of well-known documents with unknown or disputed attribution. AA has contributed to conversations about the authorship of the disputed Federalist paper [30], the Shakespearean authorship controversy [14], the author of New Testament [22] and the author of The Dark Tower [38] to name a few examples. AA has also been proven to be very effective in forensic linguistic science. Two noteworthy examples of scholarship in this vein include the use of CUSUM (Cumulative Sum Analysis) technique [29] as expert evidence in court proceedings and the use of stylometric analysis by FBI agents in solving the *unabomber* case [19].

Beyond its narrow application within literary research, AA has also paved the way for the development of several other tasks [37] such as author verification [25], plagiarism detection [13], author profiling [24], and the detection of stylistic inconsistencies [11].

Mostellar and Wallace pioneered the statistical approach to AA by using functional and non-contextual words to identify authors of disputed 'Federalist Papers', written by three American congressmen [30]. Following their success with basic lexical features, AA as a subfield of research was dominated by efforts to define stylometric features [21,34]. Over time, features used to quantify writing style, i.e., style markers [37], evolved from simple word frequency [10] and word/sentence length [28] to more sophisticated syntactic features like sentence structure and part-of-speech tags [6,7], to more contextual character/word n-grams [32,39] and finally towards semantic features like synonyms [27] and semantic dependencies [16].

As Stamatatos points out, the more detailed the text analysis for extracting stylometric features is, the less accurate and noisier produced measures get [37]. Furthermore, sophisticated and contextual features are not robust enough measures to counteract the influence of topic and genre [20].

With the widespread use of internet and social media, AA research has shifted gears in the last two decades towards texts written in everyday language. The text data from these sources are more colloquial and idiomatic and much shorter in length compared to historical texts, making stylometric features such as word and character n-grams more effective measures for AA research. Thanks to the abundance of short texts available in the form of tweets, messages, and blog posts, the most recent AA studies of shorter texts rely on various deep learning methods [8,26,33].

In this study we demonstrate that a simple linear classifier trained with word counts from the most frequently used words alone performs surprisingly well in a highly entangled AA problem involving 36 prominent and prolific writers of the Victorian era. The success of this study depended on both training and test sets including the same set of authors. We investigate the effect of vocabulary size and the size of text fragments on the overall attribution accuracy. Existing AA research on historical texts is limited for two reasons. First, studies are limited to a small number of attribution candidates, and second classification is performed in a closed-set setting. In order for computational AA to propose an alternative to costly and demanding manual author indexing methods and possibly challenge previous identifications of authorship, effective open-set classification [36] strategies are required. There are a few studies [15,35] of open-set AA in the literature dating back to early 2010's but these techniques were tested on small datasets (Federalist papers and Books of Mormons) with a few attribution candidates where stylistic features are often the most predictive ones. Our study is the first to tackle AA in an open-set classification setting with a large number of attribution candidates. Although its attribution accuracy is far from being ideal when compared to identifications from the Wellesley Index, our approach demonstrates that a linear classifier when run in an open-set classification setting can produce interesting insights about pseudonymous and anonymous contributions to Victorian periodicals that would not be readily attainable by manual indexing techniques.

3   Datasets and Methods

3.1   Datasets

Herein, we discuss the datasets used in this study. The main dataset consists of 595 novels (books) from 36 Victorian Era novelists. We also added 23 novels to the corpus from another 10 authors who have less than 4 books available in Gutenberg Project [2]. Novels from these 10 authors were not used during training phase. We used these novels as samples from unknown authors, noted as out-of-distribution (OOD) samples, in the open set classification experiments. All books are downloaded from Gutenberg Project [2]. These texts represent a wide range of genres, spanning from Gothic novels, historical fiction, satirical works, social-problem novels, detective stories, and realist novels. The distribution of genres among authors is quite erratic as some writers like Charles Dickens explored all of these genres in their novels whereas some others such as A. C. Doyle fall mainly within the purview of a single genre. Table 1 shows all 46 authors and the number of novels written by them in the dataset. This dataset is used in both closed and open-set classification experiments.

We also curated a small collection of essays from Victorian periodicals that were republished in *Bentley's Miscellaneous* (1937–1938, Volume 1 and 2 based on availability from Gutenberg Project [2]) to test our model in a realistic open-set classification setup. We picked essays that were either written anonymously or under pseudonym. There are a total of 27 essays written under pseudonyms and 3 articles published anonymously. The length of essays ranges between 50–600 sentences. For validation we used identifications from Wellesley Index as reference standard. In preparing this dataset, we ensured that all text not belonging to the author, such as footnotes, preface, author biography, commentary, etc., were all removed. Dataset and *Python Notebooks* prepared for this project are released to the public at GitHub.

Table 1. There are total of 46 authors in the corpus. Of these, the most prolific 36 authors were used during the training phase of this study. Novels from the remaining 10 authors were used as OOD samples. First column represent ranges for the number of novels. Authors whose n

# novels	Authors
41–53	Margaret Oliphant, H. Rider Haggard, Anthony Trollope
31–40	Charlotte M. Yonge, George MacDonald
21–30	Bulwer Lytton, Frederick Marryat, Mary E. Braddon, Mary A. Ward, George Gissing
11–20	Arthur C. Doyle, Geroge Meredith, Harrison Ainsworth, Mrs. Henry Wood, Ouida, Charles Dickens, Marie Corelli, Benjamin Disraeli, Wilkie Collins, Thomas Hardy, Robert L. Stevenson, Walter Besant, R. D. Blackmore, William Thackeray, Charles Reade
4–10	Bram Stoker, George Eliot, Rhoda Broughton, Elizabeth Gaskell, Frances Trollope, George Reynolds, Charlotte Brontë, Dinah M. Craik, Catherine Gore, Ford Madox, Mary Cholmondeley
1–3	Eliza L. Lynton, Lewis Caroll, Samuel Butler, Sarah Grand, Thomas Hughes, Anna Thackeray, Mona Caird, Anne Brontë, Walter Pater, Emily Brontë

3.2 Methods

Our main motivation in this study is to demonstrate that simple classification models using just word counts from the most commonly used English words can be effective even in a highly entangled AA problem as long as both training and test sets come from the same set of authors. We emphasize that a closed-set classification setting is far from being realistic and is unlikely to offer much benefit in real-world AA problems that often emerge in open-set classification settings. We draw attention to the need for more sophisticated techniques for AA by extending the highly versatile linear support vector machine (SVM) beyond its use in closed-set text classification [5, 9, 18, 23, 31, 38] to open-set classification. We show its strengths and limitations in the context of an interesting AA problem that requires identifying authorship information of anonymous and pseudonymous essays in Victorian periodicals. Although our discussion of methodology is limited to SVM, the conclusions we draw are not necessarily method specific and apply to other popular closed-set classification techniques (random forest, naive Bayes, multinomial etc.) as well.

Linear SVM optimizes a hyperplane that maximizes the margin between the positive and negative classes while minimizing the training loss incurred by samples that fall on the wrong side of the margin. We extend SVM to open-set classification as follows. For each training class (author) we train a linear SVM with a quadratic regularizer and hinge loss in a one vs. all setting by considering all text from the given author as positive and text from all other authors as negative. During the test phase, test documents are processed by these classifiers, and final prediction is rendered based on the following possibilities. If the test sample is not classified as positive by any of the classifiers, it is labeled as belonging to an unknown author. If the test sample is classified as positive by one or more classifiers, the document is attributed to the author whose corresponding classifier generates the highest confidence score.

The regularization parameter for each one vs. all classifier is tuned on the hold-out validation set to optimize F_1 score. $SGD\_classifier$ from python library $sklearn.linear\_model$ is used for SVM implementation. Class weight is set to *balanced* to deal with class imbalance and for reproducibility purposes random state is fixed at 42 (randomly chosen).

4 Experiments

In the first part of the experiment we demonstrate that usage frequency of most common words can offer important stylistic cues about the writings of renowned novelists. Towards this end, we utilized bag of words (BOW) representation to vectorize our unstructured text data. We limit the vocabulary with the most common thousand words to avoid introducing topic or genre specific bias into AA problem. One third of these words are *stop words* that would normally be removed in a standard document classification task. All special names/entities and punctuation are removed from the vocabulary. All words are converted to lower case.

Each book is divided into chunks of same number of sentences. Each of these text chunks is considered a document. Authors with less than 4 books/novels (10 of the 46) are considered as unknown authors in open-set classification experiments. These ten authors are randomly split between validation and test sets. Documents from the remaining 36 authors are split into three as train, validation, and test set using the ratio of 64/16/20. At the book level all three sets are mutually disjoint. That is, all documents from the same book are used in one of the three sets.

4.1 Varying Document Length in both Training and Test Sets

In this experiment we investigate the size of each document on the authorship attribution performance. We considered five different document lengths, noted as $|D|$: 10, 25, 50, 100 sentences or the entire book. Python's NLTK library is used for sentence and word tokenization. The length of novels in terms of the number of sentences ranges from 390 to $25,612$ and has a median of $6,801$. A document with 50 sentences has a median of $1,142$ tokens. Vocabulary size is another variable considered pairwise with document length. For vocabulary, noted as $|V|$, we considered the most frequent 100, 500, 1000, 5000, and 10000 words.

Table 2 presents mean F_1, noted as $\bar{F}_1$, scores for each pair of document length and vocabulary size. Not surprisingly the AA prediction performance improves as the document length increases. Rate of improvement is more significant for shorter document lengths. Using entire novel as a document slightly hurts the performance as the number of training samples per class dramatically decreases. The same situation is also observed with the size of the feature vector. The rate of improvement is more remarkable for smaller vocabulary sizes. These results suggest that a vocabulary size of around 1000 and a document length of about 50 sentences can be sufficient to distinguish among eminent novelists with near perfect accuracy if three or more of their books are available for training.

4.2 Varying Document Length in Test Set While Keeping It Fixed in Training

To simulate a scenario, where documents during test time may emerge with arbitrary lengths, we fixed document length to 50 sentences in the training phase and varied the number of sentences in test samples. The very first problem in this setting is the scaling problem between test and training documents due to different document lengths. To address this problem we normalized word counts in each document by dividing each count by the maximum count in that document. This scaling operation is followed by maximum absolute value scaling (*MaxAbsScaler*) on columns. MaxAbsScaler is preferred over min-max scaling to preserve sparsity as the former does not shift/center the data. Results of this experiment is reported in Table 3.

Results in Table 3 suggest that using same document length for training and test sets is not necessary. Indeed, fixing document length in the training set to a number large enough to ensure documents are sufficiently informative for the classification task at hand while small enough to provide adequate number of training samples for each class yields results comparable to those achieved by varying sizes of training document length. It is interesting to note that the larger number of training samples available in this case led to improvements in the book-level attribution accuracy as all 131 test books are correctly attributed to their true authors for vocabulary size 500 or larger.

Table 2. $\bar{F}_1$ scores from closed set experiment with 36 authors. Both training and test documents have the same length in terms of the number of sentences. Numbers of sentences considered are 10, 25, 50, 100 and entire book. Columns represent different vocabulary sizes. 100, 500, 1000, 5000 and 10000 most frequent words are considered.

$|D|/|V|$	100	500	1K	5K	10K
Whole Book	0.94	0.95	0.96	0.98	0.97
100 Sent		0.78	0.96	0.98	0.99	1.00
50 Sent		0.58	0.88	0.93	0.98	0.98
25 Sent		0.37	0.70	0.79	0.90	0.92
10 Sent		0.15	0.39	0.48	0.64	0.68

Table 3. Varying the document length in test set while it is fixed in training. Number of sentences in training documents are fixed at 50 but different lengths are considered for test documents. Numbers of sentences considered are 10, 25, 50, 100 and the entire book. Columns again represent vocabulary sizes. Reported results are $\bar{F}_1$ scores.

$|D|/|V|$	100	500	1K	5K	10K
Whole Book	0.93	1.00	1.00	1.00	1.00
100 Sent	0.74	0.97	0.98	1.00	1.00
50 Sent	0.58	0.88	0.93	0.98	0.98
25 Sent	0.40	0.71	0.79	0.90	0.92
10 Sent	0.21	0.43	0.50	0.64	0.68

4.3   Authors' Top Predictive Words

This experiment is designed to highlight each author's most distinguishing words and their impact on classification. One-vs-all SVM classifiers are trained for each author with L_1 penalty. Documents of 50 sentences length are considered along with a vocabulary containing 1000 most frequent words. Regularization constant is tuned to maximize individual F_1 scores on the validation set. L_1 penalty yields a more sparse model by pushing coefficients of some features towards zero. From each classifier 100 non-zero coefficients with the highest absolute magnitude are selected. We considered words associated with these coefficients as the author's most distinguishing words[1]. Figure 1 displays these words for six of the authors with font size for each word magnified proportional to the absolute magnitude of their corresponding coefficients. These six authors are Charles Dickens, George Eliot, George Meredith, H. Rider Haggard, Thomas Hardy and William Thackeray.

With one possible exception it is impractical, if not impossible, to identify authors based on these word clouds. George Eliot who was a critic of organized

---

[1] Scaling word counts eliminates the potential bias due to frequency.

religion in her novels is the only exception as the word 'faith' can be spotted in the corresponding word cloud. However, the nearly-perfect attribution accuracy achieved in our results demonstrates the utility of machine learning in detecting subtle patterns not easily captured by human reasoning. In addition to presence or absence of certain words and their usage frequencies in a given text SVM takes advantage of word co-occurrence in the high dimensional feature space.

These word clouds provided other interesting findings as well. For example 'summer' is the word with the largest absolute SVM coefficient for both William Thackeray (WT) and George Meredith (GM), yet the coefficient is positive for WT and negative for GM. This suggests that WT has its own characteristic way of using the word 'summer' that would set him apart from others.[2]. Beside word frequency distribution, correlation between these frequencies also carries predictive clues and SVM effectively exploits these signals.

The usage frequency of words with the largest absolute SVM coefficients significantly vary among authors. For example, George Eliot's top 3 words has a total of 200 occurrence in her novels whereas H. Rider Haggard's top 3 words occur thousands of times and includes the article 'a'.

Fig. 1. Word cloud plots for six authors. Words are magnified proportional to the magnitude of their corresponding SVM coefficients.

4.4 Open-Set Classification

We conducted two sets of experiments in the open set classification setting. The first one in a controlled and the other in a real-world setting. In both settings training data contains documents from 36 known authors (64% of author's all documents). The validation set contains documents from the first group of 5 unknown authors in addition to documents from 36 known authors (16% of author's all documents). In the controlled setting, the test set contains documents from the second group of 5 unknown authors in addition to documents

---

[2] We can draw this conclusion as features from BOW representation are always non-negative. Normalization and scaling we applied did not change this fact. Therefore, we may argue that features with the largest absolute coefficient are best individual predictors for the positive class.

from 36 known authors (remaining 20% of author's all documents). Document lengths are fixed at 100 sentences and most frequent $1,000$ words are used as vocabulary. The most frequent 1000 words are derived based on the data we collected, yet these words turned out to be not time-sensitive and generally agree with modern-day most common words. In the open-set classification setting we used two different evaluation metrics as there are documents from both known and unknown authors in the test set. In addition to mean F_1 ($\bar{F}_1$) computed for known authors, we also report detection F_1 that evaluates the performance of the model in detecting documents of unknown authors.

For the real-world experiment complete essays of various lengths from *Bentley's Miscellaneous* Volume 1 and 2 (1837–1838) were considered as test documents. Number of sentences in these essays varies between 47 and 500.

Results in the Controlled Setting. $\bar{F}_1$ score of 36 known authors in the controlled setting is 0.91 (first row in Table 4), slightly below closed set classification score of 0.98. This is expected as the number of potential false positives for each known author increases with the inclusion of documents from unknown authors. The first row in Table 5 shows the performance breakdown on OOD samples. The performance of out-of-distribution (OOD) samples detection greatly suffered from false positives (FPs), mainly because the dataset is severely unbalanced. OOD samples form only 3.9% of the test set (365 vs $8,949$). C. Brontë and D. M. Craik were two most affected authors with 70 and 38 FPs which were 50% and 55% of total test samples from them, respectively. Small training size seems to be the underlying reason for this performance as both authors had 2 books (out of 4) for training. On false negative (FN) side, 3 authors were on the spotlight: C. Brontë (16), G. Meredith (12) and M. Corelli (8) where the numbers in parenthesis stand for quantity of FNs associated with that author. First row in Table 6 displays author distribution over misclassified OOD samples. 14 authors had zero false negatives and overall 33 of them had false negatives less than or equal to five samples. Out of 365 OOD documents, our approach correctly identified 294 (80.5%) samples belonging to an unknown author.

It is interesting to note that all 16 documents that were incorrectly attributed to Charlotte Brontë are from her younger sister Emily Brontë's famous novel 'Wuthering Heights'. Although Brontë sisters had their own narrative style and originality portrayed in their novels, they may be considered homologous when it comes to the usage of most common words. Collaborative writing and imaginary story telling during their childhood might be one explanation for this phenomenon [40]. Another interesting observation is that all of the documents that were incorrectly attributed to Marie Corelli comes from Elizabeth L. Linton's book of 'Modern Woman and What is Said of Them'.

Results on Periodicals. Of the 27 essays collected from Victorian periodicals, 6 are from known authors. Three of these are written under *Boz* pseudonym, which is known to belong to Charles Dickens. The remaining three are written by C. Gore, G.W.M. Reynolds, and W. Thackeray, under their pseudonyms,

Toby Allspy, *Max*, and *Goliah Gahagan*, respectively. Remaining 21 essays are from less known writers who are not represented in our training dataset. We evaluated our results using identifications by Wellesley Index as the reference standard. Although all three articles from Charles Dickens would have been correctly classified under closed set assumption, only one was correctly attributed to him in the open set setup and the other two are misdetected as documents belonging to unknown authors. 'Adventures in Paris' under pseudonym *Toby Allspy* was accurately attributed to Catherine Gore. Articles by Reynolds and Thackeray were misdetected as documents by unknown authors.

Detection F_1 score of 0.84 is still reasonable as 18 of the 21 essays from unknown authors were correctly detected as documents by unknown authors. Out of the three that were missed 'The Marine Ghost' from Edward Howard offers an interesting case study that may challenge identification by Wellesley Index. It is attributed to Frederick Marryat, who was Howard's captain while he served in the Navy. Furthermore, Marryat chose Howard as his sub-editor [17] while he was the editor of the *Metropolitan Magazine*. This incorrect attribution suggests Marryat played a profound role in shaping Edward Howard's authorial voice, a claim supported by their shared naval experience and their long term professional relationship.

The corpus has 3 essays from periodicals that still remain unattributed in Wellesley Index. In this experiment, none of three essays were attributed to any known authors.

Table 4. Performance on samples from known authors. Reported scores are averages from 36 authors

$|V|$	Precision	Recall	$\bar{F}_1$
1K	0.97	0.87	0.91
2K	0.99	0.92	0.95

Table 5. Performance on OOD samples detection from controlled setting with 1000 and 2000 vocabulary.

$|V|$	Precision	Recall	F_1
1K	0.34	0.81	0.48
2K	0.46	0.85	0.60

Increasing Vocabulary Size. Although the most frequent $1,000$ words proved to be sufficiently informative in the closed set classification setup, they are not as effective in the open set framework. When the number of contributors in the test set is on the order of thousands more features will inevitably be required to improve attribution accuracy. To show that this is indeed the case we repeated the previous two open set classification experiments after increasing the vocabulary size to $2,000$.

In the controlled setting with the increased vocabulary size, $\bar{F}_1$ score on known authors increases to 0.95 from 0.91 while detection F_1 improves from 0.48 to 0.60 (second row in Table 4). Additional vocabulary significantly helped to decrease total number of FPs from 569 to 368. Nevertheless, these features had minimal effect on documents from C. Brontë as 40% of them are again classified as OOD samples. The distribution of false negatives is also updated in the second row of Table 6. Increasing the vocabulary size led to significant improvements in open-set classification performance yet the number of documents incorrectly attributed to Charlotte Brontë slightly increased to 18. As earlier all of these belong to Emily Brontë's "Wuthering Heights" novel. This paves the way for speculating about two possible scenarios. It is possible that children with the same upbringing develop similar unconscious daily word usage, which does not change after childhood. Considering the fact that the Brontë family wrote and shared their stories with each other from a very early age [4], this study suggests that their childhood editorial and reading practices shaped their subsequent work. The conflation of their respective authorial voices also suggests that the elder sister might have helped the younger in editing the book.

At the book level, using entire novel as a single long document produced perfect results. All 131 books from known authors are attributed to their true authors and all 12 novels from unknown authors are correctly detected as belonging to unknown authors.

Noteworthy improvements are achieved on essays from periodicals as well. In addition to two articles correctly attributed to known authors in the previous experiment, article 'The Professor - A Tale' under pseudonym G. Gahagan is now correctly attributed to William Thackeray in our model. These changes also classified a previously unindexed article to Frederick Marryat. Marryat is considered an early pioneer of sea story, and the attributed article, 'A Steam Trip to Hamburg', offers a travel narrative of a journey by sea from London to the European continent.

Table 6. False negative distribution using most frequent 2, 000 words as a feature set. Top row represents number/range of misclassified OOD samples. The second and third rows display how many classifiers correspond to the number/range in the top row, using 1000 and 2000 most frequent words, respectively.

FN ranges	0	1	2–5	6–10	$\geq$10
# authors using 1K most frequent words	14	7	12	1	2
# authors using 2K most frequent words	18	9	6	2	1

5   Conclusion and Future Work

In this paper, we took a pragmatic view of computational AA to highlight the critical role it could play in authorship attribution studies involving historical texts. We consider Victorian texts as a case study as many contemporary literary tropes and publishing strategies originate from this period. We demonstrated the strengths and weaknesses of existing computational AA paradigms. Specifically, we show that common English words are sufficient to a greater extent in distinguishing among writings of most renowned authors, especially when AA is performed in the closed-set setup. Experiments under closed-set assumption produced near perfect attribution accuracy in AA task involving 36 authors using only $1,000$ most frequent words. The performance suffered significantly as we switch to the more realistic open-set setup. Increasing the vocabulary size helped to some extent and provided some interesting insights that would challenge results of manual indexing. Open-set experiments also open interesting avenues for future research to investigate whether authors with the same upbringing may develop similar word usage habits as in the case of Brontë sisters. However, overall results from open set experiments confirm the need for a more systematic approach to open-set AA. There are several directions for future exploration following this work.

We believe that attribution accuracy in open-set configuration could improve significantly if word counts are first mapped onto attributes capturing information about themes, genres, archetypes, settings, forms, etc. rather than being directly used in the attribution task. Bayesian priors can be extremely useful to distinguish viable human-developed word usage patterns from those adversarially generated by computers. Similarly, hierarchically clustering known authors and defining meta-authors at each level of the hierarchy can help us more accurately identify writings by unknown authors. The dataset that we have compiled can be enriched with additional essays from Victorian periodicals to become a challenging benchmark dataset and an invaluable resource for evaluating future computational AA algorithms.

References

1. Gdelt project. https://www.gdeltproject.org
2. Gutenberg project. https://www.gutenberg.org
3. Wellesley index. http://wellesley.chadwyck.co.uk
4. Alexander, C.: The Early Writings of Charlotte Brontë. Wiley, Hoboken (1983)
5. Argamon, S., Levitan, S.: Measuring the usefulness of function words for authorship attribution. In: Proceedings of ACH/ALLC Conference (2005)
6. Argamon-Engelson, S., Koppel, M., Avneri, G.: Style-based text categorization: what newspaper am i reading? In: Proceedings of AAAI Workshop on Learning for Text Categorization, pp. 1–4 (1998)
7. Baayen, R., van Halteren, H., Tweedie, F.: Outside the cave of shadows: using syntactic annotation to enhance authorship attribution. Lit. Linguistic Comput. **11**, 121–131 (1996)
8. Bagnall, D.: Author identification using multi-headed recurrent neural networks. arXiv abs/1506.04891 (2015)
9. Bozkurt, I., Baglioglu, O., Uyar, E.: Authorship attribution performance of various features and classification methods. In: 22nd International Symposium on Computer and Information Sciences (2007)
10. Burrows, J.: Not unless you ask nicely: the interpretative nexus between analysis and information. Lit. Linguistic Comput. **7**, 91–109 (1992)
11. Collins, J., Kaufer, D., Vlachos, P., Butler, B., Ishizaki, S.: Detecting collaborations in text: comparing the authors' rhetorical language choices in the federalist papers. Comput. Humanit. **38**, 15–36 (2004). https://doi.org/10.1023/B:CHUM.0000009291.06947.52
12. Curran, E.: Curran index. http://curranindex.org

13. Meyer zu Eissen, S., Stein, B., Kulig, M.: Plagiarism detection without reference collections. In: Decker, R., Lenz, H.-J. (eds.) Advances in Data Analysis. SCDAKO, pp. 359–366. Springer, Heidelberg (2007). https://doi.org/10.1007/978-3-540-70981-7_40

14. Fox, N., Ehmoda, O., Charniak, E.: Statistical stylometrics and the Marlowe-Shakespeare authorship debate. M.A thesis, Brown University, providence, RI (2012)

15. Schaalje, G.B., Fields, P.J.: Open-set nearest shrunken centroid classification. Commun. Stat. Theory Methods **41**, 638–652 (2012)

16. Gamon, M.: Linguistic correlates of style: authorship classification with deep linguistic analysis features. In: Proceedings of the 20th International Conference on Computational Linguistics, pp. 611–617 (2004)

17. Goodwin, G., Howard, E.: Dictionary of National Biography, vol. 28 (1885)

18. Gungor, A.: Benchmarking authorship attribution techniques using over a thousand books by fifty Victorian era novelists. Master thesis (2018)

19. Haberfeld, M., Hassell, A.V.: A New Understanding of Terrorism: Case Studies, Trajectories and Lessons Learned. Springer, Heidelberg (2009). https://doi.org/10.1007/978-1-4419-0115-6

20. Hitschler, J., van den Berg, E., Rehbein, I.: Authorship attribution with convolutional neural networks and POS-eliding. In: Proceedings of the Workshop on Stylistic Variation, pp. 53–58 (2017)

21. Holmes, D.: The evolution of stylometry in humanities scholarship. Lit. Linguistic Comput. **13**, 111–117 (1998)

22. Hu, W.: Study of pauline epistles in the new testament using machine learning. Sociol. Mind **3**, 193–203 (2013)

23. Kim, S., Kim, H., Weninger, T., Han, J.: Authorship classification: syntactic tree mining approach. In: Proceedings of the ACM SIGKDD Workshop on Useful Patterns (2010)

24. Koppel, M., Argamon, S., Shimoni, A.: Automatically categorizing written texts by author gender. Lit. Linguistic Comput. **17**, 401–412 (2002)

25. Koppel, M., Schler, J.: Authorship verification as a one-class classification problem. In: Proceedings of the 21st International Conference on Machine Learning (2004)

26. Koppel, M., Winter, Y.: Determining if two documents are written by the same author. J. Assoc. Inf. Sci. Technol. **65**, 178–187 (2014)

27. McCarthy, P., Lewis, G., Dufty, D., McNamara, D.: Analyzing writing styles with Coh-Metrix. In: Proceedings of the Florida Artificial Intelligence Research Society International Conference, pp. 764–769 (2006)

28. Mendenhall, T.C.: The characteristic curves of composition. Science **9**, 237–249 (1887)

29. Morton, A., Michaelson, S.: The qsum plot, Technical report CSR-3-90, University of Edinburgh (1990)

30. Mosteller, F., Wallace, D.: Inference and Disputed Authorship: The Federalist. Addison-Wesley, Boston (1964)

31. Olsson, J.: Forensic Linguistics: An Introduction to Language, Crime and the Law, 2nd edn. (2008)

32. Peng, F., Shuurmans, D., Wang, S.: Augmenting Naive Bayes classifiers with statistical language models. Inf. Retr. J. **7**, 317–345 (2004)

33. Rhodes, D.: Author attribution with Cnns. Technical report (2015). http://cs224d.stanford.edu/reports/RhodesDylan.pdf

34. Rudman, J.: The state of authorship attribution studies: some problems and solutions. Comput. Humanit. **31**, 351–365 (1998)

35. Schaalje, G.B., Blades, N.J., Funai, T.: An open-set size-adjusted Bayesian classifier for authorship attribution. J. Assoc. Inf. Sci. Technol. **64**, 1815–1825 (2013)
36. Scheirer, W.J., Rocha, A.R., Sapkota, A., Boult, T.E.: Toward open set recognition. TPAMI **35** (2013)
37. Stamatatos, E.: A survey of modern authorship attribution methods. J. Am. Soc. Inf. Sci. Technol. **60**, 538–556 (2009)
38. Thompson, J.R., Rasp, J.: Did C.S. Lewis write the dark tower?: an examination of the small-sample properties of the Thisted-Efron tests of authorship. Austrian J. Stat. **38**, 71–82 (2009)
39. de Vel, O., Anderson, A., Corney, M., Mohay, G.: Mining e-mail content for author identification forensics. SIGMOD Rec. **30**, 55–64 (2001)
40. Wikipedia: Brontë family (2019). https://en.wikipedia.org/wiki/Brontë_family

A More Effective Sentence-Wise Text Segmentation Approach Using BERT

Amit Maraj[(✉)], Miguel Vargas Martin, and Masoud Makrehchi

Ontario Tech University, Oshawa, ON L1G 0C5, Canada
amit.maraj@ontariotechu.net,
{miguel.martin,masoud.makrehchi}@ontariotechu.ca

Abstract. Text Segmentation is a Natural Language Processing based task that is aimed to divide paragraphs and bodies of text into topical, semantic blocks. This plays an important role in creating structured, searchable text-based representations after digitizing paper-based documents for example. Traditionally, text segmentation has been approached with sub-optimal feature engineering efforts and heuristic modelling. We propose a novel supervised training procedure with a pre-labeled text corpus along with an improved neural Deep Learning model for improved predictions. Our results are evaluated with the P_k and WindowDiff metrics and show performance improvements beyond any public text segmentation system that exists currently. The proposed system utilizes Bidirectional Encoder Representations from Transformers (BERT) as an encoding mechanism, which feeds to several downstream layers with a final classification output layer, and even shows promise for improved results with future iterations of BERT.

Keywords: Text segmentation · Natural Language Processing · Natural language understanding

1 Introduction

Text segmentation can often be seen as a preliminary task useful for cleaning and providing meaningful text for other, more downstream Natural Language Processing (NLP) tasks, such as text summarization, relation extraction, natural language understanding, language modelling, etc. Text segmentation is the task of creating clear distinctions between contextual topics within a document. Recognizing the layout of unstructured digital documents is an important step when parsing the documents into a more structured, machine-readable format for downstream applications. Where a document could be some arbitrary length of text such as a chapter in a book or a section in a research paper, a "somewhat supervised" idea of text segments could be paragraphs. The reason this is not completely supervised is because segments within a body of text can be relatively subjective - depending on the reader. Most text segmentation datasets are very clean and include clear distinctions between segments. For example, within

© Springer Nature Switzerland AG 2021
J. Lladós et al. (Eds.): ICDAR 2021, LNCS 12824, pp. 236–250, 2021.
https://doi.org/10.1007/978-3-030-86337-1_16

a play, an act can be considered a text segment, but so can dialogue between speakers. This challenge makes it inherently difficult to appropriately identify text segments, which is usually heavily influenced by the data being used in the system.

The research field of document segmentation is very deep and as of recently with the leverage of novel deep learning methods such as Convolutional Neural Networks (CNN), has made leaps beyond what was previously capable. Although this field is of particular interest for its obvious implications, the research area of text segmentation in the realm of deep learning is still novel and relatively undiscovered.

Historically, text segmentation has proven to be a task best tackled with more unsupervised approaches. This is in part, due to a lack of available training data along with viable hardware to run more complex, resource-hungry supervised systems. Technological advancements and increased research efforts in recent years have facilitated breakthroughs in supervised learning, allowing researchers to innovate in this field with less barriers. This is not to mention the ever increasing volume of data - the essential fuel for these supervised learning techniques.

Our research leverages recent advancements in the field of NLP to improve text segmentation results. In particular, we use BERT as a rich sentence encoder and show that by including a more thoughtful text segmentation focused data augmentation technique, state-of-the-art results in this field can be achieved with minimal training. We contribute a step forward in the text segmentation space with an elementary framework, bolstered by useful propositions for future improvements.

2   Related Works

2.1   Text Segmentation

Text segmentation is a fundamental task in NLP, lending way to various implementations and approaches over the years. At a higher level, the process of text segmentation is often more aligned with the study of topic segmentation, which aims to identify more abstract, discontiguous topics within the document [10,13]. Identified topics can include a variable level of sentences, which are, at an atomic level, the building blocks of said topics. The goal in more recent years has been to adopt a more supervised approach for this task [2,4,15,41]. At a finer level, text segmentation includes the analysis of elementary discourse units, often referred to as EDU segmentation [17]. EDUs are small intent-based units within sentences that serve as building blocks to create a structured sentence. Identifying these units is called discourse parsing and are a part of Rhetorical Structure Theory [38].

There has been a myriad of unsupervised and supervised approaches towards topic segmentation, both of which have very distinct ways of understanding the document at hand. While unsupervised methods require a more globally informed context (i.e., an understanding of the entire document) before making predictions, traditional supervised techniques can make predictions with less overall contextual knowledge of the document, thus requiring less RAM and upfront processing time. Because of these limitations with unsupervised approaches, use within real-time production applications is unpractical. It is worth noting that some of these unsupervised techniques have been very effective, most notably topic modelling [12,20,25,29]. Latent Dirichlet Allocation (LDA) [6] has seen considerable success in the topic modelling domain, where the algorithm can globally analyze a document and understand topics on a sentence-wise basis successfully. Recent research has even seen LDA be more effective when combined with supervised methods [33] in an effort to provide richer features.

On the other hand, supervised techniques have shown to be more increasingly accurate in predicting segment boundaries, due to the improved capabilities of newer machine learning and deep learning propositions, thus challenges such as global document contextuality and understanding have become less important to the performance of said techniques. Recent research done by Kosharek et al. [15] shows that text segmentation can be tackled as a supervised task using Long Short-Term Memory (LSTM) [14] networks. The LSTM is great at capturing contextuality between its inputs and can understand dependencies from previous inputs. Extending upon this, a more recent approach introduces an LSTM variation, where Barrow et al. [4] layers bi-directional LSTMs to encode sentences, create a segment bound predictor, and a segment labeller.

One such model proposed by Badjatiya et al. [2] shows improvement over pre-existing benchmarks by adopting a sliding window approach to contextuality. The researchers show the effectiveness of using just three sentences (one sentence before, one target sentence, and one sentence after) as input for a binary prediction. Rather than approaching text segmentation from a global or individual context, this approach blends the two effectively. For this reason, we use this as a foundation for our work.

2.2 Data Augmentation

Rich text data is hard to come by and the lack of annotated data especially has proved to be challenging for building larger, supervised models. Data augmentation techniques for numerical data is commonplace in data science and can help to bolster the performance of models. Methods such as Synthetic Minority Over-Sampling (SMOTE) [9] can help with imbalanced data, whereas methods such as autoencoding [30] can help build new features by learning deep connections. In the space of NLP, the field of data augmentation has become vast [8].

While traditional data requires numeric-level alteration for effective augmentation and image data requires pixel-level alteration, approaches for NLP differ as augmentation required word-level alteration. Augmentation has proven to be beneficial in the world of NLP through techniques such as lexical substitution [21], vector substitution [36], back translation [40], sentence and word-wise shuffling [37,44], and embedding combinations [43].

We also prove that with thoughtful augmentation, we can boost the performance of our proposed model by preventing it from having to deal with as much class imbalance, and focus on understanding the difference between the positive and negative classes (target sentence vs. non-target sentences).

2.3 Word Embeddings

In 2013, Mikolov et al. [19] introduced Word2Vec, a word embedding method, which changed the NLP landscape for years to come. The Word2Vec architecture leveraged deep learning coupled with a large corpus to provide rich understanding of words. This discovery sparked further research and improvements in this field [7,18,22,23]. The onset of these word embedding techniques were demonstrated to retain semantic understanding between words, which can be seen in Mikolov et al.'s [19] paper. Word2Vec introduced word-level machine understanding by proving these systems with the capability of acknowledging differences between words such as "car" and "truck", but at the same time, know that those two words are somehow related.

Computer Vision has shown deep promise with the idea of transfer learning over the past decade. Devlin et al. [11] introduced BERT, which has thus revolutionized the field of NLP, similar to what ImageNet did for Computer Vision. BERT is an architecture built on top the previously proposed Transformer by Vaswani et al. [34], which provides a general NLP model, trained on roughly 40 gigabytes of diverse text curated from throughout the web. BERT has been shown to yield state-of-the-art results on a number of benchmarks including GLUE [35], MultiNLI [38], and SQuAD [27].

2.4 BERT

In contrast to ELMo, which learns sentence representations by adopting a bidirectional approach to deep learning, Bidirectional Encoder Representations from Transformers (BERT) approaches this task by reading the entire sequence of words at once. BERT accomplishes this by extending the functionality of vanilla transformers, which is an architecture built upon two fundamental mechanisms - an encoder that reads text input and a decoder that produces the prediction for the task. It has been shown in research that BERT is most useful as a preliminary embedding step that can be fine-tuned with downstream layers for specific NLP tasks.

The goal for utilizing BERT in our research was to capture semantic representation of sentences, which would provide high level understanding for further downstream layers to compute our domain-specific task. For sake of brevity, we choose to forego an in-depth technical explanation of BERT. BERT's high dimensional output provides our system with enough contextual features to fuel the downstream architecture.

2.5 DistilBERT

Due to hardware limitations, we elected to use DistilBERT [31], a smaller general purpose language representation model, based off BERT. DistilBERT is 40%

smaller than BERT-base, retains 97% of its language understanding capabilities through various tasks, and is up to 60% faster due to the significant reduction in parameters. DistilBERT proved to be a viable alternative to BERT - it has been shown to provide effective results, comparable to BERT. For brevity, all references to BERT in this paper is using the DistilBERT implementation.

3   Model Architecture

The model we propose consists of roughly 70 millions total parameters with around 3.7 million of those being trainable. This number fluctuates widely based upon chosen hyper-parameters. We go over our chosen hyper-parameters below.

We elect to use BERT as an initial sentence encoder out of the box without fine-tuning. This design choice was made as we found during preliminary experimentation that fine-tuning BERT does not seem to provide additional performance. This is perhaps due to our specific domain-related data being very variable in terms of sentence structure and understanding. For example, one segment may look normal, whereas another may begin with a target sentence such as, "Table:".

Our proposed model's architecture consists of an initial BERT-based (i.e., any BERT or BERT variant) context encoder, followed by a dense layer, a 128 cell Bi-LSTM, an attention layer, and a final dense layer. The architecture can be seen more in Fig. 1.

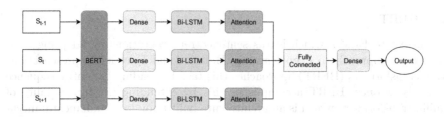

Fig. 1. Model architecture. One BERT structure is used to encode each sentence.

The first hidden dense layer acts as a feature extraction layer, mapping the highly dimensional sentence embedding to a vector with a smaller dimension. The initial dense layer is adopted from the autoencoder architecture, whereby an intermediate hidden layer handles feature extraction by reducing dimensionality. This dense layer has 128 hidden neurons, thus resulting in a reduced size of 128 to represent each sentence instead of 768. This is then sent into a Bi-directional LSTM layer, which captures longer relational structure within the sentence. The bidirectional implementation allows for the retention of important information through the sentence forwards and backwards. The hidden state hyperparameter utilized for the LSTM cells is 128. The attention layers toward the end of the network are utilized to capture the important pieces of the sentences before feeding it into the final dense layer. The attention mechanism used was proposed by Luong et al. [16] as an improvement to the originally introduced attention mechanism by Bahdanau et al. [3]. It is important to emphasize that each of the three streams are independent of each other and do not share weights. The goal is to independently achieve strong sentence encodings that are compared after the fully connected layer.

4   Dataset

We go over a few terms in this section to denote specific text segments and their contributing sentences for ease of reference within our research. We denote a text segment with S, the first sentence within a text segment (i.e., the target sentence) as s_t, and each sentence within a text segment as s_i (i.e., s_1, s_2, ..., s_n).

The dataset created by Badjatiya et al. [2] consists of samples from three very different domains, which were large and diverse enough to train their system - we elect to use these three datasets in our work. The three datasets include the following:

1. **Clinical**: A set of 227 chapters from a medical textbook. Segments are denotes as sections indicated by the author.
2. **Fiction**: A set of 85 fiction books. Each segment is a chapter break within each book.
3. **Wikipedia**: A randomly selected set of 300 documents, widely from the narrative category. The original XML dump denotes categories, which is what's used for each segment.

Due to the supervised nature of our task, careful data preparation was necessary. We decided to work with the same dataset the previous authors curated for their task, as it provided a useful spread across a variety of domains including clinical, fiction, and Wikipedia. Text segmentation as a supervised learning task can be challenging due to the inherently large segment sizes in available corpora. For example, the average segment size for the Wikipedia dataset utilized is 26 sentences long, which is a 1:26 ratio for positive to negative classes. We use a 75%–25% train to test split ratio for training and testing.

4.1   Data Preparation

All of the sentences in our dataset were reduced quite significantly, truncated to 32 words at the maximum. The intuition behind this choice was primarily focused toward prevention of overfitting along with increased performance. We tried sequence lengths of 256, 128, 64, 32, and 16. Results remained constant throughout our testing with a slight increase in precision and recall the smaller the sequence length. These findings are explained in more depth in Sect. 6

We shuffled the data segments at random. To accomplish this, we isolated each segment as its own object, collected all the those objects, and shuffled them in place. We then split the dataset into training and testing subsets. The shuffling of the dataset was necessary to represent topics evenly throughout the training and testing process. In the case where shuffling did not take place, Wikipedia entries of more recent historical events may show up in the training data, while being absent in the training data for example.

4.2 Data Augmentation

The proposed supervised learning system outlined in our research performed extremely poorly upon training with an untouched dataset. Due to the size of data needed to create an effective supervised system, we introduce a data augmentation technique specific to the text segmentation field. Our augmentation technique truncates each text segment at a certain max segment length (MSL), effectively ignoring any sentences past a certain threshold within each segment. For example, if a MSL of five was selected, every text segment within our dataset will consist of 5 or less sentences.

This step was only employed during the training phase of our system. The rationale for the introduction of this approach was to reduce the amount of imbalance between the positive and negative class (e.g., reduction from a 1:29 to a 1:5 for positive to negative classes), thus removing the amount of emphasis naturally put on the negative class. Intuitively, this augmentation approach drastically reduces the size of usable data as the removal of sentences beyond the MSL is ignored. To combat this, we fabricated new segments by associating the sentences after truncation with the previous segment (i.e., fabricated segments will have the same first sentence as the segment immediately prior). See Fig. 2 for a visual representation of how this step works. By utilizing this technique, we were able to create a more balanced dataset, while still retaining representation of sentences and topics throughout.

The inherit challenge with this technique is that it introduces a level of conflict within the training step, whereby every fabricated segment's target sentence will be immediately preceded by a sentence from the same overall text segment. We addressed this by shuffling the dataset after fabricating segments.

The chosen MSLs were arbitrary, but showed that our system could produce promising results at long and short segment sizes. We found that, without our augmented dataset approach, our system began to overfit and became more skewed toward the false prediction (i.e., 0 indicating a prediction of segment continuation, while 1 denoting a sentence being the start of a new segment). This was due to the oversampling of target sentences. For example, an original text segment consisting of 30 sentences will end up having its first sentence seen six times within training through one epoch if we choose an MSL of five. The lack of diversity in these over-sampled target sentences creates a false sense of importance, which our system tends to seek out during inference.

Although our proposed data augmentation technique provided promising results, overfitting is an eventuality. To remedy this, we also decided to explore the efficacy of utilizing our technique without the use of replicating the target sentence. In essence, instead of copying the initial target sentence at the

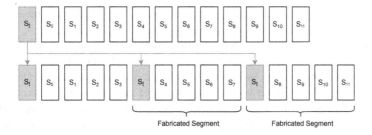

Fig. 2. The real target sentence is copied throughout the course of the segment to be used repeatedly as a fabricated target sentence for the purposes of balancing an imbalanced dataset. We shuffle the resulting segments within the global corpus to ensure sentences immediately prior to fabricated target sentences are sampled from other segments.

beginning of every subsequent segment, we decide to mark the first real sentence as the target sentence for those segments. In short, this method would take a segment of 30 sentences and make it six segments of five sentences, with each segment's first sentence being the original target sentence. This will remove our oversampling problem entirely, but could introduce a misrepresentation of target sentences throughout our dataset.

The idea of subsequently re-sampling the first sentence in every segment poses its own challenges. In a situation where our augmentation technique is used, given an average segment size of 25 sentences and a MSL of five, the beginning sentence can be re-sampled up to 5 times, introducing an obvious overfitting problem. We believe that introducing other NLP augmentation methods on the text segment's first sentence every time it is subsequently sampled (e.g., antonym replacements, synonym replacements, word embedding distances, etc.) can provide enough of a synthetic difference to reduce the overfitting challenge. Qiu et al. [26] shows the effectiveness of some of these augmentation techniques. We hope to explore this regularization technique in future works.

It is worth noting that although our data augmentation technique altered the dataset dramatically, an imbalance was still prevalent. For example, given an MSL of five, the dataset will still have a 5:1 class ratio in favor of the negative class. We chose to overcome this by using a weighted binary cross entropy loss function. Badjatiya et al. [2] showed the effectiveness of utilizing a weighted loss function in their research. This design choice introduced a faster path to overfitting due to the weighting emphasis on the positive classes, which is why we notice a performance ceiling at around 15 epochs.

5   Training

The training procedure seemed to start overfitting after roughly 15 epochs. At 30 epochs, training performance began to flat-line. However, testing performance began to see marginal improvements beyond the 5–10 epoch mark. We believe this is due to the tendency of our system to overfit on the training data over time.

BERT's parameters do not have to be trained. We used the out-of-the-box BERT configuration as our default system due to the enormous dataset and vocabulary it comes pre-trained with. The BERT authors mentioned that simple fine-tuning of 4–5 epochs on your custom dataset will give BERT the ability to understand and make meaningful embeddings. Our preliminary testing with fine-tuning BERT before our downstream model produced less accurate results by a wide margin. We suspect this is due to overfitting on larger sentences. Ultimately, we stopped this endeavor short, but would like to revisit this in future work.

Due to hardware limitations, a larger dataset could not be used. We stuck to the dataset outlined in [2] due to the variety of domain, and intra-domain topics within the respective corpora.

We also explored the use of a weighted binary cross entropy loss function, but it did not enhance the performance. We believe this is due to the lack of real target sentences. We plan on testing the possibility of using a weighted loss function along with more aggressive regularization in our future works.

5.1 Hardware Specifications

Through our tests, a variety of hyper-parameters were chosen and tested. All of the respective models were trained on a GTX 1070 graphics card with 8GB of GDDR5 memory. Each epoch took roughly 400–600 s and every model was trained for 5, 15, 30, and 40 epochs. Epoch training time decreased significantly when the MSL for sentence input was reduced. For reference, with an MSL of 32, each epoch took 100–120 s to complete.

In the future, we believe domain-specific data can be used to fine-tune our model over 5–10 epochs to make viable predictions on inference datasets. With a max sequence length of 32, fine-tuning our system with a dataset of roughly 50,000 samples would take no longer than 15 min on consumer level hardware.

6 Results

It is worth re-emphasizing that the dataset we chose to work with is extremely imbalanced, whereby roughly 95% on average (96% for Clinical, 97% for Wikipedia, and 92% for Fiction) had a ground truth class of 0. This created a large challenge in training our system to recognize and accurately predict these very infrequent text segments (i.e., predicting 1 to denote the beginning of a new segment). This is, of course, indicative of a real-world scenario, whereby a randomly sampled sentence would be much more likely to be part of a segment than the beginning of a segment.

As discussed in Sect. 4.1, the performance with varying max sequence lengths tend to lean in favor of shorter rather than longer. We elect to use 32 as the value for our testing due to the observed performance to training speed tradeoff. In addition, performance tends to degrade as the sequence length grows and training takes significantly longer.

We decided to explore the use of metrics that were more indicative of the actual performance of a text segmentation system instead of the typical accuracy metric. For reference, due to the imbalanced datasets we worked with, a system that predicts 0 for every sentence would receive an accuracy score of roughly 95%. This is of course, not reflective of the actual "learning" and performance of a supervised machine learning system. For this obvious reason, we could not use accuracy as a reliable metric for evaluating our system. We are using WindowDiff [24] and P_k [5] are representative metrics as they have shown to be reliable in the text segmentation space for the past two decades. Pk was originally introduced in 1999 and WindowDiff built upon this algorithm by providing more penalization toward false positive predictions. Due to P_k's leniency towards false positives, WindowDiff scores are usually higher by comparison (higher values indicate poorer performance).

In a task like text segmentation, classifications for the wrong class can be forgiving when it is a false positive. Some results that show our system's eagerness to over-predict can be seen in Table 1. This is untrue however for the mirror case, whereby an imperfect system should ideally be better at predicting the false positive class

We begin to see a somewhat linear growth in results as the window size grows. Suggested in the original [5] P_k paper, the optimal k-value that should be used is half the average segment size - the Wikipedia dataset has an average of 25.97 words. The larger the k-value, the more strict the penalization factor is when observing a false positive within the sliding window. For example, Figs. 3a and 3b illustrate this clearly.

Table 1. Sample results from Wikipedia prediction. The model as trained tends to over-predict. In this case, the model predicted false positives for lines 3–5. We believe this is due to no matching words in the sentences before and after. E.g., for sentence 3, the word "Duryodhana", which might be interpreted as an important feature, is not seen in the sentence before or after. $\hat{y}$ is the predicted value, whereas y is the true value.

Text	$\hat{y}$	y
In Villiputuralvar's 14th-century version, Krishna...	0	0
There is no mention of Duryodhana in this version...	1	0
In other accounts, Aravan is sacrificed in order t...	1	0
In the traditions of the village of Neppattur, in ...	1	0
So Krishna prescribes the human sacrifice of Arava...	0	0
This allows Aravan to make the initial sacrifice o...	0	0
Gattis grew up in Forney, Texas, and began playing...	1	1
His parents divorced when he was eight years old, ...	0	0
Busy playing baseball, Gattis never processed his ...	0	0
Gattis played for the Dallas Tigers, one of the pr...	0	0

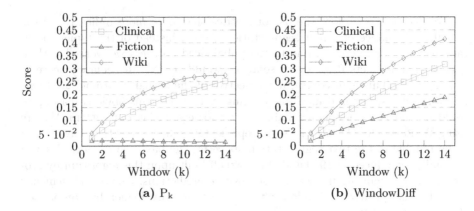

Fig. 3. P_k and WindowDiff Results.

Our results show a considerable improvement over Badjatiya et al.'s [2] work, where their attention based neural model achieved roughly 0.32, 0.34, and 0.38 for clinical, wikipedia, and fiction datasets respectively using the P_k metric. Our results show sub 0.3 results for all these datasets throughout a similar training pattern and identical window sizes.

7 Limitations

All of the models we tested were trained on an Nvidia GTX 1070 graphics card with 8 GB of GDDR5 memory. While we were capable of training the system at a reduced scale, increasing hyper-parameters for longer training stints became challenging. For example, increasing the sequence length from 64 to 128 and the dense layer's hidden neurons would significantly impact the required training time. The LSTM cell size in particular (i.e., increasing from 128 to 256 or 512) inflated the required training time. Utilizing a Gated Recurrent Unit (GRU) cell instead of LSTM was not an option as it does not return the hidden state as it is necessary for the following attention layer.

Due to hardware limitations, dataset size was critical in the training and benchmarking of our system. We elected to work with static datasets with domain-specific verbiage and terminologies. For example, the clinical dataset included medical terms which would usually be considered out-of-vocabulary (OOV) in systems with vocabulary restrictions. Because we are using BERT as our encoding strategy, most of these words are taken care of gracefully by Word-Piece, a sub-word algorithm proposed by Wu et al. [39], which builds off the Byte Pair Encoding (BPE) strategy [32]. Unfortunately, the generic dataset used to pre-train BERT does not include many utterances of these rarer words and as of such, lacks a richer domain-specific linguistic understanding. While the BERT authors have shown that fine-tuning BERT on the dataset being worked with yields favorable results, our preliminary analysis testing this resulted in little to no improvement. A larger dataset and longer training times could be beneficial for fine-tuning BERT in this setting.

8   Future Work

Although our work is very preliminary, it opens the door to a wide variety of future possibilities - most of which we hope to explore. Drawing inspiration from Badjatiya et al.'s [2] work, we kept a 3 sentence approach to our system. We hypothesize that with the increased contextual leverage, a 4 or 5 sentence approach could significantly improve the results of our system. This would deepen the contextual understanding of the system by providing more samples per record. We can see based on Table 1 that in certain situations, important words may not show up in the sentences immediately before or after, but in sentences two orders of magnitude away.

We also believe that introducing other NLP augmentation methods on the text segment's first sentence every time it is subsequently sampled (e.g., back translation, antonym replacements, synonym replacements, etc.) can provide enough of a synthetic difference to reduce the overfitting challenge. Qiu et al. [26] shows the effectiveness of some of these augmentation techniques.

In recent years, lots of breakthrough success has come from multi-modal approaches, whereby a system will take multiple types of inputs. For example, Adobe has shown that novel Document Segmentation success can be achieved by feeding both text and image based data into a system instead of one or the other [42]. We believe that including more sentence topical information as additional input features into the system can help it learn similarities and differences. For example, utilizing Topic Modelling techniques such as LDA in combination with BERT, has been shown to be an effective combination by Shoa [33].

Through a Siamese Network learning technique, more semantically-aware BERT-based sentence vectors can be generated in works done by [28]. Substituting the initial BERT encoding with these vectors not only reduces dimensionality significantly, but also has proven to demonstrate a higher level of semantic understanding. In a similar vein, Angelov et al. [1] created a topic-based implementation using BERT, which has proved to be highly accurate when it comes to identifying topics due to BERT's contextual nature.

9   Conclusion

In conclusion, our research aimed to provide a sentence-wise approach to text segmentation by utilizing modern NLP techniques. We show that BERT and BERT-based architectures (namely DistilBERT) result in measurable improvements within specific domains, given a sufficient amount of training data. Our research shows that our approach holds true for three completely different domains (fiction, wikipedia, clinical). Fiction outperformed the other two domains, giving us hope that our approach can be extended to similar structure-based documents in other domains, such as legal and accounting.

Acknowledgements. The second author thanks the support of an NSERC Discovery Grant.

References

1. Angelov, D.: Top2Vec: distributed representations of topics. arXiv:2008.09470 [cs, stat], August 2020)
2. Badjatiya, P., Kurisinkel, L.J., Gupta, M., Varma, V.: Attention-based neural text segmentation. In: Pasi, G., Piwowarski, B., Azzopardi, L., Hanbury, A. (eds.) ECIR 2018. LNCS, vol. 10772, pp. 180–193. Springer, Cham (2018). https://doi.org/10.1007/978-3-319-76941-7_14
3. Bahdanau, D., Cho, K., Bengio, Y.: Neural machine translation by jointly learning to align and translate. arXiv:1409.0473 [cs, stat], May 2016
4. Barrow, J., Jain, R., Morariu, V., Manjunatha, V., Oard, D., Resnik, P.: A joint model for document segmentation and segment labeling. In: Proceedings of the 58th Annual Meeting of the Association for Computational Linguistics, pp. 313–322. Association for Computational Linguistics, July 2020. https://doi.org/10.18653/v1/2020.acl-main.29, https://www.aclweb.org/anthology/2020.acl-main.29
5. Beeferman, D., Berger, A., Lafferty, J.: Statistical models for text segmentation. Mach. Learn. **34**(1), 177–210 (1999). https://doi.org/10.1023/A:1007506220214
6. Blei, D.M.: Latent Dirichlet Allocation, p. 30
7. Bojanowski, P., Grave, E., Joulin, A., Mikolov, T.: Enriching word vectors with subword information. Trans. Assoc. Comput. Linguist. **5**, 135–146 (2017)
8. Chaudhary, A.: A visual survey of data augmentation in NLP, May 2020. https://amitness.com/2020/05/data-augmentation-for-nlp/
9. Chawla, N.V., Bowyer, K.W., Hall, L.O., Kegelmeyer, W.P.: SMOTE: synthetic minority over-sampling technique. J. Artif. Intell. Res. **16**, 321–357 (2002)
10. Choi, F.Y.Y.: Advances in domain independent linear text segmentation. arXiv:cs/0003083, March 2000
11. Devlin, J., Chang, M.W., Lee, K., Toutanova, K.: BERT: pre-training of deep bidirectional transformers for language understanding. arXiv:1810.04805 [cs], May 2019
12. Eisenstein, J., Barzilay, R.: Bayesian unsupervised topic segmentation. In: Proceedings of the 2008 Conference on Empirical Methods in Natural Language Processing, pp. 334–343. Association for Computational Linguistics, Honolulu, October 2008. https://www.aclweb.org/anthology/D08-1035
13. Hearst, M.A.: TextTiling: a quantitative approach to discourse segmentation. Technical report (1993)
14. Hochreiter, S., Schmidhuber, J.: Long short-term memory. Neural Comput. **9**(8), 1735–1780 (1997)
15. Koshorek, O., Cohen, A., Mor, N., Rotman, M., Berant, J.: Text segmentation as a supervised learning task. In: Proceedings of the 2018 Conference of the North American Chapter of the Association for Computational Linguistics: Human Language Technologies, Volume 2 (Short Papers), pp. 469–473. Association for Computational Linguistics, New Orleans, June 2018. https://doi.org/10.18653/v1/N18-2075, https://www.aclweb.org/anthology/N18-2075
16. Luong, M.T., Pham, H., Manning, C.D.: Effective approaches to attention-based neural machine translation. arXiv:1508.04025 [cs], September 2015
17. Marcu, D.: The Theory and Practice of Discourse Parsing and Summarization. MIT Press, Cambridge (2000).Google-Books-ID: VyjED9VOn5MC

18. McCann, B., Bradbury, J., Xiong, C., Socher, R.: Learned in translation: contextu-alized word vectors. In: Guyon, I., et al. (eds.) Advances in Neural Information Processing Systems 30, pp. 6294–6305. Curran Associates, Inc. (2017). http://papers.nips.cc/paper/7209-learned-in-translation-contextualized-word-vectors.pdf

19. Mikolov, T., Chen, K., Corrado, G., Dean, J.: Efficient estimation of word representations in vector space. arXiv:1301.3781 [cs], September 2013

20. Misra, H., Yvon, F., Jose, J.M., Cappe, O.: Text segmentation via topic modeling: an analytical study. In: Proceedings of the 18th ACM Conference on Information and Knowledge Management. CIKM 2009, pp. 1553–1556. Association for Computing Machinery, New York, November 2009. https://doi.org/10.1145/1645953.1646170

21. Mueller, J., Thyagarajan, A.: Siamese recurrent architectures for learning sentence similarity. In: Proceedings of the AAAI Conference on Artificial Intelligence, vol. 30, no. 1, March 2016. https://ojs.aaai.org/index.php/AAAI/article/view/10350, number: 1

22. Pennington, J., Socher, R., Manning, C.: GloVe: global vectors for word representation. In: Proceedings of the 2014 Conference on Empirical Methods in Natural Language Processing (EMNLP), pp. 1532–1543. Association for Computational Linguistics, Doha, October 2014. https://doi.org/10.3115/v1/D14-1162, https://www.aclweb.org/anthology/D14-1162

23. Peters, M.E., et al.: Deep contextualized word representations. arXiv:1802.05365 [cs], March 2018

24. Pevzner, L., Hearst, M.A.: A critique and improvement of an evaluation metric for text segmentation. Comput. Linguist. **28**(1), 19–36 (2002)

25. Purver, M., Körding, K.P., Griffiths, T.L., Tenenbaum, J.B.: Unsupervised topic modelling for multi-party spoken discourse. In: Proceedings of the 21st International Conference on Computational Linguistics and 44th Annual Meeting of the Association for Computational Linguistics, pp. 17–24. Association for Computational Linguistics, Sydney, July 2006. https://doi.org/10.3115/1220175.1220178, https://www.aclweb.org/anthology/P06-1003

26. Qiu, S., et al.: EasyAug: an automatic textual data augmentation platform for classification tasks. In: Companion Proceedings of the Web Conference 2020. WWW 2020, pp. 249–252. Association for Computing Machinery, New York, April 2020. https://doi.org/10.1145/3366424.3383552

27. Rajpurkar, P., Zhang, J., Lopyrev, K., Liang, P.: SQuAD: 100,000+ questions for machine comprehension of text, June 2016. https://arxiv.org/abs/1606.05250v3

28. Reimers, N., Gurevych, I.: Sentence-BERT: Sentence Embeddings using Siamese BERT-Networks. arXiv:1908.10084 [cs], August 2019

29. Riedl, M., Biemann, C.: TopicTiling: a text segmentation algorithm based on LDA. In: Proceedings of ACL 2012 Student Research Workshop, pp. 37–42. Association for Computational Linguistics, Jeju Island, July 2012. https://www.aclweb.org/anthology/W12-3307

30. Rumelhart, D.E., Mcclelland, J.L.: Parallel Distributed Processing: Explorations in the Microstructure of Cognition, vol. 1. Foundations (1986)

31. Sanh, V., Debut, L., Chaumond, J., Wolf, T.: DistilBERT, a distilled version of BERT: smaller, faster, cheaper and lighter. arXiv:1910.01108 [cs], February 2020

32. Sennrich, R., Haddow, B., Birch, A.: Neural machine translation of rare words with subword units. arXiv:1508.07909 [cs], June 2016

33. Shoa, S.: Contextual Topic Identification: Identifying meaningful topics for sparse Steam reviews, March 2020. Publication Title: Medium

34. Vaswani, A., et al.: Attention is all you need. In: Guyon, I., et al. (eds.) Advances in Neural Information Processing Systems 30, pp. 5998–6008. Curran Associates, Inc. (2017). http://papers.nips.cc/paper/7181-attention-is-all-you-need.pdf

35. Wang, A., Singh, A., Michael, J., Hill, F., Levy, O., Bowman, S.R.: GLUE: a multi-task benchmark and analysis platform for natural language understanding, April 2018. https://arxiv.org/abs/1804.07461v3

36. Wang, W.Y., Yang, D.: That's so annoying!!!: a lexical and frame-semantic embedding based data augmentation approach to automatic categorization of annoying behaviors using #petpeeve tweets. In: Proceedings of the 2015 Conference on Empirical Methods in Natural Language Processing, pp. 2557–2563. Association for Computational Linguistics, Lisbon, September 2015. https://doi.org/10.18653/v1/D15-1306, https://www.aclweb.org/anthology/D15-1306

37. Wei, J., Zou, K.: EDA: easy data augmentation techniques for boosting performance on text classification tasks, January 2019. https://arxiv.org/abs/1901.11196v2

38. Williams, A., Nangia, N., Bowman, S.R.: A broad-coverage challenge corpus for sentence understanding through inference, April 2017. https://arxiv.org/abs/1704.05426v4

39. Wu, Y., et al.: Google's neural machine translation system: bridging the gap between human and machine translation. arXiv:1609.08144 [cs], October 2016

40. Xie, Q., Dai, Z., Hovy, E., Luong, M.T., Le, Q.V.: Unsupervised data augmentation for consistency training. arXiv:1904.12848 [cs, stat], November 2020

41. Yang, H.: BERT meets Chinese word segmentation, September 2019. https://arxiv.org/abs/1909.09292v1

42. Yang, X., Yumer, E., Asente, P., Kraley, M., Kifer, D., Lee Giles, C.: Learning to extract semantic structure from documents using multimodal fully convolutional neural networks, pp. 5315–5324 (2017). https://openaccess.thecvf.com/content_cvpr_2017/html/Yang_Learning_to_Extract_CVPR_2017_paper.html

43. Zhang, H., Cisse, M., Dauphin, Y.N., Lopez-Paz, D.: mixup: beyond empirical risk minimization. arXiv:1710.09412 [cs, stat], April 2018

44. Zhang, X., Zhao, J., LeCun, Y.: Character-level convolutional networks for text classification. arXiv:1509.01626 [cs], April 2016

Data Augmentation for Writer Identification Using a Cognitive Inspired Model

Fabio Pignelli[1]([✉]), Yandre M. G. Costa[1], Luiz S. Oliveira[2],
and Diego Bertolini[1,3]

[1] State University of Maringá, Maringá, PR, Brazil
yandre@din.uem.br
[2] Federal University of Paraná, Curitiba, PR, Brazil
luiz.oliveira@ufpr.br
[3] Federal Technological University of Paraná, Campo Mourão, PR, Brazil
diegobertolini@utfpr.edu.br

Abstract. Assuming that two people do not have the same handwriting and do not write twice identically, handwriting can be considered a biometric characteristic of a person, frequently used in forensic document analysis. In the writer identification and verification scenario, the number of samples available for training is not always sufficient. Thus, in this work, we investigate the impact of the increase in the number of manuscript samples on the writer identification task, using synthetic samples generation inspired on a cognitive model. To better compare the proposed approach, we also performed experiments using a Gaussian Filter to generate new samples. Experiments were accomplished on the IAM and CVL databases, containing samples collected from 301 and 310 volunteer writers, respectively. From genuine samples of these databases, we generate new samples of images using the Duplicator approach. Feature vectors obtained using some well-known texture operators were used to feed an SVM classifier. Experiments showed that increasing the number of synthetic samples in the training set can be beneficial for the writer identification task, particularly when there are very few manuscript samples available for some writers. In the best scenario, the gain increased 31.7% in the identification rate using three genuine and fifteen duplicated samples at line-level.

Keywords: Writer identification · Duplicator · Texture descriptor

1 Introduction

Writer identification and writer verification tasks have already been investigated for some time by the research community, and good performance rates have been reported [15,27]. However, there are still many open issues to be investigated in this field of research. Several works have been published, and, in the

© Springer Nature Switzerland AG 2021
J. Lladós et al. (Eds.): ICDAR 2021, LNCS 12824, pp. 251–266, 2021.
https://doi.org/10.1007/978-3-030-86337-1_17

majority of cases, the authors were pursuing improvements particularly related to specific steps of the writer identification/verification process [29]. Among the different strategies already proposed, we can highlight some well-known ideas in this research field.

The first concept concerns online and off-line methods. These methods differ from each other regarding the way how the manuscript is acquired. In online methods, additional information about the writing movement is recorded, as the direction or the pen pressure [29]. On the other hand, in the off-line mode, only the manuscript image is available [4]. Another aspect is whether the text depends or not on the writer. In text-dependent, all the volunteers who contributed to creating the database are requested to copy a predefined given text [19]. In text-independent, the volunteers have the freedom to create the text content [23,31].

In another vein, some works investigated the effects of the creation of particular zones on the manuscript image [4], the use of different descriptors to capture the content of these images [5], or also the use of data augmentation strategies, aiming to supply the lack of data [11,30]. Different classification schemes have also been studied in other works [29]. Finally, it is still worth mentioning the recent efforts accomplished towards using deep learning techniques aiming to improve these tasks' performance [15].

The challenges behind these tasks have also become bigger. To exemplify that, we can cite the multi-script approaches that have been recently proposed [3]. In the multi-script scenario, the classifier is trained using manuscripts made using one kind of script (i.e. Arabic or Roman). The test is performed using manuscripts written using a different script. In a more challenging scenario, the researchers have been investigating the effects of the availability of scarce data from the writers, like only one paragraph of the manuscript, or only one line, or even a single word [15].

By analyzing the context above, we can observe that this subject has already been exhaustively investigated, thanks to the challenging behavior inherent to it. However, we can still find several minor details inside the more general steps present in the task that can be investigated more deeply. Considering some peculiarities of the task, we decided to make additional efforts to evaluate the impact of some strategies commonly used to deal with the lack of data to perform training and test on the writer identification task. Moreover, the investigations conducted here consider using the "Document Filter" [26] protocol, which enforces that all the data taken from different parts of the same manuscript must be placed either in the training set or in the test set.

Although this seems obvious at first sight, it is important to observe that many handwritten documents available in the databases widely used in this kind of investigation are composed of more than one page. Thus, frequently researchers use different pages from the same document both on training and test sets. Therefore, it is important to emphasize that it is not fair to compare the rates obtained here with the state-of-the-art directly. Our proposal relies on techniques inspired by cognitive models aiming to increase the number of samples included in the training set. Even though we can find in the literature other works that investigated the use of data augmentation techniques to address this issue [2,35,37], none have considered the use of techniques inspired by cognitive models, like those proposed by Diaz et al. [9] in the context of signature verification.

In this work, we assess the use of cognitive models based on human characteristics to generate synthetic samples to increase the training set, aiming at alleviating the negative effects of the lack of samples from writers who contributed to the creation of the manuscript database. Very often, the number of manuscript samples available to train the models is quite small. Real cases do not work on the text-dependent mode, so we may have to deal with small amounts of handwritten content.

In this way, we intend to investigate if the addition of synthetic samples in the training set leads to the improvement of the system's general performance. Besides, we investigate to what extent it can be done favoring the achievement of good results. It is important to highlight that our main objective here is to gauge how we can circumvent the difficulties imposed by the lack of data, and we are not looking for state-of-the-art identification rates. To better compare the proposed approach, we also performed experiments using a data augmentation based on a Gaussian Filter proposed in [24].

In order to make these investigations, we chose two well-known benchmarks, i.e. IAM [23] and CVL [19]. The IAM dataset is quite challenging and has some favorable characteristics considering our investigations, as the text-independent property makes the amount of content from one manuscript to another far from uniform. CVL dataset, on the other hand, was selected since it is a database created on the text-dependent mode. Considering the constraints imposed by the "Document Filter" protocol, we have used only manuscripts from writers who have contributed with at least two manuscript samples during the creation of the database. Therefore, the number of writers available on the IAM database drops from 657 to 301. This is particularly important to avoid a document bias, which could be related to other marginal factors concerning the manuscript, as the pen used for writing, the paper texture, and so on.

This work is organized as follows: Sect. 2 briefly describes the cognitive inspired model; Sect. 3 shows the materials and the proposal itself, including the methodology; Sect. 4 presents the obtained results; and finally, the concluding remarks are presented in Sect. 5.

2   Cognitive Inspired Model - Duplicator

Data augmentation is a well-known approach widely used in several tasks to increase the dataset by synthetically generating new samples based on variations of the original ones. This technique is generally employed in problems where there are few samples per class, such as in the signature verification task, in which usually there are few signatures available for each author to perform the training step.

A large number of approaches to create synthetic samples can be found in the literature [2,11], and some of them have already been applied to signature verification [30]. The rationale behind the approach proposed by Diaz et al. [9] is similar to the basis of the data augmentation concept, that is, generate synthetic samples starting from the (few) samples originally available. However, a Cognitive Inspired Model, in this specific case the Duplicator approach, is based on cognitive models that seek to imitate human behavior to generate handwriting documents. Thus, it is expected that the samples synthetically created can contribute to the generation of more robust and realistic classifier models. As the writer identification task involves a specific type of manuscript, similar to the signature, we investigate the Duplicator approach's use with the same purposes of the typical use of data augmentation methods.

The Cognitive Inspired Model (or Duplicator approach) employed in this work is based on the neuro-motor equivalence theory, which investigates the nervous system's activity to control posture and movement of hands and arms [21]. Some studies with signatures describe the use of effector-dependent, and effector-independent aspects in the handwriting signature [36]. Once the signature and the writing are learned, they become something automatic, requiring motor control and high cognitive skills.

Based on this idea, Diaz et al. [9] presented an algorithm designed to generate new samples of signature handwriting, considering the intra-personal variability. This method was initially proposed to be used with signatures in [9,10,12]. In this work, we propose an extension of that method to be used in the writer identification. The process of generating new samples comprises five steps: segmentation, intra-component variability, component labeling, inter-component variability, and inclination. Besides, a pre-processing step previous to the segmentation is carried out using the images in grayscale. In the segmentation step, noises are removed from the image, and a binarization process is performed. The intra-component variability stage uses a sine wave transformation by distorting the inked portions of the image, aiming to obtain new resulting images with a small variation that can be seen as a variation of the writer's writing. Component labeling performs a search for connected components in the image. It generates L sets of image components, in which L corresponds to the number of regions of connected components found in the image. The inter-component variability stage introduces a spatial variation between the non-connected components. Diaz et al. [9] describe a relationship between the number of connected components and the ink stroke.

Many connected components may suggest that the strokes were drawn quickly or indicate less contact time between the pen and paper. Thus, there may be greater variability between the components. Finally, the inclination modification step simulates the slant variation intra-personal of writing. Slant is a characteristic of the signature and is also present in writing [7]. So, a method used as a parameter the slant variation of a person's writing can generate even better synthetic samples. In this way, it is possible to generate new samples from the original document varying the parameters of the steps previously described, both for signature and handwriting applications. Maruyama et al. [24] uses a search method using Particle Swarm Optimization to define six intra-component variability parameters. Thus, we use six parameters proposed in [24] and the other 24 default parameters as proposed by [9].

As aforementioned, even though the Duplicator technique's original proposal was designed to create synthetic samples of signatures. In this work, we intend to investigate its use for writer identification. Preliminary experiments were accomplished using the document as a whole (the entire letter). However, we observed that samples containing many connected components generated very large intra-personal variation using Duplicator, such a way that the distortion of the writer's writing pattern was visible. In this way, we decided to focus on experiments aiming at creating new samples of text blocks (i.e., pieces of the manuscript), lines, and words. This strategy seems to be quite suitable considering contexts in which there is a severe scarcity of samples, i.e., only one line of the manuscript or a few words.

Figure 1 shows a genuine sample (block) and examples of blocks of a document generated using Duplicator as proposed by Diaz et al. [9]. Figure 1a depicts a genuine handwriting block, and Figs. 1b and 1c are duplicates generated from the original block. We can note that the images are similar, but with slant variations slightly different.

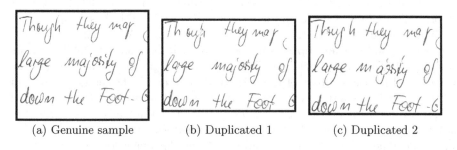

(a) Genuine sample   (b) Duplicated 1   (c) Duplicated 2

Fig. 1. Example of duplicated handwriting generated from a genuine block.

Figure 2 shows the superimposition of genuine and duplicated blocks. By looking at this figure, it is clear that there is a small variation in the samples generated.

In addition, it is worth mentioning that the use of duplicator allows to generate different samples from a given sample, once there are 30 parameters that can be ranged ensuring the creation of samples inside the intra-personal variation.

3 Proposed Methodology

In this section, we describe our proposed methodology. Subsection 3.1 describes the databases used in our experiments. In Subsect. 3.2 we introduce the feature extraction methods. A brief description of the Gaussian filter used to compare our results is presented in Subsect. 3.3. In Subsect. 3.4 the classification framework is detailed.

Fig. 2. The superimposition using one genuine document and nine synthetic samples.

3.1 Databases

Currently, there are several databases composed of offline handwritten documents, such as ICDAR 2013 dataset [22], BFL [13], and Firemaker [31]. In this work, we employ two widely-known public databases, the IAM [23] (Institut fur Informatik und angewandte Mathematik), and the CVL-Database [19].

The manuscripts of the IAM database were written in American English, and they correspond to transcriptions made by 657 writers (from 1 to 59 documents per writer). To comply with the "Document Filter" protocol requirements, we were forced to use a subset of the original database. All the manuscripts from writers who had contributed with a single sample were discarded. The final subset was then composed of two manuscript samples randomly chosen from the 301 remaining writers, who had contributed with at least two samples. Therefore, we use document one for training and the second document for testing, and the contrariwise. One of the biggest challenges on this base is the variation in the amount of written text (text-independent), since many writers have only three to five text lines in total.

The CVL database, proposed by Kleber et al. [19], is composed of text transcriptions taken from Wikipedia and literary works worldwide known, such as, "Origin of Species" and "Mac Beth". Each document in the database is an excerpt of text from these works. In this database, 310 writers contributed with five or seven documents (284 writers handed in four documents, and 27 writers handed in seven documents). The documents were scanned in color images with a resolution of 300 dpi [19]. In this work, we use two from the five samples available. The CVL is a text-dependent database. Thus, all the samples necessarily have the same content.

3.2 Feature Extraction

The feature extraction step was performed in this work considering both the handcrafted and the non-handcrafted forms. On the handcrafted scenario, a large number of features has already been experimented to address the writer identification task (e.g., LBP [4], Textural and Allographic Features [5], and SURF [32]). Here, we decided to use some features which have already demonstrated a good performance in similar scenarios: SURF [1], BSIF [17], EQP [25], LETRIST [34] and LDN [28].

Regarding the non-handcrafted features, three well-known CNN architectures were experimented: VGGNet [33], ResNet [14] and MobileNet [16]. For this purpose, pretrained CNNs were employed as feature extractors, and following these feature vectors were used to train a general classifier. More details about those descriptors and codes can be found in the respective references. In Table 1, we describe the parameter settings used here to generate each of those descriptors.

Table 1. Features dimensions and main parameters.

Feature	Parameters	Dimensions
SURF	$SurfSize = 64$	257
BSIF	$filter = ICAtextureFilters\text{-}11 \times 11\text{-}8bit$	256
EQP	$loci = ellipse$	256
LETRIST	$sigmaSet = 1, 2, 4;\ noNoise$	413
LDN	$mSize = 3;\ mask = kirsch;\ \sigma = 0.5$	56
VGG16	As done by Simonyan and Zisserman [33]	512
ResNet50	As done by He et al. [14]	2048
MobileNet	As done by Howard et al. [16]	1280

3.3 Gaussian Filter

Maruyama et al. [24] evaluated the use of Gaussian filter as a data augmentation strategy on the context of handwritten signature identification. This strategy can also be used when the quantity of samples is not enough to properly train the classification models [8,20]. Gaussian filter is based on the idea of adding random noises in the feature space aiming at creating new samples. The parameter σ determines the intensity of the Gaussian filter. As in the work of Maruyama et al. [24], we chose this value, considering the uniform distribution between 0.29 and 0.72 for σ_{min} and σ_{max}, respectively.

3.4 Classification Schema

In this section, we describe the experimental protocol used to perform the classification in this work. We adopted a protocol similar to that used in real cases, in which only genuine handwriting can be used in the test set (that is, duplicated samples must be exclusively placed into the training set). Besides, the genuine samples used to generate duplicated samples cannot be used in the testing set.

On the IAM database, we used only the manuscripts of writers who contributed with at least two documents, summing up 301 writers [26]. We randomly selected two documents per writer when there were more than two documents available. We adopted this approach because, if we were to use all 657 writers, 356 of them have only one document, making training and testing using different documents impossible. Pinhelli et al. [26] describe some impacts that may artificially boost the performance when different parts of the same manuscript are used both in training and test sets. For the CVL database, two documents in English from 310 writers were used. As it is a balanced database, that is, all writers have at least five documents, it was possible to use all writers. From each writer, we chose to use his/her first and third manuscript. Figure 3 illustrates the classification schema used in this work.

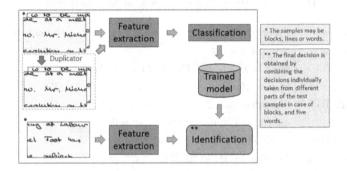

Fig. 3. Classification schema.

We have also evaluated the use of blocks of text, lines, and words from the writer's handwritten documents. For the blocks, the blank spaces at the top, bottom, left, and right margins were discarded. Thus, each manuscript was divided into nine blocks of size $(m \times n)$, where m and n are, respectively, the width and height size from preprocessed manuscript divided by 3. In the experiments using lines, we have taken the first three lines of the document, because there are many documents with only three lines on the IAM database. The same protocol was employed on the CVL database for the same reason. On the experiments conducted using words, we selected the five biggest words from the text, taking into account the area (i.e., width × height) of the region occupied by the word in terms of pixels.

To properly evaluate the impact of duplicated samples in the training set, we investigated two scenarios: (i) Using only genuine samples in the training set, as a baseline; (ii) Using both genuine and its duplicated samples in the training set. In all cases, we used only genuine samples in the test set. Thus, from each of these genuine documents, feature vectors were computed using different texture descriptors. In this case, one document was used on the training set, and another one on the test set, and the contrariwise. As aforementioned, in the second scenario (ii), genuine and synthetic samples were used for training. Our goal here is to evaluate if duplicated samples can positively contribute to performance improvement. The generation of duplicated samples from a handwritten document presented very large distortions regarding the intra-class patterns homogeneity. Because of this, we made duplicates specifically from blocks, lines, and words of text. Experiments with blocks, lines, and words showed no large intra-personal distortion.

The classification was performed using Support Vector Machine (SVM), since it has already shown good performance in related works [4,29]. We use the LibSVM implementation [6] with the configuration of hyper-parameters made by grid search. As the main goal in this work is to perform the writer identification task for a given document as a whole, it was necessary to combine the individual scores, when applicable. For this purpose, we used the Sum Rule [18], which has already demonstrated good results in the literature [4].

4   Results and Discussion

In all experiments, the SVM classifier was used with a Gaussian kernel with parameters C and γ optimized using grid search. The identification rates were obtained using accuracy. In some cases, we present results expressed in terms of TOP-1 (i.e. the same as accuracy in this context), TOP-5, and TOP-10. This means that a hit list will be considered correct if at least one version of the queried specimen appears on it.

For block-level experiments, we used nine blocks taken from each document, both for training and test sets. For line-level experiments, we always use $N = \{1..3\}$ lines for the training set and one line from another document to perform the test. In the case of word-level experiments, we use $N = \{1..5\}$ words for the training set and the test is performed with one or with five words taken from another document. In our first experiment, we are interested in assessing the impacts of the use of feature extractors starting only from genuine samples. In this case, we aim at identifying the most promising feature extractors to use in the further experiments to be described here. The results are described in Table 2.

As presented in Table 2, SURF and BSIF presented the best results. Hence, they were chosen to perform the next experiments. As far as we know, BSIF is used to address the writer identification task in this work for the first time. We also did not achieve interesting rates using pre-trained CNN. In fact, these pre-trained architectures using a generic database (ImageNet) may be inefficient for this task.

In the following set of experiments, we use both scenarios previously described in Sect. 3.4. We use SURF and BSIF descriptors in all experiments since these descriptors presented the best results in the preliminary experiments. Experiments were conducted on the IAM and CVL databases, and the manuscript images were explored on a block-level mode, with nine blocks created from each document. The results are shown in Table 3, where "1orig" stands for

Table 2. Accuracy (%) using nine genuine blocks both for training and test.

Descriptor	IAM	CVL
SURF	87.54 (±1.17)	80.17 (±6.16)
BSIF	**90.53 (±1.17)**	**85.16 (±2.28)**
EQP	79.90 (±0.24)	75.81 (±4.11)
LDN	43.19 (±1.88)	68.07 (±0.91)
LETRIST	44.85 (±2.35)	67.42 (±3.20)
VGG16	51.30 (±0.07)	73.38 (±2.96)
ResNet50	61.60 (±0.30)	66.61 (±3.42)
MobileNet	69.70 (±0.04)	74.52 (±5.47)

nine genuine blocks, and "*N*dupl" stands for $N = \{0, 1, 2, 3\}$ duplicates of each genuine block. Experiments at level-blocks employed nine blocks in the test set and the Sum Rule as a fusion scheme. The results presented were obtained by making an average between the shots performed considering the cross-validation protocol. The standard deviations are also shown in brackets.

Table 3. Identification rates (%) varying the number of blocks for training.

Feat.	#Docs	IAM				CVL		
		Top-1	Top-5	Top-10		Top-1	Top-5	Top-10
SURF	1orig0dupl	87.54 (±1.17)	95.02 (±0.47)	96.01 (±0.00)	–	80.17 (±6.16)	95.81 (±0.46)	97.42 (±0.00)
	1orig1dupl	90.04 (±0.94)	95.35 (±0.00)	**96.18 (±0.24)**	–	85.17 (±4.56)	96.46 (±0.91)	97.74 (±0.45)
	1orig2dupl	**91.20 (±4.00)**	95.68 (±0.47)	95.85 (±0.23)	–	85.97 (±3.87)	**97.10 (±0.00)**	97.91 (±0.69)
	1orig3dupl	89.70 (±2.81)	**95.85 (±0.71)**	**96.18 (±0.24)**	–	**86.29 (±3.42)**	96.78 (±0.46)	**98.07 (±0.46)**
BSIF	1orig0dupl	90.53 (±1.17)	95.35 (±0.47)	95.85 (±0.23)	–	85.16 (±2.28)	94.84 (±0.45)	96.61 (±0.23)
	1orig1dupl	92.36 (±0.47)	**95.52 (±0.70)**	**96.18 (±0.24)**	–	89.84 (±1.15)	95.48 (±0.00)	96.61 (±0.23)
	1orig2dupl	92.53 (±0.23)	**95.52 (±0.23)**	96.02 (±0.47)	–	90.81 (±0.23)	**95.97 (±0.23)**	96.77 (±0.00)
	1orig3dupl	**92.69 (±0.00)**	**95.52 (±0.23)**	96.01 (±0.00)	–	**91.46 (±0.69)**	95.81 (±0.46)	**96.94 (±0.23)**

Although slightly better rates were obtained using duplicated blocks, we do not claim that these results are better once their difference does not seem to be statistically significant.

In the next set of experiments, we assess the writer identification task on the line-level. In this way, we evaluate the impact of varying the number of lines used to feed the classifier training and the effect of duplicated lines. For this, we assessed the use of one, two, and three genuine lines, and we created five, ten, and fifteen duplicates samples from each of these lines. In all experiments at line-level, was used a single line in the test set. We used only genuine lines to create a baseline, and at least one original sample takes part in the training.

The results are described in Table 4. As we can note, the rates increased significantly when duplicates were added to the training set. We may also conclude that Duplicator's use is recommended when many synthetic samples (e.g.,

45 samples) are generated. With bigger datasets, we could obtain identification rates that probably would not be achieved using only the genuine samples. As we can observe by analyzing the results presented in Table 4, the use of duplicated samples led to better performance rates in the vast majority of cases. Considering the Top-1 identification, the accuracy increased approximately 32% points (from 38.54 to 70.27) on the IAM dataset using the BSIF descriptor and three original plus fifteen duplicated samples.

It is worth mentioning that the identification rates achieved when we use one genuine line and fifteen duplicates on IAM and CVL databases increased compared to its baselines. Still, in most cases, it did not overcome the use of three genuine lines alone. Although these results do not overcome those obtained using three genuine lines, we must consider that it is quite common to face situations like that in a real-world scenario, in which we do not have three genuine samples available. We highlighted the best rates for each descriptor in bold. Figure 4 shows, using bar graphs, the remarkable performance of duplicator using three lines in test set. In this experiment, we generated the model with three original lines and fifteen duplicated line samples. We can note that, in all cases, by increasing the number of samples in test set, we get better performance.

Table 4. Identification rates (%) varying the number of lines for training.

Feat.	#Orig	#Dupl	IAM				CVL		
			Top-1	Top-5	Top-10		Top-1	Top-5	Top-10
SURF	1orig	0dupl	21.59 (±4.23)	49.84 (±5.64)	63.96 (±7.76)	–	22.90 (±0.46)	45.48 (±0.00)	56.29 (±0.68)
		5dupl	33.55 (±3.76)	62.79 (±4.70)	74.59 (±6.81)	–	27.74 (±2.74)	57.42 (±3.20)	70.49 (±0.23)
		10dupl	35.88 (±1.88)	66.28 (±4.94)	75.92 (±4.93)	–	30.97 (±4.56)	60.00 (±4.10)	72.10 (±0.69)
		15dupl	37.87 (±2.82)	67.11 (±2.81)	76.91 (±3.99)	–	34.35 (±4.79)	62.26 (±2.74)	72.42 (±0.68)
	2orig	0dupl	32.72 (±3.05)	65.62 (±0.70)	76.91 (±1.17)	–	26.77 (±0.46)	52.26 (±4.11)	63.23 (±3.65)
		5dupl	50.50 (±0.00)	77.91 (±2.58)	83.72 (±1.41)	–	43.71 (±3.88)	72.10 (±2.96)	81.78 (±2.51)
		10dupl	52.16 (±1.88)	78.24 (±1.17)	84.89 (±0.70)	–	44.84 (±4.11)	74.52 (±3.65)	83.55 (±2.74)
		15dupl	54.65 (±1.64)	79.40 (±0.47)	86.05 (±0.94)	–	46.61 (±2.51)	74.36 (±2.96)	83.87 (±1.82)
	3orig	0dupl	47.51 (±1.41)	79.07 (±4.23)	85.88 (±2.59)	–	37.90 (±2.05)	59.68 (±0.46)	68.87 (±0.68)
		5dupl	56.64 (±2.58)	79.74 (±0.47)	84.39 (±0.94)	–	51.13 (±3.42)	77.26 (±3.87)	85.16 (±2.28)
		10dupl	58.97 (±1.17)	81.73 (±2.82)	87.54 (±1.64)	–	54.19 (±3.65)	79.68 (±2.28)	86.62 (±1.14)
		15dupl	**61.13** (±1.41)	**82.40** (±1.41)	**88.54** (±0.71)	–	**54.68** (±2.51)	**80.17** (±0.69)	**86.94** (±1.14)
BSIF	1orig	0dupl	26.58 (±1.88)	54.99 (±1.65)	64.29 (±1.65)	–	23.23 (±0.00)	46.13 (±0.91)	57.91 (±2.51)
		5dupl	39.70 (±2.11)	69.94 (±3.53)	78.91 (±1.65)	–	32.58 (±2.28)	63.71 (±2.97)	74.03 (±5.25)
		10dupl	44.02 (±3.52)	71.93 (±2.58)	81.73 (±0.47)	–	38.39 (±0.46)	64.84 (±4.10)	74.52 (±5.93)
		15dupl	44.68 (±3.52)	73.09 (±3.76)	82.23 (±1.17)	–	38.87 (±1.14)	65.16 (±5.47)	74.52 (±5.93)
	2orig	0dupl	34.22 (±0.94)	60.80 (±2.35)	72.59 (±1.64)	–	30.97 (±0.46)	53.23 (±0.46)	62.91 (±2.74)
		5dupl	55.32 (±2.11)	80.90 (±2.58)	88.04 (±0.93)	–	50.81 (±7.53)	75.49 (±5.01)	85.48 (±3.65)
		10dupl	59.30 (±2.11)	83.56 (±1.18)	90.04 (±0.94)	–	56.13 (±5.47)	80.00 (±5.93)	86.94 (±4.79)
		15dupl	59.80 (±1.41)	84.06 (±1.41)	89.37 (±0.00)	–	56.77 (±8.67)	80.16 (±6.62)	86.94 (±3.88)
	3orig	0dupl	38.54 (±2.82)	69.77 (±0.94)	79.74 (±0.47)	–	37.58 (±6.61)	66.13 (±7.75)	76.94 (±3.42)
		5dupl	64.95 (±0.70)	84.55 (±0.71)	**90.70** (±0.47)	–	61.94 (±3.19)	**84.04** (±2.51)	**90.33** (±2.28)
		10dupl	66.11 (±0.00)	85.88 (±1.64)	90.53 (±1.64)	–	**62.26** (±4.56)	83.87 (±3.20)	89.52 (±2.51)
		15dupl	**70.27** (±2.58)	**87.88** (±3.53)	91.36 (±2.35)	–	60.32 (±3.19)	84.36 (±4.33)	89.36 (±0.91)

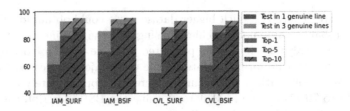

Fig. 4. Performance using one and three lines in test set.

Next, we describe the experimental results obtained using words. For that purpose, we used the biggest words of each manuscript from the IAM and CVL datasets. As done in the experiments using lines, we selected N genuine samples to define a baseline, and then we added five and ten duplicates in the training set. The test was performed using a single word and five words. The results using IAM dataset are presented in Table 5 and CVL dataset in Table 6, described using Top-1, Top-5 and Top-10. As we can see, it is possible to take advantage of duplicates created from words in some specific scenarios. Its use worths it or not depending on the availability/scarcity of manuscript content.

Table 5. Performance varying the number of words for training - IAM Database

Feat	#Orig	#Dupl	Test in 1 genuine word				Test in 5 genuine words		
			Top-1	Top-5	Top-10		Top-1	Top-5	Top-10
SURF	1orig	0dupl	5.48 (±1.17)	13.29 (±0.94)	20.93 (±0.47)	–	6.65 (±1.87)	18.60 (±0.00)	27.74 (±1.17)
		5dupl	4.99 (±0.47)	19.77 (±0.23)	30.90 (±2.35)	–	10.80 (±0.71)	32.23 (±1.41)	45.85 (±0.94)
		10dupl	5.98 (±0.93)	22.76 (±1.65)	33.72 (±5.87)	–	12.63 (±0.47)	37.21 (±4.23)	49.50 (±3.76)
	5orig	0dupl	21.43 (±4.46)	44.85 (±2.81)	59.14 (±3.75)	–	48.67 (±1.64)	78.08 (±1.41)	86.05 (±0.00)
		5dupl	25.75 (±3.05)	49.84 (±0.94)	60.14 (±0.47)	–	48.67 (±1.17)	76.58 (±0.71)	86.55 (±2.58)
		10dupl	27.41 (±4.46)	51.66 (±1.17)	61.96 (±0.24)	–	49.67 (±3.05)	80.07 (±0.47)	87.38 (±0.94)
BSIF	1orig	0dupl	12.62 (±1.88)	30.23 (±0.47)	39.70 (±3.05)	–	17.44 (±0.71)	40.70 (±1.17)	50.34 (±2.11)
		5dupl	17.44 (±1.17)	36.21 (±0.47)	46.51 (±4.23)	–	25.42 (±0.23)	50.67 (±0.23)	65.29 (±2.11)
		10dupl	18.28 (±0.94)	38.87 (±1.88)	49.34 (±1.65)	–	30.90 (±2.82)	55.15 (±2.35)	67.94 (±1.17)
	5orig	0dupl	33.22 (±6.58)	62.13 (±7.52)	71.27 (±6.34)	–	60.13 (±6.58)	85.05 (±0.47)	89.87 (±0.23)
		5dupl	46.35 (±0.70)	67.78 (±1.41)	76.41 (±1.88)	–	72.92 (±5.40)	90.37 (±0.47)	94.02 (±0.00)
		10dupl	**49.34** (±1.17)	69.11 (±0.47)	77.58 (±0.70)	–	**74.59** (±5.41)	90.87 (±0.70)	94.52 (±0.23)

To evaluate the Duplicator's performance as a data augmentation approach, we compare it to the Gaussian filter strategy. We performed the comparison considering three scenarios: block, line, and word level. These experiments' main goal is to test if the use of Duplicator pays off or not when faced with a simpler data augmentation approach.

In the block-level scenario, we used models generated with nine genuine blocks and one duplicate from each document block. For evaluating the Gaussian Filter, the model was created using the same nine genuine blocks, and one synthetic sample for each block. In the line-level scenario, the models were generated using three genuine lines and fifteen duplicates from each of these lines for

Table 6. Performance varying the number of words for training - CVL Database

Descriptor	#Orig	#Dupl	Test in 1 genuine word				Test in 5 genuine words		
			Top-1	Top-5	Top-10		Top-1	Top-5	Top-10
SURF	1orig	0dupl	5.16 (±0.45)	12.75 (±1.14)	21.94 (±0.46)	–	6.61 (±1.60)	18.39 (±3.65)	27.10 (±3.65)
		5dupl	5.97 (±0.69)	19.84 (±2.05)	29.20 (±1.14)	–	8.07 (±0.46)	26.13 (±0.45)	38.87 (±0.68)
		10dupl	7.42 (±0.45)	21.94 (±2.74)	31.61 (±3.65)	–	10.65 (±2.28)	29.03 (±1.82)	42.42 (±0.68)
	5orig	0dupl	12.26 (±0.91)	29.04 (±0.91)	37.58 (±2.05)	–	27.42 (±3.20)	46.94 (±4.33)	57.58 (±4.79)
		5dupl	15.00 (±1.15)	36.13 (±1.37)	48.39 (±1.82)	–	33.88 (±2.74)	62.10 (±2.96)	73.07 (±2.51)
		10dupl	17.10 (±0.46)	36.94 (±1.14)	50.97 (±3.20)	–	36.62 (±2.96)	61.62 (±4.11)	73.55 (±0.00)
BSIF	1orig	0dupl	8.07 (±2.74)	22.91 (±0.91)	36.13 (±0.91)	–	11.13 (±2.97)	28.87 (±5.25)	42.74 (±5.70)
		5dupl	11.46 (±0.69)	32.26 (±2.28)	43.87 (±3.65)	–	16.13 (±1.37)	41.94 (±0.46)	55.16 (±0.00)
		10dupl	12.42 (±1.60)	33.87 (±3.20)	45.65 (±4.33)	–	19.20 (±0.69)	46.61 (±0.23)	60.81 (±1.60)
	5orig	0dupl	22.42 (±3.42)	49.52 (±2.96)	60.65 (±4.56)	–	47.42 (±3.20)	77.10 (±2.28)	85.65 (±0.23)
		5dupl	28.87 (±1.15)	56.29 (±3.87)	66.29 (±3.42)	–	64.03 (±3.42)	83.71 (±2.50)	90.00 (±2.28)
		10dupl	**32.58** (±2.28)	57.10 (±5.02)	68.39 (±6.39)	–	**64.36** (±3.88)	83.87 (±2.28)	90.33 (±2.28)

the Duplicator Approach. Regarding Gaussian Filter's use, the models were created using three genuine lines and fifteen synthetic samples from each of these lines. In the word-level scenario using Duplicator Approach, the models were generated using five genuine words and five duplicates from each of these words. We compared these models to the one created using five genuine words and five Gaussian filter synthetic samples gotten from each word. In all cases, we used the test set with genuine samples from a document that had not been used in the training set. In Table 7, the column "Test" describes the number of samples used in the test set. The Sum Rule was used to combine the classifier's outputs with more than one document in test set.

We show experiments comparing the Duplicator with the Gaussian filter in Table 7. The results are presented using Top-1, Top-5, and Top-10. In most

Table 7. Comparison between Duplicator and Gaussian Filter

Scenario	Test	Feature	Data	Duplicator				Gaussian Filter		
				Top-1	Top-5	Top-10		Top-1	Top-5	Top-10
Blocks	9 blocks	SURF	IAM	90.04 (±0.94)	95.35 (±0.00)	96.18 (±0.24)	–	88.87 (±1.64)	94.85 (±0.24)	96.18 (±0.24)
			CVL	85.17 (±4.56)	96.46 (±0.91)	97.74 (±0.45)	–	83.71 (±5.70)	95.81 (±0.91)	97.10 (±0.00)
		BSIF	IAM	92.36 (±0.47)	95.52 (±0.70)	96.18 (±0.24)	–	92.53 (±0.70)	95.02 (±0.00)	96.18 (±0.24)
			CVL	89.84 (±1.15)	95.48 (±0.00)	96.61 (±0.23)	–	88.87 (±0.68)	95.81 (±0.46)	96.45 (±0.45)
Lines	1 line	SURF	IAM	61.13 (±1.41)	82.40 (±1.41)	88.54 (±0.71)	–	55.48 (±3.29)	78.91 (±4.46)	86.71 (±2.35)
			CVL	54.68 (±2.51)	80.17 (±0.69)	86.94 (±1.14)	–	44.84 (±5.02)	70.33 (±0.91)	80.81 (±0.69)
		BSIF	IAM	70.27 (±2.58)	87.88 (±3.53)	91.36 (±2.35)	–	61.79 (±1.41)	78.24 (±1.17)	84.55 (±1.17)
			CVL	60.32 (±3.20)	84.36 (±4.33)	89.36 (±0.91)	–	52.58 (±3.19)	76.13 (±2.28)	82.42 (±0.23)
	3 lines	SURF	IAM	78.74 (±2.35)	93.03 (±0.47)	95.18 (±0.71)	–	72.26 (±3.52)	91.20 (±1.17)	93.52 (±1.64)
			CVL	69.04 (±6.39)	88.39 (±3.65)	92.91 (±0.46)	–	60.33 (±8.21)	83.39 (±4.79)	89.52 (±1.60)
		BSIF	IAM	85.55 (±5.40)	94.19 (±0.23)	95.52 (±0.23)	–	80.90 (±1.65)	90.54 (±0.23)	93.19 (±0.24)
			CVL	75.17 (±2.74)	90.00 (±0.00)	92.90 (±0.00)	–	69.84 (±3.42)	86.78 (±2.74)	90.81 (±1.60)
Words	1 word	SURF	IAM	25.75 (±3.05)	49.84 (±0.94)	60.14 (±0.47)	–	19.43 (±1.65)	43.86 (±0.47)	56.15 (±1.88)
			CVL	15.00 (±1.15)	36.13 (±1.37)	48.39 (±1.82)	–	12.10 (±1.60)	30.32 (±1.37)	42.42 (±1.15)
		BSIF	IAM	46.35 (±0.70)	67.78 (±1.41)	76.41 (±1.88)	–	38.21 (±3.29)	63.96 (±1.65)	74.59 (±3.53)
			CVL	28.87 (±1.15)	56.29 (±3.87)	66.29 (±3.42)	–	23.71 (±0.23)	50.00 (±2.28)	64.03 (±3.42)
	5 words	SURF	IAM	48.67 (±1.17)	76.58 (±0.71)	86.55 (±2.58)	–	39.04 (±3.05)	69.44 (±3.29)	79.57 (±3.52)
			CVL	33.88 (±2.74)	62.10 (±2.96)	73.07 (±2.51)	–	28.55 (±1.14)	57.26 (±2.96)	68.71 (±0.91)
		BSIF	IAM	72.92 (±5.40)	90.37 (±0.47)	94.02 (±0.00)	–	67.28 (±0.70)	87.21 (±0.24)	91.70 (±0.47)
			CVL	64.03 (±3.42)	83.71 (±2.50)	90.00 (±2.28)	–	54.19 (±1.82)	79.36 (±2.28)	86.94 (±1.14)

cases, Duplicator performed better. Considering Top-1, Top-5, and Top-10, the Duplicator approach proved to be superior to other methods in the vast majority of cases. In the scenario using lines, the Duplicator showed its best contribution, with an improvement of at least 13% points. Besides that, in one case, the improvement reached 31% points.

5   Concluding Remarks and Future Directions

This work evaluates a cognitive model inspired by human behavior as a data augmentation strategy for handwritten samples in writer identification task. We investigate whether duplicated manuscripts can cooperate with genuine ones available, aiming to overcome the lack of manuscript samples.

The experiments carried out show that the Duplicator has a good capacity to simulate samples from writers. Results obtained using Duplicator performed better than the baseline and also than the Gaussian filter. In our experiments using blocks, lines, and words, we can conclude that the Duplicator approach can be an interesting method when the content per writer is limited. We also noticed that the Duplicator showed greater gains at the line-level.

The most successful use case of the Duplicator occurred when three genuine lines and fifteen duplicates were combined. The result increased by 31% points compared to the use of only the three original genuine lines using BSIF descriptor, and 13% points in the same condition using SURF descriptor. This indicates that, in cases where there are few samples per writer, the Duplicator's use may provide a significant performance improvement. In future, we intend to evaluate fine-tuning in pre-trained models and other techniques inspired by cognitive models.

Acknowledgment. We thank the Brazilian research support agencies: Coordination for the Improvement of Higher Education Personnel (CAPES), and National Council for Scientific and Technological Development (CNPq) for their financial support.

References

1. Bay, H., Tuytelaars, T., Van Gool, L.: SURF: speeded up robust features. In: Leonardis, A., Bischof, H., Pinz, A. (eds.) ECCV 2006. LNCS, vol. 3951, pp. 404–417. Springer, Heidelberg (2006). https://doi.org/10.1007/11744023_32
2. Beresneva, A., Epishkina, A.: Data augmentation for signature images in online verification systems. In: Tavares, J.M.R.S., Dey, N., Joshi, A. (eds.) BIOCOM 2018. LNCVB, vol. 32, pp. 105–112. Springer, Cham (2020). https://doi.org/10.1007/978-3-030-21726-6_10
3. Bertolini, D., Oliveira, L.S., Sabourin, R.: Multi-script writer identification using dissimilarity. In: 2016 23rd International Conference on Pattern Recognition (ICPR), pp. 3025–3030 (2016)
4. Bertolini, D., Oliveira, L.S., Justino, E., Sabourin, R.: Texture-based descriptors for writer identification and verification. Expert Syst. Appl. **40**(6), 2069–2080 (2013)

5. Bulacu, M., Schomaker, L.: Text-independent writer identification and verification using textural and allographic features. IEEE Trans. Pattern Anal. Mach. Intell. **29**(4), 701–717 (2007)
6. Chang, C.C., Lin, C.J.: LIBSVM: a library for support vector machines. ACM Trans. Intell. Syst. Technol. **2**, 27:1-27:27 (2011)
7. Daniels, Z.A., Baird, H.S.: Discriminating features for writer identification. In: 2013 12th International Conference on Document Analysis and Recognition, pp. 1385–1389 (2013)
8. DeVries, T., Taylor, G.W.: Dataset augmentation in feature space. In: 5th International Conference on Learning Representations, ICLR 2017, Toulon, France, 24–26 April 2017, Workshop Track Proceedings (2017)
9. Diaz, M., Ferrer, M.A., Eskander, G.S., Sabourin, R.: Generation of duplicated offline signature images for verification systems. IEEE Trans. Pattern Anal. Mach. Intell. **39**(5), 951–964 (2017)
10. Diaz, M., Ferrer, M.A., Sabourin, R.: Approaching the intra-class variability in multi-script static signature evaluation. In: 2016 23rd International Conference on Pattern Recognition (ICPR), pp. 1147–1152. IEEE (2016)
11. Fang, B., Leung, C., Tang, Y.Y., Kwok, P., Tse, K., Wong, Y.: Offline signature verification with generated training samples. IEE Proc.-Vis. Image Signal Process. **149**(2), 85–90 (2002)
12. Ferrer, M.A., Diaz-Cabrera, M., Morales, A.: Synthetic off-line signature image generation. In: 2013 International Conference on Biometrics (ICB), pp. 1–7, June 2013
13. Freitas, C., Oliveira, L.S., Sabourin, R., Bortolozzi, F.: Brazilian forensic letter database. In: 11th International Workshop on Frontiers on Handwriting Recognition, Montreal, Canada (2008)
14. He, K., Zhang, X., Ren, S., Sun, J.: Deep residual learning for image recognition. In: Proceedings of the IEEE Conference on Computer Vision and Pattern Recognition (2016)
15. He, S., Schomaker, L.: Deep adaptive learning for writer identification based on single handwritten word images. Pattern Recogn. **88**, 64–74 (2019)
16. Howard, A.G., et al.: MobileNets: efficient convolutional neural networks for mobile vision applications (2017)
17. Kannala, J., Rahtu, E.: BSIF: binarized statistical image features. In: Proceedings of the 21st International Conference on Pattern Recognition (ICPR2012), pp. 1363–1366 (2012)
18. Kittler, J., Hater, M., Duin, R.P.: Combining classifiers. In: Proceedings of 13th International Conference on Pattern Recognition, vol. 2, pp. 897–901. IEEE (1996)
19. Kleber, F., Fiel, S., Diem, M., Sablatnig, R.: CVL-database: an off-line database for writer retrieval, writer identification and word spotting. In: 2013 12th International Conference on Document Analysis and Recognition, pp. 560–564 (2013)
20. Kumar, V., Glaude, H., de Lichy, C., Campbell, W.: A closer look at feature space data augmentation for few-shot intent classification. In: Proceedings of the 2nd Workshop on Deep Learning Approaches for Low-Resource NLP (DeepLo 2019) (2019)
21. Lashley, K.S.: Basic neural mechanisms in behavior. Psychol. Rev. **37**(1), 1–24 (1930)
22. Louloudis, G., Gatos, B., Stamatopoulos, N., Papandreou, A.: ICDAR 2013 competition on writer identification. In: 2013 12th International Conference on Document Analysis and Recognition, pp. 1397–1401. IEEE (2013)

23. Marti, U.V., Bunke, H.: The IAM-database: an English sentence database for offline handwriting recognition. Int. J. Doc. Anal. Recogn. **5**, 39–46 (2002)
24. Maruyama, T.M., Oliveira, L.S., Britto, A.S., Sabourin, R.: Intrapersonal parameter optimization for offline handwritten signature augmentation. IEEE Trans. Inf. Forensics Secur. **16**, 1335–1350 (2021)
25. Nanni, L., Lumini, A., Brahnam, S.: Local binary patterns variants as texture descriptors for medical image analysis. Artif. Intell. Med. **49**(2), 117–125 (2010)
26. Pinhelli, F., Britto Jr., A.S., Oliveira, L.S., Costa, Y.M., Bertolini, D.: Single-sample writers-'document filter' and their impacts on writer identification. arXiv preprint arXiv:2005.08424 (2020)
27. Plamondon, R., Lorette, G.: Automatic signature verification and writer identification - the state of the art. Pattern Recogn. **22**(2), 107–131 (1989)
28. Ramirez Rivera, A., Rojas Castillo, J., Oksam Chae, O.: Local directional number pattern for face analysis: face and expression recognition. IEEE Trans. Image Process. **22**(5), 1740–1752 (2013)
29. Rehman, A., Naz, S., Razzak, M.I.: Writer identification using machine learning approaches: a comprehensive review. Multimedia Tools Appl. **78**(8), 10889–10931 (2018). https://doi.org/10.1007/s11042-018-6577-1
30. Ruiz, V., Linares, I., Sanchez, A., Velez, J.F.: Off-line handwritten signature verification using compositional synthetic generation of signatures and Siamese neural networks. Neurocomputing **374**, 30–41 (2020)
31. Schomaker, L., Vuurpijl, L.: Forensic writer identification: a benchmark data set and a comparison of two systems. Internal report for the Netherlands Forensic Institute (2000)
32. Sharma, M.K., Dhaka, V.P.: Offline scripting-free author identification based on speeded-up robust features. Int. J. Doc. Anal. Recogn. (IJDAR) **18**(4), 303–316 (2015). https://doi.org/10.1007/s10032-015-0252-0
33. Simonyan, K., Zisserman, A.: Very deep convolutional networks for large-scale image recognition (2015)
34. Song, T., Li, H., Meng, F., Wu, Q., Cai, J.: LETRIST: locally encoded transform feature histogram for rotation-invariant texture classification. IEEE Trans. Circuits Syst. Video Technol. **28**(7), 1565–1579 (2018)
35. Tang, Y., Wu, X.: Text-independent writer identification via CNN features and joint Bayesian. In: 2016 15th International Conference on Frontiers in Handwriting Recognition (ICFHR), pp. 566–571 (2016). https://doi.org/10.1109/ICFHR.2016.0109
36. Wing, A.M.: Motor control: mechanisms of motor equivalence in handwriting. Curr. Biol. **10**(6), R245–R248 (2000)
37. Xing, L., Qiao, Y.: DeepWriter: a multi-stream deep CNN for text-independent writer identification. In: 2016 15th International Conference on Frontiers in Handwriting Recognition (ICFHR), pp. 584–589, October 2016. https://doi.org/10.1109/ICFHR.2016.0112

Key-Guided Identity Document Classification Method by Graph Attention Network

Xiaojie Xia[✉], Wei Liu, Ying Zhang, Liuan Wang, and Jun Sun

Fujitsu R&D Center Co., Ltd., Beijing, China
{xiaxiaojie,liuwei,zhangying,liuan.wang,sunjun}@fujitsu.com

Abstract. Identify documents, such as personal identity cards, driving licenses, passports, and residence permits, are indispensable and necessary in human life. Various industries, like hotel accommodation, banking service, traffic service and car renting, need identify documents to extract and verify the personal information. Automatic identity document classification can help people to quickly acquire the valid information from the documents, saving the labor cost and improving the efficiency. However, due to the inconsistency and diversity of identify documents between different countries, automatic classification of these documents is still a challenging problem. We propose an identity document classification method in this paper. GAT (Graph Attention Network) and its edge-based variant are exploited to generate the graph embedding, which fuse visual features and textual features together. Moreover, the key information is learnt to guide to a better representation of the document feature. Extensive experiments on the identify document dataset have been conducted to prove the effectiveness of the proposed method.

Keywords: Identify document · Document classification · GAT

1 Introduction

Identity document, which contains rich personal information, plays an important role in human life. Typical identity document includes personal identity card, driving license, passport and residence permit. Many industries including but not limited to hotel accommodation, banking service, traffic service and car rental require identity document to confirm identity and provide services.

Automatic identity document processing can extract the valid information from identity documents automatically and effectively reduce the labor cost. However, due to the inconsistency and diversity of identity document in different countries, and the layout of identify documents are usually not discriminant, automatic classification of these documents is still a very challenging task.

Identity document classification is generally related to the document classification field. It can be divided into two categories: visual-based methods [1,2] and text-based methods [3]. The visual-based methods mainly extract the visual features or layout information from the training document images and then group

ⓒ Springer Nature Switzerland AG 2021
J. Lladós et al. (Eds.): ICDAR 2021, LNCS 12824, pp. 267–280, 2021.
https://doi.org/10.1007/978-3-030-86337-1_18

the document according to its visual representation. However, some documents have very similar visual appearance, this approach can lead to misclassification. The text-based approaches use the whole text information to construct models to represent documents and group them to different categories by the textual similarity. The classification performance depends heavily on the accuracy of the text detection and text recognition engine. Furthermore, the commonly happened texture background in identity documents impose additional difficult text-based methods. Most of the previous identity document classification methods [4–6] use handcrafted features for classification. However, the performance is still limited and not ready for the practical applications.

In this paper we propose a Graph Neural Network based framework for identity document classification. Compared with the previous methods, this method can directly predict the country (or district) and type of the input document without any redundant post-processing. Both textual and visual features are fused for generating a rich representation. The classification of the country and type is separate, and the key area of the document image and corresponding quantitative key value can be learnt, which provides more information for the final classification. The classification of the proposed system supports 88 countries (or districts) and 5 types.

Our method has the following advantages: 1) Using GAT (Graph Attention Network) [7] and edge-enhanced graph network to generate the graph embeddings for classification, integrating visual features and text features; 2) Learning key information to guide to represent the document feature; 3) Fusing the graph embedding together with the global image embedding for the final classification. The performance of our method is evaluated on a database with 198 real identity documents. The result proves the effectiveness of the proposed method.

The organization of the rest of the paper is as follows: Sect. 2 introduces the proposed method; Sect. 3 shows the experiments and result analysis; the conclusion and summary are presented in Sect. 4.

2 Method

In this section we introduce the proposed key-guided identify document classification method using the graph model. At a high level, this method is composed of three modules: 1) text reading module; 2) graph embedding module; 3) classification module. Optimization functions are as shown in Sect. 2.4. The architecture is illustrated in Fig. 1.

2.1 Text Reading Module

The text reading module is designed to locate and recognize all the texts in the document images. As shown in Fig. 1, this module has four parts: backbone, text detection, RoIAlign [9] and text recognition. The text detection part takes the original images as the network input and predicts all texts' coordinates in each image. Then the corresponding text features of the detected regions are

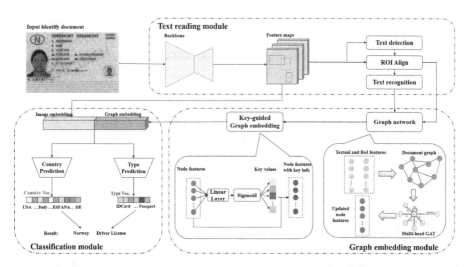

Fig. 1. The architecture of proposed method. The showed input identify document sample is public specimen from [8].

extracted by the RoIAlign part. Finally, the text recognition part based on a sequence recognizer decodes all texts from the features. For ease understanding, each part is described in detail in below.

Backbone. For an input image X, we adopt ResNet [10] as the backbone network to obtain shared features $FM \in \mathbb{R}^{h \times w \times c}$, where h, w, c are the height, width and channel of the output feature maps, respectively.

Text Detection. The text detection part takes FM as the input and predicts the text regions' locations. The detector can be any segmentation-based or anchor-based text detection methods [11–15]. The output locations of the detection part, denoted as $B = (b_0, b_1, \cdots, b_{M-1})$, is a set of M all possible text bounding boxes. And $b_i = (x_0, y_0, x_1, y_1) \in B$ denotes the coordinates of the top-left and bottom-right points of the i-th text segments.

RoIAlign. Based on the shared features FM, we extract the text region features by RoIAlign [9]. The M text region features are denoted as $A = \{a_0, a_1, \cdots, a_{M-1}\}$, where $a_i \in \mathbb{R}^{h' \times w' \times c}$, h' and w' are the spatial dimensions, and channel c is same as in FM.

Text Recognition. Finally, the text recognition part decodes the text strings from all text region features. Convolutional Recurrent Neural Network method [16] is applied to obtain the text transcriptions. Recurrent neural networks (RNNs) [17] is combined with CNN-based feature extraction method for a sequential representation of the text lines, and the final text transcriptions are obtained by a CTC-based [18] method. Similarly, the attention-based decoding method [19] can be adopted to replace the CTC decoder. The strings are denoted as $S = \{s_0, s_1, \cdots, s_{M-1}\}$, where s_i is a set of characters for the i-th text segments.

Note that there are many open-source OCR systems for examples [20,21], which can be called directly for the text detection and recognition. If there were not abundant data for model training, or the performance was not good enough, the OCR systems can be adopted to replace of the text reading module. Moreover, the specific text detection and recognition models also can be applied individually.

2.2   Graph Embedding Module

In this module, the ID documents are embedded by construct a graph, of which the detected textual regions as vertices and cooccurrences between the vertices as edges. The graph embedding is based on graph attention network, which can catch the relationship between nodes and obtain richer representations.

From the text reading module, the text locations and corresponding decoded strings are acquired. The decode text strings represent the textual information of different detected text segments. Meanwhile, the region features, containing local image patterns, can be extracted by the RoIAlign part. It is obvious that these two features belong to different modalities and are mutually corresponding with each other. Graph Attention Network (GAT) [7] is proposed here for the graph embedding.

Graph Construction. Given an input document image $\boldsymbol{X}$ with M detected text blocks, $A = \{a_0, a_1, \cdots, a_{M-1}\} \in \mathbb{R}^{M \times F_1}$ and $S = \{s_0, s_1, \cdots, s_{M-1}\} \in \mathbb{R}^{M \times F_2}$ denote the text region features and the text transcriptions, respectively. The graph is constructed as $G = (V, C)$, where $S = (s_0, s_1, \cdots, s_{N-1})$ is the set of N nodes and C is a binary matrix with shape $N \times N$. $v_i \in R_F$ is the feature of i-th node, and $\varepsilon_{ij} \in C$ has the value of 1 or 0, which represents the connection from node i to node j. F is the dimension of node feature and $F = F_1 + F_2$.

For the node feature, we combine the regional visual features (namely, RoI feature) and the textual feature together. $\hat{a}_i$ is the flatten pattern of i-th text regional feature a_i, followed by a normalization, denoted as

$$\hat{a}_i = Flatten(Norm(a_i)) \tag{1}$$

The text transcription s_i also need to be encoded to a word embedding. Here we adopt a high-dimensional embedding method Pyramidal Histogram of Characters (PHOC) [22] to generate the textual feature $\hat{s}_i$, denoted as

$$\hat{s}_i = PHOC(s_i) \tag{2}$$

Finally, we obtain the fused node set V by element-wise concatenating the two embeddings, of which the i-th v_i can be expressed as

$$v_i = concat(\hat{a}_i, \hat{s}_i) \tag{3}$$

The $N \times N$ adjacency matrix of the edges C is denoted as

$$\varepsilon_{ij} = \begin{cases} 1, i\text{-th node is connected to } j\text{-th node,} \\ 0, \text{otherwise.} \end{cases} \tag{4}$$

After generating the initial embedding of nodes and edges, the graph is constructed.

Graph Network. We formulate the graph embedding aggregation process by GAT layer and its variant style by applying the edge features. The output of the graph network is the updated node embedding set with a reasonable representation for the further classification module.

GAT Layer. We apply the GAT layer to build a graph network. Each node can receive the information a from its adjacent neighbors and then merge with its own representation to update. The input node feature set is denoted as $V = \{v_0, v_1, \cdots, v_{N-1}\} \in \mathbb{R}^{N \times F}$. A weight value matrix $W \in \mathbb{R}^{F' \times F}$ need be trained for all nodes to transform the F-dimensional input node features to F-dimensional output. Self-attention is performed on the nodes with a shared attentional mechanism. The attention coefficient α_{ij} between i-th and j-th node is defined as

$$\alpha = f(Wv_i, Wv_j) \tag{5}$$

where $f(\cdot)$ is the attentional mechanism, and only the attention coefficients between connected nodes are computed. Softmax layer is introduced to regularize the values of the connected N_i nodes of i-th node, denoted as

$$\alpha_{ij}' = softmax(LeakyReLU(\alpha_{ij})) = \frac{\exp(\alpha_{ij})}{\sum_{k \in N_i} \exp(\alpha_{ik})} \tag{6}$$

The computation for attention coefficient is showed as below, where LeakyReLU is used for activation.

$$\begin{aligned} \alpha_{ij}' &= softmax(LeakyReLU(\alpha_{ij})) \\ &= \frac{\exp(LeakyReLU(f[Wh_i, Wh_j]))}{\sum_{k \in N_i} \exp(LeakyReLU(f[Wh_i, Wh_k]))} \end{aligned} \tag{7}$$

After obtaining the attention coefficients between the nodes, we can generate the output node feature embeddings based on these weights and their own features. The output node feature $v_i' \in \mathbb{R}^{F'}$ is expressed as

$$v_i' = \sigma\left(\sum_{j \in N_i} \alpha_{ij}' W v_j\right) \tag{8}$$

where σ is a non-linear activation, and N_i represents that we only update the node by other connected neighbors.

Additionally, multi-head attention is also applied to combine the outputs of multiple individual attention operations. We regard the mean value of the node output embeddings with K-head attention as the final output, which is defined as

$$v_i' = \sigma(\frac{1}{K} \sum_{k=1}^{K} \sum_{j \in N_i} {\alpha_{ij}'}^k W^k v_j) \tag{9}$$

where ${\alpha_{ij}'}^k$ and W^k are the corresponding attention coefficient and weight matrix from k-th attention mechanism.

EGNN(A) Layer. As mentioned above, we only use the connection adjacency information of the text blocks in the document as the edge features, which are likely to be noisy and not optimal. For example, if two nodes are not connected, the attention coefficient between them are forced to zero. Edge features can be multi-dimensional vector rather than binary matrix to contain more intuitional relationship and richer information. Inspired by EGNN proposed in [23], we adopt the attention-based edge enhanced graph neural network (EGNN(A)) as the graph network.

We define the edge feature as $E \in \mathbb{R}^{N \times N \times P}$, where P is the edge feature dimension. $e_{ij} \in \mathbb{R}^{N \times N}$ is a P-dimensional feature vector of the edge connecting the i-th and j-th nodes, here we use a four-dimensional feature denoted as

$$e_{ij} = \left[\frac{x_{ij}}{W}, \frac{y_{ij}}{H}, \frac{w_i}{w_j}, \frac{h_i}{h_j} \right] \tag{10}$$

where w_i and h_i are the width and height of corresponding text region of the node, x_{ij}, y_{ij} are the center distance between the i-th and j-th node in x, y directions, W and H are the width and height of input document image. If there are no connections between nodes, the edge feature is set to zero vector.

Utilizing the edge feature, the EGNN(A) aggregation operation is defined as follows:

$$v_i' = \sigma \left(\sum_{p=1}^{P} \sum_{j \in N_i} \alpha_{..p} W v_j \right) \tag{11}$$

where $v_i' \in R_F'$ is the output node feature updated by EGNN(A) layer, p is the index of the edge feature dimension. The attention coefficient is denoted as

$$\alpha_{ijp} = f (W v_i, W v_j) e_{ijp} \tag{12}$$

where f is the attention mechanism, and we use the same function with that in GAT. Similarly, multi-head attention is adopted.

Key-Guided Graph Embedding. After updating the node features by the graph net-work, we apply a key-guided method to get a final embedding representation of the document graph. The key here is the keyword that has the most important information to influence the classification result. For example, the country name in the identify document, the type string, the MRZ code in the passport, etc. They are also the nodes in the constructed graph. It is reasonable that if the keywords nodes were highlight with higher weights, the final classification is in possession to be better.

After obtaining the node embedding set $V' = \{v_0', v_1', \cdots, v_{N-1}'\} \in \mathbb{R}^{N \times F'}$ from the graph network, we concatenate all the node embedding together following with a fully connected network and softmax function. The operation is produced as

$$\rho = sigmoid \left(FC \left(\|_{i=0}^{N-1} v_i \right) \right) \tag{13}$$

Here, $\|$ is the concatenation operator for the N node embeddings, and FC is the abbreviation for the fully connection layer, which transforms the concatenated node embedding to a N-dimensional output vector ρ after the sigmoid function. ρ is the computed key value vector, indicating the key information of each node.

Then the final output of the graph embedding can be denoted as

$$v_i' = \rho_i v_i' \tag{14}$$

where v_i' is node embedding of i-th node, ρ_i is the key value of it.

2.3 Classification Module

The classification module aims at classifying the input document into the correct countries (or districts) and types.

Firstly, it merges the image embedding and graph embedding together to generate a fused feature, which contains global morphology information and overall nodes' information. The graph embedding $\boldsymbol{GE} = \{v_0', v_1', \cdot, v_{N-1}'\} \in \mathbb{R}^{N \times F'}$ is the output of graph embedding obtained in Sect. 2.2. To generate an image embedding $\boldsymbol{IE} \in R^{Fim}$, where $\mathbb{R}^{Fim}$ is the output dimension of the image embedding, convolution layers are used for transforming the image feature map $\boldsymbol{FM}$ extracted from the backbone described in Sect. 2.1. It is denoted as

$$\boldsymbol{IE} = Conv(\boldsymbol{FM}) \tag{15}$$

Then $\boldsymbol{IE}$ and $\boldsymbol{GE}$ are flattened and concatenated to constitute a fused embedding $\boldsymbol{FE}$. The procedure above is denoted as

$$\boldsymbol{FE} = \| \left[Flatten\left(\boldsymbol{IE}\right), \|_{i=0}^{N-1} v_i' \right], v_i' \in \boldsymbol{GE} \tag{16}$$

where $\|$ is the concatenation operator.

Secondly, there are two branches for the country and type classification. Both branches use full connection networks and softmax functions for the final predictions. It is expressed as

$$p_{country} = softmax \left(FC_1 \left(\boldsymbol{FE} \right) \right) \tag{17}$$

$$p_{type} = softmax \left(FC_2 \left(\boldsymbol{FE} \right) \right) \tag{18}$$

where FC_1 and FC_2 are the fully connection network for the two kinds of classification, with output dimensions of $N_{country}$ and N_{type}, respectively. $p_{country}$ and p_{type} are the predicted classification scores of different categories. And we can get the final classification results by the argmax function.

2.4   Optimization

Among the three modules, the text reading module adopts models trained on other datasets for text detection and recognition. The main reason is that we don't have enough identity document data for training from scratch. Therefore, the text reading module is not optimized for identity document classification task. The graph embedding module and final classification module are trained with the losses from three parts,

$$L = L_{country} + \lambda_1 L_{type} + \lambda_2 L_{key} \tag{19}$$

where the hyper-parameters λ_1 and λ_2 control the trade-off between different losses.

$y_{country}$ and y_{type} are the losses of the country and type prediction branches respectively, using the cross-entropy loss formulated as

$$L_{country} = -\sum_i y_{country} \log\left((p_{country})_i\right) \tag{20}$$

$$L_{type} = -\sum_j y_{type} \log\left((p_{type})_j\right) \tag{21}$$

where $y_{country}$ and y_{type} are the one-hot country and type labels.

L_{key} is the loss of the key value prediction in final graph embedding, which is denoted as

$$L_{key} = -\sum_j y_{key} \log\left((\rho_{key})_j\right) + (1 - y_{key}) \log\left(1 - (\rho_{key})_j\right) \tag{22}$$

where y_{key} is the one-hot key value label.

Noted that if we have enough identity document samples, the proposed network also can be trained in an end-to-end manner. In that case, the text reading module needs to be optimized by adding text detection loss and text recognition loss to the total loss in Eq. (19).

3   Experiments

In this section, experiments are performed to verify the effectiveness of the proposed method.

3.1   Datasets

The identify documents contain rich personal information. It is hard to collect enough real samples for training. Therefore, we divide the whole network to two parts and train them separately to achieve a better performance.

SynthText Dataset [24]. It is a synthetic dataset which consists of 800k images synthesized from 8k background images. It is only used to pre-train our model.

ICDAR 2015 Dataset [25]. It consists of 1000 training images and 500 testing images captured by Google glasses. The text instances are labeled at the word level.

Synth ID Dataset. It is our private dataset, including over 30,000 synthetic identify document samples. The dataset varies in over 80 countries (or districts) and 5 types (ID Card, driver license, passport, residence permit and insurance card). Templates of different countries and types are generated, and personal information are filled into the image to obtain the synthetic identify document samples.

The model in the text reading module is pre-trained on SynthText and ICDAR 2015 dataset and fine-tuned on Synth ID dataset. For the graph embedding module and classification module, the model is trained on the Synth ID dataset, which is separated with the model in the text reading module.

Real ID Dataset. It contains 198 real identify documents from 88 countries (or districts) and 5 types (same with the Synth ID dataset). This dataset is private for the reason of personal privacy and only used for testing.

3.2 Implementation Details

Label Generation. For the SynthText and ICDAR 2015 dataset, we used the official labels directly for training, mainly the word-level annotations (locations and strings) of all texts in the images. As for the Synth and real ID dataset, we annotated the bounding boxes and the transcripts corresponding to the boxes. Additionally, we marked each text segment with a key value for guiding the nodes to constitute an appropriate representation of the graph embedding. The keywords, for examples the country and type names, are set with the key value as 1.0. Rest text blocks have less information on classification have the key value 0.3.

Implementation. The proposed models are implemented in TensorFlow [26] and trained on 8 NVIDIA Titan Xp GPUs with 12 GB memory. In the text reading module, the text detection model and the text recognition are trained separately with a shared backbone. We firstly pretrain the text detection model based on [15] on SynthText dataset from scratch for more than 100k iterations. Then we continue train it on ICDAR 2015 dataset for 1000 iterations. Finally, the model is fine-tuned on the Synthetic ID dataset. The training batch size is set to 12 and the crop size of the input image is 512. The text recognition model is trained on the cropped text line images with a height size of 128 from SynthText and ICDAR 2015 dataset, and we used a CTC-based decoding method. *Adam* [27] is applied as the optimizer to minimize the text detection model loss and the recognition loss.

We train the graph embedding and classification module together by adopting the trained models in the text reading module above. During each iteration, the image features are extracted by shared convolutions. Meanwhile, the text blocks are detected and transcribed. They are regarded as the input to the rest models to be trained. The node number N is set to 32. When the detected text regions are more than N, we only extract the N text regions with taller and larger text areas as the node features. Oppositely, if detected text blocks were less than the set node number, the node feature will be padded with zero and corresponding key value is set to 0, too. By setting the node number, we can filter out the nodes with small areas and have convenient batch training. *Adam* is used to optimize the loss described in Sect. 2.5 with a learning rate of 10^{-4}. For the final classification module, we use a country (or district) vocabulary with 89 classes, and the number of type categories is 6, including the negative class. The trade-off hyper-parameters λ_1 and λ_2 are set to 0.5 and 2, respectively.

In the inference phase, the model in the text reading module directly predicts each text segment with the bounding box coordinates and the decoded strings, then the final country and type categories are predicted by the model of graph embedding and classification module.

Evaluation Metric. We consider the correct classification when the country and type predictions are both right. Total accuracy represents the ratio between the correct classification number and the amount of the whole samples. Accuracy of the country and type classification is also computed independently.

3.3   Results and Analysis

We conduct several experiments on our Real ID dataset to show the effectiveness of the proposed model. Properties of the experiments are described in Table 1 and corresponding detailed experimental results are shown in Table 2, where C., T., Num. and Acc. are the abbreviations of country (or district), type, number and accuracy.

Table 1. Properties of the conducted experiments.

Experiments	Image embedding	Graph embedding				
		Textual feature	RoI feature	Key value guiding	Edge connection	Graph network
Exp.1	√					
Exp.2		√			Fully connected	GAT
Exp.3			√		Fully connected	GAT
Exp.4		√	√		Fully connected	GAT
Exp.5	√	√	√		Fully connected	GAT
Exp.6	√	√	√		Five nearest connected	GAT
Exp.7	√	√	√		Five nearest connected	EGNN(A)
Exp.8	√	√	√	√	Five nearest connected	EGNN(A)

Table 2. Results of the experiments in Table 1.

Experiments	Correct C. Num.	Country Acc.	Correct T. Num.	Type Acc.	Correct C.&T. Num.	Total Acc.
Exp.1	130	65.66%	169	85.35%	121	61.11%
Exp.2	63	31.82%	140	70.71%	58	29.29%
Exp.3	75	37.88%	145	73.23%	69	34.85%
Exp.4	107	54.04%	160	80.81%	101	51.01%
Exp.5	166	83.84%	192	96.97%	162	81.82%
Exp.6	169	85.35%	193	97.47%	166	83.84%
Exp.7	174	87.88%	193	97.47%	171	86.36%
Exp.8	182	91.92%	195	98.48%	180	90.91%

Analysis. As described in Table 2, the experiments are performed under different conditions. The results also indicate the contributions of the different modules.

Exp.1 only adopts the image embedding without graph network. We can see that the results is not good, especially the country classification. Type classification result is not so bad since the category number is smaller than the country category number.

Exp.2, exp.3 and exp.4 show the differences whether to use the multi-modality features, namely the textual feature and the RoI feature. Compared with textual feature, RoI feature seems playing a more important role in the final classification. The reason might be that the extracted textual feature contained more accumulated error from the text reading module, especially the recognition part. Fused feature with the textual and RoI feature together achieved better result than using separate feature only. It is obvious that the two multi-modality features used in the graph embedding are complementary with each other.

Exp.5 combines the image embedding and graph embedding for the classification module. The classification result outperforms than the experiments above, which verifies the effect of the fusing of the two embedding patterns.

In exp.6, we modify the edge connection in the graph network from fully connection to only connecting five nearest other nodes of each individual node. Exp.7 adopts the edge feature to replace the binary matrix of edge connection. The two experiments both achieve better results than previous experiments, which indicates that the fully connected graph aggregates redundancy node information and sparse connection between the nodes is efficient. Furthermore, adding edge features by EGNN(A) also enhances the graph embedding.

Exp.8 exploits all proposed method in this paper. Surprisingly, key-guided graph embedding shows significant improvement. This indicates that the key information can lead to generate a more reasonable graph embedding.

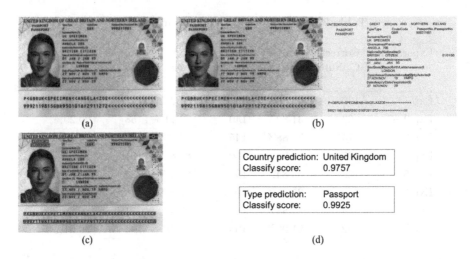

Fig. 2. Classification result of a sample based on the proposed method. (a) the source input document image; (b) the OCR results obtained by Sect. 2.1; (c) visualization of the key-guided graph embedding, where the displayed boxes and corresponding text transcriptions consist of the nodes in graph, and the thickness of bounding box refers to the key value of each node; (d) the final classification results and corresponding classify scores.

Visualization. To gain an intuition of our proposed method, we visualized a result in Fig. 2 with a public sample collected from [28]. Figure 2(b) shows the result of the text reading module. We can see that the OCR result is not perfect, and the key information of the country name and the type name is not completely recognized. Then we construct the graph model, only a set number of the detected regions and recognized texts are feed into the node features, which are illustrated in Fig. 2(c) with the red boxes. Moreover, we also show boxes with the key values above a fixed threshold, with boxes more attended having thicker lines. The key information, for example the upper country name "**UNITED KINGDOM OF GREAT BRITAIN AND NORTHERN IRELAND**" and type name "**PASSPORT**", together with the lower two *MRZ* lines, have higher key values than other node features. It demonstrates that the model has the ability to learn the key information of the nodes in constructed graph. Finally, the sample is classified into the correct country and type with high confidences, which is showed in Fig. 2(d).

4   Conclusion

In this paper, we propose a method to efficiently classify the identify documents by both image embedding and graph embedding are adopted to express the document feature. GAT and edge-enhanced GNN are introduced to generate the graph embedding and the key information leads to learn a reasonable representation. The experiment results show the superior performance in the classification of real collected identify document images.

References

1. Sidiropoulos, P., Vrochidis, S., Kompatsiaris, I.: Content-based binary image retrieval using the adaptive hierarchical density histogram. Pattern Recogn. **44**(4), 739–750 (2011)
2. Shin, C., Doermann, D.: Document image retrieval based on layout structural similarity. In: Proceedings of International Conference on Image Processing, Computer Vision and Pattern Recognition, Las Vegas, Nevada, USA, pp. 606–612 (2006)
3. Sebsatiani, F.: Machine learning in automated text categorization. J. ACM Comput. Surv. **34**(1), 147 (2002)
4. Heras, L., Terrades, O., Lladós, J., Mota, D., Cañero, C.: Use case visual Bag-of-Words techniques for camera-based identity document classification. In: Proceedings of 13th IEEE International Conference on Document Analysis and Recognition (ICDAR), Nancy, France, pp. 721–725 (2015)
5. Simon, M., Rodner, E., Denzler, J.: Fine-grained classification of identity document types with only one example. In: Proceedings of 14th IAPR International Conference on Machine Vision Applications, pp. 126–129 (2015)
6. Héroux, P., Diana, S., Ribert, A., Trupin, É.: Classification method study for automatic form class identification. In: 40th International Conference on Pattern Recognition (ICPR), Brisbane, Australia, pp. 926–928 (1998)
7. Velickovic, P., Cucurull, G., Casanova, A., Romero, A.: Graph attention networks. In: International Conference on Learning Representations (ICLR), Vancouver, BC, Canada (2018)
8. Norwegian Public Roads Administration. https://www.vegvesen.no/forerkort/har-forerkort/gyldig-forerkort-i-norge/eos-modell-3/forerkortets-side-1
9. He, K., Gkioxari, G., Dollár, P., Girshick, R.: Mask R-CNN. In: IEEE International Conference on Computer Vision and Pattern Recognition (CVPR), Venice, Italy, pp. 2980–2988. IEEE (2017)
10. He, K., Zhang, X., Ren, S., Sun, J.: Deep residual learning for image recognition. In: Proceedings of 29th IEEE Conference on Computer Vision and Pattern Recognition (CVPR), Las Vegas, NV, USA, pp. 770–778. IEEE (2016)
11. Zhou, X.: EAST: an efficient and accurate scene text detector. In: Proceedings of IEEE Conference on Computer Vision and Pattern Recognition (CVPR), Venice, Italy, pp. 2642–2651. IEEE (2017)
12. Liao, M., Shi, B., Bai, X., Wang, X., Liu, W.: TextBoxes: a fast text detector with a single deep neural network. In: Proceedings of the 31st AAAI Conference on Artificial Intelligence, San Francisco, California, USA, pp. 4161–4167 (2017)
13. Deng, D., Liu, H., Li, X., Cai, D.: Pixellink: detecting scene text via instance segmentation. In: Proceedings of the 32nd AAAI Conference on Artificial Intelligence, New Orleans, Louisiana, USA, pp. 6773–6780 (2018)
14. Baek, Y., Lee, B., Han, D., Yun, S., Lee, H.: Character region awareness for text detection. In: Proceedings of IEEE Conference on Computer Vision and Pattern Recognition (CVPR), Long Beach, CA, USA, pp. 9365–9374 (2019)
15. Liao, M., Wan, Z., Yao, C., Chen, K., Bai, X.: Real-time scene text detection with differentiable binarization. In: Proceedings of the 34th AAAI Conference on Artificial Intelligence, New York, USA, pp. 11474–11480 (2020)

16. Shi, B., Bai, X., Yao, C.: An end-to-end trainable neural network for image-based sequence recognition and its application to scene text recognition. IEEE Trans. Pattern Anal. Mach. Intell. (TPAMI) **39**(11), 2298–2304 (2017)
17. Hochreiter, S., Schmidhuber, J.: Long short-term memory. Neural Comput. **9**(8), 1735–1780 (1997)
18. Graves, A., Fernandez, S., Gomez, F., Schmidhuber, J.: Connectionist temporal classification: labelling unsegmented sequence data with recurrent neural networks. In: International Conference on Machine Learning (ICML), Pennsylvania, USA, pp. 369–376 (2006)
19. Lee, C., Osindero, S.: Recursive recurrent nets with attention modeling for OCR in the wild. In: Proceedings of the IEEE Conference on Computer Vision and Pattern Recognition (CVPR), Las Vegas, NV, USA, pp. 2231–2239. IEEE (2016)
20. Borisyuk, F., Gordo, A., Sivakumar, V.: Rosetta: large scale system for text detection and recognition in images. In: 24th ACM Conference on Knowledge Discovery and Data Mining (KDD), pp. 71–79 (2018)
21. Du, Y., et al.: PP-OCR: a practical ultra lightweight OCR system. arXiv:2009.09941 (2020)
22. Almazan, J., Gordo, A., Fornes, A., Valveny, E.: Word spotting and recognition with embedded attributes. IEEE Trans. Pattern Anal. Mach. Intell. (TPAMI) **36**(12), 2552–2566 (2014)
23. Gong, L., Cheng, Q.: Exploiting edge features for graph neural networks. In: Proceedings of IEEE Conference on Computer Vision and Pattern Recognition (CVPR), Long Beach, CA, USA, pp. 9211–9219 (2019)
24. Gupta, A., Vedaldi, A., Zisserman, A.: Synthetic data for text localisation in natural images. In: Proceedings of 29th IEEE Conference on Computer Vision and Pattern Recognition (CVPR), Las Vegas, NV, USA, pp. 2315–2324. IEEE (2016)
25. Karatzas, D., et al.: ICDAR 2015 competition on robust reading. In: Proceedings of 13th IEEE International Conference on Document Analysis and Recognition (ICDAR), Nancy, France, pp. 1156–1160 (2015)
26. Abadi, M., et al.: TensorFlow: a system for large-scale machine learning. In: Proceedings of the 12th USENIX Conference on Operating Systems Design and Implementation (OSDI), Berkeley, CA, USA, pp. 265–283 (2016)
27. Kingma, D., Ba, J.: Adam: a method for stochastic optimization. In: International Conference on Learning Representations (ICLR), San Diego, CA, USA (2015)
28. Her Majesty's Passport Office: Basic Passport Checks. https://assets.publishing. service.gov.uk/government/uploads/system/uploads/attachment_data/file/ 867550/Basic_passport_checks_1988_-2019_02.20.pdf

Document Image Quality Assessment via Explicit Blur and Text Size Estimation

Dmitry Rodin[1,2]($\boxtimes$)(iD), Vasily Loginov[1,2](iD), Ivan Zagaynov[1,2](iD), and Nikita Orlov[2](iD)

[1] R&D Department, ABBYY Production LLC, Moscow, Russia
{d.rodin,vasily.loginov,ivan.zagaynov}@abbyy.com
[2] Moscow Institute of Physics and Technology (National Research University), Moscow, Russia
nikita.orlov@picsart.com

Abstract. We introduce a novel two-stage system for document image quality assessment (DIQA). The first-stage model of our system was trained on synthetic data to explicitly extract blur and text size features. The second-stage model was trained to assess the quality of optical character recognition (OCR) based on the extracted features. The proposed system was tested on two publicly available datasets: SmartDoc-QA and SOC. The discrepancies in the results between our system and current state-of-the art methods are within statistical error. At the same time, our results are balanced for both datasets in terms of Pearson and Spearman Correlation Coefficients. In the proposed approach, features are extracted from image patches taken at different scales, thus making the system more stable and tolerant of variations in text size. Additionally, our approach results in a flexible and scalable solution that allows a trade-off between accuracy and speed. The source code is publicly available on github: https://github.com/RodinDmitry/QA-Two-Step.

Keywords: Document image · Quality assessment · Computer vision

1 Introduction

Optical character recognition (OCR) is an important part of the digitization process, as it allows information to be extracted automatically from document images. With mobile phones virtually ubiquitous nowadays, document image quality assessment (DIQA) is crucial. For example, it can be used to ensure that important information is not lost from document images captured with phone cameras. In the case of document images captured with mobile cameras, the quality of OCR is often degraded due to artifacts introduced during image acquisition, which may adversely affect subsequent business processes.

Depending on the availability of a reference document text, DIQA methods can be classified into two groups: full-reference (FR) assessment (when a reference text is available) and no-reference (NR) assessment (when no reference text

© Springer Nature Switzerland AG 2021
J. Lladós et al. (Eds.): ICDAR 2021, LNCS 12824, pp. 281–292, 2021.
https://doi.org/10.1007/978-3-030-86337-1_19

is available) [13, 20]. When images a captured with a mobile camera, no reference document text is typically available and so FR DIQA is usually impossible.

In this paper, we present a novel two-stage DIQA system, which combines the advantages of the classic "hand-crafted features" approach with the generalization ability of modern deep convolutional neural networks. The first part of the system extracts blur and text size features explicitly from the document image, while the second part assesses the quality of OCR based on the extracted features. The system is multi-scale, which allows a certain degree of variation in text size and blur. The architecture and pipeline of the proposed system is described in detail in Sect. 3, and information about the conducted experiments and the obtained results are provided in Sect. 4.

2   Related Works

To perform NR assessment for document images, several different approaches were investigated in pioneering research. The first group of methods is based on hand-crafted document features. Blando et al. [2] used the amount of white speckle, the number of white connected components, and character fragments to assess the quality of document images. Similarly, Souza et al. [18] collected statistics of connected components to assess image quality. Cannon et al. [4] analyzed background speckle and broken characters to assess the quality of a document image. Peng et al. [16] used stroke gradient and average height-to-width ratio to obtain the quality, using a Support Vector Machine for regression which was trained on normalized OCR scores. Kumar et al. [7] developed a DIQA approach based on the difference between median-filtered and original grayscale images. Bui et al. [3] used a similar approach.

Most of the approaches based on hand-crafted features require deep domain knowledge of image distortions and how they affect OCR quality. To compensate for that fact, Ye and Doermann [19] utilized localized soft-assignment and max-pooling to get a better estimate of image quality. Peng et al. [15] used latent Dirichlet allocation to obtain the final assessment by learning quality topics and mapping them to the image. Alaei et al. [1] used the bag-of-visual-words approach, extracting visual attributes from images and forming features.

Recently, convolutional neural networks (CNN) have been successfully applied to various computer vision tasks, including DIQA. For example, Li et al. [11] developed a recurrent learning system integrated into a CNN to assess the quality of image recognition, while Peng and Wang [17] proposed an approach based on a Siamese CNN, where results are obtained by comparing different image patches.

3   Proposed Method

This section describes the architectures of the proposed networks and the overall structure of the document processing pipeline. The first subsection provides information about the entire image quality assessment pipeline. The second sub-section gives details on the architecture of the first-stage feature extraction network. The third and last subsection provides information on the final prediction model.

3.1 Processing Pipeline

This subsection describes in detail the quality assessment pipeline. Since most OCR systems work with grayscale images, our system converts all input images to grayscale. In the first step, the image is resized, so that each of its dimensions becomes a multiple of 512, while approximately maintaining the aspect ratio of the original image. Then the image is separated into square patches at different scales. Patches are extracted at four different sizes: 64, 128, 256, and 512 pixels. For example, if the original size of an image is 2048 × 1536 the number of extracted patches will be 1020: 768, 192, 48, and 12 from corresponding scales. This process is presented in Fig. 1. All those patches are then resized to a fixed resolution of 128 by 128 pixels using the bilinear interpolation. The resized patches are then processed by the feature extraction model, which is explicitly trained to produce blur and text size estimates for each patch, as well as some additional features. In the third step, the extracted features are combined into tensors, three for each of the scales, and they are then fed as input to the quality assessment model. As a result, we get blur length, blur angle, and text size for each patch of the input image at four scales, and a number in the range from zero to one representing the overall image recognition quality. The full pipeline is shown in Fig. 2.

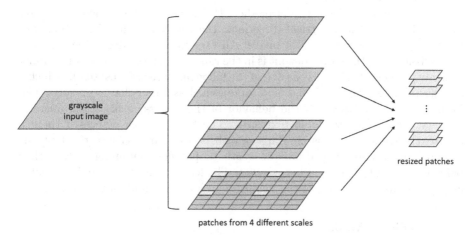

Fig. 1. Patches extraction and normalization, note that it is possible to significantly reduce the number of patches used in higher resolution scales.

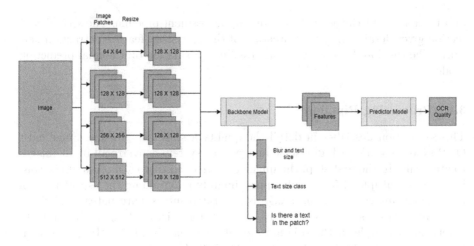

Fig. 2. Full model pipeline

3.2 Feature Extraction Network

This section describes the architecture of the feature extraction network. It accepts an image patch of 128 × 128 pixels as input. The patch is then normalized using instance normalization. The main body of the network consists of several layers of coordinate convolutions with a 3 × 3 kernel with a subsequent max-pooling layer. The regression part of the network consists of several linear layers, each branch corresponding to its own set of features. The architecture of the model is shown in Fig. 3.

The first branch of the model predicts the Gaussian blur value of the patch in both horizontal and vertical directions, the rotation angle of the blur, and the text size value measured in pixels. All targets are encoded using min-max normalization, so the model output is in the range from 0 to 1. The second branch coarsely predicts the text size as one of the following three classes: small (which is smaller than 12 pixels), normal, and large (which is larger than 60 pixels). This branch is necessary because the behavior of the regression branch is unstable on very small and very large text sizes. The third branch detects if there is any text present in the patch. This branch is necessary because the first and second branches are trained only on synthetic patches containing text, so their behaviour is undefined on other patches. The features used for the final quality assessment are the output of the penultimate layers from all of the branches.

3.3 Predictor Model

This section describes the architecture of the final predictor model. The model has a total of 12 inputs, three for each scale of the patches. Each input is first processed by a convolutional layer with a 3 × 3 kernel. The tensors for each of the scales are then concatenated, so that only four branches remain, one for

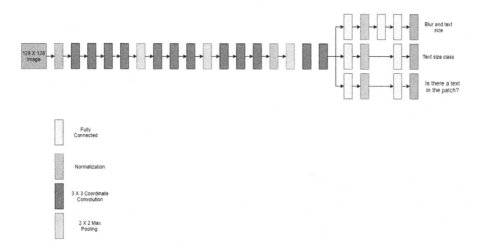

Fig. 3. Feature extraction model architecture.

each scale. Then each branch is processed by several convolutions. After that, the lowest scale branch is downsampled, concatenated with the branch of the higher scale, and combined via convolution. This process is repeated until only one branch remains. The resulting tensor is turned into a feature vector with Global Average Pooling, followed by a linear layer, which produces the output. The network architecture is shown in Fig. 4.

4   Experiments

This section provides detailed information on the training process of the models and describes the results obtained on the test datasets.

4.1   Blur Detector Model

This model was trained on fully synthetic data in three steps. In the first step, the regression branch was trained using L_2 loss with hard-negative mining and the Adam optimizer. In the second step, the weights of the first branch were frozen and the second branch was trained for the coarse text size classification task, using hard-negative mining and the cross-entropy loss function. The last branch was trained with all the other weights frozen on a different set of data, which additionally contained non-text samples. The model was trained with the Adam [6] optimizer with $\beta_1 = 0.9$, $\beta_2 = 0.999$, and $\epsilon = 1e-8$, running for 50 epochs with a learning rate of 0.001, decaying to 0.0001 starting at epoch 40.

The training data for the model was generated using a set of scanned documents in several different languages. The texts were taken from publicly available sources. Font sizes in the source documents varied from 20 to 62 pixels, but remained the same within each document. In each data generation step, a

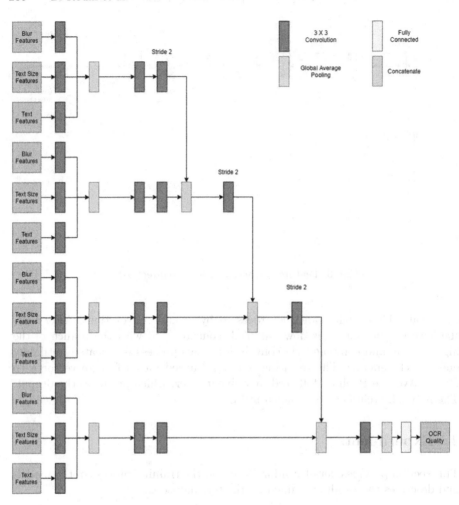

Fig. 4. Predictor model architecture.

patch was randomly cropped out of the image. The patch was then randomly defocused or blurred, with probability of defocus being 0.3. The defocus sigma varied from 1 to 5, the blur x-axis sigma varied from 0.1 to 1, and the y-axis sigma varied from 1 to 10. To apply blur at a random angle, the image was first rotated at a random angle, and once blurred, it was returned to its original orientation. In the last step, white Gaussian noise was added with sigma ranging from 1 to 10, and the JPEG compression was applied, its quality ranging from 60% to 80%. To compensate for the fact that some of the document images were in landscape orientation, the patches were rotated by 90° clockwise or counterclockwise with a probability of 0.1. Text patches used for "text"/"no text" classification accounted for one half of the training data. The "no text" patches

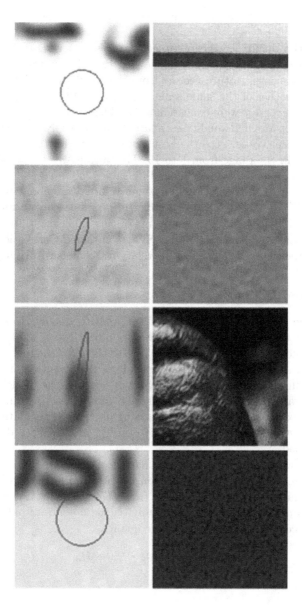

Fig. 5. Synthetic data used for training. Left column: synthetic text image patches with superimposed ellipses, equivalent to the anisotropic Gaussian blur which was applied to model distortion. The ellipses have been magnified for demonstration purposes. Note: The synthetic images were additionally modified with added white Gaussian noise and JPEG compression artifacts which are invisible here. Right column: samples of non-text image patches.

were extracted from non-document images. Examples of the synthetic data are shown in Fig. 5.

To assess the quality of the obtained features and get baseline results for the quality assessment network, the following experiment was conducted. For each image in the data, feature vectors from the backbone model were averaged among all the patches of the same scale, resulting in 12 average feature vectors per image. These vectors were concatenated and the result was used as an image descriptor. To get an OCR quality estimate, we used a k-NN-based model with Euclidean distance. We selected the best possible value for the *number of neighbors* parameter in the range from 1 to 15. The output was the weighted sum of the closest vectors. The results are listed in Table 1.

Another experiment was conducted to get information about the quality-to-speed trade-off. We used a reduced number of input patches to make our approach computationally scalable. The results of this experiment are also listed in Table 1. For feature quality metric we use Pearson Linear Correlation Coefficient (PLCC) and Spearman Rank Order Correlation Coefficient (SROCC).

4.2 Predictor Training

This subsection describes the training of the predictor model. To train the predictor, we split the publicly available datasets SmartDoc-QA [14] and SOC [8] into training and testings sets, reserving 10% of them for the test. The datasets were split on a series basis, which seemed fair as it prevented images from the same series being used both in training and in testing. The model was trained on features obtained by the feature extraction model described in Subsect. 3.2. Features were extracted only from the training portions of the datasets. For the ground truth OCR score, the average of outputs produced by different OCR systems was used.

The model was trained with a combination of losses presented in [17] with classic L_1 and L_2 losses. The loss function is presented in Eq. 1, where L_m and L_l are monotonicity and linearity losses from [17], and $\alpha = 1$, $\beta = 1$, $\gamma = 0.1$, $\delta = 1$. We also performed an ablation study on the amount of patches required to achieve the correct prediction of image quality.

$$L = \alpha L_1 + \beta L_2 + \gamma L_m + \delta L_l \tag{1}$$

Table 1. Feature quality estimation for k-NN baseline, measuring PLCC (ρ_p) and SROCC (ρ_s).

Number of samples from scale				Parameters	Metrics	
Scale 64	Scale 128	Scale 256	Scale 512	Neighbours	ρ_p	ρ_s
768	192	48	12	1	0.77	0.75
768	192	48	12	3	0.84	0.80
768	192	48	12	5	0.84	0.80
768	192	48	12	7	0.83	0.79
768	192	48	12	15	0.83	0.79
48	48	12	12	1	0.74	0.71
48	48	12	12	3	0.78	0.75
48	48	12	12	5	0.80	0.77
48	48	12	12	7	0.82	0.79
48	48	12	12	15	0.81	0.77
12	12	12	12	1	0.70	0.69
12	12	12	12	3	0.75	0.74
12	12	12	12	5	0.77	0.74
12	12	12	12	7	0.77	0.74
12	12	12	12	15	0.76	0.75

The model was trained with the Adam [6] optimizer with the learning rate of 0.001 and weight decay of 1e−6. The training was performed for 150 epochs (Table 2).

Table 2. The model performance on the SmartDoc-QA dataset with different amount of patches used.

Scale 64	Scale 128	Scale 256	Scale 512	ρ_p	ρ_s
768	192	48	12	0.914	0.848
48	48	12	12	0.893	0.833
12	12	12	12	0.890	0.840

4.3   Comparison with Other Models

We compared our model with other existing approaches to the DIQA problem on the SmartDoc-QA and SOC datasets. The results obtained on the first dataset are presented in Table 3 and the results obtained on the second dataset are presented in Table 4. The test sets for both datasets were obtained as described in Subsect. 4.2.

Table 3. Results on SmartDoc-QA dataset on PLCC(ρ_p) and SROCC(ρ_s).

Approach	ρ_p	ρ_s
RNN [11]	0.814	**0.865**
Feature-based k-NN	0.840	0.801
Li et al. [10]	0.719	0.746
X. Peng and C. Wang [17]	**0.927**	0.788
Our approach	0.914	0.848

Table 4. Results on SOC dataset on PLCC and SROCC.

Approach	PLCC	SROCC
CNN [5]	0.950	0.898
LDA [15]	N/A	0.913
Sparse representation [15]	0.935	0.928
RNN [11]	0.956	0.916
CG-DIQA [9]	0.927	0.788
Li et al. [10]	0.914	0.931
Lu et al. [12]	0.951	0.835
X. Peng and C. Wang [17]	**0.957**	**0.936**
Our approach	0.951	0.935

The results demonstrate that the accuracy of the proposed approach is comparable to that of state-of-the-art-methods within statistical error, while being well-balanced in both PLCC and SROCC. At the same time, the use of patches of multiple resolutions makes our approach more tolerant of variations in text size and, as an added benefit, yields local and interpretable features that describe the visual and OCR quality of document images.

5   Conclusion

In this paper, we have presented a novel two-stage DIQA system which combines the advantages of the classical "hand-crafted features" approach with the generalization ability of modern deep convolutional neural networks. Our model predicts document OCR quality based on interpretable features, while maintaining a prediction accuracy comparable to that of other state-of-the-art approaches. Multi-scale patch extraction makes the system more robust and tolerant of variations in text size. A trade-off between speed and accuracy is achieved by varying the number of patches used by the system.

In future work, we plan to improve our system by teaching it to localize image defects that may impair OCR. Another interesting avenue of research is obtaining more relevant local image features via a precise and robust estimate of the local point spread function (PSF), using the existing approaches to the blind deconvolution problem. We believe that with PSF estimates and OCR quality results for reconstructed images, it is possible to build a system that will obtain better OCR results on moderately degraded images.

References

1. Alaei, A., Conte, D., Martineau, M., Raveaux, R.: Blind document image quality prediction based on modification of quality aware clustering method integrating a patch selection strategy. Expert Syst. Appl. **108**, 183–192 (2018)
2. Blando, L.R., Kanai, J., Nartker, T.A.: Prediction of OCR accuracy using simple image features. In: Proceedings of 3rd International Conference on Document Analysis and Recognition, vol. 1, pp. 319–322 (1995). https://doi.org/10.1109/ICDAR. 1995.599003
3. Bui, Q.A., Molard, D., Tabbone, S.: Predicting mobile-captured document images sharpness quality. In: 2018 13th IAPR International Workshop on Document Analysis Systems (DAS), pp. 275–280 (2018). https://doi.org/10.1109/DAS.2018.32
4. Cannon, M., Hochberg, J., Kelly, P.: Quality assessment and restoration of typewritten document images. Doc. Anal. Recogn. 2 (2000). https://doi.org/10.1007/s100320050039
5. Kang, L., Ye, P., Li, Y., Doermann, D.: A deep learning approach to document image quality assessment. In: 2014 IEEE International Conference on Image Processing (ICIP), pp. 2570–2574 (2014). https://doi.org/10.1109/ICIP.2014.7025520
6. Kingma, D.P., Ba, J.: Adam: a method for stochastic optimization. In: Bengio, Y., LeCun, Y. (eds.) 3rd International Conference on Learning Representations, ICLR 2015, San Diego, CA, USA, 7–9 May 2015. Conference Track Proceedings (2015). http://arxiv.org/abs/1412.6980
7. Kumar, J., Chen, F., Doermann, D.: Sharpness estimation for document and scene images. In: Proceedings of the 21st International Conference on Pattern Recognition (ICPR 2012), pp. 3292–3295 (2012)
8. Kumar, J., Ye, P., Doermann, D.: A dataset for quality assessment of camera captured document images. In: Iwamura, M., Shafait, F. (eds.) CBDAR 2013. LNCS, vol. 8357, pp. 113–125. Springer, Cham (2014). https://doi.org/10.1007/978-3-319-05167-3_9
9. Li, H., Zhu, F., Qiu, J.: CG-DIQA: no-reference document image quality assessment based on character gradient. CoRR abs/1807.04047 (2018). http://arxiv.org/abs/1807.04047
10. Li, H., Zhu, F., Qiu, J.: Towards document image quality assessment: a text line based framework and a synthetic text line image dataset. CoRR abs/1906.01907 (2019). http://arxiv.org/abs/1906.01907
11. Li, P., Peng, L., Cai, J., Ding, X., Ge, S.: Attention based RNN model for document image quality assessment. In: 2017 14th IAPR International Conference on Document Analysis and Recognition (ICDAR), vol. 01, pp. 819–825 (2017). https://doi.org/10.1109/ICDAR.2017.139
12. Lu, T., Dooms, A.: A deep transfer learning approach to document image quality assessment. In: 2019 International Conference on Document Analysis and Recognition (ICDAR), pp. 1372–1377 (2019). https://doi.org/10.1109/ICDAR.2019.00221
13. Ma, K., Liu, W., Liu, T., Wang, Z., Tao, D.: dipIQ: blind image quality assessment by learning-to-rank discriminable image pairs. IEEE Trans. Image Process. **26**(8), 3951–3964 (2017)

14. Nayef, N., Luqman, M.M., Prum, S., Eskenazi, S., Chazalon, J., Ogier, J.: SmartDoc-QA: a dataset for quality assessment of smartphone captured document images - single and multiple distortions. In: 2015 13th International Conference on Document Analysis and Recognition (ICDAR), pp. 1231–1235 (2015). https://doi.org/10.1109/ICDAR.2015.7333960

15. Peng, X., Cao, H., Natarajan, P.: Document image OCR accuracy prediction via latent Dirichlet allocation. In: 2015 13th International Conference on Document Analysis and Recognition (ICDAR), pp. 771–775 (2015). https://doi.org/10.1109/ICDAR.2015.7333866

16. Peng, X., Cao, H., Subramanian, K., Prasad, R., Natarajan, P.: Automated image quality assessment for camera-captured OCR. In: 2011 18th IEEE International Conference on Image Processing, pp. 2621–2624 (2011). https://doi.org/10.1109/ICIP.2011.6116204

17. Peng, X., Wang, C.: Camera captured DIQA with linearity and monotonicity constraints. In: Bai, X., Karatzas, D., Lopresti, D. (eds.) DAS 2020. LNCS, vol. 12116, pp. 168–181. Springer, Cham (2020). https://doi.org/10.1007/978-3-030-57058-3_13

18. Souza, A., Cheriet, M., Naoi, S., Suen, C.Y.: Automatic filter selection using image quality assessment. In: Proceedings of the Seventh International Conference on Document Analysis and Recognition, vol. 1, pp. 508–512 (2003). https://doi.org/10.1109/ICDAR.2003.1227717

19. Ye, P., Doermann, D.: Learning features for predicting OCR accuracy. In: Proceedings of the 21st International Conference on Pattern Recognition (ICPR 2012), pp. 3204–3207 (2012)

20. Ye, P., Doermann, D.: Document image quality assessment: a brief survey. In: 2013 12th International Conference on Document Analysis and Recognition, pp. 723–727, August 2013. https://doi.org/10.1109/ICDAR.2013.148

Analyzing the Potential of Zero-Shot Recognition for Document Image Classification

Shoaib Ahmed Siddiqui[1,2]($\boxtimes$) ⓘ, Andreas Dengel[1,2] ⓘ, and Sheraz Ahmed[1,3] ⓘ

[1] German Research Center for Artificial Intelligence (DFKI),
67663 Kaiserslautern, Germany
{shoaibahmed.siddiqui,andreas.dengel,sheraz.ahmed}@dfki.de
[2] TU Kaiserslautern, 67663 Kaiserslautern, Germany
[3] DeepReader GmbH, 67663 Kaiserlautern, Germany

Abstract. Document image classification is one of the most important components in business automation workflow. Therefore, a range of different supervised image classification methods have been proposed, which rely on a large amount of labeled data, which is rarely available in practice. Furthermore, retraining of these models is necessary upon the introduction of new classes. In this paper, we analyze the potential of zero-shot document image classification based on computing the agreement between the images and the textual embeddings of class names/descriptions. This enables the deployment of document image classification models without the availability of any training data at zero training cost, alongside providing seamless integration of new classes. Our results show that using zero-shot recognition achieves significantly better than chance performance on document image classification benchmarks (49.51% accuracy on Tobacco-3482 in contrast to 10% random classifier accuracy and 39.22% on RVL-CDIP dataset in contrast to 6.25% random classifier accuracy). We also show that the representation learned by a vision transformer using image-text pairs is competitive to CNNs by training a linear SVM on top of the pre-computed representations which achieves comparable performance to state-of-the-art convolutional networks (85.74% on Tobacco-3482 dataset in contrast to 71.67% from ImageNet pretrained ResNet-50). Even though the initial results look encouraging, there is still a large gap to cover for zero-shot recognition within the domain of document images in contrast to natural scene image classification, which achieves comparable performance to fully supervised baselines. Our preliminary findings pave the way for the deployment of zero-shot document image classification in production settings.

Keywords: Document image classification · Zero-shot learning · Deep learning · Representation learning · Contrastive Language-Image Pretraining (CLIP) · Transformer · Vision transformer

This work was supported by the BMBF project DeFuseNN (Grant 01IW17002) and the NVIDIA AI Lab (NVAIL) program.

J. Lladós et al. (Eds.): ICDAR 2021, LNCS 12824, pp. 293–304, 2021.
https://doi.org/10.1007/978-3-030-86337-1_20

1   Introduction

There has been a wide-scale deployment of deep learning models in a range of different scenarios, especially in the context of visual recognition [15,16,19,30], speech recognition [4], natural language processing [32] and game playing [7]. Deep learning models particularly shine in the presence of a large amount of labeled data, where they automatically discover useful features along with learning the final output head to perform the given task. Therefore, this naturally creates a dependence on a large number of labeled training examples for training these deep models. This dependence of data is specifically concerning in sensitive industrial domains where the cost of data collection is extremely high [2].

In order to reduce this dependence on the amount of label training data required, recent progress in the domain of visual recognition has moved towards self-supervised learning. Self-supervised learning offloads the task of feature learning by employing auxiliary pretext tasks [6,13]. These tasks enable the network to learn general image structure. Once the model is pretrained, the model fine-tuned in a supervised way using the given labeled dataset. A recent attempt has been to incorporate these self-supervised pretext tasks to improve the performance of the document image classification systems [8]. Another dominant direction in the domain of document image classification is to use the potential of transfer learning to achieve competitive performance [1,2]. The model is pre-trained on a large image-classification dataset (i.e. ImageNet [26]) transforming the initial convolutional layers of the network into generic feature extractors. The model is fine-tuned after the reinitialization of the output layer for the given dataset. Although both of these approaches have the potential to drastically reduce the amount of labeled training data required, they still require some data to train. Therefore, they lack in their ability to achieve zero-shot learning. This problem is even more severe when the number of classes is dynamically adapted at run-time. Conventional approaches require retraining of at least the final classification layer, hence going through the lengthy process of retraining.

Previous work in the direction of zero-shot learning requires intermediate class representations [9,24]. Furthermore, they usually achieve zero-shot recognition only after training for a few classes on the given dataset [9,24]. Seminal work from Radford et al. (2021) [24] has achieved a major leap in this direction where they were able to achieve impressive performance on zero-shot learning benchmarks by pairing images to their respective textual descriptions. Therefore, this alleviates the requirement to obtain a difficult class representation in addition to the ability to transfer across datasets rather than just new classes on a particular dataset. They achieved this by training two different transformer encoders (one for text and one for images) in a contrastive formulation. Given randomly crawled image-text pairs from the internet, the two encoders learn to map image and textual representations in a way to maximize the agreement between the correct image-text pairs while maximizing the distance to wrong assignments.

Following this leap in performance, in this paper, we explore the potential of zero-shot recognition for the task of document image classification by leveraging these encoders trained using a contrastive loss. Therefore, the task is to compute the embeddings of the given class descriptions and the given images and identify their agreement. The class with the maximum agreement is selected as the predicted class. To summarize, the contributions of this paper are two-fold:

- This paper shows for the first time that zero-shot learning achieves significantly better performance than a random classifier on document image classification benchmarks. We also show that the image representation learned with these image-text pairs is competitive with the image representation learned by fully-supervised ImageNet models [2].
- This paper also analyzes the performance of recently proposed vision transformers [11] for the task of document image classification when trained in a fully-supervised way after pretraining on the ImageNet dataset [26].

2   Related Work

2.1   Document Image Classification

There is rich literature on the topic of document image classification. However, we restrict ourselves to some of the most recent work in this direction. For a more comprehensive treatment of the prior work, we refer readers to the survey from Chen & Blostein (2007) [5] where they explored different features, feature representations, and classification algorithms prevalent in the past for the task of document image classification.

Shin et al. (2001) [28] defined a set of hand-crafted features which were used to cluster documents together. The selected features included percentages of textual and non-textual regions, column structures, relative font sizes, content density as well as their connected components. These features were then fed to a decision tree for classification. Following their initial work, they proposed an approach to compute geometrically invariant structural similarity along with document similarity for querying document image databases in 2006 [29]. Reddy & Govindaraju (2008) [25] used low-level pixel density information from binary images to classify documents using an ensemble of K-Means clustering-based classifiers leveraging AdaBoost. Sarkar et al. (2010) [27] proposed a new method to learn image anchor templates automatically from document images. These templates could then be used for downstream tasks, e.g., document classification or data extraction.

Kumar et al. (2012) [21] proposed the use of code-books to compute document image similarity where the document was recursively partitioned into small patches. These features were then used for the retrieval of similar documents from the database. Building on their prior work, they additionally used a random forest classifier trained in an unsupervised way on top of the computed representations to obtain document images that belonged to the same category. This enabled them to achieve unsupervised document classification [20]. The following year, they achieved state-of-the-art performance on the table and tax form retrieval tasks [22] by using the same approach as [20], but evaluated the performance in the scenario of limited training data.

After the deep learning revolution, mainly attributed to the seminal paper from Krizhevsky et al. [19] in 2012 where they introduced the famous *AlexNet* architecture, a range of deep learning-powered document classification systems were introduced. Kang et al. (2014) [17] defined one of the first uses of deep Convolutional Neural Networks (CNN) for document image classification. They achieved significant gains in performance in contrast to prior hand-coded feature engineering approaches. Afzal et al. (2015) [1] introduced *DeepDocClassifier* powered by a deep CNN (AlexNet [19] in their case), and demonstrated the potential of transfer learning where the weights of the model were initialized from the pretrained model trained on the large-scale ImageNet [26] dataset comprising of 1.28 million training images belonging to a 1000 different categories. This transformed the initial convolutional layers into generic feature extractors, achieving a significant performance boost in contrast to prior methods. Noce et al. (2016) [23] additionally included textual information extracted using an off-the-shelf OCR alongside the raw images to boost classification performance. The extracted text was fed to an NLP model for projection into the feature space. Kolsch et al. (2017) [18] used extreme learning machine on top of frozen convolutional layers initialized from pretrained AlexNet [19] model. They achieved significant gains in efficiency without any major degradation in accuracy.

Afzal et al. (2017) [2] showed that using pretrained state-of-the-art image classification networks trained on a large amount of labeled data yields significant improvements on document image classification benchmarks where they evaluated the performance of VGG, ResNet, and GoogLeNet [15,30,31]. Asim et al. (2019) [3] proposed a two-stream network taking both visual and textual inputs into the network for producing the output. In contrast to prior work, they included a feature ranking algorithm to pick the top-most features from the textual stream. Similar to [23], Ferrando et al. (2020) [12] combined predictions from OCR with that of images but leveraged the representational power of BERT [10] as the NLP model to boost performance.

2.2   Self-supervised Recognition

Although a range of different self-supervised methods exists in the literature, we only consider the two most recent methods in this regard. Chen et al. (2020) [6] proposed *SimCLR* where they trained the feature backbone using a contrastive loss formed by two different random augmentations of the input. Without any labeled data, they achieved comparable performance to fully-supervised baselines when evaluating on transfer learning tasks. Grill et al. (2020) [13] proposed *BYOL* where they attempted to predict the embeddings of an offline version of the network which is a slow-moving average of the online network. With this simple formulation, they were able to achieve consistent performance improvements over SimCLR [6].

Cosma et al. (2020) [8] recently analyzed the potential of self-supervised representation learning for the task of document image classification. However, their analysis was limited to some obsolete self-supervised tasks (Jigsaw puzzles) where they showed its insufficiency to learn useful representations. They also analyzed some custom variants of this jigsaw puzzle task which they found to improve performance over this trivial baseline. The authors then moved to fuse representations in a self-supervised way for both images and text using LDA and showed significant performance gains in a semi-supervised setting.

2.3 Zero-Shot Recognition

Most methods for zero-shot learning require some mid-level semantic representation of the input which is hard to obtain in practice [9,24]. One of the most important breakthroughs in this direction is achieved by Radford et al. (2021) [24]. They performed contrastive pretraining of transformer encoders for both text and images based on random image-text pairs crawled from the internet. They called their method Contrastive Language-Image Pretraining (CLIP). One of the biggest strengths of CLIP over other zero-shot methods is readily available textual description of the classes, rather than a more intricate semantic representation which is required by all prior methods [9,24]. This enables direct classification in zero-shot settings by computing the agreement between the embeddings of the text and the image encoder. Although most zero-shot learning refers to the direct inclusion of new classes at test time, we evaluate it in a more competitive novel dataset setting where it has never been specifically trained for the document image classification task. This is similar to how CLIP was evaluated in the original paper [24].

3 Method

An overview of the method is presented in Fig. 1. The model is comprised of two different transformer encoders where one is trained to encode the textual descriptions while the other one is trained to encode images such that the agreement between the correct image-text pair is maximized. We will now discuss each of the components in detail below.

3.1 Image/Text Encoder

The field of image recognition has been dominated by convolutional networks [15, 16,19,30]. Transformers, on the other hand, has taken over the field of NLP away from recurrent networks [32] by leveraging the power of self-attention. A transformer takes in a sequence of words where the network can attend to all the other words in the sequence using multi-head self-attention modules. Dosovitskiy et al. (2020) [11] analyzed the potential of transformers for the task of visual recognition. In an attempt to minimize deviation from the original transformer architecture, they fed in small patches of the input image which is analogous to

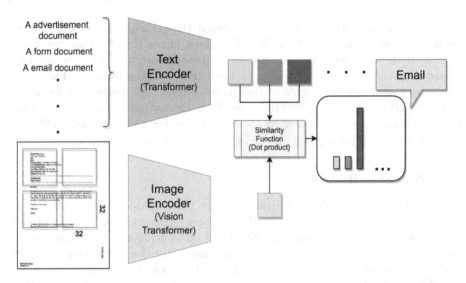

Fig. 1. Overview of the method where the images and the class labels are encoded separately using transformers which are jointly trained using a contrastive loss function to maximize agreement between the correct image-text pair. For zero-shot recognition, we encoded the classes as "A $\langle CLS \rangle$ document" where $\langle CLS \rangle$ is replaced with the actual class name, and the label with the highest agreement with the image representation is chosen as the predicted class (email in the given example).

the words in a sentence. The network can then attend to any other patch within the image during the self-attention process.

CLIP [24] leverages the classical transformer architecture for text encoding [32] and vision transformer for the task of image encoding [11]. The patches were projected using a linear projection layer. These encoders are trained together using a contrastive loss where the aim is to maximize the cosine similarity between the correct image-text pairs. We used their pretrained ViT-B/32 model which is one of the biggest encoder variants they tested. The encoder was trained on image patches of size 32×32.

3.2 Contrastive Pretraining

Let $\Phi(\cdot; \mathcal{W}_t)$ be the text encoder parameterized by $\mathcal{W}_t$ which is a transformer in our case. Furthermore, let $\Psi(\cdot; \mathcal{W}_i)$ be the image encoder parameterized by $\mathcal{W}_i$ which is a vision transformer in our case (trained on patch sizes of 32×32 – ViT-B/32). Let $(\mathbf{x}_i, \mathbf{t}_i)$ be the ith input example where $\mathbf{x}$ represents the input image while $\mathbf{t}$ represents the corresponding text. Let $\mathbf{y}_i = \mathbf{W}_t \cdot \Phi(\mathbf{t}_i; \mathcal{W}_t)$ be the text representation after the linear projection using $\mathbf{W}_t$ where $\Phi(\mathbf{t}_i; \mathcal{W}_t)$ represents the word embedding. Similarly, let $\mathbf{z}_i = \mathbf{W}_{im} \cdot \Psi(\mathbf{x}_i; \mathcal{W}_{im})$ be the image representation after the linear projection using $\mathbf{W}_{im}$ where $\Psi(\mathbf{x}_i; \mathcal{W}_i m)$ represents the patch embedding. In our case, the patch embedding is an identity operator, as more sophisticated approaches have been shown to achieve negligible gain in performance [24]. The similarity between the image and the text representation can be represented as:

$$s_{i,j} = \frac{\mathbf{y}_i^T \mathbf{z}_j}{\|\mathbf{y}_i\|\|\mathbf{z}_j\|} \tag{1}$$

The task of contrastive pretraining is to maximize the agreement between the correct image-text pairs. Finally, the negative log-likelihood between the representations of the correct image-text pairs is minimized. Therefore, the loss for the ith training example is:

$$\ell(i) = -\log \frac{s_{i,i} \cdot \exp(T)}{\sum_{k=1}^{N} s_{i,k} \cdot \exp(T)} - \log \frac{s_{i,i} \cdot \exp(T)}{\sum_{k=1}^{N} s_{k,i} \cdot \exp(T)} \tag{2}$$

where T is a learned parameter in the case of CLIP [24]. As the loss is symmetric, we compute it for both the different assignments of the images for the given text and different assignments of the text for the given image as indicated by the first term and the second term respectively. The final loss then becomes:

$$\mathcal{L} = \frac{1}{N} \sum_{i=1}^{N} \ell(i)/2 \tag{3}$$

where the division by 2 is to cater for the summation running over both the text as well as the images. The encoders are trained by collecting 400 million image-text pairs randomly crawled from the internet. This transforms the encoders into producing similar representations for input and image pairs that are aligned with each other.

4 Results

4.1 Datasets

We evaluated the performance of zero-shot learning methods on two famous document image classification benchmarks i.e. RVL-CDIP [14] and Tobacco-3482 dataset [2]. The RVL-CDIP dataset is comprised of 16 different classes: letter, form, email, handwritten, advertisement, scientific report, scientific publication, specification, file folder, news article, budget, invoice, presentation, questionnaire, resume, and memo. The dataset is divided into 320k training images, 40k validation, and 40k test images. Tobacco-3482 dataset on the other hand is comprised of 3482 images belonging to 10 different classes: advertisement, email, form, letter, memo, news, note, report, resume, and scientific publication. Since there is no predefined split of the dataset, we use randomly sampled 20% of the dataset as test set while the remaining data as train set.

Table 1. Classification accuracy using CLIP [24]. The zero-shot results include the complete dataset for evaluation including both the train as well as the test splits.

Dataset	Random classifier	Zero-shot
Tobacco-3482	10.0%	49.51%
RVL-CDIP	6.25%	39.22%

Table 2. Linear evaluation results computed by training a linear SVM on top of the obtained representations from the model on the corresponding training sets. ResNet-18/50 were pretrained on the ImageNet [26] dataset, while CLIP was pretrained on a randomly crawled 400 million image-text pairs from the internet [24].

Dataset	CLIP	ResNet-18*	ResNet-50*
Tobacco-3482	**85.74%**	70.14%	71.67%
RVL-CDIP	**78.02%**	6.72%	6.73%

It is important to note that since zero-shot classification requires no training set, we combine all different sets for both datasets to report the results for zero-shot classification. Therefore, the results for zero-shot classification on RVL-CDIP dataset are computed using all 400k images while the results for Tobacco-3482 dataset are computed using all 3482 images.

4.2   Zero-Shot Document Image Classification

The results for zero-shot image classification are summarized in Table 1. It is clear from the table that the accuracy achieved by CLIP [24] on zero-shot benchmarks is way above the performance of the random classifier. For the Tobacco-3482 dataset with 10 classes, the random classifier accuracy is just 10%. Zero-shot performance on the other hand is close to 50%. Similarly, for RVL-CDIP dataset with 16 classes, the random classifier accuracy is only 6.25% in contrast to 40% accuracy achieved by zero-shot learning.

In order to characterize the power of the computed image representations within the CLIP framework [24], we computed a linear evaluation result where the accuracy is computed by training a hard-margin linear SVM ($C = 10$) on top of the representation computed by the image encoder on the train set. These results are summarized in Table 2. It is interesting to note that the linear evaluation accuracy on both datasets is above the state-of-the-art CNNs trained on ImageNet [26], indicating that the representations computed by the CLIP image encoder are generic and sufficient to represent the class identity of document images.

Fig. 2. Average attention maps for a randomly selected image from the given class categories (left to right): advertisement, email, news, and note. Model predictions along with their corresponding probabilities are visualized at the bottom left.

Table 3. Zero-shot classification results on Tobacco3482 when using different class embeddings. $\langle CLS \rangle$ here refers to the actual class name (e.g. email, advertisement etc.). "a photo of a $\langle CLS \rangle$" is the default template recommended in the original paper [24].

Class description template	Zero-shot accuracy
$\langle CLS \rangle$	39.00%
a $\langle CLS \rangle$	44.46%
a/an $\langle CLS \rangle$	44.23%
a photo of a $\langle CLS \rangle$	35.21%
a $\langle CLS \rangle$ document	**49.51%**
a/an $\langle CLS \rangle$ document	48.94%
image of a $\langle CLS \rangle$ document	44.08%
a scanned image of a $\langle CLS \rangle$ document	44.80%

As vision transformers leverage the power of self-attention, we visualize the average attention maps from all the heads of a CLIP image encoder for a few sample images in Fig. 2. It is interesting to note that the network learns to focus on textual regions while ignoring the other context. In cases where the text is scattered throughout the document, the attention heads nearly attend to the entire document.

4.3 Comparison Between Class Descriptions for Zero-Shot Learning

It is particularly important to choose the right textual description of the classes when using the text encoder [24]. We compare the different class description templates in Table 3. The most straightforward is to use the class name itself without any additional description. Since the CLIP encoder is trained on real-world image-text pairs, there are rarely any single-word descriptions. Therefore, using this simple encoding achieves only 39% zero-shot accuracy on the Tobacco-3482 dataset. After comparing several different input variations, the one we found

Table 4. Classification results using an independently fine-tuned Vision Transformer (ViT-B/32) pretrained on the ImageNet21k+ImageNet2012 [11] as well as some prior results based on CNNs pretrained on ImageNet-2012 [2]. * results computed with only 100 examples per class in contrast to 80% of the dataset in our case.

Dataset	Model	Accuracy
Tobacco-3482	ViT-B/32	84.50%
	ResNet-50 [2]*	90.40%
	VGG-16 [2]*	90.97%
RVL-CDIP	ViT-B/32	82.35%
	ResNet-50 [2]	91.13%
	VGG-16 [2]	91.01%

to be the most useful one is "a ⟨CLS⟩ document" which achieved zero-shot accuracy of 49.51% on the Tobacco-3482 dataset. Therefore, all of our results in Table 1 are based on this class description. It is important to note that our search for the best encoding is not exhaustive. A more systematic search for new encoding can be used which can improve the performance significantly.

4.4   The Power of Vision Transformer

As CLIP [11] leverages the power of the vision transformer, it is important to characterize the power of the vision transformer itself. In order to characterize this performance, we trained a vision transformer (ViT-B/32) which is the same as the one used in CLIP but trained on the ImageNet dataset in a fully supervised way rather than the image-text pairs. The results are summarized in Table 4. Vision transformers require larger training datasets in contrast to CNNs for achieving state-of-the-art performance [11]. Therefore, despite the fact that the vision transformers were pretrained on the ImageNet dataset, and fine-tuned on the RVL dataset, the performance is still significantly below that of CNNs. However, it is quite likely that the hyperparameters chosen were not optimal as we did not particularly employ any hyperparameter search. Therefore, hyperparameter tuning can make a big impact on performance. We leave more systematic studies in this regard as future work. We believe that the difference between the two can be covered up, but achieving significantly better performance than conventional CNNs seems unlikely.

5   Conclusion

This paper explores the potential of the recently proposed image-text pairing-based zero-shot learning framework for the task of document image classification. Our results show that using zero-shot framework achieves significantly high accuracy as compared to a random classifier. However, the accuracy level is significantly below the supervised baselines in contrast to natural images where the gap is narrow. Therefore, pretraining the models on document data crawled from the internet to pair image-text inputs can significantly improve the performance of such systems in the real world, enabling the direct deployment of such systems in business workflows.

References

1. Afzal, M.Z., et al.: Deepdocclassifier: document classification with deep convolutional neural network. In: 2015 13th International Conference on Document Analysis and Recognition (ICDAR), pp. 1111–1115 (2015). https://doi.org/10.1109/ICDAR.2015.7333933
2. Afzal, M.Z., Kölsch, A., Ahmed, S., Liwicki, M.: Cutting the error by half: investigation of very deep CNN and advanced training strategies for document image classification. In: 2017 14th IAPR International Conference on Document Analysis and Recognition (ICDAR), vol. 1, pp. 883–888. IEEE (2017)
3. Asim, M.N., Khan, M.U.G., Malik, M.I., Razzaque, K., Dengel, A., Ahmed, S.: Two stream deep network for document image classification. In: 2019 International Conference on Document Analysis and Recognition (ICDAR), pp. 1410–1416. IEEE (2019)
4. Baevski, A., Zhou, H., Mohamed, A., Auli, M.: wav2vec 2.0: a framework for self-supervised learning of speech representations. arXiv preprint arXiv:2006.11477 (2020)
5. Chen, N., Blostein, D.: A survey of document image classification: problem statement, classifier architecture and performance evaluation. IJDAR **10**(1), 1–16 (2007)
6. Chen, T., Kornblith, S., Norouzi, M., Hinton, G.: A simple framework for contrastive learning of visual representations. arXiv preprint arXiv:2002.05709 (2020)
7. Chen, Y., et al.: Bayesian optimization in AlphaGo. arXiv preprint arXiv:1812.06855 (2018)
8. Cosma, A., Ghidoveanu, M., Panaitescu-Liess, M., Popescu, M.: Self-supervised representation learning on document images. arXiv preprint arXiv:2004.10605 (2020)
9. Das, D., Lee, C.G.: Zero-shot image recognition using relational matching, adaptation and calibration. In: 2019 International Joint Conference on Neural Networks (IJCNN), pp. 1–8. IEEE (2019)
10. Devlin, J., Chang, M.W., Lee, K., Toutanova, K.: BERT: pre-training of deep bidirectional transformers for language understanding. arXiv preprint arXiv:1810.04805 (2018)
11. Dosovitskiy, A., et al.: An image is worth 16x16 words: transformers for image recognition at scale. arXiv preprint arXiv:2010.11929 (2020)
12. Ferrando, J., et al.: Improving accuracy and speeding up document image classification through parallel systems. In: Krzhizhanovskaya, V.V., et al. (eds.) ICCS 2020. LNCS, vol. 12138, pp. 387–400. Springer, Cham (2020). https://doi.org/10.1007/978-3-030-50417-5_29
13. Grill, J.B., et al.: Bootstrap your own latent: a new approach to self-supervised learning. arXiv preprint arXiv:2006.07733 (2020)
14. Harley, A.W., Ufkes, A., Derpanis, K.G.: Evaluation of deep convolutional nets for document image classification and retrieval. In: International Conference on Document Analysis and Recognition (ICDAR) (2015)

15. He, K., Zhang, X., Ren, S., Sun, J.: Deep residual learning for image recognition. In: Proceedings of the IEEE Conference on Computer Vision and Pattern Recognition, pp. 770–778 (2016)
16. Huang, G., Liu, Z., Van Der Maaten, L., Weinberger, K.Q.: Densely connected convolutional networks. In: Proceedings of the IEEE Conference on Computer Vision and Pattern Recognition, pp. 4700–4708 (2017)
17. Kang, L., Kumar, J., Ye, P., Li, Y., Doermann, D.: Convolutional neural networks for document image classification. In: 2014 22nd International Conference on Pattern Recognition, pp. 3168–3172. IEEE (2014)
18. Kölsch, A., Afzal, M.Z., Ebbecke, M., Liwicki, M.: Real-time document image classification using deep CNN and extreme learning machines. In: 2017 14th IAPR International Conference on Document Analysis and Recognition (ICDAR), vol. 1, pp. 1318–1323. IEEE (2017)
19. Krizhevsky, A., Sutskever, I., Hinton, G.E.: ImageNet classification with deep convolutional neural networks. Commun. ACM 60(6), 84–90 (2017)
20. Kumar, J., Doermann, D.: Unsupervised classification of structurally similar document images. In: 2013 12th International Conference on Document Analysis and Recognition, pp. 1225–1229. IEEE (2013)
21. Kumar, J., Ye, P., Doermann, D.: Learning document structure for retrieval and classification. In: Proceedings of the 21st International Conference on Pattern Recognition (ICPR 2012), pp. 1558–1561. IEEE (2012)
22. Kumar, J., Ye, P., Doermann, D.: Structural similarity for document image classification and retrieval. Pattern Recogn. Lett. 43, 119–126 (2014)
23. Noce, L., Gallo, I., Zamberletti, A., Calefati, A.: Embedded textual content for document image classification with convolutional neural networks. In: Proceedings of the 2016 ACM Symposium on Document Engineering, pp. 165–173 (2016)
24. Radford, A., et al.: Learning transferable visual models from natural language supervision. Image 2, T2 (2021)
25. Reddy, K.V.U., Govindaraju, V.: Form classification. In: Yanikoglu, B.A., Berkner, K. (eds.) Document Recognition and Retrieval XV, vol. 6815, pp. 302–307. International Society for Optics and Photonics, SPIE (2008). https://doi.org/10.1117/12.766737
26. Russakovsky, O., et al.: ImageNet large scale visual recognition challenge. Int. J. Comput. Vision 115(3), 211–252 (2015). https://doi.org/10.1007/s11263-015-0816-y
27. Sarkar, P.: Learning image anchor templates for document classification and data extraction. In: 2010 20th International Conference on Pattern Recognition, pp. 3428–3431. IEEE (2010)
28. Shin, C., Doermann, D., Rosenfeld, A.: Classification of document pages using structure-based features. Int. J. Doc. Anal. Recogn. 3(4), 232–247 (2001)
29. Shin, C.K., Doermann, D.S.: Document image retrieval based on layout structural similarity. In: IPCV, pp. 606–612 (2006)
30. Simonyan, K., Zisserman, A.: Very deep convolutional networks for large-scale image recognition. arXiv preprint arXiv:1409.1556 (2014)
31. Szegedy, C., Ioffe, S., Vanhoucke, V., Alemi, A.: Inception-v4, inception-resnet and the impact of residual connections on learning. arXiv preprint arXiv:1602.07261 (2016)
32. Vaswani, A., et al.: Attention is all you need. arXiv preprint arXiv:1706.03762 (2017)

Gender Detection Based on Spatial Pyramid Matching

Fahimeh Alaei$^{(\boxtimes)}$ (ID) and Alireza Alaei (ID)

Southern Cross University, Gold Coast, Australia
{fahimeh.alaei,ali.alaei}@scu.edu.au

Abstract. The similarity and homogeneous visual appearance of male and female handwriting make gender detection from off-line handwritten document images a challenging research problem. In this paper, an effective method based on spatial pyramid matching is proposed for gender detection from handwritten document images. In the proposed method, the input handwritten document image is progressively divided into several sub-regions from coarse to fine levels. The weighted histograms of the sub-regions are then calculated. This process is resulting in a spatial pyramid feature set which is an extension of the orderless bag-of-features image representation. Classical classifiers, such as Support Vector Machines and ensemble classifiers, are considered for determining the gender (male and female) of individuals from their handwriting. Experiments were conducted on two benchmarks, QUWI and MSHD datasets, and the proposed method provided a promising improvement in gender detection accuracies, especially in script-dependent scenarios, compared with the results reported in the literature.

Keyword: Gender detection · Spatial pyramid matching · Handwritten document images

1 Introduction

Analysis of the patterns and physical characteristics of handwriting is an experimental field of science that refers to graphology. Handwriting is one of the traces that can be used for classifying human soft biometric. Writer identification, gender identification/detection, forensic domain, medicine, sociology, and psychological studies are instances of graphology applications. Detection of the gender of individuals from their off-line handwritten document images is a research problem, which has received great attention in the last decades [1–3]. However, it is still an attractive research problem to many researchers because of the challenging nature of analysing individuals' handwriting styles.

Different features and classifiers were used in the literature for offline handwritten gender detection [1–10]. In [4], data mining models were considered for gender detection. A list of 133 features, including contours, areas, slant, crown, arch, and triangle were extracted from Turkish handwritten document images. Two sets of rules and a decision

© Springer Nature Switzerland AG 2021
J. Lladós et al. (Eds.): ICDAR 2021, LNCS 12824, pp. 305–317, 2021.
https://doi.org/10.1007/978-3-030-86337-1_21

tree were applied for the classification of handwritten documents into two classes: male and female.

In [5], a bank of Gabor filters with different scales and orientations was considered to obtain several filtered images from each input handwritten document image. The mean and standard deviation of the filtered images were computed and saved in a matrix. The Fourier transform was then applied to the obtained matrix and the extracted values were considered as features. A feed-forward neural network was finally used to carry out the gender classification from the extracted features.

In [6], handwritten images were interpreted as textures and images, therefore, were decomposed into a series of Wavelet sub-bands to perform off-line gender detection. The features were extracted from each sub-band using probabilistic finite-state automata. Support vector machine (SVM) and artificial neural networks (ANN) were used as classifiers for dependent and text-independent, as well as script-dependent and script-independent gender detection.

In [7], visual appearance of the handwriting was used for gender classification. Different features, including direction, number of pixels occupied by text, and length of edges were calculated. SVM, logistic regression, KNN and majority voting were considered as classification techniques to detect gender from handwritten document images.

In [8], the prediction of gender, age, and nationality was carried out based on off-line handwritten document images. A set of features, including chain codes, directions, and edge-based directional, curvatures, tortuosities was extracted to characterise document images. Random forests and kernel discriminant classifiers were used as classifiers. The performance of each feature and their combinations for gender, age, and nationality were computed and compared.

In [9], various attributes of handwriting, including slant/skew, curvature, and legibility were used for the classification of handwriting into male and female. The histograms of chain codes, LBPs, and fractal dimensions were considered for quantifying the mentioned attributes. The classification was carried out using SVMs and artificial neural networks.

In [2], Cloud of line distribution and Hinge features were considered for gender classification. The features were extracted by analysing the relationship between the unique shapes of different components in handwritten document images. The classification was then conducted using SVM classifiers.

In [1], ensemble classifiers were employed for improving the handwritten-based gender classification. For this purpose, four texture features, including the histogram of oriented gradients (HoG), local binary patterns (LBP), grey-level co-occurrence matrices (GLCM), and segmentation-based fractal texture analysis were considered to characterise document images. The results of the nearest neighbour classifier, artificial neural networks, decision trees, SVMs, and random forests were compared to find the most accurate gender classification result. The classifiers were applied individually as well as in the ensemble forms, using voting, bagging, and stacking strategies.

From the literature, it can be noted that various feature extraction techniques, such as wavelet transform, Gabor filter, local binary pattern, (LBP), histogram of oriented gradients (HoG), oriented basic image features (oBIFs), and gradient features [1–6] were used to detect the gender of each individual from their off-line handwritten document images. Most of the applied techniques focused on texture and local features. However, a combination of weighted local and global features has not been used in the literature of gender detection based on handwritten document images. Moreover, though there has been a considerable amount of research in this domain, neither automatic machine operated document image analysis nor manual human perception based gender detection have achieved high accuracies results so far [11]. It is also important to mention that employing deep learning approaches for gender detection may not be feasible in many real-world scenarios, as the sizes of available gender detection datasets are relatively small, and manual labelling/tagging document images for creating large datasets is time-consuming and expensive.

Inspired by the work [12], in this research work, a set of robust and transform invariant structural features is proposed for representing the handwritten document images. To compute a set of features, a handwritten document image is iteratively divided into sub-regions at multiple scales/levels from coarse to fine. The histograms of local features, for example, Scale-Invariant Feature Transform (SIFT) [13], are computed from coarse to fine sub-regions and then combined to form a set of multiple resolution features. In addition, K-means clustering is employed on the extracted features to create a visual dictionary. A kernel-based spatial pyramid method is finally applied to provide weights for the extracted features from handwritten document images and sub-regions at multiple levels [7]. Spatial pyramids are particularly used in this research work, as they are good at capturing the organization of major pictorial elements, such as blobs, and the directionality of dominant lines and edges. As the pyramid is based on features computed at the original image resolution, even high-frequency details can be preserved. Different classical classifiers, such as SVMs, and ensemble classifiers are applied for the final gender detection purposes. It is worth noting that incorporating spatial pyramid matching in the proposed method significantly improved the performance of the proposed gender detection method especially when it is compared to the basic bag-of-features representation extracted from single resolution images.

The rest of the paper is organized as follows. In Sect. 2, the proposed gender detection approach is explained. Experimental results and comparative analysis are discussed in Sect. 3. Conclusions and future work are finally presented in Sect. 4.

2 Proposed Approach

The block diagram of the proposed method is demonstrated in Fig. 1. The proposed method includes pre-processing, feature extraction, spatial pyramid matching, and classification components. In the pre-processing step, the image is converted into the greyscale and then resized to speed up the process. The processed image is then divided into sub-regions based on the given levels from coarse to fine. Local features from each sub-region are computed and visual dictionaries are created. The histogram of local features is then weighted according to the level of subregions using spatial pyramid matching and then concatenated to form a set of features. The system is trained using different classifiers, including the SVM and ensemble learner to provide gender detection results. At the testing phase, the system detects the gender of an unseen writer using the given writer's handwritten document image. Details of each step are presented in the following subsections.

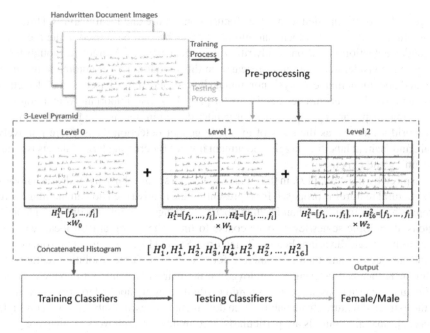

Fig. 1. Block diagram of the proposed method

2.1 Feature Extraction and Bag-of-Features

In the literature, the bag-of-features (BoF) technique has received great attention in image processing and computer vision [14, 15]. In the conventional BoF technique, the image is described as a set of orderless features, and codebooks are created based on the extracted features from entire images [16]. The descriptive power of the image representation and its level of details (local or global) are two important factors that can help to detect gender from off-line handwritten documents. Local descriptors can be obtained by dividing input images into fine sub-regions and computing local features at each sub-region. To extract detailed features from input images, finer levels of input images (smaller subregions) are more beneficial. This process allows extracting features at different levels (coarse to fine). Depending on how much detail is needed to achieve desirable results, coarse to fine division of the input image and feature extraction can be iteratively applied to the images/sub-images.

In this research work, first, the input images are subjected to a few pre-processing techniques to prepare them for feature extraction. To do so, handwritten document images are converted into greyscale and then they are resized into 1000×1300 pixels size. This size is chosen experimentally. Employing these pre-processing techniques is decreasing the computing time required to compute features and so it impacts the speed of the rest of the gender detection tasks.

After the pre-processing step, each image is characterised by a set of global and local features computed from the pre-processed images as well as sub-regions obtained by dividing the pre-processed images at different levels from coarse to fine. It is worth mentioning that the multi-level division of the image and computing features from each subregion is different from multiresolution feature extraction methods [12], which involve repeatedly subsampling an image and computing global features (for example, global histograms) at each new level. To characterise the image and sub-regions, the SIFT features, as a powerful descriptor, are extracted from the whole image, and sub-regions are obtained at different levels. For the SIFT descriptor, the patches of 16×16 pixels with one pixel gap are considered in our system.

It has been argued in the literature that locally orderless representation plays an important role in visual perception [12]. Therefore, visual vocabulary is considered in this research work. The K-means clustering is one of the options to form a codebook and cluster the feature descriptors. To form a visual vocabulary several patches from the training set are chosen and K-means clustering is applied. We consider the vocabulary sizes of $X = 200$ and $X = 400$ in our experiments, while vocabulary size was set to $X = 200$, as it provided better results compared to 400. Once the codebook is obtained, the image is represented by histograms of the visual vocabulary of the codebook.

2.2 Spatial Pyramid Matching

Spatial pyramid matching approach was used for image classification [17] and scene recognition [12] in the literature. Moreover, the spatial pyramid framework has achieved great success when the image was iteratively divided into sub-regions and the features were extracted using a visual dictionary.

To the best of our knowledge, no gender detection method in the literature has incorporated spatial pyramid matching in their gender detection pipeline. Therefore, we adapted and explored the use of this new method for gender detection in this research work.

Pyramid Match Kernels. In the pyramid match kernel, the feature sets transfer into multi-resolution histograms [18]. These multi-resolution histograms can attain co-occurring features. At each level of resolution, coarser grids increase consecutively and taking a weighted sum of the matched numbers. The points that did not match at fine resolution may match at the coarse resolution. The histogram of finer resolutions weighted more compared to the histogram of coarser resolutions. To estimate the similarity of the best partial matching between feature sets, the created histogram is compared with a weighted histogram.

Let M and N be two sets of d-dimensional vectors. The sequence of grids at $0, \ldots, L$ resolutions at level l has 2^l cells along each dimension and the total grids at each level is $D = 2^{dl}$. In each resolution, the number of the points from M and N that fall in i th cells of grids defines by $H_M^l(i)$ and $H_N^l(i)$. The histogram intersection function provides the number of matches that happened at level l and it is computed using the following equation [19].

$$\Gamma\left(H_M^l, H_N^l\right) = \sum_{i=1}^{D} min\left(H_M^l(i), H_N^l(i)\right) \tag{1}$$

The number of the points that matched at the coarse level includes the number of the points that match at the finer level, so, the number of new matches is calculated by $\Gamma^l - \Gamma^{l+1}$ where Γ^l is the brevity of Eq. (1). The matches found in larger cells increase the dissimilar features, thus, the weight associated with level l is $w_L = \frac{1}{2^{L-l}}$ given to all cells of that level. As a result, the pyramid match kernel is defined by:

$$k^L(M, N) = \Gamma^L + \sum_{l=0}^{L-1} \frac{1}{2^{L-l}} \left(\Gamma^l - \Gamma^{l+1} \right)$$

$$= \frac{1}{2^L} \Gamma^0 + \sum_{l=1}^{L} \frac{1}{2^{L-l+1}} \Gamma^l. \tag{2}$$

Spatial Matching. From the literature, we noted that spatial pyramids can capture perceptually salient features. Furthermore, locally orderless matching has been a powerful mechanism for estimating overall perceptual similarity between images [12]. These characteristics are quite in line with how human experts look at document images to find similarities between handwritings and then to distinguish the writer or the gender of the writer of a document image. Therefore, the use of this concept is proposed in our proposed method of gender detection.

Pyramid match kernel precisely matches two collections of features. However, it ignores all spatial information of the features. Thus, an orthogonal approach that quantizes all feature vectors into X discrete types has been proposed [12]. The assumption in this approach is that only features of the same type can be matched to each other. Each channel x provides us with two sets of two-dimensional vectors. The coordinates of features of type x represent by M_x and N_x, in the respective images. The final kernel obtains by summing up the separate channel kernels as follows:

$$K^L(M, N) = \sum_{x=1}^{X} k^L(M_x, N_x). \tag{3}$$

As mentioned above, the pyramid match kernel was obtained by a weighted sum of histogram intersections. By considering $c = \min(a, b) = \min(ca, cb)$, the K^L also can be obtained by concatenating the weighted histograms of all channels at all resolutions with the advantage of sustaining continuity with the popular visual vocabulary. The dimensionality of resulting vector for L levels and X channels is:

$$X \sum_{l=0}^{L} 4^l = X \frac{1}{3} (4^{L+1} - 1). \tag{4}$$

In our experiments, we considered $L = 2$ and $X = 200$ resulting in 4200-dimensional histogram intersections. The histograms were further normalized by the total weight of all features in the image.

2.3 Classification

Classification is carried out using the support vector machine (SVM) with linear kernel [20] as a common classifier used in the gender detection literature [3, 6, 27]. In addition to the SVM classifier, experiments are conducted with the AdaBoost classifier, as an ensemble of multiple decision trees [21].

In an ensemble classifier, handwritten images are classified by taking a vote on the predictions obtained from different classifiers. There are generally two ways of combining multiple classifiers [22]. The first one is a classifier selection that selects the classifiers with the most accurate results. The second one is classifier fusion, where different classifiers are applied in parallel, and each learner has a contribution to this decision. In this way, the final decision is based on the group's consent [23].

The important property of AdaBoost, which is composed of N number of decision trees [24], is its ability to reduce the training error. Consider $G = \{1, \ldots, |G|\}$ is a classifier of the form $h : B \rightarrow G$. In the boosting algorithm, weak learners adding iteratively and sustain a distribution D of weights over the learning set $J = \{(b_1, g_1), \ldots, (b_N, g_N); b_e \in B, g_e \in G\}$. The weight of this distribution denotes by $D_t(e)$ on training example e on round t. Finding a hypothesis h_t appropriate for the distribution D_t is the aim of the weak learner. This hypothesis h_t is measured by its weighted error rate.

$$\epsilon_t = \sum_e D_t(e) E(h_t(b_e) \neq g_e) \tag{5}$$

where E denotes the indicator function. The weights increase for the examples misclassified by h_t and decrease for correctly classified examples. The result is creating a composite strong learner with higher accuracy than the weak learner classifiers. In our experiments, the number of iterations was 50.

3 Experimental Results and Discussions

3.1 Databases and Evaluation Metrics

Two different diverse datasets were considered for the evaluation of the proposed method. The first one is the Multi-script Handwritten Database (MSHD) [25], which contains 1300 handwritten documents written by 100 male and female writers. Each writer copied 6 texts in Arabic, 6 texts in French, and 1 page of digits. For our experiments, the digit documents were ignored. The document image sizes are approximately 2200 × 1600 pixels. The experiments were conducted with 600 Arabic samples and 600 French samples in two scenarios, script-dependent, and script-independent. In the script-dependent experiments, training and testing samples were from the same scripts, either Arabic or French. In the script-independent experiments, training and testing samples were from different scripts. That means the system was trained with Arabic document images, but it was tested with French document images, and vice versa.

The second dataset considered for experimentation is the Qatar University Writer Identification (QUWI) dataset [26], which contains Arabic and English handwritten documents written by 475 male and female writers. From each writer 2 Arabic and 2 English texts were collected. The document image sizes are approximately 2500 × 3400 pixels. The experiments were conducted with 950 Arabic samples and 950 English samples in two scenarios similar to the MSHD dataset. A few handwritten samples from both datasets are shown in Fig. 2.

For the evaluation metrics, accuracy, as a commonly used metric in gender detection, was considered in this research work. The number of correctly detect the gender over the actual number of test handwritten images is defined as accuracy.

From each script in each round of the experiment, 300 samples were randomly used for training, 100 samples for validation, and 100 samples for testing. The training and testing data did not have any overlapping. To evaluate the performance of the proposed method, the experiments were repeated 10 times, and the statistical mode value of the obtained accuracies was reported as the results in this paper.

Fig. 2. (a1, a2) and (b1, b2) are examples of English and Arabic document samples from the QUWI dataset written by a male and female individual, respectively. (c1, c2) and (d1, d2) are examples of French and Arabic document samples from the MSHD dataset written by a male and female individual, respectively.

3.2 Results and Discussion

The results obtained from our proposed method on two datasets with two scenarios are demonstrated in Table 1. The results obtained using SVM and ensemble classifiers are listed for comparison. From Table 1 it is evident that in the script-dependent scenario by applying the proposed method on the QUWI, 82% and 81% of handwritten document images were classified correctly when training and testing scripts were Arabic and English respectively. Similarly, by applying the method on the MSHD, 90% and 87% classification rates were obtained in Arabic and French scripts. Conducting the experiments using Arabic scripts obtained a higher classification rate compared to the experiments using English scripts, this might be due to the cursive nature of the Arabic script and the effectiveness of the SIFT feature extraction. It is also noted that in this scenario, the ensemble classifier performed better than the SVM classifier.

Table 1. The obtained gender detection results using spatial pyramid.

Scenario	Dataset	Script		Classification	
		Train	Test	SVM	Ens
Script-dependent	QUWI	Arabic	Arabic	74	**82**
		English	English	76	**81**
	MSHD	Arabic	Arabic	85	**90**
		French	French	83	**87**
Script-independent	QUWI	Arabic	English	**73**	68
		English	Arabic	**70**	64
	MSHD	Arabic	French	**69**	62
		French	Arabic	**68**	66

As individuals use common characteristics (loops, strokes, etc.) in their handwriting in different scripts, in our research work the script-independent experiments were also conducted to show that gender can be detected from handwriting regardless of the script. The script-independent scenario is more challenging compared to script-dependent as training and testing documents are from different scripts. Considering the difficulty of the problem, the classification rates obtained in the script-independent scenario are lower than the classification rates obtained from the script-dependent scenario. In the QUWI dataset, when the system was trained with Arabic and tested with English, an accuracy of 73% was obtained. When the system was trained with English samples and tested with Arabic samples, the accuracy was 70%. Evaluating the system with the MSHD dataset provided a 69% classification rate when training samples were Arabic scripts and testing were from French scripts. In contrast, a 68% classification rate was obtained when the training was performed using French document images and tested with Arabic documents. It is also noted that the SVM classifier performed better in script-independent scenario compared to the ensemble classifier.

To demonstrate the significance of the proposed method for gender detection compared to the standard bag-of-features, the results obtained using only bag-of-features and our proposed approach on the same datasets are shown in Table 2. From Table 2, it is apparent that the proposed method integrating the bag-of-features extraction and the pyramid matching provided a higher classification rate compared to the method using standard bag-of-features alone for feature extraction. In the QUWI dataset, the proposed method shows up to 15% and 16% improvement in classification rate in script-dependent and script-independent, respectively. Similarly, the proposed method provided approximately 12% and 7%, better classification rate, respectively, in script-dependent and script-independent on MSHD dataset. Considering the results obtained using standard bag-of-features, the SVM classifier performed better in most of the cases compared to the ensemble classifier in both the datasets.

Table 2. The obtained gender detection results using standard bag-of-features.

Scenario	Dataset	Script		Bag-of-features		Proposed approach	
		Train	Test	SVM	Ens	SVM	Ens
Script-dependent	QUWI	Arabic	Arabic	**72**	67	74	**82**
		English	English	**74**	68	76	**81**
	MSHD	Arabic	Arabic	82	**87**	85	**90**
		French	French	**78**	75	83	**87**
Script-independent	QUWI	Arabic	English	62	52	**73**	68
		English	Arabic	62	54	**70**	64
	MSHD	Arabic	French	56	**57**	**69**	62
		French	Arabic	**63**	59	**68**	66

3.3 Comparative Analysis

To have a fair and meaningful comparison, the same dataset and the evaluation protocol considered in the ICDAR2015 competition were employed in our experimentation. The results obtained from the proposed method using the SVM and ensemble classifiers are compared with the state-of-the-art methods, and also the results reported in the ICDAR 2015 gender classification competition [27], and the comparisons are shown in Table 3. The classification rates reported from the competition include the position of the submitted systems in the parenthesis.

Tasks A and B represent the script-dependent scenario, where Arabic and English samples were only used for training and testing. Tasks C and D refer to the script-independent scenario, where Arabic samples were considered for training and English samples for testing in Task C, and English samples were used for training and Arabic samples used for testing in Task D.

From Table 3 it can be observed that the proposed method performed better than the winner system (1) in the competition in all the Tasks. The proposed method using an ensemble classifier provided a 3% higher classification rate compared to the other reported results (Ahmed et al. [1]) in Task A and promising results in Task B. The proposed method provided better accuracies in Tasks C and D compared to the competition winner and the method proposed in Mirza et al. [5]. However, the methods proposed in [1] and [3] provided better results in Tasks C and D compared to our model. This was expected as our model uses SIFT features as a feature extraction technique and this method is sensitive to the local points. Moreover, as handwritten shapes and patterns in English and French are completely different from Arabic, the type of local features (and ultimately visual dictionary) extracted from these two completely different scripts were different resulting in a lower detection accuracy in script-independent tasks. It is important noting that in most of the methods in the literature, the classification rates of script-independent are lower compared to the classification rates of the script-dependent evaluations. The classification rates of script-independent from our proposed method are in line with the results reported in the literature.

Table 3. Results of ICDAR 2015 gender classification competition and other existing methods.

Method	Classification rate			
	Task A	Task B	Task C	Task D
LISIC	60(3)	42(8)	49(5)	55(2)
ACIRS	60(3)	54(3)	53(3)	49(6)
Nuremberg	62(2)	**60(1)**	55(2)	53(3)
MCS-NUST	47(7)	51(5)	48(6)	45(8)
CVC	**65(1)**	57(2)	**63(1)**	**58(1)**
QU	44(8)	52(4)	53(3)	47(7)
UBMA	51(5)	50(6)	44(7)	50(5)
ESI-STIC	48(6)	46(7)	42(8)	53(3)
Mirza et al. [5]	70	67	69	63
Ahmed et al. [1]	79	85	79	80
Gattal et al. [3]	78	81	76	76
Proposed method (SVM)	74	76	73	70
Proposed method (Ens)	**82**	**81**	68	64

4   Conclusion and Future Work

An effective technique for detecting gender from off-line handwritten document images was presented in this paper. The proposed method is an extension and modification of orderless bag-of-features image representation. The handwritten document images were repeatedly divided into sub-regions and a histogram of features from sub-regions was computed. The experiments conducted on two diverse datasets, QUWI and MSHD, show encouraging improvement, especially in the script-dependent scenario. In the future, we plan to investigate different feature extraction techniques to demonstrate the impacts of different features for the task of script-independent gender detection.

References

1. Ahmed, M., Rasool, A.G., Afzal, H., Siddiqi, I.: Improving handwriting based gender classification using ensemble classifiers. Expert Syst. Appl. **85**, 158–168 (2017). https://doi.org/10.1016/j.eswa.2017.05.033
2. Gattal, A., Djeddi, C., Bensefia, A., Ennaji, A.: Handwriting based gender classification using COLD and hinge features. In: El Moataz, A., Mammass, D., Mansouri, A., Nouboud, F. (eds.) ICISP 2020. LNCS, vol. 12119, pp. 233–242. Springer, Cham (2020). https://doi.org/10.1007/978-3-030-51935-3_25
3. Gattal, A., Djeddi, C., Siddiqi, I., Chibani, Y.: Gender classification from offline multi-script handwriting images using oriented Basic Image Features (oBIFs). Expert Syst. Appl. **99**, 155–167 (2018). https://doi.org/10.1016/j.eswa.2018.01.038
4. Topaloglu, M., Ekmekci, S.: Gender detection and identifying one's handwriting with handwriting analysis. Expert Syst. Appl. **79**, 236–243 (2017). https://doi.org/10.1016/j.eswa.2017.03.001
5. Mirza, A., Moetesum, M., Siddiqi, I., Djeddi, C.: Gender classification from offline handwriting images using textural features. In: Proceedings of the ICFHR, pp. 395–398 (2016). https://doi.org/10.1109/ICFHR.2016.0080
6. Akbari, Y., Nouri, K., Sadri, J., Djeddi, C., Siddiqi, I.: Wavelet-based gender detection on off-line handwritten documents using probabilistic finite state automata. Image Vis. Comput. **59**, 17–30 (2017). https://doi.org/10.1016/j.imavis.2016.11.017
7. Maken, P., Gupta, A.: A method for automatic classification of gender based on text- independent handwriting. Multimed. Tools Appl. **80**(16), 24573–24602 (2021). https://doi.org/10.1007/s11042-021-10837-9
8. Al Maadeed, S., Hassaine, A.: Automatic prediction of age, gender, and nationality in offline handwriting. EURASIP J. Image Video Process. **2014**(1), 1 (2014). https://doi.org/10.1186/1687-5281-2014-10
9. Siddiqi, I., Djeddi, C., Raza, A., Souici-meslati, L.: Automatic analysis of handwriting for gender classification. Pattern Anal. Appl. **18**(4), 887–899 (2014). https://doi.org/10.1007/s10044-014-0371-0
10. Hannad, Y., Siddiqi, I., Kettani, M.E.Y.: Writer identification using texture descriptors of handwritten fragments. Expert Syst. Appl. **47**, 14–22 (2016). https://doi.org/10.1016/j.eswa.2015.11.002
11. Illouz, E., (Omid) David, E., Netanyahu, N.S.: Handwriting-based gender classification using end-to-end deep neural networks. In: Kůrková, V., Manolopoulos, Y., Hammer, B., Iliadis, L., Maglogiannis, I. (eds.) ICANN 2018. LNCS, vol. 11141, pp. 613–621. Springer, Cham (2018). https://doi.org/10.1007/978-3-030-01424-7_60
12. Lazebnik, S., Schmid, C., Ponce, J.: Beyond bags of features: spatial pyramid matching for recognizing natural scene categories. In: Proceedings of the CVPR, vol. 2, pp. 2169–2178 (2006). https://doi.org/10.1109/CVPR.2006.68
13. Lowe, D.G.: Object recognition from local scale-invariant features. In: Proceedings of the ICCV, vol. 2, pp. 1150–1157 (1999). https://doi.org/10.1109/ICCV.1999.790410
14. Zhou, L., Zhou, Z., Hu, D.: Scene classification using a multi-resolution bag-of-features model. Pattern Recogn. **46**(1), 424–433 (2013). https://doi.org/10.1016/j.patcog.2012.07.017

15. Caicedo, J.C., Cruz, A., Gonzalez, F.A.: Histopathology image classification using bag of features and kernel functions. In: Combi, C., Shahar, Y., Abu-Hanna, A. (eds.) AIME 2009. LNCS (LNAI), vol. 5651, pp. 126–135. Springer, Heidelberg (2009). https://doi.org/10.1007/978-3-642-02976-9_17

16. Yuan, X., Yu, J., Qin, Z., Wan, T.: A SIFT-LBP image retrieval model based on bag-of-features. In: Proceedings of the ICIP, pp. 1061–1064 (2011)

17. Yang, J., Yu, K., Gong, Y., Huang, T.: Linear spatial pyramid matching using sparse coding for image classification. In: Proceedings of the CVPR, pp. 1794–1801 (2009). https://doi.org/10.1109/CVPR.2009.5206757

18. Grauman, K., Darrell, T.: The pyramid match kernel: efficient learning with sets of features. J. Mach. Learn. Res. **8**, 725–760 (2007)

19. Swain, M., Ballard, D.: Color indexing. Int. J. Comput. Vis. **7**, 11–32 (2004)

20. Cortes, C., Vapnik, V.: Support-vector networks. Mach. Learn. **20**(3), 273–297 (1995). https://doi.org/10.1007/BF00994018

21. Eibl, G., Pfeiffer, K.P.: How to make AdaBoost.M1 work for weak base classifiers by changing only one line of the code. In: Elomaa, T., Mannila, H., Toivonen, H. (eds.) ECML 2002. LNCS (LNAI), vol. 2430, pp. 72–83. Springer, Heidelberg (2002). https://doi.org/10.1007/3-540-367 55-1_7

22. Ho, T.K., Hull, J.J., Srihari, S.N.: Decision combination in multiple classifier systems. IEEE Trans. Pattern Anal. Mach. Intell. **16**(1), 66–75 (1994). https://doi.org/10.1109/34.273716

23. Woods, K., Bowyer, K., Kegelmeyer, W.P.: Combination of multiple classifiers using local accuracy estimates. In: Proceedings of the CVPR, pp. 391–396 (1996). https://doi.org/10.1109/CVPR.1996.517102

24. Freund, Y., Schapire, R.E.: A decision-theoretic generalization of on-line learning and an application to boosting. J. Comput. Syst. Sci. **55**(1), 119–139 (1997). https://doi.org/10.1006/jcss.1997.1504

25. Djeddi, C., Gattal, A., Souici-Meslati, L., Siddiqi, I., Chibani, Y., Abed, H.E.: LAMIS-MSHD: a multi-script offline handwriting database. In: Proceedings of the ICFHR, pp. 93–97 (2014). https://doi.org/10.1109/ICFHR.2014.23

26. Al Maadeed, S., Ayouby, W., Hassaïne, A., Aljaam, J.M.: QUWI: an Arabic and English handwriting dataset for offline writer identification. In: Proceedings of the ICFHR, pp. 746–751 (2012). https://doi.org/10.1109/ICFHR.2012.256

27. Djeddi, C., Al Maadeed, S., Gattal, A., Siddiqi, I., Souici-Meslati, L., Abed, H.E.: ICDAR2015 competition on multi-script writer identification and gender classification using 'QUWI' database. In: Proceedings of the ICDAR, pp. 1191–1195 (2015). https://doi.org/10.1109/ICDAR.2015.7333949

EDNets: Deep Feature Learning for Document Image Classification Based on Multi-view Encoder-Decoder Neural Networks

Akrem Sellami$^{(\boxtimes)}$(iD) and Salvatore Tabbone(iD)

Université de Lorraine, CNRS, LORIA, UMR 7503, Campus Scientifique, 615 Rue du Jardin-Botanique, 54506 Vandœuvre-lès-Nancy, France
`akrem.sellami@inria.fr, salvatore.tabbone@univ-lorraine.fr`
`https://www.loria.fr/fr/`

Abstract. In document analysis, text document images classification is a challenging task in several fields of application, such as archiving old documents, administrative procedures, or security. In this context, visual appearance has been widely used for document classification and considered as a useful and relevant features for the classification. However, visual information is insufficient to achieve higher classification rates, where relevant additional features, including textual features can be leveraged to improve classification results. In this paper, we propose a multi-view deep representation learning which allows combining textual and visual-based information respectively measured through the text and visual document images. The multi-view deep representation learning is designed to find a deeply shared representation between textual and visual features by fusing them into a joint latent space where a classifier model is trained to classify the document images. Our experimental results demonstrate the ability of the proposed model to outperform competitive approaches and to produce promising results.

Keywords: Document image classification · Multi-view representation learning · Deep learning

1 Introduction

In the field of document analysis and recognition, text document image classification is still a big challenge for the scientific community [15]. Especially, analyzing and understanding textual data from document images manually is very expensive and time-consuming. Unlike classical images, document images may have very complex structures and the extraction of structured relevant information from text is a difficult task [4].

Several approaches based on deep learning have been proposed for document images classification, which are based on visual features and structural information [12,13]. Some techniques have been used to extract text from document images using optical character recognition (OCR) [14].

© Springer Nature Switzerland AG 2021
J. Lladós et al. (Eds.): ICDAR 2021, LNCS 12824, pp. 318–332, 2021.
https://doi.org/10.1007/978-3-030-86337-1_22

Tensmeyer et al. [19] proposed a convolutional neural network (CNN) trained on RVL-CDIP dataset to extract region-specific features. Muhammed et al. [1] investigated various deep neural network architectures, including GoogLeNet, VGG, and ResNet and using transfer learning techniques (from real images), validated on the Tobacco-3482 and the large-scale RVL-CDIP datasets. Furthermore, the textual content of document has been used for the text document images classification [17]. A new approach extracting code-words from different locations of text document images is developed in [12]. Also, Kumar et al. [13] proposed an approach based on the construction of code-book from a text document images to perform the classification. Chen et al. [3] focused on an approach for structured document images classification by matching the salient feature points between the reference images and the query image. Kang et al. [10] used the CNN model for document classification based on the structural similarity. Moreover, Fesseha et al. [5] used pre-trained word embedding architectures using various CNNs to predict class labels. CNNs have been constructed with a continuous bag-of-words method, a skip-gram method, and without word2vec. Harley et al. [8] presented a comparative study for document image classification and retrieval, using features learned by CNNs. They proved that the CNNs are able to learn a hierarchical process of abstraction from pixel to concise and descriptive representations. Moreover, based on extensive experiments they showed that features learned from CNNs are robust to classification, and pre-trained CNNs on non-document images are efficient to document classification tasks. Moreover, in the natural language processing context, multiple deep learning methods have been proposed also to extract semantic features using the word embedding based on transfer learning techniques [20]. Recently, deep learning-based approaches have been developed for natural language processing and image classification [6]. Based on previous works, we can notice that deep neural networks models have shown their high performance in improving document image classification by extracting effective features. Most of the previous deep learning-based methods aim to use a single view (textual or visual features). However, some features are not useful for classification and may be noisy. Considering both visual contents and textual features remain an open challenge and we believe that fusing both features may provide a better accurate documents classification.

In this perspective, for text document images classification, we expect to leverage both text and image features. We propose a novel approach based on multi-view deep representation learning that aims at extracting and fusing textual and visual features into joint latent representation in order to improve the document images classification. Therefore, the aim of the proposed model is to extract only useful features and finding a shared representation, which can be effective for the classification.

The remainder of the paper is organized as follows. In Sect. 2, we explain the proposed methodology called Multi-View Encoder-Decoder Neural Networks (EDNets) which includes the multi-view deep autoencoder (MDAE) that allows combining textual- and visual-features, and the fully connected layer model.

In Sect. 3, we detail our experimental results. We conclude in Sect. 4 and give some perspectives for future works.

2   Proposed Method

This section describes the proposed multi-view deep representation learning based on the autoencoder model for document image classification. It contains three main steps: a) Feature extraction (textual and visual), b) Multi-view representation learning, and c) Classification of the document image. In the feature extraction step, we extract textual features from document text images using an OCR tool, and visual features using a transfer learning technique. Then, we use both features, i.e. textual and visual features to learn our multi-view deep representation learning in order to find a common space. The final step seeks to classify the obtained multi-view latent representation using a fully connected network (FCN) classifier. Figure 1 report the whole architecture of the proposed method.

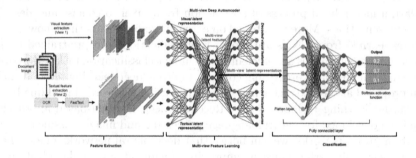

Fig. 1. The general flowchart of the proposed approach

2.1   Feature Extraction

In this section, the main goal of the feature extraction phase is to extract and build a set of textual and visual features from the document image (input), and to feed them into the multi-view deep learning model in order to improve the classification of the document image. We propose then to build a set of textual and spatial features vectors for both views.

Textual Feature Extraction X_t. This step seeks to extract textual features from text document images to feed them into the multi-view deep representation learning model. Let us consider X_t be the extracted textual features vectors. Therefore, in order to extract textual features from text document images, we used the Tesseract OCR model[1]. It is based on the Long Short Term Memory (LSTM) deep learning model and trained on a large-scale dataset. The main goal of the Tesseract model is to detect the text orientation by applying a set of rotations and transformations. Usually, it detects the black text from binary images using Otsu's thresholding. However, the outputs of Tesseract OCR are often noisy due to complex structures (see Fig. 2). Moreover, in order to get only the relevant and discriminative features from the extracted text, we use the FastText model [2], which is considered as a useful library for sentence classification and efficient learning of word representations. Finally, all extracted features are converted into a sequence of document words embedding using the variable-length document embedding technique [2].

Visual Feature Extraction X_v. In the visual feature extraction step, we use a deep learning model such as Convolutional Neural Network (CNN) [19] to extract relevant and discriminative visual features from the document image. To do so, we performed a transfer learning to enhance the visual feature extraction. Therefore, the EfficientNet model [18] is applied to extract visual features from all datasets. It uses a neural architecture search (NAS) to design a new neural network and to improve the accuracy of the classification. This model has proven its higher performance in the classification of ImageNet compared to standard CNN models [18]. Usually, it uses the mobile inverted bottleneck convolution (MBConv) and the AutoML MNAS model to optimize both efficiency and accuracy. Figure 3 reports the general architecture of the EfficientNet model.

2.2 Multi-view Feature Learning

In this section, we designed a multi-view deep autoencoder (MDAE) that aims at learning better representation and extracting relevant features from the fusion of multiple input views (X_t and X_v), from which these input views may be reconstructed ($\hat{X}_t$ and $\hat{X}_v$). We proposed three main autoencoder neural networks architectures reported respectively in Figs. 5, and 6. In the three proposed architectures, the input of the neural network model is a pair (x_t, x_v) of the two feature views (inputs), x_t for textual features, and x_v for visual features. In Fig. 5 (A), a simple AE (AutoEncoder) is designed to consider as input the concatenated features, i.e., the concatenation of textual and visual features $x = concat(x_t, x_v)$. Figure 5 (B) reports the second architecture, where obtained latent representations z_t and z_v are concatenated into a new latent space vector $z = concat(z_t, z_v)$. The last architecture in Fig. 6 is an MDAE based on shared representations, i.e., a common space between both latent representations z_t and z_v.

---

[1] https://github.com/tesseract-ocr/tesseract.

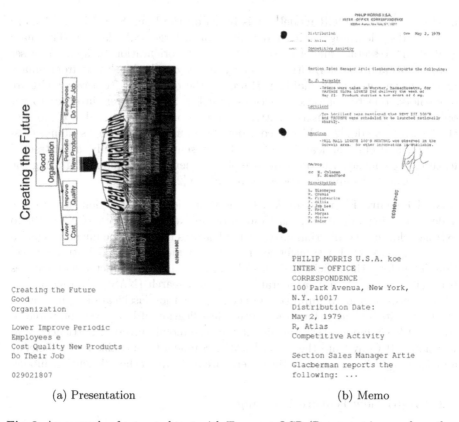

(a) Presentation

Creating the Future
Good
Organization

Lower Improve Periodic
Employees e
Cost Quality New Products
Do Their Job

029021807

(b) Memo

PHILIP MORRIS U.S.A. koe
INTER - OFFICE
CORRESPONDENCE
100 Park Avenua, New York,
N.Y. 10017
Distribution Date:
May 2, 1979
R, Atlas
Competitive Activity

Section Sales Manager Artie
Glacberman reports the
following: ...

Fig. 2. An example of extracted text with Tesseract OCR (Document images from the RVL-CDIP) [8].

Fig. 3. General architecture of the EfficientNet model [18].

Simple (Mono-View) Deep AE. A simple AE aims to reduce the high dimensionality of document image dataset X from D to d ($d \ll D$). Formally, it takes as input the initial feature data $\boldsymbol{X} \in \mathbb{R}^{N \times D}$, which includes an encoder model noted $E_\theta(\boldsymbol{X})$ and a decoder model noted $D_\phi(\boldsymbol{z})$ ($\boldsymbol{z}$ is the bottleneck layer i.e., latent space). The encoder model E_θ transforms $\boldsymbol{X} \in \mathbb{R}^{N \times D}$ into a new latent feature space $\boldsymbol{z} = E_\theta(\boldsymbol{X})$ ($\boldsymbol{z} \in \mathbb{R}^{N \times d}$). The decoder D_ϕ then aims to recover the initial data $\boldsymbol{X}$, i.e. $D_\phi(E_\theta(\boldsymbol{X})) \approx \boldsymbol{X}$. The objective function of the AE seeks to minimize the reconstruction error between the initial feature $\boldsymbol{X}$ and its output (reconstructed feature) $\hat{\boldsymbol{X}}$ using a Mean Squared Error (MSE) criterion as follows (see Fig. 4):

$$\mathcal{L}(\theta, \phi; \boldsymbol{x}) = \mathbb{E}\left[(\boldsymbol{x} - E_\theta(D_\phi(\boldsymbol{x})))^2\right] \tag{1}$$

where $E(.)$ and $D(.)$ are parameterized by θ and ϕ, respectively. The parameters (θ, ϕ) are learned together to reconstruct data $\hat{\boldsymbol{x}}$ same as the initial input $\boldsymbol{x}$.

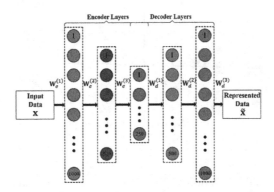

Fig. 4. The architecture of the deep AE. In this example, the sizes of first and second layers are 1000 and 500, respectively. The size of the bottleneck layer is of 250 units.

Multi-view Deep Autoencoder (MDAE). We designed an MDAE model that aims at extracting relevant and useful features from the combination of textual and visual features, from which these input features can be reconstructed. Our assumption is that textual and visual features are complementary for document images classification. The main goal is to find a shared multi-view representation from which one may get accurate classification results. The MDAE model contains one encoder model per view (X_t and X_v) noted E_t and E_v for textual and visual features inputs. Each of the two encoders models is a multi-layer neural network that projects the input feature (view) into a new latent feature space. We note $\boldsymbol{z}_t$ and $\boldsymbol{z}_v$ the corresponding relevant features extracted by the two encoders models, $\boldsymbol{z}_t = E_t(\boldsymbol{x}_t)$ and $\boldsymbol{z}_v = E_v(\boldsymbol{x}_v)$. A multi-view latent representation, $\boldsymbol{z}$, is extracted from the encoding of the two input features $\boldsymbol{X}_t$ and $\boldsymbol{X}_v$, then this multi-view latent representation $\boldsymbol{z}$ is feed to two decoders D_t and D_v that reconstruct both input features, $\hat{\boldsymbol{x}}_t = D_t(\boldsymbol{z})$ and $\hat{\boldsymbol{x}}_v = D_v(\boldsymbol{z})$. D_t and D_v are deep neural networks including three hidden dense layers. The optimization criterion of the MDAE model is the sum of the MSE criterion of both input features, defined as follows

$$\mathcal{L}(\boldsymbol{x}_t, \boldsymbol{x}_v; \theta) = \sqrt{\frac{1}{D}||\boldsymbol{x}_t - \hat{\boldsymbol{x}}_t||^2 + \frac{1}{w^2 d}||\boldsymbol{x}_v - \hat{\boldsymbol{x}}_v||^2} \tag{2}$$

The main goal of the MDAVE model is to find a common space from the two latent representations $\boldsymbol{z}_t$ and $\boldsymbol{z}_v$ using a merging layer, which aims to concatenate

z_t and z_v as: $z = concat(z_t, z_v)$ (see Fig. 5 (B)). We use another architecture based on shared weights and define the multi-view latent representation using a specific fusion dense layer as: $z = E(W \times z_t + W \times z_v)$. The aim is to find a common latent feature space, i.e., shared representation between two views: textual and visual features (see Fig. 6).

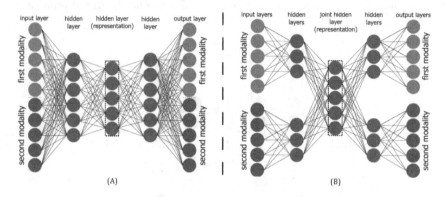

Fig. 5. Two typical multi-view deep AE architectures. (A) concatenated inputs (textual and visual features); (B) separated inputs, outputs and hidden layers, and concatenated latent representations

2.3   Classification with the Fully Connected Neural Network (FCN)

The obtained multi-view latent representation Z will be fed into the multi-layer perception (MLP) to perform the classification of document images. The general architecture of the MLP contains an input layer with n input-neurons, a set of hidden layers, and an output layer with m outputs (labels). Figure 7 reports the general architecture of the MLP model. In the output layer, the output of each neuron in the network, i.e., the label is obtained based on the probability theory using the *softmax* activation function as follows:

$$\mathcal{L}(z, \theta_l) = \frac{expf_l(z, \theta_l)}{\sum_{j=1}^{J} expf_j(z, \theta_j)}; \ 1 \leq l \leq J \tag{3}$$

where $\mathcal{L}(z, \Theta)$ is the probability that neuron z belongs to class l; $\theta_l = (\beta_0^l, ..., \beta_d^l; W_1, ..., W_d)$ is the vector of learned parameters, i.e., weights and bias of the output node l; $W_j = (W_{j0}, ..., W_{jK})$ is the vector of weights of the node j and $f_j(z, \theta_j)$, and $f_l(z, \theta_l)$ is the output value of node l defined as follows

$$f_l(z, \theta_l) = \beta_0^l + \sum_{j=1}^{d} \beta_j^l \sigma \left(w_{j0} + \sum_{i=1}^{K} w_{ji} z_i \right) \tag{4}$$

where $\sigma(.)$ is the *softmax* activation function. The sample z with a maximum probability will be assigned to a class l as follows

$$Y(z) = \arg \max_{l} \mathcal{L}(z, \theta_l); \ l = 1, ..., J$$

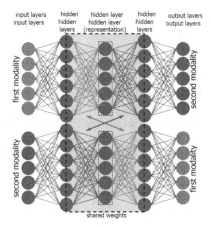

Fig. 6. Multi-view deep autoencoder based on shared weights

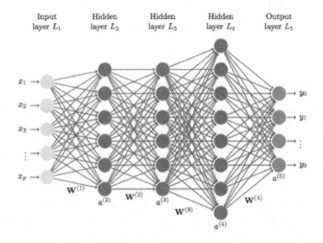

Fig. 7. The architecture of the MLP model

3   Experimental Results

3.1   Datasets and Preprocessing

To assess the performance and the effectiveness of the proposed MDAE model, we conduct our experiments on two real document images dataset respectively Tobacco3482 and RVL-CDIP.

- The Tobacco3482 image-based dataset [13] consists of 3482 white and black documents images. It is a subset from the Truth Tobacco Industry Documents archives[2]. There are ten true classes of documents (e.g. AVDE, email, forum, letter, memo,...).

---

[2] https://www.industrydocuments.ucsf.edu/tobacco/.

- The RVL-CDIP dataset [8] contains 400000 grayscale digitized documents with 16 labeled classes of documents (e.g. eadvertisement, email, form, handwritten, letter, memo, news article, ...). Each class contains 25000 images.

The Tobacco3482 and RVL-CDIP have been used to assess our multi-view deep representation learning model. Therefore, we use firstly the Tesseract OCR tool to extract text data from the grayscales document images. This operation has been performed on both images datasets. Table 1 reports the statistical information, including the number of documents, classes, and features of the two datasets Tobacco-3482 and RVL-CDIP (the last column is the number of extracted features).

Table 1. Statistical information of Tobbacco-3482, and RVL-CDIP datasets

Dataset	Documents	Classes	Features
Tobacco-3482	3482	10	130320
RVL-CDIP	399743	16	4257214

3.2 Multi-view Deep Autoencoders Implementations

We developed different types of deep autoencoders: mono-view (simple) autoencoder, and multi-view deep autoencoder model (MDAE). For all deep neural network models, we variate the size of the bottleneck layer from 2 to 100 features ($z \in [2, ..., 100]$). Furthermore, we tested multiple pairs of activation functions for the hidden layers and output layer, respectively: (*relu, linear*), (*relu, relu*), and (*relu, sigmoid*). All different deep autoencoders models were developed using the Keras API and Tensorflow. In the training stage, we use ADAM as an optimizer, with a learning rate $lr = 0.001$ over 400 epochs and a batch size of 200 training samples. We performed also regularization methods such as Dropout and Batch Normalization to leverage the overfitting problem. Finally, we tested three hidden layers with different sizes and we reported the best deep learning architecture.

3.3 Benchmarked Representation Learning Techniques

A comparative study has been performed between different representation learning methods, including principal component analysis (PCA) [16], independent component analysis (ICA) [9], and the AE model with different activation functions. Figure 8 reports the obtained average MSE of reconstruction of the visual features x_v using different representation learning (PCA, ICA, AE(*relu, linear*), AE(*relu, relu*), and AE(*relu, sigmoid*)). We use the concatenated textual and visual features as input for the different methods of learning representations. We can see in Fig. 8a that the best MSE is obtained with the AE(*relu, linear*) model for the Tobacco3482 Dataset, which is near to 0.04. Moreover, for the RVL-CDIP dataset, the same model gives the best results than other representation learning methods (see Fig. 8b). This improves the efficiency of the proposed deep learning model.

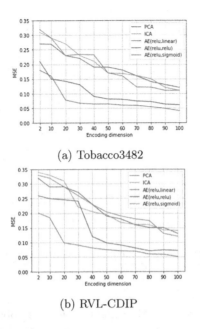

(a) Tobacco3482

(b) RVL-CDIP

Fig. 8. Average MSE of reconstruction versus encoding dimension

Classification Performance. In this section, we compare and evaluate the classification rates of our proposed multi-view deep representation learning MDAE with PCA, ICA, CNN, multi-channel CNN [21], AlexNet [11], and DAE. The Multi-channel CNN consists of 3 channels where each channel starts with an embedding layer. The classification task is performed using the FCN model. Figure 9 reports the obtained Overall Accuracies (OAs) using the multi-view representation learning models when the number of encoding dimensions varies from 2 to 100. For the Tobacco3482, the best OA (96.95%) is reached for $MDAE(Z_t + Z_v)$ when the number of encoding dimension equals 20 features. However, using mono-view models, i.e., $DAE(X_t)$ and $DAE(X_v)$, the obtained OAs are at most 91.35% and 95.21% respectively. For the remaining multi-view deep learning models, the highest values of OAs are between 94% and 95% (see Fig. 9a). In Fig. 9b, the best OA for the RVL-CDIP dataset is obtained also with the $MDAE(Z_t + Z_v)$ where the number of encoding dimension is equal to 97.81%, whereas with the $MDAE(X_t + X_v)$ the best OA is equal to 96.42%. This proves the added value and the effectiveness of the concatenated latent representation Z_t and Z_v.

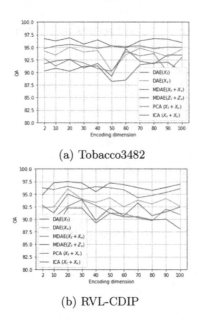

(a) Tobacco3482

(b) RVL-CDIP

Fig. 9. OA of classification versus encoding dimension using DAE(X_t), DAE(X_v), MDAE($X_t + X_v$), and MDAE($Z_t + Z_v$)

Table 2 reports the obtained MSE and OAs for the Tobacco3482 dataset. Following these results, we can notice that the proposed model MDAE (*relu, linear*) based on the concatenated latent representations ($Z_t + Z_v$) gives better OAs, compared to other representation learning models. In fact, the obtained MSE is 0.058, and OA is 96.95%. However, in the MDAE (*relu, linear*) model based on the concatenated latent representations ($X_t + X_v$), the best MSE and OA are equal to 0.083 and 95.75, respectively. In Table 3, best performances are obtained using the multi-view architecture where $MSE = 0.049$ and $OA = 97.81\%$ for the MDAE(*relu, linear*) model. This clearly proves again the superiority of the concatenation of latent representations of textual and visual features to get a higher classification of document images. The multi-channel CNN, ICA, and PCA give low classification accuracy compared to the MDAE model which their OAs are equal to 95.40%, 94.98% and 94.11%, respectively. Figures 10 and 11 show the confusion matrix of our multi-view deep representation learning network with the concatenated latent representations X_t and X_v. For the Tobacco342, 8 OAs of 9 classes are ≥94% (see Fig. 10). Moreover, in Fig. 11, we can see that most OAs are >95%, which demonstrates the effectiveness and the higher performance of the proposed MDAE model in the classification task. Based on the obtained results (the multi-view representation and the classification) and the comparative study, the proposed model MDAE gives better classification performances

Table 2. Best average MSE and OA ($\pm$ standard error) using textual and visual features on Tobbaco3482 dataset

	Model	Textual features (x_t)		Visual features (x_v)	
		Average MSE	Average OA	Average MSE	Average OA
Mono-view	PCA/FCN	0.135 ($\pm$ 0.023)	87.14 ($\pm$ 0.039)	0.129 ($\pm$ 0.072)	88.44 ($\pm$ 0.074)
	ICA/FCN	0.111 ($\pm$ 0.041)	89.56 ($\pm$ 0.026)	0.109 ($\pm$ 0.084)	91.23 ($\pm$ 0.062)
	CNN	N/A	73.90	N/A	79.91
	Multi-Channel CNN	N/A	87.10	N/A	93.20
	AlexNet	N/A	79.23	N/A	89.12
	DAE/FCN (*relu, relu*)	0.091 ($\pm$ 0.024)	90.04 ($\pm$ 0.066)	0.088 ($\pm$ 0.095)	94.92 ($\pm$ 0.035)
	DAE/FCN (*relu, lin.*)	**0.075** ($\pm$ 0.094)	**91.35** ($\pm$ 0.035)	**0.074** ($\pm$ 0.012)	**95.25** ($\pm$ 0.009)
	DAE/FCN (*relu, sig.*)	0.097 ($\pm$ 0.049)	90.03 ($\pm$ 0.028)	0.094 ($\pm$ 0.023)	94.12 ($\pm$ 0.052)
	Model	Concatenated inputs ($x_t + x_v$)		Fused latent rep. ($z_t + z_v$)	
		Average MSE	Average OA	Average MSE	Average OA
Multi-view	PCA/FCN	0.127 ($\pm$ 0.092)	93.85 ($\pm$ 0.047)	N/A	N/A
	ICA/FCN	0.113 ($\pm$ 0.056)	94.34 ($\pm$ 0.092)	N/A	N/A
	Multi-Channel CNN	N/A	87.10	N/A	93.20
	MDAE/FCN (*relu, relu*)	0.091 ($\pm$ 0.046)	95.12 ($\pm$ 0.034)	0.085 ($\pm$ 0.038)	95.24 ($\pm$ 0.039)
	MDAE/FCN (*relu, lin.*)	**0.083** ($\pm$ 0.016)	**95.75** ($\pm$ 0.013)	**0.058** ($\pm$ 0.063)	**96.95** ($\pm$ 0.026)
	MDAE/FCN (*relu, sig.*)	0.096 ($\pm$ 0.056)	94.97 ($\pm$ 0.031)	0.090 ($\pm$ 0.036)	95.06 ($\pm$ 0.044)

Table 3. Best average MSE and OA ($\pm$ standard error) using textual and visual features on RVL-CDIP dataset

	Model	Textual features (x_t)		Visual features (x_v)	
		Average MSE	Average OA	Average MSE	Average OA
Mono-view	PCA/FCN	0.152 ($\pm$ 0.032)	86.22 ($\pm$ 0.081)	0.134 ($\pm$ 0.058)	90.02 ($\pm$ 0.074)
	ICA/FCN	0.129 ($\pm$ 0.076)	90.11 ($\pm$ 0.038)	0.103 ($\pm$ 0.033)	92.02 ($\pm$ 0.023)
	CNN	N/A	65.35	N/A	89.8
	Multi-Channel CNN	N/A	88.67	N/A	92.81
	AlexNet	N/A	77.6	N/A	90.97
	DAE/FCN (*relu, relu*)	0.097 ($\pm$ 0.036)	91.62 ($\pm$ 0.002)	0.089 ($\pm$ 0.031)	94.82 ($\pm$ 0.029)
	DAE/FCN (*relu, lin.*)	**0.075** ($\pm$ 0.094)	**92.74** ($\pm$ 0.007)	**0.074** ($\pm$ 0.041)	**95.89** ($\pm$ 0.016)
	DAE/FCN (*relu, sig.*)	0.095 ($\pm$ 0.022)	91.12 ($\pm$ 0.062)	0.091 ($\pm$ 0.049)	94.59 ($\pm$ 0.067)
	Model	Concatenated inputs ($x_t + x_v$)		Fused latent rep. ($z_t + z_v$)	
		Average MSE	Average OA	Average MSE	Average OA
Multi-view	PCA/FCN	0.112 ($\pm$ 0.102)	94.11 ($\pm$ 0.072)	N/A	N/A
	ICA/FCN	0.109 ($\pm$ 0.066)	94.98 ($\pm$ 0.059)	N/A	N/A
	Multi-Channel CNN	N/A	95.40	N/A	N/A
	MDAE/FCN (*relu, relu*)	0.083 ($\pm$ 0.052)	95.32 ($\pm$ 0.046)	0.082 ($\pm$ 0.042)	95.47 ($\pm$ 0.066)
	MDAE/FCN (*relu, lin.*)	**0.069** ($\pm$ 0.084)	**96.64** ($\pm$ 0.058)	**0.049** ($\pm$ 0.036)	**97.81** ($\pm$ 0.057)
	MDAE/FCN (*relu, sig.*)	0.093 ($\pm$ 0.060)	95.03 ($\pm$ 0.032)	0.089 ($\pm$ 0.016)	95.43 ($\pm$ 0.026)

compared to the other representation learning and classification methods. Furthermore, MDAE allows finding a multi-view latent representation of document images while combining textual and visual features in order to improve the classification accuracy.

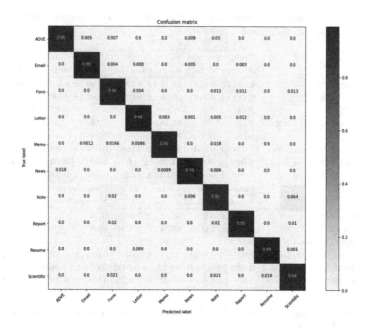

Fig. 10. Confusion matrix using the MDAE model (Tobacco3482 dataset)

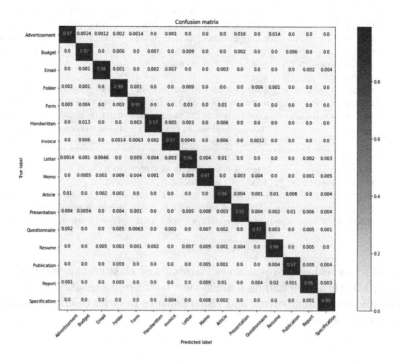

Fig. 11. Confusion matrix using the MDAE model (RVL-CDIP dataset)

4   Conclusion

In this paper, we propose a multi-view deep representation learning based on autoencoder (MDAE) that learns simultaneously from the visual information extracted from document images and textual features to perform document images classification. We show that, combining two views with different multi-view learning schemes improves the performance of the classification comparatively to mono-view networks. Furthermore, our proposed multi-view deep representation learning outperforms the current state-of-the-art methods. The fusion of textual and visual features has proved its efficiency to get a better classification of document images. In the future, we will investigate graph deep representation learning with different views [7], exploring new strategies that may improve the performance of the classification of documents.

References

1. Afzal, M.Z., Kölsch, A., Ahmed, S., Liwicki, M.: Cutting the error by half: investigation of very deep CNN and advanced training strategies for document image classification. In: 14th IAPR International Conference on Document Analysis and Recognition (ICDAR), vol. 1, pp. 883–888. IEEE (2017)
2. Bojanowski, P., Grave, E., Joulin, A., Mikolov, T.: Enriching word vectors with subword information. Trans. Assoc. Computat. Linguist. **5**, 135–146 (2017)
3. Chen, S., He, Y., Sun, J., Naoi, S.: Structured document classification by matching local salient features. In: Proceedings of the 21st International Conference on Pattern Recognition (ICPR), pp. 653–656. IEEE (2012)
4. Do, T.H., Ramos Terrades, O., Tabbone, S.: DSD: document sparse-based denoising algorithm. Pattern Anal. Appl. **22**(1), 177–186 (2018). https://doi.org/10.1007/s10044-018-0714-3
5. Fesseha, A., Xiong, S., Emiru, E.D., Diallo, M., Dahou, A.: Text classification based on convolutional neural networks and word embedding for low-resource languages: Tigrinya. Information **12**(2), 52 (2021)
6. Guo, J., et al.: GluonCV and GluonNLP: deep learning in computer vision and natural language processing. J. Mach. Learn. Res. **21**(23), 1–7 (2020)
7. Hanachi, R., Sellami, A., Farah, I.R.: Interpretation of human behavior from multimodal brain MRI images based on graph deep neural networks and attention mechanism. In: 16th International Joint Conference on Computer Vision, Imaging and Computer Graphics Theory and Applications (VISIGRAPP 2021), vol. 12, pp. 56–66. SCITEPRESS (2021)
8. Harley, A.W., Ufkes, A., Derpanis, K.G.: Evaluation of deep convolutional nets for document image classification and retrieval. In: 13th International Conference on Document Analysis and Recognition (ICDAR), pp. 991–995. IEEE (2015)
9. Jayasanthi, M., Rajendran, G., Vidhyakar, R.: Independent component analysis with learning algorithm for electrocardiogram feature extraction and classification. Signal Image Video Process. **15**, 391–399 (2021)
10. Kang, L., Kumar, J., Ye, P., Li, Y., Doermann, D.: Convolutional neural networks for document image classification. In: 22nd International Conference on Pattern Recognition (ICPR), pp. 3168–3172. IEEE (2014)

11. Krizhevsky, A., Sutskever, I., Hinton, G.E.: ImageNet classification with deep convolutional neural networks. Adv. Neural. Inf. Process. Syst. **25**, 1097–1105 (2012)
12. Kumar, J., Ye, P., Doermann, D.: Learning document structure for retrieval and classification. In: Proceedings of the 21st International Conference on Pattern Recognition (ICPR), pp. 1558–1561. IEEE (2012)
13. Kumar, J., Ye, P., Doermann, D.: Structural similarity for document image classification and retrieval. Pattern Recogn. Lett. **43**, 119–126 (2014)
14. Noce, L., Gallo, I., Zamberletti, A., Calefati, A.: Embedded textual content for document image classification with convolutional neural networks. In: Proceedings of the 2016 ACM Symposium on Document Engineering, pp. 165–173 (2016)
15. Patil, P.B., Ijeri, D.M.: Classification of text documents. In: Chiplunkar, N.N., Fukao, T. (eds.) Advances in Artificial Intelligence and Data Engineering. AISC, vol. 1133, pp. 675–685. Springer, Singapore (2021). https://doi.org/10.1007/978-981-15-3514-7_51
16. Shah, A., Chauhan, Y., Chaudhury, B.: Principal component analysis based construction and evaluation of cryptocurrency index. Expert Syst. Appl. **163**, 113796 (2021)
17. Shin, C., Doermann, D., Rosenfeld, A.: Classification of document pages using structure-based features. Int. J. Doc. Anal. Recogn. (IJDAR) **3**(4), 232–247 (2001)
18. Tan, M., Le, Q.: EfficientNet: rethinking model scaling for convolutional neural networks. arXiv 2019. arXiv preprint arXiv:1905.11946 (2020)
19. Tensmeyer, C., Martinez, T.: Analysis of convolutional neural networks for document image classification. In: 14th IAPR International Conference on Document Analysis and Recognition (ICDAR), vol. 1, pp. 388–393. IEEE (2017)
20. Yang, X., Yumer, E., Asente, P., Kraley, M., Kifer, D., Lee Giles, C.: Learning to extract semantic structure from documents using multimodal fully convolutional neural networks. In: Proceedings of the IEEE Conference on Computer Vision and Pattern Recognition (CVPR), pp. 5315–5324 (2017)
21. Zhang, Y., Roller, S., Wallace, B.: MGNC-CNN: a simple approach to exploiting multiple word embeddings for sentence classification. arXiv preprint arXiv:1603.00968 (2016)

Fast End-to-End Deep Learning Identity Document Detection, Classification and Cropping

Guillaume Chiron[(✉)], Florian Arrestier, and Ahmad Montaser Awal

Research Department AriadNEXT, Cesson-Sévigné, France
{guillaume.chiron,florian.arrestier,montaser.awal}@ariadnext.com

Abstract. The growing use of Know Your Customer online services generates a massive flow of dematerialised personal Identity Documents under variable capturing conditions and qualities (e.g. webcam, smartphone, scan, or even handcrafted pdfs). IDs are designed, depending on their issuing country/model, with a specific layout (i.e. background, photo(s), fixed/variable text fields) along with various anti-fraud features (e.g. checksums, Optical Variable Devices) which are non-trivial to analyse. This paper tackles the problem of detecting, classifying, and aligning captured documents onto their reference model. This task is essential in the process of document reading and fraud verification. However, due to the high variation of capture conditions and models' layout, classical handcrafted approaches require deep knowledge of documents and hence are hard to maintain. A modular approach using a fully multi-stage deep learning based approach is proposed in this work. The proposed approach allows to accurately classify the document and estimates its quadrilateral (localization). As opposed to approaches relying on a single end-to-end network, the proposed modular framework offers more flexibility and a potential for future incremental learning. All networks used in this work are derivatives of recent state-of-the-art ones. Experiments show the superiority of the proposed approach in terms of speed while maintaining good accuracy, both on the MIDV-500 academic dataset and on an industrial based dataset compared to hand crafted solutions.

Keywords: ID documents segmentation · Deeplearning · Classification

1 Introduction

Registration to critical services (e.g. bank account, online gambling, mobile phone subscription, etc.) is regulated and follow KYC[1] principles. Rules specify that customers' IDs have to be verified and assessed for conformity. These procedures are more and more done remotely, and the COVID-19 pandemic has

---

[1] Know Your Customer [16]: set of laws, certifications and regulations preventing criminal from either impersonating other people or forging false IDs.

© Springer Nature Switzerland AG 2021
J. Lladós et al. (Eds.): ICDAR 2021, LNCS 12824, pp. 333–347, 2021.
https://doi.org/10.1007/978-3-030-86337-1_23

accelerated this tendency to dematerialization. Automatized verification using computed vision can surly not yet replace the eye of qualified employees, however it can perform some obvious checks and alleviate the redundant manual labour. This paper settles in this context of need for fast automatic and accurate solutions for IDs processing.

Commercial solutions do exist for IDs recognition and reading, but the lack of transparency (e.g. eventual manual post treatment) makes them hard to benchmark. The community has shown a growing interest for this problematic and has recently released public datasets of IDs to play with. A major challenge was identified around the classification of the input document and its alignment onto its corresponding reference model (i.e. image registration problem). These specific early steps are essential to perform any further contextual analysis such as security checks and fields decoding.

To the best of our knowledge, only few solutions [4, 9, 12] published in the literature are viable in an industrial context (i.e. accurate, scalable and maintainable). They rely for the most on handcrafted features along with more or less advanced filtering and matching techniques (e.g. SIFT/FLANN/RANSAC). Whereas being quite robust, these approaches hardly run on mobile devices and especially suffer from performance issues with a growing number of models to support. Designing a solution to such a problem is challenging and requires to address various difficulties, as detailed here below:

- **Uncontrolled capturing conditions:** Scale (A4 scans, adjusted smartphone shots), definition (100 DPI to 4K), pose (orientation, perspective), light (glares, shadows), occlusions (overlaps), coloration (B&W, RGB).
- **Intrinsic variability of the models:** Multi-pass printing with unstable offsets, Fixed labels slight variations (e.g. language adaptation), Layout alteration (e.g. stamps added afterwards, frauds "collage").
- **Physical support:** Material (semi-rigid plastic, foldable paper booklets), alterations (effect of time, handmade plasticized), fuzzy bounds (i.e. unreliable borders and corners).
- **Models in circulation:** Constant evolution (models released/disposed), models number is the product of Countries * Document types (e.g. id, passport, residence permit, driving licence) * Versions (temporal updates, language variations), unbalanced samples (e.g. rare diplomatic passports).

Recent advances in deep-learning have brought new opportunities, especially regarding local features and matching (e.g. SuperPoint [11] and SuperGlue [20], respectively). In this context, we propose in this paper a modular, scalable, and maintainable end-to-end deep learning based approach for ID detection, classification and alignment. The Fig. 1 gives an overview of the pipeline, composed of a "Model agnostic" part generic to any documents, and a "Model specific" part requiring some specific training with multiple document samples (only for the classifier). This two-part split allows the system to detect and localize documents previously unseen by the classifier, and therefore they can be proceeded differently with an eventually less optimized fallback approach (e.g. [4]). The final alignment step requires a unique reference sample for each supported model.

Our contributions are twofold. The first innovative aspect is the pipeline itself, being fast, modular and able to rely 100% on deep-learning modules (i.e. spare from complex parameterization of some hand-crafted approaches). The second innovation specifically involves the combined localization and classification. A custom lightweight EfficientDet-like [25] is trained to detect and roughly localize any kind of IDs, followed by an even lighter version of MNASNet-A1 [26] dedicated to classification. As explained after, this specific configuration provides a pre-cropped document and paves the way for the use of deep-learning based fine alignment approaches (e.g. SuperPoint/SuperGlue).

To the best of our knowledge, it is the first time that such networks are applied to the task of ID localization and classification. When compared to hand-crafted approaches [4,12,24], the proposed networks are faster and more accurate on both the MIDV-500 academic dataset and on an industrial one. On-shelf local deep features were used as a proof of concept, but it would certainly benefit from retraining from scratch using a IDs dataset. The modularity of our framework is highlighted by an extensive study of different combination of classifiers, local features and matching methods.

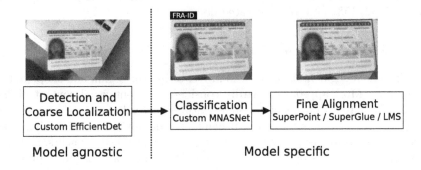

Fig. 1. Overview of the proposed end-to-end deep-learning pipeline.

The paper is organized as follows. Related approaches and techniques are presented in Sect. 2. Then, Sect. 3 details the proposed end-to-end multi-stage framework for ID detection, classification and fine alignment. Section 4 shows the main results obtained by the proposed approaches compared to existing and hybrid approaches. Finally, a conclusion is given with some perspectives.

2 Related Work

This section is organized as follows. First some existing ID datasets are presented, then an overview of previous works related to document segmentation and image alignment is proposed. Finally, end-to-end solutions for IDs detection, classification, and cropping are detailed.

2.1   Identity Document Segmentation Reference Datasets

Although public datasets of (IDs) are rare due to obvious privacy restrictions, some efforts have been made by the community to build up and share some reference data to benchmark on. MIDV-500 [2] is an annotated dataset composed of 15 000 ID images spread across 50 classes (issued from annotated video clips). Capturing conditions were limited (e.g. distortions and lighting conditions) and thus struggle to represent well a real industrial data flow. MIDV-2019 [7], an upgraded version of MIDV-500, did address some of these issues. However, the main drawback of both MIDV-500 and MIDV-2019 datasets is the use of an unique source document per class to generate all the image samples, hence missing the variable nature of IDs.

SmartDoc [8] dataset also consists of fully annotated video clips (i.e. 25 000 frames) of documents captured by a handheld devices. Compared to MIDV-500, SmartDoc dataset is composed of only 10 classes with only 2 classes being ID classes. The SmartDoc competition results shows a relatively low margin for improvement and indeed highlights the need for more challenging datasets.

Closely related works also refers to unpublished datasets: Brazilian [15], Colombian [9], Vietnamese [29] IDs and mixed industrial one [4,12,23] (composed of 5982, 1587 and 375 annotated documents respectively).

2.2   Blind Document Segmentation

Blind document segmentation consists in recovering the region and shape of the document(s). Most approaches hardly considers the contextual content of the document and relies on generic visible features instead (i.e. borders, corners, strong area contrasts). As a result, it can't provide any information about the orientation and also, it's prone to errors on partially visible documents, multiple overlapping documents or barely contrasted grayscale scans. Blind approaches can be split in the two categories detailed below: contour based and region based.

Contour based approaches relies mainly on borders, corners or any common features (e.g. text). Hough Transforms, salient lines detection and custom filtering are usual steps for extracting document boundaries. Various derivatives were proposed [3,18,27] and seems to perform well on academic datasets (i.e. SmartDoc and MIDV-500). All these approaches rely on the assumption that the document is visually separable form the background which is not always the case, especially in industrial dataset. Deep learning based solutions seem to outperform the traditional approaches and U-Net like network have become references in terms of contour detection. Gated and Bifurcated Stacked U-Net [5] were used for document border detection and dewarping. Also, designed to be faster than U-Nets, the Hough Encoder [22] (including Fast Hough Transform embeded layers) have shown goods results on MIDV-500/2019 datasets. Regarding corner detection based methods, deep learning techniques have been widely used. It seems that a CNN can be trained to regress corners positions [1], although two-step approaches [14,32] appear to be more robust. In the latest, corners are first roughly detected using an attentional map and then their positions are individually refined using a corner-specific deep network.

Region based approaches can be seen as a pixel wise segmentation problem. As for contour detection, U-Net like networks offer interesting results in terms of document segmentation, namely for background removal [9]. A U-Net variant using octave convolutions [15] was also proposed as fast alternative to classical convolutions based networks for document and text segmentation.

2.3   Generic Content Based Alignment

Content based alignment (or image registration) consists in aligning a query image onto a reference model which is supposed to be known a priori. This section do not exclusively focus on ID applications.

Sparse alignment refers here to the classic 5 steps process of keypoints detection, descriptors extraction, pairwise matching, filtering and homography estimation. Many combinations of detector and descriptors (e.g. SIFT, SURF, ORB, BRISK, LF-Net, SuperPoint)[2] have been exhaustively compared in the literature [6]. In terms of robustness to variations of illumination and geometry, SIFT often catches a place on the podium. The ORB binary descriptor also gets an honorable mention for its speed. More recently, deeplearning based solutions have been proposed and seem to outperform traditional methods on certain tasks. LF-Net and SuperPoint [11] performs well on HPatches benchmark with an advantage for SuperPoint regarding illumination changes. Authors of SuperPoint later introduced SuperGlue [20], a deeplearning sparse local features matcher that performs particularly well with SuperPoint features.

Dense alignment has also largely benefited from deep-learning in the past few years, first for short displacement applications (e.g. optical flow with FlowNet 2.0 [13]) and then for the larger problem of arbitrary image registration. HomographyNet [10] was first introduced as a generic solution able to estimate, after a supervised training, 4 displacement vectors yielding to the homography between a pair of images. An unsupervised approach [17], supposedly more generic, uses Spatial Transform Networks (STNs) and additionally relies on depth information. In STNs, one limitation is the photometric loss computed on the whole image intensities. This limitation has been overcame by learning content aware mask [30]. Finally, GLU-Net [28] proposes a universal solution based in attention mechanisms for dense alignment. All the previous work seems to still have limitations in terms of maximum displacement, scale and rotation they can handle. Modular solutions, as the one proposed in our paper, do not suffer from the same limitations such as RANSAC-Flow [21] which combines off-shelf deep features with RANSAC for coarse alignment and refinement with unsupervisedly trained local flow prediction. These latest solutions are promising, but relies on heavy networks (hundreds of MB) and are still too slow to fully replace traditional sparse approaches. Regarding document specific solutions, it can be noted some

---

[2] References for local features are not exhaustively provided and can be found in [6].

attempts to make a CNN directly regress the transformation matrix values from pair of images [31].

2.4 End-to-End Document Detection, Classification, and Cropping

On the one hand, blind document segmentation approaches listed in Sect. 2.2 do not bring satisfaction in regard of the difficulties mentioned in the introduction. On the other hand, whether they are sparse or dense, content based alignment methods listed in Sect. 2.3 do not bring any on-shelf solution for document cropping as the reference model (corresponding to the query) has to be known in advance, which is not the case in our current context.

To fill this gap, all-in-one document detection, classification and alignment solutions have been proposed in the literature [4,12,24]. Those approaches rely on sparse handcrafted keypoints/descriptors and perform a "1 query" to "N models" matching (combining FLANN and inverse-FLANN). It performs the classification and the quadrilateral estimation simultaneously and has the advantage of requiring only one sample per class. However, it hardly runs on mobile devices and do not necessarily scale well with a growing number of models to support.

3 Proposed End-to-End Pipeline for Detection, Classification and Fine-Grained Alignment

In this paper, we propose an end-to-end ID detection, classification and fine alignment pipeline, illustrated in Fig. 2. This pipeline is inspired from and built upon on the work of Awal et al. [4]. The pipeline in [4] performs the detection, classification and fine alignment in one shot using local hand-crafted features and a direct and inverse match with known models using FLANN. We made a first attempt to improve this pipeline was to replace the hand-crafted features with more modern and learned features extracted with the SuperPoint [11] network. Secondly, we tried to replace FLANN matching with the SuperGlue [20] network. However both of these attempt revealed to be impractical for different reasons. Firstly, replacing hand-crafted features with SuperPoint increases significantly the processing time (around 5 times experimented with SIFT) at equal resolution. This first problem can be mitigated by down scaling images to lower resolutions (e.g. 340×240 pixels) but it reduces SuperPoint performances due to signal loss. Secondly, using SuperGlue for matching is impractical in the same context as in [4] due to the number of keypoints in input.

However, we still think that deep learned features and matcher have a potential for improving detection, classification and fine alignment of IDs. To resolve the aforementioned issues, we propose to split the problem into 4 individual and distinct steps and designed a pipeline aiming at maximizing speed, accuracy, and maintainability. This novel pipeline is shown in Fig. 2 and described hereafter:

1. **Detection:** A customized EfficientDet-like [25] network is proposed for pre-localizing the document(s) in the image. The network is used without its

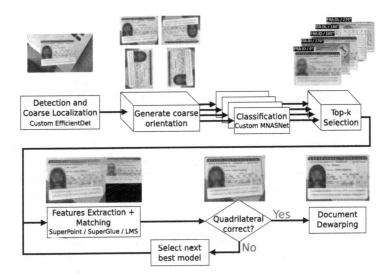

Fig. 2. Multi step end-to-end identity document processing chain.

classification ability, i.e. the training is done on various document samples but using one unique generic label, i.e. whether the box contains a document or not. As mentioned in the Fig. 1, this first part is "Model agnostic" and sufficiently generic to detect any IDs.

2. **Classification:** After the document has been roughly localized, a custom MNASNet-like [26] classifier is proposed to estimate the proper document class model. This part is "Model specific" and thus requires a specific training with multiple document samples for every classes to be recognized. The classification is performed along four different orientations $(0°, 90°, 180°, 270°)$ in order to reduce the rotation sensitivity of the classifier. The first to top-k outputs will possibly be explored in the next steps until a satisfying results is obtained.

3. **Features extraction:** Taking advantage of the precrop, the SuperPoint [11] network is used for computing keypoints and extracting local descriptors on both the detected document and its reference model (know by the classifier) at a unified scale.

4. **Fine-grain alignment:** In the last step, the SuperGlue [20] network performs the feature matching. The transformation between the cropped document and its normalized reference model is estimated by a Least Mean Square (LMS) regression. In case the quality of the resulting homography is not sufficient, then the current model is rejected and the next best i^{th} model of Step 2 is used as input of Step 3 until either the last of the best k-models has been tried or an homography satisfying the quality criteria is found.

The following highlights the benefits of using 2 distinct networks (EfficientDet for detection and MNASNet for classification) rather than an all-in-one detection network eventually conceivable with EfficientDet alone.

The advantage of using of EfficientDet as a generic solution for coarse pre-localization is twofold. First it makes the pipeline able to detect and localize document models that were not part of the training set, as EfficientDet generalizes well for this specific application. Those unseen documents can eventually be rejected by the classifier and therefore be processed by another less optimised classic fallback (e.g. such as [4]). This could be helpful to collect new training samples to further upgrade the classifier. Second, localizing the useful content (i.e. documents) solves the problem of variable DPIs and resolutions encountered in our dataset. Knowing the document box enables some kind of normalization (e.g. 320px width) of the query image to proceed down the line and in the classifier and makes the problem much easier for both the classifier and the matcher.

The MNASNet-like network performs a rough top-k classification of the pre-localized document. This way, the exploration of alignment combinations can be limited to only a reduced number of models (e.g. "1 query to top-5 models" rather "1 query to all supported models"). This is especially useful as SuperPoint descriptors are heavy (e.g. twice the size of SIFT ones) and would otherwise be too slow in any exhaustive matching process (e.g. direct/inverse-FLANN already being a bottleneck in [4]).

In Step 4, the rejection mechanism requires an estimation of the homography quality (or the quality of the transformed quad Q). The assumption made is that bad homographies lead to hazardous quads hardly explainable by any legit camera pose (translation and rotation). As the reference model width/height ratio is known a priori thanks to the classification step, it is possible to estimate, the most probable camera pose that would lead to the quad Q. This is done with a dicotomic search on the basis of arbitrary defined calibration parameters. Therefore, the quality metric is the Intersection over Union (IoU) between the quad Q and the theoretical quad obtained by transforming the reference model quad with the estimated camera pose.

Special considerations for runtime optimization were made for the design of the pipeline. The original backbone of EfficientDet (i.e. B0, the smaller configuration) was changed by a MNASNet-A1 [26] network which makes the detector run 4 times faster on CPU. No significant drop of box detection accuracy were observed. Also, MNASNet-like networks are by essence very light. But as it classified pre-localised ID with an excellent accuracy by default, we proposed to use an even lighter design (half the number of channels per layer) compared to MNASNet-A1 (the lightest version of MNASNet [26]) for the classification. Theses network optimisations made our pipeline more mobile compliant.

4 Experiments and Evaluation

4.1 Experimental Setup

The different experiments are conducted on two datasets, the academic MIDV500 dataset [2] (publicly available) and a private industrial dataset which is supposedly more challenging and representative of a real life application, both illustrated in Fig. 3. All experiments are performed on a laptop with an INTEL

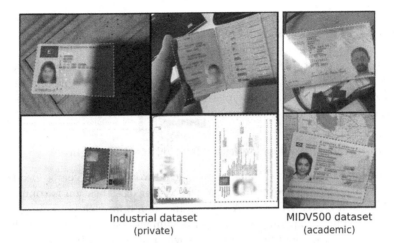

Industrial dataset   MIDV500 dataset
(private)   (academic)

Fig. 3. Samples of an industrial dataset of Identity Documents (on the left) and academic MIDV-500 dataset (on the right).

i7-7820HQ CPU @ 2.90 GHz processor and equipped with 32 Go of RAM. Importantly, all experiments involving deep neural networks were not run on GPUs to present comparable results with traditional handcrafted approaches.

The MIDV500 academic dataset, which is described in Sect. 2.1, provides a reference baseline but suffers from several limitations. The main limitation being the lack of diversity in the number of unique documents per class. Indeed, with only 1 unique document per class, classification tasks performed on this dataset are strongly biased. Secondly, in order to provide comparable results, we use the same protocol as the one proposed in [12], that is documents having an out-of-image-bounds area exceeding 10% are removed from the dataset, which correspond to roughly a 10% pruning of the dataset. When training Efficient-Det [25] and MNASNet [26] networks on this dataset for detection and classification tasks, the pruned dataset, composed of roughly 13500 samples distributed over 50 classes, is split in half for training and evaluation.

Similarly, the industrial dataset is composed of about 14k sample images uniformly distributed over 67 different classes, composed of passports, National ID Cards (NICs), health cards, driving licences, etc. from 14 different countries. The industrial dataset offers variability on multiple levels: each document is unique (i.e. no two samples have the same name), capturing conditions are non standardized and physical supports are diverse. Figure 3 illustrates some of these challenges with partially shadowed documents, paper based documents or non color captured documents. Contrarily to the MIDV500 dataset, the ground truth of document quadrilaterals was not available off-the-shelf and it was generated using a semi-automated process. A first detection is performed using the SIFT + FLANN approach such as [4], the generated quads are then manually verified and corrected when needed. We find that this semi-automated annotation process do generate better reference quads than fully manual annotations due to sometime a lack of clear, or out-of-bound, boundaries on documents.

The next two sections present results of two main experiments. Section 4.2 present the effect of the coarse localization step on the classification speed and accuracy. Then, Sect. 4.3 details the results of the complete processing chain and, specifically, the validity rate of the detected quads in both MIDV500 and Industrial datasets.

4.2   Detection and Classification: Faster, Better, Lighter

In this section, the use of the EfficientDet-like network [25] for coarse localization combined with the MNASNet-like network [26] for classification is compared to a non detection based SIFT+FLANN classification approach [4]. Following recommendations in [4,12], variable fields of documents (such as names or dates of birth) are masked in the reference models when using the SIFT+FLANN approach as it provides better classification results. Additionally, different, arbitrary chosen, resolutions of input images are tested for the SIFT+FLANN approach in order to evaluate the impact on classification. Indeed, due to the multi-scale nature of the SIFT features, there is a trade-off to be found between classification accuracy and processing time. Finally, we compare 3 approaches using the EfficientDet network: an EfficientDet (Multi) only method with classification output in addition to box detection, an EfficientDet (Mono) + MNASNet pipeline performing box detection only and classification only tasks, respectively, and finally, EfficientDet (Mono) combined with a SIFT + FLANN classifier.

Table 1 shows the results obtained for both approaches on both datasets. The "Loc." column directly refers to the accuracy of the EfficientDet coarse localization, and thus is not provided for SIFT+FLANN approach as it performs not proper pre-localization. The value represents the percentage of documents with an IoU value computed from the ground truth quad over the detected box > 0.5. The "Top_1" and "Top_5" columns refer to the percentage of correct classification when taking only the best result and the best 5 results, respectively.

From the results of Table 1, on the Industrial dataset, the pipelines based on EfficienDet only or "EfficientDet + MNASNet" networks appear to be orders of magnitude faster than the SIFT+FLANN pipelines while achieving highest Top_1 and Top_5 classification accuracy. Interestingly, the EfficienDet only method is outperformed by both the SIFT+FLANN method for an 800 pixel input resolution and the MNASNet approaches when looking at the Top_1 classification accuracy. However, it achieves the highest Top_5 classification compared to every other approaches. Finally, results of Table 1 show higher accuracy on estimated bounding boxes when training EfficientDet with only one generic document label for all samples (Mono configuration), rather than training it with specific labels corresponding to each model (Multi configuration).

On the MIDV500 dataset, results show a clear superiority in Top_1 classification accuracy of SIFT+FLANN approaches over deep learning based ones. This superiority is almost reduced to none when looking at the Top_5 classification accuracy and is explained by the biased nature of the MIDV500 dataset. Indeed,

Table 1. Classification accuracy and speed for the SIFT + FLANN approach for different resolutions VS the proposed EffDet + MNASNet based approach.

Method/Resolution	Industrial				MIDV500			
	Loc. (%)	Classif (%)		Time (s)	Loc. (%)	Classif (%)		Time (s)
		Top$_1$	Top$_5$			Top$_1$	Top$_5$	
SIFT + FLANN/320px	–	71.38	74.44	1.38	–	98.61	98.74	1.18
SIFT + FLANN/480px	–	86.44	89.42	1.45	–	98.61	98.88	1.26
SIFT + FLANN/640px	–	90.67	93.53	1.54	–	98.46	98.74	1.34
SIFT + FLANN/800px	–	91.27	94.31	1.62	–	97.30	98.19	1.41
SIFT + FLANN/1024px	–	89.93	93.78	1.85	–	95.90	97.11	1.53
EffDet (Multi)/512px	97.47	90.85	**96.13**	**0.04**	99.3	96.71	99.09	0.04
EffDet (Mono)/512px + MNASNet/240px	**98.95**	**94.98**	95.29	0.06	**99.43**	93.91	98.87	**0.03**
EffDet (Mono)/512px + SIFT + FLANN/640px	**98.95**	90.21	94.17	1.3	**99.43**	**98.97**	**99.10**	0.7

as there is only one document, albeit seen from different viewpoints and capture conditions, keypoints extracted during the learning phase and during the test part are most likely to be very close whereas this lack of diversity seems to alter the quality of the learned features of the MNASNet network.

Figure 4 shows a visualization of confusion matrices for the "EfficientDet (Mono) + MNASNet" and the SIFT+FLANN at 640px resolution approaches. The clear diagonal and the sparse distribution of the errors indicate a healthy classification. It's important to note that no rejection class was included in these experiments. Additionally, the left column on the matrices give an estimation of the good coarse localization of the detection with EfficientDet, thus not available for the SIFT+FLANN approach.

Overall, the EfficientDet + MNASNet method seem to be the best choice as it provides highly accurate coarse localized bounding boxes of documents while maintaining good classification results and low processing times.

4.3   Fine Alignment: Learned vs Handcrafted Features

This second group of experiments aims at validating our main contribution which is our overall end-to-end deep-learning pipeline for ID classification and fine grain alignment. In this section, all configurations explored rely on our robust "EfficientDet + MNASNet" preprocessing as proven efficient in Sect. 4.2.

As explained in Sect. 3, once the detection and classification steps are performed, the last step before obtaining an exploitable ID unwarped for security checks and fields decoding is the fine grained alignment step. The image alignment process consists in computing local features onto both the query and the reference model images, followed by descriptors matching and homography estimation. In this section, performances of different local features extraction methods and features matcher algorithms, both deep learning and hand-crafted based, are compared.

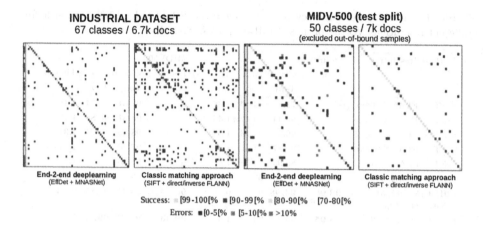

Fig. 4. Confusion matrices obtained with our end-2-end deeplearning pipeline versus the a SIFT/FLANN classification approach (as in [4]) on both MIVD500 and our industrial dataset.

Table 2. Crop results for our EfficientDet-like (detector)/MNASNet-like (classifier) combined with various local features and matchers. (*) LMS estimator is used for SuperGlue rather than USAC as recommended in [20].

Dataset	Features	Matcher	Valid crop(%) ↑	Avg Rank	Time (s) ↓
MIDV500	SIFT	FLANN/BF	97.78/97.67	1.10/1.11	0.74/0.40
	SURF	FLANN/BF	96.23/97.68	1.11/1.11	0.46/**0.36**
	SuperPoint	FLANN/BF	97.09/**98.28**	1.08/1.10	0.95/0.93
		SuperGlue*	97.99	1.10	1.39
Industrial	SIFT	FLANN/BF	80.96/86.41	1.07/1.10	0.42/0.20
	SURF	FLANN/BF	78.66/85.79	1.08/1.09	0.24/**0.17**
	SuperPoint	FLANN/BF	78.10/90.17	1.00/1.07	0.59/0.46
		SuperGlue*	**90.43**	1.07	0.74

The main metric used for the evaluation and comparison of the different methods is the IoU value of the estimated quadrilateral, a.k.a crop, over the ground truth quadrilateral of an ID. A crop is considered as accepted if and only if the resulting IoU value w.r.t the ground truth is >0.9.

Table 2 shows the results obtained on both the Industrial and MIDV500 datasets in terms of accepted crop rate and processing time. For each matching algorithm, the USAC [19] framework is used for the homography estimation, apart for the SuperGlue [20] matcher. Indeed, following recommendations of authors, the SuperGlue network produces good enough matches to be used with a simple, non-robust, LMS based estimator while yielding better results than when used with RANSAC. Importantly, results reported in Table 2 are obtained using the Top$_5$ classification result of Sect. 4.2, meaning that we loop onto the top 5 best classification result as long as no satisfying crop is found.

The results of Table 2 show that the deep-learning based features extractor and matcher composed of SuperPoint [11] and SuperGlue [20] networks outperforms hand-crafted alternative in term of pure accepted crop percentage, especially on the Industrial dataset. On MIDV500 however, SuperGlue seems not necessary and a simpler Brute Force (BF) matcher seems to make the job. It's important to note that both SuperPoint and SuperGlue networks were taken pretrained off-the-shelf and do not show their full potential as they were not specifically trained for the current task.

5   Conclusion

In this paper, we proposed a novel fast end-to-end framework to classify and accurately localize Identity Documents. Our proposed approach leverage the recent advances in deep learning and uses a fast custom derivative of the recent EfficientDet-B0 network to perform a coarse detection and localization of the IDs in the image. Following this coarse detection step, we use again a custom light derivative of the efficient MNASNet-A1 classification network to identify the document model with a relatively high accuracy. This proposed scheme of detection then classification showed orders of magnitude faster processing time compared to classical approach (e.g. SIFT + inverse/direct-FLANN) while improving the results on both test datasets (academic MIDV500 and our Industrial). Our solution has the advantage of being able to detect previously unseen ID models, which can afterward be rejected by the classifier and thus be sent to another less optimised classic fallback approach (e.g. such as [4]).

Secondly, we showed that using off-the-shelves learned local features and descriptors such as the one of the SuperPoint network provide higher quality fine alignment on IDs. Similarly, results showed that using a pre-trained version of the SuperGlue network for features matching almost always yield the best performances, albeit not by a large margin. However, although a bit disappointing as is, these preliminary results lead us to believe that retraining SuperPoint and SuperGlue networks on our specific datasets could lead to state of the art performances.

As future perspective, we intend to perform these retraining of SuperPoint and SuperGlue networks, as well as focus on the scalability of our current solution. Indeed, using a deep learning network for classification reduce the potential for adding new classes with few samples, which is a constraint often met in industrial identity document verification applications. An existing work [23] (tackling a very similar problem), has shown the possibility of using the already trained features of a CNN network to perform one-shot learning and thus facilitating the support of new models.

References

1. Abbas, S.A., ul Hussain, S.: Recovering homography from camera captured documents using convolutional neural networks. arXiv preprint arXiv:1709.03524 (2017)
2. Arlazarov, V.V., et al.: MIDV-500: a dataset for identity documents analysis and recognition on mobile devices in video stream. CoRR (2018)
3. Attivissimo, F., et al.: An automatic reader of identity documents. In: Systems, Man and Cybernetics (SMC). IEEE (2019)
4. Awal, A.M., et al.: Complex document classification and localization application on identity document images. In: 14th IAPR International Conference on Document Analysis and Recognition, pp. 426–431 (2017)
5. Bandyopadhyay, H., et al.: A gated and bifurcated stacked U-Net module for document image dewarping (2020). arXiv:2007.09824 [cs.CV]
6. Bojanić, D., et al.: On the comparison of classic and deep keypoint detector and descriptor methods. In: 11th International Symposium on Image and Signal Processing and Analysis (ISPA), pp. 64–69. IEEE (2019)
7. Bulatov, K., et al.: MIDV-2019: challenges of the modern mobile based document OCR. In: ICMV 2019, vol. 11433 (2020)
8. Burie, J.-C., et al.: ICDAR2015 competition on smartphone document capture and OCR (SmartDoc). In: 13th International Conference on Document Analysis and Recognition, pp. 1161–1165. IEEE (2015)
9. Castelblanco, A., Solano, J., Lopez, C., Rivera, E., Tengana, L., Ochoa, M.: Machine learning techniques for identity document verification in uncontrolled environments: a case study. In: Figueroa Mora, K.M., Anzurez Marín, J., Cerda, J., Carrasco-Ochoa, J.A., Martínez-Trinidad, J.F., Olvera-López, J.A. (eds.) MCPR 2020. LNCS, vol. 12088, pp. 271–281. Springer, Cham (2020). https://doi.org/10.1007/978-3-030-49076-8_26
10. DeTone, D., et al.: Deep image homography estimation. arXiv preprint arXiv:1606.03798 (2016)
11. DeTone, D., et al.: Superpoint: self-supervised interest point detection and description. In: Proceedings of the IEEE Conference on Computer Vision and Pattern Recognition Workshops, pp. 224–236 (2018)
12. Chiron, G., et al.: ID documents matching and localization with multi-hypothesis constraints. In: 25th International Conference on Pattern Recognition (ICPR). IEEE (2020)
13. Ilg, E., et al.: FlowNet 2.0: evolution of optical ow estimation with deep networks. In: Proceedings of the IEEE Conference on Computer Vision and Pattern Recognition, pp. 2462–2470 (2017)
14. Javed, K., Shafait, F.: Real-time document localization in natural images by recursive application of a CNN. In: 14th IAPR International Conference on Document Analysis and Recognition (ICDAR), vol. 1, pp. 105–110. IEEE (2017)
15. das Neves Junior, R.B., et al.: A fast fully octave convolutional neural network for document image segmentation. arXiv preprint arXiv:2004.01317 (2020)
16. Mullins, R.R., et al.: Know your customer: how salesperson perceptions of customer relationship quality form and influence account profitability. J. Mark. 78(6), 38–58 (2014)
17. Nguyen, T., et al.: Unsupervised deep homography: a fast and robust homography estimation model. IEEE Rob. Autom. Lett. 3(3), 2346–2353 (2018)
18. Puybareau, É., Géraud, T.: Real-time document detection in smartphone videos. In: 25th IEEE International Conference on Image Processing, pp. 1498–1502 (2018)

19. Raguram, R., et al.: USAC: a universal framework for random sample consensus. IEEE Trans. Pattern Anal. Mach. Intell. **35**(8), 2022–2038 (2012)

20. Sarlin, P.-E., et al.: Superglue: learning feature matching with graph neural networks. In: Proceedings of the IEEE/CVF Conference on Computer Vision and Pattern Recognition, pp. 4938–4947 (2020)

21. Shen, X., et al.: RANSAC-flow: generic two-stage image alignment. arXiv preprint arXiv:2004.01526 (2020)

22. Sheshkus, A., et al.: Houghencoder: neural network architecture for document image semantic segmentation. In: IEEE International Conference on Image Processing (ICIP), pp. 1946–1950 (2020)

23. Simon, M., et al.: Fine-grained classification of identity document types with only one example. In: 2015 14th IAPR International Conference on Machine Vision Applications (MVA), pp. 126–129. IEEE (2015)

24. Skoryukina, N., et al.: Fast method of ID documents location and type identification for mobile and server application. In: International Conference on Document Analysis and Recognition, pp. 850–857 (2019)

25. Tan, M., et al.: EfficientDet: scalable and efficient object detection. In: Proceedings of the IEEE/CVF Conference on Computer Vision and Pattern Recognition, pp. 10781–10790 (2020)

26. Tan, M., et al.: MnasNET: platform-aware neural architecture search for mobile. In: IEEE/CVPR, pp. 2820–2828 (2019)

27. Tropin, D.V., et al.: Approach for document detection by contours and contrasts. arXiv preprint arXiv:2008.02615 (2020)

28. Truong, P., et al.: GLU-Net: global-local universal network for dense flow and correspondences. In: IEEE/CVPR (2020)

29. Viet, H.T., et al.: A robust end-to-end information extraction system for Vietnamese identity cards. In: NAFOSTED (2019)

30. Zhang, J., et al.: Content-aware unsupervised deep homography estimation. In: Vedaldi, A., Bischof, H., Brox, T., Frahm, J.-M. (eds.) ECCV 2020. LNCS, vol. 12346, pp. 653–669. Springer, Cham (2020). https://doi.org/10.1007/978-3-030-58452-8_38

31. Zhou, Q., Li, X.: STN-homography: estimate homography parameters directly. arXiv preprint arXiv:1906.02539 (2019)

32. Zhu, A., Zhang, C., Li, Z., Xiong, S.: Coarse-to-fine document localization in natural scene image with regional attention and recursive corner refinement. Int. J. Doc. Anal. Recogn. (IJDAR) **22**(3), 351–360 (2019). https://doi.org/10.1007/s10032-019-00341-0

Gold-Standard Benchmarks and Data Sets

Image Collation: Matching Illustrations in Manuscripts

Ryad Kaoua[1], Xi Shen[1], Alexandra Durr[2], Stavros Lazaris[3], David Picard[1], and Mathieu Aubry[1(✉)]

[1] LIGM, Ecole des Ponts, Université Gustave Eiffel, CNRS,
Marne-la-Vallée, France
mathieu.aubry@enpc.fr
[2] Université de Versailles-Saint-Quentin-en-Yvelines, Versailles, France
[3] CNRS (UMR 8167), Ivry-sur-Seine, France

Abstract. Illustrations are an essential transmission instrument. For an historian, the first step in studying their evolution in a corpus of similar manuscripts is to identify which ones correspond to each other. This image collation task is daunting for manuscripts separated by many lost copies, spreading over centuries, which might have been completely re-organized and greatly modified to adapt to novel knowledge or belief and include hundreds of illustrations. Our contributions in this paper are threefold. First, we introduce the task of illustration collation and a large annotated public dataset to evaluate solutions, including 6 manuscripts of 2 different texts with more than 2 000 illustrations and 1 200 annotated correspondences. Second, we analyze state of the art similarity measures for this task and show that they succeed in simple cases but struggle for large manuscripts when the illustrations have undergone very significant changes and are discriminated only by fine details. Finally, we show clear evidence that significant performance boosts can be expected by exploiting cycle-consistent correspondences. Our code and data are available on http://imagine.enpc.fr/~shenx/ImageCollation.

1 Introduction

Most research on the automatic analysis of manuscripts and particularly their alignment, also known as collation, has focused on text. However, illustrations are a crucial part of some documents, hinting the copyist values, knowledge and beliefs and are thus of major interest to historians. One might naively think that these illustrations are much easier to align than text and that a specialist can identify them in a matter of seconds. This is only true in the simplest of cases, where the order of the illustrations is preserved and their content relatively similar. In harder cases however, the task becomes daunting and is one of the important limiting factor for a large scale analysis.

As an example, the "De materia Medica" of Dioscorides, a Greek pharmacologist from the first century, has been widely distributed and copied between the 6th and the 16th century. Depending on the versions, it includes up to 800

© Springer Nature Switzerland AG 2021
J. Lladós et al. (Eds.): ICDAR 2021, LNCS 12824, pp. 351–366, 2021.
https://doi.org/10.1007/978-3-030-86337-1_24

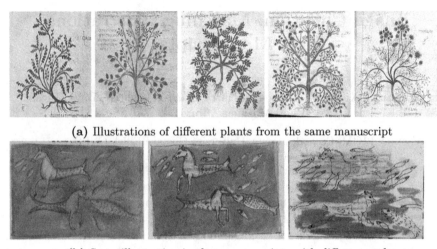

(a) Illustrations of different plants from the same manuscript

(b) Same illustration in three manuscripts with different styles

Fig. 1. The illustration alignment task we tackle is challenging for several reasons. It requires fine-grained separation between images with similar content (a), while being invariant to strong appearance changes related to style and content modifications (b). The examples presented in this figure are extracted from the two groups of manuscript we include in our dataset: (a) the *De Materia Medica* of Dioscoride; (b) the *Physiologus*.

depictions of natural substances. In particular the manuscripts we study in this paper contain around 400 illustrations of plants. They have been re-organized in different orders in the 17 different known illustrated versions of the text, for example alphabetically or depending on their therapeutic properties. The changes in the illustrations and their organizations hints both at the tradition from which each manuscript originates and at the evolution of scientific knowledge. However, the important shifts both in the illustrations appearance and in the order in which they appear makes identifying them extremely time consuming. While the text could help, it is not always readable and it is sometime not next to the illustrations.

From a Computer Vision perspective, the task of retrieving corresponding illustrations in different versions of the manuscripts present several interesting challenges, illustrated in Fig. 1. First, we are faced with a fine-grained problem, since many illustrations correspond to similar content, such as different plants (Fig. 1a). Second, the style, content and level of details vary greatly between different versions of the same text (Fig. 1b). Third, we cannot expect relevant supervision but can leverage many constraints. On the one hand the annotation cost is prohibitive and the style and content of the illustrations vary greatly depending on the manuscripts and topics. On the other hand the structure of the correspondences graph is not random and could be exploited by a learning or optimization algorithm. For example, correspondences should mainly be one on one, local order is often preserved, and if three or more versions of the same

Table 1. The manuscripts in our dataset come from two different texts, have diverse number of illustrations and come from diverse digitisations. In total, it includes more than 2000 illustrations and 1 200 annotated correspondences.

Name	Code	Number of folios	Folios' resolution	Number of illustrations	Annotated correspondences
Physiologus	P1	109	1515 × 2045	51	P2: 50 - P3: 50
	P2	176	755 × 1068	51	P1: 50 - P3: 51
	P3	188	792 × 976	52	P1: 50 - P2: 51
De Materia Medica	D1	557	392 × 555	816	D2: 295 - D3: 524
	D2	351	1024 × 1150	405	D1: 295 - D3: 353
	D3	511	763 × 1023	839	D1: 524 - D2: 353

text are available correspondences between the different versions should be cycle consistent.

In this paper, we first introduce a dataset for the task of identifying correspondences between illustrations of collections of manuscripts with more than 2 000 extracted illustrations and 1 200 annotated correspondences.

Second, we propose approaches to extract such correspondences automatically, outlining the crucial importance both of the image similarity and its nontrivial use to obtain correspondences. Third, we present and analyze results on our dataset, validating the benefits of exploiting the problem specificity and outlining limitations and promising directions for future works.

2   Related Work

Text Collation. The use of mechanical tools to compare different versions of a text can be dated back to Hinman's collator, an opto-mechanical device which Hinman designed at the end of the 1940s to visually compare early impressions of Shaekspeare's works [30]. More recently, computer tools such as CollateX [13] have been developed to automatically compare digitised versions of a text. The core idea is to explain variants using the minimum number of edits, or block move [5], to produce a variant graph [26]. Most text alignement methods rely on a transcription and tokenization step, which is not adapted to align images. Methods which locally align texts and their transcriptions e.g., [10,14,16,17,25], are also related to our task. Similar to text specific approaches, we will show that leveraging local consistency in the alignments has the potential to improve results for our image collation task.

Image Retrieval in Historical Documents. Given a query image, image retrieval aims at finding images with similar content in a database. Classic approaches such as Video-Google [29] first look for images with similar SIFT [20] features then filter out those which features cannot be aligned by a simple spatial transformation. Similar bag of words approaches have been tested for pattern spotting

Fig. 2. Structure of the correspondences for the "De Materia Medica". From left to right we show crops of the full correspondence matrices for D1-D2, D1-D3 and D2-D3. The black dots are the ground truth annotations. While the order is not completely random, the illustrations have been significantly re-ordered. Best viewed in electronic version.

in manuscripts [9]. However, handcrafted features such as SIFTs fail in the case of strong style changes [27] which are characteristic of our problem.

Recent studies [12,23] suggest that directly employing global image features obtained with a network trained on a large dataset such as ImageNet [7] is a strong baseline for image retrieval. Similarly, [31] leverages features from a RetinaNet [18] network trained on MS-Coco [19] for pattern spotting in historical manuscripts and shows they improve over local features. If annotations are available, the representation can also to be learned specifically for the retrieval task using a metric learning approach, i.e., by learning to map similar samples close to each other and dissimilar ones far apart [12,22,24]. Annotations are however rare in the case for historical data.

Recently, two papers have revisited the Video-Google approach for artistic and historical data using pre-trained deep local features densely matched in images to define an image similarity: [27] attempts to discover repeated details in artworks collections and [28] performs fine-grained historical watermark recognition. Both papers propose approaches to fine-tune the features in a self-supervised or weakly supervised fashion, but report good results with out-of-the box features.

3   Dataset and Task

We designed a dataset to evaluate image collation, i.e., the recovery of corresponding images in sets of manuscripts. Our final goal is to provide a tool that helps historians analyze sets of documents by automatically extracting candidate correspondences. We considered two examples of such sets originating from different libraries with online text access to digitized manuscripts [1–4], which characteristics are summarized in Table 1 and which are visualized in Fig. 1 and 3:

– The "Physiologus" is a christian didactic zoological text compiled in Greek during the 2nd century AD. The three manuscripts we have selected contain illuminations depicting real and fantastic animals and contain respectively 51,

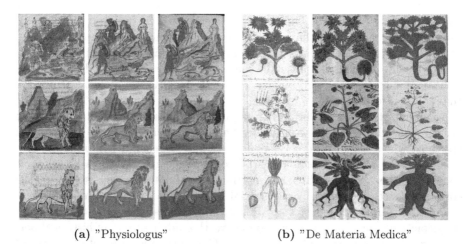

(a) "Physiologus"   (b) "De Materia Medica"

Fig. 3. Examples of annotated image triplets in our two sets of manuscripts. Note how the depiction can vary significantly both in style and content.

51 and 52 illustrations which could almost all be matched between versions. In the three versions we used the order of the illustrations is preserved. The content of the illuminations is similar enough that the correspondences could be easily identified by a human annotator. The style of the depictions however varied a lot as can be seen Figs. 1b and 3a.

– "De Materia Medica" was originally written in greek by Pedanius Dioscorides. The illustrations in the manuscript are mainly plants drawings. We consider three versions of the text with 816, 405 and 839 illustrations and annotated 295, 353 and 524 correspondences in the three associated pairs. Finding corresponding images in this set is extremely challenging due to three difficulties: many plants are visually similar to each other (Fig. 1b), the appearance can vary greatly in a matching image pair (Fig. 3b) and the illustrations are ordered differently in different manuscripts (Fig. 2).

Note that we included three manuscripts in both sets so that algorithms and annotations could leverage cycle-consistency.

Illustrations Annotations. We ran an automatic illustration extraction algorithm [21] and found it obtained good results, but that some bounding boxes were inaccurate and that some different but overlapping illustrations were merged. To focus on the difficulty of finding correspondences rather than extracting the illustrations, we manually annotated the bounding boxes of the illustrations in each manuscript using the VGG Image Annotator [8]. The study of joint detection and correspondence estimation is left for future work.

Correspondences Annotations. For the manuscripts of the Physiologus, annotating the corresponding illustration was time-consuming but did not present any significant difficulties. For the De Materia Medica however, the annotation presented significant challenge. Indeed, as explained above and illustrated in Fig. 2, the illustrations have been significantly re-ordered, modified, and are often visually ambiguous. Since the manuscripts contain hundreds of illustrations, manually finding correspondences one by one was simply not feasible. We thus followed a three step procedure. First, for each illustration we used the image similarity described in Sect. 4.1 to obtain its 5 nearest neighbors in each other manuscript. Second, we provided these neighbors and their context to a specialist who selected valid correspondences and searched neighboring illustrations and text to identify other nearby correspondences. Third, we used cycle consistency between the three manuscripts to validate the consistency of the correspondences identified by the specialist and propose new correspondences. Interestingly, during the last step we noticed 51 cases where the captions and the depictions were not consistent. While worth studying from an historical perspective, these cases are ambiguous from a Computer Vision point of view, and we removed all correspondences leading to such inconsistencies from our annotations.

Evaluation Metric. We believe our annotations to be relatively exhaustive, however the difficulty of the annotation task made this hard to ensure. We thus focused our evaluation metric on precision rather than recall. More precisely, we expect algorithms to return a correspondence in each manuscript for each reference image and we compute the average accuracy on annotated correspondences only. In our tables, we report performances on pairs of manuscripts $M_1 - M_2$ by finding correspondences in both directions (finding a correspondence in M_2 for each image of M_1, then a correspondence in M_1 for each image in M_2) and averaging performances. Note that there is a bias in our annotations in the De Materia Medica since we initially provided the annotator with the top correspondences using our similarity. However, while it may slightly over-estimate the performance of our algorithm, qualitative analysis of the benefits brought by our additional processing remains valid.

4   Approach

In this section, we present the key elements of our image collation pipeline, visualized in Fig. 4. Except when explicitly mentioned otherwise, we focus on studying correspondences in a pair of manuscripts. First, we discuss image similarities adapted to the task. Second, we introduce different normalizations of the similarity matrix associated to a pair of manuscripts. Third, we present a method to propagate information from confident correspondences to improve results. Finally, we give some implementation details.

4.1   Image Similarities

We focus on similarities based on deep features, following consistent observations in recent works that they improve over their classical counterparts for historical

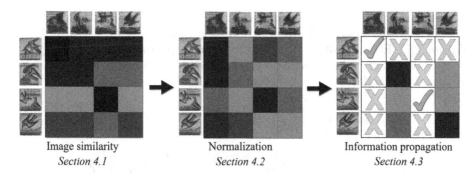

Image similarity	Normalization	Information propagation
Section 4.1	*Section 4.2*	*Section 4.3*

Fig. 4. Overview of our approach. We first compute a similarity score between each pair of image, which we visualize using darker colors for higher similarity. We then normalize the similarity matrix to account for images that are similar to many other, such as the first line of our example. Finally, we propagate signal from confident correspondences which are maxima in both directions (green marks) to the rest of the matrix. (Color figure online)

image recognition [27,31]. Since we want our approach to be directly applicable to new sets of images, with potentially very different characteristics, we use off-the-shelf features, without any specific fine-tuning. More precisely we used ResNet-50 [15] features trained for image classification on ImageNet [7], which we found to lead to better performances (see Sect. 5).

Raw Features. Directly using raw features to compute image similarity is a strong baseline. Similar to other works [27,27], we found that using *conv4* features and averaging cosine similarity of these features at the same location consistently performed best. More formally, given two images I_1 and I_2 we consider their *conv4* features $f_k = (f_k^i)_{i \in \{1, ..., N\}}$, where $k = 1$ or 2 is the image ID, i is the index of the spatial location in feature map and N is the size of the feature map. We define the feature image similarity as:

$$S_{features}(I_1, I_2) = \frac{1}{N} \sum_{i=1}^{N} \frac{f_1^i}{\|f_1^i\|} \cdot \frac{f_2^i}{\|f_2^i\|} \qquad (1)$$

where $\cdot$ is the scalar product. Note that the normalization is performed for each local feature independently and this similarity can only be defined if the two images are resized at a constant size. We used 256×256 in our implementation.

Matching-Based Similarity. The feature similarity introduced in the previous paragraph only considers the similarity of local features at the same spatial location and scale, and not their similarity with other features at other locations in the image. To leverage this information, [28] proposed to use a local matching score. Each feature f_k^i of a source image I_k is matched with the features extracted at several scales in a target images I_l. Then, each of the features of the target image is matched back in the source image and kept only if it matches back to the original feature, i.e. if it is a cycle consistent match. Finally, the best cycle consistent match among all scales of the target image $m_{k,l}(f_k^i)$ is identified. Writing $x_k^i \in \mathbb{R}^2$ the position of the feature f_k^i in the feature map and $x_{k,l}(f_k^i) \in \mathbb{R}^2$ the position of its best match $m_{k,l}(f_k^i)$ (which might be at a different scale), we define the similarity between I_1 and I_2 as:

$$S_{matching}(I_1, I_2) = \frac{1}{2N} \sum_{i=1}^{N} e^{-\frac{\|x_1^i - x_{1,2}(f_1^i)\|^2}{2\sigma^2}} \frac{f_1^i}{\|f_1^i\|} \cdot \frac{m_{1,2}(f_1^i)}{\|m_{1,2}(f_1^i)\|}$$

$$+ \frac{1}{2N} \sum_{i=1}^{N} e^{-\frac{\|x_2^i - x_{2,1}(f_2^i)\|^2}{2\sigma^2}} \frac{f_2^i}{\|f_2^i\|} \cdot \frac{m_{2,1}(f_2^i)}{\|m_{2,1}(f_2^i)\|}$$

where $\cdot$ is the scalar product and σ is a real hyperparameter. This score implicitly removes any contribution for non-discriminative regions and for details that are only visible in one of the depictions, since they will likely match to a different spatial location and thus have a very small contribution to the score. It will also be insensitive to local scale changes. Note that [28] considered only the first term of the sum, resulting in a non-symmetric score. On the contrary, our problem is completely symmetric and we thus symmetrized the score.

Transformation Dependent Similarity. While the score above has some robustness to local scale changes, it assumes the images are coarsely aligned. To increase robustness to alignment errors, we follow [27] and use RANSAC [11] to estimate a 2D affine transformation between the two images. More precisely, keeping the notations from the previous paragraph, we use RANSAC to find an optimal affine transformation $T_{k,l}$ between image I_k and I_l:

$$T_{k,l} = \arg\max \sum_{i=1}^{N} e^{-\frac{\|T_{k,l}x_k^i - x_{k,l}(f_k^i)\|^2}{2\sigma^2}} \frac{f_k^i}{\|f_k^i\|} \cdot \frac{m_{k,l}(f_k^i)}{\|m_{k,l}(f_k^i)\|} \qquad (2)$$

Note this is slightly different from [27] which only uses the RANSAC to minimize the residual error in the matches to optimize the transformation. We found that maximizing the score instead of the number of inliers significantly improved the performances. Considering again the symmetry of the problem, this leads to the following score:

$$S_{trans}(I_1, I_2) = \frac{1}{2N} \sum_{i=1}^{N} e^{-\frac{\|T_{1,2}x_1^i - x_{1,2}(f_1^i)\|^2}{2\sigma^2}} \frac{f_1^i}{\|f_1^i\|} \cdot \frac{m_{1,2}(f_1^i)}{\|m_{1,2}(f_1^i)\|}$$

$$+ \frac{1}{2N} \sum_{i=1}^{N} e^{-\frac{\|T_{2,1}x_2^i - x_{2,1}(f_2^i)\|^2}{2\sigma^2}} \frac{f_2^i}{\|f_2^i\|} \cdot \frac{m_{2,1}(f_2^i)}{\|m_{2,1}(f_2^i)\|}$$

This score focuses on discriminative regions, is robust to local scale changes and affine transformations. We found it consistently performed best in our experiments, outperforming the direct use of deep features by a large margin.

4.2   Normalization

Let us call S the similarity matrix between all pairs of images in the two manuscripts, $S(i,j)$ being a similarity such as the ones defined in the previous section between the ith image of the first manuscript and the jth image of

Table 2. Row-wise and Column-wise normalizations.

Normalization	$R(i, j)$	$C(i, j)$
$sm(\lambda S)$	$\exp(\lambda S(i, j))/\sum_k \exp(\lambda S(i, k))$	$\exp(\lambda S(i, j))/\sum_k \exp(\lambda S(k, j))$
$S/\mathrm{avg}(S)$	$R_{\mathrm{avg}} = S(i, j)/\mathrm{avg}_k S(i, k)$	$C_{\mathrm{avg}} = S(i, j)/\mathrm{avg}_k S(k, i)$
$S/\max(S)$	$R_{\max} = S(i, j)/\max_k S(i, k)$	$C_{\max} = S(i, j)/\max_k S(k, i)$
$sm(\lambda S/\mathrm{avg}(S))$	$\exp(\lambda R_{\mathrm{avg}}(i, j))/\sum_k \exp(\lambda R_{\mathrm{avg}}(i, k))$	$\exp(\lambda C_{\mathrm{avg}}(i, j))/\sum_k \exp(\lambda C_{\mathrm{avg}}(k, j))$
$sm(\lambda S/\max(s))$	$\exp(\lambda R_{\max}(i, j))/\sum_k \exp(\lambda R_{\max}(i, k))$	$\exp(\lambda C_{\max}(i, j))/\sum_k \exp(\lambda C_{\max}(k, j))$

the second manuscript. For each image in the first manuscript, one can simply predict the most similar image in the second one as a correspondence, i.e. take the maximum over each row of the similarity matrix. This approach has however two strong limitations. First, it does not take into account that some images tend to have higher similarity scores than other, resulting in rows or columns with higher values in the similarity matrix. Second, it does not consider the symmetry of the problem, i.e., that one could also match images in the second manuscript to images in the first one.

To account for these two effects, we propose to normalize the similarity matrix S along each row and each column resulting in two matrices R and C.

We experimented with five different normalization operations using softmax (sm), maximum (max) and average (avg) operations either along the rows (leading to R) or the columns (leading to C), as shown in Table 2.

We then combine the two matrices R and C into a final score: we experimented with summing them or using element-wise (Hadamard) multiplication. Both performed similarly, with a small advantage from the sum, we thus only report those results. We found in our experiments that the max normalization performed best, without requiring an hyper-parameter. As such, our final normalized similarity matrix N_S is defined as:

$$N_S(i, j) = \frac{S(i, j)}{\max_k S(i, k)} + \frac{S(i, j)}{\max_k S(k, j)} \tag{3}$$

4.3 Information Propagation

While the normalized score N_S obtained in the previous section includes information about both directions of matching in a pair of manuscripts, it does not ensure that correspondences are 2-cycle consistent, i.e. that the maxima in the rows of N_S correspond to maxima in the columns. If one has access to more than 2 manuscripts, one can also check consistency between triplets of manuscripts and identify correspondences that are 2 and 3-cycle consistent. Correspondences that verify such cycle-consistency are intuitively more reliable, as we validated in our experiments, and thus can be used as anchors to look for other correspondences in nearby images. Indeed, while the order of the images is not strictly preserved in the different versions, there is still a clear locally consistent structure as can be seen in the ground truth correspondence matrices visualized in Fig. 2.

(a) Query and 5 nearest neighbors according to our similarity score (Section 4.1)

(b) Query and 5 nearest neighbors according to our normalized score (Section 4.2)

(c) Query and 5 nearest neighbors after information propagation (Section 4.3)

Fig. 5. Query (in blue) and 5 nearest neighbors (ground truth in green) after the different steps of our method. Despite not being in the top-5 using the similarity, the correct correspondence is finally identified after the information propagation step. (Color figure online)

Many approaches could be considered to propagate information from confident correspondences and an exhaustive study is beyond the scope of our work. We considered a simple baseline as a proof of concept. Starting from an initial score N_S and a set of confident correspondences $\mathcal{C}^*$ as seeds (e.g., correspondences that verify 2 or 3 cycle consistency constraints), we define a new score after information propagation P_S as:

$$P_S(i,j) = N_S(i,j) \prod_{(k,l)\in\mathcal{C}^*} \left(1 + \alpha \exp\left(\frac{-||(i,j)-(k,l)||^2}{2\sigma_p^2}\right)\right) \tag{4}$$

where σ_p and α are hyperparameters. Note that this formula can be applied with any definition of $\mathcal{C}^*$, and thus could leverage sparse correspondence annotations.

Implementation Details. In all the experiments, we extract *conv4* features of a ResNet50 architecture [15] pre-trained on ImageNet [7]. To match illustrations between different scales, we keep the original aspect ratios and resize the source image to have 20 features in the largest dimension and the target image to five scales such that the numbers of features of the largest dimension are 18, 19, 20, 21, 22. We set σ in Eq. 4.1 and 4.1 to $\frac{1}{\sqrt{50}}$ times the size of the image and the number of iterations in the RANSAC to 100. For the information propagation, we find that $\sigma_p = 5$ and $\alpha = 0.25$ performs best. With our naive Pytorch

Table 3. Percentage of accuracy of the correspondences obtained using $S_{features}$ with different conv4 features for all manuscripts pairs.

Pairs	ResNet18	MoCo-v2	ArtMiner	ResNet50	Pairs	ResNet18	MoCo-v2	ArtMiner	ResNet50
P1-P2	78.0	75.0	**92.0**	84.0	D1-D2	31.9	31.2	35.3	**35.4**
P1-P3	75.0	62.0	**78.0**	73.0	D1-D3	42.0	35.6	43.7	**46.1**
P2-P3	99.0	99.0	98.0	100.0	D2-D3	27.6	26.6	31.7	**34.1**

Table 4. Accuracy of the correspondences obtained using the different similarities explained in Sect. 4.1, as well as the similarity used in [27], which is similar to S_{trans} but uses the number of inliers instead of our score to select the best transformation.

Pairs	$S_{features}$	$S_{matching}$	[27]	S_{trans}	Pairs	$S_{features}$	$S_{matching}$	[27]	S_{trans}
P1-P2	84.0	98.0	99.0	**100.0**	D1-D2	35.4	54.6	56.3	**61.7**
P1-P3	73.0	94.0	**98.0**	**98.0**	D1-D3	46.1	69.8	71.3	**77.7**
P2-P3	100.0	98.0	**100.0**	**100.0**	D2-D3	34.1	51.8	51.7	**60.1**

implementation, computing correspondences between D1 (816 illustrations) and D2 (405 illustrations) takes approximately 80 min, 98% being spent to compute similarities between all the 330,480 pairs of images.

5   Results

In this section, we present our results. In 5.1 we compare different features similarities. In 5.2 we show the performance boost by the different normalizations. In 5.3 we demonstrate that the results can be improved by leveraging the structure of the correspondences. Finally, in 5.4 we discuss the failure cases and limitations.

To measure the performance, as described in Sect. 3, we compute both the accuracy a_1 obtained by associating to each illustration of the first manuscript the illustration of the second manuscript which maximizes the score and the accuracy a_2 by associating illustrations of the second manuscript to illustrations of the first manuscript. We then report the average of these two accuracies $\frac{a_1+a_2}{2}$. Because using a good image similarity already led to almost perfect results on the Physiologus, we focus our analysis on the more challenging case of the De Materia Medica. The benefits of the three steps of our approach are illustrated in Fig. 5, where one can also assess the difficulty of the task.

5.1   Feature Similarity

We first compare in Table 3 the accuracy we obtained using the baseline score $S_{features}$ with different conv4 features: ResNet18 and ResNet50 trained on ImageNet, MoCo v2 [6], and the ResNet18 features fine-tunned by [27]. The feature from [27] achieve the best results on Physiologus manuscripts. However, on the challenging De Materia Medica manuscripts, the ResNet50 features perform best, and we thus use them in the rest of the paper.

In Table 4, we compare the accuracy using the different similarities explained in Sect. 4.1. The results obtained using S_{trans} leads to the best performances

Table 5. Accuracy of the correspondences obtained using the different normalizations explained in Sect. 4.2 and in Table 2.

Pairs	S	$sm(\lambda S/\mathrm{avg}(S))$	$sm(\lambda S/\max(s))$	$sm(\lambda S)$	$S/\mathrm{avg}(S)$	$S/\max(S)$
D1-D2	61.7	68.3	**70.5**	67.8	67.1	**70.5**
D1-D3	77.7	83.5	**85.3**	83.1	82.5	**85.3**
D2-D3	60.1	66.3	66.7	66.1	65.3	**69.0**

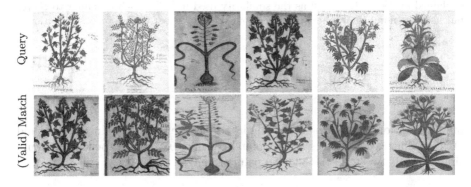

Fig. 6. Examples of correspondences recovered only after the information propagation. These examples are with many local appearance changes.

on all pairs, and we consider only this score in the rest of the paper. Note in particular that optimizing the function of Eq. (2) leads to clearly better result than using the number of inliers as in [27]. Since results on the Physiologus, where illustrations are fewer and more clearly different, are almost perfect, we only report the quantitative evaluation on the more challenging De Materia Merdica in the following sections.

5.2   Normalization

In Table 5, we compare the accuracy we obtained using the different normalizations presented in Sect. 4.2 and in Table 2. For the softmax-based normalizations that include an hyperparamter, we optimized it directly on the test data, so the associated performance should be interpreted as an upper bound. Interestingly, a simple normalization by the maximum value $S/max(S)$ outperforms these more complexe normalization without requiring any hyper-parameter tuning. It is also interesting that all the normalization schemes we tested provide a clear boost over the raw similarity score, outlining the importance of considering the symmetry of the correspondence problem.

5.3   Information Propagation

We analyze the potential of information propagation in Table 6. Using the normalized similarity score, we can compute correspondences (column 'all'). Some

Table 6. The left part of the table details the accuracy of the correspondences with the normalized score N_S on different subsets of the annotated correspondences: all (the measure used in the rest of the paper), the correspondences obtained with N_S are 2-cycle consistent and and the correspondences obtained with N_S are 2 and 3-cycle consistent. The number in parenthesis is the number of correspondences. The right part of the table present the average accuracy obtained when performing information propagation from either the 2-cycle or the 3-cycle consistent correspondences.

Pairs	N_S			P_S - $\mathcal{C}^*$:2-cycles	P_S - $\mathcal{C}^*$:3-cycles
	All	Only 2-cycle	Only 3-cycle	All	All
D1-D2	70.5 *(295)*	83.5 *(224)*	99.2 *(118)*	**82.5**	82.0
D1-D3	85.3 *(524)*	90.2 *(457)*	98.3 *(118)*	88.5	**88.6**
D2-D3	69.0 *(353)*	78.5 *(279)*	96.6 *(118)*	**81.7**	79.3

of these correspondences will be 2-cycle or 3-cycle consistent. These correspondences will be more reliable, as can be seen in the 'only 2-cycle' and 'only 3-cycle' columns, but there will be fewer (the number of images among the annotated ones for which such a cycle consistent correspondence is found given in parenthesis). In particular, the accuracy restricted to the 3-cycle consistent correspondences is close to 100%. Because the accuracy of these correspondences is higher, one can use them as a set of confident correspondences $\mathcal{C}*$ to compute a new score P_S as explained in Sect. 4.3. The results, on all annotations, can be seen in the last two column. The results are similar when using either 2 or 3-cycle consistent correspondences for $\mathcal{C}^*$ and the improvement over the normalized scores is significant.

This result is a strong evidence that important performance boost can be obtained by leveraging consistency. Qualitatively, the correspondences that are recovered are difficult cases, where the depictions have undergone significant changes, as shown in Fig. 6.

5.4   Failure Cases, Limitations and Perspectives

Figure 7 shows some typical examples of our failure cases. As expected they correspond to cases where the content of the image has been significantly altered and where very similar images are present. For such cases, it is necessary to leverage the text to actually be able to discriminate between the images. Extracting the text and performing HTR in historical manuscripts such as ours is extremely challenging, and the text also differs considerably between the different versions. However, a joint approach considering both the text and the images could be considered and our dataset could be used for such a purpose since the full folios are available.

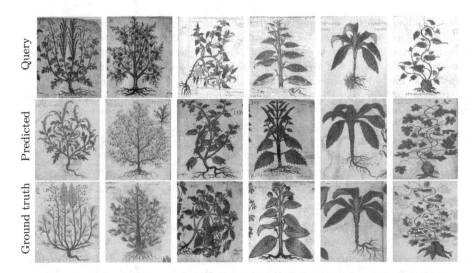

Fig. 7. Examples of failure cases. We show the queries, predicted matches and ground truth correspondences in the first, second and third line respectively.

6   Conclusion

We have introduced the new task of image collation and an associated dataset. This task is challenging and would enable to study at scale the evolution of illustrations in illuminated manuscripts. We studied how different image similarity measures perform, demonstrating that direct deep feature similarity is outperformed by a large margin by leveraging matches between local features and modeling image transformations. We also demonstrated the strong benefits of adapting the scores to the specificity of the problem and propagating information between correspondences. While our results are not perfect, they could still speed-up considerably the manual collation work, and are of practical interest.

Acknowledgements. This work was supported by ANR project EnHerit ANR-17-CE23-0008, project Rapid Tabasco, and gifts from Adobe. We thank Alexandre Guilbaud for fruitful discussions.

References

1. https://www.wdl.org
2. https://www.themorgan.org
3. https://digi.vatlib.it
4. http://www.internetculturale.it
5. Bourdaillet, J., Ganascia, J.G.: Practical block sequence alignment with moves. In: LATA (2007)

6. Chen, X., Fan, H., Girshick, R., He, K.: Improved baselines with momentum contrastive learning. arXiv (2020)
7. Deng, J., Dong, W., Socher, R., Li, L. J., Li, K., Fei-Fei, L.: ImageNet: a large-scale hierarchical image database. In: CVPR (2009)
8. Dutta, A., Zisserman, A.: The VIA annotation software for images, audio and video. In: ACM Multimedia (2019)
9. En, S., Petitjean, C., Nicolas, S., Heutte, L.: A scalable pattern spotting system for historical documents. Pattern Recognit. **54**, 149–161 (2016)
10. Ezra, D.S.B., Brown-DeVost, B., Dershowitz, N., Pechorin, A., Kiessling, B.: The dead sea scrolls. In: ICFHR, Transcription alignment for highly fragmentary historical manuscripts (2020)
11. Fischler, M.A., Bolles, R.C.: Random sample consensus: a paradigm for model fitting with applications to image analysis and automated cartography. Commun. ACM **24**(6), 381–395 (1981)
12. Gordo, A., Almazan, J., Revaud, J., Larlus, D.: End-to-end learning of deep visual representations for image retrieval. IJCV **124**(2), 237–254 (2017)
13. Haentjens Dekker, R., Van Hulle, D., Middell, G., Neyt, V., Van Zundert, J.: Computer-supported collation of modern manuscripts: collatex and the Beckett Digital Manuscript Project. DSH **30**(3), 452–470 (2015)
14. Hassner, T., Wolf, L., Dershowitz, N.: OCR-free transcript alignment. In: ICDAR (2013)
15. He, K., Zhang, X., Ren, S., Sun, J.: Deep residual learning for image recognition. In: CVPR (2016)
16. Hobby, J.D.: Matching document images with ground truth. IJDAR **1**(1), 52–61 (1998)
17. Kornfield, E.M., Manmatha, R. and Allan, J.: Text alignment with handwritten documents. In: DIAL (2004)
18. Lin, T.Y., Goyal, P., Girshick, R., He, K., Dollár, P.: Focal loss for dense object detection. In: ICCV (2017)
19. Lin, T.Y., et al.: Microsoft COCO: common objects in context. In: Fleet, D., Pajdla, T., Schiele, B., Tuytelaars, T. (eds.) ECCV 2014. LNCS, vol. 8693, pp. 740–755. Springer, Cham (2014). https://doi.org/10.1007/978-3-319-10602-1_48
20. Lowe, D.G.: Distinctive image features from scale-invariant keypoints. IJCV **60**(2), 91–110 (2004)
21. Monnier, T., Aubry, M.: docExtractor: an off-the-shelf historical document element extraction. In: ICFHR (2020)
22. Radenović, F., Tolias, G., Chum, O.: Fine-tuning CNN image retrieval with no human annotation. TPAMI **41**(7), 1655–1668 (2018)
23. Razavian, A.S., Sullivan, J., Carlsson, S., Maki, A.: Visual instance retrieval with deep convolutional networks. MTA **4**(3), 251–258 (2016)
24. Revaud, J., Almazán, J., Rezende, R.S., Souza, C.R.D.: Learning with average precision: training image retrieval with a listwise loss. In: ICCV (2019)
25. Sadeh, G., Wolf, L., Hassner, T., Dershowitz, N., Ben-Ezra, D.S.: Viral transcript alignment. In: ICDAR (2015)
26. Schmidt, D., Colomb, R.: A data structure for representing multi-version texts online. Int. J. Hum.-Comput. Stud. **67**(6), 497–514 (2009)
27. Shen, X., Efros, A.A., Aubry, M.: Discovering visual patterns in art collections with spatially-consistent feature learning. In: CVPR (2019)
28. Shen, X., et al.: Large-scale historical watermark recognition: dataset and a new consistency-based approach. In: ICPR (2020)

29. Sivic, J., Zisserman, A.: Video google: a text retrieval approach to object matching in videos. In: ICCV (2003)
30. Smith, S.E.: The eternal verities verified: Charlton Hinman and the roots of mechanical collation. Stud. Bibliogr. **53**, 129–161 (2000)
31. Úbeda, I., Saavedra, J.M., Stéphane, N., Caroline, P., Heutte, L.: Pattern spotting in historical documents using convolutional models. In: HIP (2019)

Revisiting the Coco Panoptic Metric to Enable Visual and Qualitative Analysis of Historical Map Instance Segmentation

Joseph Chazalon[(✉)] and Edwin Carlinet

EPITA Research and Development Laboratory (LRDE),
14-16 rue Voltaire, 94270 Le Kremlin-Bicêtre, France
{joseph.chazalon,edwin.carlinet}@lrde.epita.fr

Abstract. Segmentation is an important task. It is so important that there exist tens of metrics trying to score and rank segmentation systems. It is so important that each topic has its own metric because their problem is too specific. Does it? What are the fundamental differences with the ZoneMap metric used for page segmentation, the COCO Panoptic metric used in computer vision and metrics used to rank hierarchical segmentations? In this paper, while assessing segmentation accuracy for historical maps, we explain, compare and demystify some the most used segmentation evaluation protocols. In particular, we focus on an alternative view of the COCO Panoptic metric as a classification evaluation; we show its soundness and propose extensions with more "shape-oriented" metrics. Beyond a quantitative metric, this paper aims also at providing qualitative measures through *precision-recall maps* that enable visualizing the success and the failures of a segmentation method.

Keywords: Evaluation · Historical map · Panoptic segmentation

1 Introduction

The massive digitization of historical data by the national institutions have lead to a huge volume of data to analyze. These data are a source of fundamental knowledge for historians and archaeological research. In this paper, we focus on the processing of a specific type of documents: the historical maps. Historical maps are very rich resources that can identify archaeological sites [15], can track the social evolution of places [6], can help study the urban mobility by analyzing historical changes performed on a road system [14]. To exploit effectively these data in geographical applications, one need to extract the interesting features from the map automatically. The automation is required as the time needed to extract them by hand is not tractable. The process, *i.e.* the image processing pipeline required to exploit the information from the maps is application dependent but most of them require the following items: *1.* spotting the elements of interest in the map such as texts, legends, geometric patterns; *2.* segmentation and classification of each of these elements; *3.* georeferencing the extracted

© Springer Nature Switzerland AG 2021
J. Lladós et al. (Eds.): ICDAR 2021, LNCS 12824, pp. 367–382, 2021.
https://doi.org/10.1007/978-3-030-86337-1_25

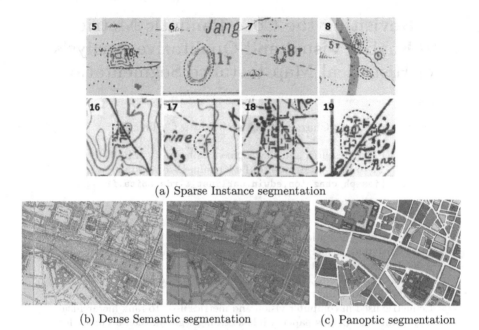

(a) Sparse Instance segmentation

(b) Dense Semantic segmentation   (c) Panoptic segmentation

Fig. 1. Instance segmentation vs Panoptic segmentation. In (a), the instance segmentation of the mounds (1st row) of the Sol map and the ruins (2nd row) on the French Levant maps (images from [10]). Each object is described by its region (usually a binary mask) and its class. In (b), the semantic segmentation of a 1925 urban topographic map: building blocks (orange), parks (green), roads (gray) and rivers (blue); each pixel is associated to a class label. In (c) the panoptic segmentation of the same 1925 urban topographic map. A pixel-wise annotation that combines object instances (in false colors) and classes [5]. (Color figure online)

elements. The automation of historical map processing has been studied for a long time. For instance, in [1], the authors apply well-known morphological approaches for segmenting maps. With the advances of the deep learning techniques over the last decade, these techniques have been further developed and are getting more and more efficient. In [5], the authors combine the morphological methods from [1] with deep edge detectors and [10] rely on the U-Net deep-network to spot and segment features on the map.

In this context, we are interested in **evaluating and comparing different approaches for historical map processing**.

Segmentation is itself very content and application specific, even when considering just maps. Indeed, one may expect from a segmentation to have strong guaranties such as being a total partition of the space or just delimiting interesting areas. Other may require the segmentation to have thin separators between regions... In Fig. 1 (a) and (b), we show two classes of segmentation tasks related to historical map processing: object detection (or instance segmentation) and semantic segmentation. Recently, it has been suggested in [11] that both types of segmentation should be joint in a unique framework and should be evaluated with the same metrics. The work of [11] has resulted in the creation of a new segmentation task, namely the *panoptic segmentation*, that encompasses both the semantic and the instance segmentation. Moreover, they have proposed a parameter-less metric, the *panoptic quality*, that renders the quality of the segmentation and is becoming a standard for the evaluation.

In this paper, we aim at providing a set of tools that enables to visualize, compare and score a (panoptic) segmentation in the context of a challenge [4]. The task consists in providing an instance segmentation of the building blocks of the maps. In others words, there are several classes in the groundtruth but only the *block* class is of interest. Each pixel must be assigned to an instance id (of class *block*) or to the *void* class (with a unique instance). The blocks are **closed** 4-connected shapes and should be separated by 8-connected boundaries (pixels of class *void*). The boundary requirement is task specific as it enables vectorizing and georeferencing the blocks afterward. The set of *blocks* and *void* segments forms a partition of the image.

The contributions of this paper lies in three points that supply simple and fast tools for visualizing, evaluating and comparing segmentation methods. First, we propose another point of view of the COCO Panoptic metric where it has solid foundations in terms of prediction theory and is closely related to the well-known precision, recall and Area Under the Curve. It follows a quantitative evaluation of the segmentation systems and a ranking of the systems with respect to a shape pairing score. Second, we provide a meaningful way to visualize the segmentation results in a qualitative way. We introduce the *precision* and *recall* maps based the metrics previously defined, that highlights the locations where a system succeeds and where it fails. Third, we provide an insight about how to extend these metrics by studying the algebraic requirements of the metrics in bipartite graph formalism. It follows that the panoptic metric can be easily customized with another metric that may make more sense in some specific domains like document processing.

The paper is organized as follows. First, we have a short review of the evaluation protocols for image segmentation and their differences in Sect. 2 with an emphasis on the COCO Panoptic score. In Sect. 3, we demystify, explain and extend the COCO metric within a bipartite graph framework and propose a qualitative evaluation of a segmentation through the *prediction* and *recall* maps. In Sect. 4, we compare our approach with other usual segmentation scoring and highlight their pitfalls. Last, we conclude in Sect. 5.

2   State of the Art

Here we focus on assessing which existing approaches are suitable for evaluating the quality of dense instance segmentation (of map images). Classification (in the sense of semantic segmentation) is not our priority here, as it can be handled quite easily on top of an instance segmentation evaluation framework. Hence, the challenge consists in combining two indicators: a measure of detection performance and a measure of segmentation quality. Among the methods currently used for object detection, the accuracy of the segmentation usually is of little importance. An approximate location is usually sufficient to count a detection as successful. Indeed, most of the approach reported in a recent survey [12] show that an IoU of 0.5 is often used, which is far from being acceptable from a segmentation point of view. Pure segmentation metrics, on the other hand, often focus on pixel classification and do not consider shapes as consistent objects. The ability to accurately delineate objects in an image therefore requires dedicated metrics which can be either *contour-based*, i.e., measuring the quality of the detection of the boundary of each shape, or *region-based*, i.e., measuring the quality of the matching and coverage of each shape [2]. Contour-based approach often rely on an estimation of the fraction of pixels accurately detected, which does not incur any performance regarding actual shape detection because the removal of a single pixel may prevent a shape from being detected. Region-based approaches combine a matching stage between expected and detected shapes, followed by the computation of a segmentation quality.

Several metrics have been proposed, both for natural images and document images, with differences in mind. In both cases, the matching step is performed by measuring a spatial consistency using a coverage indicator, i.e., the intersection of two shapes normalized by the area of either one or the union of the shapes. For natural images, evaluation metrics used to assume that objects to detect were sparse and non-overlapping. As a result, segmentation quality used to be reported as a rough indication of the distribution of matching scores: average of maximum pair-wise IoUs [2] or average of IoUs over 0.5 to consider only one-to-one matches [7]. Such information is usually insufficient in the context of document processing, especially when it comes to manual annotation: a measure of error costs is necessary to assess whether a system will provide a gain over manual work, and a simple global accuracy measure is not sufficient.

Early document segmentation metrics [17] focused on classifying error cases to enable the identification of correct, missed, false alarms, over- and under-segmentation cases, but relied on many thresholds and did not provide a normalized scoring; it was possible to rank systems but not to know whether their performance was acceptable or not. Further work like DetEval [20] proposed an improved formulation with a normalized score leading to precision and recall indicators blending detection and segmentation measures, with the possibility to include over- and under-segmentation costs. However, this approach requires the calibration of several thresholds and setting a minimal segmentation quality is hard to tune. The ZoneMap [9] metric improved DetEval to enable the support of overlapping shapes thanks to a greedy matching strategy to identify matching shapes, but did not make the usage cost of a system easier to assess as the metric is not normalized.

The F_{ob} [16] also performs a greedy strategy while reporting normalized precision, recall and F-measures while detecting under- and over-segmentation, but depends on two thresholds which prevent an easy calibration. The COCO Panoptic metric [11], finally, combines most of the features we are looking for: a single threshold on the minimum IoU to consider shapes as matching, a measure based on a F-score combining detection performance and segmentation accuracy, and can be evaluated class-wise if needed. This metric is both a solid and simple foundation to build our evaluation protocol on.

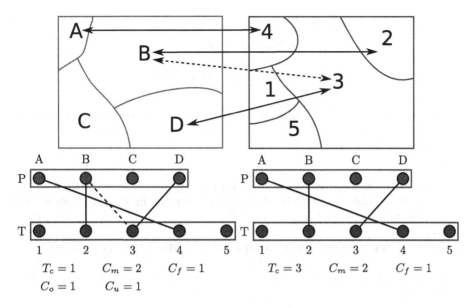

Fig. 2. Bipartite matching graph

3 Extending and Visualizing the COCO Metrics

3.1 Metric on Pairings and Bipartite Graph

While aiming at providing a measure of the quality of two segmentations, our measure relies on a metric on bipartite graphs. Let $G = (P, T, E)$ denote a bipartite graph of a pairing relation between partitions P (the *prediction*) and T (the *target*). Each node of P (resp. T) actually represents a component of the predicted (resp. ground truth) segmentation. This graph is said to be a *matching* when no edge share the same endpoints (in Sect. 3.3, we provide a way to build such a graph from two segmentations). In other words, a *matching* is a 1-1 relation between components of P and T. Note that this relation does not need to be total (or serial), some nodes of P or T may not have any match in the other set. Figure 2 shows an example of such a bipartite graph and a matching. In [9,17], the authors deduce several metrics from these graphs:

Number of matched components (T_c) The number of nodes (in P or T) that have a one-to-one match.

Number of missed components (type-I C_m**)** is the number of nodes in T that do not match any component in P, i.e., the number of components of the groundtruth that our segmentation has not been able to recover.

Number of false alarms (type-II C_f**)** is the number of nodes in P that do not match any component in T, i.e., the number of components in our segmentation that are false detection.

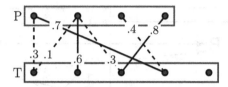

Fig. 3. Bipartite graph G with edges weighted by a matching score. Plain edges have weights above $\alpha = 0.5$ and form a one-to-one relation. Dashed edges have scores below α and exhibit a many-to-many relation.

Number of over-segmented components - Splits (C_o) is the number of nodes in T that match more than one component in P, *i.e.*, involved in a many-to-one association. We note T_{split} the subset of T involved in this relation.

Number of under-segmented components - Merges (C_u) is the number of nodes in P that match more than one component in T, *i.e.*, involved in one-to-many association. We note P_{merge} the subset of P involved in this relation.

It is worth noticing that counting the numbers of over/under-segmented components does not make sense on *matchings* as they are non-null on non-one-to-one relations only. **The type of relation actually characterizes of the features needed to assess the quality of the pairing.** If the relation is always one-to-one, an evaluation protocol would rely on T_c, C_m, C_f only. It does not mean that over-segmentations or under-segmentations are not possible in protocols handling only one-to-one relation but rather than they are converted to C_m or C_f errors in these frameworks. Therefore, *matchings* can be likened to a binary classification where matched components are *true positives*, missed components are *false negatives* and false alarms are *false positives*. It follows the definitions of the *recall* and the *precision* as:

$$precision = \frac{T_c}{T_c + C_f} = \frac{T_c}{|P|} \tag{1}$$

$$recall = \frac{T_c}{T_c + C_m} = \frac{T_c}{|T|} \tag{2}$$

$$F\text{-}score = \frac{T_c}{T_c + \frac{1}{2}(C_m + C_f)} = \frac{2.T_c}{|P| + |T|} \tag{3}$$

In the context of a segmentation evaluation, the *recall* stands for the ability to recover the components of the ground truth, while the *precision* stands for the ability to match the components of our segmentation with those of the ground truth.

3.2   Metrics on Weighted Pairings

We are now focusing on the case where edges are weighted by a *matching score* (the higher, the better). These scores are typically base on shape matching

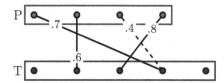 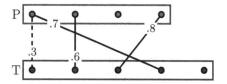

Fig. 4. Precision $G_{\text{precision}}$ (left) and recall G_{recall} (right) sub-graphs of G.

metrics like the DICE, the IoU, the Hausdorff distance.... We note $w(X,Y)$ the similarity score between any two components X and Y normalized in $[0,1]$.

Moreover, the similarity score must have the following property:

$$\exists \alpha \in [0,1] \text{ s.t. } \{(X,Y) \mid w(X,Y) > \alpha\} \text{ is a bijection.} \tag{4}$$

In other words, there exists a threshold α for which the subset of edges whose weights are above α forms a one-to-one relation in $P \times T$ as illustrated in Fig. 3. It follows that by considering any subgraph of G formed by edges with weights greater than $\alpha < t \leq 1$, we have a one-to-one relation for which we can compute the *precision* and *recall* with Eq. (3).

If the scoring function does not feature the property Eq. (4), it is still possible to extract a bipartite graph using a maximum weighted bipartite matching algorithm as in [19] or any greedy algorithm [16].

Qualitative Assessment: The Recall and Precision Maps. We propose the *precision* and *recall* maps that provide a *meaningful* way to *visualize and summarize* **locally** the performance of the segmentation.

The *precision* map reports how well a component of the *predicted* segmentation matches with the *ground truth*. It is built upon the subgraph of G with the best incident edge of each node of the prediction. Then, the pixels of the components of the prediction partition are rendered with the corresponding incident edge value. On the other hand, the *recall* map reports how well the ground truth has been match by the prediction. It is built upon the subgraph of G with the best incident edge of each node of the ground truth. The selected edge weights are then rendered pixel-wise just as before.

$$G_{\text{precision}} = (P,T,E_p) \text{ with } E_p = \{(X,Y), X \in P, Y = \arg\max_Y w(X,Y)\}$$

$$G_{\text{recall}} = (P,T,E_r) \text{ with } E_r = \{(X,Y), Y \in T, X = \arg\max_X w(X,Y)\}$$

The precision and recall graphs are depicted in Fig. 4 while precision and recall maps deduced from them are illustrated on Fig. 5. The interpretation of the color of a component depends on the "section" it is located (green when $> \alpha$, red when $< \alpha$). The green section reports the minimum score that a pairing must have so that a component is counted as a *matched component* (T_c). Below this score, the two components of the pairing would be considered as a *missed component* $(T_m+ = 1)$ and a *false alarm* $(T_f+ = 1)$. Components in the red section have to be interpreted quite differently. Those components will never be

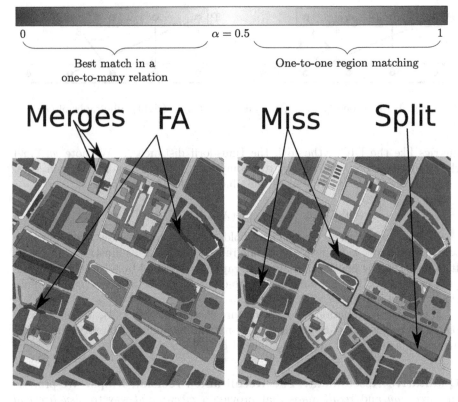

Fig. 5. Precision (left) - recall (right) maps. (Color figure online)

part of *match* as they are included in a one-to-many relation in G. Therefore, they are always counted as *missed components* (if a red region the recall map) and as *false alarms* (if a red region of the precision map).

Quantitative Assessment: Precision, Recall and F-Measures Curves on Matchings. When assessing the segmentation quality, most protocols (especially those used in detection in computer vision) start with defining a minimum overlap score to define the *matched components*. It makes sense to have such threshold as a mean to dismiss regions that are not good enough to be "usable" (*usable* meaning is application dependent). This is equivalent to taking the pairing graph G, removing edges below a given score, getting a one-to-one relation, and computing the precision/recall/F-measure as in Sect. 3.1. Afterward, edges values are not considered anymore.

Instead of relying on an arguably given overlap threshold and forgetting the edges weights, one can compute the precision/recall/F-score as a function of the edges weights. Indeed, we have seen that for any threshold t, $\alpha < t \leq 1$, the sub-graph is a one-to-one relation. When t increases, the number of matched components $T_c(t)$ decreases, and the number of missed components $C_m(t)$ and

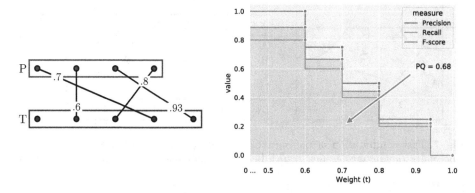

Fig. 6. Precision/recall/F-score curves from a one-to-one weighted pairing and the Panoptic quality got as the AUC of the F-score.

false alarms $C_f(t)$ increases. It allows to plot the precision, recall and f-score curves as a function of the matching score threshold t as shown on Fig. 6.

Quantitative Assessment: The Panoptic Quality Score. While providing a visual insight of the performance of a segmentation method when being more and more strict on the segmentation quality, the previous curves does not allow comparing and rank a set of methods. We need a single, simple and informative metric for this purpose. When applied on predictors, the Area Under the Curve (AUC) of the PR (or the ROC) curves is a widely used metric that summarizes how well a predictor perform for the whole space of prediction thresholds [8]. Here, the AUC of the F-score curve reflects how well the segmentation performs, both in-term of recall and precision, without committing to a particular threshold on the matching quality.

$$\text{NPQ} = (1 - \alpha)^{-1} \int_{t=\alpha}^{1} \textit{f-score}(t) \tag{5}$$

The *Normalized Panoptic Quality* depicted in Eq. (5) is named after the *Panoptic Quality (PQ)* metric described in [11]. It is an unacknowledged rewriting of the Area under the F-score Curve as shown with Eq. (8) (with $\alpha = 0$). Interpreted this way, the PQ metric matches the well-established practices in prediction evaluation, more specifically, averaging the performance over all the thresholds. Also, splitting the formula into the terms *segmentation quality* and *recognition quality* (as in [11]) has a straightforward interpretation with this formalism. The *recognition quality* is related to the initial number of edges in the graph (one-to-one matches) and corresponds to the f-score of the most permissive segmentation. On the other hand, the *segmentation quality* integrates the weights of the edges and renders the loss of matched regions while being less and less permissive with the region matching quality.

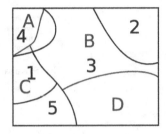

$X \cap Y$	1	2	3	4	5	$|Y|$
A				11		11
B		15	31	12		58
C	20			3	22	45
D			29			29
$|X|$	20	15	40	36	22	

Fig. 7. Fast computation of pairing scores based on the 2D-histogram of the image label pairs *(prediction, ground truth)*. Left: label-pair maps, right: 2d-histogram with marginal sums.

$$\mathrm{PQ} = \int_{t=0}^{1} \text{f-score}(t) \tag{6}$$

$$= \text{f-score}(\alpha) . \int_{0}^{1} \frac{T_c(t)}{T_c(\alpha)} \tag{7}$$

$$= \underbrace{\text{f-score}(\alpha)}_{\text{recognition quality}} . \underbrace{\sum_{(X,Y)\in E} \frac{w(X,Y)}{T_c(\alpha)}}_{\text{segmentation quality}} \tag{8}$$

The *Normalized Panoptic Quality* ensures that the score is distributed in [0-1] and removes a constant offset of the PQ due to the parameter α which is minimal matching quality (the "usable" region quality threshold). This normalization enables comparing PQ scores with several "minimum matching quality".

3.3   Pairing Strategies

In the previous section, we have defined the algebraic requirements of a weighted bipartite graph so that we can compute a parameter-less measure of the quality of a segmentation. In this section, we provide hints to build a graph out of a segmentation.

Symmetric Pairing Metrics. We first need a shape matching metric that renders the matching of two regions. In Mathematical Morphology, distances between any set of points have been largely studied [1,3,18]. While the Jaccard distance (one minus the Jaccard index) and the Hausdorff distance are some well-known examples of such distances, more advanced metrics based on skeletons and median sets allow for a more relevant analysis of the shape families.

The *Intersection over Union (IoU)* (or *Jaccard index*) defined in Eq. (10) has many advantages and is a *de facto* standard to compute similarity between two regions [2,7]. Whereas advanced morphological distances are computationally expensive, the IoU is fast to compute between any pair of regions of a segmentation as it is based the matrix of the intersections (see Fig. 7). Second, it features

$$\underline{\text{C}}\text{ov}_B(A) = \frac{|A \cap B|}{|\mathbf{B}|} \tag{9}$$

$$\underline{\text{J}}\text{accard}(A, B) = \frac{|A \cap B|}{|\mathbf{A} \cup \mathbf{B}|} \tag{10}$$

$$\underline{\text{D}}\text{ice}(A, B) = \frac{|A \cap B|}{|\mathbf{A}| + |\mathbf{B}|} \tag{11}$$

$$\underline{\text{F}}\text{orce}(A, B) = \text{Cov}_A(B)^2 + \text{Cov}_B(A)^2 \tag{12}$$

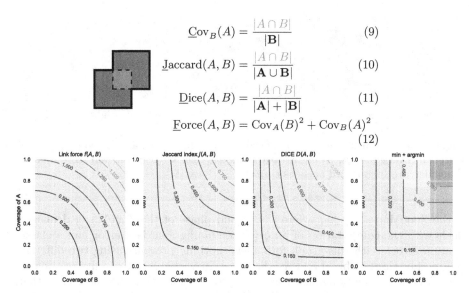

Fig. 8. Standard shape similarity for scoring edges. Link force [9], IoU (*Jaccard*), F1-score (*DICE*), and min(Cov$_A$, Cov$_B$) with edge categorization [16] (the color zone defines the category of the edge).

the property required in Sect. 3.2 as shown in [11]. Indeed, the authors have proven that *each ground-truth region can have at most one corresponding predicted region with IoU strictly greater than 0.5 and vice-versa*. Other popular pairing metrics includes the DICE coefficient D (or F1 score) which is closely related to the Jaccard index and so ensures a one-to-one pairing when the $D > \frac{2}{3}$. In the ZoneMap metric used for page segmentation[9], the regions' coverage are combined with a euclidean distance. All these metrics are related and can be expressed as a function of the coverage of the groundtruth's shapes by the prediction and the coverage of the prediction's shapes by the ground truth as shown Fig. 8. The Zonemap pairing score differs as it allows good score as soon as one of the coverage score is high. Roughly speaking, *min*, the DICE and the IoU are "logical ands" while the Zonemap score is a "logical or". In [16], the score is also associated with a category depending on its "zone"—it will be further detailed in Sect. 4.

The question when choosing a pairing function is "what is a match?" In our opinion, a *match* has a practical definition. Two objects/regions should be considered as *matching* if there is no need for human manual intervention to correct the result produced by a system. In the case of a segmentation, it means that the predicted region should perfectly fit with the ground truth and vice verse. As a consequence, the IoU of the DICE score should be favored.

Different Pairing and Edge Weighting Functions. Most evaluation framework use the same scoring function for pairing, *i.e.* building the association graph, and *evaluating*, *i.e.* getting a score out-of-the graph. A notable exception

(a) Precision. Left: IoU, right: $\mathrm{Cov}_Y(X)$ (b) Recall. Left: IoU, right: $\mathrm{Cov}_X(Y)$

Fig. 9. Comparison of the precision/recall values when edges are valued with IoU vs marginal coverage scores. Marginal coverages are clearly optimistic and score some of the under/over-segmented areas as correct.

is the ZoneMap score [9], where the second phase of the ZoneMap calculation deals with the computation of an error combining a surface error and a classification error. With this graph formalism, nothing prevents using two different scoring functions to build and weight edges. In particular, it is quite reasonable to build the associations based on a fast-to-compute and sensible metric that exhibits the properties in Eq. (4) (like the IoU) and keep only edges that provide a matching. On this new graph, we can either keep the original edge weight and compute the statistics described in Sect. 3.1 – this yields the COCO Panoptic metric – or weight edges with another function more sensible for the application. Since this metric is only computed between regions which are one-to-one match, it does not matter if it is more computationally expensive. In the context of map segmentation and block partitioning, the 95% Hausdorff has shown to be closer to the human perception of distance between shapes as it penalizes non-regular boundaries. Yet, once the graph re-weighted (with normalized distances), the evaluation protocol described in Sect. 3.1 becomes available.

4   Comparison with Other Metrics and Limitations

Non-symmetric Metrics for Precision-Recall. In [16], the authors have an alternate definition of the precision, recall, F-measure based on partition coverage and refinement (Eq. (9)). Explained in a bipartite graph framework, given two shapes X and Y from the predicted segmentation P and the groundtruth T, an edge is added to E if $\min(\mathrm{Cov}_X(Y), \mathrm{Cov}_Y(X)) > \gamma_1$ and $\max(\mathrm{Cov}_X(Y), \mathrm{Cov}_Y(X)) > \gamma_2$ (with $\gamma_1 < \gamma_2$) where γ_1 represents a *candidate* threshold and γ_2 a *match* threshold. These thresholds define the *match* category areas as shown in Fig. 8. When both $\mathrm{Cov}_X(Y)$, $\mathrm{Cov}_Y(X)$ pass the match threshold, we have an *object match* (simply a *match* in our terminology). If only one of them passes the *match* threshold, we get a *partial* match that is either a *merge* or a *split* depending on which of $\mathrm{Cov}_X(Y)$ and $\mathrm{Cov}_Y(X)$ is higher. It follows definitions of the precision/recall based on the these classes. They have been rewritten to highlight the relations with the Panoptic metric.

$$precision[16] = \underbrace{precision(1)}_{\text{P match rate}} + \underbrace{\beta.\frac{C_u}{|P|}}_{\text{under-segmentation rate}} + \underbrace{\sum_{(x,y)\in E}\frac{\text{Cov}_Y(x)}{|P|}}_{\text{fragmentation quality}} \qquad (13)$$

$$recall[16] = \underbrace{recall(2)}_{\text{T match rate}} + \underbrace{\beta.\frac{C_o}{|T|}}_{\text{over-segmentation rate}} + \underbrace{\sum_{(X,Y)\in E}\frac{\text{Cov}_X(Y)}{|T|}}_{\text{fragmentation quality}} \qquad (14)$$

These metrics are actually over-scoring the quality of the segmentation. Indeed, *recall* and *precision* are related to measuring "correct" things, *i.e.*, *matches*. Here the precision is maximal whenever $T \sqsubseteq P$ where $\sqsubseteq$ denotes the partition refinement and the recall is maximal whenever $P \sqsubseteq T$. However, neither an over-segmentation, nor an under-segmentation are good as they may have no "correct" regions (regions that would need no human correction). This statement is motivated by the experiment in Fig. 9 where the precision and recall maps are valued with the IoU, $\text{Cov}_Y(X)$ and $\text{Cov}_Y(X)$ and shows that some over/under-segmented regions appear (and are counted) as "correct".

Nevertheless, the metrics are sensible to measure the quality of an over-/under-segmentation. For instance, it may be used to measure the ability to recover a correct segmentation by only merging regions. The *recall* and *precision* from [16] should not be used for assessing a fully-automated system and might rather be named "Coarse Segmentation Score" and "Fine Segmentation Score". These metrics also make sense when evaluating hierarchies of segmentation to allow over- and under-segmented regions as soon as the boundaries match [13].

Comparison with DETVAL and ZoneMap Document-Oriented Metrics. The DETVAL metric [20] shares many features with [16]. The pairing function between two regions depends on the marginal coverage $\text{Cov}_X(Y)$ and $\text{Cov}_Y(X)$ that must pass two detection quality thresholds. It also uses a fragmentation score in the precision/recall that allows to consider a one-to-many and many-to-one (splits and merges) as partial match. For document processing, it may make sense to consider such situation where it is "better have an over/under segmentation than nothing" since documents are structured hierarchically. In [20], precision/recall are defined as follows (again rewritten to highlight the relations with the Panoptic metric):

$$precision[20] = \underbrace{precision(1)}_{\text{P match rate}} + \underbrace{\sum_{X\in E_{merge}}\frac{fs(X)}{|P|}}_{\text{fragmentation quality}} \qquad (15)$$

$$recall[20] = \underbrace{recall(2)}_{\text{T match rate}} + \underbrace{\sum_{Y\in E_{split}}\frac{fs(Y)}{|P|}}_{\text{fragmentation quality}} \qquad (16)$$

where $fs(X) = (1 + \ln(degree(X)))^{-1}$ is the fragmentation score that decreases as the number of incident edges in X increases. At the end, the drawbacks of this metric are: 1. it considers *fragmentation* as an acceptable error, 2. it has two detection quality thresholds (vs one for the PQ), 3, it does not average the quality of the segmentation once a match has been accepted.

The ZoneMap[9] is generalization of the DETVAL protocol that supports zone overlapping. As a consequence, it takes account of *splits* and *merges* by decomposing partial overlaps in sub-zones that are scored and contribute to the final score. Contrary to DETVAL, the score penalizes all "imperfect" matches since the pixels that are not in the intersection participate in the error. ZoneMap also introduces a classification error that is of prime interest as part of a panoptic segmentation where instances are associated to classes. However, it has some drawbacks for our application. 1. There are (almost) no one-to-one association when computing the graph with all non-zero edges over two map partitions, we need to filter out edges with a detection quality threshold. 2. The error is based on the global surface and not the instance surface, *i.e.* imprecise small regions are not as important as imprecise big regions. 3. the score, not normalized between [0-1], is hard to interpret.

5   Limitations, Conclusions and Perspectives

The COCO Panoptic metric shines for the segmentation task: it is fast, simple and intelligible. We have shown that it is theoretically sound as it can be seen as classical prediction evaluation for which a strong background exists and is widely accepted. Also, expressed in a bipartite graph framework, it is extendable as we can separate the metric used for pairing regions (that needs strong properties and has to be fast to compute) and the one assessing the segmentation quality. As a matter of reproducible research, the metric implementation and a python package are available as a tool set on Github[1]. We have also highlighted the similarities and differences with some other metrics used in the document and natural image segmentation. Especially, considering an over-/under-segmentation as "half"-success is debatable for map segmentation, but may make sense for page segmentation. We believe that a generalization of the COCO Panoptic metric will lead to a unification of the evaluation segmentation protocol with just an application-dependent customization. Once the actual limitations are addressed, we will be able to quantify the differences between a unified-COCO Panoptic metric and task-specific metrics on various dataset. The following table summarizes the current limitations and the domain accessible once addressed:

Limitation	Application – Domain
1. Region Confidence Score	A. Evaluation of hierarchies [13]
2. Region overlapping	B. Page/Multi-layer map segmentation
3. Over/under-segmentation scoring	A + B

---

[1] https://github.com/icdar21-mapseg/icdar21-mapseg-eval.

Acknowledgements. This work was partially funded by the French National Research Agency (ANR): Project SoDuCo, grant ANR-18-CE38-0013.

References

1. Angulo, J., Meyer, F.: Morphological exploration of shape spaces. In: Wilkinson, M.H.F., Roerdink, J.B.T.M. (eds.) ISMM 2009. LNCS, vol. 5720, pp. 226–237. Springer, Heidelberg (2009). https://doi.org/10.1007/978-3-642-03613-2_21
2. Arbelaez, P., Maire, M., Fowlkes, C., Malik, J.: Contour detection and hierarchical image segmentation. IEEE Trans. Pattern Anal. Mach. Intell. **33**(5), 898–916 (2010)
3. Charpiat, G., Faugeras, O., Keriven, R., Maurel, P.: Distance-based shape statistics. In: IEEE International Conference on Acoustics Speech and Signal Processing Proceedings, vol. 5, p. V (2006)
4. Chazalon, J., et al.: ICDAR 2021 competition on historical map segmentation. In: IEEE International Conference on Document Analysis and Recognition, September (2021, to appear). https://icdar21-mapseg.github.io/
5. Chen, Y., Carlinet, E., Chazalon, J., Mallet, C., Duménieu, B., Perret, J.: Combining deep learning and mathematical morphology for historical map segmentation. In: Lindblad, J., Malmberg, F., Sladoje, N. (eds.) DGMM 2021. LNCS, vol. 12708, pp. 79–92. Springer, Cham (2021). https://doi.org/10.1007/978-3-030-76657-3_5
6. Dietzel, C., Herold, M., Hemphill, J.J., Clarke, K.C.: Spatio-temporal dynamics in California's central valley: empirical links to urban theory. Int. J. Geogr. Inf. Sci. **19**(2), 175–195 (2005)
7. Everingham, M., Van Gool, L., Williams, C.K., Winn, J., Zisserman, A.: The pascal visual object classes (VOC) challenge. Int. J. Comput. Vis. **88**(2), 303–338 (2010). https://doi.org/10.1007/s11263-009-0275-4
8. Flach, P.A., Hernández-Orallo, J., Ramirez, C.F.: A coherent interpretation of AUC as a measure of aggregated classification performance. In: ICML (2011)
9. Galibert, O., Kahn, J., Oparin, I.: The zonemap metric for page segmentation and area classification in scanned documents. In: IEEE International Conference on Image Processing, pp. 2594–2598 (2014)
10. Garcia-Molsosa, A., Orengo, H.A., Lawrence, D., Philip, G., Hopper, K., Petrie, C.A.: Potential of deep learning segmentation for the extraction of archaeological features from historical map series. Archaeol. Prospect. **28**, 187–199 (2021)
11. Kirillov, A., He, K., Girshick, R., Rother, C., Dollár, P.: Panoptic segmentation. In: Proceedings of the IEEE/CVF Conference on Computer Vision and Pattern Recognition, pp. 9404–9413 (2019)
12. Padilla, R., Passos, W.L., Dias, T.L.B., Netto, S.L., da Silva, E.A.B.: A comparative analysis of object detection metrics with a companion open-source toolkit. Electronics **10**(3), 279 (2021)
13. Perret, B., Cousty, J., Guimaraes, S.J.F., Maia, D.S.: Evaluation of hierarchical watersheds. IEEE Trans. Image Process. **27**(4), 1676–1688 (2018)
14. Perret, J., Gribaudi, M., Barthelemy, M.: Roads and cities of 18th century France. Sci. Data **2**(1), 1–7 (2015)
15. Petrie, C.A., Orengo, H.A., Green, A.S., Walker, J.R., et al.: Mapping archaeology while mapping an empire: using historical maps to reconstruct ancient settlement landscapes in modern India and Pakistan. Geosciences **9**(1), 11 (2019)
16. Pont-Tuset, J., Marques, F.: Supervised evaluation of image segmentation and object proposal techniques. IEEE Trans. Pattern Anal. Mach. Intell. **38**(7), 1465–1478 (2015)

17. Shafait, F., Keysers, D., Breuel, T.: Performance evaluation and benchmarking of six-page segmentation algorithms. IEEE Trans. Pattern Anal. Mach. Intell. **30**(6), 941–954 (2008)
18. Vidal, J., Crespo, J.: Sets matching in binary images using mathematical morphology. In: 2008 International Conference of the Chilean Computer Science Society, pp. 110–115 (2008)
19. West, D.B., et al.: Introduction to Graph Theory, vol. 2 (2001)
20. Wolf, C., Jolion, J.M.: Object count/area graphs for the evaluation of object detection and segmentation algorithms. Int. J. Doc. Anal. Recogn. **8**(4), 280–296 (2006)

A Large Multi-target Dataset of Common Bengali Handwritten Graphemes

Samiul Alam[1,2](✉) ⓘ, Tahsin Reasat[1], Asif Shahriyar Sushmit[1],
Sadi Mohammad Siddique[1], Fuad Rahman[3]ⓘ, Mahady Hasan[4]ⓘ,
and Ahmed Imtiaz Humayun[1]ⓘ

[1] Bengali.AI, Dhaka, Bangladesh
salam10@uh.edu
[2] Bangladesh University of Engineering and Technology, Dhaka, Bangladesh
[3] Apurba Technologies Inc., Dhaka, Bangladesh
[4] Independent University, Bangladesh, Dhaka, Bangladesh

Abstract. Latin has historically led the state-of-the-art in handwritten optical character recognition (OCR) research. Adapting existing systems from Latin to *alpha-syllabary* languages is particularly challenging due to a sharp contrast between their orthographies. Due to a cursive writing system and frequent use of diacritics, the segmentation and/or alignment of graphical constituents with corresponding characters becomes significantly convoluted. We propose a labeling scheme based on *graphemes* (linguistic segments of word formation) that makes segmentation inside alpha-syllabary words linear and present the first dataset of Bengali handwritten graphemes that are commonly used in everyday context. The dataset contains $411k$ curated samples of 1295 unique commonly used Bengali graphemes. Additionally, the test set contains 900 uncommon Bengali graphemes for out of dictionary performance evaluation. The dataset is open-sourced as a part of a public Handwritten Grapheme Classification Challenge on Kaggle to benchmark vision algorithms for multi-target grapheme classification. The unique graphemes present in this dataset are selected based on commonality in the Google Bengali ASR corpus. From competition proceedings, we see that deep learning methods can generalize to a large span of out of dictionary graphemes which are absent during training (Kaggle Competition kaggle.com/c/bengaliai-cv19, Supplementary materials and Appendix https://github.com/AhmedImtiazPrio/ICDAR2021supplementary).

1 Introduction

Speakers of languages from the alpha-syllabary or *Abugida* family comprise of up to 1.3 billion people across India, Bangladesh, and Thailand alone. There is significant academic and commercial interest in developing systems that can optically recognize handwritten text for such languages with numerous applications in e-commerce, security, digitization, and e-learning. In the alpha-syllabary writing system, each word is comprised of segments made of character units that

© Springer Nature Switzerland AG 2021
J. Lladós et al. (Eds.): ICDAR 2021, LNCS 12824, pp. 383–398, 2021.
https://doi.org/10.1007/978-3-030-86337-1_26

Fig. 1. Orthographic components in a Bangla (Bengali) word compared to English and Devnagari (Hindi). The word 'Proton' in both Bengali and Hindi is equivalent to its transliteration. Characters are color-coded according to phonemic correspondence. Unlike English, characters are not arranged horizontally according to phonemic sequence in Hindi and Bengali.

are in phonemic sequence. These segments act as the smallest written unit in alpha-syllabary languages and are termed as *Graphemes* [12]; the term alpha-syllabary itself originates from the alphabet and syllabary qualities of graphemes [7]. Each grapheme comprises of a *grapheme root*, which can be one character or several characters combined as a conjunct. The term character is used interchangeably with unicode character throughout this text. Root characters may be accompanied by *vowel* or *consonant diacritics-* demarcations which correspond to phonemic extensions. To better understand the orthography, we can compare the English word *Proton* to its Bengali transliteration প্রোটন (Fig. 1). While in English the characters are horizontally arranged according to phonemic sequence, the first grapheme for both Bengali and Devanagari scripts have a sequence of glyphs that do not correspond to the linear arrangement of unicode characters or phonemes. As most OCR systems make a linear pass through a written line, we believe this non-linear positioning is important to consider when designing such systems for Bengali as well as other alpha-syllabary languages.

We propose a labeling scheme based on grapheme segments of *Abugida* languages as a proxy for character based OCR systems; grapheme recognition instead of character recognition bypasses the complexities of character segmentation inside handwritten alpha-syllabary words. We have curated the first *Hand-*

written Grapheme Dataset of Bengali as a candidate alpha-syllabary language, containing 411882 images of 1295 unique commonly used handwritten graphemes and ∼900 uncommon graphemes (exact numbers are excluded for the integrity of the test set). While the cardinality of graphemes is significantly larger than that of characters, through competition proceedings we show that the classification task is tractable even with a small number of graphemes- deep learning models can generalize to a large span of unique graphemes even if they are trained with a smaller set. Furthermore, the scope of this dataset is not limited to the domain of OCR, it also creates an opportunity to evaluate multi-target classification algorithms based on root and diacritic annotations of graphemes. Compared to Multi-Mnist [25] which is a frequently used synthetic dataset used for multi-target benchmarking, our dataset provides natural data of a multi-target task comprising of three target variables.

The rest of the paper is organized as follows. Section 2 discuses previous works. Section 3 shows the different challenges that arise due to the orthography of the Bengali language, which is analogous to other *Abugida* languages. Section 4 formally defines the dataset objective, goes into the motivation behind a grapheme based labeling scheme, and discusses briefly the methodology followed to gather, extract, and standardize the data. Section 5 discusses some insights gained from our Bengali Grapheme Classification Competition (2019–2020) on Kaggle along with solutions developed by the top-ranking participants. Finally Sect. 6 presents our conclusions. For ease of reading, we have included the IPA standard English transliteration of Bengali characters in {.} throughout the text.

2   Related Work

Handwritten OCR methods in recent times have systematically moved towards recognition at word or sentence levels instead of characters, facilitated by word recognition datasets like the IAM-Database [21] that contains 82, 227 words for English with a lexicon of 10, 841 words. Bengali on the other hand, is a low-resource language considering the volume of the word level recognition datasets currently available [24,26] e.g. Roy et al. [24] introduced a dataset of alpha-syllabary words comprising of 17, 091 Bengali and 16, 128 Devnagari words, which is significantly low compared to their English counterparts. The absence of a standardized corpus for Bengali, analogous to the Lancaster-Oslo/Bergen Corpus (LOB) [20], inhibits the collection of rich datasets of Bengali handwritten words. In recent times, several datasets [1,6,23] have been made for Bengali handwritten isolated characters and their effectiveness has been limited. Apart from the absence of datasets, adopting state-of-the-art methods from Latin to alpha-syllabary languages face significant challenges since the sequence of glyphs don't always follow the sequence of characters in the word due to the use of diacritics. This introduces a disconnect between the order of graphical states in the word image with the order of states in the ground truth string, which is key to alignment methods [3] and also the alignment free forward-backward algorithm of connectionist temporal classification (CTC) models [13] for word level recognition. Languages with different writing systems therefore require language specific design of the recognition pipeline and need more understanding of how it affects performance. For Bengali, grapheme represents the smallest unit of a word encapsulated by the relevant glyphs (Fig. 1). Representing handwritten words as graphemes would allow the detection model to bypass glyph sequence based complexities- graphemes (as collection of character glyphs) are sequentially arranged in the same order for both the word image and the word string. Therefore grapheme recognition models trained on our dataset can be employed as a pretrained front-end for a CTC pipeline, that does word level recognition with graphemes as units. To the best of our knowledge, this is the first work that proposes grapheme recognition for alpha-syllabary OCR. This dataset is also meant to be a potential stepping stone for word-level datasets. The lack of large datasets is a serious restriction for word-level OCR studies on Bengali. The only word level dataset available is from [22, 26] which are not large enough for accurate inference. Large grapheme datasets can be used to create synthetic dataset of handwritten words independently or with existing word level datasets to attenuate or even redress this problem.

3   Challenges of Bengali Orthography

As mentioned before in Sect. 1, each Bengali word is comprised of segmental units called graphemes. Bengali has 48 characters in its alphabet- 11 vowels and 38 consonants (including special characters 'ৎ'{ṭ},'ং' {ṁ},'ঃ'{ḥ}). Out of the 11 vowels, 10 vowels have diacritic forms. There are also four consonant diacritics, '্য' (from consonant য {ya}), '্র' (from consonant র {ra}), '্র' (also from consonant র {ra}) and '়'. We follow the convention of considering 'ং'{ṁ},'ঃ' {ḥ} as standalone consonants since they are always present at the end of a grapheme and can be considered a separate root character.

3.1   Grapheme Roots and Diacritics

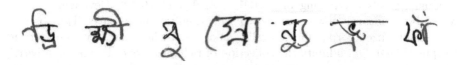

Fig. 2. Different vowel diacritics (green) and consonant diacritics (red) used in Bengali orthography. The placement of the diacritics are not dependent on the grapheme root. (Color figure online)

Graphemes in Bengali consist of a root character which may be a vowel or a consonant or a consonant conjunct along with vowel and consonant diacritics whose occurrence is optional. These three symbols together make a grapheme in Bengali. The consonant and vowel diacritics can occur horizontally, vertically adjacent to the root or even surrounding the root (Fig. 2). These roots and diacritics cannot be identified in written text by parsing horizontally and detecting each glyph separately. Instead, one must look at the whole grapheme and identify them as separate targets. In light of this, our dataset labels individual graphemes with root characters, consonant and vowel diacritics as separate targets.

3.2   Consonant Conjuncts or Ligatures

Consonant conjuncts in Bengali are analogous to ligatures in Latin where multiple consonants combine together to form glyphs which may or may not contain characteristics from the standalone consonant glyphs. In Bengali, up to three consonants can combine to form consonant conjuncts. Consonant conjuncts may have two (second order conjuncts, e.g. ষ্ট = শ + ট {sta = śa + ta }) or three (third order conjuncts, e.g. ক্ষ্ণ = ক + ষ + ন {kṣṇa = ka + ṣa + na}) consonants in the cluster. Changes in the order of consonants in a conjunct may result in complete or partial changes in the glyph. The glyphs for conjuncts can get very complex and can even be hard for human subjects to discern (See Section C of Appendix in supplementary materials).

3.3   Allographs

(a)   (b)   (c)   (d)

Fig. 3. Examples of allograph pairs for the same consonant conjunct 'ঙ্গ' {ṅga} (3a and 3b) and the same vowel diacritic 'ু' {u} (3c and 3d) marked in green. 3b and 3d follows an orthodox writing style. (Color figure online)

It is also possible for the same grapheme to have multiple styles of writing, called allographs. Although they are indistinguishable both phonemically and in their unicode forms, allographs may in fact appear to be significantly different in their handwritten guise (Fig. 3). Allograph pairs are sometimes formed due to simplification or modernization of handwritten typographies, i.e. instead of using the orthodox form for the consonant conjunct ঙ্গ = ঙ + গ as in Fig. 3b, a simplified more explicit form is written in Fig. 3a. The same can be seen for diacritics in Fig. 3c and Fig. 3d. It can be argued that allographs portray the linguistic plasticity of handwritten Bengali.

3.4 Unique Grapheme Combinations

One challenge posed by grapheme recognition is the huge number of unique graphemes possible. Taking into account the 38 consonants (n_c) including three special characters, 11 vowels (n_v) and ($n_c^3 + n_c^2$) possible consonant conjuncts (considering 2nd and 3rd order), there can be $((n_c - 3)^3 + (n_c - 3)^2 + (n_c - 3)) + 3$ different grapheme roots possible in Bengali. Grapheme roots can have any of the $10 + 1$ vowel diacritics (n_{vd}) and $7 + 1$ consonant diacritics (n_{cd}). So the approximate number of possible graphemes will be $n_v + 3 + ((n_c - 3)^3 + (n_c - 3)^2 + (n_c - 3)) \cdot n_{vd} \cdot n_{cd}$ or 3883894 unique graphemes. While this is a big number, not all of these combinations are viable or are used in practice.

4 The Dataset

Of all the possible grapheme combinations, only a small amount is prevalent in modern Bengali. In this section we discuss how we select the candidates for data collection and formalize the grapheme recognition problem.

4.1 Grapheme Selection

To find the popular graphemes, we use the text transcriptions for the Google Bengali ASR dataset [17] as our reference corpus. The ASR dataset contains a large volume of transcribed Bengali speech data. It consists of 127565 utterances comprising 609510 words and 2111256 graphemes. Out of these graphemes, 1295 commonly used Bengali graphemes in everyday vocabulary is selected. Each candidate grapheme had to occur more than twice in the entire corpus or used in at least two unique words to be selected in our pool. Graphemes from highly frequent transliterations and misspelled words were also considered. The uncommon graphemes were synthesized by uniformly sampling from all the possible combinations and verifying their legibility. (See Section B of Appendix in supplementary materials for a full list of grapheme roots and diacritics)

4.2 Labeling Scheme

Bengali graphemes can have multiple characters depending on the number of consonants, vowels or diacritics forming the grapheme. We split the characters of a Bengali grapheme into three target variables based on their co-occurrence:

1. Vowel Diacritics, i.e. আ, ি, ী, ু, ূ, ৃ, ে, ৈ, ো, ৌ. If the grapheme consists of a vowel diacritic, it is generally the final character in the unicode string. Graphemes cannot contain multiple vowel diacritics. The vowel diacritic target variable has 11 (N_{vd}) orthogonal classes including a null diacritic denoting absence.

2. Consonant Diacritics, i.e. ্র, ্য, ঁ, ্. Graphemes can have a combination of consonant diacritic characters e.g. '্যু' = '্য' + '্র'. We consider each combination to be a unique diacritic in our scheme for ease of analysis. The consonant diacritic target variable has 8 (N_{cd}) orthogonal classes including combinations and a null diacritic.

3. Grapheme roots, which can be comprised of vowels, consonants or conjuncts. In unicode these are placed as the first characters of a grapheme string. An alternative way of defining grapheme roots would be considering all the characters apart from diacritics as root characters in a grapheme. While possible orthogonal classes under this target variable can be a very big number (see Sect. 3.4), we limit the number of consonant conjuncts based on commonality in everyday context. There are in total 168 (N_r) roots in the dataset.

Grapheme recognition thus becomes a multi-target classification task for handwritten Bengali, rather than the traditional multi class classification adopted by previous datasets of this kind [1,2,23,26]. Here, a vision algorithm would have to separately recognize grapheme roots, vowel diacritics and consonant diacritics as three target variables. Formally, we consider our dataset $\mathcal{D} = \{s_1, s_2, ..., s_N\}$ as composed of N data points $s_i = \{x_i, y_i^r, y_i^{vd}, y_i^{cd}\}$. Each datum s_i consists of an image, $x_i \in \mathbb{R}_{H \times W}$ and a subset of three target variables y_i^r, y_i^{vd} and y_i^{cd} denoting the ground truth for roots, vowel diacritics and consonant diacritics respectively, i.e., $y_i^r \in \mathcal{R}$, $y_i^{vd} \in \mathcal{V}$, and $y_i^{cd} \in \mathcal{C}$. Here, $\mathcal{R}, \mathcal{C}$, and $\mathcal{V}$ is the set of roots, vowel diacritics and consonant diacritics, respectively, where $|\mathcal{R}| = N_r$, $|\mathcal{C}| = N_{cd}$, and $|\mathcal{V}| = N_{vd}$. The multi-target classification task, thus will consist of generating a classifier h which, given an image x_i, is capable of accurately predicting its corresponding components, i.e., $h(x_i) = \{y_i^r, y_i^{vd}, y_i^{cd}\}$.

Although we have formulated the grapheme recognition challenge as a multi-target classification task, it is only one way of defining the grapheme recognition problem. In fact, we will see in Sect. 5.2, that the problem can also be defined as a multi-label and a metric learning task.

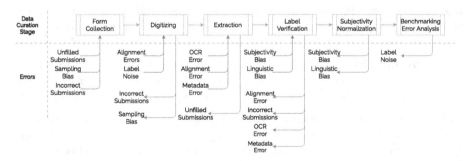

Fig. 4. Overview of dataset creation process. Green arrows refer to the bias/errors removed in each step and red refers to the ones inevitably introduced. (Color figure online)

4.3   Dataset Collection and Standardization

The data was obtained from Bengali speaking volunteers in schools, colleges and universities. A standardized form (See Section A of Appendix in supplementary materials) with alignment markers were printed and distributed. A total of 2896 volunteers participated in the project. Each subject could be uniquely identified through their institutional identification number submitted through the forms, which was later de-identified and replaced with a unique identifier for each subject. The dataset curation pipeline is illustrated in Fig. 4.

Collection. Contributors were given one of 16 different template forms with prompts for graphemes. The templates were automatically generated using Adobe Photoshop scripts. Each template had a unique set of graphemes compared to the others. Since the 16 different templates were not dispersed uniformly every time minor sampling bias was introduced during collection.

Pruning and Scanning. The forms were scanned and analysed carefully to remove invalid submissions and reduce sampling bias. In this step additional errors were introduced due to misalignment during scanning. Unfilled or improperly filled samples were still retained. All the forms were scanned using the same device at 300 dpi.

Extraction. An OCR algorithm was used to automatically detect the template ID. The template identifier asserted which ground truth graphemes where present in which boxes of the form. In this step, OCR extraction errors introduced label noise. The scanned forms were registered with digital templates to extract handwritten data, which sometimes introduced alignment errors or errors while extracting metadata. Unfilled boxes were removed automatically in this step.

Preliminary Label Verification. The extracted graphemes were compiled into batches and sent to 22 native Bengali volunteers who analysed each image and matched them to their corresponding ground truth annotation. In this step OCR errors and label noise was minimised. However additional error was introduced in the form of conformity bias, linguistic bias (i.e. allograph not recognized), grapheme bias (i.e. particular grapheme has significantly lesser number of samples) and annotator subjectivity. Samples selected as erroneous by each annotator was stored for further inspection instead of being discarded.

Label Verification. Each batch from the previous step, was sent to one of two curators who validated erroneous samples submitted by annotators and re-checked unique graphemes which had a higher frequency of mislabeled samples.

Subjectivity Normalization. A fixed guideline is decided upon by all curators that specifies how much and the nature of deformity a sample can contain. Based on this, subjectivity errors were minimized for unique graphemes with high frequency mislabeled samples.

4.4   Training Set Metadata

The metadata collected through forms are compiled together for further studies on dependency of handwriting with each of the meta domains. Since the data

Table 1. Number of samples present in each subset of the dataset. Null diacritics are ignored.

Targets	Sub-targets	Classes	Samples	Training Set	Public Test Set	Private Test Set
Roots	Vowel Roots	11	5315	2672	1270	1398
	Consonant Roots	38	215001	107103	52185	57787
	Conjunct Roots	119	184050	91065	45206	53196
	Total	168	404366	200840	98661	112381
Diacritics	Vowel Diacritics	10	320896	159332	78503	89891
	Consonant Diacritics	7	152010	75562	37301	44649

was crowd-sourced, the distribution with respect to factors such as age and education is subject to interest and availability of contributors. This could possibly introduce demographic biases so we provide metadata on the age group, gender, handedness and education for every sample to allow future ablation studies. Only the training set metadata is made public; the test set metadata will be made available upon request. The training set contains handwriting from 1448 individuals, each individual contributing 138.8 graphemes on average; 1037 of the contributors identified as male, 383 as female, 4 as non-binary and 24 declined to identify. The medium of instruction during primary education for 1196 contributors was Bengali, for 214 English and for 12 Madrasha (Bengali and Arabic); 33 are left-handed while 1192 are right handed. Of all the contributors, 93 fall in the age group between $0 - 12$, 245 in $13-17$, 1057 in $18-24$, 22 in $25-35$ and 2 in ages between $36-50$.

4.5 Dataset Summary

A breakdown of the composition of the train and test sets of the dataset is given in Table 1. Additionally, a breakdown of the roots into vowels, consonants and conjuncts along with the number of unique classes and samples for each target is also shown. Note that the absence of a diacritic which is labeled as the null diacritic '0' is not considered when counting the total samples as the glyph for the diacritic is not present in such samples. The final dataset contains a total of 411882 handwritten graphemes of size 137 by 236 pixels. See supplementary materials and Appendix A for dataset collection forms, tools and protocols.

4.6 Class Imbalance in Dataset

We divide the roots into three groups- vowels, consonants, and consonant conjuncts- and inspect class imbalance within each. There are linguistic rules which constrict the number of diacritics that may occur with each of these roots, e.g. vowel roots never have added diacritics. Although imbalance in vowel roots is not major, it must be noted because the relatively infrequent vowel roots 'ঈ', 'ঊ' {ī, ū} and 'ঐ' {ai} share a close resemblance to the more frequent roots 'ই', 'উ' {i, u} and 'এ' {ē} respectively.

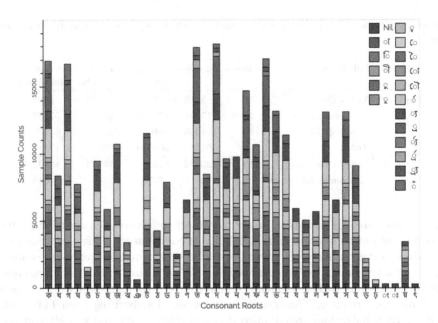

Fig. 5. Number of samples per consonant root. Each bar represents the number of samples which contain a particular consonant and the divisions in each bar represent the distribution of diacritics in the samples containing that consonant.

The imbalance in consonant roots is however much more striking as we can see in Fig. 5. The imbalance here is twofold- in the number of total sample images of consonant roots and the imbalances in the distribution of the vowel and consonant diacritics that can occur with each consonant.

The consonant conjuncts demonstrate imbalance similar to the consonant roots but with an added degree of complexity. We can visualize this imbalance much better via the chord diagram in Fig. 6. The consonant conjuncts are made up of multiple consonant characters and since the glyph of a consonant conjunct often shares some resemblance with its constituent consonants, highly frequent consonants may increase estimation bias for less frequent conjuncts containing them. This phenomenon is indeed visible and further discussed in Sect. 5.3.

5   The Challenge

The dataset is open-sourced as part of a public Handwritten Grapheme Recognition Kaggle competition. Of all the samples present in the dataset, 200840 were placed in the training set, 98661 in the public test set, and 112381 in the private test set; while making sure there is no overlap in contributors between sets. Most of the uncommon graphemes were placed in the private test set and none in the training subset. Throughout the length of the competition, the participants try to improve their standings based on the public test set results. The

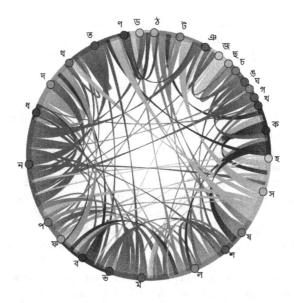

Fig. 6. Connectivity graph between consonants forming second-order conjuncts. The length of each arc shows how often the consonant occurs as a consonant conjunct. Higher frequency of consonants (e.g. ক {ka}) may bias lower frequency conjuncts towards the constituent.

private test set result on the other hand, is kept hidden for each submission and is only published after the competition is over. Of the OOD graphemes, 88.4% were placed in the private test set to prevent over-fitting models based on public standings. This motivated the participants to build methods that have the capacity to classify out of dictionary graphemes by recognizing the target variables independently.

5.1 Competition Metric

The metric for the challenge is a hierarchical macro-averaged recall. First, a standard macro-averaged recall is calculated for each component. Let the macro-averaged recall for grapheme root, vowel diacritic, and consonant diacritic be denoted by R_r, R_{vd}, and R_{cd} respectively. The final score R is the weighted average

$$R = \frac{1}{4}(2R_r + R_{vd} + R_{cd}). \tag{1}$$

5.2 Top Scoring Methods

The Kaggle competition resulted in $31,002$ submissions from $2,059$ teams consisting of $2,623$ competitors. The participants have explored a diverse set of algorithms throughout the competition (See Section E in Appendix for more details);

the most popular being state of the art image augmentation methods such as cutout [11], mixup [30], cutmix [29], mixmatch [5] and fmix [14]. Ensemble methods incorporating snapshots of the same or different network architectures were also common.

The *winner* took a grapheme classification approach rather than component recognition. The input images were classified into 14784 (168 x 11 x 8) classes, that is, all the possibles graphemes that can be created using the available grapheme roots and diacritics. An EfficientNet [28] model is initially used to classify the graphemes. However, if the network is not confident about its prediction then the sample is considered as an OOD grapheme and it is passed on to the OOD grapheme classification pipeline. The pipeline consists of a Cycle-Gan [31] that is trained to convert handwritten graphemes into typeface rendered graphemes. An EfficientNet classifier trained on a 14784 class synthetic grapheme dataset is used as an OOD grapheme classifier.

The *second* place team built their solution upon a multi-label grapheme classifier, but the number of classes were limited to the 1295 unique graphemes in training. A post processing heuristic is employed to generalize for any OOD graphemes. For each class in a target variable (e.g. consonant diacritic 'ে'), probabilities of all the graphemes containing the target (e.g., স্ক, ব, থি etc.) are averaged. This is repeated for every class for a target and the class with the highest average probability is selected for that target variable. Architectures used are different variants of SE-ResNeXt [15].

The *third* placed team dealt with the OOD graphemes by using metric learning. They used Arcface [10] to determine if an input test sample is present or absent in the train dataset. If the grapheme is present, a single EfficientNet model is used that detects the grapheme components in a multi-target setting. Otherwise, a different EfficientNet model is used to recognize each grapheme component.

A brief outline of the different approaches used by the participants and how they handled In Dictionary (ID) and Out of dictionary (OOD) graphemes are given in Table 2. For comparison, we also listed two simple preliminary baseline models based on VGG16 and DenseNet121 to showcase where improvements were or could be made. More then half of the teams achieved a decent score without considering any specialized method for detecting OOD graphemes, proving the hypothesis that it is possible to create a generalized model based on frequently occurring graphemes.

5.3   Submission Insights

For exploratory analysis on the competition submissions, we take the top 20 teams from both the private and public test set leaderboards and categorize their submissions according to quantile intervals of their private set scores as *Tier 1* ($> .976$), *Tier 2* ($> .933, < .976$), *Tier 3* ($> .925, < .933$) and *Tier 4* ($> .88, < .925$). It is seen that the *Tier 4* submissions have low discrepancy between public and private test set metrics; suggesting these to be high bias - low variance estimators. The *Tier 3* submissions were the total opposite and had high discrepancy on average, indicating fine-tuning of the model on the public test set. This discrepancy increases as we go to *Tier 2* submissions but then decreases for *Tier 1* submissions. In fact if we observe the progression of the *Tier 2* teams, many of them were once near the top of the private leaderboard but later fine-tuned for the public test set. Contrary to intuition, error rate doesn't significantly vary depending on the frequency of the unique grapheme in the training set, within each of these quantile groups. However, the error rate is higher by 1.31% on average for OOD graphemes, with *Tier 1* error rate at $4.4(\pm0.07)\%$.

Considering the size of the challenge test set, possible reasons behind test set error could be label noise, class imbalance or general challenges surrounding the task. We start by inspecting the samples that were misclassified by all the *Tier 1* submissions and find that only 34.8% had label noise. Significant error can be seen due to the misclassification of consonant diacritic classes which are binary combinations of {'ে', 'ো' or 'ৌ'}, with high false positives for the individual constituents. This can be attributed to the class imbalance in the dataset since combinations are less frequent that their primitives; separately recognizing the primitives in a multi-label manner could be a possible way to reduce such error. The vowel diacritic component has the highest macro-averaged recall, proving to be the easiest among the three tasks.

The false negatives of different grapheme roots give us significant insights on the challenges present in Bengali orthography. A pair of grapheme roots can have high similarity due to common characters or even similarity between glyphs of constituents. Probing the *Tier 1* false negatives, we find that 56.5% of the error is between roots that share at least one character. Misclassification between consonant conjuncts with the same first and second characters accounts for 28.8%

Table 2. Top 10 results on grapheme recognition challenge. Number of teams used separate models for in dictionary (ID) and out of dictionary (OOD) classification. MC = multi-class, ML = multi-label; MT = multi-target; SM = similarity metric learning. Note that the first two rows are preliminary baselines added for reference.

Rank	Augmentation	Problem Transform ID	OOD	OOD Detection	Model Architecture ID (M1)	OOD (M2)	Public Score	Private Score
–	Cutout [11]	MT	–	–	VGG16 [19]	–	0.946	0.876
–	Cutout	MT	–	–	DenseNet121 [16]	–	0.915	0.885
1	Auto-Augment [9]	MC	MC	M1 confidence below threshold and M2 output is OOD	EfficientNet-b7 [28]	CycleGan [31], EfficientNet-b7	0.995	0.976
2	Fmix [14]	ML	–	–	SeResNeXt [15]	–	0.996	0.969
3	CutMix [29]	SM	MT	Arcface [10]	EfficientNet, SeResNeXt	EfficientNet	0.995	0.965
4	Cutout	SM	MT	Arcface	Arcface	EfficientNet	0.994	0.962
5	CutMix	MT	–	–	SeResNeXt	–	0.994	0.958
6	CutMix, MixUp [30], Cutout, GridMask [8]	MT	–	–	SeResNeXt, InceptionResNetV2 [27]	–	0.987	0.956
7	CutMix, Cutout	MT	–	–	PNASNet-5 [18]	–	0.994	0.955
8	Cutout [11]	MT	MT	Arcface	EfficientNet-b7	EfficientNet	0.994	0.955
9	CutMix, Grid-Mix[4]	MT	–	–	SeResNeXt, Arcface	–	0.992	0.954
10	Cutout	MT	–	–	SeResNeXt	–	0.984	0.954

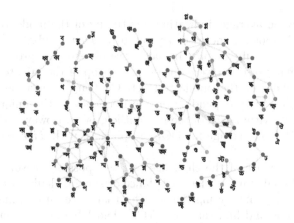

Fig. 7. Similarity between handwritten grapheme roots based on *Tier 1* confusion. Nodes are color coded according to the first character in each root. Edges correspond to sum of false negative rates between nodes, higher edge width corresponds to higher similarity between handwritten grapheme roots. (Color figure online)

and 21.5% of the error. Confusion between roots by *Tier 1* submissions highly correlate with similarity between glyphs and is visualized in Fig. 7. Edges correspond to the sum of false negative rates between nodes. Edges are pruned if the sum is below .5%. Sub-networks are formed by groups that are similar to each other. Class imbalance also plays an interesting role in such cases; misclassification of roots with high similarity between their handwritten glyphs can also be biased towards one class due to higher frequency, e.g. roots with 'ণ' {ṇa} are more frequently misclassified as roots with 'ন' {na} because of 'ন' {na} being more frequent. For more insight, see Section D and F in Appendix for confusion between roots.

6   Conclusion

In this paper, we outlined the challenges of recognizing Bengali handwritten text and explained why a character based labeling scheme- that has been widely successful for English characters- does not transfer well to Bengali. To rectify this, we propose a novel labeling scheme based on graphemes and present a dataset based on this scheme. Crowd sourced benchmarking on Kaggle shows that algorithms trained on this dataset can generalize on out of dictionary graphemes. This proves that it is possible to summarize the entire cohort of unique graphemes through some representative samples. One limitation of grapheme level recognition is that graphemes written in isolation may shift away from natural handwriting. Our proposed grapheme labeling scheme could be used as a stepping stone for future word level datasets. Grapheme recognition models could also be used as a front-end to solve OCR related tasks in not only Bengali but also other related languages in the alpha-syllabary family.

Acknowledgment. We would like to thank Kaggle for their generous support through their platform. We would also like to thank Md. Asif Bin Khaled for his leadership in coordinating the dataset collection in IUB. We are also grateful to our volunteers who participated and helped organize collection of data for our dataset. We would like to thank the late Dr. Ali Shihab Sabbir for providing invaluable help and guidance to Bengali.AI since its inception. Full list of contributions available at https://rb.gy/qd3ur2.

References

1. Alam, S., Reasat, T., Doha, R.M., Humayun, A.I.: NumtaDB - assembled Bengali handwritten digits. arXiv preprint arXiv:1806.02452 (2018)
2. AlKhateeb, J.H.: A database for Arabic handwritten character recognition. Procedia Comput. Sci. **65**, 556–561 (2015)
3. Arora, A., et al.: Using ASR methods for OCR. In: In Proceedings of the ICDAR (2019)
4. Baek, K., Bang, D., Shim, H.: GridMix: strong regularization through local context mapping. Pattern Recogn. **109**, 107594 (2020)
5. Berthelot, D., Carlini, N., Goodfellow, I., Papernot, N., Oliver, A., Raffel, C.: Mixmatch: a holistic approach to semi-supervised learning. In: 2019 Proceedings of the NeurIPS (2019)
6. Biswas, M., et al.: Banglalekha-isolated: a comprehensive bangla handwritten character dataset. arXiv preprint arXiv:1703.10661 (2017)
7. Bright, W.: A matter of typology: alphasyllabaries and abugidas. Writ. Lang. Lit. **2**(1), 45–65 (1999)
8. Chen, P.: Gridmask data augmentation. arXiv preprint arXiv:2001.04086 (2020)
9. Cubuk, E.D., Zoph, B., Mane, D., Vasudevan, V., Le, Q.V.: Autoaugment: learning augmentation policies from data. arXiv preprint arXiv:1805.09501 (2018)
10. Deng, J., Guo, J., Xue, N., Zafeiriou, S.: ArcFace: additive angular margin loss for deep face recognition. In: Proceedings of the 2019 IEEE/CVF Conference on CVPR, June 2019
11. DeVries, T., Taylor, G.W.: Improved regularization of convolutional neural networks with cutout. arXiv preprint arXiv:1708.04552 (2017)
12. Fedorova, L.: The development of graphic representation in abugida writing: the Akshara's grammar. Lingua Posnaniensis **55**(2), 49–66 (2013)
13. Graves, A., Fernández, S., Gomez, F., Schmidhuber, J.: Connectionist temporal classification: labelling unsegmented sequence data with recurrent neural networks. In: Proceedings of the ICML, pp. 369–376 (2006)
14. Harris, E., Marcu, A., Painter, M., Niranjan, M., Prügel-Bennett, A., Hare, J.: FMix: enhancing mixed sample data augmentation. arXiv e-prints arXiv:2002.12047, February 2020
15. Hu, J., Shen, L., Sun, G.: Squeeze-and-excitation networks. In: 2018 Proceedings of the IEEE/CVF Conference on CVPR, pp. 7132–7141, June 2018
16. Huang, G., Liu, Z., Pleiss, G., Van Der Maaten, L., Weinberger, K.: Convolutional networks with dense connectivity. IEEE Trans. Pattern Anal. Mach. Intell., 1–1 (2019). https://doi.org/10.1109/TPAMI.2019.2918284
17. Kjartansson, O., Sarin, S., Pipatsrisawat, K., Jansche, M., Ha, L.: Crowd-sourced speech corpora for Javanese, Sundanese, Sinhala, Nepali, and Bangladeshi Bengali. In: Proceedings of the 6th International Workshop on SLTU, pp. 52–55, August 2018

18. Liu, C., et al.: Progressive neural architecture search. In: Ferrari, V., Hebert, M., Sminchisescu, C., Weiss, Y. (eds.) ECCV 2018. LNCS, vol. 11205, pp. 19–35. Springer, Cham (2018). https://doi.org/10.1007/978-3-030-01246-5_2

19. Liu, S., Deng, W.: Very deep convolutional neural network based image classification using small training sample size. In: 2015 3rd IAPR Asian Conference on Pattern Recognition (ACPR), pp. 730–734 (2015). https://doi.org/10.1109/ACPR.2015.7486599

20. de Marcken, C.: Parsing the LOB corpus. In: 28th Annual Meeting of the Association for Computational Linguistics, pp. 243–251 (1990)

21. Marti, U.V., Bunke, H.: The IAM-database: an English sentence database for offline handwriting recognition. Int. J. Doc. Anal. Recogn. **5**(1), 39–46 (2002). https://doi.org/10.1007/s100320200071

22. Mridha, M., Ohi, A.Q., Ali, M.A., Emon, M.I., Kabir, M.M.: BanglaWriting: a multi-purpose offline Bangla handwriting dataset. Data Brief **34**, 106633 (2021)

23. Rabby, A.K.M.S.A., Haque, S., Islam, M.S., Abujar, S., Hossain, S.A.: Ekush: a multipurpose and multitype comprehensive database for online off-line Bangla handwritten characters. In: Santosh, K.C., Hegadi, R.S. (eds.) RTIP2R 2018. CCIS, vol. 1037, pp. 149–158. Springer, Singapore (2019). https://doi.org/10.1007/978-981-13-9187-3_14

24. Roy, P.P., Bhunia, A.K., Das, A., Dey, P., Pal, U.: HMM-based Indic handwritten word recognition using zone segmentation. Pattern Recogn. **60**, 1057–1075 (2016)

25. Sabour, S., Frosst, N., Hinton, G.E.: Dynamic routing between capsules. In: Advances in Neural Information Processing Systems, pp. 3856–3866 (2017)

26. Sarkar, R., Das, N., Basu, S., Kundu, M., Nasipuri, M., Basu, D.K.: CMATERdb1: a database of unconstrained handwritten Bangla and Bangla-English mixed script document image. IJDAR **15**(1), 71–83 (2012). https://doi.org/10.1007/s10032-011-0148-6

27. Szegedy, C., Ioffe, S., Vanhoucke, V., Alemi, A.: Inception-v4, inception-resnet and the impact of residual connections on learning. arXiv preprint arXiv:1602.07261 (2016)

28. Tan, M., Le, Q.: EfficientNet: Rethinking model scaling for convolutional neural networks. In: PMLR, vol. 97, pp. 6105–6114 (2019)

29. Yun, S., Han, D., Chun, S., Oh, S.J., Yoo, Y., Choe, J.: CutMix: regularization strategy to train strong classifiers with localizable features. In: 2019 Proceedings of the ICCV, pp. 6022–6031 (2019)

30. Zhang, H., Cisse, M., Dauphin, Y.N., Lopez-Paz, D.: Mixup: beyond empirical risk minimization. arXiv preprint arXiv:1710.09412 (2017)

31. Zhu, J., Park, T., Isola, P., Efros, A.A.: Unpaired image-to-image translation using cycle-consistent adversarial networks. In: 2017 Proceedings of the ICCV, pp. 2242–2251 (2017)

GNHK: A Dataset for English Handwriting in the Wild

Alex W. C. Lee[1], Jonathan Chung[2(✉)], and Marco Lee[1]

[1] GoodNotes, Hong Kong, Hong Kong
{alex,marco}@goodnotes.com
[2] Amazon Web Services, Vancouver, Canada
jonchung@amazon.com

Abstract. In this paper, we present the GoodNotes Handwriting Kollection (GNHK) dataset. The GNHK dataset includes unconstrained camera-captured images of English handwritten text sourced from different regions around the world. The dataset is modeled after scene text datasets allowing researchers to investigate new localisation and text recognition techniques. We presented benchmark text localisation and recognition results with well-studied frameworks. The dataset and benchmark results are available at https://github.com/GoodNotes/GNHK-dataset.

Keywords: Handwriting recognition · Benchmark dataset · Scene text

1 Introduction

Understanding handwriting from images has long been a challenging research problem in the document analysis community. The MNIST dataset, introduced by LeCun et al. [15] in 1998, was a widely known dataset containing 70,000 grayscale images with 10 classes of handwritten digits. In the age of deep learning, classifying individual digits is no longer a difficult task [2]. However, in reality, we see that handwriting can vary immensely in both style and layout (e.g. orientation of handwritten text lines compared with printed text lines). As such, detecting and recognizing sequences of handwritten texts in images still pose a significant challenge despite the power of modern learning algorithms.

With researchers moving from using Hidden Markov Models (HMM) [21] to using variants of Recurrent Neural Networks (RNN) (e.g. BiLSTM [11], MDL-STM [32] and CRNN [29]) with Connectionist Temporal Classification (CTC) loss, problems like word recognition and line recognition have quickly become less difficult. More recent challenges focus on trying to recognize a full page of handwriting without explicit segmentation. Two datasets that are commonly used in research are IAM Handwriting Database [22] and RIMES Database [3], consisting of distinct and mostly horizontally oriented text lines from scanned documents. Numerous attempts have been made to solve the problem, such as using

---

A. W. C. Lee and J. Chung—Denotes equal contribution.

© Springer Nature Switzerland AG 2021
J. Lladós et al. (Eds.): ICDAR 2021, LNCS 12824, pp. 399–412, 2021.
https://doi.org/10.1007/978-3-030-86337-1_27

attention mechanisms with encoder-decoder architecture [6,7,9] or approaches that exploit the 2D layout of the document images [28,35].

While research in handwriting recognition is making great progress, outside of the document analysis community, text localisation and recognition in the natural scene is becoming a more active research area [19]. As texts are the medium of communication between people, the ubiquity of cameras makes it possible for us to capture the text in the wild and store them in pixels. The existing scene-text datasets that are commonly used as benchmarks are all in printed text. Recently, Zhang et al. released SCUT-HCCDoc [37], an unconstrained camera-captured documents dataset containing Chinese handwritten texts, which is novel to the research community, as there has been a lack of data for handwritten text in the wild. However, such a dataset does not exist for offline English handwritten images and we believe there is a necessity to create a similar one.

Therefore, we created a dataset for offline English handwriting in the wild called GoodNotes Handwriting Kollection (GNHK), which contains 687 document images with 172,936 characters, 39,026 texts and 9,363 lines. Note that "texts" defined in this paper include words, ASCII symbols and math expressions (see Sect. 3.2 for a detailed explanation). As English is the leading global lingua franca, handwriting styles and lexicons can vary across regions. For example, words like "Dudu-Osun" will more likely appear in documents from Nigeria and "Sambal" may more likely be written in Malaysia. To capture the diversity of handwriting styles from different regions, we collected the images across Europe, North America, Asia and Africa. We also attempted to create a more representative "in the wild" dataset by including different types of document images such as shopping lists, sticky notes, diaries, etc.

In this paper, we make the following contributions to the field of document analysis:

1. We provide a benchmark dataset, GNHK, for offline English handwriting in the wild.
2. We present a text localisation and a text recognition model, both serving as baseline models for our provided dataset.

2   Related Work

Detecting and recognizing offline handwriting remains a challenging task due to the variety of handwriting and pen styles. Therefore, many researchers have created datasets that allow the document analysis community to compare against their results. For Latin scripts, the most popular datasets are IAM Handwriting Database (English) [22], the IUPR Dataset (English) [8], the IRONOFF dataset (English) [31], and RIMES Database (French) [3]. For other scripts, there are KHATT Database of Arabic texts [20], IFN/ENIT Database of Arabic words [24], CASIA Offline Chinese Handwriting Database [18], HIT-MW Database of Chinese text [30] and SCUT-EPT of Chinese text [38]. One common feature among the aforementioned datasets is that the images were scanned using flatbed scanners, so they do not have much noise and distortion compared to images captured using cameras. Recently, a Chinese dataset called SCUT-HCCDoc [37] was released, which has a set of unconstrained camera-captured documents taken at different angles.

Table 1. Overview of different offline Latin-based handwriting datasets.

Dataset	# Texts	# Lines	Image Type
IAM	115,320	13,353	Flatbed-scanned
RIMES	66,982	12,093	Flatbed-scanned
Ours	**39,026**	**9,363**	**Camera-captured**

Our contribution in GNHK is similar to SCUT-HCCDoc, but for English texts. To the best of our knowledge, much work in offline Latin-based handwriting has been done using flatbed-scanned images. Table 1 shows a comparison between IAM, RIMES, and our dataset. We believe our dataset will bring in diversity and a modern take to the researchers working on offline English handwriting recognition.

3   Dataset Overview

3.1   Data Collection

The images were collected by a data-labeling firm [1] upon our request. Since penmanship can vary from country to country [5], to make sure we have a diverse set of English handwriting, we sourced the images across Europe, North America, Africa and Asia (see Table 2). In addition, no more than 5 images were written by the same writer in the collection.

Table 2. Number of images per region in the dataset.

Region	# Images
Europe (EU)	330
North America (NA)	146
Africa (AF)	117
Asia (AS)	94
TOTAL	**687**

The dataset consists of 687 images containing different types of handwritten text, such as shopping lists, sticky notes, diaries, etc. This type of text tends to favor handwriting over typing because people found them to be more reliable when capturing fleeting thoughts [27]. Images were captured by mobile phone cameras under unconstrained settings, as shown in Fig. 1.

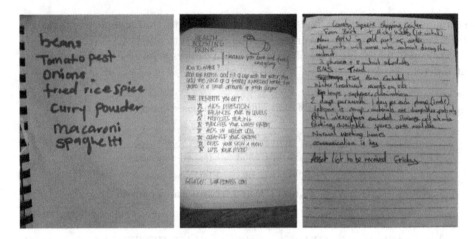

Fig. 1. Examples of the handwriting images captured under unconstrained settings.

3.2   Data Annotation and Format

For each handwritten-text image, there is a corresponding JSON file that contains the annotations of the image. Each JSON file contains a list of objects and each of them has a list of key-value pairs that correspond to a particular text, its bounding polygon and other associated information (see Fig. 2 for example):

Text. A sequence of characters belonging to a set of ASCII printable characters plus the British pound sign (£). Note that no whitespace characters are included in the value. In addition, instead of having a sequence of characters as value, sometimes we use one of the three special tokens for polygon annotations that meet some criteria (see Fig. 3 for examples):

- %math%: it includes math expressions that cannot be represented by ASCII printable characters. For example, ∞ and $\sum$.
- %SC%: it considered as illegible scribbles.
- %NA%: it does not contain characters or math symbols.

Polygon. Each polygon is a quadrilateral and it is represented by four (x, y) coordinates in the image. They are listed in clockwise order, with the top-left point (i.e. "x0", "y0") being the starting point.

Line Index. Texts belonging to the same line have the same index number.

Type. Either H or P, indicating whether the label is handwritten or printed.

```
{
    "text": "Proof",
    "polygon": {
        "x0": 845, "y0": 1592, "x1":  859, "y1": 1809,
        "x2: 1188, "y2": 1888, "x3": 1300, "y3": 1588
    },
    "line_idx": 4,
    "type": "H",
}, ...
```

Fig. 2. A sample object from one of the JSON files.

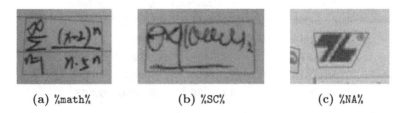

| (a) %math% | (b) %SC% | (c) %NA% |

Fig. 3. Examples of the three special tokens for the values of `"text"`. For (a), some of the math symbols in the polygon cannot be represented by ASCII characters. For (b), we can tell it is a sequence of characters inside the polygon, but it is hardly legible. For (c), the annotator labelled it as text, but it is not.

4 Dataset Statistics

Among the 687 images in the dataset, there are a total of 39,026 texts with 12,341 of them being unique. The median number of texts per image is 57 and the mean is 44. If we look at the text statistics by region, their distributions are roughly the same, as shown in Fig. 4. In Fig. 5, it shows the top 40 frequently used text in the dataset.

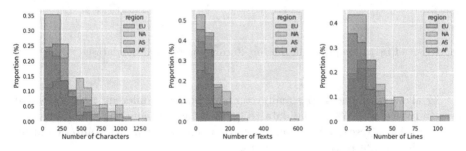

Fig. 4. Distributions of characters, texts, and lines for each region. Each plot includes four histograms with bin width of 100, 50, 10 for number of characters, texts, and lines respectively. Each histogram in a plot represents a region denoted by the color. For each histogram the counts are normalized so that the sum of the bar height is 1.0 for each region. (Color figure online)

Figure 6 illustrates the top 40 frequently used characters in the dataset. The dataset has 96 unique characters, with each of them showing up 3 to 17887 times. The median count is 486 and the mean is 1801.

In terms of total number of characters and lines, there are 172,936 and 9,363 respectively. Similar to texts, the distributions of character and lines are fairly similar, except for the EU region for characters (see Fig. 4). Detailed statistics are summarized in Table 3.

Table 3. Statistics of characters, texts and lines for each region.

Region	# Characters	# Texts	# Lines
Europe (EU)	58,982	13,592	3,306
North America (NA)	47,361	10,967	2,099
Asia (AS)	39,593	8,586	2,780
Africa (AF)	27,000	5,881	1,178
TOTAL	**172,936**	**39,026**	**9,363**

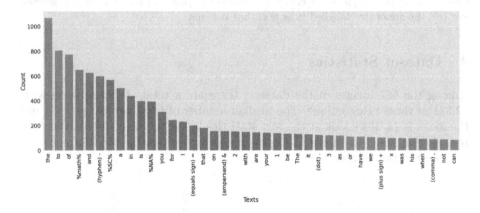

Fig. 5. Frequency of the top 40 common texts in the dataset.

To allow researchers to equally compare between different techniques, we randomly selected 75% of the data to be training data and the remaining 25% to be test data. Researchers are expected to determine their own train/validation split with the 75% of the training data.

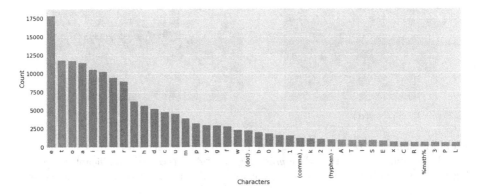

Fig. 6. Frequency of the top 40 common characters in the dataset.

5 Benchmark

Building upon works in scene text, we evaluated the dataset with different state-of-the-art methods in localisation and recognition. Specifically, instance segmentation with Mask R-CNN [12] frameworks were used to localise handwritten text and the Clova AI deep text recognition framework [4] to perform text recognition. The benchmark and implementation details are provided at https://github.com/GoodNotes/GNHK-dataset.

5.1 Text Localisation Benchmark

Recent handwriting recognition frameworks investigating on the IAM Handwriting database typically forgo the text localisation steps [9,35]. More similar to a scene text dataset (e.g., [14,23]), the GNHK dataset poses several challenges: 1) the GNHK dataset includes camera captured images which are not limited to handwritten text (i.e., there are printed text, images, etc.), 2) the images are not perfectly aligned with varying degrees of brightness, luminosity, and noise, and 3) there are no set lines guides/spacing, and image size/resolution restrictions.

Unlike printed text, handwritten allographs widely differ between individuals. Especially with the absence of line guides, individuals' writing style on character ascenders and/or descenders (i.e., the portion of a character that extends above or below the baseline) complicate the assignment of text boxes (see Fig. 7 for examples).

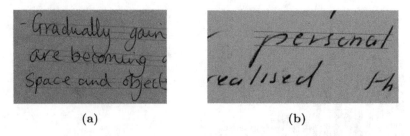

(a)   (b)

Fig. 7. Examples of the handwriting with problematic character ascenders and descenders. For example in (a), the "y" in *gradually* and "g" in *becoming* significantly increasing the bounding box size eventually overlapping with the regions of the words below.

We modelled the text localisation task using instance segmentation problem where the Mask R-CNN [12] framework was used as a baseline. Our dataset does not contain pixel-level segmentation masks separate between word vs non-word. To circumvent this issue, given the polygon around each word, the maximum and minimum points in the x and y direction of each polygon were used as the bounding box for the R-CNN component of the Mask R-CNN framework. The polygon itself was then used as the segmentation mask.

The text localisation task was built upon the detectron2 [34] framework. The network consisted of a backbone of a ResNet-50 [13] with FPN [16] pretrained on ImageNet. Furthermore, the whole network was pretrained on the MS COCO [17] dataset for segmentation. We compared the results of the Mask R-CNN to Faster R-CNN [26] within the same detectron2 framework.

5.2   Text Recognition Benchmark

We modelled the text recognition benchmark as a segmented offline handwriting recognition problem [10, 25]. That is, the ground truth bounding boxes for each word are used to crop word images and the word images are fed into a neural network to predict the corresponding text. Recent works have demonstrated segmentation free offline handwriting recognition [9, 33, 35, 36], however considering that our dataset includes content other than handwritten text, we opted with a segmented offline handwriting recognition framework.

We leveraged Clova AI's deep text recognition framework for scene text to assess the handwriting recognition component. Clova AI's deep text recognition framework consists of four major components: transformation, feature extraction, sequence modelling, and prediction. Table 4 shows the settings that we considered for our benchmark (eight possible configurations in total). Note that we did not consider using the VGG image features as the performance was lower compared to ResNet in [4].

Table 4. Considered configurations of the text recognition benchmark. TPS - *thin-plate spline*, BiLSTM - *Bidirectional Long short-term memory*, CTC - *Connectionist temporal classification*

Component	Configurations
Transformations	{None, TPS}
Sequence	{None, BiLSTM}
Prediction	{CTC, Attention}

It is important to note that it was essential to pre-process the data that was fed into the network. Specifically, words that 1) are unknown, 2) contained scribbles, 3) contained mathematical symbols, and 4) contained only punctuation were removed. Please see Sect. 3.2 for more details.

6   Benchmark Results

6.1   Text Localisation Results

To evaluate text localisation, we used the criteria: recall, precision, and f-measure with an intersection-over-union (IOU) > 0.5 ratio of the segmentation mask or bounding boxes (Shown in Fig. 8 and Table 5).

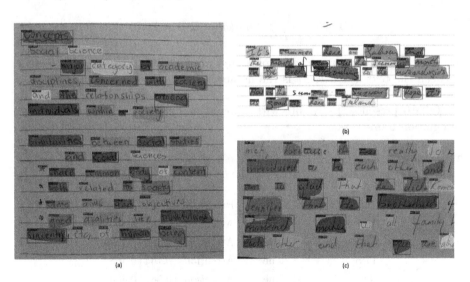

Fig. 8. Examples of the Mask R-CNN results.

Figure 8 presents three examples of the instance segmentation results. We can observe that the majority of text are detected, with the exception of several shorter texts (e.g., * in Fig. 8(a), *of* in Fig. 8(b)). Furthermore, in Fig. 8(c) even though the *f* in "Jennifer" overlaps with "mankind" in the line below, the network managed to differentiate between the two words. Table 5 presents the quantitative results of Mask and Faster R-CNN. Both Mask R-CNN and Faster R-CNN have greater than 0.86 f-measure scores. It is clear that the f-measure score is a result of high precision suggesting that the network has a smaller tendency to detect false positives.

Table 5. Text localisation results with IOU > 0.5.

Frameworks	Recall	Precision	f-measure
Mask R-CNN	0.8237	0.9079	0.864
Faster R-CNN	0.8077	0.9215	0.860

6.2   Text Recognition Results

Text recognition was performed on the individually cropped images from the ground truth bounding boxes and the results are shown in Table 6. The main assessment criteria are the character accuracy rate (CAR) and word accuracy rate (WAR):

$$CAR = \frac{1}{M} \sum_{i}^{M} 1 - \frac{dist(gt_i, preds_i)}{N_i} \tag{1}$$

$$WAR = \frac{1}{M} \sum_{i}^{M} [dist(gt_i, preds_i) == 0] \tag{2}$$

where gt = ground truth, $preds$ = predicted words, $dist(a, b)$ is the edit distance between the text a and b, M is the number of words in the dataset and N_i is the length of the word i.

Table 6 presents the results with different configurations of the deep text recognition framework:

Table 6. Text recognition results. TPS - *thin-plate spline*, BiLSTM - *Bidirectional Long short-term memory*, CTC - *Connectionist temporal classification*

Transformation	Sequence	Prediction	CAR	WAR
TPS	BiLSTM	Attention	0.861	0.502
None	BiLSTM	Attention	0.808	0.377
TPS	None	Attention	0.764	0.453
None	None	Attention	0.587	0.430
TPS	BiLSTM	CTC	0.416	≈0.000
None	BiLSTM	CTC	0.403	≈0.000
TPS	None	CTC	0.399	≈0.000
None	None	CTC	0.405	≈0.000

The benchmark recognition method achieved the highest CAR of 0.861 and WAR of 0.502. This configuration uses attention-based decoding, TPS for image transformation, and BiLSTM for sequential modelling. We can see that the attention-based decoding significantly outperforms CTC based decoding. The results of [4] also showed that attention-based decoders perform better than CTC based decoders, but by a much lesser extent. When comparing the performance with and without TPS, we can see that TPS improves the CAR and WAR. For example, setting the sequence modelling to BiLSTM and prediction to attention decoding, the performance TPS improves the CAR by 5.3% and WAR by 12.5%. Similar results are found when the sequence modelling is not used (CAR increase = 17.7%, WAR increase = 2.3%). The results are likely suggesting that text alignment and input image normalisation assists in the recognition. Similarly, Table 6 shows that the BiLSTM consistently improves the CAR. When investigating the performance of including the BiLSTM without TPS, it is clear that the CAR substantially improves. However, the performance measured by the WAR decreases with the BiLSTM (0.377 vs 0.430).

7   Conclusion

To fill the gap in the literature for a new handwriting recognition dataset, we presented the GNHK dataset. The GNHK dataset consists of unconstrained camera-captured images of English handwritten text sourced from different regions around the world. The dataset is modelled after scene text dataset providing opportunities to novel investigate localisation with instance segmentation or object detection frameworks and text recognition techniques. In this paper, we demonstrated benchmark results of text localisation and recognition with well-studied architectures. Future works could explore end-to-end approaches to perform localisation and recognition sequentially within the same framework. The dataset and benchmark are available at https://github.com/GoodNotes/GNHK-dataset and we look forward to contributions from the researchers and developers to build new models with this dataset.

Acknowledgements. Thanks to Paco Wong, Felix Kok, David Cai, Anh Duc Le and Eric Pan for their feedback and helpful discussions. We also thank Elizabeth Ching for proofreading the first draft of the paper.

We thank Steven Chan and GoodNotes for providing a product-driven research environment that allows us to get our creativity ideas to the hands of millions of users.

References

1. Basicfinder. https://www.basicfinder.com/. Accessed 20 Jan 2021
2. Image classification on MNIST. https://paperswithcode.com/sota/image-classification-on-mnist. Accessed 20 Jan 2021
3. Augustin, E., Carré, M., Grosicki, E., Brodin, J.M., Geoffrois, E., Prêteux, F.: Rimes evaluation campaign for handwritten mail processing (2006)

4. Baek, J., et al.: What is wrong with scene text recognition model comparisons? Dataset and model analysis. In: Proceedings of the IEEE International Conference on Computer Vision, pp. 4715–4723 (2019)

5. Bernhard, A.: What your handwriting says about you, June 2017. https://www.bbc.com/culture/article/20170502-what-your-handwriting-says-about-you. Accessed 20 Jan 2021

6. Bluche, T.: Joint line segmentation and transcription for end-to-end handwritten paragraph recognition. In: Advances in Neural Information Processing Systems, pp. 838–846 (2016)

7. Bluche, T., Louradour, J., Messina, R.: Scan, attend and read: end-to-end handwritten paragraph recognition with MDLSTM attention. In: 2017 14th IAPR International Conference on Document Analysis and Recognition (ICDAR), vol. 1, pp. 1050–1055. IEEE (2017)

8. Bukhari, S.S., Shafait, F., Breuel, T.M.: The IUPR dataset of camera-captured document images. In: Iwamura, M., Shafait, F. (eds.) CBDAR 2011. LNCS, vol. 7139, pp. 164–171. Springer, Heidelberg (2012). https://doi.org/10.1007/978-3-642-29364-1_13

9. Coquenet, D., Chatelain, C., Paquet, T.: End-to-end handwritten paragraph text recognition using a vertical attention network. arXiv preprint arXiv:2012.03868 (2020)

10. Dutta, K., Krishnan, P., Mathew, M., Jawahar, C.: Improving CNN-RNN hybrid networks for handwriting recognition. In: 2018 16th International Conference on Frontiers in Handwriting Recognition (ICFHR), pp. 80–85. IEEE (2018)

11. Graves, A., Liwicki, M., Fernández, S., Bertolami, R., Bunke, H., Schmidhuber, J.: A novel connectionist system for unconstrained handwriting recognition. IEEE Trans. Pattern Anal. Mach. Intell. **31**(5), 855–868 (2008)

12. He, K., Gkioxari, G., Dollár, P., Girshick, R.: Mask R-CNN. In: Proceedings of the IEEE International Conference on Computer Vision, pp. 2961–2969 (2017)

13. He, K., Zhang, X., Ren, S., Sun, J.: Deep residual learning for image recognition. In: Proceedings of the IEEE Conference on Computer Vision and Pattern Recognition, pp. 770–778 (2016)

14. Karatzas, D., et al.: ICDAR 2015 competition on robust reading. In: 2015 13th International Conference on Document Analysis and Recognition (ICDAR), pp. 1156–1160. IEEE (2015)

15. LeCun, Y., Bottou, L., Bengio, Y., Haffner, P.: Gradient-based learning applied to document recognition. Proc. IEEE **86**(11), 2278–2324 (1998)

16. Lin, T.Y., Dollár, P., Girshick, R., He, K., Hariharan, B., Belongie, S.: Feature pyramid networks for object detection. In: Proceedings of the IEEE Conference on Computer Vision and Pattern Recognition, pp. 2117–2125 (2017)

17. Lin, T.-Y., et al.: Microsoft COCO: common objects in context. In: Fleet, D., Pajdla, T., Schiele, B., Tuytelaars, T. (eds.) ECCV 2014. LNCS, vol. 8693, pp. 740–755. Springer, Cham (2014). https://doi.org/10.1007/978-3-319-10602-1_48

18. Liu, C.L., Yin, F., Wang, D.H., Wang, Q.F.: CASIA online and offline Chinese handwriting databases. In: 2011 International Conference on Document Analysis and Recognition, pp. 37–41. IEEE (2011)

19. Long, S., He, X., Yao, C.: Scene text detection and recognition: the deep learning era. Int. J. Comput. Vis. **129**(1), 161–184 (2020). https://doi.org/10.1007/s11263-020-01369-0

20. Mahmoud, S.A., et al.: KHATT: an open Arabic offline handwritten text database. Pattern Recogn. **47**(3), 1096–1112 (2014)

21. Marti, U.V., Bunke, H.: Using a statistical language model to improve the performance of an hmm-based cursive handwriting recognition system. In: Hidden Markov Models: Applications in Computer Vision, pp. 65–90. World Scientific (2001)

22. Marti, U.V., Bunke, H.: The IAM-database: an English sentence database for offline handwriting recognition. Int. J. Doc. Anal. Recogn. **5**(1), 39–46 (2002). https://doi.org/10.1007/s100320200071

23. Mishra, A., Alahari, K., Jawahar, C.: Scene text recognition using higher order language priors (2012)

24. Pechwitz, M., Maddouri, S.S., Märgner, V., Ellouze, N., Amiri, H., et al.: IFN/ENIT-database of handwritten Arabic words. In: Proceedings of CIFED, vol. 2, pp. 127–136. Citeseer (2002)

25. Puigcerver, J.: Are multidimensional recurrent layers really necessary for handwritten text recognition? In: 2017 14th IAPR International Conference on Document Analysis and Recognition (ICDAR), vol. 1, pp. 67–72. IEEE (2017)

26. Ren, S., He, K., Girshick, R., Sun, J.: Faster R-CNN: towards real-time object detection with region proposal networks. IEEE Trans. Pattern Anal. Mach. Intell. **39**(6), 1137–1149 (2016)

27. Riche, Y., Riche, N.H., Hinckley, K., Panabaker, S., Fuelling, S., Williams, S.: As we may ink?: learning from everyday analog pen use to improve digital ink experiences. In: CHI, pp. 3241–3253 (2017)

28. Schall, M., Schambach, M.P., Franz, M.O.: Multi-dimensional connectionist classification: reading text in one step. In: 2018 13th IAPR International Workshop on Document Analysis Systems (DAS), pp. 405–410. IEEE (2018)

29. Shi, B., Bai, X., Yao, C.: An end-to-end trainable neural network for image-based sequence recognition and its application to scene text recognition. IEEE Trans. Pattern Anal. Mach. Intell. **39**(11), 2298–2304 (2016)

30. Su, T., Zhang, T., Guan, D.: Corpus-based hit-mw database for offline recognition of general-purpose Chinese handwritten text. IJDAR **10**(1), 27 (2007). https://doi.org/10.1007/s10032-006-0037-6

31. Viard-Gaudin, C., Lallican, P.M., Knerr, S., Binter, P.: The IRESTE On/Off (IRONOFF) dual handwriting database. In: Proceedings of the Fifth International Conference on Document Analysis and Recognition. ICDAR 1999 (Cat. No. PR00318), pp. 455–458. IEEE (1999)

32. Voigtlaender, P., Doetsch, P., Ney, H.: Handwriting recognition with large multidimensional long short-term memory recurrent neural networks. In: 2016 15th International Conference on Frontiers in Handwriting Recognition (ICFHR), pp. 228–233. IEEE (2016)

33. Wigington, C., Tensmeyer, C., Davis, B., Barrett, W., Price, B., Cohen, S.: Start, follow, read: end-to-end full-page handwriting recognition. In: Ferrari, V., Hebert, M., Sminchisescu, C., Weiss, Y. (eds.) ECCV 2018. LNCS, vol. 11210, pp. 372–388. Springer, Cham (2018). https://doi.org/10.1007/978-3-030-01231-1_23

34. Wu, Y., Kirillov, A., Massa, F., Lo, W.Y., Girshick, R.: Detectron2 (2019). https://github.com/facebookresearch/detectron2

35. Yousef, M., Bishop, T.E.: OrigamiNet: weakly-supervised, segmentation-free, one-step, full page text recognition by learning to unfold. In: Proceedings of the IEEE/CVF Conference on Computer Vision and Pattern Recognition, pp. 14710–14719 (2020)

36. Yousef, M., Hussain, K.F., Mohammed, U.S.: Accurate, data-efficient, unconstrained text recognition with convolutional neural networks. Pattern Recogn. **108**, 107482 (2020)

37. Zhang, H., Liang, L., Jin, L.: SCUT-HCCDoc: a new benchmark dataset of handwritten Chinese text in unconstrained camera-captured documents. Pattern Recogn. **108**, 107559 (2020)
38. Zhu, Y., Xie, Z., Jin, L., Chen, X., Huang, Y., Zhang, M.: SCUT-EPT: a new dataset and benchmark for offline Chinese text recognition in examination paper. IEEE Access **7**, 370–382 (2018)

Personalizing Handwriting Recognition Systems with Limited User-Specific Samples

Christian Gold[(✉)], Dario van den Boom, and Torsten Zesch

Language Technology Lab, University of Duisburg and Essen, Duisburg, Germany
{christian.gold,dario.van-den-boom,torsten.zesch}@uni-due.de

Abstract. Personalization of handwriting recognition is still an understudied area due to the lack of a comprehensive dataset. We collect a dataset of 37,000 words handwritten by 40 writers that we make publicly available. We investigate the impact of personalization on recognition by training a baseline recognition model and retraining it using our dataset. After controlling that our model really adapts to the personal handwriting style and not just to the overall domain, we show that personalization in general requires several hundred samples to be effective. However, we show that the choice of transfer samples is important and that we can quickly personalize a model with a limited number of samples. We also examine whether we can detect adversarial behavior trying to reduce recognition performance.

Keywords: Handwriting recognition · Personalization · Dataset

1 Introduction

Handwriting is a relatively understudied input modality and handwriting recognition has mainly been of interest for historical texts [11,19,21]. Another area where handwriting is still widespread is education. Learning to write by hand in general - especially for cursive writing - is considered beneficial for memory [15] and reading skills [6]. Still, most university students write notes electronically [15], but while doing so often just transcribe what was said in the lecture [12,15].

Even if (in a university context) almost all writing is in an electronic form, exams are often still written by hand. Teachers then need to decipher answers by students not being used to handwriting anymore. Automatic handwriting recognition (HWR) has been proposed as a means for supporting teachers in that task. For example, automatic or assisted scoring systems [4,20,23,25] usually only work with electronic input, but could now also be used for handwritten exams. As bad handwriting can induce a negative grading bias, having teachers only look at the digitized version might improve fairness [9,13,22].

Variance in Writing. Anyone who has ever graded handwritten exams knows that there are some challenges to reliable recognition. One is *intra-writer*

© Springer Nature Switzerland AG 2021
J. Lladós et al. (Eds.): ICDAR 2021, LNCS 12824, pp. 413–428, 2021.
https://doi.org/10.1007/978-3-030-86337-1_28

variance, i.e. the same letter, letter-combination, or word is written differently by the same writer. We show here an example for the variance when writing *the* several times (by the same writer):

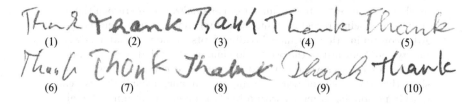

Another challenge is the *inter-writer* variance, whose origin may lie in factors such as sex, age, first learned alphabet, etc. The figure below shows *Thank* written by various writers. The letter T can look similar to a J (8) or a B when combined with the next letter h (3). Or the letter k which might be easily mistaken for an h (3, 6 and 9).

The more the target writing style deviates from the writing styles the classifier was trained on, the worse the recognition rate becomes. An extreme example for this was given by [7] who trained a model on the 20th century writing IAM database [14] that was able to recognize less than 20% of all characters correctly when evaluating it on the handwriting of 18th century George Washington.

Personalization. In this paper, the term *personalization* describes the process of transforming a generic model that works well for a wide variety of writers into a writer-specific model. Hence, we explore the personalization of handwriting recognition as a way to deal with inter-writer variance in particular.

If we knew about the idiosyncrasies of a certain handwriting style, we could try to adapt the model to it. Contrary to historical applications, where we usually already have handwriting samples of the authors we are interested in, there is no collection of handwritings from all students attending a course. Instead, the students could copy a short text as part of the exam that can then be used as a dataset for personalization to improve the model's recognition performance. This could e.g. happen on the first page of the exam as a kind of warm-up exercise.

In this paper, we analyze how well an HWR system can be personalized to individual handwriting when only limited data for personalization is available (like in our educational setting). For this purpose, we collect a handwriting dataset of 40 writers and 926 words per writer. We make the dataset of personalized handwriting samples publicly available to foster future research on this topic.

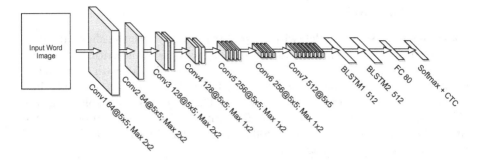

Fig. 1. Architecture of the recognizer.

2   Handwriting Recognition System

Handwriting recognition systems (HWR) can be divided into two different categories, the first of which being online handwriting recognition and primarily used for electronic devices that are able to provide information about the pen tip's path (including timestamps) on the display. Offline handwriting recognition systems on the other hand, deal with images of handwritten text with the only information available being the raw pixel values [17]. As in the educational area, writing is usually 'pen and paper' based, we are focusing on offline approaches.

In most traditional offline handwriting recognition systems, the first step involves splitting the handwritten text into lines and - in the case of word recognition - words. Then, a machine learning classifier is used to map the resulting segments to sequences of characters.

The most popular architecture in recent years, the convolutional and recurrent neural network (CNN-RNN) hybrid architecture, uses convolutional layers to transform the input images into sequences of feature vectors [2,24]. A major breakthrough for neural networks in the field of HWR was the introduction of connectionist temporal classification (CTC) [5]. This facilitated the training of HWR systems without having to provide the exact position of each character in an image. The feature vectors can be classified using the CTC as an output layer. A type of recurrent layer commonly used in this context is the Long Short-Term Memory (LSTM) layer, which is able to exploit more context during classification [26].

2.1   Architecture

The architecture used in our experiments consists of seven convolutional layers, two bidirectional LSTM layers with 512 units each, followed by a single softmax layer for the prediction. The input to the network are grayscale images of handwritten words with a width of 256 and a height of 64 pixels. The network architecture with the individual layer setup is shown in Fig. 1. Afterwards the predicted sequence is decoded using CTC.

2.2   Baseline Model

For training a baseline model (without personalization) with the above architecture, we need a comprehensive database of transcribed handwriting. A widely used dataset is called IAM [14]. We use the current version 3.0 with over 115,000 isolated labeled word images. On the word level, there is no separate split into training, evaluation and test sets provided. Thus, we chose the split from the "Large Writer Independent Text Line Recognition Task" as stated on the website[1]. The split is structured in the following way: 69,120 word images for training, 13,513 for evaluation, and 13,754 for testing.

We apply basic transformations for data augmentation including translation, rotation, shearing, scaling, and elastic distortions within realistic factors randomly during training. Additionally, we do not apply more advanced techniques like using fonts that mimic handwriting [10] as this might interfere with the personalization experiments.

Our baseline model was trained on the training dataset and evaluated on the validation data after each epoch. The learning rate was initially set to .001 and divided by 10 after five epochs without improvements in validation loss. Furthermore, early stopping was used instead of determining a fixed number of epochs.

Baseline Model Performance. To compare new results with other research in this field, we exclusively use the character error rate (CER) due to it being more representative for smaller datasets than the word error rate (WER).

For the IAM dataset modern models like the one presented in [2] typically show a CER of 5%. Our baseline model performed a CER of 9.44%. Despite being far from reaching state-of-the-art performance, the model's architecture is nevertheless comparable to modern models, thus making it a decent baseline model for our experiments with personalization. We choose a standard model over a highly-optimized model to focus on the effect of personalization instead of hunting state-of-the-art numbers.

3   Personalizing the Model

The term *personalization* as we use it in this paper describes the process of transforming a generic model that works for a wide variety of users into a user-specific model which shows improved performance for a target user only. We will use *Transfer Learning* to personalize the baseline model.

3.1   Transfer Learning

Transfer Learning aims at leveraging the knowledge of a model trained on a source dataset by retraining on data from the target domain, thus making it possible to obtain a high performance with very little additional training data.

---

[1] http://www.fki.inf.unibe.ch/databases/iam-handwriting-database.

Due to transferability of low level features and syntactic knowledge, writer adaptation can be viewed as a special case of *Transfer Learning*. By retraining a generic model on the target handwriting, the model can be adapted to the writing style's characteristics. This allows for the creation of a writer-specific model without having to train it from scratch.

Regarding the number of writer-specific training examples required for effective personalization only very little research has been done. [7] trained a model on the IAM dataset and found 325 handwritten lines to be sufficient for reducing the model's CER for the Washington Database from 82% to 5.3%. When decreasing the number of lines from 325 to 250 and 150 the CER were slightly higher (7.6% and 11.9%), but still far better than the non-personalized results. This suggests that few personalization samples are sufficient for personalization.

We assume that this trend also exists for word recognition due to the similarity of line and word recognizer. Further, it is necessary to identify properties that make personalized data more sample-efficient in order to decrease the amount of training data to a minimum.

3.2 Setup

In this work, we follow the architecture of [7,16] and use standard transfer learning by fine-tuning the pretrained model (as described in Sect. 2.2) on a writer-specific dataset. In order to work with a minimal training set we use a batch size of 10 samples for retraining.

As a result, we had to replace the automatic learning rate schedule and early stopping with a fixed number of epochs and a constant learning rate of .001. In our experiments, we found three epochs to be sufficient to retrain the model on any of our datasets without overfitting. Contrary to [7] we chose not to lock layers during retraining. Furthermore, we do not apply any transformations for data augmentation.

4   Dataset for Handwriting Personalization

We now describe how we collected the handwriting data used in our personalization experiments. Afterwards, we present the statistics of our dataset, *GoBo*. GoBo is made publicly available at GitHub[2].

4.1 Data Sources

As we want to investigate the effect of personalization based on different types of data, we include words from different sources:

Random As a baseline, we select random words from the Brown Corpus [3].
Nonword Here, we use pseudowords that are similar to the English language but carry no meaning. Our nonwords were selected from the ARC Nonword Database[18], using the attribute *polymorphemic only syllables*. Exemplary nonwords are *plawgns*, *fuphths*, and *ghrelphed*.

---

[2] https://github.com/ltl-ude/GoBo.

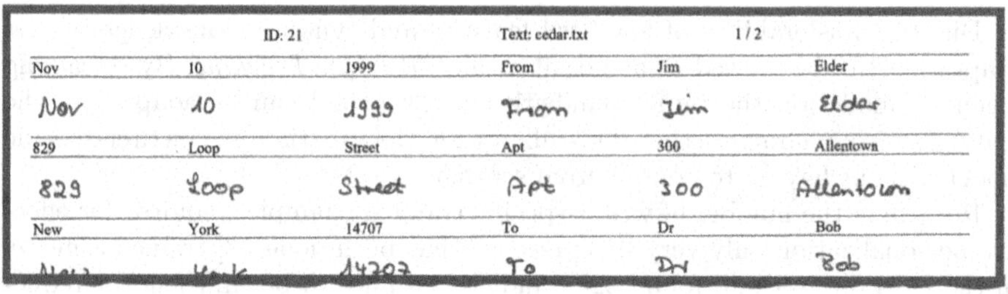

ID: 21			Text: cedar.txt		1 / 2
Nov	10	1999	From	Jim	Elder
Nov	*10*	*1999*	*From*	*Jim*	*Elder*
829	Loop	Street	Apt	300	Allentown
829	*Loop*	*Street*	*Apt*	*300*	*Allentown*
New	York	14707	To	Dr	Bob
New	*York*	*14707*	*To*	*Dr*	*Bob*

Fig. 2. Example extract of a data acquisition sheet with 2 of 16 lines.

CEDAR Letter This letter is specifically crafted to contain all English characters and common character combinations [1]. All characters appear at least once at the beginning and the middle of a word. Moreover, all letters appear in upper and lower case as well as all numbers.

Domain-specific Finally, we use domain-specific words from two topics *academic conferences* and *artificial intelligence*. For the first topic, we use (part of) the Wikipedia article "Academic Conference"[3]. For the second topic, we use the summary of an AI lecture. As the topic itself is not important, we refer to these as DOMAIN A and DOMAIN B.

In the experimental setup, we differentiate between handwriting samples used for personalizing the model and test data. The exact same sample may not appear in personalization data and test data, but of course common words like *the* may appear in both sets (but not written exactly the same due to intra-writer variance as discussed above).

As test data, we use the domain specific sets A and B as well as their combination A+B. For personalization, we use all of the sources mentioned above, including domain-specific words, but have each individual participant write the words again. For the domain-specific personalization sets we do not let them copy the whole text, but only a small number of selected words. The rationale for using such a small domain-specific personalization set is that in a real-life exam situation examinees can write these words on the front page of their exam just before the exam starts in one or two minutes. For example, the most frequent content words for Domain A are *academic, conferences, panel, presentation*, etc.

4.2 Dataset Statistics

All texts were written on white paper and were color-scanned at a resolution of 300 dpi but were later used in gray-scale only (see Fig. 2). In contrast to the IAM dataset, we do not include punctuation marks. Punctuation marks are usually ignored in educational content scoring, so we can safely ignore them for our purposes while putting more emphasis on improving the recognition rates for words.

---

[3] https://en.wikipedia.org/wiki/Academic_conference.

We collected handwriting samples from a group of 40 writers. They are quite diverse with respect to a wide range of factors e.g. alphabet of first handwriting, native language, origin of birth, etc. The gender and age of the writers splits as follows:

Total	Gender		Age					
#	Female	Male	20–29	30–39	40–49	50–59	60–99	ø
40	17	23	22	10	1	5	2	34

Our full dataset comprises 5 different personalization sets and 2 test sets (see Table 1). A total of 926 words were collected from each writer which sums up to 37,000 words for the full dataset.

Table 1. Statistics of our handwriting personalization dataset.

	source	words	unique words	chars	chars per word
PERSONALIZATION	RANDOM	155	120	751	4.8
	NONWORD	156	156	1117	7.2
	CEDAR	156	122	700	4.5
	DOMAIN A	22	22	201	9.1
	DOMAIN B	39	39	304	7.8
	COMBINED	528	437	3073	5.8
TEST	DOMAIN A	200	125	1182	5.9
	DOMAIN B	198	142	1230	6.2

4.3   Comparison with Other Datasets

Our dataset GoBo was specifically designed for the personalization task. However, some published datasets could have been partly used as well.

The CEDAR database [1] includes 1568 individual copies of the CEDAR Letter. To use its balanced character distribution for personalization, the full letter must be used without having a test set. However, we liked the idea of the character distribution and thus integrated the letter as one domain.

The CVL-DATABASE [8] consists of 5 English and German text pieces of a length between 50–90 words written by 311 writers. Although, stating a high type-token ratio on word level, the character balance as provided by CEDAR is not given. Furthermore, a personalization and test set of the same domain was not provided.

Additionally, IAM [14] can be used for personalization as writers are exclusively partitioned into training, validation and test set using the split of the

Large Writer Independent Text Line Recognition Task. Although, several writer wrote multiple pages, the texts differ thus making comparisons difficult.

We think that with our dataset GoBo, where every author wrote the same text, analyses are more consistent. A further characteristic of our dataset is the selection of several domains (especially random words and nonwords) which lead to a broader usage.

5   Experiments and Results

We now take our newly collected dataset (see Sect. 4) and the baseline HWR model (see Sect. 2.2) to investigate the potential improvement of personalization. Due to a potential large variance in performance between randomly selected subsets (see Sect. 5.5), we repeated the experiments 10 times for each writer and averaged the final results.

5.1   Impact of Personalization

To evaluate the overall effect, we combined all training sets for each writer. We then retrained our baseline model with these samples and the setup described in Sect. 3.2. The individual models were tested on the combined test set *Domain A+B*. This procedure was repeated 10 times for each writer and values were finally averaged. Figure 3 shows the results. The average CER of the baseline model is 14.1% and drops to 8.0% with personalization, which shows that our personalization strategy works in general. As the boxplot on the left shows, the average CER values are 'hiding' a lot of individual variance and personalization significantly reduces the variance in model performance. The learning curve (right part of the figure) shows that we need quite a lot of samples for personalization. We further explore both issues in the next section.

5.2   Required Amount of Personalized Handwriting Samples

In our second experiment, we analyzed the relationship between the amount of training samples used for personalization and the resulting performance on the target writing style. For this, we combined all training sets and randomly selected subsets of increasing size. Starting with a subset of 10 samples, the training set was gradually increased by 10 samples up until it included all 528 training examples. After each training step, the model was tested using the test sets *Domain A*, *Domain B* and the combination of these *Domain A+B*. The retraining with a different shuffled training set was repeated 10 times and averaged for each individual writer.

The averaged CER across all writers is shown on the right side in Fig. 3 with 0 denoting the results from the baseline model. Our results indicate that the amount of training samples unsurprisingly correlates with an improvement of CER. In total, an average improvement of roughly 6% is achieved. Further, our personalization approach improves the recognition rate even with only few handwritten samples. For the examination example, 50 handwritten samples, which can be quickly produced by the examinee, would decrease the CER by more than 1%.

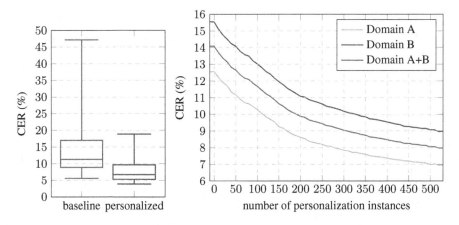

Fig. 3. Left: comparison of baseline and personalized model. Right: learning curve averaged over all writers.

Table 2. Examples for easy and hard to recognize handwritings.

writer	CER(%) initial	CER(%) retrained	industrial	later	unconference
17	47.18	18.87	*industrial*	*later*	*unconference*
26	33.58	17.45	*industrial*	*later*	*unconference*
34	5.80	3.93	*industrial*	*later*	*unconference*

Inter-Writer Variance. As we are so far only looking at averages, this might mask the variance in performance between single writers. We thus show in Fig. 4 a breakdown by writer for the 'Domain A+B' setting.

Obviously, handwriting styles already covered by the generic model do not benefit much from personalization (c.f. Writer 34 in Table 2 who has an improvement of 1.87%). In contrast, for the writer for which the generic model performs worst (Writer 17) the CER drops from 47% to 19%. Bad performing handwritings lack in completion and execution of the characters as can be seen in Table 2 (Writer 17 and 26). This might be due to a fast and speedy handwriting which one would deal with in our example of handwriting recognition from exams. However, the recognition of these handwritings profit the most using personalization.

5.3   Personalization or Just Task Adaptation

The previous results indicate that personalization improves recognition, especially as the size of the effect strongly depends on the individual writer.

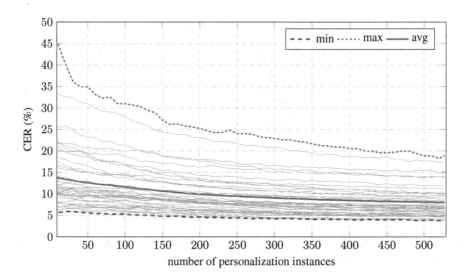

Fig. 4. Individual recognition results by increasing number of personalization instances. (Evaluated on test set Domain A+B)

However, the properties of our personalization dataset differ a bit from the IAM training data, which has continuous text, a different segmentation scheme, and more background noise. Therefore, it is possible that our personalization only bridges between the dataset styles.

We test this by applying the personalized model for each writer to: (i) the test data for the same writer (personalization), (ii) the test data of all other writers (no personalization, but possible style transfer), and (iii) the IAM test set (no personalization, no style transfer). We compare all results against the baseline system trained on the IAM dataset. The results averaged over all writers are presented in Table 3.

Table 3. CER (%) for different validation datasets before and after personalization.

Same writer (our dataset)		Other writers (our dataset)		IAM dataset	
IAM baseline	Retrained	IAM baseline	Retrained	IAM baseline	Retrained
14.07	**7.99**	14.07	16.60	9.44	13.87

The only performance gain over the baseline can be seen when retraining the model on data from the same writer. In both other cases, average performance decreases which shows that the performance gain can be solely attributed to personalization, not general adaptation to our dataset style.

5.4   Domain-Specific Results

Based on the finding that the model can be adapted to a writer, we evaluate the impact of the domain. Thus, we train our model with each training set individually. From Fig. 5 it can be inferred that the domain-specific training words reduce the CER faster than any other training sets. Using only 20 individual words we reach the same improvement as *Random* and *CEDAR* at about 100 words.

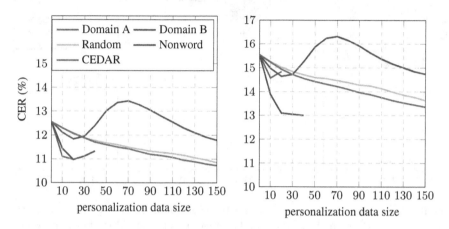

Fig. 5. Influence of the domain on the CER (%) performance for *Domain A* (left) and *Domain B* (right).

However, *Nonword* performed worse and increases the error rate interim. Presumably, this might be due to an unnatural behaviour of the writers which had to write uncommon words and thus forces the writer to consciously think about how to write them. Alternatively, this behaviour could change the writing style from cursive to print or vice versa. Thus, the model needs to adapt to both, print and cursive.

Overall, we conclude from this experiment that already a small amount of domain-specific words improves the recognition performance on the corresponding domain. This might be due to the combination of two factors— personalization and domain adaptation.

5.5   Best Performing Samples

While the previous results suggest that the domain-specific personalization sets are the most efficient, our intuition is that the other personalization sets also contain words that are well suited for personalization.

To compare potentially better suited words with the results of the domain-specific ones, we use the same test setting in terms of test set and sample size. We retrain the baseline model 1,000 times with 20 (for *Domain A*) or respectively

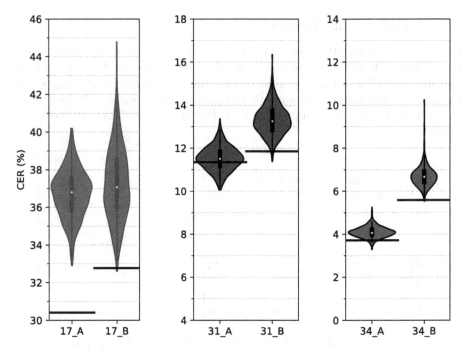

Fig. 6. Distribution of the performance for 1000 personalization runs tested on *Domain A* and *Domain B* for the best, worst and average performing writers in comparison to the domain-specific samples (lines).

40 (for *Domain B*) random samples. As these calculations are time and resource consuming, we perform this experiment only for the best, worst, and average performing writers.

In Fig. 6 we display the performance distribution of the 1000 runs for the Writers 17 (worst), 31 (average) and 34 (best). In all cases, except for ID 17 on *Domain A*, there is a slight performance improvement over the domain-specific personalization sets (black lines). However, the best performing sets are mainly not composed of domain-specific samples, but reflect the normal distribution of the overall personalization set. Furthermore, those samples do not overlap between the writers. Also, there is no conspicuousness in the word length which varies between 4.9 and 6.6 characters, whereas the domain-specific personalization sets have an average length of 9.1 and 7.8.

Highlighting the effect that all domain-specific personalization sets perform close to the best performing sets and due to their simplicity in creation, we recommend to use them in the exam scenario.

5.6 Adversarial Attacks

Staying with the example of exams, a test taker might want to sabotage the personalization. This is easily achieved by not writing the words one is asked to

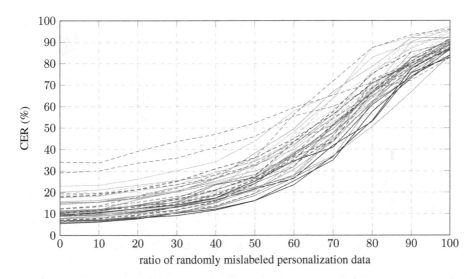

Fig. 7. Recognition performance on test set *Domain A+B* under varying amounts of adversarial training data.

write, e.g. when prompted to write 'handwriting' one could write 'love'. Understandably, the resulting personalization set will negatively influence the personalization, correlating with the number of words that have been intentionally written incorrectly. Although a performance decrease is certain, we evaluate the impact of such an attack. Therefore, we retrain our baseline model with a total of 100 words and gradually increase the number of randomly mislabeled words by 10% from 0 to 100%.

A performance decrease can already be noted with about 20% of changed labels (see Fig. 7) when testing on *Domain A+B*. If we want to automatically detect adversarial attacks, one could of course compare the performance of the model before and after personalization on a test data set. However, such test data is not available in a realistic setting. Hence, only an analysis of the potentially invalid training set can be performed.

One possible solution to detect adversarial attacks could be performed using the recognition result of the baseline model. The predicted word could be collated with the expected word using the Levenshtein Distance and rejected when exceeding a set threshold. However, this validation will fail when the baseline model performance is insufficient. A more reliable indicator could be the word length, which should be similar even in case of wrongly recognized characters.

Additionally, we assume that a personalization using incorrectly labeled samples will result in the personalized model deviating significantly from the baseline model. To analyse this hypothesis, we re-run the previous experiment, but instead of calculating the CER after retraining, we measure the reduction in loss using 10-fold cross validation. It can be seen in Fig. 8 having a larger share of mislabeled samples results in greater loss changes. However, using a low

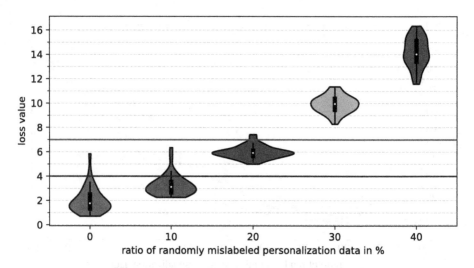

Fig. 8. Loss value distribution by increasing attack rate using 10-fold cross validation. $2 * \sigma$ is depicted by the blue line. A manual threshold is depicted by the green line. (Color figure online)

threshold (blue line denoting $2 * \sigma$) on the loss value for detecting attacked training sets is critical as we cannot distinguish between a challenging handwriting and a partial attack. When setting up the threshold to high, writers could have produced several malicious samples without the system noticing (green line denoting a manually set threshold).

Instead, we propose to use a combination of the presented analyses to detect adversarial attacks. In the first step, a verification of the word length should be made. The writer could then be left out if several occurrences of attacked samples were detected. If however, only a few samples were detected, these could be ignored in the following personalization process. In the second step, the writers should be filtered out according to a pre-set loss value threshold.

6 Summary

We collected a new, publicly available dataset, *GoBo* for personalization with a total of 37,000 words from 40 writers. Using our dataset, we find that by retraining a HWR system on the personalization data, we are really adapting to the individual handwriting style, not just to the overall domain. Error rate reduction varies heavily between individual writers. Learning curve experiments show that several hundred samples are needed for personalization to yield its full effect. However, the required amount of personalization samples can be heavily reduced by selecting the right kind of transfer samples. In our educational setting, we already know which words are likely to appear in an answer and adding those to the personalization reduces the number of required samples. We also explore how adversarial behavior by students (who want to impede the personalization process) would influence recognition. We show that severe harmful behavior can be detected by looking for unusual loss values during personalization.

References

1. Cha, S.H., Srihari, S.N.: Handwritten document image database construction and retrieval system. Document Recognition and Retrieval VIII, vol. 4307, pp. 13–21 (2000)
2. Dutta, K., Krishnan, P., Mathew, M., Jawahar, C.: Improving CNN-RNN hybrid networks for handwriting recognition. In: ICFHR, pp. 80–85 (2018)
3. Francis, W.N., Kucera, H.: Brown corpus manual. Lett. Editor **5**(2), 7 (1979)
4. Gold, C., Zesch, T.: Exploring the impact of handwriting recognition on the automated scoring of handwritten student answers. In: ICFHR, pp. 252–257. IEEE (2020)
5. Graves, A., Liwicki, M., Fernández, S., Bertolami, R., Bunke, H., Schmidhuber, J.: A novel connectionist system for unconstrained handwriting recognition. TPAMI **31**(5), 855–868 (2008)
6. James, K.H., Engelhardt, L.: The effects of handwriting experience on functional brain development in pre-literate children. Trends Neurosci. Educ. **1**(1), 32–42 (2012)
7. Jaramillo, J.C.A., Murillo-Fuentes, J.J., Olmos, P.M.: Boosting handwriting text recognition in small databases with transfer learning. In: ICFHR, pp. 429–434 (2018)
8. Kleber, F., Fiel, S., Diem, M., Sablatnig, R.: CVL-database: an off-line database for writer retrieval, writer identification and word spotting. In: ICDAR, pp. 560–564. IEEE (2013)
9. Klein, J., Taub, D.: The effect of variations in handwriting and print on evaluation of student essays. Assess. Writ. **10**(2), 134–148 (2005)
10. Krishnan, P., Jawahar, C.: Generating synthetic data for text recognition. arXiv preprint arXiv:1608.04224 (2016)
11. Lavrenko, V., Rath, T.M., Manmatha, R.: Holistic word recognition for handwritten historical documents. In: DIAL, pp. 278–287 (2004)
12. Mangen, A., Anda, L.G., Oxborough, G.H., Brønnick, K.: Handwriting versus keyboard writing: effect on word recall. J. Writ. Res. **7**(2), 227–247 (2015)
13. Markham, L.R.: Influences of handwriting quality on teacher evaluation of written work. Am. Educ. Res. J. **13**(4), 277–283 (1976)
14. Marti, U.V., Bunke, H.: The IAM-database: an English sentence database for offline handwriting recognition. IJDAR **5**(1), 39–46 (2002)
15. Mueller, P.A., Oppenheimer, D.M.: The pen is mightier than the keyboard: advantages of longhand over laptop note taking. Psychol. Sci. **25**(6), 1159–1168 (2014)
16. Nair, R.R., Sankaran, N., Kota, B.U., Tulyakov, S., Setlur, S., Govindaraju, V.: Knowledge transfer using neural network based approach for handwritten text recognition. In: DAS, pp. 441–446 (2018)
17. Plamondon, R., Srihari, S.N.: Online and off-line handwriting recognition: a comprehensive survey. TPAMI **22**(1), 63–84 (2000)
18. Rastle, K., Harrington, J., Coltheart, M.: 358,534 nonwords: the arc nonword database. QJEP **55**(4), 1339–1362 (2002)

19. Rothacker, L., Sudholt, S., Rusakov, E., Kasperidus, M., Fink, G.A.: Word hypotheses for segmentation-free word spotting in historic document images. In: ICDAR, vol. 1, pp. 1174–1179 (2017)
20. Rowtula, V., Oota, S.R., Jawahar, C.: Towards automated evaluation of handwritten assessments. In: ICDAR, pp. 426–433 (2019)
21. Rusiñol, M., Aldavert, D., Toledo, R., Lladós, J.: Efficient segmentation-free keyword spotting in historical document collections. Pattern Recogn. **48**(2), 545–555 (2015)
22. Russell, M., Tao, W.: The influence of computer-print on rater scores. Pract. Assess. Res. Eval. **9**(1), 10 (2004)
23. Sharma, A., Jayagopi, D.B.: Automated grading of handwritten essays. In: ICFHR, pp. 279–284 (2018)
24. Shi, B., Bai, X., Yao, C.: An end-to-end trainable neural network for image-based sequence recognition and its application to scene text recognition. TPAMI **39**(11), 2298–2304 (2016)
25. Srihari, S.N., Srihari, R.K., Babu, P., Srinivasan, H.: On the automatic scoring of handwritten essays. In: IJCAI, pp. 2880–2884 (2007)
26. Voigtlaender, P., Doetsch, P., Ney, H.: Handwriting recognition with large multidimensional long short-term memory recurrent neural networks. In: ICFHR, pp. 228–233 (2016)

An Efficient Local Word Augment Approach for Mongolian Handwritten Script Recognition

Haoran Zhang, Wei Chen, Xiangdong Su$^{(\boxtimes)}$, Hui Guo, and Huali Xu

Inner Mongolia Key Laboratory of Mongolian Information Processing Technology, College of Computer Science, Inner Mongolia University, Hohhot, China
`cssxd@imu.edu.cn`

Abstract. The scarcity problem of Mongolian handwritten data greatly limits the accuracy of Mongolian handwritten script recognition. Most existing augmentation methods generate new samples by making holistic transformation on the existing samples, which cannot fully reflect the variation of the Mongolian handwritten words. According to the characteristics of Mongolian words, this paper proposes a local word augment approach for Mongolian handwriting data, effectively improving the diversity of the augmented samples. We make local variation on the strokes by moving the endpoints and the out-stroke control point of the strokes and reconstructing the strokes with Bezier splines. The overall generation process is flexible and controllable. Experiment on few-shot Mongolian handwritten OCR demonstrates that our approach significantly improves the recognition accuracy and outperforms the holistic augmentation methods.

Keywords: Data augmentation · Mongolian handwritten words · Word stroke · Few-shot OCR

1 Introduction

Good performance of deep neural networks has been reported in optical character recognition (OCR). The success partly depends on the large scale training dataset to cope with the problem of overfitting. With the increase of the training samples, the generalization error bound will be minimized and the recognition accuracy will be improved. However, the scarcity of Mongolian handwritten data greatly limits the accuracy of Mongolian handwritten recognition, since Mongolian is an agglutinative language which results in a very huge vocabulary size. When collecting enough Mongolian handwritten samples is a costly work, it is necessary to automatic augment the Mongolian handwritten dataset for Mongolian handwritten script recognition.

There are many universal data augmentation methods for image recognition task, such as rotating, flipping, scaling, clip, mask and translation. These methods can be used in Mongolian handwritten word augmentation and advance the

© Springer Nature Switzerland AG 2021
J. Lladós et al. (Eds.): ICDAR 2021, LNCS 12824, pp. 429–443, 2021.
https://doi.org/10.1007/978-3-030-86337-1_29

recognition performance in some extent. However, these methods are not specifically designed for Mongolian OCR. They generated new samples by making transformation on the holistic word images. This cannot fully reflect the variation of the characters and limited their effect on the recognition of Mongolian handwriting words. Even part of the generated samples may have negative effects on model training.

Recently, the GAN-based neural augmentation approach [1] takes data from a source domain and learns to take any data item and generalize it to generate other within-class data items. Another neural augmentation approach [2] learns to make the sample generation flexible and controllable by using a set of custom fiducial points. It bridges the gap between the isolated processes of data augmentation and network optimization by joint learning. However, it takes a lot of time to train the deep neural network that generates the data. Although some of these methods have shown good augmentation effects, a lot of comparative experiments are needed to find the optimal hyper-parameters for Mongolian handwritten script.

To avoid the problems related to the above global augmentation approaches, this paper proposes a local word augment approach for Mongolian handwritten dataset. The motivation of our method is to generate word samples with various appearances according to Mongolian writing styles. The different ways of writing word strokes make the words present a variety of writing styles. That is, the deformation of strokes is the key to changing the writing style. We make local variation on the strokes by moving the endpoints and the out-stroke control point of the strokes and reconstructing the strokes with Bezier splines. Our method possesses great flexibilities by controlling the moving range and the number of strokes. Experiment on few-shot Mongolian handwritten OCR demonstrates that our approach outperforms the holistic augmentation methods and significantly improves the recognition accuracy.

The contributions of this paper are as follows:

- This paper proposes a local word augmentation method for few-shot Mongolian handwritten word recognition, which generated new Mongolian word samples through local stroke deformation. It effectively improve the recognition accuracy of Mongolian handwritten words.
- This paper set an example for handwritten word image augmentation based on local strokes, which can generate more various word samples than the holistic augmentation methods.

2   Related Work

The existing data augmentation methods can be summarized into two types: white-box based and black-box based. The white-box based data augmentation methods are usually realized by geometric transformation and color transformation. And the black-box based data augmentation methods are implemented with deep learning [3].

White-box based data augmentation methods usually performs affine transformation [3–6] on images, including rotation, flipping, and scaling and so forth.

It can effectively improve the generalization ability of deep neural networks for most tasks, which has strong versatility. Engstrom et al. [4] proved that even simple rotation and translation of the image can make the network misclassify the image. This also shows the effectiveness of affine transformation as a data augmentation method. Jin et al. [5] added stochastic additive noise to the input image to improve the robustness of the model in the presence of adversarial noise. Mikołajczyk et al. [3] combined the input image with the appearance of another image to generate a high perceptual quality image. Moreover, Krishna et al. [6] force the network to find other relevant content in the image by removing the interference information from the input image by randomly hiding patches. All data augmentation methods mentioned above are used in image classification task. They usually cannot obtain the good effects in OCR task, especially for Mongolian OCR. Wei et al. [7] augment Mongolian text images using the simple affine transformation method. All the above works use static augmentation strategy, and the transformation of each sample in dataset is treated equally. Therefore, these methods cannot meet the needs of sample characteristic augmentation.

Black-box based methods usually augment sample using the Generative Adversarial Network (GAN) [8]. Antoniou et al. [1] proposed Data Augmentation Generative Adversarial Network (DAGAN), which generalizes the input data by learning the characteristics of it to generate more reasonable intra-class data. Cubuk et al. [9] proposed a search-based data augmentation method based on the reinforcement learning. In addition, [10] is improved to select all augmented sub-strategies with equal probability. In the handwriting text recognition task, text images with different writing styles can improve the generalization ability of the model. Luo et al. [2] proposed a hybrid augmentation method to deform word images. Fogel et al. [11] proposed a semi-supervised method ScrabbleGAN, which can generate handwritten words in different fonts. However, it takes a lot of time and comparative experiments to train the deep neural network well to generate the data.

Our method regenerate the strokes of partial Mongolian handwritten characters, and increase the data scale on the basis of increasing the writing style. Therefore, it provides a new way to solve the problem of character recognition with fewer samples.

3   Methodology

Mongolian is an agglutinative language, and its letters present different visual forms in the initial, middle and final positions [12]. To generate word samples with various appearances according to Mongolian writing styles, we make local variations on the strokes by moving the endpoints and the out-stroke control point of the strokes and reconstructing the strokes with Bezier splines. The augmentation algorithm consists of four steps, including word skeleton extraction, stroke segmentation, moving area computing and augmented sample generation, as shown in Algorithm 1.

The strokes of the handwritten Mongolian words are wider than 1-pixel, which makes the detection of key points difficult. Thus, we first extract the skeletons of handwritten characters. Then, we locate the corner points based on the skeletons and divide each Mongolian word into multiple strokes using these corner points as the boundary. Next, we compute the out-stroke control points according to the strokes and calculate the moving areas for the corner points and the out-stroke control point. At this point, each stroke is considered as a Bezier curve. Finally, we reconstruct each stroke with Bezier spline by moving its control points. We explain each step in detail in the following four subsections.

Algorithm 1: Local Word Augment Algorithm

Input: Mongolian handwritten word image I.

Output: augmented sample set $\{S_1, S_2, \ldots, S_N\}$.

Step 1: Word skeleton extraction

   Extract the skeleton I_{ske} for each word image I

Step 2: Stroke segmentation

   (1) Detecting the corner point on the word skeleton I_{ske}

   (2) Segment the word into strokes according to the corner point

Step 3: Moving area computing

   (1) Computing the out-stroke control point for each stroke

   (2) Computing the moving area for the corner points

   (3) Computing the moving area for out-stroke control point

Step 4: Augmented sample generation

   (1) Moving the corner points and out-stroke control point

   (2) Generating the new strokes with Bezier curves according to the position of the corner points and out-stroke control point

   (3) Generating new word sample by combining the new strokes

---

3.1   Word Skeleton Extration

Mongolian word skeleton detection uses the method of Zhang et al. [13] to detect the 1-pixel framework of each word image. The core idea is removing redundant pixels according to specific rules. Whether the point P_k on the stroke is a skeleton point are determined by the eight neighborhood points. Suppose the pixel vertical above the point P_k is P_1, and the rest neighborhood points taken in the clockwise direction are $P_2, P_3 \ldots P_8$. Each iteration of this method is divided into two subiterations. The judgment conditions of the first subiteration are as follows:

$$P_1 + P_2 + P_3 + P_4 + P_5 + P_6 + P_7 + P_8 \in [2, 6] \qquad (1)$$

$$\sum_{i=1}^{N=8} J\left(P_i, P_{(i \bmod 8)+1}\right) = 1 \qquad (2)$$

$$P_1 \times P_3 \times P_5 = 0 \qquad (3)$$

$$P_3 \times P_5 \times P_7 = 0 \qquad (4)$$

where $J(P_i, P_{(i \bmod 8)+1})$ is a judgment formula, if and only if $P_i = 0$ and $P_{(i \bmod 8)+1} = 1$, $J = 1$. In other cases, $J = 0$. In the second subiteration, we keep the conditions (1) and (2) and change the rest conditions (3) and (4) as follows:

$$P_1 \times P_3 \times P_7 = 0 \tag{5}$$

$$P_1 \times P_5 \times P_7 = 0 \tag{6}$$

When the eight neighborhood values of the point P_k to be determined meet one of the above conditions, the point P_k is set to 0, meaning it is not a skeleton point. According to the above rules, we check each pixel on the Mongolian word image. The remaining pixels form the word skeleton. Figure 1 shows the Mongolian word "ᠮᠣᠩ" and its skeleton.

word image   word skeleton

Fig. 1. The Mongolian word "ᠮᠣᠩ" and its skeleton.

3.2 Stroke Segmentation

The different ways of writing word strokes make the words present a variety of writing styles. That is, the deformation of strokes is the key to changing the writing style. The proposed augmentation approach makes local variation on the strokes to produce new Mongolian word samples. Thus, we need to segment each word image into strokes. We use the corner point as the boundary of two strokes. These corner points and the out-stroke control points determine the deformation of strokes together. We named these corner points as the endpoints of the strokes.

In Mongolian words, the corner points are the turning points where there is an obvious turn at the handwriting tracking. We employ Shi-Tomas [14] to detect the corner in Mongolian handwritten word images. Shi-Tomas corner detector takes the following corner selection criteria. A score is calculated for each pixel, and if the score is above a certain value, the pixel is marked as a corner. We first calculate the following matrix M for each pixel:

$$M = \sum_{(x,y)} w(x,y) \begin{bmatrix} I_x^2 & I_x I_y \\ I_x I_y & I_y^2 \end{bmatrix} \tag{7}$$

where $w(x, y)$ denotes the weights for the window centered at position (x, y). I_x and I_y represent the gradient of pixels (x, y) in the horizontal and vertical directions. Next, we calculate the eigenvalues λ_1, λ_2 of M. The score is calculated using two eigenvalues as:

$$R = \min\left(\lambda_1, \lambda_2\right) \tag{8}$$

When all the corner points are available, we obtain the strokes by removing the corner points. We can use connected component analysis then to obtain these strokes. Note that some corner points in this stage may be redundantly added in rare cases. The segmented components also can be treated as strokes to produce the varied words. The corner points and segmented strokes of the Mongolian word "ᠮᠣᠷ" are shown in Fig. 2. The red points denote corner points. The word is divided into six strokes with different color.

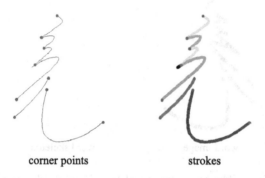

corner points   strokes

Fig. 2. The corner points and segmented strokes of the Mongolian word "ᠮᠣᠷ" .

3.3   Moving Area Computing

The out-stroke control point is another key point to control stroke deformation. It can control the convexity of the associated strokes. According to the definition of Bezier curve [15], the farthest point on the curve from the straight line connecting two endpoints represents the turning point of Bezier curve. It is also the middle point between the out-stroke control point and the middle point of two endpoints. We use the following formula to locate this farthest point.

$$x^*, y^* = \arg\max_{x, y}\left(\frac{(X_2 - X_1) \times y + (Y_2 - Y_1) \times x + X_2 Y_1 - X_1 Y_2}{(X_2 - X_1)^2 + (Y_2 - Y_1)^2}\right) \tag{9}$$

where X_1, Y_1, X_2, Y_2 are the coordinate values of the two endpoints of the strokes, respectively. x, y represent the coordinate value of each point on the curve. x^* and y^* are coordinate values of the farthest point mentioned above. The out-stroke control point can be determined by the following formula:

$$P\left(x_{out}, y_{out}\right) = 2P\left(x^{*}, y^{*}\right) - \frac{1}{2}\left(P\left(X_1, Y_1\right) + P\left(X_2, Y_2\right)\right) \tag{10}$$

where $P(x, y)$ represents a point whose x-axis is x and y-axis is y in the word image.

When the two endpoints (corner points) and the out-stroke control point of each stroke are available, we compute the moving areas of these control points. For the corner points, the moving areas should ensure that there are no other corner points in such area. Suppose corner point $P\left(X_c, Y_c\right)$ and its nearest corner point $P\left(X_n, Y_n\right)$, the width w_{area} and the height h_{area} of the moving area are calculated as follows:

$$w_{area} = 2\alpha \times |X_c - X_n| \tag{11}$$

$$h_{area} = 2\alpha \times |Y_c - Y_n| \tag{12}$$

where $\alpha \in [0, 1]$ is a coefficient to control the maximum moving area of the corner point.

For the out-stroke control point, we should ensure that it do not change the concavity of the stroke after moving it. Meanwhile, we move it to produce new stroke styles as many as possible. Therefore, we define the movement area of the out-stroke control point as a square area. Suppose out-stroke control point $P\left(X_o, Y_o\right)$, the side length s_{area} of this area can be calculated by the following formula:

$$s_{area} = \beta \times \left(\frac{(X_2 - X_1) \times Y_o + (Y_2 - Y_1) \times X_o + X_2 Y_1 - X_1 Y_2}{(X_2 - X_1)^2 + (Y_2 - Y_1)^2} \right) \tag{13}$$

where $\beta \in [0, 1]$ is a coefficient to control the maximum moving area of the out-stroke control point. For the Mongolian word "ᠮᠠᠷ" , its movement areas for the out-stroke control points and the corner points are shown in Fig. 3.

3.4   Augmented Sample Generation

The key to reconstructing strokes is to make them continuous and smooth. Simply moving the points will lead to discontinuous strokes and bring the extra computation of the moving distance of the stroke. This will lead to an increase in system overhead and a decrease in augmentation efficiency. From Mongolian handwritten word images, we found that the strokes can be approximately fitted to a curve with a certain degree of convexity. Therefore, we used the Bezier spline [16] to approximate the fitting method for stroke reconstruction.

Based on the three control points of each stroke, we use De Casteljau's algorithm [15] to construct a second-order Bezier curve. Suppose the two corner points are P_A and P_B, and the out-stroke control point is P_C, we moved them to P'_A, P'_B and P'_C within their movement areas, respectively. The points on the deformed new stroke curve after moving the control points are computed as follows:

$$P_t = (1 - t)^2 P'_A + 2t\left(1 - t\right) P'_C + t^2 P'_B \tag{14}$$

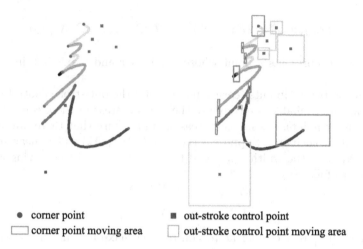

● corner point   ■ out-stroke control point
▭ corner point moving area   ▢ out-stroke control point moving area

Fig. 3. Out-stroke control point and moving areas of the Mongolian word "ᠮᠣᠷᠢ" .

where $t \in [0, 1]$ represents the time step. By randomly selecting different P'_A, P'_B and P'_C, we can generate multiple word samples.

After completing the deformation of all the strokes, we combine the newly generated strokes to form an augmented sample. The new handwriting constructed in this way can make the handwritten characters appear more different writing styles. The augmented sample of the Mongolian word "ᠮᠣᠷᠢ" generated in this step is shown in Fig. 4. In the broken line box, the light color denotes the stroke before deformation and the deep color denote the new stroke. The sample on the right is formed with all the deformed strokes.

4   Experiment

4.1   Dataset

The experiment was performed on Mongolian handwritten word set provided by Inner Mongolia University. The word images in this dataset are collected from the handwriting function of the Mongolian input method on the mobile device. After proofreading, we selected $10,000$ commonly used words form the dataset. Each word contains two samples. For each word, we used one sample as the training sample, and the other as the test sample. This means the training dataset before augmentation includes $10,000$ word samples, and the testing dataset also contains $10,000$ word samples. The width of all sample images is 300 pixels, and the height varies with the word. We used our method to augment the training dataset, and used the test dataset to evaluate the effect of data augmentation in Mongolian handwritten script recognition.

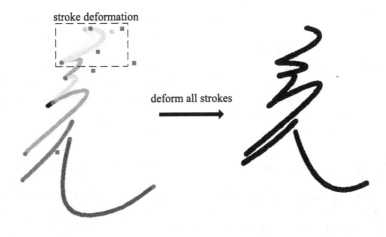

Fig. 4. An Illustration of stroke deformation and an augmented sample of the Mongolian word "ᠮᠤᠷ" .

4.2   Baselines

To prove the effectiveness of the proposed method, we select six widely used data augmentation methods for comparison. These methods are listed as follows:

- **Affine transformation** It is currently the most widely applicable data augmentation method. Its commonly used transformation operations include translation, zoom, flip, rotation, and shear. Each operation has some hyperparameters to control the magnitude of the transformation. When we use it, we usually combine all operations randomly to transform the sample.
- **Elastic transformation** [17] This is an augmentation method proposed for handwritten numbers. It first generates random standard deviations in the interval $[-1, 1]$ for each dimension of the pixel. After that, the mean value is θ and the standard deviation is σ Gaussian kernel to filter the deviation matrix of each dimension. Finally, an amplification factor is used to control the deviation range.
- **Advanced transformation** [5,18] This augmentation method mixes random elimination and noise addition. It is often used in target detection tasks, which make the model focus on more features through random elimination. At the same time, light effects such as salt and pepper noise, Gaussian noise, rain and fog is added to the sample.
- **Hide and Seek** [6] This method divides the input image into multiple grids of the same size, and each grid generates a mask with a certain probability. In this way, the network is forced to make good use of all parts of the global information.

- **ScrabbleGAN** [11] This is a semi-supervised approach to generate handwritten word images. It can manipulate the resulting text styles with the help of a generative model.
- **Learn to Augment** [2] This data augmentation method can be used for scene text and handwritten text. It divides the text image into N patches and gives each patch a control point. The text is deformed by moving these control points within the radius of R. This method can be optimized with neural networks to find the most suitable N and R.

4.3   Evaluation Method

Because our approach is used to augment the Mongolian handwritten samples to enhance OCR performance, this study employs word recognition accuracy to evaluate the quality of the augmented samples by our approach. We compared the word accuracies on the test images before and after word augmentation. The deep neural network for the Mongolian handwriting recognition is Convolutional Recurrent Neural Network (CRNN), which was proposed in [19]. It is composed of 7 convolution modules to form a convolutional neural network part, and 2 bidirectional LSTMs form a recurrent neural network part. It is a lightweight model that can be trained quickly. Since the text recognition task is sequence to sequence, we use CTC loss [20] to optimize the parameters of the neural network. The output vector of the model is the predicted letter sequence. After the CTC method is processed, the final prediction sequence is obtained. The model is optimized by Adam optimizer.

We adopt the data preprocessing method in [21] for Mongolian word recognition. Specifically, each word image is divided into frames of equal size, and the height of each frame is set to 300 pixels. There is a half-frame overlapping between each pair of adjacent frames. That is, when the next frame is taken, the interception starts at the half height of the previous frame. The contents of half frames in two adjacent frames overlap. Next, we make the word images have the same aspect ratio. To ensure the aspect ratio is 32 : 280, we use the pixels of zero to pad the word images. Finally, we scale the filled image to a size of 32 × 160, and then rotate it 90 degrees counterclockwise as the input of the recognition model.

5   Results and Discussion

5.1   Ablation Experiment

Our method is to deform the strokes. It cannot determine the moving range of each key point directly. If we used a fixed movement area for all the key points, it may produce wrong deformation on short strokes. Thus, we consider the relationship between key points and their surrounding points, and adaptively determine the moving areas according to criterion in Sect. 3.3. Here, α donate the moving area control coefficient of corner point, and β denote the moving area control coefficient of out-stroke control point. This section conducts an ablation experiment to select the optimal α and β for Mongolian word augmentation.

According to our observation, we limit α and β in $[0.3, 0.5]$ to obtain word samples with rich styles. We compare the recognition accuracy with different α and β when the dataset augmentation times is 20. That is, we augment the training dataset into $200,000$, and use the augment dataset to training the word recognizer. The testing result is shown in Table 1.

Table 1. The recognition accuracy with different α and β (the dataset augmentation times is 20).

β / α	0.3	0.4	0.5
0.3	0.2980	0.3118	0.3181
0.4	0.3212	0.3243	0.3254
0.5	0.3211	0.3359	0.3246

The results in Table 1 show that setting α to 0.5 and β to 0.4 achieves the best recognition performance. The samples generated with this setting possess rich style variation while maintains the correct presentations of the original words. Although the samples generated by setting lower α and β than the optimal parameters can also maintain the correct word patterns, their styles will decrease and cannot cover more writing styles. On the contrary, the samples generated by higher α and β are various in styles, but may introducing wrong samples. Therefore, the recognition network trained with the samples generated by the optimal parameters achieves the best recognition performance.

5.2 Performance on Recognition Model

This section compares the effectiveness of baseline augmentation methods and our method on different augmentation times. Our method employs the optimal α and β obtained from the ablation study. The evaluation metric is also the word accuracy.

For fair comparison, we adopt the best parameter settings for each baseline according to their works. The training dataset are augmented according to the augmentation time from 2 to 200. The recognizers trained on different augmented datasets are evaluated with the same test dataset. The results are shown in Table 2.

From Table 2, we found that our method obtains the recognition accuracy of 0.3729 with 120 times of augmentation. This is also the highest word among the above methods on all the augmentation times. Learn to Augment [2] ranks the second, who obtains 0.3693 word accuracy when the augmentation time is 150. Our method significantly outperforms Affine Transformation, Elastic Transformation [17], Advanced transformation [5,18], Hide and Seek [6] and Scrabble-GAN [11] on each augmentation scale. This indicates that our method is the best

Table 2. Word accuracy comparison between our method and the baselines on different augmentation times.

Method	Training data size (augmentation times)					
	2	5	10	15	20	25
Affine Transformation	0.1201	0.1977	0.2460	0.2608	0.2840	0.2825
Elastic Transformation	0.0724	0.0779	0.0864	0.0894	0.0934	0.0929
Advanced Transformation	0.0754	0.0802	0.0876	0.0820	0.0855	0.0797
Hide and Seek	0.0655	0.0706	0.0734	0.0736	0.0709	0.0619
ScrabbleGAN	0.0481	0.0602	0.0704	0.0747	0.0797	0.0784
Learn to Augment	**0.1433**	**0.2146**	0.2595	0.2906	0.3015	0.3143
Ours	0.1389	0.2139	**0.2713**	**0.3022**	**0.3359**	**0.3459**
Method	Training data size (augmentation times)					
	50	80	100	120	150	200
Affine Transformation	0.2887	0.2984	0.2912	0.2988	0.3108	0.3005
Elastic Transformation	0.1065	0.1104	0.1088	0.1057	0.1067	0.1126
Advanced transformation	0.0670	0.0695	0.0670	0.0659	0.0624	0.0667
Hide and Seek	0.0536	0.0474	0.0456	0.0450	0.0525	0.0360
ScrabbleGAN	0.0806	0.0748	0.0725	0.0658	0.0705	0.0663
Learn to Augment	0.3345	0.3545	0.3538	0.3692	**0.3693**	0.3684
Ours	**0.3629**	**0.3656**	**0.3707**	**0.3729**	0.3663	**0.3689**

augmentation for Mongolian handwritten script. This is because our method is a stroke based augmentation method which can introduce more various writing styles than the above holistic augmentation baselines. This suggests that the key to improve the accuracy of word recognition is to cover the diversity of word appearance.

Except that the augmentation times are 2, 5, 150, the proposed augmentation method obviously performs better than Learn to Augment [2]. On these three cases, the performance of our method approximates to that of Learn to Augment [2]. Actually, when the augmentation time is very small, the variation of each augmentation method cannot be well reflected. Our method achieves optimal effects with 120 dataset augmentation times, while Learn to Augment [2] and affine transformation achieve best effects with 150 augmentation times. This shows our method achieves the best performance with the smallest augmentation times, which reduce the operating cost of the equipment and the training time of the recognizer.

The results also show that Elastic Transformation [17], Advanced transformation [5,18] and Hide and Seek [6] do not work well on Mongolian handwritten word augmentation, because these methods are unable to extend the handwriting style of the sample. The results of ScrabbleGAN [11] are also not well, since Mongolian is an agglutinative language and the length of each word is inconsistent which makes it impossible to obtain accurate letter styles. Besides, it is difficult for ScrabbleGAN [11] to learn more writing styles from a single Mongolian sample.

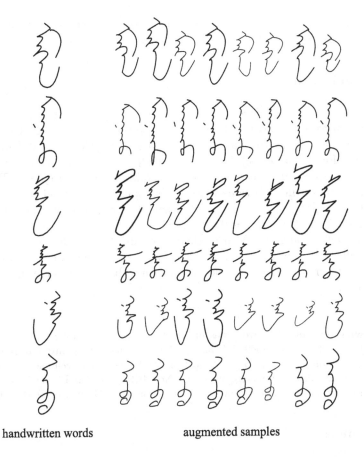

handwritten words   augmented samples

Fig. 5. Mongolian handwritten words and their augmented samples.

For each augmentation method, with the increase of augmentation time, the word accuracies first sharply increase. Then the increase speeds become slow. After reaching the peak, the word accuracies begin to decrease. The reason is as follows. At first, with the increase of augmentation time, the variations of training samples are increase. This brings benefit to the word recognizer. When the augmentation is excessive, there are some error augmented samples. This will hurt the word recognizer on the contrary. Therefore, it is necessary to select an ideal augmentation scale.

5.3   A Case Study

We take a case study on the augmented Mongolian word samples, as shown in Fig. 5. The style of samples generated by our method is obviously different from the existing handwritten ones, which exhibit local variations. This makes the data augmented by our method improve the recognizer significantly. Meanwhile, our method is less likely to introduce error augmented samples.

6   Conclusion

This paper proposes an efficient local word augmentation method for Mongolian handwritten words to deal with the scarcity problem of Mongolian handwritten data in OCR task. To generate new word samples, the method deforms the strokes in Mongolian words by moving the endpoints and the out-stroke control point of the strokes and reconstructing the strokes with Bezier splines. The augmented samples exhibit obvious writing variations. Our method also can avoid introducing useless or error augmented samples. Experiment shows that our method significantly improves the Mongolian word recognizer and outperforms the other holistic augmentation methods. We also found that word augmentation should be kept in a reasonable range. Excessive augmentation may hurt the word recognizer since error or useless samples may be introduced in the excessive augmentation.

Acknowledgement. This work was funded by National Natural Science Foundation of China (Grant No. 61762069, 61773224), Natural Science Foundation of Inner Mongolia Autonomous Region (Grant No. 2019ZD14), Science and Technology Program of Inner Mongolia Autonomous Region (Grant No. 2019GG281) and Inner Mongolia Discipline Inspection and Supervision Big Data Laboratory (Grant No. 21500-5206043).

References

1. Antoniou, A., Storkey, A., Edwards, H.: Data augmentation generative adversarial networks. arXiv preprint arXiv: 1711.04340 (2017)
2. Luo, C., Zhu, Y., Jin, L., Wang, Y.: Learn to augment: joint data augmentation and network optimization for text recognition. In: IEEE Computer Society Conference on Computer Vision and Pattern Recognition (CVPR), pp. 13743–13752. IEEE, USA (2020)
3. Mikolajczyk, A., Grochowski, M.: Data augmentation for improving deep learning in image classification problem. In: 2018 International Interdisciplinary PhD Workshop (IIPhDW), pp. 117–122. IEEE, Swinoujscie (2018)
4. Engstrom, L., Tran, B., Tsipras, D., Schmidt, L., Madry, A.: A rotation and a translation suffice: fooling CNNs with simple transformations. arXiv preprint arXiv: 1712.02779 (2017)
5. Jin, J., Dundar, A., Culurciello, E.: Robust convolutional neural networks under adversarial noise. In: 4th International Conference on Learning Representations (ICLR), pp. 1–8. Computational and Biological Learning Society (2016)
6. Singh, K.-K., Yu, H., Sarmasi, A., Pradeep, G., Lee, Y.-J.: Hide-and-seek: a data augmentation technique for weakly-supervised localization and beyond. arXiv preprint arXiv:1811.02545 (2018)

7. Wei, H., Liu, C., Zhang, H., Bao, F., Gao, G.: End-to-end model for offline handwritten mongolian word recognition. In: Tang, J., Kan, M.-Y., Zhao, D., Li, S., Zan, H. (eds.) NLPCC 2019. LNCS (LNAI), vol. 11839, pp. 220–230. Springer, Cham (2019). https://doi.org/10.1007/978-3-030-32236-6_19

8. Goodfellow, I.-J., et al.: Generative adversarial nets. In: 28th Conference on Neural Information Processing Systems (NIPS), Montreal, QC, Canada, pp. 2672–2680 (2014)

9. Cubuk, E.-D. Zoph, B., Mane, D., Vasudevan, V., Le, Q.-V.: AutoAugment: learning augmentation strategies from data. In: IEEE Computer Society Conference on Computer Vision and Pattern Recognition (CVPR), pp. 113–123. IEEE, Long Beach (2019)

10. Cubuk, E.-D., Zoph, B., Shlens, J., Le, Q.-V.: RandAugment: practical automated data augmentation with a reduced search space. In: IEEE Computer Society Conference on Computer Vision and Pattern Recognition Workshops, pp. 3008–3017. IEEE, USA (2020)

11. Fogel, S., Averbuch-Elor, H., Cohen, S., Mazor, S., Litman, R.: ScrabbleGAN: semi-supervised varying length handwritten text generation. In: IEEE Computer Society Conference on Computer Vision and Pattern Recognition (CVPR), pp. 4323–4332. IEEE, USA (2020)

12. Wei, H., Gao, G.: A keyword retrieval system for historical Mongolian document images. Int. J. Doc. Anal. Recogn. (IJDAR) 17(1), 33–45 (2013). https://doi.org/10.1007/s10032-013-0203-6

13. Zhang, T.-Y., Suen, C.-Y.: A fast parallel algorithm for thinning digital patterns. Commun. ACM 27(3), 236–239 (1984)

14. Shi, J., Tomasi, C.: Good features to track. In: IEEE Computer Society Conference on Computer Vision and Pattern Recognition (CVPR), pp. 593–600. IEEE, Seattle (1994)

15. Farin, G., Hansford, D.: The Essentials of CAGD. CRC Press, USA (2000)

16. Forrest, A.-R.: Interactive interpolation and approximation by Bézier polynomials. Comput. J. 15(1), 71–79 (1972)

17. Simard, P., Steinkraus, D., Platt, J.: Best practices for convolutional neural networks applied to visual document analysis. In: 7th International Conference on Document Analysis and Recognition (ICDAR), pp. 958–963. IEEE, Edinburgh (2003)

18. Zhong, Z., Zheng, L., Kang, G., Li, S., Yang, Y.: Random erasing data augmentation. In: AAAI Conference on Artificial Intelligence, pp. 13001–13008 (2020)

19. Shi, B., Bai, X., Yao, C.: An End-to-end trainable neural network for image-based sequence recognition and its application to scene text recognition. IEEE Trans. Pattern Anal. Mach. Intell. 39(11), 2298–2304 (2017)

20. Graves, A., Fernandez, S., Gomez, F., Schmidhuber, J.: Connectionist temporal classification: labelling unsegmented sequence data with recurrent neural networks. In: 23rd international conference on Machine learning, Pittsburgh, PA, USA, pp. 369–376 (2006)

21. Zhang, H., Wei, H., Bao, F., Gao, G.: Segmentation-free printed traditional mongolian ocr using sequence to sequence with attention model. In: 14th International Conference on Document Analysis and Recognition (ICDAR), pp. 585–590. IEEE, Kyoto (2017)

IIIT-INDIC-HW-WORDS: A Dataset for Indic Handwritten Text Recognition

Santhoshini Gongidi$^{(\boxtimes)}$ and C. V. Jawahar

Centre for Visual Information Technology, IIIT Hyderabad, Hyderabad, India
santhoshini.gongidi@research.iiit.ac.in, jawahar@iiit.ac.in

Abstract. Handwritten text recognition (HTR) for Indian languages is not yet a well-studied problem. This is primarily due to the unavailability of large annotated datasets in the associated scripts. Existing datasets are small in size. They also use small lexicons. Such datasets are not sufficient to build robust solutions to HTR using modern machine learning techniques. In this work, we introduce a large-scale handwritten dataset for Indic scripts containing 868K handwritten instances written by 135 writers in 8 widely-used scripts. A comprehensive dataset of ten Indic scripts are derived by combining the newly introduced dataset with the earlier datasets developed for Devanagari (IIIT-HW-DEV) and Telugu (IIIT-HW-TELUGU), referred to as the IIIT-INDIC-HW-WORDS.

We further establish a high baseline for text recognition in eight Indic scripts. Our recognition scheme follows the contemporary design principles from other recognition literature, and yields competitive results on English. IIIT-INDIC-HW-WORDS along with the recognizers are available publicly. We further (i) study the reasons for changes in HTR performance across scripts (ii) explore the utility of pre-training for Indic HTRs. We hope our efforts will catalyze research and fuel applications related to handwritten document understanding in Indic scripts.

Keywords: Indic scripts · Handwritten text recognition · Pre-training

1 Introduction

In the last two decades, large digitization projects converted paper documents as well as ancient historical manuscripts into the digital forms. However, they remain often inaccessible due to the unavailability of robust handwritten text recognition (HTR) solutions. The capability to recognize handwritten text is fundamental to any modern document analysis systems. There is an immediate need to provide content-level access to the millions of manuscripts and personal journals, large court proceedings and also develop HTR applications to automate processing of medical transcripts, handwritten assessments etc. In recent years, efforts towards developing text recognition systems have advanced due to the success of deep neural networks [15,27] and the availability of annotated datasets. This is especially true for Latin scripts [4,5,21]. The IAM [20] handwritten dataset

© Springer Nature Switzerland AG 2021
J. Lladós et al. (Eds.): ICDAR 2021, LNCS 12824, pp. 444–459, 2021.
https://doi.org/10.1007/978-3-030-86337-1_30

introduced over two decades ago, the historic George Washington [10] dataset and the RIMES [14] dataset are some of the popularly known datasets for handwritten text recognition. These public datasets enabled research for Latin HTR. Even today, these datasets are still being utilized to study handwritten data.

Compared to Latin HTR, Indic HTR remains understudied due to a severe deficit of annotated resources in Indian languages. Unlike many other parts of the world, a wide variety of languages and scripts are used in India. Therefore, collecting sizeable handwritten datasets for multiple Indic scripts becomes challenging and expensive. Existing annotated datasets for Indic HTR are limited in size and scope. They are approximately 5× smaller than the Latin counterparts.

Recent progress in text recognition is mainly credited to the easy access to large annotated datasets. Through this work, we make an effort to bridge the gap between the state of the arts in Latin and Indian languages by introducing a handwritten dataset written in 8 Indian scripts. We introduce a dataset for Bengali, Gujarati, Gurumukhi, Kannada, Odia, Malayalam, Tamil, and Urdu scripts. This dataset along with the datasets IIIT-HW-DEV, IIIT-HW-TELUGU introduced by Dutta et al. [8,9] for Devanagari and Telugu scripts provide handwritten datasets for text recognition in all ten prominent scripts used in India. We refer to this collective dataset as the IIIT-INDIC-HW-WORDS[1]. Figure 1 gives a glimpse into the new dataset and the writing style in eight different Indic scripts. This new dataset contains 868K word instances written in 8 prominent Indic scripts by 135 writers. Possibly this is the first attempt in even attempting offline handwriting in some of these scripts.

We hope that our dataset provides the much-needed resources to the research community to develop Indic HTR and to build valuable applications around Indian languages. The diversity of the IIIT-INDIC-HW-WORDS dataset makes it possible to utilize this dataset for other document analysis problems also. Script identification, handwriting analysis, and synthesis for Indian languages are examples of other problems that can be enabled using this dataset. This dataset could also be beneficial for study of script-independent recognition architectures for HTR.

In this work, we also present a simple and effective text recognition architectures for Indian HTR. We establish a high baseline on scripts presented in the IIIT-INDIC-HW-WORDS. The results and related discussion are in Sect. 4. We also study the benefit of using architectures pre-trained on other scripts. Building robust recognizers in various scripts requires a large amount of data for each script. With transfer learning from other scripts, the excessive requirements of large dataset and training time can be reduced. We also investigate the relation between script similarity and pre-training. We explore both these ideas in Sect. 5.

The major contribution of this work is as follows:

i. We introduce a large dataset—IIIT-INDIC-HW-WORDS, consisting of annotated handwritten words written in 8 Indic scripts.
ii. We establish a high baseline recognizer for Indic HTR. Our recognition algorithm is highly script independent. In this process, we identify the appro-

---

[1] http://cvit.iiit.ac.in/research/projects/cvit-projects/iiit-indic-hw-words.

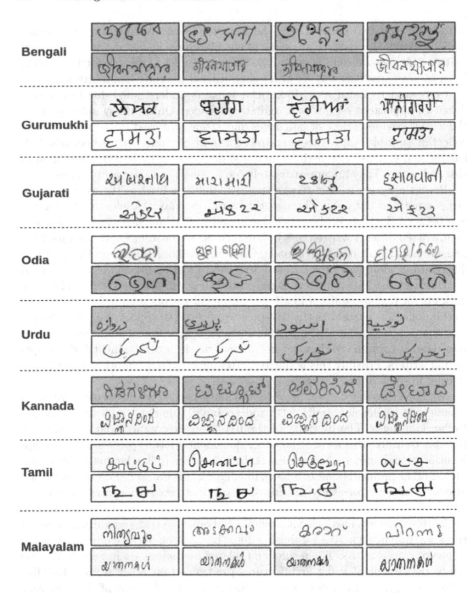

Fig. 1. Word instances from our IIIT-INDIC-HW-WORDS dataset. Out of the 10 major Indic scripts used, we present datasets for 8 scripts: Bengali, Gujarati, Gurumukhi, Kannada, Malayalam, Odia, Tamil, Urdu and complement the recent efforts in Dutta et al. [8] and [9] for Hindi and Telugu. For each script, row 1 shows writing style variations for a specific writer across different words. Row 2 images presented in a specific script block show four writing variations for a particular word.

priate architecture for Indic scripts and establish benchmarks on the IIIT-INDIC-HW-WORDS dataset for future research.

iii. We explore the possibility of transferability across scripts with a set of systematic experiments.

2 Datasets for HTR

Approaches for recognizing natural offline handwriting are heavily data-driven today. Most of them use machine learning methods and learn from annotated examples. This demands for script/language-specific collection of examples. For machine learning methods to be effective, the annotated datasets should be substantial and voluminous in size and diverse in nature. For HTR, this implies that the dataset should have (i) many writers, (ii) extensive vocabulary, and (iii) huge number of samples.

Public datasets such as IAM [20] and George Washington [10] have catalyzed the research in handwriting recognition and retrieval in the past. Datasets for Latin scripts such as IAM [20] and Bentham collection [24] have over 100K running words with large lexicons. The IAM dataset is one of the most commonly used dataset in HTR for performance evaluation even today. This dataset contains 115,320 word instances written by 657 writers, and has a lexicon size of 10,841. This dataset provides annotations at word-level, line-level, and page-level. As the performance on IAM dataset has started to saturate, more challenging datasets have started to surface. The historical datasets such as Bentham collection [24] and the READ dataset [25] are associated with line-level and page-level transcriptions. Annotations at the line, word, and document level provide opportunities for developing methods that use language models and higher order cues. Indian languages are still in their infancy as far as HTRs are concerned, and we limit our attention to the creation of word-level annotations in this work.

Lack of large handwritten datasets remains as a major hurdle for the development of robust solutions to Indic HTR. Table 1 presents the list of publicly available datasets in Indian languages. We also contrast our newly introduced IIIT-INDIC-HW-WORDS dataset with the existing ones in size, vocabulary and the number of writers in Table 1. Most of the Indic handwritten datasets are smaller than the IAM in the number of word instances, writing styles, or lexicon size. For example, the lexicon used to build the CMATER2.1 [3] and the TAMIL-DB datasets [28] consist of city names only. The CENPARMI-U dataset provides only 57 different financial terms in the entire dataset. The small and restrictive nature of the lexicon for these datasets limits the utility while building a generic HTR to recognize text from a large corpus of words. ROYDB [22] and LAW datasets [19] use a larger lexicon. However, the size of these dataset is small, and possibly insufficient to capture the natural variability in handwriting. The PBOK [1] dataset provides the page-level transcriptions for 558 text pages written in three Indic scripts. Page-level text recognizers require excessive amount of data in comparison to word-level recognizers. On the IAM and READ datasets the state-of-the-art full-page text recognizer [31] has high error rates due to additional challenges like extremely skewed text, overlapping lines and inconsistent

Table 1. Publicly available handwritten datasets for Indic scripts and comparison to the dataset introduced in this work. All the datasets provide word-level transcriptions, except for the PBOK dataset.

Name	Script	#Writers	#Word instances	#Lexicon
PBOK [1]	Bengali	199	21K	925
	Kannada	57	29K	889
	Odia	140	27K	1040
ROYDB [22]	Bengali	60	17K	525
	Devanagari	60	16K	1030
CMATERDB2.1 [3]	Bengali	300	18K	120
CENPARMI-U [23]	Urdu	51	19K	57
LAW [19]	Devanagari	10	27K	220
TAMIL-DB [28]	Tamil	50	25K	265
IIIT-HW-DEV [8]	Devanagari	12	95K	11,030
IIIT-HW-TELUGU [9]	Telugu	11	120K	12,945
IIIT-INDIC-HW-WORDS (this work)	Bengali	24	113K	11,295
	Gujarati	17	116K	10,963
	Gurumukhi	22	112K	11,093
	Kannada	11	103K	11,766
	Odia	10	101K	13,314
	Malayalam	27	116K	13,401
	Tamil	16	103K	13,292
	Urdu	8	100K	11,936
	8 scripts	**135**	**868K**	**97,060**

gaps between lines. Note that the IAM dataset is 3× bigger than PBOK dataset. Therefore, solving Indic HTR for full page text recognition requires a major effort and is beyond the scope of this work.

IIIT-HW-DEV [8] and IIIT-HW-TELUGU [9] are possibly the only Indic datasets that are comparable to the IAM dataset at word level. Both these datasets have around 100K word instances each with a lexicon size of over 10K unique words. In this work, we complement this effort and extend similar datasets in many other languages so as to cover 10 prominent Indic scripts. We introduce a unified database for handwritten datasets that are comparable to the IAM dataset in size and diversity. This dataset for Indic scripts is referred to as the IIIT-INDIC-HW-WORDS dataset. We compare our dataset to existing datasets in Table 1. We describe the dataset and discuss further details in the next section.

3   IIIT-INDIC-HW-WORDS

This section introduces the IIIT-INDIC-HW-WORDS dataset consisting of handwritten image instances in 8 different Indic scripts. The scripts present in the

Table 2. Statistics of IIIT-INDIC-HW-WORDS dataset and IAM dataset.

Dataset	Script	Word length (avg)	#Instances		
			Train set	Val set	Test set
IAM	English	5	53,838	15,797	17,615
IIIT-INDIC-HW-WORDS	Bengali	7	82,554	12,947	17,574
	Gujarati	6	82,563	17,643	16,490
	Gurumukhi	5	81,042	13,627	17,947
	Kannada	9	73,517	13,752	15,730
	Odia	7	73,400	11,217	16,850
	Malayalam	11	85,270	11,878	19,635
	Tamil	9	75,736	11,597	16,184
	Urdu	5	71,207	13,906	15,517

collection are Bengali, Gujarati, Gurumukhi, Kannada, Malayalam, Odia, Tamil and Urdu. Indian scripts belong to the Brahmic writing system. The scripts later evolved into two distinct linguistic groups: Indo-Aryan languages in the Northern India and Dravidian languages in the South India. Out of the 8 scripts discussed in this work, 5 scripts are of Indo-Aryan descent, and 3 are of Dravidian descent. Bengali, Gujarati, Gurumukhi, Odia, and Urdu belong to the Indo-Aryan family. Kannada, Malayalam, and Tamil are Dravidian languages. Indian scripts run from left to right, except for Urdu. Table 2 shows the statistics of this dataset. In the dataset, more than 100K handwritten word instances are provided for each script. The entire dataset is written by 135 natural writers aged between 18 and 70. More details about the dataset are mentioned on the dataset web-page.

Dutta et al. [9] propose an effective pipeline for the annotation of word-level datasets. We employ the same approach for collection and annotation generation for IIIT-INDIC-HW-WORDS. A large and appropriate vocabulary is selected to cover the language adequately. Participants are asked to write in a large coded space on an A4 paper. The written pages are scanned at 600 DPI using a flatbed scanner. Word images and associated annotations are automatically extracted from the scanned forms using image processing techniques. The average image height and width in the dataset is 288 and 1120. Samples that are wrongly segmented or those containing printed text, QR codes, and box borders are eliminated from the dataset. Such samples accounted for 4% of the total extracted samples. Except for removing wrongly segmented words, we do not perform any other kind of image pre-processing. Figure 1 shows word instances for each of these scripts. We describe some of our observations here. The words found in Dravidian languages are longer than those present in Indo-Aryan languages. We plot the differences in word lengths between these two groups in Fig. 2. We also show the inter-class and intra-class variability in handwriting styles across Indic

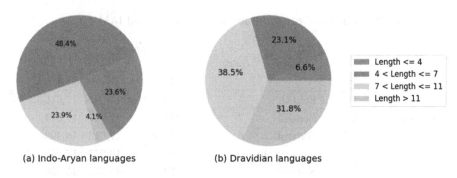

(a) Indo-Aryan languages   (b) Dravidian languages

Fig. 2. Plots showing the variation in distribution of word lengths in IIIT-INDIC-HW-WORDS dataset for two linguistic groups: Indo-Aryan languages and Dravidian languages.

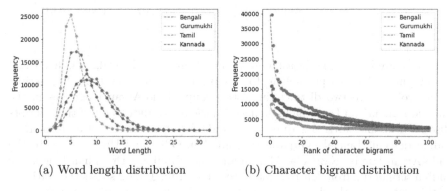

(a) Word length distribution   (b) Character bigram distribution

Fig. 3. Distribution plots for four scripts: Bengali, Gurumukhi, Tamil, and Kannada. We observe that above distributions for Gujarati, Odia are similar to that of Bengali script. Distributions for Urdu and Malayalam script are comparable to those of Gurumukhi and Tamil script respectively.

scripts in Fig. 1. The variable handwriting styles, noisy backgrounds, and choice of the unwanted pen makes it a diverse and challenging collection. Many samples in the dataset contain overwritten characters or poorly visible characters due to pen use. The varying background noise and paper quality also adds to the complexity of the dataset.

The text lexicon for the datasets is sampled from the Leipzig corpora collection [11]. It consists of text files with content from newspaper articles, Wikipedia links, and random websites. More than 10K unique words are sampled from these collections in each language to generate the coded pages for participants to write. The number of unique characters per script includes the basic characters in the respective script's Unicode block and some special characters. The statistics of lexicon size for the IIIT-INDIC-HW-WORDS collection are listed in Table 1. Figure 3a shows the distribution of word length across different datasets. The plot 3b shows that the character bigrams follow Zipf's law. We observe that the law holds true for character n-grams as well with $n = 3, 4, 5, 6$.

The annotated dataset consists of image files along with a text file containing the corresponding label information. The text labels are encoded using Unicode. We also release the train, test, and validation splits for the word recognition task. Around 70% of the instances are added to the training set, 15% to the test set, and the remaining instances comprise the validation set. We ensure that 9–12% of the test labels are out-of-vocabulary (OOV) words. Test sets with a high rate of OOV samples are challenging. The evaluated metrics on such sets inform whether a proposed solution is biased towards the vocabulary of training and validation sets. The total number of samples having out-of-vocabulary text labels in the test sets varies from 35% to 40% per script.

4   Baseline for Text Recognition

Unconstrained offline handwritten text recognition (HTR) is the task of identifying the written characters in an image without using any external dictionaries for a language. Previous works in text recognition [2,4,21,26] show that deep neural networks are very good at solving this problem due to their generalization capability and representational power. Several deep neural network architectures have been proposed for recognition of scene text, printed text, handwritten content. Baek et al. [2] propose a four-stage text recognition framework derived from existing scene text recognition architectures. The recognition flow at different stages of the framework is demonstrated in Fig. 4. This framework can be applied to the HTR task as well. In this section, we discuss and study the different stages of the pipeline. We identify an appropriate architecture for HTR and establish a baseline on the IIIT-INDIC-HW-WORDS dataset.

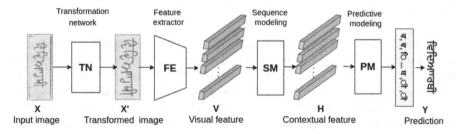

Fig. 4. Generic pipeline for text recognition. Here, we demonstrate the flow of a sample image **X** through the pipeline to generate a text prediction **Y**. The sample shown here is written in Gurumukhi script.

Transformation Network (TN): Diverse styles observed in handwriting data are a significant challenge for HTR. A transformation block learns to apply input-specific geometric transformations such that the end goal of text recognition is simplified. In other words, this module reduces the burden of the later stages of the pipeline. Spatial Transformer Network (STN) [17] and its variants are commonly used to rectify the input images. In this work, we experiment with two types of rectification, affine transformation (ATN), and thin-plate spline transformation (TPS). Affine module applies a transformation to rectify the scale, translation, and shear. TPS applies non-rigid transformation by identifying a set of fiducial points along the upper and the bottom edges of the word region.

Feature Extractor (FE): The transformed image is forwarded to a convolutional neural network (CNN) followed by a map-to-sequence operation to extract feature maps. This visual feature sequence v has a corresponding distinguishable receptive field along the horizontal line of the input image. Words are recognized by predicting the characters at every step of the feature sequence v. We study two commonly used CNN architecture styles: VGG [27] and RESNET [15]. VGG-style architecture comprises multiple convolutional layers followed by a few fully connected layers. RESNET-style architecture uses residual connections and eases the training of deep CNNs.

Shi et al. [26] tweak the original VGG architecture so that the generated feature maps have larger width to accommodate for recognition of longer words. Dutta et al. [8] introduce a deeper architecture with residual connections. We study both these architectures in this work. We refer to these as HW−VGG and HW−RESNET.

Sequence Modeling (SM): The computed visual sequence v lacks contextual information, which is necessary to recognize characters across the sequence. Therefore, the feature sequence is forwarded to a stack of recurrent neural networks to capture the contextual information. Bidirectional LSTM (BLSTM) is the preferred block to compute the new feature sequence H as it enables context modeling from both directions. Due to its success [2,7,21,26], we use a 2 layer BLSTM architecture with 256 hidden neurons in each layer as the SM module in our experiments.

Predictive Modeling (PM): This module is responsible for decoding a character sequence from the contextual feature H. To decode and recognize the characters, this block learns the alignment between the feature sequence H and the target character sequence. One of the commonly used methods to achieve this is Connectionist Temporal Classification (CTC) [13]. It works by predicting a character for every frame in the sequence and removing recurring characters and blanks.

Data Augmentation: Training deep networks to learn generalized features is crucial for the task of handwriting recognition. Work done in [7,9,29] shows that data augmentation can improve HTR performance on Latin and Indic datasets. It enables the architecture to learn invariant features for the given task and prevents the networks from overfitting. In this work, we apply affine and elastic transformations. This is done to imitate the natural distortions and variations observed in handwriting data. We also apply brightness and contrast augmentation to learn invariant features for text and background.

Table 3. Results on Bengali and Tamil script in IIIT-INDIC-HW-WORDS dataset for common HTR architectures. TN-FE-SM-PM refers to the specific modules used for text recognition. BLSTM-CTC is used for SM-PM stages in the experiments.

#	TN	FE	CER		Params $\times 10^6$
			Bengali	Tamil	
M1	None	HW–VGG	11.48	7.25	8.35
M2	ATN	HW–VGG	9.41	5.99	8.38
M3	ATN	HW–RESNET	5.47	1.47	16.54
M4	TPS	HW–RESNET	5.22	1.38	17.15

Results: In this section, we evaluate and discuss four architecture combinations in the HTR pipeline. Through this setup, we aim to determine the best model for the Indic HTR task. The model architectures are listed in Table 3. For every upgrade done from M1 to M4, we introduce better alternatives in a single stage of the HTR pipeline. We observe the improvement introduced by a specific module. The best performing architecture is identified from the combinations while considering the trade-off between evaluation metrics and computational complexity. We evaluate the performance of these architectures on two datasets: Tamil and Bengali. The Tamil dataset belongs to the Dravidian linguistic group, and the Bengali dataset belongs to the Indo-Aryan group.

We compare the alternatives in TN and FE stages of the HTR pipeline against character error rate (CER) and total number of parameters associated with the architecture. The results obtained are presented in Table 3. For feature extraction stage, replacing HW–VGG with HW–RESNET architecture(M2 → M3) increases the number of parameters by 2×. However, architecture M3 significantly improves the error rate by at least 4%. As the number of layers are increased, architecture M3 benefits from the network's increased complexity and representation power.

Changing the transformation network from ATN to TPS (M3 → M4) reduces the error rate by minor margin only. We conduct another experiment to understand the effectiveness of TN. For this study, we train two architectures with and without TN for different sizes of training data. We train the architectures with 10K, 30K, 50K, and 82K samples for Bengali script. The remaining stages of the pipeline are chosen as HW–RESNET-BLSTM-CTC. The comparison plot in Fig. 5 shows that the WER for the TPS variant reduces significantly when the training datasets have limited samples. With limited training data, the TN stage is crucial to build robust HTR. The introduced changes in TN and FE stages increases the architecture's complexity resulting in smaller error rates.

With the HTR pipeline as TPS-HW–RESNET-BLSTM-CTC, we train and evaluate its performance on all the scripts in the IIIT-INDIC-HW-WORDS dataset. The samples are augmented randomly with affine, elastic, and color transformations. The model is fine-tuned on the pre-trained weights from the IAM dataset. To this end, we train the architecture on IAM dataset as well and the WER and CER obtained are 13.17 and 5.03 respectively. The obtained performance metrics are reported in Table 4. We report error rates on two sets; the test set and its

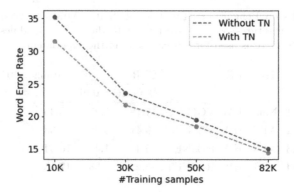

Fig. 5. Word error rate (WER) vs. training size for two architectures: with TPS TN and without TN. The FE, SM, PM stages are fixed as HW–RESNET, BLSTM, CTC. The results are shown for Bengali script in IIIT-INDIC-HW-WORDS dataset.

Table 4. Results on IIIT-INDIC-HW-WORDS dataset. Metrics listed are computed on the test set and out-of-vocabulary (OOV) test set. CER is character error rate and WER is word error rate. Characters refers to number of unique Unicode symbols in each dataset.

Script	Characters	Test		OOV Test	
		CER	WER	CER	WER
Bengali	91	4.85	14.77	3.71	16.65
Gujarati	79	2.39	11.39	5.27	25.8
Gurumukhi	84	3.42	12.78	6.85	23.97
Odia	81	3.00	14.97	4.97	23.40
Kannada	83	1.03	5.90	1.59	9.39
Malayalam	95	1.92	9.85	2.06	10.6
Tamil	72	1.28	7.38	1.88	8.25
Urdu	81	3.67	15.00	9.33	33.23

subset containing only OOV words. From the table, the error rates observed for Kannada script is the lowest and the error rates for Urdu script are the highest. It is interesting to observe similar performances within the Indo-Aryan group and Dravidian group. We also note that the Urdu dataset is more challenging than other datasets due to high error rates, despite having fewer writing styles. Figure 6 shows the predicted text for selective samples from the IIIT-INDIC-HW-WORDS dataset. The model fails to recognize the samples presented in the third column by less than two characters. The recognized text for these samples is incorrect due to confusing character pairs or predicting an extra character at the end of the text string. For the samples presented in the fourth column, the longer length of words causes the recognizer to make mistakes and generate shorter predictions.

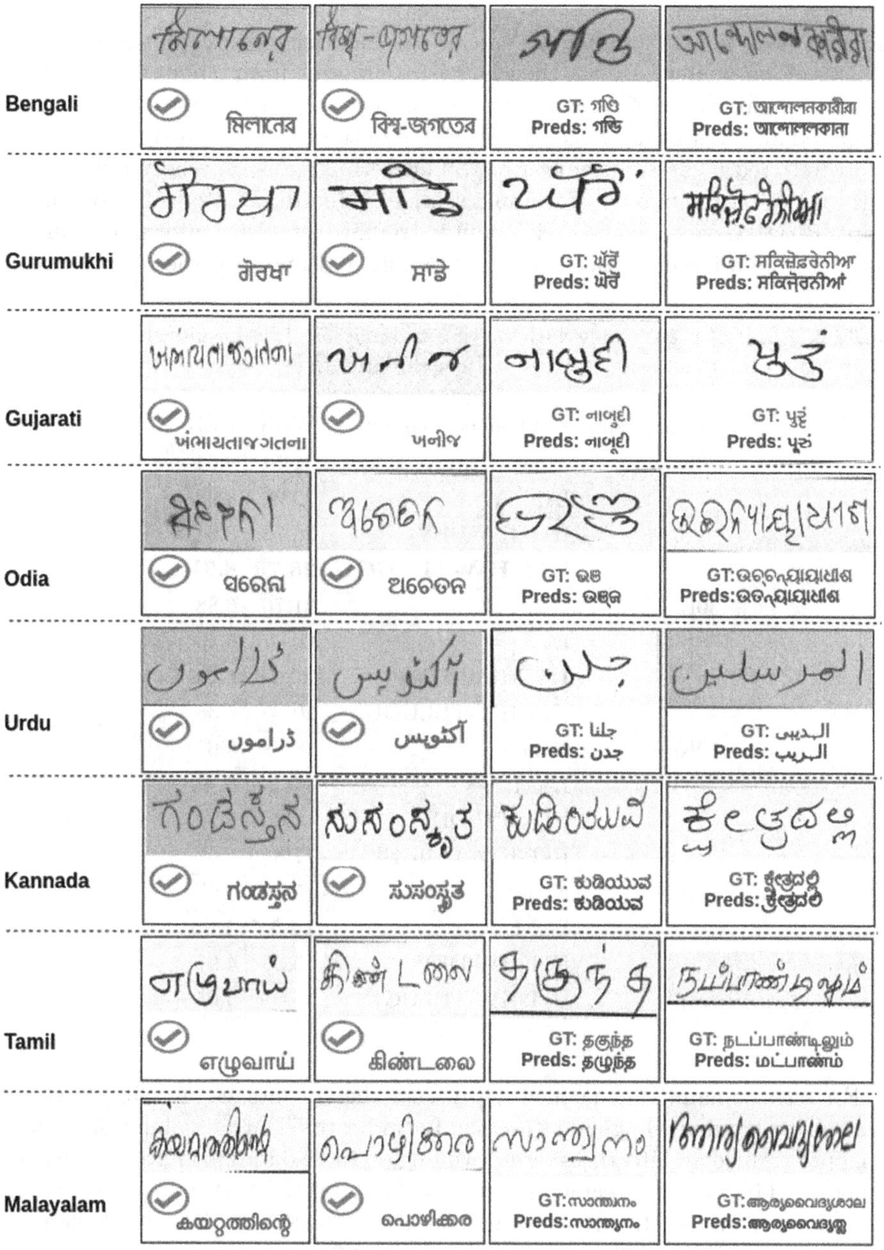

Fig. 6. Qualitative results on the IIIT-INDIC-HW-WORDS dataset. Word images are converted to grayscale and forwarded to the recognizer. Column 1 & 2 present correctly predicted samples. Column 3 & 4 shows incorrect predictions. GT and Preds refer to ground truth and predicted label respectively.

5   Analysis

Training deep neural architectures for a specific task requires optimization of millions of parameters. Due to their large size, training from randomly initialized weights requires a lot of time and hyper-parameter tuning. Pre-trained weights computed for other tasks are generally used to initialize new networks [6,30]. This is also referred to as transfer learning. This technique is especially meaningful in HTR where the number of training samples is limited in a specific domain or script [12,18]. Given a dataset with limited training samples, performing transfer learning from a large dataset eases the training and improves the performance.

Table 5. Effect of pre-training with varying training size. The results shown here are computed on the IIIT-INDIC-HW-WORDS Bengali dataset.

Training size	Pre-trained weights	WER	CER
10K	None	31.54	9.64
	IAM	29.28	8.98
	IIIT-HW-DEV	28.73	8.94
	IIIT-HW-TELUGU	**28.79**	**8.71**
30K	None	21.76	6.88
	IAM	20.37	6.43
	IIIT-HW-DEV	**20.10**	**6.36**
	IIIT-HW-TELUGU	20.35	6.38
50K	None	18.08	5.67
	IAM	**17.23**	**5.49**
	IIIT-HW-DEV	18.30	5.72
	IIIT-HW-TELUGU	17.8	5.67
82K	None	15.34	4.97
	IAM	**14.77**	**4.85**
	IIIT-HW-DEV	15.35	4.97
	IIIT-HW-TELUGU	15.0	5.04

For HTR domain, Bluche et al. [4] discuss the advantages of training recognizers using big multi-lingual datasets in Latin scripts. Dutta et al. [9] show that using pre-trained architectures for fine tuning on Indic datasets improves the performance. They utilize weight parameters learnt on the IAM dataset to reduce the error rate on the IIIT-HW-DEV and the IIIT-HW-TELUGU datasets. Whereas, recent research [16] points out a network trained from scratch for long intervals is not worse than a network finetuned on pre-trained weights for detection and segmentation task. In this study, we explore whether pre-training is useful within the context of handwritten data across multiple scripts. We use three scripts to conduct these experiments: English, Devanagari and Telugu. We also explore the utility of pretrained architectures with varying number of training samples.

In this study, we explore the training architectures using randomly initialized weights, and pre-trained weights from IAM, IIIT-HW-DEV and IIIT-HW-TELUGU datasets. We also explore these pre-training strategies for varying sizes of training data. We present the results on the Bengali data in the IIIT-INDIC-HW-WORDS dataset. The networks are trained for same number of iterations using Adadelta optimization method. The final M4 architecture is trained with data augmentation in this experiment and the results obtained are presented in Table 5. The benefit of using pre-trained architectures for smaller training datasets is notable and gives an improvement of 2.75% in word error rate (WER). As the available training data increases, the reduction in error rates is not apparent. We observe that a randomly initialized network is not necessarily worse at recognition than the network using pre-trained weights for initialization.

Through the above experiment, we also conclude that language similarity is not a major contributing factor to performance improvement for HTR task. The datasets used in the study are written in Latin, Devanagari, and Telugu script. The corresponding datasets for these scripts are IAM, IIIT-HW-DEV, IIIT-HW-TELUGU. Bengali and Devanagari scripts share similarities and both the scripts belong to the Indo-Aryan group. However, results show that pre-training from similar language outperforms other pre-training techniques only for one of the cases and also by very small margin. Interestingly, the pretrained weights from the IAM dataset provide best results as the training data increases to 82K samples.

From this brief study, we conclude that pre-training is especially meaningful for handwritten tasks when the available training data is limited. Therefore, pre-trained architectures can be utilized to build recognizer for historical data or regional scripts. Transcriptions for a few samples are sufficient to fine-tune the recognizers for such target tasks. We also observe that pretraining from similar scripts does not supplement the training process for Indic scripts.

6 Summary

In this work, we extend our earlier efforts in creating Devanagari [8] and Telugu [9] datasets. We introduce a collective handwritten dataset for 8 new Indian scripts: Bengali, Gujarati, Gurumukhi, Kannada, Malayalam, Odia, Tamil, and Urdu. We hope the scale and the diversity of our dataset in all 10 prominent Indic scripts will encourage research on enhancing and building robust HTRs for Indian languages. It is essential to continue improving Indic datasets to enable Indic HTR development, and therefore, future work will include enriching the dataset with more writers and natural variations. Another important direction is to create handwritten data at line, paragraph and page level.

For the introduced dataset, we establish a high baseline on the 8 scripts present in the IIIT-INDIC-HW-WORDS dataset and discuss the effectiveness of the HTR modules. We also conduct a brief study to explore the utility of pre-training recognizers on other scripts.

Acknowledgments. The authors would like to acknowledge the funding support received through IMPRINT project, Govt. of India to accomplish this project. We would like to thank all the volunteers who contributed to this dataset. We would also like to express our gratitude to Aradhana, Mahender, Rohitha and annotation team for their support in data collection. We also thank the reviewers for their detailed comments.

References

1. Alaei, A., Pal, U., Nagabhushan, P.: Dataset and ground truth for handwritten text in four different scripts. IJPRAI **26**, 1253001 (2012)
2. Baek, J., et al.: What is wrong with scene text recognition model comparisons? Dataset and model analysis. In: ICCV (2019)
3. Bhowmik, S., Malakar, S., Sarkar, R., Basu, S., Kundu, M., Nasipuri, M.: Off-line Bangla handwritten word recognition: a holistic approach. Neural Comput. Appl. **31**, 5783–5798 (2019)
4. Bluche, T., Messina, R.: Gated convolutional recurrent neural networks for multi-lingual handwriting recognition. In: ICDAR (2017)
5. Bluche, T., Ney, H., Kermorvant, C.: Feature extraction with convolutional neural networks for handwritten word recognition. In: ICDAR (2013)
6. Donahue, J., et al.: DeCAF: a deep convolutional activation feature for generic visual recognition. In: ICML (2014)
7. Dutta, K., Krishnan, P., Mathew, M., Jawahar, C.V.: Improving CNN-RNN hybrid networks for handwriting recognition. In: ICFHR (2018)
8. Dutta, K., Krishnan, P., Mathew, M., Jawahar, C.: Offline handwriting recognition on devanagari using a new benchmark dataset. In: DAS (2018)
9. Dutta, K., Krishnan, P., Mathew, M., Jawahar, C.: Towards spotting and recognition of handwritten words in indic scripts. In: ICFHR (2018)
10. Fischer, A., Keller, A., Frinken, V., Bunke, H.: Lexicon-free handwritten word spotting using character HMMs. PR **33**, 934–942 (2012)
11. Goldhahn, D., Eckart, T., Quasthoff, U.: Building large monolingual dictionaries at the leipzig corpora collection: from 100 to 200 languages. In: LREC (2012)
12. Granet, A., Morin, E., Mouchère, H., Quiniou, S., Viard-Gaudin, C.: Transfer learning for handwriting recognition on historical documents. In: ICPRAM (2018)
13. Graves, A., Fernández, S., Gomez, F., Schmidhuber, J.: Connectionist temporal classification: labelling unsegmented sequence data with recurrent neural networks. In: ICML (2006)
14. Grosicki, E., El-Abed, H.: ICDAR 2011-French handwriting recognition competition. In: ICDAR (2011)
15. He, K., Zhang, X., Ren, S., Sun, J.: Deep residual learning for image recognition. In: CVPR (2016)
16. He, K., Girshick, R., Dollár, P.: Rethinking ImageNet pre-training. In: CVPR (2019)
17. Jaderberg, M., Simonyan, K., Zisserman, A., Kavukcuoglu, K.: Spatial transformer networks. In: NIPS (2015)
18. Jaramillo, J.C.A., Murillo-Fuentes, J.J., Olmos, P.M.: Boosting handwriting text recognition in small databases with transfer learning. In: ICFHR (2018)
19. Jayadevan, R., Kolhe, S.R., Patil, P.M., Pal, U.: Database development and recognition of handwritten devanagari legal amount words. In: ICDAR (2011)

20. Marti, U.V., Bunke, H.: The IAM-database: an English sentence database for offline handwriting recognition. IJDAR **5**, 39–46 (2002)
21. Puigcerver, J.: Are multidimensional recurrent layers really necessary for handwritten text recognition? In: ICDAR (2017)
22. Roy, P.P., Bhunia, A.K., Das, A., Dey, P., Pal, U.: HMM-based Indic handwritten word recognition using zone segmentation. Pattern Recognit. **60**, 1057–1075 (2016)
23. Sagheer, M.W., He, C.L., Nobile, N., Suen, C.Y.: A new large Urdu database for off-line handwriting recognition. In: Foggia, P., Sansone, C., Vento, M. (eds.) ICIAP 2009. LNCS, vol. 5716, pp. 538–546. Springer, Heidelberg (2009). https://doi.org/10.1007/978-3-642-04146-4_58
24. Sánchez, J.A., Romero, V., Toselli, A.H., Vidal, E.: ICFHR2014 competition on handwritten text recognition on transcriptorium datasets (HTRtS). In: ICFHR (2014)
25. Sanchez, J.A., Romero, V., Toselli, A.H., Villegas, M., Vidal, E.: ICDAR2017 competition on handwritten text recognition on the READ dataset. In: ICDAR (2017)
26. Shi, B., Bai, X., Yao, C.: An end-to-end trainable neural network for image-based sequence recognition and its application to scene text recognition. PAMI **39**, 2298–2304 (2016)
27. Simonyan, K., Zisserman, A.: Very deep convolutional networks for large-scale image recognition. In: ICLR (2015)
28. Thadchanamoorthy, S., Kodikara, N., Premaretne, H., Pal, U., Kimura, F.: Tamil handwritten city name database development and recognition for postal automation. In: ICDAR (2013)
29. Wigington, C., Stewart, S., Davis, B., Barrett, B., Price, B., Cohen, S.: Data augmentation for recognition of handwritten words and lines using a CNN-LSTM network. In: ICDAR (2017)
30. Yosinski, J., Clune, J., Bengio, Y., Lipson, H.: How transferable are features in deep neural networks? In: NIPS (2014)
31. Yousef, M., Bishop, T.E.: OrigamiNet: weakly-supervised, segmentation-free, one-step, full page text recognition by learning to unfold. In: CVPR (2020)

Historical Document Analysis

AT-ST: Self-training Adaptation Strategy for OCR in Domains with Limited Transcriptions

Martin Kišš[(✉)] [iD], Karel Beneš [iD], and Michal Hradiš [iD]

Brno University of Technology, Brno, Czechia
{ikiss,ibenes,hradis}@fit.vutbr.cz

Abstract. This paper addresses text recognition for domains with limited manual annotations by a simple self-training strategy. Our approach should reduce human annotation effort when target domain data is plentiful, such as when transcribing a collection of single person's correspondence or a large manuscript. We propose to train a seed system on large scale data from related domains mixed with available annotated data from the target domain. The seed system transcribes the unannotated data from the target domain which is then used to train a better system. We study several confidence measures and eventually decide to use the posterior probability of a transcription for data selection. Additionally, we propose to augment the data using an aggressive masking scheme. By self-training, we achieve up to 55% reduction in character error rate for handwritten data and up to 38% on printed data. The masking augmentation itself reduces the error rate by about 10% and its effect is better pronounced in case of difficult handwritten data.

Keywords: Self-training · Text recognition · Language model · Unlabelled data · Confidence measures · Data augmentation

1 Introduction

When transcribing documents from a specific *target domain*, e.g. correspondence of a group of people or historical issues of some periodical, the accuracy of pre-trained optical character recognition (OCR) systems is often insufficient and the automatic transcriptions need to be manually corrected. The manual effort should naturally be minimized. One possibility is to improve the OCR system through adapting it using the manually corrected transcriptions. In this paper, we propose to improve the OCR system more efficiently by utilizing the whole target domain, not only the manually corrected part.

We explore the scenario where a small part of the target domain is manually transcribed (annotated) and a large annotated dataset from *related domains* is available (e.g. large-scale public datasets, such as READ [15] or IMPACT [14]). Our pipeline starts by training a *seed* system on this data. The seed system transcribes all unannotated text lines in the target domain, and the most confident transcriptions are then used as a ground truth to retrain and adapt the system.

© Springer Nature Switzerland AG 2021
J. Lladós et al. (Eds.): ICDAR 2021, LNCS 12824, pp. 463–477, 2021.
https://doi.org/10.1007/978-3-030-86337-1_31

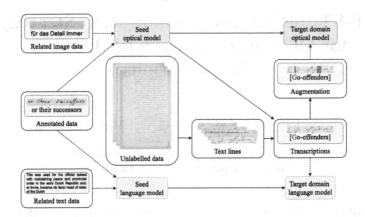

Fig. 1. Proposed OCR system adaptation pipeline to a target domain.

While our approach does not depend on the actual implementation of the OCR system and should be applicable to a broad range of methods, we limit our experimentation to a combination of a CTC-based [9] optical model with an explicit language model. We show that:

1. The posterior probability of a transcription produced by prefix search decoder on top of a CTC optical model is a well-behaved OCR confidence measure that allows to select lines with low error rate.
2. A masking data augmentation scheme helps the optical model to utilize the machine-annotated data.
3. Even when well pre-trained on a related domain, both the optical and the language model benefit significantly from machine annotated data from the target domain and the improvements are consistent over a range of overall error-rates, languages, and scripts.

2   Related Work

Semi-supervised learning approaches in OCR are based on *Pseudo-Labeling* [13] or *Self-Training* [3,17]. They differ in that pseudo-labeling continues the training with the seed system, whereas in self-training, a new model is trained on the extended data [21]. However, these terms are often mixed at authors' liberty.

Automatic transcriptions naturally contain errors, but consistent data is desirable for training recognition models. Thus, the goal is to either correct these errors or select only text lines that do not contain these errors. If multiple models are trained on annotated data, it is possible to combine the outputs of these models to obtain a more reliable transcript [7,13]. If there is only one trained model, transcriptions can be checked against a language model [11], a lexicon [17], or page-level annotations [11] if available.

Semi-supervised training of language models is not explored very well in literature. It has been shown that small gains can indeed be achieved with count based n-gram LMs trained on transcripts produced by a speech recognition system [6]. Machine annotations are a bit more common in discriminative language modeling, where the LM is trained to distinguish a hypothesis with low error rate among others. This work typically focuses on getting a diverse enough training data for the training [2]. While recent LM literature focuses on training universal models on vast amounts of data, with the hope that such LM can adapt to anything [1], a careful selection of sentence pairs has been shown to provide improvement for a count based machine translation system [8].

In a recent paper, an approach similar to ours was proposed [3]. The authors propose an iterative self-training scheme in combination with extensive fine-tuning to improve a CTC-based OCR. However, their experiments are limited to printed text, they recognize individual words and they achieve negligible improvements from the self-training itself. Also, no attention is paid to adaptation of language models.

Data augmentation is a traditional approach to reduce overfitting of neural networks. In computer vision, the following are commonly used [3,5,20]: affine transformations, change of colors, geometric distortions, adding noise, etc. A technique similar to our proposed masking is *cutout* [4], which replaces content of random rectangle areas in the image by gray color. To the best of our knowledge, it has never been used for text recognition.

3 AT-sT: Adapting Transcription System by Self-Training

Our goal is to obtain a well performing OCR system for the *taget domain* with little human effort and when there is the following available: (1) *Related domain data*, which loosely matches the style, language or overall condition of the target domain; (2) *annotated target domain data*, which has human annotations available. Together, these two provide the *seed data*. The rest of the target domain constitutes (3) *unannotated target domain data*, which we want to utilize without additional human input. Eventually, we measure the performance of our systems by Character Error Rate (CER) on a held out *validation* and *test* portion of the annotated target domain data.

As summarized in Fig. 1, our pipeline progresses in four steps: (1) A seed OCR system is trained on the seed data. (2) This OCR is used to process all unannotated lines from the target domain. (3) For each of these lines, confidence score is computed using a suitable OCR confidence measure. Then, a best-scoring portion of a defined size is taken as the *machine annotated* (MA) data. (4) MA data is then merged with the seed data and a new OCR is trained on it. In general, steps 2–4 can be repeated, which might lead to further adaptation of the OCR to the target domain.

3.1 Implementing OCR System

Our pipeline could in principle be implemented with any type of optical and language model. In our implementation, the optical model is a neural network based on the CRNN architecture [16] trained with the CTC loss function [9]. This means that the optical model transforms a 2-D image of a text line ℓ into a series of *frames* $\mathbf{f}_1, \ldots, \mathbf{f}_T$, where every frame $\mathbf{f}_t$ is a vector of probabilities over an alphabet including a special *blank* symbol representing empty output from the frame.

It is viable to obtain the transcription by simply taking the most probable character from each frame (greedy decoding). More accurate results can be obtained by *prefix search decoding* [9], which accommodates the fact that the probability of a character may be spread over several frames. The prefix search decoding keeps a pool of the most probable partial transcripts (prefixes) and updates their probabilities frame by frame. All possible single-character prefix extensions are considered at every frame and only the most probable prefixes are retained.

This approach can be readily extended by introduction of a language model that estimates the probability of a sequence of characters. The total score of prefix a is then computed as:

$$\log S(a) = \log S_O(a) + \alpha \log S_L(a) + \beta |a| \tag{1}$$

where $S_O(a)$ is the score of a as given by the optical model, $S_L(a)$ is the language model probability of a, α is the weight of the language model, $|a|$ is the length of a in characters, and β is the empirically tuned insertion bonus. Since we use an autoregressive language model that estimates probability of every character given the previous ones, we introduce the additional terms on-the-fly during the search. A total of K best scoring prefixes is kept during decoding.

3.2   OCR Confidence Measures

To identify the most accurately transcribed lines, we propose several confidence measures M. The output of a measure M is a score in range $[0, 1]$. The score serves as the predictor of error rate of the greedy hypothesis h_g.

CTC loss is the probability $P(h_g|\ell)$ as defined by the CTC training criterion, i.e. the posterior probability of the hypothesis h_g marginalized over all possible alignments of its letters to the frames. We normalize this probability by the length of the hypothesis in characters.

Transcription posterior probability is estimated using prefix decoding algorithm using only the optical scores, i.e. with $\alpha = \beta = 0$. For each[1] line ℓ, the decoding produces a set of hypotheses $\mathcal{H} = \{h_1, \ldots, h_K\}$ together with their associated logarithmic unnormalized probabilities $C(h_i)$. We normalize these scores to obtain posterior probabilities of those hypotheses:

$$M_{\text{posterior}}(\ell) = P(h_g|\ell) = \frac{exp(C(h_g))}{\sum_{i=1} exp(C(h_i))} \tag{2}$$

---

[1] Except lines with only one frame, where there may be fewer prefixes considered.

Fig. 2. Example of the proposed masking augmentation

We assign $P(h|\ell) = 0$ to hypotheses outside of $\mathcal{H}$.

Additionally, we propose four measures that focus on the maximal probabilities in each frame. This should reflect the fact that greedy decoding – the source of the hypothesis – also begins by taking these maxima. Denoting these maxima as m_t, we define:

- *probs mean* as the average of m_t for all $t \in \{1, \ldots, T\}$.
- *char probs mean* as the average of m_t for those t where blank is not the most probable output.
- *Inliers Rate* is based on the distribution of the maxima from an examined data set. This is modelled by a maximum likelihood Gaussian. Then, the confidence score of each text line is computed as the ratio of t such, that m_t lies within 2σ.
- *Worst best* finds the Viterbi alignment of the hypothesis to the output probabilities. Then for each character in the hypothesis, it takes a corresponding frame with the highest m_t. Finally, the confidence score is the minimum of the values.

3.3   Masking Augmentation

To provide a more challenging learning environment, we propose to mask parts of the input image out. Examples are shown in Fig. 2. Since the transcription of the image remains unchanged, we believe this technique might especially improve the implicit language model inside the optical network.

The masking augmentation replaces random parts of the input image by noise with uniform distribution. The number of masked regions is sampled from binomial distribution. The masked regions span full height of a text line and their width is sampled from another uniform distribution.

We chose the random noise, instead of for example constant color, to clearly mark the masked regions. We believe that conveying this information to the model should avoid ambiguity between empty and masked regions, reduce the complexity of the training task, and consequently improve convergence and the final accuracy.

4   Datasets

We performed experiments on printed and handwritten data. The description of individual datasets is in the following paragraphs and the sizes of the datasets

Table 1. Size of datasets in lines.

	Handwritten datasets			Printed datasets		
	Training	Validation	Test	Training	Validation	Test
Related domain	189 805	–	700	1 280 000	–	12 000
Target d. annotated	9 198	1 415	860	32 860	6 546	1 048
Target d. unannotated	1 141 566	–	–	2 673 626	–	–

are summarized in Table 1 and text line examples from each dataset are depicted in Fig. 3.

As we used several datasets in the experiments, we needed to ensure consistency in terms of baseline positioning and text line heights. We used text line detection model to detect text lines in all datasets and mapped the produced text lines to the original transcriptions based on their content and location if available. The detection model is based on ParseNet architecture [10] and was trained on printed and handwritten documents from various sources.

Handwritten Datasets. For experiments with handwritten data, we used the ICDAR 2017 READ Dataset [15], the ICFHR 2014 Bentham Dataset [19], and additional unannotated pages (*Unannotated Bentham Dataset*) obtained from the Bentham Project[2] as the related domain, target domain annotated and target domain unannotated datasets respectively. The READ Dataset contains pages written in German, Italian, and French, while the Bentham Dataset and the Unannotated Bentham Dataset consist of pages written in early 19th-century English. In contrast to the other datasets with plentiful writer, majority of the dataset has been written by J. Bentham himself or his secretarial staff.

Printed Datasets. For experiments with printed data, we used IMPACT Dataset [14] as the related domain data. As target domain, we used historical printings of Czech newspapers[3]. This data is generally of relatively low quality as it was scanned from microfilms. Approximately 2000 pages of it were partially transcribed by volunteers, these lines serve as the annotated target domain data. In addition, we checked the test set and removed text lines with ambiguous transcriptions. The rest of the pages was used as the target domain unannotated data. From the IMPACT dataset, we randomly sampled 1.3M text lines. The most common languages represented in the IMPACT dataset are Spanish, English, and Dutch, while the Czech newspapers dataset contains pages written in Czech.

Related Domain Data for Language Models

It is reasonable to expect availability of contemporary text data for any language with writing system. To reflect that in our experiments, we provide pre-training

---

[2] https://www.ucl.ac.uk/bentham-project.
[3] https://www.digitalniknihovna.cz/mzk.

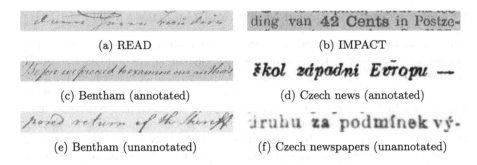

(a) READ

(b) IMPACT

(c) Bentham (annotated)

(d) Czech news (annotated)

(e) Bentham (unannotated)

(f) Czech newspapers (unannotated)

Fig. 3. Example text lines by datasets.

corpora to our language models. This way, the overall structure of data for the optical model and the language model is the same. We do not attempt to use the training transcriptions from the related domain OCR data, as those come in various languages and are orders of magnitude smaller than available text data.

For English, we use the raw version of Wikitext-103 [12], which contains ca. 530M characters (103 M words). Since the dataset is originally prepared for word-level language modeling based on whitespace-delimited tokenization, we adjusted the spacing around punctuation to follow the standard rules.

For Czech, we use an in-house corpus consisting mainly of news from internet publishing houses and wikipedia articles. The total size is approx. 2.2 B characters (320 M words).

5   Experiments

Experimentally, we demonstrate the effectiveness of both the proposed self-training scheme and the masking augmentation. The first step of our self-training pipeline is training seed optical models on human annotated data. Using their outputs, we explore the predictive power of the proposed confidence measures and estimate the error rate of the untranscribed data in Sect. 5.2. Then in Sect. 5.3, we show the effect of adding this machine annotated data into the training process of both the optical model and the language model. As all the above mentioned optical models are trained with our proposed masking augmentation, we finally explore its effect separately in Sect. 5.4.

To simulate the effect of investing more or less effort into manually transcribing the target domain, we consider two conditions for each dataset: *Big*, which is equal to using all of the annotated target domain data, and *small*, which is simulated by randomly taking 10% of it[4]. When training the optical model in the big setup, we give extra weight to the text lines from the target domain so that the model adapts to it more. As a result, we have four different setups of seed data—*handwritten small, handwritten big, printed small*, and *printed big*.

---

[4] Specifically, we do the subsampling on the level of pages, to emulate the possible effect of reduced variability in data.

5.1   Specification of Optical Models and Language Models

The optical model is a neural network based on the CRNN architecture [16], where convolutional blocks are followed by recurrent layers and the network is optimized using the CTC loss [9]. To speed up the training process, we initialize the convolutional layers from a pretrained VGG16[5]. The recurrent part consists of six parallel BLSTM layers processing differently downsampled representations of the input. Their outputs are upsampled back to the original time resolution, summed up, and processed by an additional BLSTM layer to produce the final output. In all experiments, the input of the optical model is an image $W \times 40$, where W is its width and the height is fixed to 40 pixels, and the output is matrix of dimensions $W/4 \times |V|$, where $|V|$ is the size of the character vocabulary including the blank symbol. The exact definition of the architecture is public[6].

Each optical model was optimized using Adam optimizer for 250k iterations. The initial learning rate was set to 2×10^{-4} and was halved after 150k and 200k iterations. Except for the experiments in Sect. 5.4, the masking augmentation was configured as follows: The number of masked regions was sampled from a binomial distribution with the number of trials equal to the width of the input image in pixels, and the success probability equal to 5×10^{-3}. The width of the masked region was sampled uniformly from interval [5, 40], making the largest possible region a square.

All language models (LMs) were implemented as LSTMs [18], with 2 layers of 1500 units. Input characters are encoded into embeddings of length 50. We optimize the LMs using plain SGD without momentum, with initial learning rate 2.0, halving it when validation perplexity does not improve. For validation, we always use the reference transcriptions of the respective validation data. We train all our LMs using BrnoLM toolkit[7].

When doing the prefix decoding, we tune the LM weight α in range 0.0–1.5 and the maximal number of active prefixes K up to 16. To keep the hyperparameter search feasible, the character insertion bonus β is kept at 1.0. This value is a result of preliminary experiments: lower values lead invariably to increased number of deletions, whereas higher values resulted in a trade-off between insertions and deletions, and a slow but steady increase in total error rate.

5.2   Predictive Power of Confidence Measures

Having the seed optical models, we use them to evaluate the confidence measures proposed in Sect. 3.2. We do so by sorting the validation lines by their confidence scores and then calculating the CER on the most confident subsets of increasing size. The resulting progress of the CER is visualized in Fig. 4. To assess the quality of the measures quantitatively, we calculate the area-under-curve (AUC) for each confidence measure. We report the AUCs in Table 2. The *Transcription*

---

[5] From PyTorch module `torchvision.models.vgg16`.
[6] https://github.com/DCGM/pero-ocr.
[7] https://github.com/BUTSpeechFIT/BrnoLM.

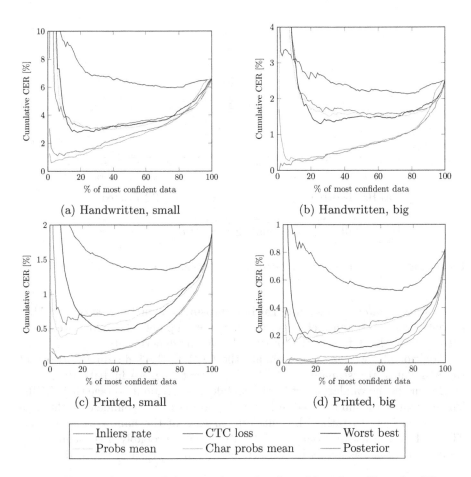

(a) Handwritten, small

(b) Handwritten, big

(c) Printed, small

(d) Printed, big

| —— Inliers rate | —— CTC loss | —— Worst best |
| —— Probs mean | —— Char probs mean | —— Posterior |

Fig. 4. CER as a function of the size of considered confident data. Note the different scale of the error rate.

Table 2. AUCs for individual confidence measures.

Confidence measure	Handwritten		Printed	
	Small	Big	Small	Big
CTC loss	7.082	2.453	1.557	0.656
Posterior	2.798	**0.754**	**0.429**	**0.109**
Probs mean	4.218	2.035	0.747	0.283
Char probs mean	**2.657**	0.844	0.431	0.131
Inliers rate	4.053	2.277	0.896	0.305
Worst best	5.443	2.428	1.189	0.260

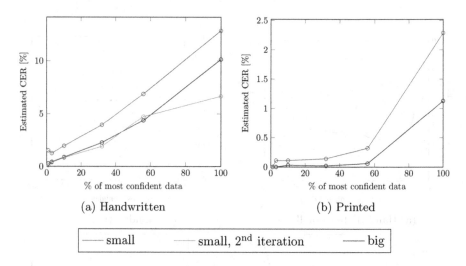

(a) Handwritten   (b) Printed

———— small   ———— small, 2nd iteration   ———— big

Fig. 5. CER of portions of unannotated data, as estimated by the confidence measure.

posterior probability comes out as the best measure, having the best overall AUC as well as very consistent behaviour for all operating points. Therefore, we picked it to perform the data selection in the self-training experiments.

Additionally, we use it to estimate the CER of the individual portions of the MA data. To estimate the CER of any text line, we find its 10 nearest neighbours in the validation set by the confidence score, and average their CERs. The estimates are summarized in Fig. 5. According to the confidences, the data varies from easy to rather difficult, esp. in case of the handwritten dataset. The slightly inconsistent output of 2nd iteration is probably a consequence of confirmation bias, as model has already been trained on parts of this data.

5.3 Effect of Introducing Machine-Annotated Data

The core of our contribution is in demonstrating the improvements from adapting the text recognition system to the unannotated part of the target domain. We report how much the OCR improves when additionally presented 1, 3, 10, 32, 56, and 100% of the most confident MA data. We perform these experiments in two separate branches.

The first branch focuses on the optical model (OM) only, where output is obtained using greedy decoding. In this case, the MA data is mixed with the seed data and a new OM is trained from scratch. Results are shown in Figs. 6a and 6c.

The second branch keeps the seed OM fixed and studies the effect of adding the MA data to the language model. Training LMs is done in stages: (1) The LM is pretrained on the related domain data, (2) it is further trained on the given amount of MA data, and (3) it is fine-tuned to the target domain annotations.

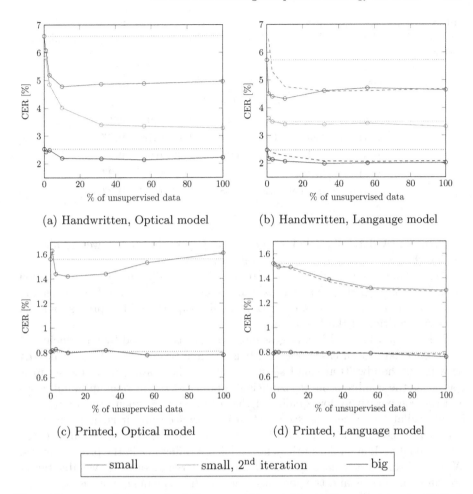

(a) Handwritten, Optical model   (b) Handwritten, Langauge model

(c) Printed, Optical model   (d) Printed, Language model

| —— small | —— small, 2nd iteration | —— big |

Fig. 6. System performance in different experimental branches and datasets. Dotted line shows the corresponding baseline. Dashed line in the LM branch denotes decoding with LM trained on target domain only.

The LM is introduced into prefix search decoding as shown in Eq. (1). Results are shown in Figs. 6a and 6d.

Additionally, we demonstrate the possibility to obtain further gains from a second iteration on the handwritten small setup, which is the most challenging one. In this case, we take the fusion of the best performing OM with best performing LM as a new seed system to transcribe the unannotated data. This setup is referred to as *small, 2$_{nd}$ iteration*.

From the validation results in Fig. 6, we see that, with the exception of the printed big setup, adding the MA data provides significant improvement to the performance of the system and this improvement is smooth with respect to the

Table 3. Final comparison of results on test datasets, reported as CER [%]. Optimal size of MA data as well as other hyperparameters are selected as per validation performance.

OM	LM	Handwritten			Printed	
		Small	Small-2	Big	Small	Big
Seed	None	6.43	4.41	2.58	1.03	0.29
Optimal	None	4.41	2.92	2.07	0.78	**0.27**
Seed	Seed	5.34	3.27	2.53	0.99	**0.27**
Seed	Optimal	4.17	3.21	2.02	0.76	**0.27**
Optimal	Optimal	**3.27**	**2.88**	**1.94**	**0.64**	**0.27**

amount of MA data. We assume that in the printed big setup, we are already hitting the level of errors in the annotations.

When adapting the OMs, it is more beneficial to only take a smaller portion of the MA data in the small setups. On the other hand, in the big setups and the second iteration of the handwritten small setup, the OM improves more on the larger portion of the MA data.

When adapting the LMs, largest improvement is achieved by introducing all of the MA data in all cases but one: It is optimal to only take 10% of the MA data in the handwritten small setup (Fig. 6b). The reason, why this setup is special, is the combination of insufficient amount of target domain annotations and the comparatively low quality of the seed system – when pretraining data is not available or when the seed system is more accurate, it is again the most beneficial to consider all of the available MA data. In contrast to the handwritten setups, related domain data brings no significant difference in the printed setups. We expect this to be a consequence of relatively large amount of available target domain annotations and the good accuracy of the machine annotations.

Comparing LM and OM, we hypothesize the difference in behaviour in the small setups comes from the fact that while LM is a generative model which is expected to produce a less sharp distribution in the face of noisy data, the discriminative OM is prone to learning a wrong input-output mapping. This is in line with the observation that the OM benefits much more from the second iteration, where the error rate of the machine annotations is lower. Additionally, it probably benefits from the different knowledge brought in by the LM, which may break the limit of confirmation bias. On the other hand, the LM is simply exposed to less noisy data, but not to a new source of knowledge.

Finally, we take the best performing amounts of MA data and compare all the relevant combinations of models on the test data, as summarized in Table 3. In line with the findings on the validation set, the fusion of the best OM with the best LM performs the best in all setups, providing massive gains over the baseline.

Table 4. Augmentations comparison in CER [%]

Augmentation	READ	IMPACT
None	5.28	0.66
Masking	3.91	0.65
Traditional	3.69	**0.58**
Both	**3.20**	**0.58**

5.4 Masking Augmentation Effect

Besides the experiments with unannotated data, we also explored the effect of the masking augmentation. The aim of the first experiment is to compare the effect of the masking augmentation with the traditional augmentations (affine transformations, change of colors, geometric distortions, etc.) We conducted this experiment on the related domain only, results are in Table 4. The results show that the masking does improve the recognition accuracy, however the impact is smaller than that of a combination of traditional approaches and is not significant in the case of the easy printed data.

The second experiment aims to determine the effect of distribution of the masking across the text line. We did so by training three models with different settings of the success probability and the maximal width of the masked region. In order to isolate the effect of frequency of masking, we designed the settings so that the expected volume of the masking remains the same. This experiment was performed on the combination of the related domain, the annotated target domain, and 32 % of MA data in the first iteration of our self-training pipeline. The settings and the results are presented in Table 5. Overall, the impact of the masking augmentation is very robust to the frequency of the masking.

Table 5. Results of different masking augmentation settings in CER [%]. The success probability and the width of the region are denoted as p and W respectively. Base-probability refers to the setting used in self-training experiments.

	Settings		Handwritten		Printed	
	p	W	Small	Big	Small	Big
Without masking	0	–	5.30	2.27	1.51	0.86
Half-probability	2.5×10^{-3}	$[5, 80]$	4.98	**2.13**	1.49	0.85
Base-probability	5×10^{-3}	$[5, 40]$	**4.87**	2.18	**1.44**	**0.82**
Double-probability	10×10^{-3}	$[5, 20]$	4.89	2.15	1.50	0.83

6   Conclusion

In this paper, we have described and experimentally validated a self-training adaptation strategy for OCR. We carefully examined several confidence measures for data selection. Also, we experimented with image masking as an augmentation scheme.

Our self-training approach led to reduction in character error rate from 6.43 % to 2.88 % and 1.03 % to 0.64 % on handwritten and printed data respectively. As the confidence measure, the well-motivated transcription posterior probability was identified as the best performing one. Finally, we showed that the proposed masking augmentation improves the recognition accuracy by up to 10 %.

Our results open way to efficient utilization of large-scale unlabelled target domain data. This allows to reduce the amount of manual transcription needed while keeping the accuracy of the final system high. As a straight-forward extension of this work, our self-training method can be applied to sequence-to-sequnce models. Regarding the masking augmentation, it would be interesting to specifically assess its impact on learning from the machine annotated data and to test whether it truly pushes the CTC optical model towards learning a better language model. Finally, data selection method considering also novelty of the line could lead to better utilization of the unannotated target domain data.

Acknowledgement. This work has been supported by the Ministry of Culture Czech Republic in NAKI II project PERO (DG18P02OVV055), by the EC's CEF Telecom programme in project OCCAM (2018-EU-IA-0052), and by Czech National Science Foundation (GACR) project "NEUREM3" No. 19-26934X.

References

1. Brown, T., et al.: Language models are few-shot learners. In: Larochelle, H., Ranzato, M., Hadsell, R., Balcan, M.F., Lin, H. (eds.) Advances in Neural Information Processing Systems, vol. 33, pp. 1877–1901. Curran Associates, Inc. (2020)
2. Celebi, A., et al.: Semi-supervised discriminative language modeling for Turkish ASR. In: 1988 International Conference on Acoustics, Speech, and Signal Processing, ICASSP-88, pp. 5025–5028, March 2012
3. Das, D., Jawahar, C.V.: Adapting OCR with limited supervision. In: Bai, X., Karatzas, D., Lopresti, D. (eds.) DAS 2020. LNCS, vol. 12116, pp. 30–44. Springer, Cham (2020). https://doi.org/10.1007/978-3-030-57058-3_3
4. DeVries, T., Taylor, G.W.: Improved regularization of convolutional neural networks with cutout. arXiv:1708.04552 [cs], November 2017.
5. Dutta, K., Krishnan, P., Mathew, M., Jawahar, C.: Improving CNN-RNN hybrid networks for handwriting recognition. In: 2018 16th International Conference on Frontiers in Handwriting Recognition (ICFHR), pp. 80–85. IEEE, Niagara Falls, August 2018
6. Egorova, E., Serrano, J.L.: Semi-supervised training of language model on Spanish conversational telephone speech data. In: SLTU-2016 5th Workshop on Spoken Language Technologies for Under-Resourced Languages, 09–12 May 2016, Yogyakarta, Indonesia, vol. 81, pp. 114–120 (2016)

7. Frinken, V., Bunke, H.: Evaluating retraining rules for semi-supervised learning in neural network based cursive word recognition. In: 2009 10th International Conference on Document Analysis and Recognition, pp. 31–35 (2009)
8. Gascó, G., et al.: Does more data always yield better translations? In: Proceedings of the 13th Conference of the European Chapter of the Association for Computational Linguistics, pp. 152–161. Association for Computational Linguistics, Avignon, April 2012
9. Graves, A., Fernández, S., Gomez, F., Schmidhuber, J.: Connectionist temporal classification: labelling unsegmented sequence data with recurrent neural networks. In: Proceedings of the 23rd International Conference on Machine Learning, ICML 2006, New York, NY, USA, pp. 369–376 (2006)
10. Kodym, O., Hradiš, M.: Page layout analysis system for unconstrained historic documents (2021)
11. Leifert, G., Labahn, R., Sánchez, J.A.: Two semi-supervised training approaches for automated text recognition. In: 2020 17th International Conference on Frontiers in Handwriting Recognition (ICFHR), pp. 145–150, September 2020
12. Merity, S., Xiong, C., Bradbury, J., Socher, R.: Pointer sentinel mixture models. In: 5th International Conference on Learning Representations, ICLR 2017, Toulon, France, 24–26 April 2017, Conference Track Proceedings. OpenReview.net (2017)
13. Nagai, A.: Recognizing Japanese historical cursive with pseudo-labeling-aided CRNN as an application of semi-supervised learning to sequence labeling. In: 2020 17th International Conference on Frontiers in Handwriting Recognition (ICFHR), pp. 97–102, September 2020
14. Papadopoulos, C., Pletschacher, S., Clausner, C., Antonacopoulos, A.: The IMPACT dataset of historical document images. In: Proceedings of the 2nd International Workshop on Historical Document Imaging and Processing, HIP 2013, pp. 123–130. Association for Computing Machinery, New York, August 2013
15. Sanchez, J.A., Romero, V., Toselli, A.H., Villegas, M., Vidal, E.: ICDAR2017 competition on handwritten text recognition on the READ dataset. In: 2017 14th IAPR International Conference on Document Analysis and Recognition (ICDAR), pp. 1383–1388. IEEE, Kyoto, November 2017
16. Shi, B., et al.: An end-to-end trainable neural network for image-based sequence recognition and its application to scene text recognition. IEEE Trans. Pattern Anal. Mach. Intell. **39**(11), 2298–2304 (2017)
17. Stuner, B., Chatelain, C., Paquet, T.: Self-training of BLSTM with lexicon verification for handwriting recognition. In: 2017 14th IAPR International Conference on Document Analysis and Recognition (ICDAR), vol. 01, pp. 633–638, November 2017. iSSN 2379–2140
18. Sundermeyer, M., Schlüter, R., Ney, H.: LSTM neural networks for language modeling. In: INTERSPEECH (2012)
19. Sánchez, J.A., Romero, V., Toselli, A.H., Vidal, E.: ICFHR2014 competition on handwritten text recognition on transcriptorium datasets (HTRtS). In: 2014 14th International Conference on Frontiers in Handwriting Recognition, pp. 785–790, September 2014. iSSN 2167–6445
20. Wigington, C., et al.: Data augmentation for recognition of handwritten words and lines using a CNN-LSTM network. In: 2017 14th IAPR International Conference on Document Analysis and Recognition (ICDAR), vol. 01, pp. 639–645, November 2017. iSSN 2379–2140
21. Xie, Q., Luong, M.T., Hovy, E., Le, Q.V.: Self-training with noisy student improves ImageNet classification. In: Proceedings of the IEEE/CVF Conference on Computer Vision and Pattern Recognition (CVPR), June 2020

TS-Net: OCR Trained to Switch Between Text Transcription Styles

Jan Kohút[(✉)] and Michal Hradiš

Faculty of Information Technology, Brno University of Technology,
Brno, Czech Republic
{ikohut,ihradis}@fit.vutbr.cz

Abstract. Users of OCR systems, from different institutions and scientific disciplines, prefer and produce different transcription styles. This presents a problem for training of consistent text recognition neural networks on real-world data. We propose to extend existing text recognition networks with a Transcription Style Block (TSB) which can learn from data to switch between multiple transcription styles without any explicit knowledge of transcription rules. TSB is an adaptive instance normalization conditioned by identifiers representing consistently transcribed documents (e.g. single document, documents by a single transcriber, or an institution). We show that TSB is able to learn completely different transcription styles in controlled experiments on artificial data, it improves text recognition accuracy on large-scale real-world data, and it learns semantically meaningful transcription style embeddings. We also show how TSB can efficiently adapt to transcription styles of new documents from transcriptions of only a few text lines.

Keywords: Transcription styles · Adaptive instance normalization · Text recognition · Neural networks · CTC

1 Introduction

When multiple users transcribe the same document the results might differ. The difference might not be given only by errors, but also by the diversity in

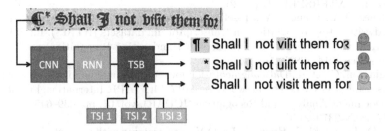

Fig. 1. Our proposed Transcription Style Block (TSB) learns to switch between various transcription styles based on transcription style identifiers (TSI).

J. Lladós et al. (Eds.): ICDAR 2021, LNCS 12824, pp. 478–493, 2021.
https://doi.org/10.1007/978-3-030-86337-1_32

transcription styles e.g., a historic character called 'long s' can be transcribed with the respective UTF-8 symbol or with the standard 's'. As it is not desirable to enforce a specific transcription style on transcribers, especially when it comes to open online systems, the presence of inconsistent transcriptions is inevitable.

The obvious solutions would be to unite the transcription styles of the transcribers or define mappings between the styles. Uniting transcription styles might be realistic for a small and closed group of transcribers. The second solution would require knowledge about the transcription styles in a form of translation tables, which is usually not available. If we had a reliable neural network, we could derive the tables in a semi-automatic manner. However, such a neural network would require a large amount of consistently transcribed data, resulting in a chicken and egg problem. Moreover, a transcription style mapping with 1 to n character relationships cannot be unambiguously represented. A common disadvantage of both options is the inability to automatically preserve personalized transcription styles.

OCR neural networks [2,8,25,28] trained with CTC loss function [11] learn to average probabilities of inconsistently transcribed characters if the transcription style can not be estimated from the input image, resulting in mostly random and independent transcription choices for such characters. The Seq2Seq approaches [4,21] can produce locally consistent transcriptions because of the autoregressive decoding process, but the choice of a transcription style may be arbitrary.

Figure 1 shows our proposed solution. Apart from the input, the neural network takes a transcription style identifier (TSI) and outputs a transcription in the respective style. A TSI identifies data with the same transcription style, it might be an identifier of a transcriber, of a document, or of any other transcription consistent entity. There is no need for any explicit knowledge (translation tables/rules) about the transcription styles (TS). TS are automatically learned by Transcription Style Block (TSB), which can be added near the end of any standard neural network (in our case convolutional layers (CNN) followed by BLSTM [12] layers (RNN) with the CTC loss function). TSB is an adaptive instance normalization layer with an auxiliary network, which learns to generate scales and offsets for each SI. It is trained jointly with the rest of the neural network.

To test if TSB is able to handle extreme scenarios, we evaluated our system on two datasets with synthetic TS. A synthetic TS is a full permutation of the character set. The first dataset consists of synthetic text line images with random text, the second is the IMPACT dataset [24]. To show practical benefits, we experimented with documents from Deutsches Textarchiv (DTA) where the transcribers are inconsistent among a handful of characters. We show that based on text line images, transcriptions, and TSI, the neural network is able to learn TS automatically. The adaptation to a new TS is fast and can be done with a couple of transcribed informative text line images.

2   Related Work

To our knowledge, no existing text recognition approach can automatically handle multiple transcription styles both during training and inference. Usually, there is an effort to avoid any inconsistencies in transcriptions from the start by introducing a transcription policy and resolving conflicts manually [3,24].

As our approach is based on adaptation to transcription style, it is closely related to the neural networks adaptation field. In text recognition, several authors explored various techniques for domain adaptation, transfer learning, and similar [14,31,36], but these techniques produce a unique model for each target domain and the individual domains have to be clearly defined. The idea of model adaptation was more extensively explored in speech recognition [1], mainly in the form of structured transforms approaches, which allow models to efficiently adapt to a wide range of domains or even handle multiple data domains simultaneously.

Model adaptation in speech recognition targets mostly the variability of individual speakers and communication channels. The main idea of structured transforms approaches is to adapt a subset of network parameters to speaker identity (speaker-dependent, SD) and keep the rest of the network speaker-independent (SI). The published methods adapt input layer (linear input network, LIN [23]), hidden layer (linear hidden network, LHN [9]), or output layer (linear output network, LON [19]). The main disadvantage of these methods is the large number of adapted parameters for each speaker. Adaptation is therefore slow and it is prone to overfitting without strong regularization.

Many methods strive to decrease the number of speaker-dependent parameters to reduce overfitting and improve adaptation speed. Learning Hidden Unit Contributions (LHUC) [32] adds an SD scale parameter after every neuron (kernel). Zhang et al. [35] parametrized activation functions (ReLU and Sigmoid) while making some of these parameters SD. Wang et al. [33] and Mana et al. [20] repurposed scales and offsets of batch normalization as SD parameters. Zhao et al. [37,38] found that most of the information in SD FC layers is stored in diagonals of the weight matrices and they proposed Low-Rank Plus Diagonal (LRPD, eLRPD) approaches which decompose (factorize) the original weight matrix of an FC layer into its diagonal and several smaller matrices. Factorized Hidden Layer (FHL) [26] offers a similar solution.

Another way to restrict the number of parameters is to use an SI auxiliary network that generates SD parameters based on a small SD input. The SD input representing the speaker can be an i-vector, learned speaker embedding, or similar features. Delcroix et al. [6] aggregate outputs of several network branches using weights computed from the SD input. The assumption is that the branches would learn to specialize in different types of speakers. Cui et al. [5] used an auxiliary network to generate SD scales and offsets applied to hidden activations.

Some methods do not require explicit information about speaker identity and learn to extract such SD information from a longer utterance and distribute it locally. L. Sarı et al. [27] trained an auxiliary network to predict SD single layer local activation offsets by pulling information from an entire utterance.

Similarly, Kim et al. [16] predict activation offsets and scales of multiple layers. Xie et al. [34] use an auxiliary network to generate the LHUC parameters in an online fashion based on acoustic features.

When the number of SD parameters is relatively small, the structure transforms approaches allow efficient network adaptation which is resistant to overfitting and requires only a small amount of adaptation data. As the speaker correlated information generally vanishes in deeper layers of a network [22], the SD parameters are usually located in the early layers of the network (in the acoustic part).

Our approach shares many ideas with the existing structured transforms approaches as the Transcription Style Block (TSB) uses learned domain embeddings transformed by an auxiliary network to scale and offset activations. However, this adaptation is done near the network output as it aims to change the output encoding, not to adapt to different styles of inputs.

The architecture of TSB is inspired by style transfer approaches and style-dependent Generative Adversarial Networks. These methods often use an adaptive instance normalization (AdaIN) to broadcast information about the desired output style across a whole image [7,10,13,15]. The published results demonstrate that this network architecture can efficiently generate a large range of output styles while keeping the style consistent across the whole output.

3   Transcription Style Block

Our neural network architecture (see Fig. 2) is similar to the current state of the art text recognition networks [2,8,25,28] which use CTC loss [11]. It consists of three main parts, convolutional part (CNN, blue), recurrent part (RNN, purple), and the Transcription Style Block (TSB, red). TSB allows the network to change the transcription style of the output (orange, green, yellow, etc.) based on the style transcription identifier (TSI). The transcription style switching is realized by adaptive instance normalization (AdaIN) [13]:

$$\text{AdaIN}(\mathbf{X}_c, \gamma_c, \beta_c) = \gamma_c \left(\frac{\mathbf{X}_c - \mu(\mathbf{X}_c)}{\sigma(\mathbf{X}_c) + \epsilon} \right) + \beta_c, \tag{1}$$

where $\mu(\mathbf{X}_c)$, $\sigma(\mathbf{X}_c)$ are mean and standard deviation of a channel c in a activation map $\mathbf{X}$ (ϵ is a small positive number). A transcription style is changed by scaling and offsetting the normalized channels by γ_c and β_c, respectively. Vectors of scales and offsets γ, β are given by affine transformations of a transcription style embedding e:

$$\gamma = \mathbf{W}^\gamma e + b^\gamma, \beta = \mathbf{W}^\beta e + b^\beta, \tag{2}$$

where $\mathbf{W}^\gamma$, $\mathbf{W}^\beta$ are the transformation matrices and b^γ, b^β biases. Every TSI is therefore represented by one transcription style embedding e. The output of the recurrent part is transcription style independent feature vectors that are turned into the transcription style dependent by the TSB. We decided to place the TSB near the end of the architecture as there should be no need for transcription style dependent processing in the visual and context part.

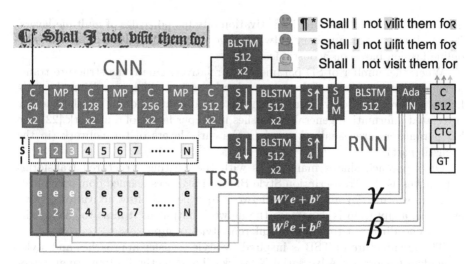

Fig. 2. Our proposed neural network takes a text line image (top left) and a transcription style identifier (TSI) and outputs a transcription in the respective style (top right). It consists of three main parts, convolutional part (CNN, blue), recurrent part (RNN, purple), and the Transcription Style Block (TSB, red). Each TSI is represented by a transcription style embedding e. Affine transformations of an e results in the scales and offsets, γ and β, for the adaptive instance normalization (AdaIN). (Color figure online)

Training. Our datasets consist of triplets of text line images, transcriptions, and TSI, which are integer numbers representing transcription consistent entities. The whole neural network is trained jointly with Adam optimizer [17]. The only transcription style dependant parameters are the embeddings, the rest of the parameters, including $\boldsymbol{W}^{\gamma}$, $\boldsymbol{W}^{\beta}$, $\boldsymbol{b}^{\gamma}$, and $\boldsymbol{b}^{\beta}$ are updated with each training sample. We initialize each embedding from standard normal distribution $\mathcal{N}(0, 1)$. The transformation matrices $\boldsymbol{W}^{\gamma}$, $\boldsymbol{W}^{\beta}$ are initialized from uniform distribution $\mathcal{U}(-k, k)$, where k is a small positive number (more details in Sect. 4, Table 4). The biases $\boldsymbol{b}^{\gamma}$ are set to ones, the biases $\boldsymbol{b}^{\beta}$ are set to zeros.

We use data augmentation including color changes, geometric deformations, and local masking which replaces the input text line image with noise at random locations and with random width limited to roughly two text characters. The idea of masking is to force the network to model language and not to rely only on visual cues. Further details about training and datasets are given in Sects. 4 and 5.

Architecture Details. The convolutional part (blue) is similar to the first 4 blocks of VGG 16 [29] with the exception that only the first two pooling layers reduce horizontal resolution and Filter Response Normalization (FRN) [30] is placed at the output of each block. It consists of four convolutional blocks (C) interlaid with max pooling layers (MP) with kernel and step size of 2. Each convolutional block consists of two (x2) convolutional layers with 64, 128, 256, and 512 output channels, respectively. The last convolutional layer aggregates all remaining vertical pixels by setting the kernel height equal to the height of its input tensor.

The subsequent recurrent part (purple) consists mainly of BLSTM layers [12]. It processes the input by three parallel BLSTM blocks (each block has two BLSTM layers). The first block processes directly the convolutional output, the second and third work at twice (S2↓) and four (S4↓) times reduced horizontal resolution, respectively. Outputs of all three branches are converted back to the same resolution by nearest neighbor upsampling and are summed together and processed by the final BLSTM layer. The reason for this multi-scale processing is to enlarge the receptive field (context) of the output hidden features.

4   Synthetic Transcription Styles Experiments

To test if the proposed TSB is indeed able to learn different transcription styles by itself and that this functionality is not learned by the whole network relying on correlations between the transcription style and visual inputs, we evaluated the whole network in extreme artificial scenarios. These experiments also give information about the limits and scalability of this approach. We evaluated our system on two datasets (SYNTHETIC and IMPACT) with 10 synthetic transcription styles (TS). A synthetic TS is a full permutation of the character set, e.g. if (a, b, c) is the visual character set then the TS (c, a, b) transcribes the character 'a' as the character 'c', etc. In real-life datasets, the difference between TS would be limited to a couple of characters. To test the robustness of the system, we represented each TS by multiple transcription style identifiers TSI (up to 10k) and we investigated the correlation between the respective learned transcription style embeddings and TS.

In all following experiments, the transcription style agnostic baseline network replaces TSB with FRN normalization. Also as we show the results for augmented training data (TRN), the results are worse than for the testing data (TST).

Datasets. The SYNTHETIC dataset contains generated images of random text lines rendered with various fonts, including historic fonts (see Fig. 3). The images were degraded by simulated whitening, noise, stains, and other imperfections, and slight geometric deformations were introduced in the text (character warping, rotation, and translation). It should be noted that the visual complexity of this dataset is mostly irrelevant because the experiments focus on TS. We generated 130k text lines for each of 10 synthetic TS. All the 10 TS are unique mappings for all characters, except for the white space. The size of the character set is 67.

The IMPACT dataset [24] contains page scans and manually corrected transcriptions collected as a representative sample of historical document collections from European libraries. The documents contain 9 languages and 10 font families. We randomly sampled 1.3M text line images from the original collection with ParseNet [18] and we translated their transcriptions into additional nine

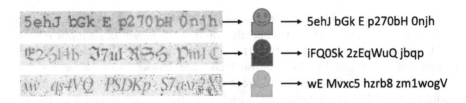

Fig. 3. Synthetic text line images with random text and random transcription styles.

Table 1. CER (in %) on the SYNTHETIC and the IMPACT datasets with 10 synthetic transcription styles and for various embedding dimensions (ED). NA is the baseline network trained and tested only on a single TS.

ED	1	2	4	8	16	32	64	NA
TST SYNTHETIC	2.95	0.58	0.66	0.50	0.67	**0.32**	0.38	0.35
TRN SYNTHETIC	5.72	1.70	1.81	1.59	1.77	**1.06**	1.29	0.91
TST IMPACT	10.87	2.57	1.13	1.09	1.09	**0.89**	0.99	0.74
TRN IMPACT	12.04	3.46	2.00	1.94	1.93	**1.53**	1.71	1.27

synthetic TS. Thus, each text line has ten transcription versions where one is the original one. The size of the character set is 399.

Transcription Style Embedding Dimension. The choice of the embedding dimension (ED) influences the ability of the network to transcribe the input in various TS. Fewer tunable parameters should provide higher overfitting resistance but may not be able to capture all the TS differences. The goal is to find the smallest ED, which will not degrade the text recognition accuracy. In this experiment, each TS is uniformly represented by multiple TSI (TSI per TS). Table 1 shows the test and the train character error rates (CER) for various ED. 10 TSI per TS was used for the SYNTHETIC dataset and 100 TSI per TS for the IMPACT dataset (as the IMPACT dataset contains 10 times more data, each embedding is trained on approximately 13k text line images). The networks were trained for 66k iterations. To estimate an upper bound on transcription accuracy, the baseline network (NA) was trained on a single TS.

The ED 32 brought the best results. Surprisingly, the network with ED 1 is already able to switch the transcription style to a limited degree, especially considering that all AdaIN scales and all offsets in this case differ just by a scale constant. The results of the ED 32 and these of the baseline network NA are comparable, which means that the TSB learned to switch between the transcription styles without lowering the performance of the text recognition.

Number of Transcription Style Identifiers per Transcription Style. The goal of the following experiment is to show if the network is capable to learn TS from a

Table 2. CER (in %) on the SYNTHETIC dataset for various TSI per TS, ED 32.

TSI per TS	10	100	1000
TST	0.32	0.35	0.29
TRN	1.06	1.13	1.01
Lines per E	13k	1.3k	130
TRN iterations	66k	132k	330k

Table 3. CER (in %) on the IMPACT dataset for various TSI per TS, ED 32.

TSI per TS	100	1000	10000
TST	0.89	1.40	1.40
TRN	1.53	2.50	2.00
Lines per E	13k	1.3k	130
TRN iterations	66k	100k	378k

Table 4. CER (in %) for various initialization energy of W^γ, W^β matrices. The weights were initialized by sampling from uniform distribution $\mathcal{U}(-k, k)$.

k	0.03	0.075	0.15	0.3	0.6	1.2
TST	0.91	0.90	0.89	1.01	0.96	1.48
TRN	1.55	1.57	1.53	1.70	1.70	2.59

Fig. 4. 2D projection (PCA) of the AdaIN scales and offsets for all embeddings learned on the SYNTHETIC dataset with 100 TSI per TS.

relatively small number of lines, and also if the network is able to handle a large amount of TSI. We can explore this by increasing the number of TSI per TS. We note that the network is not aware of which TSI represent the same TS in any way and because all embeddings are randomly initialized, it must learn the TS of TSI only from the respective text line images and transcriptions. Tables 2 and 3 show that the proposed approach scales well with the number of TSI and the final achieved transcription accuracy does not seem to depend on the number of TSI. However, the network needs more training iterations to converge with an increasing number of TSI. Consider that for the highest explored number of TSI, only 130 text line images were available for each TSI.

Figure 4 shows 2D projection (PCA) of the AdaIN scales and offsets for all embeddings learned on the SYNTHETIC dataset with 100 TSI per TS. The projections of different TS are well separated and it is apparent that the network learned a semantically meaningful embedding space. We observed that the separation increases during training proportionally to the error rate reduction.

Additionally we experimented with initialization of the matrices W^γ and W^β. We trained the network on the IMPACT dataset for 66k iterations, with 100 TSI per TS and ED 32. Table 4 shows that the network is very robust to the initial weight distribution energy represented as k, and that it converges to

Table 5. CER (in %) on the DTA dataset for various ED and the baseline network NA, 200k training iterations. TST-STSI refers to the TST set with randomly shuffled TSI.

ED	1	2	4	8	16	32	64	NA
TST	0.62	0.59	0.58	0.54	0.52	**0.51**	0.51	0.66
TST-STSI	0.86	0.86	1.22	1.57	1.73	**1.74**	1.81	–
TRN	0.79	0.80	0.70	0.70	0.71	**0.72**	0.70	0.79

similar error rates until the energy exceeds a specific threshold. We observed slightly faster convergence for lower initial energies.

5 Real Transcription Styles Experiments

To demonstrate the benefits of our system in a practical application, we experimented with a collection of documents from Deutsches Textarchiv (DTA)[1]. DTA collects German printed documents from various fields, genres, and time periods. Most documents date from the early 16th to the early 20th century and are typeset mostly in Fraktur fonts. We chose this collection because it contains several transcription styles related to archaic character variants. For example, a character called 'long s' is often transcribed as the standard 's', also the German umlauted vowels in old notation are sometimes transcribed with their modern versions.

DTA Dataset. Our DTA dataset is a subset of the original DTA collection. We represent a document by a single TSI, as we assume that each document was transcribed by a single organization using a consistent transcription style, but we have no information on how or where the documents were transcribed. We randomly sampled 1.08M text lines from 2445 randomly picked DTA documents with ParseNet [18]. The number of lines per document ranges from 200 to 3029. The size of the character set is 271. The test set (TST) contains 12.5k random samples, the train set (TRN) contains the rest. We also created two additional held-out test sets of documents that are not present in the TRN/TST partitions. These two sets allowed us to assess the behavior of the model on unknown documents. The first set (TST-HO) contains 12.5k lines from 7905 documents. The second set (TST-C) contains only 1355 lines, each line is from a different document of the TST-HO set. We transcribed TST-C consistently and accurately with the original graphemes.

TSB Performance on Real Data. We trained our system with different embedding dimensions (ED), and the baseline network, for 200k iterations. The results are shown in Table 5. In comparison to the baseline network (NA), our system

---

[1] https://www.deutschestextarchiv.de.

brought 22.7% relative decrease in character error rate on the TST set and 11.4% on the augmented TRN set. There is a noticeable drop in accuracy when we randomly shuffle TSI which means TSB learned to change the transcriptions and did not degrade into standard instance normalization. The performance for larger ED improves, but gradually worsens when TSI are shuffled. This behavior indicates that larger embeddings are better in capturing the transcription styles. As the ED 64 did not bring any significant improvement, we chose the ED 32 for the following experiments (it was also the best ED for the experiments in Sect. 4).

TSB reduces CER on the DTA dataset even if we allow the output to match any transcription style. We verified this by transforming transcriptions of the DTA TST set and the network outputs to a consistent modern style by a hand-designed look up table (reflecting substitution statistics, see Table 6). In this setting network with TSB (ED 32) provides 0.504% CER and network without TSB 0.568% CER.

Analysis of DTA Transcription Style Inconsistencies. Characterizing the transcription styles learned by the network is not straightforward. We analyzed the produced transcriptions, their differences, and relations. First, we explore which characters are substituted with changing TSI. Average statistics of the substations should clearly indicate if the changes are meaningful and correspond to expected transcription styles or if they are random and correspond only to errors or uncertain characters.

We gathered character substitution statistics on held-out documents (TST-HO) by comparing outputs of all TSI pairs using standard string alignment (as used for example in Levenshtein distance). We aggregated the statistics over all TSI pairs to identify globally the most frequently substituted character pairs.

The most frequently substituted (inconsistent) characters in relative and absolute terms are shown in Table 6. The substitutions are mostly unambiguous and match our expectations of possible transcription styles of historic German texts, i.e. substitutions of archaic and modern character variants. Other substitutions represent more technical aspects of transcription styles such as transcription of different dashes and hyphenation characters. Perhaps most interestingly, the network learned with some TSI to differentiate between 'I' and 'J' even though these modern characters are represented by the same grapheme in most documents.

As expected, the most inconsistent pair is 's' and 'long s' (ſ). The substitutions of the character 'sharp s' (ß) are meaningful but rare. The next group represents inconsistent transcriptions of historic umlauted vowels 'a̐', 'o̐', 'u̐'. As the umlauted vowels in the old notation were transcribed with two separate characters, sometimes the network outputs just the historic umlaut without the base character (represented as 'e̥'). Other significant inconsistent groups are 'rotunda r' (ꝛ) and 'r', single quotes, and various forms of hyphen or other characters that were used for word splitting ('¬', '='). When interpreting the substitution table, keep in mind that the substitutions are not with respect to the real graphemes, instead, they are substitutions between the different learned transcription styles.

Table 6. The most substituted characters of the learned transcription styles. The statistics were gathered on the DTA TST-HO set. Each column describes how the character in the header is substituted. The 'top n' rows show with which characters it is substituted together with the relative frequency of those substitutions. R is a ratio of substituted character occurrences. All numerical values are percentages. Some characters in the table represent similar historic characters or special substitutions. The mapping is: 'ſ' - long s; 'ꝛ' - rotunda r; '␣' - white space; 'ẹ' - historic umlaut without the base vowel; ∅ - character deletion; '<' stands for negligible values lower than 1.

	ſ	s	ß	å	ä	o̊	ö	ů	ü	I	J	ꝛ	r	'	'	¬	=	—	-
top 1	s	ſ	s	ä	å	ö	o̊	ü	ů	J	I	r	ꝛ	'	'	-	-	∅	¬
	97	97	92	82	77	89	85	83	81	96	97	73	34	93	93	97	84	52	50
top 2	∅	ß	z	ẹ	∅	o	o	ẹ	∅	∅	∅	e	∅	∅	∅	∅	∅	-	∅
	2	2	3	11	11	6	8	9	9	1	1	16	20	4	5	2	15	37	20
top 3	f	∅	ſ	a	a	ẹ	∅	u	u	1	F	∅	t	°	e	.	—	␣	=
	1	<	2	7	10	5	6	5	8	1	<	3	14	1	<	<	2	3	11
top 4	l	f	∅	á	ü	∅	h	ů	ä	l	l	&	x	"	␣	r	␣	„	—
	<	<	2	<	1	<	<	2	1	1	<	2	7	1	<	<	<	2	6
R	12	13	2	10	7	10	9	9	9	17	11	11	<	45	42	85	44	7	6

Structure of Learned Transcription Style Embeddings. Without a surprise, we observed that more similar transcription style embeddings produce more similar outputs. The Euclidean distances between embeddings and Levenshtein distances between the respective outputs are strongly correlated. We measured an almost perfect correlation for training documents with a high number of text lines. The correlation decreases with a lower number of training lines, but it plateaus approximately at 800 training lines and the Pearson correlation coefficient does not drop below 0.55.

Figure 5 shows a t-SNE projection of all learned transcription style embeddings. The distance metric we used for the projection is an average Levenshtein distance of respective output transcriptions on the DTA held out set TST-HO. We also categorized the embeddings based on how they transcribe 'long s', historic umlauted vowels and 'J'. We were able to do this categorization automatically thanks to the grapheme-accurate transcriptions of the TST-C dataset. The categories form a meaningful structure in the projection.

We also found embeddings that produced inconsistent transcriptions (red points in Fig. 5). Closer examination revealed that these embeddings represent documents with a high amount of annotation errors. Interestingly, most of these embedding are grouped together.

Adaptation to a New Transcription Style. The proposed system provides two possibilities of how to choose a transcription style for users. A straightforward way is to provide transcription style prototypes with descriptions (e.g. by clustering learned embeddings and manually creating their text descriptions). Alter-

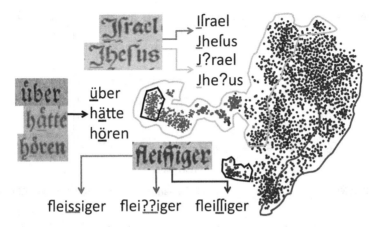

Fig. 5. Learned transcription style embeddings projected to 2D by t-SNE using output text edit distances on DTA TST-HO. Colors represent different transcription styles for certain characters. Blue and green dots differentiate between modern and historic transcriptions of 'long s'. The black polygons represent transcription styles that transcribe historic umlauts by their modern forms. Embeddings inside the brown polygon distinguish between I and J, while those inside the yellow polygon do not. Embeddings producing inconsistent transcriptions are shown in red color. (Color figure online)

natively, it is possible to estimate a transcription style embedding from a small set of exemplary text line transcriptions. The transcription style embedding can be estimated by similar optimization as when training the whole network. The loss function is the same, but the network remains fixed and only the single embedding is optimized. We use the LBFGS optimization algorithm which tries to better approximate local loss function curvature and it generally converges faster on problems with a low number of optimized variables. We fixed the number of LBFGS iterations to 150 and we initialized the new embeddings as the mean of all training document embeddings. The optimalization takes approximately 1 min on GeForce RTX 2080 Ti. We augmented the exemplary text lines in the same way as during full network training and did not use any additional regularization. The augmentation was necessary to avoid overfitting on small sets of exemplary text lines.

We tested the transcription style adaptation on held-out dataset TST-C for which we generated three additional transcription styles from the grapheme-accurate transcriptions. The four transcription styles are grapheme-accurate, modern s, modern umlaut, modern s+umlaut, where modern s replaces all 'long s' characters with the modern 's' character and normal umlaut replaces the historic umlauts with their respective modern variants.

We repeated all optimizations 50× on randomly sampled text lines from TST-C. We tested how the adaptation behaves with 1, 4, 12, 20, 50, and 100 exemplar text lines. Figure 6 shows that adaptation quality stabilizes and plateaus with the randomly sampled 100 exemplar text lines. In real use-cases, it is reasonable

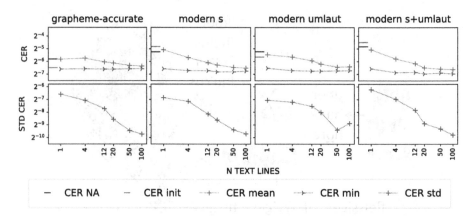

Fig. 6. Adaptation of transcription style embeddings on unseen documents with LBFGS optimization. Mean (red), standard deviation (green), and minimum (blue) test character error rate (CER) of the proposed system with adapted embeddings over 50 randomized text line sets. Black line marks the mean CER of the baseline network and the purple line marks the mean CER of our system with the initial embedding. The horizontal axis represents the number of text lines used for adaptation. Columns represent four different transcription styles. (Color figure online)

to expect that even fewer text lines would be needed as a user would probably pick representative text lines instead of sampling randomly (i.e. text lines which contain characters defining the transcription style). This expectation is supported by the fact that the adaptation achieves the same quality regardless of the number of exemplar lines for specific line sets (represented by the minimal error rates over the 50 runs).

Transcription accuracy after the adaptation is consistently better compared to the transcription style agnostic baseline network (NA). The reaching best relative improvements 42%, 66%, 65%, and 77% on the individual transcription styles. Interestingly, the initial mean embedding produces the grapheme-accurate transcriptions and it provides comparable accuracy as the baseline network on the other styles. The adaptations are able to achieve the same accuracy regardless of the target transcription style.

6   Conclusion

We proposed the Transcription Style Block (TSB) to solve the problems caused by inconsistent transcription styles present in real-world large-scale training datasets, and to allow OCR system users to easily choose their preferred transcription style. By connecting TSB near the end of any standard text recognition network, the network learns to output transcriptions in different styles from the data, without any explicit knowledge about the transcription rules. We showed that our approach can handle extreme synthetic scenarios, where the transcription styles are random permutations of the character set, without any degradation of text recognition. Our approach outperforms transcription style agnostic networks on a real-world large-scale dataset by a large margin. The adaptation to a new transcription style is fast and can be done with a couple of representative text line transcriptions.

Acknowledgment. This work has been supported by the Ministry of Culture Czech Republic in NAKI II project PERO (DG18P02OVV055).

References

1. Bell, P., Fainberg, J., Klejch, O., Li, J., Renals, S., Swietojanski, P.: Adaptation algorithms for speech recognition: an overview (2020)
2. Bluche, T., Messina, R.: Gated convolutional recurrent neural networks for multilingual handwriting recognition. In: ICDAR 2017, vol. 01, pp. 646–651 (2017)
3. Causer, T., Grint, K., Sichani, A.M., Terras, M.: 'Making such bargain': transcribe Bentham and the quality and cost-effectiveness of crowdsourced transcription. Digit. Sch. Hum. **33**(3), 467–487 (2018)
4. Chowdhury, A., Vig, L.: An efficient end-to-end neural model for handwritten text recognition. CoRR abs/1807.07965 (2018)
5. Cui, X., Goel, V., Saon, G.: Embedding-based speaker adaptive training of deep neural networks. CoRR abs/1710.06937 (2017)
6. Delcroix, M., Kinoshita, K., Ogawa, A., Huemmer, C., Nakatani, T.: Context adaptive neural network based acoustic models for rapid adaptation. IEEE/ACM Trans. Audio Speech Lang. Process. **26**(5), 895–908 (2018)
7. Dumoulin, V., Shlens, J., Kudlur, M.: A learned representation for artistic style. CoRR abs/1610.07629 (2016)
8. Dutta, K., Krishnan, P., Mathew, M., Jawahar, C.V.: Improving CNN-RNN hybrid networks for handwriting recognition. In: ICFHR 2018, pp. 80–85 (2018)
9. Gemello, R., Mana, F., Scanzio, S., Laface, P., De Mori, R.: Linear hidden transformations for adaptation of hybrid ANN/HMM models. Speech Commun. **49**(10–11), 827–835 (2007)
10. Ghiasi, G., Lee, H., Kudlur, M., Dumoulin, V., Shlens, J.: Exploring the structure of a real-time, arbitrary neural artistic stylization network. CoRR abs/1705.06830 (2017)
11. Graves, A., Fernández, S., Gomez, F., Schmidhuber, J.: Connectionist temporal classification: labelling unsegmented sequence data with recurrent neural networks. In: ICML 2006, pp. 369–376 (2006)
12. Hochreiter, S., Schmidhuber, J.: Long short-term memory. Neural Comput. **9**(8), 1735–1780 (1997)
13. Huang, X., Belongie, S.J.: Arbitrary style transfer in real-time with adaptive instance normalization. CoRR abs/1703.06868 (2017)
14. Kang, L., Rusiñol, M., Fornés, A., Riba, P., Villegas, M.: Unsupervised writer adaptation for synthetic-to-real handwritten word recognition. CoRR abs/1909.08473 (2019)
15. Karras, T., Laine, S., Aila, T.: A style-based generator architecture for generative adversarial networks. CoRR abs/1812.04948 (2018)
16. Kim, T., Song, I., Bengio, Y.: Dynamic layer normalization for adaptive neural acoustic modeling in speech recognition. CoRR abs/1707.06065 (2017)

17. Kingma, D.P., Ba, J.: Adam: a method for stochastic optimization. In: Bengio, Y., LeCun, Y. (eds.) ICLR 2015, San Diego, CA, USA, 7–9 May 2015, Conference Track Proceedings (2015)
18. Kodym, O., Hradiš, M.: Page layout analysis system for unconstrained historic documents (2021)
19. Li, B., Sim, K.C.: Comparison of discriminative input and output transformations for speaker adaptation in the hybrid NN/HMM systems. In: Eleventh Annual Conference of the International Speech Communication Association (2010)
20. Mana, F., Weninger, F., Gemello, R., Zhan, P.: Online batch normalization adaptation for automatic speech recognition. In: IEEE ASRU 2019, pp. 875–880. IEEE (2019)
21. Michael, J., Labahn, R., Grüning, T., Zöllner, J.: Evaluating sequence-to-sequence models for handwritten text recognition. In: ICDAR 2019, Sydney, Australia, 20–25 September 2019, pp. 1286–1293. IEEE (2019)
22. Mohamed, A.R., Hinton, G., Penn, G.: Understanding how deep belief networks perform acoustic modelling. In: IEEE ICASSP 2012, pp. 4273–4276. IEEE (2012)
23. Neto, J., et al.: Speaker-adaptation for hybrid HMM-ANN continuous speech recognition system (1995)
24. Papadopoulos, C., Pletschacher, S., Clausner, C., Antonacopoulos, A.: The impact dataset of historical document images. In: Proceedings of the 2nd International Workshop on Historical Document Imaging and Processing, pp. 123–130 (2013)
25. Puigcerver, J.: Are multidimensional recurrent layers really necessary for handwritten text recognition? In: ICDAR 2017, vol. 01, pp. 67–72 (2017)
26. Samarakoon, L., Sim, K.C.: Factorized hidden layer adaptation for deep neural network based acoustic modeling. IEEE/ACM Trans. Audio Speech Lang. Process. **24**(12), 2241–2250 (2016)
27. Sarı, L., Thomas, S., Hasegawa-Johnson, M., Picheny, M.: Speaker adaptation of neural networks with learning speaker aware offsets. Interspeech (2019)
28. Shi, B., Bai, X., Yao, C.: An end-to-end trainable neural network for image-based sequence recognition and its application to scene text recognition. CoRR abs/1507.05717 (2015)
29. Simonyan, K., Zisserman, A.: Very deep convolutional networks for large-scale image recognition. In: ICLR (2015)
30. Singh, S., Krishnan, S.: Filter response normalization layer: eliminating batch dependence in the training of deep neural networks. CoRR abs/1911.09737 (2019)
31. Soullard, Y., Swaileh, W., Tranouez, P., Paquet, T., Chatelain, C.: Improving text recognition using optical and language model writer adaptation. In: ICDAR 2019, pp. 1175–1180 (2019)
32. Swietojanski, P., Li, J., Renals, S.: Learning hidden unit contributions for unsupervised acoustic model adaptation. CoRR abs/1601.02828 (2016)
33. Wang, Z.Q., Wang, D.: Unsupervised speaker adaptation of batch normalized acoustic models for robust ASR. In: IEEE ICASSP 2017, pp. 4890–4894. IEEE (2017)
34. Xie, X., Liu, X., Lee, T., Wang, L.: Fast DNN acoustic model speaker adaptation by learning hidden unit contribution features. In: INTERSPEECH, pp. 759–763 (2019)
35. Zhang, C., Woodland, P.C.: Parameterised sigmoid and ReLU hidden activation functions for DNN acoustic modelling. In: Sixteenth Annual Conference of the International Speech Communication Association (2015)

36. Zhang, Y., Nie, S., Liu, W., Xu, X., Zhang, D., Shen, H.T.: Sequence-to-sequence domain adaptation network for robust text image recognition. In: Proceedings of the IEEE/CVF Conference on Computer Vision and Pattern Recognition, pp. 2740–2749 (2019)
37. Zhao, Y., Li, J., Gong, Y.: Low-rank plus diagonal adaptation for deep neural networks. In: IEEE ICASSP 2016, pp. 5005–5009. IEEE (2016)
38. Zhao, Y., Li, J., Kumar, K., Gong, Y.: Extended low-rank plus diagonal adaptation for deep and recurrent neural networks. In: IEEE ICASSP 2017, pp. 5040–5044. IEEE (2017)

Handwriting Recognition with Novelty

Derek S. Prijatelj[1], Samuel Grieggs[1], Futoshi Yumoto[2],
Eric Robertson[2], and Walter J. Scheirer[1]($\boxtimes$)

[1] University of Notre Dame, Notre Dame, IN 46556, USA
{dprijate,sgrieggs,walter.scheirer}@nd.edu
[2] PAR Government, 421 Ridge St, Rome, NY 13440, USA
{futoshi_yumoto,eric_robertson}@partech.com

Abstract. This paper introduces an agent-centric approach to handle novelty in the visual recognition domain of handwriting recognition (HWR). An ideal transcription agent would rival or surpass human perception, being able to recognize known and new characters in an image, and detect any stylistic changes that may occur within or across documents. A key confound is the presence of novelty, which has continued to stymie even the best machine learning-based algorithms for these tasks. In handwritten documents, novelty can be a change in writer, character attributes, writing attributes, or overall document appearance, among other things. Instead of looking at each aspect independently, we suggest that an integrated agent that can process known characters and novelties simultaneously is a better strategy. This paper formalizes the domain of handwriting recognition with novelty, describes a baseline agent, introduces an evaluation protocol with benchmark data, and provides experimentation to set the state-of-the-art. Results show feasibility for the agent-centric approach, but more work is needed to approach human-levels of reading ability, giving the HWR community a formal basis to build upon as they solve this challenging problem.

Keywords: Handwriting recognition · Novelty · Agents · Writer identification · Style recognition

1 Introduction

Reading comprehension is a complex human activity that requires symbol acquisition and manipulation, the perception of salient information, and an

This research was sponsored by the Defense Advanced Research Projects Agency (DARPA) and the Army Research Office (ARO) under multiple contracts/agreements including HR001120C0055, W911NF-20-2-0005,W911NF-20-2- 0004,HQ0034-19-D-0001, W911NF2020009. The views contained in this document are those of the authors and should not be interpreted as representing the official policies, either expressed or implied, of the DARPA or ARO, or the U.S. Government.

Electronic supplementary material The online version of this chapter (https://doi.org/10.1007/978-3-030-86337-1_33) contains supplementary material, which is available to authorized users.

© Springer Nature Switzerland AG 2021
J. Lladós et al. (Eds.): ICDAR 2021, LNCS 12824, pp. 494–509, 2021.
https://doi.org/10.1007/978-3-030-86337-1_33

understanding of what is known and what is novel on a page [16]. Why has it been reduced to simple optical character recognition (OCR) within the field of machine learning? While this has decreased the complexity of the domain to make it more tractable for standard data-driven approaches, it has also steered researchers away from the core of one of the most important competencies of natural intelligence: handling novelty. Consequently, OCR algorithms are not effective on handwritten documents that exhibit a wide degree of variation in appearance [24]. Yet such documents can be effortlessly read by humans, even children who have recently become literate. Fundamentally, this task is made difficult by the presence of novelties that are unknown at training time, which must be expected when a writer can do essentially whatever they want on a page. In this paper we introduce an agent-centric approach to the handwriting recognition (HWR) domain with novelty to address this challenge.

Novelty, of course, is not unique to the HWR domain. In general, the ability to act appropriately and effectively in novel situations that occur in open worlds, as opposed to closed datasets, has been singled out as a crucial challenge in AI that is inhibiting progress in multiple domains [6]. Recent theoretical work has sought to understand what implications the unknown has in the context of activity and perceptual domains, and how it should be treated by an agent operating within them. This includes risk formulations that account for the unknown [7,22], generative models of novelty [11], and enumerations of the different possible types of novelty that can occur in practice [4].

As Boult *et al.* have noted, a universal theory of novelty is not possible to construct. This is because each domain will require its own definitions for a world state, dissimilarity (*i.e.*, how far away from known data something must be to be considered novel), regret (*i.e.*, the consequence of not detecting novelty) and other constructs needed to solve a given task within a specific domain. Accordingly, researchers must formalize and attempt to solve tasks within a variety of domains to test the generalization capabilities of core novelty processing algorithms. Handwritten documents are particularly well suited for the study of novelty because of the human creativity that goes into making them. Thus we propose this domain as a challenge problem for researchers not only studying HWR, but also novelty in AI.

Beyond new characters, novelties may occur in the stylistic attributes of the written text. The discriminating markers of a writer's style are the (often subtle) differences in the way characters are drawn, as seen in Fig. 1. Global changes to the appearance of an image are also important, such as changes to the background or the application of filters (*e.g.*, Photoshop's vintage photo filter), which may have been made after a page has been acquired to improve human readability or to stylize to match the overall look of a digital text edition. Not only are these elements important for managing novelty in ordinary transcription tasks, but they are also important to scholars who are interested in identifying stylistic markers of historical significance [12,26].

An agent-centric approach reflects the real human behavior of reading. Individual algorithms can be applied to specific tasks in HWR, such as transcription or visual style recognition, but information can't easily be passed between

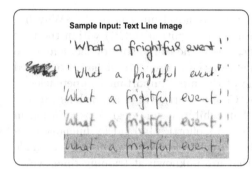

Fig. 1. Examples from the IAM offline handwriting dataset [14] depicting the difference in writing style between writers, unknown characters in the scratched out entry represented as "#" in the transcription, and the difference in appearance between the same samples with new modifications to simulate real-world situations. Novelty may occur in any of these labels, such as novel characters, writer, or appearance. HWR agents should be able to handle such novelty.

them, nor can it be used in an informed joint manner. Here we suggest that an agent with integrated task-specific modules and novelty-specific modules can autonomously process multiple information streams, understanding what it has seen before and what it hasn't, while it performs a transcription task. Further, the novelty-specific modules should not exclusively rely on simple thresholding applied over standard classifiers, and instead consist of open world classifiers optimized with a risk model of the unknown [17]. In this paper, we introduce an evaluation protocol and baseline agent for this agent-centric approach to HWR.

Related Work. Work related to HWR with novelty can be found in the fields of machine learning and computer vision. There is a strong foundation in deep learning-based approaches to HWR, which have yielded good performance in closed world data set evaluations. State-of-the-art approaches for diverse document sets [20,27] are based on the Convolutional Recurrent Neural Network (CRNN) [23] in combination with a Connectionist Temporal Classification (CTC) loss [9].

Beyond anomaly detection [18], machine learning work on classifiers has started to look at other ways in which novelty can be handled. Promising work in this direction relies on statistical modeling using extreme value theory, which more accurately accounts for the samples in the tails of distributions, which is consequential for decision boundaries in classifiers [2,17,29]. HWR in human biometrics is a mature area of research, having demonstrated that reliable person-specific features exist and can be learned for different languages [13,28]. It is an open question as to how well such features work for the characterization of novelty.

Similar to HWR at large, the problem of writer identification is known to be an unsolved problem, especially in the context of historical documents [8]. Some works have tried to use ImageNet pre-trained deep neural networks for writer identification as well as other HWR tasks [25] and have found improved performance in doing so. These approaches are built specifically for their task in mind, and tend to not focus on the sharing of information between other HWR tasks as we do in this paper.

Contributions. There are four primary contributions this work makes towards introducing a new challenge problem. (1) The formalization of HWR with novelty to standardize the domain. (2) A baseline agent integrating a deep learning-based transcription network and the Extreme Value Machine (EVM) [17] for novelty detection. (3) An evaluation protocol with benchmark data for this domain, including a fully implemented distributed software package for large-scale evaluations[1]. (4) A comprehensive set of experiments including results from over 55,000 experimental tests, setting the state-of-the-art for this challenge problem.

2   Formalization of HWR with Novelty

The formalization of the HWR domain with novelty includes two parts. The first is a series of definitions that represent a theory of novelty for the domain. The second is an ontological specification that characterizes the space of novelty and facilitates measurement of novelty detection difficulty.

2.1   A Theory of Novelty for HWR

The HWR domain, as defined for this paper's proposed benchmark, consists of two high-level tasks: (1) text transcription and (2) style recognition. The transcription task involves an agent taking a digital image of a handwritten document as input and processing it to recognize the individual characters to produce a plaintext output. The style recognition task involves the agent identifying known and unknown aspects of visual appearance for both the text (*e.g.*, how are individual characters stylized?) and page (*i.e.*, what does the page look like holistically?). Two subtasks for style recognition are considered: (2a) writer identification and (2b) overall document appearance identification (ODAI). The former involves multi-class classification to distinguish between individual known writers and new writers unseen at training time, while the latter involves multi-class classification to distinguish between known global appearances of handwritten documents and appearances unseen at training time. Any type of novelty can occur in both tasks, thus an important objective of the domain is to detect and manage it. Note that many other tasks can be defined for transcription and style recognition, and the subsequent theory is general enough to cover those as well.

A theory of novelty for the HWR domain can be constructed using the recently introduced framework of Boult et al. [4], which provides a common

---

[1] The code for this paper will be made publicly available after publication at https://github.com/prijatelj/handwriting_recognition_with_novelty.

basis to define and compare models of novelty across different domains. In this framework, an agent accesses the world indirectly through a perceptual operator, updating its internal state and acting on the world state as is necessary and possible. In that regime, the following must be defined: a world, an observational space, perceptual operators, a task, dissimilarity functions to assess potential novelties with respect to the task, and regret functions to determine the impact of incorrect assessments of potential novelties with respect to the task. A task may consist of multiple subtasks weighted by some priority.

The theory of novelty for HWR extends the image classification theory of novelty, defined in the extended version of the paper by Boult et al. [3]. As is the case for standard image classification, the samples do not necessarily have any meaningful order. However, there is a sequential relation of the characters and words within a sample image of a handwritten document. Below, time step t is in reference to the point in time when a sample image is considered, rather than a character or word within that image. The following is the specification of the key components that form the theory:

- In this paper, an HWR task $\mathcal{T}$ can be text transcription, writer identification or ODAI.
- A world $\mathcal{W}$ in HWR consists of a d'-dimensional space of pages of handwritten documents.
- An observation space $\mathcal{O}$ that is accessible to the agent is the d-dimensional space that can encode all possible images of handwritten documents. This space serves as the agent's feature space, extracted from the image.
- The family of perceptual operators $\mathcal{P}_t$ in HWR are optical sensors, such as cameras, that capture a visible region of the world $\mathcal{W}_t$, where a time-step t results in a single image of a handwritten document in the case of still image HWR. The perceptual operator may continue with feature extraction on the captured image to represent the image as a feature vector E_t of arbitrary dimensions to the agent in the observational space.
- The task-dependent world dissimilarity functions $\mathcal{D}_{w,\mathcal{T};E_t}$ and associated novelty threshold δ_w are determined by ground truth labels associated with the images from the sampling process, where the complete datasets used serve as an oracle (See Supp. Mat. Sec. 2.1.1[2]). The measurement of dissimilarity uses a distance measure (e.g., Euclidean distance) with a threshold determined by the probability distribution of the data.
- The task-dependent observational space dissimilarity functions $\mathcal{D}_{o,\mathcal{T};E_t}$ and novelty threshold δ_o are determined by the agent's knowledge and design for the task. In a learning-based agent, this is typically done via generalizing from the ground truth in any available training or validation data. These functions could make use of Euclidean distance or whichever distance measure suits the observational space.
- The world regret function $\mathcal{R}_{w,\mathcal{T}}$ is based on the error as measured in the world space given ground truth labels. For transcription, this could be

---

[2] The Supplemental Material is publicly available at https://arxiv.org/abs/2105.06582.

Levenshtein Edit Distance, Character Error Rate, or Word Error Rate. For the nominal tasks of writer identification and ODAI, the regret is captured by the confusion matrix from which measures of regret derive, such as the normalized mutual information (NMI). In this work, NMI is the focus due to ease of interpretability where values near zero indicate poor performance and thus high regret, and values approaching one indicate perfect predictions. Of course, any measure, such as NMI, whose value is inversely correlated to regret may have their inverse taken to form a proper measure of regret to be minimized. NMI also has a strong information theoretic backing and may be theoretically compared to the NMI of random variables with differing sample spaces, such as continuous random variables.

– The observational space regret function $\mathcal{R}_{o,\mathcal{T}}$ defines what the agent deems important to the task. This is embodied by the agent's internal model for the task, such as the loss function of a neural network or the likelihood calculation in a probabilistic model.

Given the above specification, novelty in HWR is deemed to occur in the world when the world $\mathcal{D}_{w,\mathcal{T};E_t}$ exceeds the novelty threshold δ_w. E_t serves as the history of experience of the agent and plays a key role when a change in the world state is considered novel at a time step. This paper's baseline agent's E_t is simply the training set and indirect information from the validation set, which informs the agent about how it should set its internal threshold δ_o. A world novelty may or may not effect the agent's performance on the task and the novelty's effect may vary in impact to task performance. This is reflected in the world regret $\mathcal{R}_{w,\mathcal{T}}$. The world novelty in any domain, including HWR, must be properly defined in an experiment to assess how an agent performs in its presence.

In transcription, the sampled world state includes the image and the oracle knows the correct transcription. World novelty thus occurs in the transcription task whenever a new character, word, or phrase appears in the sequence never before seen in the sampled world at that time-step, given the current label set for each. For writer identification, world novelty occurs within the handwritten text's style. In ODAI, world novelty includes a change in document material, backgrounds, perceptual sensor changes, or similar unseen changes in appearance of the document. World novelty is known in this paper's experiments by the data's labels and dissimilarity is calculated using the confusion matrix or its related measures NMI and accuracy. The term novelty in this paper is typically in reference to world novelty.

World Novelty is actual novelty that exists in an environment. For the HWR domain, World Novelty (*e.g.*, novel characters, novel writers, novel backgrounds) is the novelty an agent should be most focused on detecting and managing, because it is actual novelty in the world. An Observation Novelty occurs when the observation from the perceptual operators of the agent is sufficiently dissimilar from every past observation in the agent's stored experience. These novelties can only be detected in the HWR domain if the camera acquiring the image of a document has sensed the novelty that is present in the world. Finally, Agent

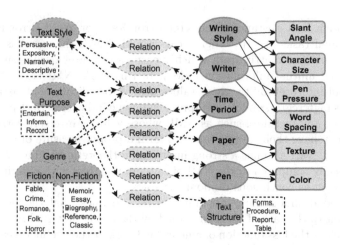

Fig. 2. HWR Novelty Ontology adapted from [15]. Entities are represented as blue ellipses. Entity attributes are represented as purple boxes. Examples of entities are dashed, white boxes overlapping the entity node. Nodes and edges with solid lines and bold text are the focus of this study.

Novelty occurs when an agent's internal processing cannot map an image to a known state. In HWR, as described in this paper, Agent Novelty is equivalent to Perceptual Novelty, because we treat state as the predicted label from the perceptual operators.

Secondary types of novelty exist that are combinations of the above three types. Unanimous Novelty is the presence of World Novelty, Observation Novelty, and Agent Novelty, and represents a valid transcription or style novelty. Imperceptible Novelty is novelty that cannot be sensed by the perceptual operators. In HWR, this can be novelty in the microscopic composition of the material that forms a document page, the historical context in which a document was discovered, the provenance of the document, or any other novelty a camera cannot capture, but a human examiner can determine via other means. Faux Novelty is a false positive determination of novelty. False positives can occur if the perceptual operators encounter noise at acquisition time, injecting novelty into the resulting image that does not exist in the environment. Further, in perceptual operators for HWR, imperfect machine learning models for transcription and style recognition have error rates, which can also create a false positive situation.

Regret factors into two additional types of novelty. Managed Novelty is any novelty that has a minimal impact on agent performance. In HWR, this could be a change in language expressed by a known character set, which does not impact character transcription performance in any meaningful way. Nuisance Novelty is a novelty whose world regret and observational space regret significantly differ. In a document, this could be stain, tear, or other physical artifact on a page that an agent consistently mistakes for a character (thus negatively impacting its error rate), but has little bearing on the environment of the page from the perspective of the world.

Given this formalization, experiments of HWR with novelty may be designed with a consistent understanding of novelty and its variations. This provides algorithm designers with an outline for implementing an algorithm given these theoretic constructs. The formalization establishes a common language to understand how different HWR tasks and their proposed solutions relate to each other. This allows for ablation studies of agents and understanding how different sources of novelty in HWR images affect performance, as done in the large-scale experiment described in Supp. Mat. Sec. 5.1.

2.2 Ontological Specification of HWR with Novelty

An ontological specification serves to describe the knowledge of the world held by the oracle and agent. Functionally, the ontology provides terms and structures to reason about and characterize actual and perceived novelty. We interpret the differences between an agent's task-dependent knowledge of the world and a newly experienced change in the world as a measurable dissimilarity between the world knowledge of the agent and the oracle. The degree of dissimilarity forms a basis for assessing the difficulty an agent has in both detecting novelty and performing its task within that novelty space, which is reflected in the expected world regret $\mathcal{R}_{w,\mathcal{T}}$. For example, in the HWR domain, writing samples from a novel writer with a similar style to a known writer are both difficult to detect as being novel and to identify as being written by an unknown writer.

The ontology's components consist of entities, attributes, actions, relations, interactions (passive) and rules often associated with a specific context or domain. An agent that can detect novelty maintains knowledge elements of the world as described by the ontology. In closed world supervised learning systems, these knowledge elements are provided through meta-data in the training sets.

The HWR ontology focuses on those components where novelty occurs. We characterize novelty in terms of text elements including writing style and pen selection, as well as in terms of background elements, i.e., those novelties not specific to the text. Writing style corresponds to the writer while the last two correspond to ODAI. The intent of ontological specification is to describe all observable features that may contain novelty including environmental novelties (e.g. water damage to the writing medium), temporal and locale novelties (e.g. date and time representations and document structures), and text-related novelties, such as copyedit marks.

The foundational ontology for transcription and style recognition is shown in Fig. 2 (adapted from [15]). We focus on a small group of core attributes representative of each ontological entity that best characterizes the set of novelty in the experiments. We excluded latent attributes from the ontology since they are difficult to qualify in the ontology and beyond the scope of the current study. However, they may play a critical role in novelty detection and characterization.

The HWR ontology defines these entities and attributes:

- Four attributes of writer style: (1) pen pressure, (2) slant angle, (3) character size, and (4) word space.
- The writer with associations to each style attribute.
- The image of a handwriting sample.
- Writing medium (background) categories including types of background noise, textures, and colors.
- Pen categories including textures and colors.

Each component specified by the ontology is associated with a measurement function $\mathcal{F}_{\mathcal{O},c}$ applied to each writing sample in the observation space, where c is a category of novelty. For example, the measurement function for pen pressure is the mean pixel intensity of the written text: $\left(\sum_{i=1}^{\mathcal{N}} \text{pixel}[i]\right) /\mathcal{N}$ where $\mathcal{N}$ is the number pixels in the written text and $\text{pixel}[i]$ is the intensity of a pixel i in the written text. The complete set of measures can be found in Table 1 of the Supp. Mat. The collection of normalized measures of each component composes a feature vector for use in distance functions (e.g., cosine similarity) to measure the similarity of a novel writing sample to the body of known non-novel writing samples.

We can also represent the writing style attributes graphically by first creating discrete attributes through binning the component measures and assigning each style and bin to a node in the world knowledge graph (see Supp. Mat. Sec. 2.2). Similarity of writing styles is represented within the knowledge graph by the shared style relations. The graph supports the application of graph metrics such as isomorphism between two writing style sub-graphs, where a higher number of shared discrete attributes between writing styles indicates a higher similarity. We hypothesize that the degree of dissimilarity inversely impacts the ability of an agent to detect and characterize novelty. However, the choice of entities, fidelity for measurement for each ontological entity and weight of significance for each entity impact the utility of dissimilarity measures (see discussion on writer similarity in novel writer discovery in Supp. Mat. Sec. 5.2).

3   Baseline Open World HWR Agent

To show that HWR tasks are feasible in the presence of novelty and to indicate the room for improvement, a baseline open world HWR agent was designed and evaluated on the transcription task and two style recognition subtasks: writer identification and ODAI. The agent consists of a CRNN [23], specifically for the transcription task, and an EVM [17] for each style subtask.

For transcription, the CRNN is trained in a supervised learning fashion on given cropped text lines from documents with the transcribed text as the labels. The CRNN's architecture consists of a sequence of CNNs leading to a bidirectional LSTM (details are in Supp. Mat. Sec. 3.1). The CRNN covers the feature extraction and transcript predictor modules of the agent as shown in Fig. 3.

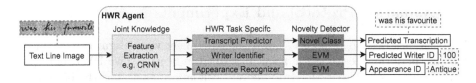

Fig. 3. The baseline open world HWR agent that is able to detect novelty consists of modules that leverage joint and task-specific knowledge of the HWR domain. Thus, it performs the tasks as desired while managing novelty. The "Appearance Recognizer" is the module for identifying the global appearance of the document, *e.g.*, white background with black foreground, noisy, or antique. The "Novel Class" novelty detector for transcription indicates that the model is trained with a class that represents novelty to isolate those instances from known classes.

Other feature extractors can be used in place of, or in addition to, the CRNN features. The CRNN was trained with a single novel character class to manage novel characters and glyphs in the images.

The writer identification and ODAI style tasks are handled by separate EVMs whose supervised training consists of the input of the feature space of a text line image and whose output consists of the labels respective to their task. For writer identification, the labels are the writer identifiers, and for ODAI the labels are the nominal labels of the overall document image appearance, which is further detailed in Sect. 4. The EVMs are classifiers trained to manage novelty in classification tasks and serve as the HWR style task modules and novelty detector modules in Fig. 3. The EVM predicts a probability vector of the known classes along with the probability of novel class association, given training data. To further determine if the class is novel, a threshold is applied to the maximum known labels' probabilities. When the max known class probability is below this threshold, the sample is deemed novel. The threshold is set by finding the minimum probability that is closest to an equal error rate between the Type I and Type II errors on the training and validation data.

Multiple feature spaces were examined for the style tasks using different feature extraction methods. These included: (1) the mean of the Histogram of Oriented Gradients (HOG) [5], referred to as Mean HOG; (2) multiple, evenly spaced, sequential means of HOG over contiguous sections of the image, which is referred to as M-Mean HOG where M is the number of sequential means; (3) the penultimate layer of ResNet50 [10] pre-trained on ImageNet [19]; (4) the penultimate layer of the CRNN's RNN trained on the transcription task, with PCA applied to reduce dimensionality. We apply this only to the writer identification subtask, as the training of the CRNN is domain specific.

For HOG feature extraction, the resulting values were flattened into a single feature vector. For M-Mean HOG, $M=10$ sequential means were obtained and the Mean HOG vector was appended to the beginning of the 10-Mean HOG feature vector. These feature extractions were compared to determine which performed best as the image feature representations for the EVMs handling the style subtasks. See Supp. Mat. Sec. 3.1 for more details on this.

4   Evaluation Protocol and Experiments

For each task described in Sect. 2, a basic experiment was designed to establish baseline performance using the agent described in Sect. 3. Two HWR datasets were used to provide data for training and validation, as well as the base data to be transformed for novelty injection. These were IAM [14] and RIMES [1]. For each task, IAM and RIMES were split into training, validation, and testing sets. The data was used in 5-fold cross validation fashion to obtain error bars of the measures.

For 5-fold cross validation, IAM was split by first randomly partitioning the writers into two groups with no intersection of writers, irrespective of sample count. One half is then split into 5 folds, randomly stratified by the writers to ensure a proper balance of samples representative of those writers' sample frequency. The remaining half was split into 5 folds of separate writers with no intersection of writers across the folds. The two halves' folds are paired up, forming the 5-fold split for IAM. RIMES was simply split randomly into 5 folds because RIMES has no writer identifiers. This process ensured that every cross validation experiment both shared a subset of IAM writers across all splits, as well as guaranteeing that the validation and testing splits have unique writers unseen during training. See Supp. Mat. Sec. 4 for details.

In addition and separate to the above data setup, a large-scale evaluation of the baseline open world HWR agent consisting of approximately 55K tests was conducted using synthetic modifications to the IAM dataset. For further details on the synthetic modifications, see Supp. Mat. Sec. 5.

Novel Characters in Text Transcription. To evaluate the agent's performance on text transcription, the baseline agent's CRNN was trained on the given text line images with the transcription text as the labels. The CRNN handled novelty through the addition of a novel character class. The novel characters in RIMES served as the known unknowns (*i.e.*, labeled negative examples [21]) for the CRNN to partially learn the novel character class. This included characters common in French that are not common in English (*e.g.*, vowels with diacritics). See Supp. Mat. Sec. 4.1 for details on the distribution of characters across the data splits.

Character accuracy and word accuracy were used to assess the performance of the agent's transcription. To assess the novelty detection of the transcription predictor, the presence of novelty was assessed per sample line image. If a character existed in the ground truth transcription that was not known to the agent, *e.g.*, scratched out characters or vowels with diacritics, then that image was determined to contain novelty. If the predicted transcript contained a novel character than the line image was predicted to contain novelty. With this, novelty detection was able to be assessed by a binary confusion matrix from which NMI and accuracy were calculated.

The results in Table 1 indicate that the CRNN performs well at text transcription in an evaluation regime that includes novelty. The most notable results reside in the novelty detection performance. The inclusion of the novel charac-

Table 1. The mean 5-fold results with standard error for the test split of all three experiments. "NMI" stands for Normalized Mutual Information where random guess is 0 and perfect correlation is 1. Results for the three tasks indicate that solving the tasks is feasible, but there is still substantial room for improvement. Perhaps surprisingly, the Mean HOG dominates across the board for style tasks, except for accuracy for writer identification, but NMI is a more reliable measure of correlation between the labels and predictions. All measures reported here are found after selecting the maximum probable class as predicted by the classifier after thresholding the maximum probability to determine if an input is novel.

Task	Multi-class Classif. w/ Novel Class		Binary Novelty Detection	
Model	NMI	Accuracy	NMI	Accuracy
Writer ID				
Mean HOG EVM	**0.6462 ± 1.02e-2**	0.6524 ± 6.35e-3	**0.6652 ± 1.22e-2**	**0.8748 ± 6.28e-3**
10-Mean HOG EVM	0.4127 ± 7.54e-3	0.6932 ± 4.99e-3	0.2199 ± 5.02e-3	0.4425 ± 6.53e-3
ResNet50 EVM	0.3853 ± 9.82e-3	0.6940 ± 4.87e-3	0.2126 ± 6.39e-3	0.4233 ± 8.12e-3
CRNN-PCA EVM	0.4013 ± 1.17e-3	**0.7058 ± 2.73e-3**	0.2235 ± 7.42e-3	0.4379 ± 8.91e-3
Appearances (ODAI)				
Mean HOG EVM	**0.5713 ± 3.63e-3**	**0.7865 ± 3.59e-3**	**0.3262 ± 7.16e-3**	**0.6392 ± 6.48e-3**
10-Mean HOG EVM	0.5065 ± 5.10e-3	0.7383 ± 8.70e-3	0.2955 ± 5.13e-3	0.5747 ± 6.46e-3
ResNet50 EVM	0.0140 ± 7.45e-4	0.4057 ± 3.01e-4	0.0085 ± 4.43e-3	0.0628 ± 1.71e-3
CRNN-PCA EVM	0.0551 ± 1.87e-3	0.4092 ± 5.65e-3	0.0345 ± 2.62e-2	0.4354 ± 4.26e-3
	Character Acc.	Word Acc.	NMI	Accuracy
Transcription				
CRNN	0.9494 ± 4.81e-3	0.8696 ± 2.85e-3	0.8777 ± 5.48e-3	0.9660 ± 1.86e-3

ter class in the model for the training data facilitates high scores for NMI and accuracy for novelty detection. However, there is still room for improvement, as these are baseline results.

Novelty in Writer Identification. For the writer identification subtask of the style recognition task, all RIMES documents were treated to have an unknown writer because the dataset contains no writer identifiers. Using the aforementioned data splits, novel writers from the IAM dataset exist across all data splits. See Supp. Mat. Sec. 4 for more details on the data breakdown for writer identification. Based on NMI, Table 1 shows that Mean HOG is the best feature extractor for use with the EVM for writer identification, followed by 10-Mean HOG. 10-Mean HOG includes the same features as Mean HOG in the beginning of its feature vector, so this indicates that the extra 10 sequential means were detrimental to the EVM for writer identification due to too many input variables. Following closely in third are the features extracted from the CRNN's penultimate layer after principal component analysis (PCA) of 1000 components. PCA was necessary given memory constraints and due to this a lot of useful information was probably lost, affecting the performance of the MEVM. See Supp. Mat. Sec. 3.1 for details. This is a somewhat surprising result, in that handcrafted features exceed the performance of deep learning in this case. We analyze this outcome in more detail below. Overall, the performance for writer identification in the face of novel writers is fair, with rather consistent NMI scores when used as a novelty detector as well.

Novelty in Overall Image Appearance. Given the above data splits, ODAI involved transforming a subset of the data into different document appearances. All data splits included the appearances of clean documents, Gaussian noise, antique background, horizontal reflection, and Gaussian blur. Horizontal reflection and Gaussian blur served as known unknowns to the agent in the training set to give the model some notion of novelty. To include novelties in the evaluation splits, vertical reflection was added to both validation and test data splits, and inverted color was added only to the test split. The differing writers and text in the images introduce nuisance novelty (irrelevant novelty that need not be detected [4]) into this subtask. See Supp. Mat. Sec. 4 for more details on the data split. ODAI was only performed on the IAM and RIMES datasets, not the synthetically modified IAM dataset. The agent used an EVM similar to the writer identification task and was assessed using the different feature extractors. Mean HOG features were again the best.

Feature Extraction Assessment for Style Recognition. Mean HOG for feature extraction for the EVM outperformed all other feature extraction methods in Table 1. And the second best feature extractor was the 10-mean HOG approach. However, one might expect that the CRNN, which was trained on the HWR-specific data, would be able to effectively transfer information from the transcription task to the style tasks. This transfer information may have been limited by the use of PCA, but PCA was necessary due constraints on resources. Furthermore, the style information is mostly noise for the transcription task, but it is a common source of noise that HWR agents need to manage to generalize onto novel styles. The CRNN's penultimate layer was probably beyond the point of a useful encoding of style information to being more biased towards the transcription task.

We do see evidence that the CRNN's features provided transfer information for style tasks in its performance relative to ResNet50. The pretrained ResNet50's penultimate layer as a feature extractor performed the worst out of the batch and this is probably due to limited sharing of information between the ImageNet classification task and the HWR style recognition task. We suggest that exploiting domain-specific information in a joint manner is still the best strategy in the long run for HWR with novelty. The key to improving performance is likely a hyperparameter search over many model configurations. That said, one can still achieve good performance using a combination of handcrafted features and deep learning, with far less computational effort.

Large-Scale 55K Evaluation. The large-scale experiment on synthetically modified IAM data adopted a formulation of novelty detection that provides $K+1$-way writer identification. The agent was trained on a closed set of K writers, identifying newly encountered writers with a single additional class. Besides writer identification, the agent provides a novelty prediction for each line of text, indicating if the text was produced from a novel or non-novel distribution of data.

The evaluation was composed of a series of approximately 55K tests. Each test simulates real world conditions where limited types of novelty are

Table 2. Results for the large-scale 55K evaluation. All measures reported here are found after selecting the maximum probable class as predicted by the classifier. Novelty Detection is the detection of any novelty in either transcription or writer identification. The stark difference between the NMI and accuracy of Novelty Detection indicates the well-known fact that NMI is a better summary measure than accuracy for correlation between the labels and predictions.

Task	Novelty Present	NMI	Accuracy
Writer ID			
Mean HOG	True	0.1298 ± 0.0513	0.2199 ± 0.4141
	False	0.7768 ± 0.0345	0.7268 ± 0.4456
Transcription		**Character Accuracy**	**Word Accuracy**
CRNN	True	0.6230 ± 0.2034	0.5260 ± 0.2580
	False	0.8267 ± 0.1073	0.7305 ± 0.2258
Novelty Detection		**NMI**	**Accuracy**
CRNN & Mean HOG	True	0.1300 ± 0.0999	0.8029 ± 0.0981

encountered at a specific rate and proportion after a period of no novelty. The specification of experimental design, including six independent variables, is described in Table 11 of Supp. Mat. Sec. 5.1.

A test is a stream of lines of text sampled from an unrevealed distribution. In the pre-novelty phase of the test, the agent is presented with small batches of lines written by a subset of K writers. In the post-novelty phase of the test, the agent is presented with small batches of lines from a new distribution over novel and non-novel writing samples. To eliminate early false positive novelty indications, the agent establishes a prior distribution composed of non-novelty lines presented in a pre-novelty phase of the test. The agent weights instance-level novelty predictions based on a distribution shift of presented lines, signaling entry into a post-novelty phase.

In each test, novel writing examples are sampled from collections containing one type of novelty—either unknown writers, novel pens or novel backgrounds. The novel elements of the writing samples exemplify open world variations in choice of writing instrument and medium. A high-level summary of the results can be found in Table 2, and a more detailed analysis in Supp. Mat. Sec. 5. Results indicate that solving the tasks is feasible, but given the much larger scale of evaluation compared to the experiments in Table 1, there is much room for improvement. This leaves an opening for new work on this problem.

5   Conclusion

This paper introduced an agent-centric approach to handling novelty in the HWR domain. This domain is attractive for the study of novelty, as it consists of a key challenge problem within AI: reading in a more human-like way. The HWR domain with novelty was formalized, an evaluation protocol with benchmark data was introduced, and comprehensive results from a baseline agent were presented to provide the research community with a starting point to build upon. Beyond incremental improvements in transcription performance and style recognition in the presence of novelty, we suggest that adaptation via incremental learning is the next step. Agents that can properly react to and manage novelty, as opposed to merely detecting novelty, will perform better on the task over time. With additions to the evaluation protocol supporting this, we expect a new class of agents to appear for a number of document processing applications.

References

1. Augustin, E., Carré, M., Grosicki, E., Brodin, J.M., Geoffrois, E., Prêteux, F.: RIMES evaluation campaign for handwritten mail processing. In: IWFHR (2006)
2. Bendale, A., Boult, T.E.: Towards open set deep networks. In: IEEE CVPR (2016)
3. Boult, T.E., et al.: A Unifying Framework for Formal Theories of Novelty:Framework, Examples and Discussion. arXiv:2012.04226 [cs], December 2020. http://arxiv.org/abs/2012.04226
4. Boult, T.E., et al.: A unifying framework for formal theories of novelty: framework, examples and discussion. In: AAAI (2021)
5. Dalal, N., Triggs, B.: Histograms of oriented gradients for human detection. In: IEEE CVPR (2005)
6. DARPA: Teaching AI systems to adapt to dynamic environments (2019). https://www.darpa.mil/news-events/2019-02-14. Accessed 11 Jan 2021
7. Fei, G., Liu, B.: Breaking the closed world assumption in text classification. In: NAACL-HLT (2016)
8. Fiel, S., Kleber, F., Diem, M., Christlein, V., Louloudis, G., Nikos, S., Gatos, B.: ICDAR2017 competition on historical document writer identification (Historical-WI). In: 2017 14th IAPR International Conference on Document Analysis and Recognition (ICDAR). vol. 01, pp. 1377–1382 (November 2017). https://doi.org/10.1109/ICDAR.2017.225. iSSN: 2379-2140
9. Graves, A., Fernández, S., Gomez, F., Schmidhuber, J.: Connectionist temporal classification: labelling unsegmented sequence data with recurrent neural networks. In: ICML (2006)
10. He, K., Zhang, X., Ren, S., Sun, J.: Deep residual learning for image recognition. In: IEEE CVPR, pp. 770–778 (2016)
11. Langley, P.: Open-world learning for radically autonomous agents. In: AAAI (2020)
12. van Lit, L.: Paleography: between erudition and computation. In: Among Digitized Manuscripts. Philology, Codicology, Paleography in a Digital World, pp. 102–131. Brill (2019)
13. Lorigo, L.M., Govindaraju, V.: Offline Arabic handwriting recognition: a survey. IEEE T-PAMI **28**(5), 712–724 (2006)
14. Marti, U.V., Bunke, H.: The IAM-database: an English sentence database for offline handwriting recognition. Int. J. Doc. Anal. Recognit. **5**(1), 39–46 (2002)
15. Ontario Ministry of Education: A Guide to Effective Literacy Instruction, Grades 4 to 6: A Multivolume Resource from the Ministry of Education. Volume one, Foundations of literacy instruction for the junior learner, p. 37. Ontario Ministry of Education (2006). https://books.google.com/books?id=Y8F4oAEACAAJ

16. Rayner, K., Pollatsek, A., Ashby, J., Clifton Jr, C.: Psychology of Reading. Psychology Press, London (2012)
17. Rudd, E.M., Jain, L.P., Scheirer, W.J., Boult, T.E.: The extreme value machine. IEEE T-PAMI **40**(3), 762–768 (2017)
18. Ruff, L., et al.: A unifying review of deep and shallow anomaly detection. arXiv preprint arXiv:2009.11732 (2020)
19. Russakovsky, O.: Imagenet large scale visual recognition challenge. IJCV **115**(3), 211–252 (2015)
20. Sanchez, J.A., Romero, V., Toselli, A.H., Vidal, E.: ICFHR2016 competition on handwritten text recognition on the read dataset. In: ICFHR (2016)
21. Scheirer, W.J., Jain, L.P., Boult, T.E.: Probability models for open set recognition. IEEE Trans. Pattern Anal. Mach. Intell. **36**(11), 2317–2324 (2014)
22. Scheirer, W.J., de Rezende Rocha, A., Sapkota, A., Boult, T.E.: Toward open set recognition. IEEE T-PAMI **35**(7), 1757–1772 (2012)
23. Shi, B., Bai, X., Yao, C.: An end-to-end trainable neural network for image-based sequence recognition and its application to scene text recognition. IEEE T-PAMI **39**(11), 2298–2304 (2017)
24. Smith, D.A., Cordell, R.: A research agenda for historical and multilingual optical character recognition. Technical report, Northeastern University (2018)
25. Studer, L., et al.: A comprehensive study of imagenet pre-training for historical document image analysis. In: 2019 International Conference on Document Analysis and Recognition (ICDAR), pp. 720–725 (September 2019). https://doi.org/10.1109/ICDAR.2019.00120. iSSN: 2379-2140
26. Stutzmann, D.: Clustering of medieval scripts through computer image analysis: towards an evaluation protocol. Digit. Mediev. **10** (2016). https://doi.org/10.16995/dm.61. http://journal.digitalmedievalist.org//articles/10.16995/dm.61/. ISSN: 1715-0736
27. Wigington, C., Stewart, S., Davis, B., Barrett, B., Price, B., Cohen, S.: Data augmentation for recognition of handwritten words and lines using a CNN-LSTM network. In: ICDAR (2017)
28. Yampolskiy, R.V., Govindaraju, V.: Behavioural biometrics: a survey and classification. Int. J. Biom. **1**(1), 81–113 (2008)
29. Zhang, H., Patel, V.M.: Sparse representation-based open set recognition. IEEE T-PAMI **39**(8), 1690–1696 (2016)

Vectorization of Historical Maps Using Deep Edge Filtering and Closed Shape Extraction

Yizi Chen[1,2], Edwin Carlinet[1], Joseph Chazalon[1(✉)], Clément Mallet[2], Bertrand Duménieu[3], and Julien Perret[2,3]

[1] EPITA Research and Development Lab. (LRDE), EPITA,
Le Kremlin-Bicêtre, France
yizi.chen@ign.fr, joseph.chazalon@lrde.epita.f
[2] Univ. Gustave Eiffel, IGN-ENSG, LaSTIG, Saint-Mandé, France
[3] LaDéHiS, CRH, EHESS, Paris, France

Abstract. Maps have been a unique source of knowledge for centuries. Such historical documents provide invaluable information for analyzing the complex spatial transformation of landscapes over important time frames. This is particularly true for urban areas that encompass multiple interleaved research domains (social sciences, economy, etc.). The large amount and significant diversity of map sources call for automatic image processing techniques in order to extract the relevant objects under a vectorial shape. The complexity of maps (text, noise, digitization artifacts, etc.) has hindered the capacity of proposing a versatile and efficient raster-to-vector approaches for decades. We propose a learnable, reproducible, and reusable solution for the automatic transformation of raster maps into vector objects (building blocks, streets, rivers). It is built upon the complementary strength of mathematical morphology and convolutional neural networks through efficient edge filtering. Evenmore, we modify ConnNet and combine with deep edge filtering architecture to make use of pixel connectivity information and built an end-to-end system without requiring any post-processing techniques. In this paper, we focus on the comprehensive benchmark on various architectures on multiple datasets coupled with a novel vectorization step. Our experimental results on a new public dataset using COCO Panoptic metric exhibit very encouraging results confirmed by a qualitative analysis of the success and failure cases of our approach. Code, dataset, results and extra illustrations are freely available at https://github.com/soduco/ICDAR-2021-Vectorization.

Keywords: Image processing · Deep learning · Maps · Vectorization · Semantics · Edge filtering · Shape extraction

This work was partially funded by the French National Research Agency (ANR): Project SoDuCo, grant ANR-18-CE38-0013. We thank the City of Paris for granting us with the permission to use and reproduce the atlases used in this work.

J. Lladós et al. (Eds.): ICDAR 2021, LNCS 12824, pp. 510–525, 2021.
https://doi.org/10.1007/978-3-030-86337-1_34

Fig. 1. Example (cropped) of a raster map used as input (top) and output of our vectorization approach (bottom): a set of individual polygons in vector format ready for further annotation in any GIS software. Some shapes are still undetected.

1   Introduction

Historical maps are unique and powerful tools for understanding the transformations of the geographical space over unique time spans. They are invaluable inputs in historical sciences, architecture, and urban planning. The massive digitization of archival collection resources carried out by heritage institutions dramatically increases the amount of geospatial information available for certain areas of the world. In the Western world, the rapid development of geodesy and cartography from the 18th century resulted in massive production of topographic maps at various scales. City maps are of utter interest. They contain rich, detailed, and often geometrically accurate representations of numerous geographical entities. Maps also document the distribution in space and the topological relationship of the depicted entities, while legends and text labels provide semantic information, in particular about their functions [10, 22]. Recovering spatial and semantic information represented in old maps requires a so-called *vectorization* process.

Vectorizing maps consists in transforming rasterized graphical representations of geographic entities (often maps) into instanced geographic data (or vector data), that can be subsequently manipulated (using Geographic Information Systems, GIS). This is a key challenge today to better preserve, analyze and disseminate content for numerous spatial and spatio-temporal analysis purposes.

From an image processing and a document analysis perspective, vectorization can be cast by the following, often interleaved, problems: (i) isolate the map content on pictures of map sheets (leave out the legend, in particular); (ii) detect and separate the various layers of graphical content: points, lines, and shape objects, as well as symbols and text; (iii) classify/recognize each graphical object of interest (including text), while ensuring a topologically and geometrically consistent result (often considered as an *instance segmentation* issue); (iv) georeference the extracted elements.

In this paper, we focus on closed shape detection in the 19th and early 20th-century historical map atlases of Paris (France). Currently, shape detection is usually manually performed, using GIS software. Such a costly and tedious process leads to heterogeneous data quality. The latest methodological developments in image processing leverage the ability to automatically build a significant number of geo-historical databases, which eventually benefit to multiple research areas. Unfortunately, unlike computer-generated maps, roughly following the same semiotic rules, historical maps steadily vary in terms of legend, level of generalization, type of geographic features, and text fonts. Even more, objects in historical maps exhibit very limited color and texture information. This creates ambiguities in interpretation, leading to the failure of main texture-based semantic [3,24,31]/instance segmentation [4,6,30] and raster-to-vector [23] approaches. Such an issue is exacerbated by the frequent overlap between map objects and occlusion with the carto-geodetic information (vertical and horizontal lines). Their instantiation become more complex, with often non continuous boundary retrieval. Last, ink+paper aging and damage in the historical image such as erased lines, bad contrast, noisy content, tearing, folding might cause gaps in the cartographic information and lead to poor object detection. To tackle those challenges, we are therefore interested in developing a versatile shape detection approach, loosely coupled with the style of a given map. We aim to accelerate the detection of core city structures (building blocks, rivers, street networks), as well as the georeferencing process while keeping both very accurate.

We recently found that extracting closed shapes from historical maps [8] was feasible, using a deep learning architecture as an edge filter (rather than trying to predict instances directly), connected with mathematical morphology tools to extract closed shapes from an edge map. The current paper improves over these preliminary findings by exploring more combinations of deep edge filters, followed by closed shape extraction (Sect. 3), and exploring all these potential architectures in an extensive benchmark (Sect. 4).

2   Related Work

Extracting the content of historical maps is an active topic in Geographic Information Sciences, in an effort to provide tools and methods to build large spatio-temporal datasets and historical gazetteers. This is a special case of the broader topic of digital map processing where most challenges are exacerbated by the higher complexity and heterogeneity of ancient maps compared to the more homogeneous modern maps [9,11]. Three main categories of extraction strategies can be distinguished: manual vectorization with GIS tools, automatic solutions, and hybrid frameworks. They either focus on structured geographical entities (roads, buildings) or textual information (named places mainly).

Manually extracting the content of historical maps with GIS tools is still the main strategy in digital humanities. This solution is conceivable when small-size datasets are handled, in space and time, for peculiar case studies [26]. Collaborative approaches have been proposed to handle larger map corpora, ranging from a limited number of contributors [27] to large-scale crowdsourcing experiments [5,34]. Although manual extraction creates "data-intimacy" conducive to reflexive work, the process is tedious, time-consuming, and leads to non-reproducible results.

Automatic solutions are dedicated either to specific symbols and local features extraction [14] or to map vectorization, i.e., (semantic) shape retrieval using raster-to-vector techniques. Many image vectorization solutions exist [18] but cannot be applied to complex historical maps: design, multiplicity of objects, noise, limited graphical quality (no color nor texture) hinder their performance. Approaches dedicated to old maps rely on the following unsupervised workflow: color-based segmentation of objects, binary (shape) filtering, cleaning, and vectorization. They exhibit two main drawbacks: a priori knowledge on color and object shapes narrow down the versatility power of the methods [21], and no focus is made on extracting multiple closed shapes, under a learning paradigm.

A suitable trade-off is found with hybrid methods that would target to learn shape extraction from crowdsourcing data. First attempts have shown their relevance [5] but closer interaction between extraction and annotation is desired to really benefit from both fields.

In Mathematical Morphology, the watershed transform is a *de facto* standard approach for image segmentation, and it has been widely studied in terms of topological properties [13,29] and computational efficiency [12,13]. While being efficient on rather clean images, the watershed is hard to apply in real, complex images. This hinders wider applications related to closed shape extractions. For natural images, Hanbury et al. [16] use a watershed algorithm to segment closed shapes from the learned gradient image extracted, while Arbelaez et al. [2] introduce oriented watershed to extract closed shapes in global probability boundaries created by several image features and cues. Unfortunately, there are very few applications that adopt watershed transform in spatial data, especially cartographic images. To our best knowledge, only one approach uses watershed transform to extract spatial semantic objects through color and geometrical features in color cartographic images [1]. However, this method is hard to apply in our historical map image, which has very limited colors and textual information.

Detecting edges from images is a widely studied subject. Firstly, it has been achieved with handcrafted features such as brightness, color, and textures [25]. After that, those features are efficiently grouped by mono-scale and multi-scale attributes retrieving micro-structures such as textons and salient examples [37]. This led to methods focusing on combining all existing features [2]. The weights are learned to efficiently combine the relevant cues such as gradients and textons in multiscale images [2] to estimate a probability of boundary. Then, an ultrametric contour map (UCM) is created by using probability boundary as input to extract closed shapes to represent image objects, turning soft assignments into hard ones. Recently, Convolutional Neural Networks (CNNs) have proved their relevance to extract and combine meaningful image features to delineate semantic objects. A large amount of research has focused on semantic edge detection. The most famous deep edge detector is called holistically edge detector(HED) [35] which is an end-to-end deep learning multi-scale architecture, built upon a VGG-16 backbone network [33]. The novelty lies in adopting skip-connections to combine multiple levels of features, while different losses are measured in the intermediate outputs of VGG-16 to filter out useful edge features at each stage of the network. It allows to better learn and recover multi-scale representations of image features. Recently, He et al. [17] proposed a so-called Bidirectional cascade network (BDCN) by designing a scale-enhancement module (SEM) and adding to every intermediate output of the HED to learn diversity edge features and enhance spatial contexts: traditional convolution kernels are substituted by dilated convolution kernels [36]. Most of the deep edge detectors are encoder-decoder structures (U-Net fashion [31]), built based on VGG-16 architecture which can use ImageNet [32] pretrained features for limiting the discrimination task to a transfer learning process: convergence is accelerated and accuracy is increased.

Very few papers are fully dedicated to historical map processing. Petit-pierre et al. [28] use U-Net [31] and SegNet [3] architectures to perform semantic segmentation on historical maps of various kinds and locations. They focus on few classes and do not propose a vectorization step, required to extract semantized instances. To tackle this issue, we provide a complete, trainable, raster-to-vector solution to extract closed shapes from historical maps with little to no texture or color information by combining the power of deep edge detector as a filtering tool with strong guarantees of closed shape extraction using mathematical morphology. To our best knowledge, to automate this process as far, we are among the only approaches with a strong commitment to being reproducible and reusable; all our code and data is open source and public.

3   Raster-to-Vector Pipeline

Our process for the raster-to-vector conversion of historical map images is divided into 3 main stages, as illustrated by Fig. 2. We introduce several possibilities for each stage, detailed hereafter. Stage 1 (Deep Edge Filtering) can be composed of either HED, BDCN, U-Net, or ConnNet standard architectures. It consists in producing an Edge Probability Map (EPM) from the raster input. Stage 2 (Closed Shape Extraction) can be composed of either a connected component (CC) labeling or a watershed (WS). We generate a Label Map (LM), which assigns a shape identifier to each pixel. Stage 3 (Vectorization) is performed using an off-the-shelf vectorization tool of the GDAL open-source project [15] as a proof of concept, leading to results like the one illustrated in Fig. 1.

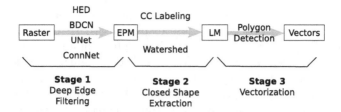

Fig. 2. Overview of our raster-to-vector pipeline and of the evaluated alternative techniques for each stage. Stage 1 produces an Edge Probability Map (EPM), which estimates how likely a pixel belongs to a semantic edge, and Stage 2 produces a Label Map (LM), where each pixel receives the label of the shape it belongs to.

3.1 Deep Edge Detectors

HED. The Holistically edge detector (HED) is a state-of-the-art deep edge detection architecture, with a so-called *skip-layer network training* procedure: different levels of intermediate features and merge into the final stage of the output layer [35]. HED is built based on VGG-16 architecture. Intermediate outputs are chosen from VGG-16 to select multi-level features from several scales. Those outputs are upsampled and concatenated, followed by 1×1 convolution to fuse multiple channel outputs into an EPM image. We will consider both a complete training and a fine-tuning from weights pre-trained using natural images (ImageNet).

BDCN. BDCN is another reference network, built on VGG-16, with some modifications compared to HED. Scale enhancement module (SEM) [17] modules are added to the intermediate outputs. The SEM modules are created by using dilated convolution [7], which enlarges the receptive field of kernels without downgrading the resolution of the feature maps [17]. In the end, all the intermediate output are similarly upsampled, concatenated, and fused into one EPM image. We will consider both a complete training and a fine-tuning from weights pre-trained using natural images (ImageNet).

U-Net. The U-Net architecture [31] is a U-shape architecture inspired by Fully convolution networks [24]. It contains two different paths to combine image features. The first path is called the contracting path, which is using downsampling to get deep and semantic features. The other one is the expansive path, which concatenates high resolution features with spatial information from downsampling, using the outputs of the up-convolution features. U-Net has a symmetrical architecture that can retain well spatial information of pixels leading to very accurate localizations. This is an advantage on boundary detection task.

ConnNet. ConnNet [19] architecture is invented by using pixel connectivity information to predict salient objects. In this paper, we decided to encode and predict pixel connectivity information for every pixel in the output. It is a learnable way to binarize EPM, making use of pixel-based spatial information at training time. The original ConnNet exhibits a significant amount of parameters and may not be efficient in our data. We modify the original ConnNet architecture by simply concatenating two networks, a deep edge detector (BDCN in this paper but any deep edge detector can be replaced here) with an encoder-decoder architecture (U-Net, but, again any full convolution architecture is relevant). The output of ConnNet is a 4- or 8-channel probability map which predicts for each pixel its 4- or 8-connectivity with each of its neighbors.

3.2   Training HED, BDCN and U-Net

Our input image is $x \in \mathbb{R}^{H \cdot W}$ and the ground truth label us $y \in 0, 1^{H \cdot W}$. The output of the predicted image is $\hat{y} = f(x, w) \in 0, 1^{H \cdot W}$ and every element of $\hat{y}$ is interpreted as the probability of pixel i having label 1: $\hat{y} \equiv p(Y_i = 1|x, w)$. Binary cross entropy loss is used as loss function, measuring the difference between edge predictions and labels. Due to the highly imbalanced nature between non-edge (97.5% in training and validation sets) and edge pixels (only 2.5%), α, β are used to re-balance the binary cross entropy loss. It is formulated as:

$$\mathcal{L}_{BCE} = -\alpha \sum_{j \in Y_-} log(1 - \hat{y}_j) - \beta \sum_{j \in Y_+} log(\hat{y}_j), \qquad (1)$$

where Y_+ is the set of indices of edge pixels, Y_- is the set of indices of non-edge pixels, $\alpha = (\lambda \cdot |Y_-|/(|Y_+| + |Y_-|))$ is the percentage of edge pixels in each batch of the historical map image, and $\beta = (|Y_+|/(|Y_+| + |Y_-|)))$ is percentage of non-edge pixels. An extra $\lambda = 1.1$ factor is used to enhance the percentage of edge pixels in order to give extra weights for edge responses.

3.3   Training ConnNet

We do detail the ConnNet training procedure. Please refer to [19]. Here we only mention the process of measuring the loss function of ConnNet, which is a combination of an edge loss and a connectivity loss:

$$\mathcal{L}_{Connnet} = \mathcal{L}_{edge} + \mathcal{L}_{pixcon}. \qquad (2)$$

$\mathcal{L}_{edge}$ is the binary cross entropy loss from Eq. 1 for edge detection, and $\mathcal{L}_{pixcon}$ is the pixel connectivity loss. L_{pixcon} is measured by the binary cross-entropy loss between connectivity labels and predictions [19]. The connectivity label y^c is created by encoding surrounding location information of the pixel. It results in eight channels, every channel represents the connectivity information in a specific direction. Then, the pixel connectivity loss L_{pixcon} is measured in Eq. 3 by using the binary cross entropy loss (Eq. 1) between connectivity labels and predictions [19]. In Eq. 3, $\hat{y}^c$ is the probability of the predicted connectivity, y^c represents the connectivity labels, N is the number of pixels in images, and C

is the number of connected pixels, respectively. $\frac{1}{(N \times C)}$ is the normalizing term for the loss.

$$\mathcal{L}_{pixcon} = \frac{1}{N \times C} \sum_{c=1}^{C} \sum_{i=1}^{N} L_{BCE}(\hat{y}_i^c, y_i^c). \tag{3}$$

3.4 Closed Shape Extraction

Closed shapes can be extracted from EPMs either through connected component labeling or watershed transform.

Connected component labeling requires a binarized edge map as input and subsequently a binarization threshold θ in order to provide a label map.

The watershed transform directly necessitates the EPM as input and returns 1-pixel thin edges and a label map. It offers a strong guarantee of closed shape extraction with efficient implementation. The strength of the watershed is to recover the boundaries of objects even on weak edge responses that would be lost by EPM thresholding. The filtering parameters (dynamic δ and minimum area σ) are important to control the trade-off between the fact we want to recover small/leaking regions (somewhat related to the recall) and the false-detection of boundaries (somewhat related to the precision). The h-minima characterize the importance of each local minimum through their **dynamic**. When flooding a basin, it refers to the water elevation required to merge with another basin. The calibration of this parameter depends on the intensity of the edges detected by the deep edge detector, and needs to be tuned accordingly. The other parameter is the **minimum area** of the components which can be easily known as prior knowledge in our historical maps, for example, objects with a size that below $100 \, m^2$ do not appear in our map. Even if the edge response is low (i.e., weak gradient), the watershed can consider this weak response and closes the contour of the region.

4 Benchmark

Here, we conduct an extensive comparison of the performance of the different modules for Stages 1 (6 variants) and 2 (2 variants) of our pipeline.

4.1 Datasets and Training Data Generation

The dataset used for training and testing the networks is one of the collections of Paris atlases. Two particular sheets in the years 1898 and 1925 are selected, focusing on the central area of the city, which exhibits high landscape diversity. These large maps were digitized with high resolution resulting in also large image sizes. Those two sheets share similar graphical content but have content, contrast, and preservation differences.

We annotated all objects in the map image by creating line vector information for each boundary to represent and label every object of interest in the map. From this vector information, we created the target outputs for each of our processing stages. We rasterized the borders of the polygons to produce a binary edge image used as the target for the Edge Probability Map deep edge filters of Stage 1 must produce. Borders were dilated to 3 pixels to match the average border thickness in the original images. Then, we created a raster label map where an integer label is assigned to every pixel of the image depending on the shapes it falls into. We assigned a special zero label to the parts of the image which did not contain map content (borders and titles); this was used to filter the outputs of Stages 1 and 2 to ensure only relevant areas were processed.

Finally, the 1925 map sheet was split into two distinct parts and used as a training and validation set, and the 1898 map sheet was used as a test set. The resulting training image has a $4,500 \times 9,000$ pixel size, with 3,343 objects, while the validation image has a size of $3,000 \times 9,000$ pixels with 2,183 shapes, and the test image has a size of $6,000 \times 5,500$ pixels with 2,836 shapes.

4.2   Protocol

Our evaluation protocol aims to identify the best deep edge detector for Stage 1 (first experiment), as well as to validate the relevance of the watershed for Stage 2 under optimal parameters settings (second experiment). The final polygon vectorization step is not used in this quantitative analysis; it was used as a pre-defined post-processing step to demonstrate that our approach effectively provides a suitable solution in the challenging sections of the raster-to-vector conversion of historical maps. The first experiment consists in comparing the performance of each deep edge detector, either trained from scratch or fine-tuned from pre-trained weights. For each of them, we check their performance on the validation set using the final metric for each training epoch. To this end, for each epoch of each detector, we compute the set of detected shapes using a connected component labeling computed on a binarized edge probability map (EPM) with a fixed threshold of 0.5. This creates a small bias in favor of our baseline for the second stage (CC labeling module). For each deep edge filter, we retain the model with the highest metric value to produce the EPMs for the second experiment. The second experiment consists in comparing the performance of the watershed module against a simpler baseline (connected component extraction) under the best set of parameters of each of them, given the best possible input from the previous stage. Using the EPMs produced by the best performing model for each variant in the first stage, we proceed in two steps. First, we calibrate the parameters for each closed shape extractor using the performance measured on the validation set. Secondly, we evaluate on the test set the performance of each combination of the best model for Stage 1 with the best calibration for each variant of Stage 2. The qualitative results illustrated in Fig. 1 are produced by vectorizing the output of the best performing combination on the test image.

Deep Edge Filtering Variants and Parameters. We consider six network variants: HED trained from scratch (**HED-scratch**), and fine-tuned using ImageNet pre-trained weights (**HED-pretrain**), BDCN trained from scratch (**BDCN-scratch**) and fine-tuned using ImageNet pre-trained weights (**BDCN-pretrain**), U-Net trained from scratch (**U-Net scratch**), and ConnNet trained from scratch (**ConnNet scratch**). As historical maps are scanned documents that have different image features compared to natural images, assessing the benefits of transfer learning is an important question. This explains why we test both HED and BDCN with and without a pre-training model on natural images. Such a pre-training was not available for U-Net and ConnNet and was discarded. During the training phase, we use the same settings for all variants: ADAM optimizer with an initial learning rate of $5e^{-5}$, a momentum of 0.9, and a weight decay of 0.002. We use an early-stopping scheme that limits the number of total epochs to consider, setting an upper limit to 60 epochs for each network.

Closed Shape Extraction Variants and Parameters. We consider two variants for the second stage of our pipeline: A connected component labeling (**CC labeling**) module used as a baseline, and a watershed (**Watershed**) module. As the CC labeling requires a binary image as input, it is necessary to binarize the Edge Probability Map produced beforehand, except in the case of ConnNet which already generates a binary image. We perform a grid search on 10 different thresholds ranging from 0.1 to 0.9 on the EPM prior to the labeling procedure. The watershed module is tuned through two parameters: a threshold on the EPM dynamic (δ) and a minimum area threshold (σ). We perform a grid search on 10 different thresholds for δ ranging from 0 to 0.04 (based on a study of the distribution of predicted edge probabilities) and with the following minimum areas for σ: 50, 100, 200, 300, 400, 500 pixels (roughly corresponding to objects with a real surface between 25 and 250 square meters—this could be set manually). The watershed does not bring any advantage over connected component labeling in the case of a binary EPM. Therefore, we do not report results for the **ConnNet scratch + Watershed** combination. Finally, large-image processing both for validation and test sets requires a tilling step: we first reconstruct the global images before running closed shape extraction to preserve the topological properties of the image and leverage a much larger spatial context than the one used by the deep edge detectors.

4.3 Metrics

To assess the performance of our system, we report the Panoptic Quality (PQ) values for each variant under test. The PQ metric [20] was introduced to simultaneously measure the detection, segmentation, and classification (hence the *panoptic* term) performance of a set of systems and rank them in the context COCO Panoptic challenge. In this work, we consider only one class of object. Wee do not make use of the multi-class capabilities of the metric and focus the joint detection and segmentation evaluation. The COCO Panoptic PQ indicator is computed as follows. First, for each pair of shapes (t_i, p_j) in the target

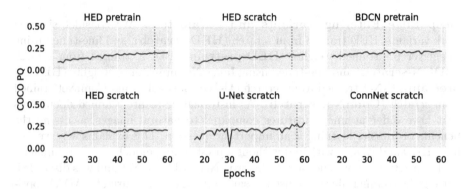

Fig. 3. COCO-PQ score of the networks under test for Stage 1 (Deep Edge Filtering), computed on the validation set, for each training epoch. The red vertical bar indicates the best model we retained to compute the results on the test set. (Color figure online)

and in the predicted segmentation, the Jaccard index (or IoU) is computed as $IoU(t_i, p_j) = \frac{t_i \cap p_j}{t_i \cup p_j}$. Then, the set of uniquely matching pairs (or *true positives*) TP is defined as the set all pairs (t_i, p_j) such as $IoU(t_i, p_j) > 0.5$, leading to the definition of PQ:

$$PQ = \frac{\sum_{(t_i, p_j) \in TP} IoU(t_i, p_j)}{|TP| + \frac{1}{2}|FP| + \frac{1}{2}|FN|}, \tag{4}$$

where FP is the set of *false positives* (the set of *predicted* shapes which do not belong to any pair TP), and FN is the set of *false negatives* (the set of *target* shapes which do not belong to any pair in TP). While this measure summarizes the segmentation and the detection quality into a single indicator, two additional metrics provide additional insights: the Segmentation Quality (SQ) and the Recognition Quality (RQ) defined such as $PQ = SQ \times RQ$ where:

$$SQ = \frac{\sum_{(t_i, p_j) \in TP} IoU(t_i, p_j)}{|TP|}, \quad RQ = \frac{|TP|}{|TP| + \frac{1}{2}|FP| + \frac{1}{2}|FN|}. \tag{5}$$

4.4 Results and Discussion

The validation of the performance of each deep edge filter using the shape detection metric on the validation set enables the selection of the best model for each of the variants. Indeed, instead of selected the best model based on the validation loss used for the training, we adopt a more complex evaluation procedure, more representative of the final goal. We report in Fig. 3 the profile of the COCO PQ indicator during the training for each variant, and identify the best model retained. It is worth noting that BDCN and ConnNet seem to plateau quite rapidly (best model obtained around epoch 40), while HED and U-Net keep progressing (best model obtained close to the limit of 60 epochs).

Table 1. Global COCO Panoptic results (in %) of our evaluation, for each combination of deep edge detector (Stage 1) and closed shape extractor (Stage 2). Best results on validation and test sets are indicated **in bold**.

Stage 1	Stage 2: Connected Component Labeling						
	Parameters	Validation			Test		
	θ	PQ	SQ	RQ	PQ	SQ	RQ
HED pretrain	0.7	27.6	77.6	35.6	**16.2**	76.1	21.3
HED scratch	0.3	23.2	76.5	30.3	14.0	74.8	18.8
BDCN pretrain	0.8	27.6	82.1	33.7	8.9	82.8	10.7
BDCN scratch	0.9	27.7	80.6	34.4	10.2	80.4	12.7
U-Net scratch	0.9	**34.8**	80.5	43.3	8.1	78.2	10.4
ConNet scratch	*none*				14.2	73.6	19.3

Stage 1	Stage 2: Watershed							
	Parameters		Validation			Test		
	δ	σ	PQ	SQ	RQ	PQ	SQ	RQ
HED pretrain	0.031	300	52.8	87.6	60.3	**38.4**	85.5	44.9
HED scratch	0.040	200	50.5	87.2	57.9	35.6	84.6	42.0
BDCN pretrain	0.027	300	53.0	88.1	60.1	37.8	86.4	43.8
BDCN scratch	0.031	200	52.5	87.8	59.8	34.9	85.8	40.6
U-Net scratch	0.000	50	**56.6**	87.7	64.5	18.3	85.2	21.4

Then, using those best models for deep edge filtering, we are able to perform a grid search on the set of parameters for the baseline (CC labeling) and our proposed watershed module for closed shape extraction. The best parameters obtained are then used to compute the final performance of each combination of a deep edge filter with a closed shape extractor on the test set. Table 1 reports these results and the value of the best set of parameters for each combination. From these results we can draw several observations.

The first observation is that **the watershed constantly improves shape detection.** The joint ability of the watershed to filter out small noisy shapes, to recover weak edges, and to produce thin edges of one pixel explain the systematic gain obtained, no matter which deep edge filter is used. This very simple, yet effective technique, is a great complement to convolutional neural network and a key enabler in our application. This confirms the potential of the global pipeline we propose, i.e., combining a deep edge filter with a shape extraction using Mathematical Morphology tools.

The second observation is that **pre-trained HED is the best deep edge filter.** The Holistically Edge Detector outperforms the other architectures both when used with a naive closed shape extraction or with a more advanced one such as our watershed module. This seems to be mainly due to the lower complexity of this model compared to BDCN, U-Net and ConnNet; given a limited training

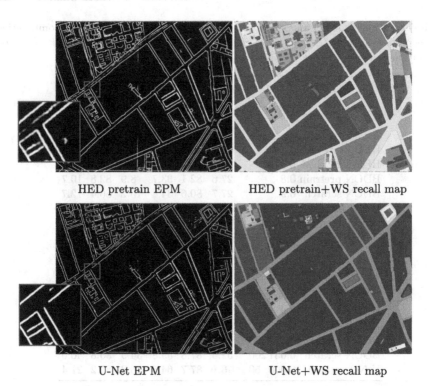

HED pretrain EPM   HED pretrain+WS recall map

U-Net EPM   U-Net+WS recall map

Fig. 4. Comparison of intermediate (EPM, *left*) and final shape prediction (recall map, *right*) results for pre-trained HED (best model, *top*) and U-Net. We can see on the left that many discontinuities in U-Net EPM prevent the detection of closed shaped, as indicated by zoomed part (red box). The recall map on the right indicates the quality of the segmentation of each shape – green shapes are correctly detected shapes and the red one are missed. (Color figure online)

set, it generalizes better, and produces less strict edges (as seen in Fig. 4), which are easier to recover in the second stage.

This is highly related to the observation that **deep edge filters tend to overfit**. This is rather obvious that deep networks can easily overfit, but in our application, where annotation comes with a great cost (annotation took approximately 200 h here), the resilience of HED is very valuable.

A further observation is that **transfer learning can improve results**: HED and BDCN benefit from features learned on natural images (ImageNet), when used with the watershed. Surprisingly, the performance is degraded for BDCN with plain component labeling for the pre-trained version.

This is unfortunate that no pre-training is available in the case of **U-Net which seems prone to overfit**. Despite reaching the best performance on the validation set, this filter performs poorly on the test set. The study of the output Edge Probability Map reveals very strong edges and equally strong cuts in them, as illustrated in Fig. 4: edges too clear and many gaps prevent a proper shape detection. Such edges cannot be recovered by the watershed, even with a dynamic threshold set to zero.

Finally, it should be noted that **the ConnNet architecture we introduced here for deep edge detection produces encouraging results.** Despite its heavier architecture, it reaches the second rank for the baseline shape detection while integrating the threshold calibration directly in its training process. Further work will be needed to enable its compatibility with the watershed.

5   Conclusion

We presented a learnable framework which enables the automatic vectorization of closed shaped from challenging historical maps. To our knowledge, this is the first approach which can accelerate manual annotation on such data. We leverage powerful convolutional neural networks as deep edge filters combined with watershed to effectively and efficiently detect closed shapes. We performed an extensive comparison on large images of several deep architectures which revealed the superiority of HED, and the relevance of the watershed transform we designed. The introduction of ConnNet architecture as a deep edge detector also opens some promising directions. Finally, our code, dataset, results and extra illustrations are freely available at https://github.com/soduco/ICDAR-2021-Vectorization.

References

1. Angulo, J., Serra, J.: Mathematical morphology in color spaces applied to the analysis of cartographic images. Proc. GEOPRO **3**, 59–66 (2003)
2. Arbelaez, P., Maire, M., Fowlkes, C., Malik, J.: Contour detection and hierarchical image segmentation. IEEE Trans. Pattern Anal. Mach. Intell. **33**(5), 898–916 (2010)
3. Badrinarayanan, V., Kendall, A., Cipolla, R.: SegNet: a deep convolutional encoder-decoder architecture for image segmentation. IEEE Trans. Pattern Anal. Mach. Intell. **39**(12), 2481–2495 (2017)
4. Bai, M., Urtasun, R.: Deep watershed transform for instance segmentation. In: Proceedings of Conference on Computer Vision and Pattern Recognition, pp. 5221–5229 (2017)
5. Budig, B., van Dijk, T.C., Feitsch, F., Arteaga, M.G.: Polygon consensus: smart crowdsourcing for extracting building footprints from historical maps. In: Proceedings of the 24th ACM SIGSPATIAL International Conference on Advances in Geographic Information Systems, pp. 1–4 (2016)
6. Chen, K., et al.: Hybrid task cascade for instance segmentation. In: Proceedings of Conference on Computer Vision and Pattern Recognition, pp. 4974–4983 (2019)
7. Chen, L.C., Papandreou, G., Kokkinos, I., Murphy, K., Yuille, A.L.: DeepLab: semantic image segmentation with deep convolutional nets, atrous convolution, and fully connected CRFs. IEEE Trans. Pattern Anal. Mach. Intell. **40**(4), 834–848 (2017)

8. Chen, Y., Carlinet, E., Chazalon, J., Mallet, C., Duménieu, B., Perret, J.: Combining deep learning and mathematical morphology for historical map segmentation. In: Proceedings of International Conference on Discrete Geometry and Mathematical Morphology (DGMM) (2021, accepted paper)
9. Chiang, Y.-Y., Duan, W., Leyk, S., Uhl, J.H., Knoblock, C.A.: Historical map applications and processing technologies. In: Chiang, Y.-Y., Duan, W., Leyk, S., Uhl, J.H., Knoblock, C.A., et al. (eds.) Using Historical Maps in Scientific Studies. SG, pp. 9–36. Springer, Cham (2020). https://doi.org/10.1007/978-3-319-66908-3_2
10. Chiang, Y.-Y., Leyk, S., Knoblock, C.A.: Efficient and robust graphics recognition from historical maps. In: Kwon, Y.-B., Ogier, J.-M. (eds.) GREC 2011. LNCS, vol. 7423, pp. 25–35. Springer, Heidelberg (2013). https://doi.org/10.1007/978-3-642-36824-0_3
11. Chiang, Y.Y., Leyk, S., Knoblock, C.A.: A survey of digital map processing techniques. ACM Comput. Surv. (CSUR) 47(1), 1–44 (2014)
12. Couprie, M., Najman, L., Bertrand, G.: Quasi-linear algorithms for the topological watershed. J. Math. Imaging Vis. 22(2), 231–249 (2005)
13. Cousty, J., Bertrand, G., Najman, L., Couprie, M.: Watershed cuts: thinnings, shortest path forests, and topological watersheds. IEEE Trans. Pattern Anal. Mach. Intell. 32(5), 925–939 (2009)
14. Dhar, D., Chanda, B.: Extraction and recognition of geographical features from paper maps. Int. J. Doc. Anal. Recogn. 8, 890–904 (2006)
15. GDAL/OGR contributors: GDAL/OGR Geospatial Data Abstraction software Library. Open Source Geospatial Foundation (2021). https://gdal.org
16. Hanbury, A., Marcotegui, B.: Morphological segmentation on learned boundaries. Image Vis. Comput. 27(4), 480–488 (2009)
17. He, J., Zhang, S., Yang, M., Shan, Y., Huang, T.: BDCN: bi-directional cascade network for perceptual edge detection. IEEE Trans. Pattern Anal. Mach. Intell. (2020)
18. Hilaire, X., Tombre, K.: Robust and accurate vectorization of line drawings. IEEE Trans. Pattern Anal. Mach. Intell. 28(6), 890–904 (2006)
19. Kampffmeyer, M., Dong, N., Liang, X., Zhang, Y., Xing, E.P.: ConnNet: a long-range relation-aware pixel-connectivity network for salient segmentation. IEEE Trans. Image Process. 28(5), 2518–2529 (2018)
20. Kirillov, A., He, K., Girshick, R., Rother, C., Dollár, P.: Panoptic segmentation. In: Proceedings of Conference on Computer Vision and Pattern Recognition, pp. 9404–9413 (2019)
21. Leyk, S., Boesch, R.: Colors of the past: color image segmentation in historical topographic maps based on homogeneity. GeoInformatica 14(1), 953–968 (2010)
22. Leyk, S., Boesch, R., Weibel, R.: Saliency and semantic processing: extracting forest cover from historical topographic maps. Pattern Recognit. 39(5), 953–968 (2006)
23. Liu, C., Wu, J., Kohli, P., Furukawa, Y.: Raster-to-vector: revisiting floorplan transformation. In: Proceedings of International Conference of Computer Vision (ICCV) (2017)
24. Long, J., Shelhamer, E., Darrell, T.: Fully convolutional networks for semantic segmentation. In: Proceedings of Conference on Computer Vision and Pattern Recognition, pp. 3431–3440 (2015)
25. Martin, D., Fowlkes, C., Tal, D., Malik, J.: A database of human segmented natural images and its application to evaluating segmentation algorithms and measur-

ing ecological statistics. In: Proceedings of International Conference of Computer Vision (ICCV), vol. 2, pp. 416–423. IEEE (2001)

26. Ostafin, K., Kaim, D., Siwek, T., Miklar, A.: Historical dataset of administrative units with social-economic attributes for Austrian Silesia 1837–1910. Sci. Data **7**(1), 1–14 (2020)

27. Perret, J., Gribaudi, M., Barthelemy, M.: Roads and cities of 18th century France. Sci. Data **2**(1), 1–7 (2015)

28. Petitpierre, R.: Neural networks for semantic segmentation of historical city maps: cross-cultural performance and the impact of figurative diversity. arXiv preprint arXiv:2101.12478 (2021)

29. Roerdink, J.B., Meijster, A.: The watershed transform: definitions, algorithms and parallelization strategies. Fundamenta Informaticae **41**(1, 2), 187–228 (2000)

30. Romera-Paredes, B., Torr, P.H.S.: Recurrent instance segmentation. In: Leibe, B., Matas, J., Sebe, N., Welling, M. (eds.) ECCV 2016. LNCS, vol. 9910, pp. 312–329. Springer, Cham (2016). https://doi.org/10.1007/978-3-319-46466-4_19

31. Ronneberger, O., Fischer, P., Brox, T.: U-Net: convolutional networks for biomedical image segmentation. In: Proceedings of Medical Image Computing and Computer Assisted Intervention (MICCAI), pp. 234–241 (2015)

32. Russakovsky, O., et al.: ImageNet large scale visual recognition challenge. Int. J. Comput. Vis. **115**(3), 211–252 (2015)

33. Simonyan, K., Zisserman, A.: Very deep convolutional networks for large-scale image recognition. arXiv preprint arXiv:1409.1556 (2014)

34. Southall, H., Aucott, P., Fleet, C., Pert, T., Stoner, M.: Gb1900: engaging the public in very large scale gazetteer construction from the ordnance survey "county series" 1: 10,560 mapping of great Britain. J. Map Geogr. Libr. **13**(1), 7–28 (2017)

35. Xie, S., Tu, Z.: Holistically-nested edge detection. In: Proceedings of International Conference on Computer Vision (ICCV), pp. 1395–1403 (2015)

36. Yu, F., Koltun, V.: Multi-scale context aggregation by dilated convolutions. arXiv preprint arXiv:1511.07122 (2015)

37. Zhu, S.C., Guo, C.E., Wang, Y., Xu, Z.: What are textons? Int. J. Comput. Vis. **62**(1), 121–143 (2005)

Data Augmentation Based on CycleGAN for Improving Woodblock-Printing Mongolian Words Recognition

Hongxi Wei[1,2,3]([✉]), Kexin Liu[1,2,3], Jing Zhang[1,2,3], and Daoerji Fan[4]

[1] School of Computer Science, Inner Mongolia University, Hohhot 010021, China
cswhx@imu.edu.cn
[2] Provincial Key Laboratory of Mongolian Information Processing Technology, Hohhot, China
[3] National and Local Joint Engineering Research Center of Mongolian Information Processing Technology, Hohhot, China
[4] School of Electronic Information Engineering, Inner Mongolia University, Hohhot, China

Abstract. In order to improve the performance of woodblock printing Mongolian words recognition, a method based on cycle-consistent generative adversarial network (CycleGAN) has been proposed for data augmentation. A well-trained CycleGAN model can learn image-to-image translation without paired examples. To be specific, the style of machine printing word images can be transformed into the corresponding word images with the style of woodblock printing by utilizing a CycleGAN, and vice versa. In this way, new instances of woodblock printing Mongolian word images are able to be generated by using the two generative models of CycleGAN. Thus, the aim of data augmentation could be attained. Given a dataset of woodblock printing Mongolian word images, experimental results demonstrate that the performance of woodblock printing Mongolian words recognition can be improved through such the data augmentation.

Keywords: Generative adversarial network · Data augmentation · Cycle consistent · Segmentation-free recognition

1 Introduction

In recent years, a number of ancient Mongolian books have been scanned into images for protecting as long as possible. The most of these ancient books were made by means of woodblock printing. The process of the woodblock printing is as follows [1]. At first, Mongolian words were engraved in woodblocks by artisans. Secondly, vermillion ink was put on the woodblocks. Finally, papers were put on the woodblocks, and then the words were printed on the papers. Due to aging, the scanned images are degraded with various noises, stains and fading. Moreover, the corresponding Mongolian words are equivalent to a kind of offline handwritten style. Hence, it is still a challenging task for woodblock printing Mongolian words recognition.

Mongolian is an alphabetic language. All letters of one Mongolian word are conglutinated together in the vertical direction and letters have initial, medial or final visual

© Springer Nature Switzerland AG 2021
J. Lladós et al. (Eds.): ICDAR 2021, LNCS 12824, pp. 526–537, 2021.
https://doi.org/10.1007/978-3-030-86337-1_35

forms according to their positions within a word. Blank space is used to separate two neighboring words. Meanwhile, the Mongolian language has a very special writing system, which is quite different from English, Arabic and other Latin languages. Its writing order is vertical from top to bottom and the column order is from left to right. A fragment of one woodblock printing Mongolian document is illustrated in Fig. 1.

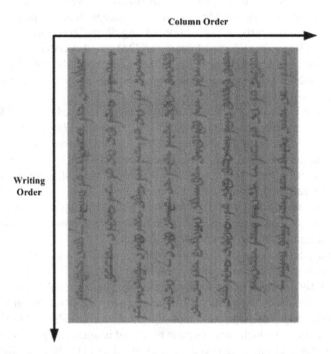

Fig. 1. A fragment of one woodblock printing Mongolian document.

In the literature, the existing approaches for solving the problem of woodblock printing Mongolian words recognition can be generally divided into two categories: segmentation based methods and segmentation free methods. In the segmentation-based methods, each Mongolian word should be segmented into several glyphs which are taken as recognition objects. In this way, the recognition result of one Mongolian word can be obtained from the results of its glyphs. Gao et al. [2] firstly proposed an approach for segmenting the woodblock printing Mongolian words into the corresponding glyphs. And a multi-layer perceptron network was utilized to accomplish the aim of distinguishing glyphs. The recognition rate is only 71% on a dataset composed of 5,500 words with good image quality. After that, Su et al. [3, 4] focused on adjusting the segmentation scheme to improve the accuracy of glyph segmentation. As a result, the recognition rate has been increased from 71% to 81% on the same dataset.

However, most of woodblock printing Mongolian words cannot be segmented into glyphs correctly because of serious degradation. Thus, the segmentation-free (i.e. holistic recognition) methods can be considered as an alternative. Recently, a convolutional neural network (CNN) based holistic recognition approach has been proposed in [5].

Therein, the task of the woodblock printing Mongolian words recognition was regarded as a special form of image classification. A suitable structure of CNN was designed for attaining the above aim. Nevertheless, the CNN based holistic recognition approach cannot solve the problem of out-of-vocabulary.

For this reason, Kang et al. presented another holistic recognition approach in [6]. A sequence to sequence (seq2seq) model with attention mechanism has been utilized to realize the woodblock printing Mongolian words recognition. The proposed seq2seq model consists of three parts: a bi-directional Long Short-Term Memory (Bi-LSTM) based encoder, a Long Short-Term Memory (LSTM) based decoder and a multi-layer perceptron based attention network between the encoder and the decoder. As long as the decoder is able to generate the whole alphabets, the seq2seq model can solve the problem of out-of-vocabulary completely.

In the ancient Mongolian books, the used Mongolian is called *classical Mongolian*. It is a standardized written language used roughly from the 17th to 19th century [7], which is quite different from the *traditional Mongolian* being used in Inner Mongolia Autonomous Region of China now. Under the circumstance, it is time-consuming and expensive to collect a large set of woodblock printing Mongolian word images with annotations. Therefore, this study mainly concentrates on how to generate much more samples from a small collection of the woodblock printing Mongolian word images. By this way, the aim of data augmentation can be realized to improve the performance of the woodblock printing Mongolian words recognition.

In this study, the procedure of generating samples could be regarded as an image-to-image translation task. Generative adversarial network (GAN) [8] is a powerful generative model firstly proposed by Goodfellow in 2014. After that, GAN was widely used in the field of computer vision especially for image-to-image translation [9]. Pix2pix [10] is a generic image-to-image translation algorithm using conditional GAN (cGAN). Providing a training set which contains pairs of related images, pix2pix could learn how to convert an image of one type into another type. Zi2zi [11] also uses GAN to realize image-to-image translation in an end-to-end fashion. The network structure of zi2zi is based on pix2pix with the addition of category embedding for multiple styles. This enables zi2zi to transform images into several different styles with one trained model. Pix2px and zi2zi require paired images of the source style and the target style as the training data. However, since it is impractical to obtain a large set of paired training examples such as in our study.

Cycle-consistent generative adversarial network (CycleGAN) learns the image translation without paired examples [12, 13], in which trains two generative models cycle-wise between the input and output images. In addition to the adversarial losses, cycle consistency loss is adopted for preventing the two generative models from contradicting each other. After that, CycleGAN has been applied for the task of data augmentation [14–16]. In recent years, CycleGAN has also been used to handwritten characters generation [17], font style transfer [18], keyword spotting [19] and so forth.

Inspired by CycleGAN based data augmentation, the style of the woodblock printing Mongolian can be transformed into the style of the machine printing Mongolian, and vice vasa. Thus, there are two manners to generate instances of the woodblock printing Mongolian word images. In the first manner, when one word image in style of machine printing is given, the corresponding word image in style of woodblock printing can be generated by one generative model. In the second manner, when one word image in style of woodblock printing is provided, a new instance of the word image can be generated by one generative model after another generative model as well. In this paper, samples will be generated by the second manner, since the alphabets between the classical Mongolian the traditional Mongolian are quite different. By this means, the aim of data augmentation can be accomplished.

The rest of the paper is organized as follows. The proposed method for generating samples is described detailedly in Sect. 2. The experimental results are shown in Sect. 3. Section 4 provides the conclusions.

2 The Proposed Method

Our aim is to generate new instances of woodblock printing Mongolian word images using unpaired source and target word images. It can be formulated as a mapping G from the source word images X to the target word images Y given training samples $\{x_i | i = 1, 2, ..., N\}$ where $x_i \in X$ and $\{y_j | j = 1, 2, ..., M\}$ where $y_j \in Y$. An example of unpaired data is shown in Fig. 2.

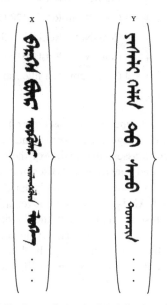

Fig. 2. A set of unpaired training samples.

When the source and the target images represent the woodblock printing Mongolian words and the machine printing Mongolian words respectively, the mapping G can be regarded as the generator of the GAN. It can learn the styles of the woodblock printing Mongolian words from the source images, and then the styles will be transferred into the target images. In our study, the generator G consists of three parts: an encoder, a transfer module and a decoder. The architecture of the generator is presented in Table 1. Besides, an adversarial discriminator (denoted by D_G) in the GAN is adopted to assess the quality of the generated target images. Its structure is given as well as in Table 1.

Table 1. The architectures of the generator and the discriminator.

Module	Structure
Encoder (Generator)	7×7 Conv-Norm-ReLU, 64 filters, stride 1
	3×3 Conv-Norm-ReLU, 128 filters, stride 2
	3×3 Conv-Norm-ReLU, 256 filters, stride 2
Transfer (Generator)	9 Residual blocks, each block contains two layers:
	3×3 Conv, 256 filters, stride 1
	3×3 Conv, 256 filters, stride 1
Decorder (Generator)	3×3 Deconv-Norm-ReLU, 128 filters, stride 1/2
	3×3 Deconv-Norm-ReLU, 64 filters, stride 1/2
	7×7 Deconv-Norm-tanh, 3 filter, stride 1
Discriminator	4×4 Conv-Norm-ReLU, 64 filter, stride 2
	4×4 Conv-Norm-ReLU, 128 filter, stride 2
	4×4 Conv-Norm-ReLU, 256 filter, stride 2
	4×4 Conv-Norm-ReLU, 512 filter, stride 2
	4×4 Conv-sigmoid, 1 filter, stride 1

In the CycleGAN, the second generator (denoted by F) is a converse mapping from the target images to the source images. The corresponding adversarial discriminator (denoted by D_F) is employed to assess the quality of the generated source images as well. The structures of the second generator and discriminator are the same as the first one. An example of the architecture of the utilized CycleGAN is illustrated in Fig. 3.

The loss function of CycleGAN is composed of adversarial losses and cycle consistency losses. The adversarial loss is able to match the distribution of generated images to the data distribution in the target domain. The cycle consistency loss ensures that the cyclic transformation is able to bring the image back to the original state, which is considered as regularization controlled by a parameter. The adversarial losses of the two GANs are defined as the following formulas.

$$L_{GAN,G}(G, D_G) = \mathbb{E}_{x\sim p(x)}[\log(1 - D_G(G(x)))] \\ + \mathbb{E}_{y\sim p(y)}[\log(D_G(y))] \tag{1}$$

$$L_{GAN,F}(F, D_F) = \mathbb{E}_{x\sim p(x)}[\log(1 - D_F(F(x)))] \\ + \mathbb{E}_{y\sim p(y)}[\log(D_F(y))] \tag{2}$$

Where G and D_G are representing the generator and discriminator in the same GAN, separately. F and D_F are the generator and discriminator of the second GAN. x and y

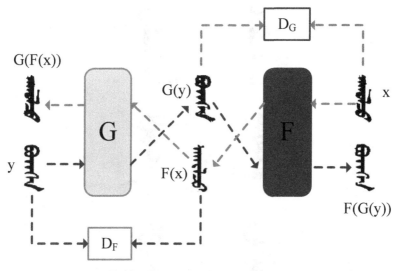

Fig. 3. The architecture of CycleGAN.

denote a source image and a target image. The cycle consistency loss is defined by the following equation.

$$L_{CYCLE}(G, F) = E_{x \sim p(x)}[\|F(G(x)) - x\|_1]$$
$$+ E_{y \sim p(y)}[\|G(F(y)) - y\|_1]$$

(3)

Thus, the whole objective of CycleGAN is the summation of the adversarial losses and the cycle consistency loss:

$$L_{Total}(G, D_G, F, D_F) = L_{GAN,G}(G, D_G)$$
$$+ L_{GAN,F}(F, D_F) + L_{CYCLE}(G, F)$$

(4)

For a CycleGAN, the Eq. (4) can be solved by the following formula:

$$G^*, F^* = \arg\min_{G,F} \max_{D_G, D_F} L_{Total}(G, D_G, F, D_F)$$

(5)

In this study, Adam was chosen as optimizer to the all modules for training a Cycle-GAN. In addition, the batch size and learning rate are set to 1 and 0.0002, severally. Moreover, a set of word images with the style of machine printing has been obtain in advance, which is composed of 2,999 word images in white font [20, 21]. These word images are formed the target images. After training a CycleGAN, we can obtain new instances of woodblock printing Mongolian word images. To be specific, each word image in the style of the woodblock printing is fed into the generator F to generate an intermediate result. And then, the intermediate result is fed into the generator G to form a new instance. Several generated instances by the proposed CycleGAN are depicted in Fig. 4.

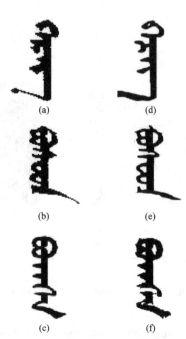

Fig. 4. The generated instances by the proposed CycleGAN. (a)-(c) Three original Mongolian word images of woodblock printing; (d)-(f) The generated Mongolian word images of woodblock printing.

3   Experimental Results

3.1   Dataset and Baseline

To evaluate the performance of the proposed representation method for word images, a collection of woodblock printing Mongolian word images has been collected as the same as in [6]. The total number of word images is 20,948 with annotations. By analyzing the annotations, the amount of vocabulary is 1,423. These word images are formed the source images for training the CycleGAN. In our experiment, *4-fold cross validation* is used for evaluating performance. Thus, the dataset has been divided into four folds randomly, where each fold contains 5,237 word images. The amount of vocabularies in each fold is listed in Table 2.

In our experiment, a holistic recognition method based on the seq2seq model with attention mechanism presented in [6], which is taken as the baseline method for comparison. The baseline method has been implemented with high level python library Keras using Theano as backend and trained on Tesla K80. Therein, RMSProp was selected as optimizer. Learning rate and batch size are set to 0.001 and 128, respectively. The amount of training epochs is 200.

The original performance (i.e. recognition accuracy) of the baseline method is listed in Table 3. Therein, various frame heights and overlap sizes have been tested. To be specific, the frame heights are decreased from 8 pixels to 2 pixels with an interval of 2 pixels and the overlapping sizes are half a frame consistently. **FxHy** denotes that the

Table 2. The detailed divisions for 4-fold cross validation.

Fold	#Vocabulary	#Word Images
Fold1	874	5237
Fold2	878	5237
Fold3	883	5237
Fold4	874	5237

Table 3. The performance of the baseline method.

Fold	Fold1	Fold2	Fold3	Fold4	Avg
F8H4	85.24%	87.19%	86.98%	87.00%	86.60%
F6H3	88.10%	87.07%	91.03%	86.94%	**88.29%**
F4H2	85.72%	87.56%	86.10%	88.01%	86.85%
F2H1	86.44%	85.7%	87.03%	85.35%	86.13%

frame height is x pixels and y pixels are overlapped between adjacent frames. From Table 3, the best performance of the baseline method is attained to **88.29%** when the height of frames and the overlapping size are set to 6 pixels and 3 pixels, severally.

In our experiment, each word image of the dataset was used for generating a new instance. As a result, the dataset were augmented one times as much. Herein, the number of samples in each fold has been doubled. After that, the same experiment settings were used for training the baseline model via the 4-fold cross validation. We will test the performance of the proposed approach for data augmentation. Furthermore, a traditional data augmentation algorithm that is synthetic minority oversampling technique (SMOTE) [22] will be compared with our proposed approach as well.

3.2 The Performance of the Proposed Approach for Data Augmentation

In this experiment, we have tested the performance of the proposed approach and the corresponding results are given in Table 4. From Table 4, we can see that the best performance is attained to **91.68%** when the height of frames and the overlapping size are set to 4 pixels and 2 pixels, separately. It comes to a conclusion that the generated samples by the proposed approach are able to improve the performance of the baseline method indeed.

3.3 The Comparison Between the Proposed Method and SMOTE

In this experiment, we have tested the performance of the SMOTE for data augmentation. In the SMOTE, the handling object is each word (i.e. one class) in vocabulary. To be specific, the synthetic samples are generated by the original samples and their k

Table 4. The performance of the proposed approach for data augmentation.

Fold	Fold1	Fold2	Fold3	Fold4	Avg
F8H4	90.24%	90.97%	90.89%	91.08%	90.80%
F6H3	90.53%	91.77%	92.48%	91.81%	91.65%
F4H2	90.83%	92.08%	91.50%	92.32%	**91.68%**
F2H1	89.65%	88.16%	88.70%	90.22%	89.18%

nearest neighbors in the corresponding class. The detailed process can be described in the following formula:

$$W_{synthetic}[i] = W_{original} + random(0, 1) \times (L[i] - W_{original}) \tag{6}$$

Where $W_{synthetic}[i]$ ($i = 1, 2, ..., n$) denotes the ith generated sample based on the original sample $W_{original}$ and its n ($n < k$) neighbors $L[i]$ ($i = 1, 2, ..., n$) randomly selected from the k nearest neighbors in the same class.

The SMOTE technique requires that the number of samples per class must be more than one. When a certain class only contains one single sample, a noise-added scheme is utilized for creating a new sample before using SMOTE. Detailedly, a new sample can be obtained by adding Gaussian noise, in which Gaussian coefficient is set to a random number between 0 and 1. By this manner, we can assure that the amount of samples per class is more than one before using SMOTE.

To be specific, all samples in the dataset with the same annotation were taken as one class in SMOTE. As far as one class, the number of synthesized instances by SMOTE equals to the number of original samples in the class. In this way, the dataset can be also augmented one times as much. Consequently, the number of samples in each fold has been doubled as well. Several instances synthesized by SMOTE are shown in Fig. 5. We can see that the synthesized instances have lots of noise. And then, the performance of the 4-fold cross validation has been tested. The corresponding results are presented in Table 5.

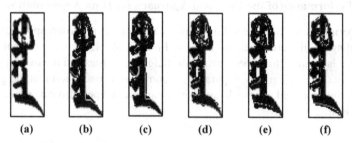

(a)   (b)   (c)   (d)   (e)   (f)

Fig. 5. Several synthetized instances by SMOTE. (a) An original woodblock printing Mongolian word image; (b)-(f) five synthetized instances using SMOTE.

Table 5. The performance of SMOTE for data augmentation.

Fold	Fold1	Fold2	Fold3	Fold4	Avg
F8H4	90.17%	90.05%	90.36%	90.89%	90.37%
F6H3	91.96%	91.24%	90.26%	89.78%	90.81%
F4H2	90.80%	92.00%	91.16%	90.38%	**91.09%**
F2H1	90.43%	89.21%	90.03%	89.67%	89.84%

From Table 5, we can see that the best performance of SMOTE algorithm is attained to **91.09%** when the height of frames and the overlapping size are also set to 4 pixels and 2 pixels. This conclusion is as the same as the proposed approach.

In addition, the comparative results (i.e. the average performance of 4-fold cross validation) among the baseline, SMOTE and the proposed approach have been listed in Table 6. We can see that the performance of SMOTE algorithm is consistently inferior to our proposed approach except for **F2H1**. Therefore, we can conclude that the proposed approach is much more suited for the task of data augmentation.

Table 6. The comparative results among the baseline, SMOTE and the proposed approach.

Methods	Baseline [6]	SMOTE [22]	Our Proposed
F8H4	86.60%	90.37%	90.80%
F6H3	88.29%	90.81%	91.65%
F4H2	86.85%	91.09%	**91.68%**
F2H1	86.13%	89.84%	89.18%

4   Conclusions

In this paper, we have proposed an approach for generating samples by CycleGAN using unpaired training data. To be specific, the source and the target images are the woodblock printing Mongolian words and the machine printing Mongolian words, respectively. After training a CycleGAN, when a Mongolian word image with the style of the woodblock printing is fed into the CycleGAN, a new instance of this word can be generated. The generated instance has the certain style of the machine printing, which will be easy to be recognized. Thus, the aim of data augmentation can be realized by this way. As a result, the performance of the woodblock printing words recognition has been improved.

Acknowledgments. This study is supported by the Project for Science and Technology of Inner Mongolia Autonomous Region under Grant 2019GG281, the Natural Science Foundation of Inner

Mongolia Autonomous Region under Grant 2019ZD14, the Program for Young Talents of Science and Technology in Universities of Inner Mongolia Autonomous Region under Grant NJYT-20-A05, and the Natural Science Foundation of China under Grant 61463038 and 61763034.

References

1. Wei, H., Gao, G.: A keyword retrieval system for historical Mongolian document images. Int. J. Doc. Anal. Recognit. (IJDAR) **17**(1), 33–45 (2014). https://doi.org/10.1007/s10032-013-0203-6
2. Gao, G., Su, X., Wei, H., Gong, Y.: Classical Mongolian words recognition in historical document. In: Proceedings of the 11th International Conference on Document Analysis and Recognition, pp. 692–697. IEEE (2011)
3. Su, X., Gao, G., Wei, H., Bao, F.: Enhancing the mongolian historical document recognition system with multiple knowledge-based strategies. In: Arik, S., Huang, T., Lai, W., Liu, Q. (eds.) Neural Information Processing. ICONIP 2015. Lecture Notes in Computer Science, vol. 9490, pp. 536–544. Springer, Cham (2015). https://doi.org/10.1007/978-3-319-26535-3_61
4. Su, X., Gao, G., Wei, H., Bao, F.: A knowledge-based recognition system for historical Mongolian documents. Int. J. Doc. Anal. Recognit. (IJDAR) **19**(3), 221–235 (2016). https://doi.org/10.1007/s10032-016-0267-1
5. Wei, H., Gao, G.: A holistic recognition approach for woodblock-print Mongolian words based on convolutional neural network. In: Proceedings of the 26th IEEE International Conference on Image Processing, pp. 2726–2730. IEEE (2019)
6. Kang, Y., Wei, H., Zhang, H., Gao, G.: Woodblock-printing Mongolian words recognition by BI-LSTM with attention mechanism. In: Proceedings of the 15th International Conference on Document Analysis and Recognition, pp. 910–915. IEEE (2019)
7. Wei, H., Gao, G., Bao, Y.: A method for removing inflectional suffixes in word spotting of Mongolian Kanjur. In: Proceedings of the 11th International Conference on Document Analysis and Recognition, pp. 88–92. IEEE (2011)
8. Goodfellow, I., et al.: Generative adversarial nets. In: Proceedings of Advances in Neural Information Processing Systems, pp. 2672–2680 (2014)
9. Sangkloy, P., Lu, J., Fang, C., Yu, F., Hays, J.: Scribbler: controlling deep image synthesis with sketch and color. In: Proceedings of 2017 IEEE International Conference on Computer Vision and Pattern Recognition, pp. 5400–5409. IEEE (2017)
10. Isola, P., Zhu, J.Y., Zhou, T., Efros, A.A.: Image-to-image translation with conditional adversarial networks. In: Proceedings of 2017 IEEE International Conference on Computer Vision and Pattern Recognition, pp. 1125–1134. IEEE (2017)
11. Tian, Y.: Zi2zi: Master Chinese calligraphy with conditional adversarial networks (2017). https://github.com/kaonashi-tyc/zi2zi
12. Zhu, J.Y., Park, T., Isola, P., Efros, A.A.: Unpaired image-to-image translation using cycle-consistent adversarial networks. In: Proceedings of 2017 IEEE International Conference on Computer Vision, pp. 2223–2232. IEEE (2017)
13. Li, M., Huang, H., Ma, L., Liu, W., Zhang, T., Jiang, Y.: Unsupervised image-to-image translation with stacked cycle-consistent adversarial networks. In: Proceedings of 2018 European Conference on Computer Vision, pp. 184–199. IEEE (2018)
14. Zhu, X., Liu, Y., Li, J., Wan, T., Qin, Z.: Emotion classification with data augmentation using generative adversarial networks. In: Phung, D., Tseng, V., Webb, G., Ho, B., Ganji, M., Rashidi, L. (eds.) Advances in Knowledge Discovery and Data Mining. PAKDD 2018. Lecture Notes in Computer Science, vol. 10939, pp. 349–360. Springer, Cham (2018). https://doi.org/10.1007/978-3-319-93040-4_28

15. Shi, Z., Liu, M., Cao, Q., Ren, H., Luo, T.: A data augmentation method based on cycle-consistent adversarial networks for fluorescence encoded microsphere image analysis. Signal Process. **161**, 195–202 (2019)

16. Hammami, M., Friboulet, D., Kechichian, R.: Cycle GAN-based data augmentation for multi-organ detection in CT images via Yolo. In: Proceedings of the 28th International Conference on Image Processing, pp. 390–393. IEEE (2020)

17. Chang, B., Zhang, Q., Pan, S., Meng, L.: Generating handwritten Chinese characters using CycleGAN. In: Proceedings of 2018 IEEE Winter Conference on Applications of Computer Vision, pp. 199–207. IEEE (2018)

18. Wu, L., Chen, X., Meng, L., Meng, X.: Multitask adversarial learning for Chinese font style transfer. In: Proceedings of 2020 International Joint Conference on Neural Networks, pp. 1–8. IEEE (2020)

19. Farooqui, F.F., Hassan, M., Younis, M.S., Siddhu, M.K.: Offline hand written Urdu word spotting using random data generation. IEEE Access **8**, 131119–131136 (2020)

20. Zhang, H., Wei, H., Bao, F., Gao, G.: Segmentation-free printed traditional Mongolian OCR using sequence to sequence with attention model. In: Proceedings of the 14th International Conference on Document Analysis and Recognition, pp. 585–590. IEEE (2017)

21. Wei, H., Zhang, H., Zhang, J., Liu, K.: Multi-task learning based traditional Mongolian words recognition. In: Proceedings of the 25th International Conference on Pattern Recognition, pp. 1275–1281. IEEE (2021)

22. Chawla, N.V., Bowyer, K.W., Hall, L.O., Kegelmeyer, W.P.: SMOTE: synthetic minority over-sampling technique. J. Artif. Intell. Res. **16**, 321–357 (2002)

SauvolaNet: Learning Adaptive Sauvola Network for Degraded Document Binarization

Deng Li[1], Yue Wu[2], and Yicong Zhou[1(✉)] ⓘ

[1] Department of Computer and Information Science, University of Macau, Macau, China
{mb85511,yicongzhou}@um.edu.com
[2] Amazon Alexa Natural Understanding, Manhattan Beach, CA, USA
wuayue@amazon.com

Abstract. Inspired by the classic *Sauvola* local image thresholding approach, we systematically study it from the deep neural network (DNN) perspective and propose a new solution called `SauvolaNet` for degraded document binarization (DDB). It is composed of three explainable modules, namely, Multi-Window Sauvola (MWS), Pixelwise Window Attention (PWA), and Adaptive Sauolva Threshold (AST). The MWS module honestly reflects the classic *Sauvola* but with trainable parameters and multi-window settings. The PWA module estimates the preferred window sizes for each pixel location. The AST module further consolidates the outputs from MWS and PWA and predicts the final adaptive threshold for each pixel location. As a result, `SauvolaNet` becomes end-to-end trainable and significantly reduces the number of required network parameters to 40K – it is only 1% of `MobileNetV2`. In the meantime, it achieves the State-of-The-Art (SoTA) performance for the DDB task – `SauvolaNet` is at least comparable to, if not better than, SoTA binarization solutions in our extensive studies on the 13 public document binarization datasets. Our source code is available at https://github.com/Leedeng/SauvolaNet.

Keywords: Binarization · Sauvola · Document processing

1 Introduction

Document binarization typically refers to the process of taking a gray-scale image and converting it to black-and-white. Formally, it seeks a decision function $f_{\text{binarize}}(\cdot)$ for a document image $\mathbf{D}$ of width W and height H, such that the resulting image $\hat{\mathbf{B}}$ of the same size only contains binary values while the overall document readability is at least maintained if not enhanced.

$$\hat{\mathbf{B}} = f_{\text{binarize}}(\mathbf{D}) \tag{1}$$

This work was done prior to Amazon involvement of the authors.
This work was funded by The Science and Technology Development Fund, Macau SAR (File no. 189/2017/A3), and by University of Macau (File no. MYRG2018-00136-FST).

© Springer Nature Switzerland AG 2021
J. Lladós et al. (Eds.): ICDAR 2021, LNCS 12824, pp. 538–553, 2021.
https://doi.org/10.1007/978-3-030-86337-1_36

Document binarization plays a crucial role in many document analysis and recognition tasks. It is the prerequisite for many low-level tasks like connected component analysis, maximally stable extremal regions, and high-level tasks like text line detection, word spotting, and optical character recognition (OCR).

Instead of directly constructing the decision function $f_{\text{binarize}}(\cdot)$, classic binarization algorithms [17,28] typically first construct an auxiliary function $g(\cdot)$ to estimate the required thresholds $\mathbf{T}$ as follows.

$$\mathbf{T} = g_{\text{classic}}(\mathbf{D}) \tag{2}$$

In global thresholding approaches [17], this threshold $\mathbf{T}$ is a scalar, *i.e.* all pixel locations use the same threshold value. In contrast, this threshold $\mathbf{T}$ is a tensor with different values for different pixel locations in local thresholding approache [28]. Regardless of global or local thresholding, the actual binarization decision function can be written as

$$\hat{\mathbf{B}}_{\text{classic}} = f_{\text{classic}}(\mathbf{D}) = th(\mathbf{D}, \mathbf{T}) = th(\mathbf{D}, g_{\text{classic}}(\mathbf{D})) \tag{3}$$

where $th(x, y)$ is the simple thresholding function and the binary state for a pixel located at i-th row and j-th column is determined as in Eq. (4).

$$\hat{B}_{\text{classic}}[i,j] = th(D[i,j], T[i,j]) = \begin{cases} +1, & \text{if } D[i,j] \geq T[i,j] \\ -1, & \text{otherwise} \end{cases} \tag{4}$$

Classic binarization algorithms are very efficient in general because of using simple heuristics like intensity histogram [17] and local contrast histogram [31]. The speed of classic binarization algorithms typical of the millisecond level, even on a mediocre CPU. However, simple heuristics also means that they are sensitive to potential variations [31] (image noise, illumination, bleed-through, paper materials, *etc.*), especially when the relied heuristics fail to hold. In order to improve the binarization robustness, data-driven approaches like [33] learn the decision function $f_{\text{binarize}}(\cdot)$ from data rather than heuristics. However, these approaches typically achieve better robustness by using much more complicated features, and thus work relatively slow in practice, *e.g.* on the second level [33].

Like in many computer vision and image processing fields, the deep learning-based approaches outperform the classic approaches by a large margin in degraded document binarization tasks. The state-of-the-art (SoTA) binarization approaches are now all based on deep neural networks (DNN) [22,27]. Most of SoTA document binarization approaches [2,19,32] treat the degraded binarization task as a binary semantic segmentation task (namely, foreground and background classes) or a sequence-to-sequence learning task [1], both of which can effectively learn $f_{\text{binarize}}(\cdot)$ as a DNN from data.

Recent efforts [2,5,9,19,30,32,34] focus more on improving robustness and generalizability. In particular, the SAE approach [2] suggests estimating the pixel memberships not from a DNN's raw output but the DNN's activation map, and thus generalizes well even for out-of-domain samples with a weak activation map. The MRAtt approach [19] further improves the attention mechanism

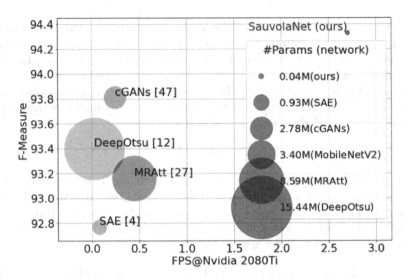

Fig. 1. Comparisons of the SoTA DNN-based methods on the DIBCO 2011 dataset (average resolution 574×1104). The SauvolaNet is of 1% of MobileNetV2's parameter size, while attaining superior performance in terms of both speed and F-Measure.

in multi-resolution analysis and enhances the DNN's robustness to font sizes. DSN [32] apply multi-scale architecture to predict foreground pixel at multi-features levels. The DeepOtsu method [9] learns a DNN that iteratively enhances a degraded image to a uniform image and then binarized via the classic Otsu approach. Finally, generative adversarial networks (GAN) based approaches like cGANs [34] and DD-GAN [5] rely on the adversarial training to improve the model's robustness against local noises by penalizing those problematic local pixel locations that the discriminator uses in differentiating real and fake results.

As one may notice, both classic and deep binarization approaches have pros and cons: 1) the classic binarization approaches are extremely fast, while the DNN solutions are not; 2) the DNN solutions can be end-to-end trainable, while the classic approaches can not. In this paper, we propose a novel document binarization solution called SauvolaNet – it is an end-to-end trainable DNN solution but analogous to a multi-window *Sauvola* algorithm. More precisely, we re-implement the *Sauvola* idea as an algorithmic DNN layer, which helps SauvolaNet attain highly effective feature representations at an extremely low cost – only two *Sauvola* parameters are needed. We also introduce an attention mechanism to automatically estimate the required *Sauvola* window sizes for each pixel location and thus could effectively and efficiently estimate the *Sauvola* threshold. In this way, the SauvolaNet significantly reduces the total number of DNN parameters to 40K, only 1% of the MobileNetV2, while attaining comparable performance of SoTA on public DIBCO datasets. Figure 1 gives the high-level comparisons of the proposed SauvolaNet to the SoTA DNN solutions.

The rest of the paper is organized as follows: Sect. 2 briefly reviews the classic *Sauvola* method and its variants; Sect. 3 proposes the `SauvolaNet` solution for degraded document binarization; Sect. 4 presents *Sauvola* ablation studies results and comparisons to SoTA methods; and we conclude the paper in Sect. 5.

2   Related *Sauvola* Approaches

The *Sauvola* binarization algorithm [28] is widely used in main stream image and document processing libraries and systems like `OpenCV`[1] and `Scikit-Image`[2]. As aforementioned, it constructs the binarization decision function (3) via the auxiliary threshold estimation function g_{Sauvola}, which has three hyper-parameters, namely, 1) w: the square window size (typically an odd positive integer [4]) for computing local intensity statistics; 2) k: the user estimated level of document degradation; and 3) r: the dynamic range of input image's intensity variation.

$$\mathbf{T}_{\text{Sauvola}} = g_{\text{Sauvola}|\theta}(\mathbf{D}). \tag{5}$$

where $\theta = \{w, k, r\}$ indicates the used hyper-parameters. Each local threshold is computed *w.r.t.* the 1st- and 2nd-order intensity statistics as shown in Eq. (6),

$$T_{\text{Sauvola}|\theta}[i,j] = \mu[i,j] \cdot \left(1 + k \cdot \left(\frac{\sigma[i,j]}{r} - 1\right)\right) \tag{6}$$

where $\mu[i,j]$ and $\sigma[i,j]$ respectively indicate the mean and standard deviation of intensity values within the local window as follows.

$$\mu[i,j] = \sum_{\delta_i=-\lfloor w/2 \rfloor}^{\lfloor w/2 \rfloor} \sum_{\delta_j=-\lfloor w/2 \rfloor}^{\lfloor w/2 \rfloor} \frac{D[i+\delta_i, j+\delta_j]}{w^2} \tag{7}$$

$$\sigma^2[i,j] = \sum_{\delta_i=-\lfloor w/2 \rfloor}^{\lfloor w/2 \rfloor} \sum_{\delta_j=-\lfloor w/2 \rfloor}^{\lfloor w/2 \rfloor} \frac{(D[i+\delta_i, j+\delta_j] - \mu[i,j])^2}{w^2} \tag{8}$$

It is well known that heuristic binarization methods with hyper-parameters could rarely achieve their upper-bound performance unless the method hyper-parameters are individually tuned for each input document image [12], and this is also the main pain point of *Sauvola* approach.

Many efforts have been made to mitigate this pain point. For example, [14] introduces a multi-grid *Sauvola* variant that analyzes multiple scales in the recursive way; [13] proposes a hyper-parameter free multi-scale binarization solution called Sauvola MS [2] by combining *Sauvola* results of a fixed set of window sizes, each with its own empirical k and r values; [8] improves the classic *Sauvola* by using contrast information obtained from pre-processing to refine *Sauvola*'s binarization; [12] estimates the required window size w in *Sauvola* by using the stroke width transform matrix. Table 1 compares these approaches with the proposed `SauvolaNet`, and it is clear that only `SauvolaNet` is end-to-end trainable.

---

[1] https://docs.opencv.org/4.5.1/.
[2] https://scikit-image.org/.

Table 1. Comparisons of various *Sauvola* document binarization approaches.

Related work	End-to-end trainable?	*Sauvola* Params.			#Used scales	Without	
		w	k	r		Preproc?	Postproc?
Sauvola [28]	✗	User specified			Single	✓	✓
[14]	✗	Auto	User specified	Auto	Multiple	✗	✗
[13]	✗	Fixed			multiple	✓	✓
[8]	✗	Fixed			Single	✗	✗
[12]	✗	Auto	Fixed	Fixed	Multiple	✗	✓
SauvolaNet	✓	Auto	Learned	Learned	Multiple	✓	✓

3   The SauvolaNet Solution

Figure 2 describes the proposed SauvolaNet solution. It learns an auxiliary threshold estimation function from data by using a dual-branch design with three main modules, namely Multi-Window Sauvola (MWS), Pixelwise Window Attention (PWA), and Adaptive Sauvola Threshold (AST).

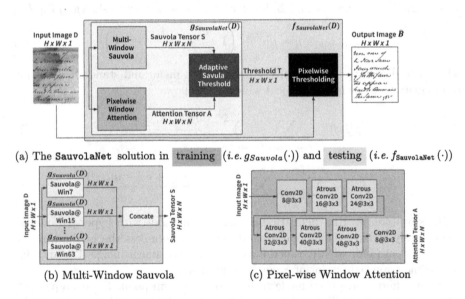

(a) The SauvolaNet solution in training (*i.e.* $g_{Sauvola}(\cdot)$) and testing (*i.e.* $f_{SauvolaNet}(\cdot)$)

(b) Multi-Window Sauvola   (c) Pixel-wise Window Attention

Fig. 2. The overview of SauvolaNet solution and its trainable modules. $g_{Sauvola}$ and $g_{SauvolaNet}$ indicate the customized *Sauvola* layer and SauvolaNet, respectively; Conv2D and AtrousConv2D indicate the traditional atrous (w/ dilation rate 2) convolution layers, respectively; each Conv2D/AtrousConv2D are denoted of format filters@ksize×ksize and followed by InstanceNorm and ReLU layers; the last Conv2D in window attention uses the Softmax activation (denoted w/ borders); and Pixelwise Thresholding indicates the binarization process (4).

Specifically, the MWS module takes a gray-scale input image **D** and leverages on the *Sauvola* to compute the local thresholds for different window sizes. The

PSA module also takes $\mathbf{D}$ as the input but estimates the attentive window sizes for each pixel location. The AST module predicts the final threshold for each pixel location $\mathbf{T}$ by fusing the thresholds of different window sizes from MWS using the attentive weights from PWA. As a result, the proposed SauvolaNet is analogous to a multi-window *Sauvola*, and models an auxiliary threshold estimation function $g_{\text{SauvolaNet}}$ between the input $\mathbf{D}$ and the output $\mathbf{T}$ as follows,

$$\mathbf{T} = g_{\text{SauvolaNet}}(\mathbf{D}) \tag{9}$$

Unlike in the classic *Sauvola*'s threshold estimation function (5), SauvolaNet is end-to-end trainable and doesn't require any hyper-parameter. Similar to Eq. (3), the binarization decision function $f_{\text{SauvolaNet}}$ used in testing as shown below

$$\hat{\mathbf{B}} = f_{\text{SauvolaNet}}(\mathbf{D}) = th(\mathbf{D}, \mathbf{T}) = th(\mathbf{D}, g_{\text{SauvolaNet}}(\mathbf{D})) \tag{10}$$

and the extra thresholding process (*i.e.* (4)) is denoted as the Pixelwise Thresholding (PT) in Fig. 2. Details about these modules are discussed in later sections.

3.1 Multi-window Sauvola

The MWS module can be considered as a re-implementation of the classic multi-window *Sauvola* analysis in the DNN context. More precisely, we first introduce a new DNN layer called Sauvola (denoted as $g_{\text{Sauvola}}(\cdot)$ in the function form), which has the *Sauvola* window size as input argument and *Sauvola*'s hyper-parameter s and r as trainable parameters. To enable multi-window analysis, we use a set of Sauvola layers, each corresponding to one window size in (11). The selection of window are verified in Sect. 4.2.

$$\mathbb{W} = \{w \,|\, w \in [7, 15, 23, 31, 39, 47, 55, 63]\} \tag{11}$$

Figure 3 visualizes all intermediate outputs of SauvolaNet, and Fig. 3-(b*) show predicted *Sauvola* thresholds based on these window sizes, and Fig. 3-(c*) further binarize the input image using corresponding thresholds. These results again confirm that satisfactory binarization performance can be achieved by *Sauvola* when the appropriate window size is used.

It is worthy to emphasize that Sauvola threshold computing window-wise mean and the standard deviation (see (6)) is very time-consuming when using the traditional DNN layers (*e.g.*, AveragePooling2D), especially for a big window size (*e.g.*, 31 or above). Fortunately, we implement our Sauvola layer by using integral image solution [29] to reduce the computational complexity to $O(1)$.

3.2 Pixelwise Window Attention

As mentioned in many works [12,13], one disadvantage when using *Sauvola* algorithm is the tuning of hyper-parameters. Among all three hyperparameters, namely, the window size w, the degradation level k, and the input deviation r, w is the most important. Existing works typically decompose an input image into

Fig. 3. Intermediate results of `SauvolaNet`. Please note: 1) window attention (e) visualizes the most preferred window size of **A** for each pixel locations (*i.e.* argmax (**A**, axis = −1)), and the 8 used colors correspond to those put before (b*); 2) binarized images (c*) are not used in `SauvolaNet` but for visualization only; and 3) $g_{\text{Sauvola}}(\cdot)$ and $th(\cdot, \cdot)$ indicates the `Sauvola` layer and the pixelwise thresholding function (4), respectively.

non-overlapping regions [13,18] or grids [14] and apply each a different window size. However, existing solutions are not suitable for DNN implementation for two reasons: 1) non-overlapping decomposition is not a differentiable operation; and 2) processing regions/grids of different sizes are hard to parallelize.

Alternatively, we adopt the widely-accepted attention mechanism to remove the dependency on the user-specified window sizes. Specifically, the proposed PWA module is a sub-network that takes an input document image **D** and predicts the pixel-wise attention on all predefined window sizes. It conceptually follows the multi-grid method introduced by DeepLabv3 [3] while using a fixed dilation rate at 2. Also, we use the InstanceNormalization instead of the common BatchNormalization to mitigate the overfitting risk caused by a small training dataset. The detailed network architecture is shown in Fig. 2.

Sample result of PWA can be found in Fig. 3-(e). As one can see, the proposed PWA successfully predicts different window sizes for different pixels. More precisely, it prefers $w = 39$ (see Fig. 3-(b5)) and $w = 15$ (see Fig. 3-(b2)) for background and foreground pixels, respectively; and uses very large window sizes, e.g., $w = 63$ (i.e. Fig. 3-(b8)) for those pixels on text borders.

3.3 Adaptive Sauvola Threshold

As one can see from Fig. 2, the MWS outputs a Sauvola tensor **S** of size $H \times W \times N$, where N is the number of used window sizes (and we use $N = 8$, see Eq. (11)), the PWA outputs an attention tensor **A** of the same size as **S** and the attention sum for all window sizes on each pixel location is always 1, namely,

$$\sum_{k=1}^{N} A[i,j,k] = 1, \ \forall 1 \leq i \leq H, 1 \leq j \leq W. \tag{12}$$

The AST applies the window attention tensor **A** to the window-wise initial Sauvola threshold tensor **S** and compute the pixel-wise threshold **T** as below

$$T[i,j] = \sum_{k=1}^{N} A[i,j,k] \cdot S[i,j,k] \tag{13}$$

Fig. 3-(g) shows the adaptive threshold **T** when using the sample input Fig. 3-(a). By comparing the corresponding binarized result (i.e. Fig. 3-(h)) with those of single window's results (i.e. Fig. 3-(c*)), one can easily verify that the adaptive threshold **T** outperforms any individual threshold result in **S**.

3.4 Training, Inference, and Discussions

In order to train SauvolaNet, we normalize the input **D** to the range of (0, 1) (by dividing 255 for uint8 image), and employ a modified hinge loss, namely

$$loss[i,j] = \max(1 - \alpha \cdot (D[i,j] - T[i,j]) \cdot B[i,j], 0) \qquad (14)$$

where $\mathbf{B}$ is the corresponding binarization ground truth map with binary values
-1 (foreground) and $+1$ (background); $\mathbf{T}$ is `SauvolaNet`'s predicted thresholds
as shown in Eq. (9); and α is a parameter to empirically control the margin of
decision boundary and only those pixels close to the decision boundary will be
used in gradient back-propagation. Throughout the paper, we always use $\alpha = 16$.

We implement `SauvolaNet` in the `TensorFlow` framework. The used training
patch size is 256×256, and the data augmentations are random crop and random
flip. The training batch size is set to 32, and we use `Adam` optimizer with the initial
learning rate of $1e-3$. During inference, we use $f_{\texttt{SauvolaNet}}$ instead of $g_{\texttt{SauvolaNet}}$
as shown in Fig. 2-(a). It only differs from the training in terms of one extra
the thresholding step (4) to compare `SauvolaNet` predicted thresholds with the
original input to obtain the final binarized output.

Unlike in most DNNs, each module in `SauvolaNet` is explainable: the MWS
module leverages the *Sauvola* algorithm to reduce the number of required network parameters significantly, and the PWA module employs the attention idea
to get rid of the *Sauvola*'s disadvantage of window size selection, and finally
two branches are fused in the AST module to predict the pixel-wise threshold.
Sample results in Fig. 3 further confirm that these modules work as expected.

4 Experimental Results

4.1 Dataset and Metrics

In total, 13 document binarization datasets are used in experiments, and they are
{(H-)DIBCO 2009 [7] (10), 2010 [23] (10), 2011 [20] (16), 2012 [24] (14), 2013 [25]
(16), 2014 [16] (10), 2016 [26] (10), 2017 [27] (20), 2018 [21] (10); PHIDB [15]
(15), Bickely-diary dataset [6] (7), Synchromedia Multispectral dataset [10] (31),
and Monk Cuper Set [9] (25)}. The braced numbers after each dataset indicates
its sample size, and detailed partitions for training and testing will be specified
in each study. For evaluation, we adopt the DIBCO metrics [16, 20, 21, 24–27]
namely, F-Measure (FM), psedudo F-Measure (F_{ps}), Distance Reciprocal Distortion metric (DRD) and Peak Signal-to-Noise Ratio (PSNR).

4.2 Ablation Studies

To simplify discussion, let θ be the set of parameter settings related to a studied
Sauvola approach f. Unless otherwise noted, we always repeat a study about
f and θ on all datasets in the leave-one-out manner. More precisely, each score
reported in ablation studies is obtained as follows

$$score(\theta, f) = \frac{1}{\|\mathbb{D}\|} \sum_{x \in \mathbb{X}} \left\{ \sum_{(\mathbf{D}, \mathbf{B}) \in x} \frac{m(\hat{\mathbf{B}}_\theta^x, \mathbf{B})}{\|x\|} \right\} \qquad (15)$$

where $m(\cdot)$ indicates a binarization metric, *e.g.* FM; and $\hat{\mathbf{B}}_\theta^x = f_\theta^{\mathbb{X}-x}(\mathbf{D})$ indicates
the predicted binarized result for a given image $\mathbf{D}$ using the solution $f_\theta^{\mathbb{X}-x}$ that is

trained on dataset $\mathbb{X} - x$ using the setting θ. More precisely, the inner summation of Eq. (15) represents the average score for the model $f_\theta^{\mathbb{X}-x}$ over all testing samples in x, and that the outer summation of Eq. (15) further aggregates all leave-one-out average scores, and thus leaves the resulting score only dependent on the used method f with setting θ.

Table 2. Trainable v.s. non-trainable *Sauvola* .

WinSize w	Train?		Converged value		Binarization scores			
	k	r	k	r	FM (%) ↑	F_{ps} (%) ↑	PSNR(db) ↑	DRD ↓
OpenCV parameter configuration								
11	✗	✗	0.50	0.50	50.09	59.95	13.44	13.73
	✗	✓	0.50	0.23 ± 0.003	77.31	81.25	15.86	8.17
	✓	✗	0.20 ± 0.005	0.50	77.92	85.44	15.99	8.09
	✓	✓	0.25 ± 0.024	0.24 ± 0.004	**80.47**	**85.49**	**16.05**	**8.01**
Pythreshold parameter configuration								
15	✗	✗	0.35	0.50	67.37	76.83	14.84	9.96
	✗	✓	0.35	0.26 ± 0.004	79.23	84.01	15.72	8.71
	✓	✗	0.22 ± 0.011	0.50	79.87	86.31	15.84	8.17
	✓	✓	0.29 ± 0.027	0.28 ± 0.009	**81.40**	**86.41**	**16.35**	**7.50**
Scikit-image parameter configuration								
15	✗	✗	0.20	0.50	77.23	85.15	15.55	8.92
	✗	✓	0.20	0.25 ± 0.003	79.51	85.34	15.67	8.43
	✓	✗	0.22 ± 0.008	0.50	79.94	86.37	15.92	8.10
	✓	✓	0.29 ± 0.023	0.28 ± 0.007	**81.46**	**86.47**	**16.38**	**7.46**

Does *Sauvola* With Learnable Parameters Work Better? Before discussing SauvolaNet , one must-answer question is whether or not re-implement the classic *Sauvola* algorithm as an algorithmic DNN layer is the right choice, or equivalently, whether or not *Sauvola* hyper-parameters learned from data could generalize better in practice. If not, we should leverage on existing heuristic *Sauvola* parameter settings and use them in SauvolaNet as non-trainable.

To answer the question, we start from one set of *Sauvola* hyper-parameters, *i.e.* $\theta = \{w, k, r\}$, and evaluate the corresponding performance of single window *Sauvola* , *i.e.* $g_{\text{Sauvola}|\theta}$ under four different conditions, namely, 1) non-trainable k and r; 2) non-trainable k but trainable r; 3) trainable k but non-trainable r; and 4) trainable k and r. We further repeat the same experiments for three well-known *Sauvola* hyper-parameter settings OpenCV (see footnote 1) ($w = 11$, $k = 0.5$, $r = 0.5$), Scikit-Image (see footnote 2) ($w = 15$, $k = 0.2$, $r = 0.5$) and Pythreshold[3] ($w = 15$, $k = 0.35$, $r = 0.5$).

Table 2 summarizes the performance scores for single-window *Sauvola* with different parameter settings. Each row is about one $score(\theta, f_{\text{Sauvola}})$, and the

---
[3] https://github.com/manuelaguadomtz/pythreshold/blob/master/pythreshold/.

three mega rows represent the three initial θ settings. As one can see, three prominent trends are: 1) the heuristic *Sauvola* hyper-parameters (*i.e.* the non-trainable k and r setting) from the three open-sourced libraries don't work well for DIBCO-like dataset; 2) allowing trainable k or r leads to better performance, and allowing both trainable gives even better performance; 3) the converged values of trainable k and r are different for different window sizes. We therefore use trainable k and r for each window size in the Sauvola layer (see Sec. 3.1).

Does Multiple-Window Help? Though it seems that having multiple window sizes for *Sauvola* analysis is beneficial, it is still unclear that 1) how effective it is comparing to a single-window *Sauvola*, and 2) what window sizes should be used. We, therefore, conduct ablation studies to answer both questions.

More precisely, we first conduct the leave-one-out experiments for the single-window *Sauvola* algorithms for different window sizes with trainable k and r. The resulting $score(w, f_{\text{Sauvola}})$ are presented in the upper-half of Table 3. Comparing to the best heuristic *Sauvola* performance attained by Scikit-Image in Table 2, these results again confirm that *Sauvola* with trainable k and r works much better. Furthermore, it is clear that f_{Sauvola} with different window sizes (except for $w = 7$) attain similar scores, possibly because there is no single dominant font size in the 13 public datasets.

Table 3. Ablation study on *Sauvola* window sizes

\multicolumn WinSize								Binarization scores			
7	15	23	31	39	47	55	63	FM (%) ↑	F_{ps} (%) ↑	PSNR (db) ↑	DRD ↓
Single-window Sauvola											
✓								77.62	79.60	15.36	9.76
	✓							81.47	86.51	16.41	7.47
		✓						82.51	87.29	16.43	7.70
			✓					82.41	57.09	16.39	7.82
				✓				82.23	86.90	16.34	7.93
					✓			82.10	86.68	16.30	8.11
						✓		82.01	86.54	16.28	8.17
							✓	81.92	86.42	16.25	8.25
Multi-window Sauvola											
✓	✓							81.55	84.41	17.09	6.62
✓	✓	✓						82.88	85.29	17.32	6.41
✓	✓	✓	✓					84.71	87.34	17.72	6.04
✓	✓	✓	✓	✓				87.83	90.70	18.47	5.30
✓	✓	✓	✓	✓	✓			89.87	92.31	18.87	4.13
✓	✓	✓	✓	✓	✓	✓		91.36	95.55	19.09	3.73
✓	✓	✓	✓	✓	✓	✓	✓	**91.42**	**95.67**	**19.15**	**3.67**

Finally, we conduct the ablation studies of using multiple window sizes in SauvolaNet in the incremental way, and report the resulting $score(\mathbb{W}, f_{\text{SauvolaNet}})$s in the lower-half of Table 3. It is now clear that 1) multi-window does help in SauvolaNet; and 2) the more window sizes, the better performance scores. As a result, we use all eight window sizes in SauvolaNet (see Eq. (11)).

4.3 Comparisons to Classic and SoTA Binarization Approaches

It is worthy to emphasize that different works [2,9,11,13,19,34] use different protocols for document binarization evaluation. In this section, we mainly follow the evaluation protocol used in [9], and its dataset partitions are: 1) training: (H)-DIBCO 2009 [7], 2010 [23], 2012 [24]; Bickely-diary dataset [6]; and Synchromedia Multispectral dataset [10], and for testing: (H)-DIBCO 2011 [20], 2014 [16], and 2016 [26]. We train all approaches using the same evaluation protocol for fairly comparison. As a result, we focus on those recent and open-sourced DNN based methods, and they are SAE [2], DeepOtsu [9], cGANs [34] and MRAtt [19]. In addition, heuristic document binarization approaches Otsu [17], Sauvola [28] and Howe [11] are also included. Finally, Sauvola MS [13], a classic multi-window Sauvola solution is evaluated to better gauge the performance improvement from the heuristic multi-window analysis to the proposed learnable analysis.

Table 4. Comparison of SauvolaNet and SoTA approaches DIBCO 2011.

Dataset	Methods	FM (%) ↑	F_{ps} (%) ↑	PSNR (db) ↑	DRD ↓
DIBCO 2011	Otsu [17]	82.10	84.80	15.70	9.00
	Howe [11]	91.70	92.00	19.30	3.40
	MRAtt [19]	93.16	95.23	19.78	2.20
	DeepOtsu [9]	93.40	95.80	19.90	**1.90**
	SAE [2]	92.77	95.68	19.55	2.52
	DSN [32]	93.30	**96.40**	20.10	2.00
	cGANs [34]	93.81	95.26	20.30	1.82
	Sauvola [28]	82.10	87.70	15.60	8.50
	Sauvola MS [13]	79.70	81.78	14.91	11.67
	SauvolaNet	**94.32**	**96.40**	**20.55**	1.97

Table 4, 5 and 6 reports the average performance scores of the four evaluation metrics for all images in each testing dataset. When comparing the three Sauvola based approaches, namely, Sauvola, Sauvola MS, and SauvolaNet, one may easily notice that the heuristic multi-window solution Sauvola MS does not necessarily outperform the classic Sauvola. However, the SauvolaNet, again a

Table 5. Comparison of `SauvolaNet` and SoTA approaches in H-DIBCO 2014.

Dataset	Methods	FM (%) ↑	F_{ps} (%) ↑	PSNR (db) ↑	DRD ↓
H-DIBCO 2014	Otsu [17]	91.70	95.70	18.70	2.70
	Howe [11]	96.50	97.40	22.20	1.10
	MRAtt [19]	94.90	95.98	21.09	1.85
	DeepOtsu [9]	95.90	97.20	22.10	0.90
	SAE [2]	95.81	96.78	21.26	1.00
	DSN [32]	96.70	97.60	23.20	0.70
	DD-GAN [5]	96.27	97.66	22.60	1.27
	cGANs [34]	96.41	97.55	22.12	1.07
	Sauvola [28]	84.70	87.80	17.80	2.60
	Sauvola MS [13]	85.83	86.83	17.81	4.88
	`SauvolaNet`	**97.83**	**98.74**	**24.13**	**0.65**

Table 6. Comparison of `SauvolaNet` and SoTA approaches DIBCO 2016.

Dataset	Methods	FM (%) ↑	F_{ps} (%) ↑	PSNR (db) ↑	DRD ↓
DIBCO 2016	Otsu [17]	86.60	89.90	17.80	5.60
	Howe [11]	87.50	82.30	18.10	5.40
	MRAtt [19]	**91.68**	94.71	19.59	2.93
	DeepOtsu [9]	91.40	94.30	19.60	2.90
	SAE [2]	90.72	92.62	18.79	3.28
	DSN [32]	90.10	83.60	19.00	3.50
	DD-GAN [5]	89.98	85.23	18.83	3.61
	cGANs [34]	91.66	94.58	**19.64**	**2.82**
	Sauvola [28]	84.60	88.40	17.10	6.30
	Sauvola MS [13]	79.84	81.61	14.76	11.50
	`SauvolaNet`	90.25	**95.26**	18.97	3.51

multi-window solution but with all trainable weights, clearly beat both by large margins for all four evaluation metrics. Moreover, the proposed `SauvolaNet` solution outperforms the rest of the classic and SoTA DNN approaches in DIBCO 2011. And `SauvolaNet` is comparable to the SoTA solutions in H-DIBCO 2014 and DIBCO 2016. Sample results are shown in Fig. 4. More importantly, the `SauvolaNet` is super lightweight and only contains 40K parameters. It is much smaller and runs much faster than other DNN solutions as shown in Fig. 1.

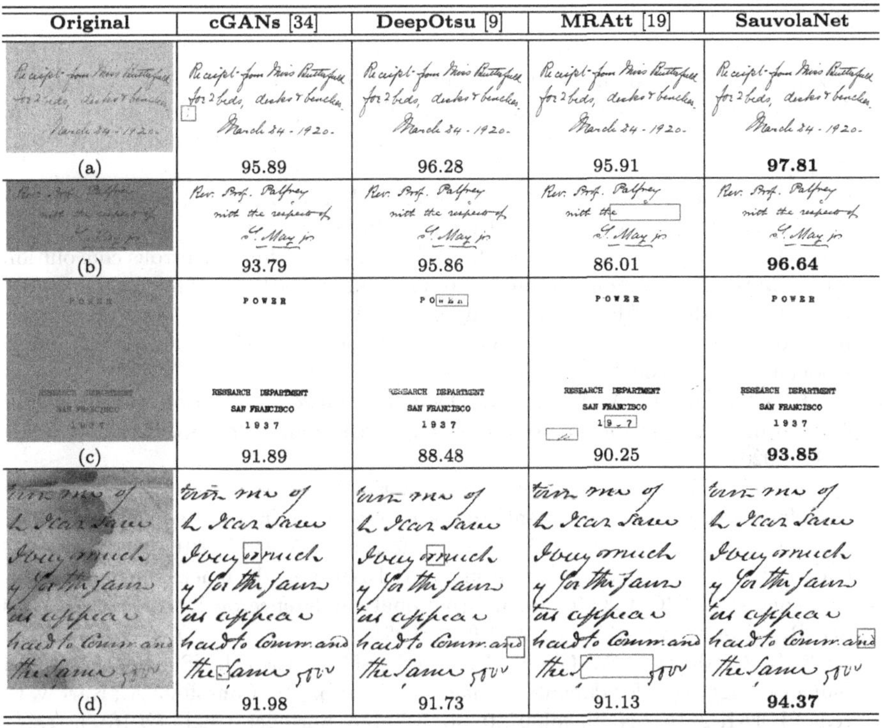

Original	cGANs [34]	DeepOtsu [9]	MRAtt [19]	SauvolaNet
(a)	95.89	96.28	95.91	**97.81**
(b)	93.79	95.86	86.01	**96.64**
(c)	91.89	88.48	90.25	**93.85**
(d)	91.98	91.73	91.13	**94.37**

Fig. 4. Qualitative comparison of SauvolaNet with SoTA document binarization approaches. Problematic binarization regions are denoted in red boxes, and the FM score for each binarization result is also included below a result. (Color figure online)

5 Conclusion

In this paper, we systematically studied the classic *Sauvola* document binarization algorithm from the deep learning perspective and proposed a multi-window *Sauvola* solution called SauvolaNet. Our ablation studies showed that the *Sauvola* algorithm with learnable parameters from data significantly outperforms various heuristic parameter settings (see Table 2). Furthermore, we proposed the SauvolaNet solution, a *Sauvola*-based DNN with all trainable parameters. The experimental result confirmed that this end-to-end solution attains consistently better binarization performance than non-trainable ones, and that the multi-window *Sauvola* idea works even better in the DNN context with the help of attention (see Table 3). Finally, we compared the proposed SauvolaNet with the SoTA methods on three public document binarization datasets. The result showed that SauvolaNet has achieved or surpassed the SoTA performance while using a significantly fewer number of parameters (1% of MobileNetV2) and running at least 5x faster than SoTA DNN-based approaches.

References

1. Afzal, M.Z., Pastor-Pellicer, J., Shafait, F., Breuel, T.M., Dengel, A., Liwicki, M.: Document image binarization using LSTM: a sequence learning approach. In: International Workshop on Historical document Imaging and Processing, pp. 79–84 (2015)
2. Calvo-Zaragoza, J., Gallego, A.J.: A selectional auto-encoder approach for document image binarization. Pattern Recognit. **86**, 37–47 (2019)
3. Chen, L.C., Papandreou, G., Schroff, F., Adam, H.: Rethinking atrous convolution for semantic image segmentation. arXiv preprint arXiv:1706.05587 (2017)
4. Cheriet, M., Moghaddam, R.F., Hedjam, R.: A learning framework for the optimization and automation of document binarization methods. Comput. Vis. Image Underst. **117**(3), 269–280 (2013)
5. De, R., Chakraborty, A., Sarkar, R.: Document image binarization using dual discriminator generative adversarial networks. IEEE Signal Process. Lett. **27**, 1090–1094 (2020)
6. Deng, F., Wu, Z., Lu, Z., Brown, M.S.: BinarizationShop: a user-assisted software suite for converting old documents to black-and-white. In: Proceedings of the 10th Annual Joint Conference on Digital Libraries, pp. 255–258 (2010)
7. Gatos, B., Ntirogiannis, K., Pratikakis, I.: ICDAR 2009 document image binarization contest (DIBCO 2009). In: International Conference on Document Analysis and Recognition, pp. 1375–1382. IEEE (2009)
8. Hadjadj, Z., Meziane, A., Cherfa, Y., Cheriet, M., Setitra, I.: ISauvola: improved Sauvola's algorithm for document image binarization. In: Campilho, A., Karray, F. (eds.) ICIAR 2016. LNCS, vol. 9730, pp. 737–745. Springer, Cham (2016). https://doi.org/10.1007/978-3-319-41501-7_82
9. He, S., Schomaker, L.: DeepOtsu: document enhancement and binarization using iterative deep learning. Pattern Recognit. **91**, 379–390 (2019)
10. Hedjam, R., Nafchi, H.Z., Moghaddam, R.F., Kalacska, M., Cheriet, M.: ICDAR 2015 contest on multispectral text extraction (MS-TEx 2015). In: International Conference on Document Analysis and Recognition, pp. 1181–1185. IEEE (2015)
11. Howe, N.R.: Document binarization with automatic parameter tuning. Int. J. Doc. Anal. Recognit. **16**(3), 247–258 (2013)
12. Kaur, A., Rani, U., Josan, G.S.: Modified Sauvola binarization for degraded document images. Eng. Appl. Artif. Intell. **92**, 103672 (2020)
13. Lazzara, G., Géraud, T.: Efficient multiscale Sauvola's binarization. Int. J. Doc. Anal. Recognit. **17**(2), 105–123 (2014)
14. Moghaddam, R.F., Cheriet, M.: A multi-scale framework for adaptive binarization of degraded document images. Pattern Recognit. **43**(6), 2186–2198 (2010)
15. Nafchi, H.Z., Ayatollahi, S.M., Moghaddam, R.F., Cheriet, M.: An efficient ground truthing tool for binarization of historical manuscripts. In: International Conference on Document Analysis and Recognition, pp. 807–811. IEEE (2013)
16. Ntirogiannis, K., Gatos, B., Pratikakis, I.: ICFHR2014 competition on handwritten document image binarization (H-DIBCO 2014). In: International Conference on Frontiers in Handwriting Recognition, pp. 809–813. IEEE (2014)
17. Otsu, N.: A threshold selection method from gray-level histograms. IEEE Trans. Syst. Man Cybern. **9**(1), 62–66 (1979)
18. Pai, Y.T., Chang, Y.F., Ruan, S.J.: Adaptive thresholding algorithm: efficient computation technique based on intelligent block detection for degraded document images. Pattern Recognit. **43**(9), 3177–3187 (2010)

19. Peng, X., Wang, C., Cao, H.: Document binarization via multi-resolutional attention model with DRD loss. In: International Conference on Document Analysis and Recognition, pp. 45–50. IEEE (2019)
20. Pratikakis, I., Gatos, B., Ntirogiannis, K.: ICDAR 2011 document image binarization contest (DIBCO 2011). In: International Conference on Document Analysis and Recognition, pp. 1506–1510 (2011)
21. Pratikakis, I., Zagori, K., Kaddas, P., Gatos, B.: ICFHR 2018 competition on handwritten document image binarization (H-DIBCO 2018). In: International Conference on Frontiers in Handwriting Recognition, pp. 489–493 (2018)
22. Pratikakis, I., Zagoris, K., Karagiannis, X., Tsochatzidis, L., Mondal, T., Marthot-Santaniello, I.: ICDAR 2019 competition on document image binarization (DIBCO 2019). In: International Conference on Document Analysis and Recognition, pp. 1547–1556 (2019)
23. Pratikakis, I., Gatos, B., Ntirogiannis, K.: H-DIBCO 2010-handwritten document image binarization competition. In: International Conference on Frontiers in Handwriting Recognition, pp. 727–732. IEEE (2010)
24. Pratikakis, I., Gatos, B., Ntirogiannis, K.: ICFHR 2012 competition on handwritten document image binarization (H-DIBCO 2012). In: International Conference on Frontiers in Handwriting Recognition, pp. 817–822. IEEE (2012)
25. Pratikakis, I., Gatos, B., Ntirogiannis, K.: ICDAR 2013 document image binarization contest (DIBCO 2013). In: International Conference on Document Analysis and Recognition, pp. 1471–1476. IEEE (2013)
26. Pratikakis, I., Zagoris, K., Barlas, G., Gatos, B.: ICFHR2016 handwritten document image binarization contest (H-DIBCO 2016). In: International Conference on Frontiers in Handwriting Recognition, pp. 619–623. IEEE (2016)
27. Pratikakis, I., Zagoris, K., Barlas, G., Gatos, B.: ICDAR2017 competition on document image binarization (DIBCO 2017). In: International Conference on Document Analysis and Recognition, vol. 1, pp. 1395–1403. IEEE (2017)
28. Sauvola, J., Pietikäinen, M.: Adaptive document image binarization. Pattern Recognit. 33(2), 225–236 (2000)
29. Shafait, F., Keysers, D., Breuel, T.M.: Efficient implementation of local adaptive thresholding techniques using integral images. In: Document recognition and retrieval XV, vol. 6815, p. 681510. International Society for Optics and Photonics (2008)
30. Souibgui, M.A., Kessentini, Y.: De-gan: A conditional generative adversarial network for document enhancement. IEEE Trans. Pattern Anal. Mach. Intell. (2020)
31. Su, B., Lu, S., Tan, C.L.: Robust document image binarization technique for degraded document images. IEEE Trans. Image Process. 22(4), 1408–1417 (2012)
32. Vo, Q.N., Kim, S.H., Yang, H.J., Lee, G.: Binarization of degraded document images based on hierarchical deep supervised network. Pattern Recognit. 74, 568–586 (2018)
33. Wu, Y., Natarajan, P., Rawls, S., AbdAlmageed, W.: Learning document image binarization from data. In: International Conference on Image Processing, pp. 3763–3767 (2016)
34. Zhao, J., Shi, C., Jia, F., Wang, Y., Xiao, B.: Document image binarization with cascaded generators of conditional generative adversarial networks. Pattern Recognit. 96, 106968 (2019)

Handwriting Recognition

Recognizing Handwritten Chinese Texts with Insertion and Swapping Using a Structural Attention Network

Shi Yan[1], Jin-Wen Wu[2,3], Fei Yin[2,3], and Cheng-Lin Liu[2,3(✉)]

[1] School of Computer Science and Technology, Anhui University, Hefei, China
`shi.yan@nlpr.ia.ac.cn`
[2] National Laboratory of Pattern Recognition, Institute of Automation of Chinese Academy of Sciences, Beijing 100190, China
`{jinwen.wu,fyin,liucl}@nlpr.ia.ac.cn`
[3] School of Artificial Intelligence, University of Chinese Academy of Sciences, Beijing 100049, China

Abstract. It happens in handwritten documents that text lines distort beyond sequential structure because of in-writing editions such as insertion and swapping of text. This kind of irregularity can not be handled using existing text line recognition methods that assume regular character sequences. In this paper, we regard this irregular text recognition as a two-dimensional (2D) problem and propose a structural attention network (SAN) for recognizing texts with insertion and swapping. Particularly, we present a novel structural representation to help SAN learn these irregular structures. With the guidance of the structural representation, SAN can correctly recognize texts with insertion and swapping. To validate the effectiveness of our method, we chose the public SCUT-EPT dataset which contains some samples of text with insertion and swapping. Due to the scarcity of text images with text insertion and swapping, we generate a specialized dataset which only consists of these irregular texts. Experiments show that SAN promises the recognition of inserted and swapped texts and achieves state-of-the-art performance on the SCUT-EPT dataset.

Keywords: Handwritten Chinese text recognition · Insertion · Swapping · 2D attention · Structural representation

1 Introduction

Handwritten text recognition is a main task in document analysis and has achieved huge progress attributed to the research efforts. However, there are still many technical challenges, such as writing style variation, large character sets, character erasure, and irregular in-writing editions. Most studies focus on documents with regular texts of sequential structure, but the non-sequential structure of irregular in-writing editions in some documents like examination

© Springer Nature Switzerland AG 2021
J. Lladós et al. (Eds.): ICDAR 2021, LNCS 12824, pp. 557–571, 2021.
https://doi.org/10.1007/978-3-030-86337-1_37

paper is also widespread in daily life. Different from other handwriting datasets such as CASIA-HWDB [1] and HIT-MW [2] dataset, the SCUT-EPT dataset [3] introduces other challenges like character erasure and irregular in-writing editions (text with insertion and swapping). Character erasure has been studied by previous works [4–6], while few works have aimed at the recognition of text with insertion and swapping. Figure 1(d) (e) show text lines that distort beyond sequential structure because of insertion and swapping of text. Existing text recognition methods which treat text images as character sequences can not handle such 2D non-sequential structure effectively.

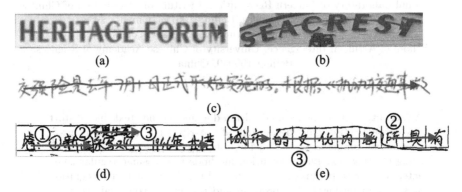

Fig. 1. Reading order of different text lines in different documents. (a) and (b) are scene texts, (c) is regular handwritten text, (d) and (e) are irregular texts with insertion and swapping. The red solid line represents the reading order of characters. Circled numbers represent the reading order of different parts of the text line. (Color figure online)

Compared with scene text or paragraph text [7], the structure of text with insertion and swapping is more complicated due to the non-sequential reading order. Specifically, for text with insertion, the inserted text is read before the text after the insertion position. While for text with swapping, the second sub-string after the swapping location is read prior to the sub-string before the location. Although many scene text recognition methods like CTC-based [8] and attention-based [9] methods have been applied to handwritten text recognition as well, they are designed for sequential structure and do not perform well for non-sequential texts with insertion and swapping. To deal with such 2D structure of texts with insertion, we propose a structural attention network and a novel structural representation for texts with insertion and swapping. The contributions of this paper are summarized as follows:

(1) We propose a structural attention network (SAN) for recognizing irregular texts with insertion and swapping. The network encodes text images into 2D feature maps and uses structural attention to enhance the perception of spatial location and structure for the decoder.

(2) We propose a novel structural representation that decomposes the irregular text into original characters and semantic structure symbols. With the representation, SAN can efficiently learn the structure of texts with insertion and swapping.
(3) Based on the scarcity of inserted and swapped texts in the SCUT-EPT dataset, we generate a specialized dataset which only consists of these irregular texts.
(4) Experimental results show that the proposed SAN promises the recognition of irregular texts with insertion and swapping, and has yielded state-of-the-art performance on the SCUT-EPT dataset.

2 Related Work

2.1 Scene Text Recognition

Existing scene text recognition methods can be roughly divided into three categories: character-based [10,11], word-based [12] and sequence-to-sequence methods [8,13]. Compared with the character-based methods, sequence-to-sequence methods do not need character-level annotations. In addition, different from word-based methods, they can predict words out of the dictionary. Hence, this category has attracted increasing attention, and these methods can be further categorized into 1D-based methods [8,9,14–16] and 2D-based methods [13,17,18]. Compared with 1D-based methods, 2D-based methods can handle texts with more complex structures like curved scene texts but need more computation costs.

According to the decoder, sequence-to-sequence methods can also be classified as CTC-based [8,14,18] and attention-based methods [9,15–17]. In general, attention-based methods achieve better performance for irregular texts, while CTC-based methods offer faster reasoning speed. Besides, attention-based methods perform poorly on contextless texts (e.g. random character sequences) and suffer from attention drift [19,20].

2.2 Handwritten Chinese Text Recognition

Previous handwritten Chinese text recognition methods [21–24] often take a pre-segmentation step, while segmentation-free methods [25,26] based on RNN can achieve higher performance. Due to the advances of scene text recognition, many methods for scene text have been applied to handwritten Chinese text recognition. However, these methods are aiming at regular shape (such as curved and slanted text), and can not cope with irregular order such as insertion and swapping of text. Specifically, the existing methods usually adopt CTC-based or 1D attention-based decoder and are aiming at solving single-line and sequentially structured text, while inserted and swapped texts are beyond 1D structure. To cope with 2D structures of texts, we adopt 2D attention-based decoder. And for non-sequential structure, we propose a novel structural representation scheme.

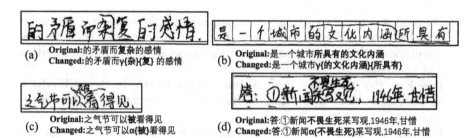

Fig. 2. Comparison between the original label and the label of our structural representation. (a) and (b) are samples of swapping text, (c) and (d) are samples of insertion text. γ and α represent swapping and insertion-up, respectively.

3 Structural Representation of Irregular Text

To explore the 2D non-sequential structure of irregular offline Chinese text, we introduce the structural representation scheme. Specifically, a structural representation consists of three key components: original characters, semantic structures, and a pair of braces (e.g. "{" and "}"). Original characters represent the characters present in the text image. The semantic structures and descriptions are illustrated as follows:

- *insertion-up* ($\alpha\{string\}$): text with insertion up structure
- *insertion-down* ($\beta\{string\}$): text with insertion down structure
- *swapping* ($\gamma\{string1\}\{string2\}$): text with swapping structure

After analyzing different types of irregular text, we use one or two pairs of braces to describe the structure of a text line. The pair of braces is used to determine which characters are inserted in the location of the insertion symbol (α or β) or which characters (sub-strings) are swapped at the location of the swapping symbol (γ). The format of insertion-up is the same as that of insertion-down, except that the edition symbols indicating the insertion direction are different. Some examples of structural representation for texts with insertion/swapping are shown in Fig. 2. With the structural representation, the non-sequential structure can be converted into the sequential problem of simultaneously predicting structural symbols and original characters.

There are three benefits of the structural representation: first, with this representation, existing methods can be directly applied to recognize handwritten texts with insertion and swapping. Second, the recognizer can learn the relationship between multiple text lines with insertion more efficiently. Third, the network can predict characters sequentially for swapped sub-strings of text with a post-processor.

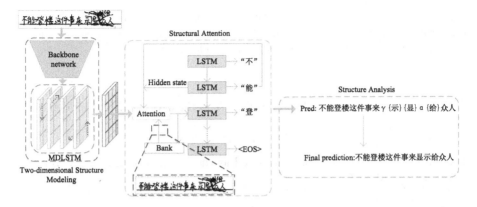

Fig. 3. Schematic illustration of the SAN framework. "Bank" means memory bank which stores the sum of past attention scores. In the third prediction step, the record vector has stored the two past attention scores. For the sake of illustration, we use the input image to show the vector.

4 Structural Attention Network

The proposed SAN framework is diagrammed in Fig. 3. The framework incorporates the structural representation scheme and consists of three parts: 2D structure modeling, structural attention, and structure analysis.

4.1 Two-Dimensional Structure Modeling

Previous methods of text recognition usually reduce the 2D features extracted from the text image into 1D sequence, which is not suitable for irregular texts which have 2D spatial structures. To cope with this structure, we keep the 2D structure of features in sequence modeling for increasing contextual information and enlarging receptive fields.

BiLSTM [27] has been frequently used for the sequence modeling. It works for 1D structure issue, but cannot satisfy texts with insertion and swapping, which could be associated with long sequences distributed in both horizontal and vertical directions. For this sake, we use multi-dimensional LSTM (MDLSTM) [7] instead of BiLSTM. MDLSTM can be formulated as:

$$(a_{(i,j)}, q_{(i,j)}) = LSTM(x_{(i,j)}, a_{(i,j\pm1)}, a_{(i\pm1,j)}, q_{(i,j\pm1)}, q_{(i\pm1,j)}), \qquad (1)$$

where $x_{(i,j)}$ is the input feature vector at position (i,j), and a and q represent the output and inner state of the LSTM cell, respectively. The ±1 choices in this recurrence depend on which of the four scanning directions is considered. Besides, we add DropBlock [28] for 2D features regularization.

4.2 Structural Attention

Due to the long sequence of handwritten Chinese texts and the non-sequential structure of irregular texts, the classic attention mechanism could not handle this long-distance dependent structural relationship well. To better illustrate our proposed structural attention, we first show the details of classic attention as follows: at t-step, the output y_t of the decoder is obtained as:

$$y_t = softmax(W_0 h_t + b_0), \tag{2}$$

where W_0 and b_0 are trainable parameters, h_t is the decoder LSTM hidden state at time t:

$$h_t = LSTM(y_{t-1}, c_t, h_{t-1}), \tag{3}$$

here c_t is a context vector and computed as the weighted sum of $a_{(i,j)}$ which is the output of MDLSTM.

$$c_t = \sum_{i=1}^{H} \sum_{j=1}^{W} \alpha_{t,(i,j)} a_{(i,j)}, \tag{4}$$

where H, W are the height and width of the feature maps obtained by CNN, and $\alpha_{t,(i,j)}$ is called attention weight and computed by

$$\alpha_{t,(i,j)} = \frac{\exp e_{t,(i,j)}}{\sum_{i=1}^{H} \sum_{j=1}^{W} \exp e_{t,(i,j)}}, \tag{5}$$

where

$$e_{t,(i,j)} = v_a^\mathsf{T} tanh(W_a h_{t-1} + U_a a_{(i,j)} + b), \tag{6}$$

and v_a, W_a, U_a and b are trainable parameters.

To handle the long-distance dependent structural relationship, we leverage all the historical reading areas of the recognizer to enhance its perception of spatial structure. In detail, it is realized by utilizing a historical memory bank which stores the information of past attended areas:

$$\beta_t = \sum_{l=1}^{t-1} \alpha_l, \tag{7}$$

$$F = Q * \beta_t, \tag{8}$$

$$e_{t,(i,j)} = v_a^\mathsf{T} tanh(W_a h_{t-1} + U_a a_{(i,j)} + U_f f_{(i,j)} + b). \tag{9}$$

Here β_t is the memory bank of decoding step t, α_l is the past attention score, $f_{(i,j)}$ is the record vector of annotation $a_{(i,j)}$ and is initialized as zero tensor. At each decoding step t, the historical memory bank serves as an additional input to the attention model and provides complementary past alignment information about whether a local region of source images has been attended to. The record vector is produced through a convolution layer for better fusing its adjacent attention probabilities. With the guide of structured representation, our structural attention-based decoder could effectively parse non-sequential structures.

Table 1. Statistics of text swapping and text insertion samples from the SCUT-EPT dataset and our synthetic dataset. "Swapping" means text with swapping and "Insertion" means text with insertion.

Dataset	Swapping	Insertion
SCUT-EPT training	86	93
SCUT-EPT test	92	235
Synthesized data	850	1,200

4.3 Objective Function

The SAN is trained by minimizing the negative log-likelihood of the conditional probability:

$$L = -\sum_{t=1}^{T} \log p(Y_i|X_i), \tag{10}$$

where X_i is the input irregular offline Chinese text image, and Y_i is the corresponding structural representation. Since the structural representations are in the form of a sequence, the vanilla beam search algorithm [29,30] can still be used to search for the optimal path during prediction.

4.4 Structure Analysis

In recognition, the SAN first outputs the structural representation via structural attention decoding. Then we analyze the structural representation in the recognition result to obtain the original format sequence. Specifically, we transpose string enclosed by two pairs of braces following the swapping symbol, then delete the swapping symbol and two pairs of braces. For text insertion, we keep the order of the inserted text in the sequence and delete the insertion symbol and a pair of braces. Among cases of texts with both insertion and swapping, only legal cases are processed. The legal format of text swapping is defined as $\gamma\{string1\}\{string2\}$. γ represents a swapping symbol, the first brace must follow the swapping symbol, and there are no other characters between the second brace and the third brace. For text with insertion, i.e. $\alpha/\beta\{string\}$, the first brace must follow the insertion symbol. After all legal cases are processed, the remaining braces, swapping/insertion symbols will be deleted.

4.5 Synthetic Data for Texts with Insertion and Swapping

To better understand the characteristics of irregular texts, we count the number of samples of texts with insertion and swapping in the SCUT-EPT dataset [3]. As shown in Table 1, there are few samples of texts with insertion and swapping. Furthermore, the average length of inserted and swapped text in the test set is longer than that in the training set. Since these samples are too few for training

(a)label:肇事推责，冷了人心   (b) label:到日后会发生这样的事情，但我们面对这样的事情,是要

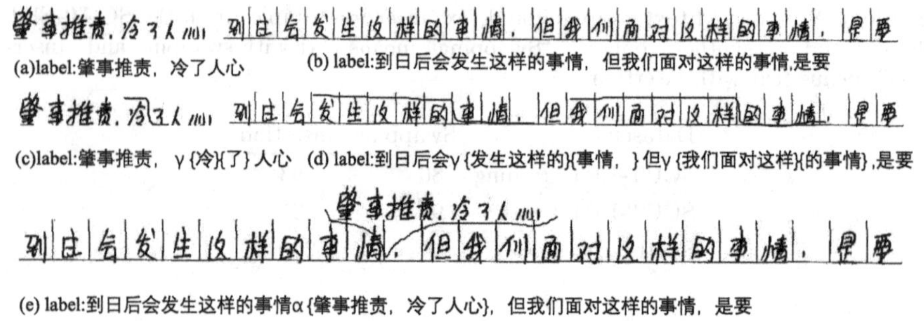

(c)label:肇事推责，γ{冷}{了}人心 (d) label:到日后会γ{发生这样的}{事情，}但γ{我们面对这样}{的事情},是要

(e) label:到日后会发生这样的事情α{肇事推责，冷了人心}，但我们面对这样的事情，是要

Fig. 4. Original training samples and synthetic samples. (a) and (b) are samples from the training set, (c) and (d) are synthetic samples of texts with swapping. (e) is a synthetic sample of text with insertion.

and evaluating our model, we synthesize some data of texts with swapping and insertion.

Comparing with samples of regular Chinese texts, we have to draw the insertion/swapping symbols in all synthetic samples by hand to obtain the image that contains texts of insertion and swapping. All synthetic samples are based on images of training set from SCUT-EPT or other synthetic Chinese handwriting images, and are only used for training. When synthesizing text swapping samples from a regular text line, we randomly select a location, pick sub-strings of random length before and after this location, and change the label in the way as in Fig. 2. For text insertion, we first select two different training text lines and insert the shorter text line into the longer one. We set a maximum length for the inserted text line and synthesize text lines with insertion of different directions according to a certain ratio (insertion-up/insertion-down = 3/1). Examples of synthesizing are shown in Fig. 4.

5 Experiments

In this section, we first introduce the datasets, implementation details, and evaluation metrics. Then, extensive experiments and visual analysis are presented to verify the effectiveness of our method.

5.1 Datasets

We conducted experiments on one public dataset and two datasets consisting irregular texts selected from the test set of the public dataset.

SCUT-EPT. This dataset was constructed from examination papers of 2,986 high school students. It contains 50,000 text line images, including 40,000 text lines for training and 10,000 text lines for testing.

SWAPPING. The samples of this dataset are the text lines with swapping selected from the test set of SCUT-EPT. It contains 92 images.

INSERTION. This dataset contains the text lines with insertion from the test set of SCUT-EPT. There are 235 samples in this set.

5.2   Implementation Details

Consistent with previous works [3], we take the character set (7,356 characters) [24] as a base, add the begin and end symbols, and set the rest characters as outlier. For the proposed model SAN, two insertion symbols, a swapping symbol, and two braces are added. During the training process, we normalize the height of the text line image to 96 pixels and then pad the width to 1,440 pixels while keeping the aspect ratio. For fair comparison, we select the ResNet [31] as backbone like previous scene text recognition methods [32]. To get suitable size of feature maps, we add two extra max-pooling layers. The dimensionality of the MDLSTM hidden state is set as 256. For evaluation, we use the correct rate (CR) and accuracy rate (AR) used in ICDAR2013 competition [24] as metrics of performance. They are given by:

$$CR = (N - D_e - S_e)/N, \tag{11}$$

$$AR = (N - D_e - S_e - I_e)/N, \tag{12}$$

where N is the total number of characters in the test set, D_e, S_e and I_e are the numbers of deletion errors, substitution errors, and insertion errors, respectively. All the experiments were implemented on PyTorch with 4 GeForce Titan-X GPUs.

5.3   Ablation Study

We performed a series of ablation experiments to validate the effectiveness of each part of the proposed SAN model.

Table 2. Ablation Study. "Baseline" means the 1D attention-based method; "2D" means 2D attention; "MD" represents replacing BiLSTM with MDLSTM; "SA" means structural attention; "SR" means structural representation and "Syn" refers to synthesized data.

	SCUT-EPT		SWAPPING		INSERTION	
	CR	AR	CR	AR	CR	AR
Baseline	76.91	72.31	78.96	75.11	68.23	65.53
+2D	79.08	74.93	81.36	77.72	70.10	68.55
+MD	79.15	75.14	81.65	77.68	70.26	68.67
+SA	80.00	75.92	82.10	78.59	71.44	69.72
+SR	80.42	76.23	81.52	77.10	70.51	68.75
+Syn	**80.83**	**76.42**	**87.23**	**84.50**	**74.52**	**70.58**

Comparing 1D Attention and 2D Attention. Compared to the baseline method using 1D attention, 2D attention-based method keeps the 2D structure of feature maps and preserves more contextual information, which is very significant for the recognition of texts with swapping and insertion due to the complex structure. As the results in Table 2 show, the 2D attention-based method outperforms the 1D attention-based method on three datasets. This validates the necessity of taking irregular offline Chinese text recognition as a 2D structure analysis problem.

Comparing BiLSTM and MDLSTM. As shown in Table 2, the improvements of MDLSTM compared to BiLSTM on three datasets from more context information and larger receptive fields are evident. This improvement originates from the fact that texts with insertion and swapping may be associated with long sequences distributed in both horizontal and vertical directions.

Comparing Classic Attention and Structural Attention. The classic attention mechanism could not cope with long-distance dependent structural relationships well. After appending the structural attention to previous system, performances on three datasets have been further improved.

Validating Structural Representation. We propose the structural representation for texts with insertion and swapping. The improvement brought by "+SR" on SCUT-EPT validates that structural representation is helpful. The main reason for slight performance degradation on SWAPPING and INSERTION datasets is the scarcity of texts with swapping and insertion in the training set, thus the SAN is not trained sufficiently to correctly predict the samples of text swapping and insertion. This also reflects the need to synthesize samples of texts with swapping and insertion.

Effects of Synthesized Data. As mentioned before, the percentage of texts with swapping and insertion in the SCUT-EPT dataset is very small. Besides, the average length of samples of irregular texts in training set and test set are different. As shown in Table 2, by synthesizing training samples with swapping and insertion, higher performances are achieved not only on SWAPPING and INSERTION datasets but also on SCUT-EPT test set.

5.4   Visualization of Structural Attention

To better illustrate the superiority of SAN, we take a sample of text with swapping as an example and show the visualization of attention in each prediction step. In Eq. 5, $\alpha_{t,(i,j)}$ is called attention weight and reflects the correlation between the feature at each position and the currently predicted character. We display the pixel in attention map in red only when the corresponding $\alpha_{t,(i,j)}$ is above a certain threshold, otherwise in white, as Fig. 5 shows. As we see, SAN

Fig. 5. Visualization of attention in every prediction step. γ means the swapping symbol.

Table 3. Results of State-of-the-Art methods and SAN on SCUT-EPT, where "Enrich" indicates whether an additional synthetic data set is used in training. "Samples" means the number of samples used in training, and "Classes" is the corresponding number of character classes.

	Enrich	Samples	Classes	CR	AR
CTC [3]	N	40,000	4,058	78.60	75.37
	Y	228,014	7,356	80.26	75.97
Attention [3]	N	40,000	4,058	69.83	64.78
	Y	228,014	7,356	73.10	67.04
Cascaded	N	40,000	4,058	54.09	48.98
Attention-CTC [3]	Y	228,014	7,356	60.20	55.64
SCATTER [33]	N	40,000	4,058	79.36	75.20
Baseline	N	40,000	4,058	76.91	72.31
SAN	N	40,000	4,058	80.42	76.23
	Y	42,050	4,058	**80.83**	**76.42**

just read the text line from left to right with our structural representation, which can be converted in post-processing to the original normal text. This shows that SAN can handle irregular offline Chinese texts properly. In the example, we can also see that the position of both the swapping symbol and the pairs of braces can be correctly predicted even with long swapped text.

5.5 Comparisons with State-of-the-Art Methods

Some transcription methods, such as CTC, attention mechanism, and cascaded attention-CTC decoder were tested in [3]. It was found that the attention mechanism has much worse performance than that of CTC. But in our experiment, we found that the performance of 1D attention-based method is not that bad.

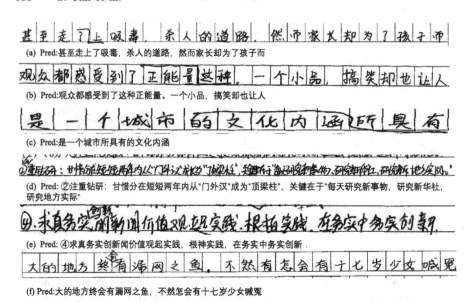

(a) Pred:甚至走上了吸毒，杀人的道路，然而家长却为了孩子而

(b) Pred:观众都感受到了这种正能量。一个小品，搞笑却也让人

(c) Pred:是一个城市所具有的文化内涵

(d) Pred: ②注重钻研: 甘惜分在短短两年内从"门外汉"成为"顶梁柱"，关键在于"每天研究新事物，研究新华社，研究地方实际"

(e) Pred: ④求真务实创新闻价值观起实践，根神实践，在务实中务实创新.

(f) Pred:大的地方终会有漏网之鱼，不然怎会有十七岁少女喊冤

Fig. 6. Recognition results on SWAPPING and INSERTION datasets. (a), (b) and (c) are samples in SWAPPING, and (d), (e) and (f) are samples in INSERTION.

As shown in Table 3, the baseline method, which is also 1D attention-based but adopts ResNet as backbone and 2-layers BiLSTM for sequence modeling, has much better performance than the 1D attention-based method [3]. Besides, we reproduced another state-of-the-art attention-based method SCATTER [33]. Its CR and AR, 79.36% and 75.20%, respectively, and are lower than the results of SAN without additional synthetic data. We believe that two-dimensional based semantic modeling and recognition is the key to our performance leadership because of the complexity of text structure resulting from the editing of text during writing. Moreover, it is worth noting that, with additional synthetic data, the training set of the CTC-based method contains 228,014 text images. Even so, the SAN trained with original training dataset outperforms CTC-based methods trained with extra data. When training with extra 2,050 samples, the SAN achieves 80.83% and 76.42% in CR and AR, respectively, outperforming all the previous methods.

5.6   Recognition Examples

We show some samples and corresponding predictions of SAN in Fig. 6. As it depicts, our method can handle short and long swapped texts. For offline Chinese text images with short text inserted, e.g. (e) and (f) in Fig. 6, our method can also effectively parse the structure. Nevertheless, as shown in Fig. 7, when the inserted text is too long, the SAN might fail to recognize all the characters completely correctly. The main reason is that samples with very long inserted texts are scarce in the training set. Though we have synthesized some samples

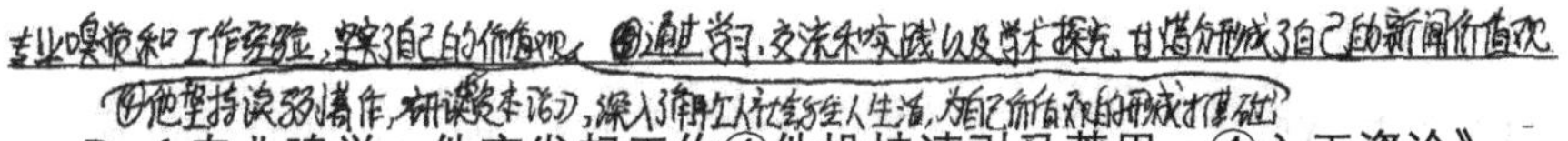

(a) Pred:专业嗅觉．他宽发起工作④他投持读引马著思，④入工资论》，深入了解工人生活，为自己后作的形成研础。自己的自己的行为自己的新闻价值观

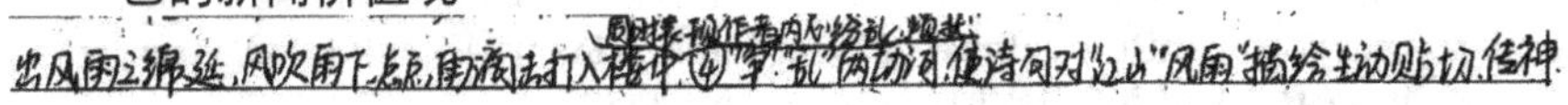

(b) Pred:出风雨之绵延，风吹雨下，点点雨滴去打入楼中中的时是一定是一个社会①"争"字乱，两句动，使诗句对，"江山""风雨"描绘生动贴切，传神

(c) Pred:诗人的无奈、郁的；颔联运用没途景象，表明奔劳之苦，渴望安定，颈联却用朱离反衬回程的欣悦，与尾联

Fig. 7. Some examples of wrongly recognized by the SAN. Most failure examples have long inserted texts. The characters in red represent the wrong prediction, characters that are not predicted are not displayed. (Color figure online)

with long text inserted, it is very different from the samples of the test set. Figure 7(b) and (c) are samples whose inserted text is very close to the remaining text, these samples are hardly to synthesize.

6   Conclusion

In this paper, we take irregular offline Chinese text recognition as a 2D problem and propose a structural attention network (SAN) for recognizing texts with insertion and swapping. Besides, we present a novel structural representation to handle the non-sequential structure of text with swapping and insertion. Due to the scarcity of samples with insertion and swapping, we synthesize a dataset consisting of irregular offline Chinese texts. Experimental results and attention visualization demonstrate the effectiveness of SAN, especially on texts with insertion and swapping. Moreover, The proposed method yielded state-of-the-art performance on the SCUT-EPT dataset. In the future, we will focus on the recognition of long inserted text, and also consider some transformer-based methods.

Acknowledgements. This work has been supported by the National Key Research and Development Program Grant 2020AAA0109702, the National Natural Science Foundation of China (NSFC) grants 61733007, 61721004.

References

1. Liu, C.L., Yin, F., Wang, D.H., Wang, Q.F.: Casia online and offline Chinese handwriting databases. In: Proceedings of 11th International Conference on Document Analysis and Recognition (ICDAR), pp. 37–41 (2011)
2. Su, T., Zhang, T., Guan, D.: HIT-MW dataset for offline Chinese handwritten text recognition. In: Proceedings of 10th International Workshop on Frontiers in Handwriting Recognition (IWFHR), pp. 1–5 (2006)
3. Zhu, Y., Xie, Z., Jin, L., Chen, X., Huang, Y., Zhang, M.: SCUT-EPT: new dataset and benchmark for offline Chinese text recognition in examination paper. IEEE Access **7**, 370–382 (2019)
4. Bhattacharya, N., Frinken, V., Pal, U., Roy, P.P.: Overwriting repetition and crossing-out detection in online handwritten text. In: Proceedings of Asian Conference on Pattern Recognition (ACPR), pp. 680–684 (2015)
5. Chaudhuri, B.B., Adak, C.: An approach for detecting and cleaning of struck-out handwritten text. Pattern Recogn. **61**, 282–294 (2017)
6. Adak, C., Chaudhuri, B.B.: An approach of strike-through text identification from handwritten documents. In: Proceedings of Nineth International Conference on Frontiers in Handwriting Recognition, pp. 643–648 (2014)
7. Graves, A., Schmidhuber, J.: Offline handwriting recognition with multidimensional recurrent neural networks. In: Advances in Neural Information Processing Systems, pp. 545–552 (2009)
8. Shi, B., Bai, X., Yao, C.: An end-to-end trainable neural network for image-based sequence recognition and its application to scene text recognition. IEEE Trans. Pattern Anal. Mach. Intell. **39**(11), 2298–2304 (2017)
9. Shi, B., Wang, X., Lyu, P., Yao, C., Bai, X.: Robust scene text recognition with automatic rectification. In: Proceedings of IEEE Conference on Computer Vision and Pattern Recognition (CVPR), pp. 4168–4176 (2016)
10. Liao, M., et al.: Scene text recognition from two-dimensional perspective. In: Proceedings of the AAAI Conference on Artificial Intelligence, vol. 33, pp. 8714–8721 (2019)
11. Lyu, P., Liao, M., Yao, C., Wu, W., Bai, X.: Mask textspotter: an end-to-end trainable neural network for spotting text with arbitrary shapes. In: Proceedings of European Conference on Computer Vision (ECCV), pp. 67–83 (2018)
12. Jaderberg, M., Simonyan, K., Vedaldi, A., Zisserman, A.: Reading text in the wild with convolutional neural networks. Int. J. Comput. Vis. **116**(1), 1–20 (2016)
13. Cheng, Z., Xu, Y., Bai, F., Niu, Y., Pu, S., Zhou, S.: AON: towards arbitrarily-oriented text recognition. In: Proceedings of IEEE Conference on Computer Vision and Pattern Recognition (CVPR), pp. 5571–5579 (2018)
14. Liu, H., Jin, S., Zhang, C.: Connectionist temporal classification with maximum entropy regularization. In: Advances in Neural Information Processing Systems, pp. 831–841 (2018)
15. Shi, B., Yang, M., Wang, X., Lyu, P., Yao, C., Bai, X.: ASTER: an attentional scene text recognizer with flexible rectification. IEEE Trans. Pattern Anal. Mach. Intell. **41**(9), 2035–2048 (2019)
16. Yang, M., et al.: Symmetry-constrained rectification network for scene text recognition. In: Proceedings of International Conference on Computer Vision (ICCV), pp. 9147–9156 (2019)
17. Li, H., Wang, P., Shen, C., Zhang, G.: Show, attend and read: a simple and strong baseline for irregular text recognition. In: Proceedings of the AAAI Conference on Artificial Intelligence, vol. 33, pp. 8610–8617 (2019)

18. Wan, Z., Xie, F., Liu, Y., Bai, X., Yao, C.: 2D-CTC for scene text recognition. arXiv preprint arXiv:1907.09705 (2019)

19. Cheng, Z., Bai, F., Xu, Y., Zheng, G., Pu, S., Zhou, S.: Focusing attention: towards accurate text recognition in natural images. In: Proceedings of International Conference on Computer Vision (ICCV), pp. 5076–5084 (2017)

20. Yue, X., Kuang, Z., Lin, C., Sun, H., Zhang, W.: RobustScanner: dynamically enhancing positional clues for robust text recognition. In: Vedaldi, A., Bischof, H., Brox, T., Frahm, J.-M. (eds.) ECCV 2020. LNCS, vol. 12364, pp. 135–151. Springer, Cham (2020). https://doi.org/10.1007/978-3-030-58529-7_9

21. Hong, C., Loudon, G., Wu, Y., Zitserman, R.: Segmentation and recognition of continuous handwriting Chinese text. Int. J. Pattern Recogn. Artif. Intell. **12**(02), 223–232 (1998)

22. Srihari, S.N., Yang, X., Ball, G.R.: Offline Chinese handwriting recognition: an assessment of current technology. Front. Comput. Sci. China **1**(2), 137–155 (2007)

23. Wang, Q., Yin, F., Liu, C.: Handwritten Chinese text recognition by integrating multiple contexts. IEEE Trans. Pattern Anal. Mach. Intell. **34**(8), 1469–1481 (2012)

24. Yin, F., Wang, Q.F., Zhang, X.Y., Liu, C.L.: ICDAR 2013 Chinese handwriting recognition competition. In: Proceedings of 12th International Conference on Document Analysis and Recognition (ICDAR), pp. 1464–1470 (2013)

25. Messina, R., Louradour, J.: Segmentation-free handwritten Chinese text recognition with LSTM-RNN. In: Proceedings of 13th International Conference on Document Analysis and Recognition (ICDAR), pp. 171–175 (2015)

26. Xie, Z., Sun, Z., Jin, L., Feng, Z., Zhang, S.: Fully convolutional recurrent network for handwritten Chinese text recognition. In: Proceedings of 23th International Conference on Pattern Recognition (ICPR), pp. 4011–4016 (2016)

27. Hochreiter, S., Schmidhuber, J.: Long short-term memory. Neural Comput. **9**(8), 1735–1780 (1997)

28. Ghiasi, G., Lin, T.Y., Le, Q.V.: DropBlock: a regularization method for convolutional networks. In: Advances in Neural Information Processing Systems, pp. 10727–10737 (2018)

29. Ney, H., Haeb-Umbach, R., Tran, B., Oerder, M.: Improvements in beam search for 10000-word continuous speech recognition. In: Proceedings of International Conference on Acoustics, Speech and Signal Processing (ICASSP), vol. 1, pp. 9–12 (1992)

30. Ow, P.S., Morton, T.E.: Filtered beam search in scheduling. Int. J. Prod. Res. **26**(1), 35–62 (1988)

31. He, K., Zhang, X., Ren, S., Sun, J.: Deep residual learning for image recognition. In: Proceedings of IEEE Conference on Computer Vision and Pattern Recognition (CVPR), pp. 770–778 (2016)

32. Baek, J., et al.: What is wrong with scene text recognition model comparisons? Dataset and model analysis. In: Proceedings of International Conference on Computer Vision (ICCV), pp. 4715–4723 (2019)

33. Litman, R., Anschel, O., Tsiper, S., Litman, R., Mazor, S., Manmatha, R.: SCATTER: selective context attentional scene text recognizer. In: Proceedings of IEEE Conference on Computer Vision and Pattern Recognition (CVPR), pp. 11962–11972 (2020)

Strikethrough Removal from Handwritten Words Using CycleGANs

Raphaela Heil[(✉)][iD], Ekta Vats[iD], and Anders Hast[iD]

Division for Visual Information and Interaction, Department of Information
Technology, Uppsala University, Uppsala, Sweden
{raphaela.heil,ekta.vats,anders.hast}@it.uu.se

Abstract. Obtaining the original, clean forms of struck-through handwritten words can be of interest to literary scholars, focusing on tasks
such as genetic criticism. In addition to this, replacing struck-through
words can also have a positive impact on text recognition tasks. This
work presents a novel unsupervised approach for strikethrough removal
from handwritten words, employing cycle-consistent generative adversarial networks (CycleGANs). The removal performance is improved upon
by extending the network with an attribute-guided approach. Furthermore, two new datasets, a synthetic multi-writer set, based on the IAM
database, and a genuine single-writer dataset, are introduced for the
training and evaluation of the models. The experimental results demonstrate the efficacy of the proposed method, where the examined attribute-
guided models achieve F_1 scores above 0.8 on the synthetic test set,
improving upon the performance of the regular CycleGAN. Despite being
trained exclusively on the synthetic dataset, the examined models even
produce convincing cleaned images for genuine struck-through words.

Keywords: Strikethrough removal · CycleGAN · Handwritten words ·
Document image processing

1 Introduction

The presence of struck-through words in handwritten manuscripts can affect the
performance of document image processing tasks, such as text recognition or
writer identification [1,3,12,15]. In addition to this, it may also be of interest
to researchers in the humanities, who examine the evolution of a manuscript,
for example in the form of genetic criticism (i.e., how literature scholars study
how a text has changed over time), and who may therefore be interested in the
original word that was struck through [9]. The removal of strikethrough strokes,
restoring a clean, original version of a given word is therefore of relevance to
several research areas.

In this work, we treat the removal of strikethrough from handwritten words
as an unpaired image-to-image translation problem, as the ground truth of a
word cannot be obtained directly from a manuscript, once it has been struck

© Springer Nature Switzerland AG 2021
J. Lladós et al. (Eds.): ICDAR 2021, LNCS 12824, pp. 572–586, 2021.
https://doi.org/10.1007/978-3-030-86337-1_38

through. We propose the use of cycle-consistent generative adversarial networks (CycleGANs [21]) in this regard.

This work makes four main contributions. Firstly, a new approach towards the removal of strikethrough is proposed, based on CycleGANs and extended with an attribute-guiding feature [13]. Secondly, we present an approach for the generation of synthetic strikethrough and apply it to the IAM database [14] to create a synthetic multi-writer dataset. Thirdly, we present a new single-writer dataset, consisting of genuine samples of struck-through handwritten words. Lastly, we also address the two upstream tasks of differentiating clean from struck words, and classifying the applied strikethrough into one of seven categories (cf. Fig. 1), using DenseNets [8]. To the best of our knowledge, this is the first work that utilises neural networks for the identification, classification and cleaning of struck-through words.

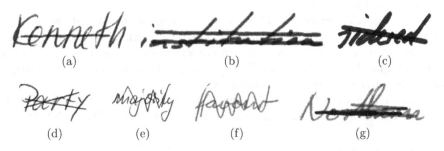

(a)   (b)   (c)

(d)   (e)   (f)   (g)

Fig. 1. Examples of different strikethrough types considered in this work. a) Single, b) Double, c) Diagonal, d) Cross, e) Zig Zag, f) Wave, g) Scratch. All strikethrough strokes were generated using the approach presented in Subsect. 3.4.

2   Related Works

2.1   Handling of Struck Through Text

Only a small number of works in the literature have concerned themselves with the impact and handling of struck-through text. Most recently, Chaudhuri and Adak [6] presented a strikethrough removal approach based on constrained shortest paths on graphs of word skeletons. They evaluate their method on words written and struck-through by volunteers (i.e. genuine strikethrough). The empirical results presented in the paper highlight the efficacy of their proposed approach. To the best of our knowledge, [6] is the only approach that deals with the removal of unwanted strokes, returning a clean version of the original word. At the time of writing, the complete dataset used in [6] was not fully available. It was therefore not possibly to compare our approach with the one presented in that work. In addition to this, it should be noted that our work also considers the additional stroke type, denoted by us as *scratch*, which Chaudhuri and Adak discard during preprocessing.

Besides the above, a number of studies have explored the impact of struck-through text on writer identification and text recognition. Brink et al. [3] used a decision tree with handcrafted features to detect struck-through words, which were then cut from the document image. The authors report that the impact of removing struck-through words is not statistically significant.

In [1], Adak et al. employed a hybrid CNN-SVM classifier for the detection of struck-through text and reported an increase in writer identification performance by three to five percentage points (p.p.) after removing the detected words.

Likforman-Sulem and Vinciarelli [12] generated a synthetic dataset by superimposing waves and single, double or triple straight lines over genuine handwritten words. They use a Hidden Markov Model to recognise the words, noting a maximum decrease of 57.6 p.p. and 64.7 p.p., respectively, in recognition rate on the two evaluated data sets.

Nisa et al. [15] trained a CRNN on two versions of the IAM database [14], with one containing a large amount of synthetic, superimposed strikethrough. They report a decrease in recognition accuracy when testing a model, trained on the original IAM, on their modified version, indicating that struck-through words impact the recognition performance.

Lastly, Dheemanth Urs and Chethan [7] recently presented a brief survey of methods handling struck-through text, noting a dominance of English texts.

2.2 CycleGANs

Cycle-Consistent Adversarial Networks (CycleGANs) were proposed by Zhu et al. [21] in 2017 and have since been used for a variety of tasks centred around unpaired image-to-image translation. In the context of handwritten text generation and document image processing, CycleGANs have for example been used to generate handwritten Chinese characters from printed material [4]. Furthermore, Sharma et al. have applied CycleGANs to the task of removing noise, such as blurring, stains and watermarks from printed texts [18]. Although strikethrough could be considered a type of noise, there is a clear distinction between the degradations handled in the aforementioned paper and struck-through words. In contrast to the types of noise considered by Sharma et al., which are typically created by a source which differs from the one that created the underlying text (e.g. coffee stains), strikethrough is typically applied by the same writer, using the same writing implement, therefore making the character and strikethrough strokes virtually indistinguishable from each other.

3   Method

3.1   Strikethrough Identification

When considering the application of a strikethrough removal approach, the first step, following the segmentation of words, will be to differentiate between clean and struck words. In this regard, we propose to use a standard DenseNet121 [8], modified to receive single-channel (greyscale) images as input and to provide a binary classification (clean vs struck) as output.

3.2 Strikethrough Stroke Classification

In addition to identifying struck-through words, it may be relevant for downstream tasks, such as the one presented in Subsect. 3.3, to distinguish which type of strikethrough has been applied to a given struck word. Similar to the previous task, we propose to use a single-channel DenseNet121 [8], in this case classifying an image as one out of seven stroke types, examples of which are pictured in Fig. 1.

3.3 Strikethrough Removal

As mentioned above, we consider the removal of strikethrough strokes from handwritten words as an unpaired image-to-image translation problem. We therefore base our approach on the work of Zhu et al., who introduced the concept of cycle-consistent generative adversarial networks (CycleGANs) [21], dealing specifically with image translation tasks as the one we are concerned with. Figure 2 shows the general CycleGAN structure, which consists of two conditional GANs, where each generator G is conditioned on images from one domain, with the goal of producing images from the respectively other domain, i.e. in our case, clean to struck and struck to clean. In order to improve the networks' translation capabilities between domains, Zhu et al. introduced a cycle consistency loss. This loss is applied in conjunction with the common adversarial loss and is based on the idea that when using the output of one generator as input to the other, the second output should be close to the original input of the generation cycle [21].

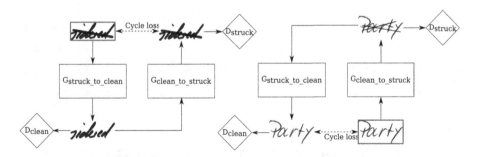

Fig. 2. General structure of a CycleGAN. Input images to the network marked with a red border. Left: forward cycle, generating a clean word from a struck one and returning it to its struck form. Right: backward cycle, generating a struck word from a clean one and returning it to its clean form. The difference between input and output images of each cycle should be minimised.

In contrast to the one-to-one translation tasks examined by Zhu et al. [21], the translation from clean words to struck ones describes a one-to-many relationship. Any given clean word may be struck-through with an arbitrary stroke, with varying control parameters, such as start and end points. Several approaches have been proposed in this regard, such as [2] and [5], however in this work, we focus on attribute-guided CycleGANs, presented by Lu et al. [13]. They propose to extend the input image that is provided to the generator which is tasked with the one-to-many translation, and the respective discriminator, with several channels, each encoding a separate feature that should be present in the output, in their case facial features. In addition to this, the authors employ an auxiliary discriminator, comparing the feature outputs of a pre-trained face verification network for a genuine and a related, generated face image [13].

Based on the aforementioned approach, we propose to use an attribute-guided generator for the task of generating struck images from clean ones and a regular generator for the opposite translation direction. In our case, the additional feature input to the generator $G_{clean \rightarrow struck}$ and discriminator D_{struck} encodes the stroke type that should be applied to a given image. In addition to this, we investigate the use of an auxiliary discriminator which is based on the feature extraction part of the DenseNet [8], pre-trained on the task of classifying the stroke type from Subsect. 3.2. We base our CycleGAN implementation on the code provided by Zhu et al. [21], replacing the ResNet blocks with dense blocks [8], followed by dense transition layers without the original average pooling. Figure 3 shows the generator architecture used in our experiments. Besides this, we use the same PatchGAN discriminators as proposed by [21] and an L_1-loss for the auxiliary discriminator.

In the context of this work, we focus our investigations on the performance of the struck-to-clean generator, essentially discarding the generator for the opposite direction after training.

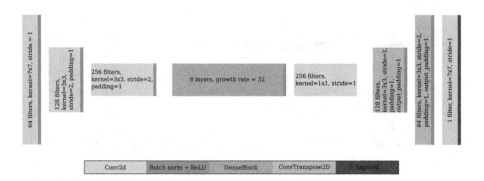

Fig. 3. Modified generator architecture, based on [21].

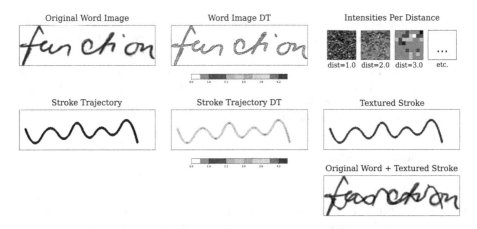

Fig. 4. Major steps of the strikethrough generation pipeline. First row: original word image, distance transform of the binarised word and pixel intensities from the original word at a given distance. Note that only three distances are pictured for illustration purposes. Second row: Randomly generated wave stroke trajectory, its distance transform and the textured stroke. Final row: outcome of the generation process

3.4 Synthetic Strikethrough Generation

Inspired by Likforman-Sulem and Vinciarelli [12], we introduce a novel approach for applying synthetic strikethrough to a given handwritten word in order to address the lack of a fully available strikethrough dataset. For this, we consider the six types handled by [6] (examples shown in Fig. 1(a)–(f)), as well as a new seventh stroke type, denoted as *scratch* (Fig. 1(g)). Each stroke is created based on features of the given input word image, namely the core region and stroke width, both calculated based on [16], the average and maximum stroke intensities, and the pixel intensity distribution. Figure 4 shows the outcomes of the main steps of our stroke generation pipeline.

In the first step, the control points for the stroke, such as the start and end points, are randomly generated within a range that ensures that a large portion of the stroke will cross the core of the word. Together with the stroke width, these points are then used to draw the trajectory. Afterwards the Euclidean distance transforms are calculated for the stroke and the (binarised) word image. Combining the original word image and its distance transform, the intensities for pixels with the same distance are collected in separate pools. Subsequently, the stroke texture is generated by sampling each pixel's intensity from the word intensity pool with the same distance. This is based on the observation that strokes tend to be of lighter colour towards the edges. To further increase the realism of a given stroke, the 'straightness' is reduced by applying an elastic transformation (following [19], implemented by [10], with α and σ drawn randomly from $[0, 20.0]$ and $[4.0, 6.0]$, respectively). Finally, the obtained stroke image is superimposed on the original word image. In order to reduce stark gradients that often occur when simply summing the images, areas in which both the original word and the generated stroke exceed the original's average stroke intensity are combined according to: $new\_pixel\_intensity = original\_word\_intensity + \omega * stroke\_intensity$, where $\omega = \frac{average\_stroke\_instensity}{max\_stroke\_intensity}$ (calculated based on the original word image). In addition to this, the newly created image is smoothed with a Gaussian blur (3×3 mask, $\sigma = 0.8$) over the area of the original strikethrough stroke. Lastly, the intensities are clipped to [0, 255]. For further details regarding the implementation, please see Sect. A.

4   Experiments and Results

4.1   Datasets

Synthetic Dataset. In order to quantitatively evaluate our work and provide a large corpus of training data, we opted to generate a new database specifically for the task of strikethrough removal. It is based on handwritten words taken from the *IAM* database [14]. In its latest version (3.0), this database contains 1539 pages of English text snippets, written by 657 different writers. Marti and Bunke provide datasplits that separate the documents into train, test, validation1 and validation2 subsets, with each writer being represented in only one of these splits [14]. We base our dataset generation on this distribution, thus preventing the same handwriting style (i.e. writer) to appear in more than one set. In addition to this, all duplicate documents (i.e. containing the same text snippets) are removed, resulting in one document per writer. The subset denoted as validation2 was not used for this work, hence validation1 will simply be referred to as validation set for the remainder of this work.

We randomly select a number of clean words with width and height in the range of [60, 600] from each writer. It should be noted here that while we ensured uniqueness of writer and source text, we did not limit the image selection with respect to the content, i.e. each actual word may occur multiple times across datasets, written and possibly struck-through in different styles. Furthermore, due to the size restrictions, a number of writers were excluded from the process. Overall, we selected 28 word images for the train and 14 for each of the other two splits, resulting in 6132 words for the train, 546 for the validation and 1638 words for the test split.

In order to generate our final database, we firstly remove the background from all images using [20]. Subsequently, we randomly divide the images for each writer into 50% of clean words and 50% of words to be artificially struck-through, using the approach outlined in Subsect. 3.4 (cf. Fig. 1 for examples of generated strokes). During the stroke generation, we ensure that the distribution of stroke types remains balanced for each writer and thus for the whole datasplit. We discard the ground truth for the struck words of the train set but retain it for the other two sets, in order to use it to quantitatively evaluate our strikethrough removal approach.

Please note that unless otherwise stated, all word images shown in this work are taken from [14] and, where applicable, were altered using the strikethrough generation approach described above.

Genuine Dataset. In order to further test the performance of the proposed method with real-world strikethrough data, we created a dataset that consists of a section of Bram Stroker's Dracula. It encompasses 756 words, handwritten by a single writer on white printer paper (A4, $80\,g/m^2$), using a blue ballpoint pen. After writing and scanning (at 600 dpi) the clean version of the text, each word was systematically struck through using one of the seven presented types and scanned again. The clean and struck scans of the pages were aligned using [22] and segmented by using the bounding boxes of the connected components of the struck version of the document. Lastly, the images were manually annotated with their respective stroke type and split into type-balanced subsets. The train set consists of 126 clean and 126 struck words with no overlap between subsets (i.e. 252 separate words in total). For the validation and test set, we use 126 and 378 struck words respectively, without providing separate clean sets. Instead, for the latter two splits, the clean ground truth of the struck words is provided. It should be noted that we opted for not providing additional clean subsets for validation and testing. This was done in order to avoid further splitting, and thus reducing, the set sizes, considering the limited word count. In addition, as is the case for the synthetic dataset, we only balanced the data splits in regard to the word images per stroke type but with respect to word content or other statistics like character count and image size.

4.2 General Neural Network Training Protocol

All word images are preprocessed by converting them to greyscale and then resizing them to a height of 128 pixels, maintaining the ratio and padding them with the background colour (white) to a maximum width of 512 pixels. Images that, after scaling, exceed the above width are squeezed accordingly to achieve uniform dimensions over the whole input set. Lastly, all images are inverted and scaled to an intensity range of $[0.0, 1.0]$.

All of the experiments below are implemented in PyTorch [17] version 1.7.1 and trained using the Adam optimiser [11] with a learning rate of 0.0002 and betas of $(0.5, 0.999)$. We train the models for a fixed number of epochs, evaluating their performance on the synthetic validation set after each epoch. For each model, we retain its state from the epoch providing the best validation performance.

4.3 Strikethrough Identification

We train the aforementioned DenseNet121 for 15 epochs on the full synthetic train set with a batch size of 64, using a cross entropy loss. The training is repeated five times, yielding five separate model states, achieving a mean F_1 score on the synthetic test set of **0.9899** (±0.0031). The mean identification performance on the genuine test dataset reaches an F_1 score of **0.7759** (±0.0243). While there is a noticeable drop in performance applying the models to the unseen, genuine dataset, it can also be noted that they do not collapse entirely and still retain a good amount of identification capability.

4.4   Strikethrough Stroke Classification

As above, we train the proposed DenseNet121 for 15 epochs on the full synthetic train set with a batch size of 64, using a cross entropy loss, with a repetition factor of five. For this task the macro F_1 score on the synthetic test set is **0.9238** (± 0.0060) and for the genuine test set **0.6170** (± 0.0405). Figure 5 shows the normalised confusion matrices, summarising the performances of the five models for the (a) synthetic and (b) genuine test sets.

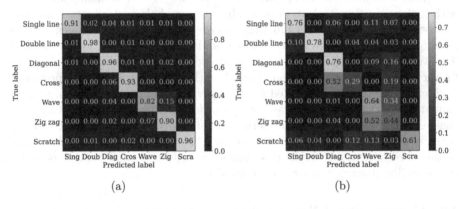

(a)   (b)

Fig. 5. Normalised confusion matrices for (a) the synthetic and (b) the genuine test sets, averaged over the five models.

As can be seen from Fig. 5(a), most of the stroke types are identified correctly and there is little to no confusion with other types. Although still comparably small, there is a noticeable misclassification between wave and zigzag strokes. Considering the examples of these two stroke types, e.g. in Fig. 1, it can be observed that, they share a lot of features in their stroke trajectories. It is therefore not surprising, that these two are misclassified to some extent. While Fig. 5(b) shows similar trends, the misclassifications between wave and zigzag are considerably more pronounced and there is a large increase in cross strokes being misclassified as diagonal ones. With regard to the first observation, it can be noted that in genuine handwriting, especially when writing quickly, corners (e.g., of zigzag strokes) may become more rounded and the softer, rounder trajectories, such as of wave strokes, may become more pointed. It may thus be advisable to review this part of the synthetic dataset for future works. One straightforward solution may be to collapse the two subsets for zigzag and wave into a single, more general wave category. Regarding the second observation and taking the concrete, misclassified images into account, we have not yet been able to identify the source of the confusion of crosses as diagonal strokes, although the former could be considered a 'double diagonal' stroke.

4.5   Strikethrough Removal

We examine a number of different model configurations, all of which are trained for 30 epochs on the full synthetic train set, with a batch size of four, repeating the process five times for each model. As a quantitative metric, we use the harmonic mean between the detection rate (DR) and the recognition accuracy (RA), i.e. the F_1-score, as defined by [6].

$$DR = O2O/N, \quad RA = O2O/M \tag{1}$$

where M is the number of ink pixels (i.e. black) in the image cleaned by the proposed method, N is the number of ink pixels in the ground truth and $O2O$ is the number of matching pixels between the two images.

Prior experiments indicated that following the same feature scheme as [13] yielded the most promising results. We therefore append seven zero-filled channels to the clean input image, switching the channel that corresponds to the desired stroke type to ones. During every training epoch, each clean image is paired with a randomly drawn struck-through word to be passed to the two generating cycles. We use the ground truth stroke type from the struck word to initialise the channel input for the clean image. In order to evaluate the impact of using the seven channels as additional inputs to $G_{clean \rightarrow struck}$ and D_{struck}, as well as that of the auxiliary discriminator and its weighting factor (considered values: 0.1, 1.0, 10.0), we exhaustively examined a combination of these parameters and compare the performances against the model without any of these modifications, i.e. the original CycleGAN structure, which we therefore denote as *Original* for the remainder of this paper.

Overall, it can be noted that all of the models are capable of removing strikethrough to some degree, achieving average F_1 scores larger than 0.75. Despite this, not all of the models improve upon the *Original*'s performance. Additionally, we observed better performances for models which were trained with a loss weighting factor of 0.1 for the auxiliary discriminator, as compared to those trained with weights of 1 and 10. In the following, we focus our evaluation on the three models that yielded the highest average F_1 scores and compare them with the *Original* one. Specifically, these models are: a) the original setup augmented with the auxiliary discriminator with a loss weight of 0.1, denoted as *Original with aux. discriminator*, b) the original attribute-guided CycleGAN, i.e. using the auxiliary discriminator (loss weight of 0.1) and augmenting the clean input with the channel-encoded stroke type, denoted as *Attribute-guided* and c) the same attribute-guided model but without the auxiliary discriminator, denoted as *Attribute-guided without aux. discriminator*.

Figure 6 shows the box plot summarising the performances of these four models on the synthetic test set for the seven stroke types, the mean performance over all seven types ('overall mean'), as well as the mean performance excluding the *scratch* type ('without scratch'). We include the latter to offer some grounds for comparison with [6], where heavily blacked-out strokes, such as the one that we denote as *scratch*, were removed during the preprocessing phase. As can be seen from the plot, the full attribute-guided model tends to achieve the highest

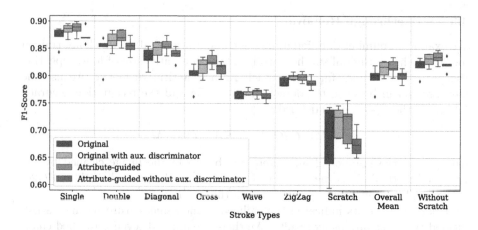

Fig. 6. Box plot showing the F_1 scores per stroke type, their overall mean and their overall mean without scratch strokes, for the four models, evaluated on the synthetic test set. Note the clipped y-axis.

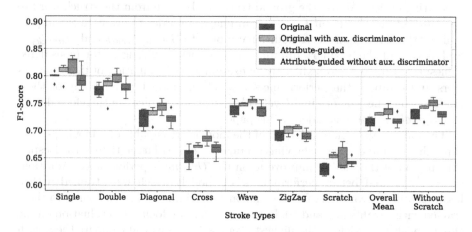

Fig. 7. Box plot showing the F_1 scores per stroke type, their overall mean and their overall mean without scratch strokes, for the four models trained on synthetic data, evaluated on the genuine test set. Note the clipped y-axis.

Table 1. Mean F_1 scores (higher better) and RMSEs (lower better) for the four models, evaluated on the *synthetic* test set. Standard deviation over five runs given in parentheses. Best model marked in bold.

Model	Overall F_1	Overall RMSE
Original	0.7966 ($\pm$ 0.0212)	0.0937 ($\pm$ 0.0123)
Original with aux. discr	0.8140 ($\pm$ 0.0140)	0.0839 ($\pm$ 0.0045)
Attribute-guided	**0.8172** ($\pm$ 0.0145)	**0.0833** ($\pm$ 0.0045)
Attribute-guided without aux. discr.	0.8005 ($\pm$ 0.0116)	0.0862 ($\pm$ 0.0028)

Table 2. Mean F_1 scores (higher better) and RMSEs (lower better) for the four models, evaluated on the *genuine* test set. Standard deviation over five runs given in parentheses. Best model marked in bold.

Model	Overall F_1	Overall RMSE
Original	0.7169 (± 0.0109)	0.0630 (± 0.0031)
Original with aux. discr	0.7263 (± 0.0134)	0.0585 (± 0.0039)
Attribute-guided	**0.7376** (± 0.0107)	**0.0576** (± 0.0013)
Attribute-guided without aux. discr	0.7199 (± 0.0111)	0.0588 (± 0.0022)

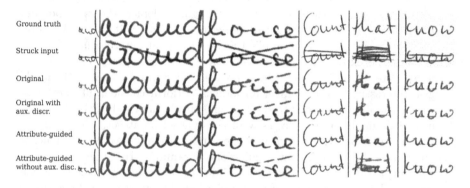

Fig. 8. Cherry-picked examples for the four models. First three examples taken from the synthetic, others from the genuine dataset.

Fig. 9. Lemon-picked examples for the four models. First three examples taken from the synthetic, others from the genuine dataset.

F_1 scores. Removing either of the additional parts, i.e. the auxiliary discriminator or stroke type encoding, decreases performance, with the lack of the former having a higher impact than the latter. Overall, a decrease in performance can be observed as the complexity of the stroke types increases. In that regard, the largest decrease in F_1 score and the highest increase in variation between instances of the same model can be observed for the *scratch*. This stroke type produces challenging struck-through word images, which are generally altered with more than one or two strokes and are sometimes heavily occluded. To summarise the model performances, Table 1 presents the mean overall F_1 score per model. As our models produce greyscale images, which have to be binarised, in order to calculate the F_1 score, we also report the intensity reconstruction capabilities, measured via the root-mean-square error (RMSE) in the same table.

Similar to the two upstream tasks presented in earlier sections, we also evaluate the CycleGANs on the genuine test set. Figure 7 depicts the corresponding box plot and Table 2 summarises the model performances. It should be noted that, as before, the models are applied as-is without an intermediate transfer-learning stage. As expected, there is a noticeable decrease in performance regarding the F_1 score, which is reduced by approximately 8 p.p. for each of the model averages. Interestingly, on average, all models perform considerably better on the genuine than the synthetic test set, when considering the RMSE. An explanation for this might be the diversity of the synthetic dataset, which consists of a large number of writers, writing styles and especially writing implements, as compared to the single-writer genuine one. Examining the individual stroke types, it can be observed that *scratch* remains to be the most challenging one, however the F_1 score for *cross* decreased considerably as well, reaching almost the same level as the former. This result requires further investigations but is likely due to a discrepancy between the creation of synthetic and genuine cross strokes.

Lastly, Fig. 8 and Fig. 9 demonstrate cherry and lemon-picked examples, i.e. convincing outputs and failure cases, respectively. While some differences in quality, particularly residual strokes, can be observed for the cherry-picked outputs, it appears that all of the models encounter similar difficulties on the lemon-picked examples.

5   Conclusions

This work presented an effective approach towards solving the problem of strikethrough removal by successfully applying a DenseNet-based attribute-guided CycleGAN to seven types of strikethrough, including one, termed *scratch*, which had not been considered by the existing literature. In addition to this, we demonstrated that regular DenseNet121s are suitable networks for the tasks of strikethrough identification and classification and can thus be used as preprocessing steps before the strikethrough removal. In order to train and subsequently evaluate these methods quantitatively, we have introduced a novel approach for strikethrough generation and used it to create a synthetic multi-writer dataset.

Furthermore, we introduced a new genuine single-writer dataset with struck-through handwritten words. Despite being exclusively trained on the former, all approaches produce convincing results when evaluated on genuine strikethrough data. This underlines both the efficacy of the different neural networks for the different tasks, as well as the authenticity and applicability of our synthetic dataset.

In the future, we intend to further examine the applicability of our strike-through generation approach as an augmentation technique for both paired and unpaired approaches. Furthermore, we plan to extend our corpus of genuine strikethrough to include more writers, writing styles and writing implements.

Acknowledgements. R.Heil would like to thank Nicolas Pielawski, Håkan Wieslander, Johan Öfverstedt and Anders Brun for their helpful comments and fruitful discussions. The computations were enabled by resources provided by the Swedish National Infrastructure for Computing (SNIC) at the High Performance Computing Center North (HPC2N) partially funded by the Swedish Research Council through grant agreement no. 2018-05973. This work is partially supported by the Riksbankens Jubileumsfond (RJ) (Dnr P19-0103:1).

A   Dataset and Code Availability

Synthetic strikethrough dataset: https://doi.org/10.5281/zenodo.4767094. Strikethrough generation code: https://doi.org/10.5281/zenodo.4767063. Genuine strikethrough dataset: https://doi.org/10.5281/zenodo.4765062. Deep learning code and checkpoints: https://doi.org/10.5281/zenodo.4767168.

References

1. Adak, C., Chaudhuri, B.B., Blumenstein, M.: Impact of struck-out text on writer identification. In: IJCNN, pp. 1465–1471 (2017)
2. Almahairi, A., Rajeshwar, S., Sordoni, A., Bachman, P., Courville, A.: Augmented CycleGAN: learning many-to-many mappings from unpaired data. In: ICML, pp. 195–204 (2018). http://proceedings.mlr.press/v80/almahairi18a.html
3. Brink, A., van der Klauw, H., Schomaker, L.: Automatic removal of crossed-out handwritten text and the effect on writer verification and identification. In: Document Recognition and Retrieval XV, vol. 6815, pp. 79–88. SPIE (2008). https://doi.org/10.1117/12.766466
4. Chang, B., Zhang, Q., Pan, S., Meng, L.: Generating Handwritten Chinese Characters Using CycleGAN. In: WACV, pp. 199–207 (2018)
5. Chang, H., Lu, J., Yu, F., Finkelstein, A.: PairedCycleGAN: asymmetric style transfer for applying and removing makeup. In: CVPR, June 2018
6. Chaudhuri, B.B., Adak, C.: An approach for detecting and cleaning of struck-out handwritten text. Pattern Recogn. **61**, 282–294 (2017). https://doi.org/10.1016/j.patcog.2016.07.032
7. Dheemanth Urs, R., Chethan, H.K.: A study on identification and cleaning of struck-out words in handwritten documents. In: Jeena Jacob, I., Kolandapalayam Shanmugam, S., Piramuthu, S., Falkowski-Gilski, P. (eds.) Data Intelligence and Cognitive Informatics. AIS, pp. 87–95. Springer, Singapore (2021). https://doi.org/10.1007/978-981-15-8530-2_6

8. Huang, G., Liu, Z., van der Maaten, L., Weinberger, K.Q.: Densely connected convolutional networks. In: CVPR, July 2017
9. Hulle, D.V.: The stuff of fiction: digital editing, multiple drafts and the extended mind. Text. Cult. **8**(1), 23–37 (2013). http://www.jstor.org/stable/10.2979/text cult.8.1.23
10. Jung, A.B., et al.: imgaug (2020). https://github.com/aleju/imgaug. Accessed 15 May 2021
11. Kingma, D.P., Ba, J.: Adam: a method for stochastic optimization. arXiv preprint arXiv:1412.6980 (2014)
12. Likforman-Sulem, L., Vinciarelli, A.: Hmm-based offline recognition of handwritten words crossed out with different kinds of strokes (2008). http://eprints.gla.ac.uk/59027/
13. Lu, Y., Tai, Y.-W., Tang, C.-K.: Attribute-guided face generation using conditional CycleGAN. In: Ferrari, V., Hebert, M., Sminchisescu, C., Weiss, Y. (eds.) ECCV 2018. LNCS, vol. 11216, pp. 293–308. Springer, Cham (2018). https://doi.org/10.1007/978-3-030-01258-8_18
14. Marti, U.V., Bunke, H.: The IAM-database: an English sentence database for offline handwriting recognition. IJDAR **5**(1), 39–46 (2002). https://doi.org/10.1007/s100320200071
15. Nisa, H., Thom, J.A., Ciesielski, V., Tennakoon, R.: A deep learning approach to handwritten text recognition in the presence of struck-out text. In: IVCNZ, pp. 1–6 (2019)
16. Papandreou, A., Gatos, B.: Slant estimation and core-region detection for handwritten Latin words. Pattern Recogn. Lett. **35**, 16–22 (2014). https://doi.org/10.1016/j.patrec.2012.08.005
17. Paszke, A., et al.: PyTorch: an imperative style, high-performance deep learning library. In: NeurIPS, pp. 8024–8035. Curran Associates, Inc. (2019)
18. Sharma, M., Verma, A., Vig, L.: Learning to clean: A GAN perspective. In: Carneiro, G., You, S. (eds.) ACCV 2018. LNCS, vol. 11367, pp. 174–185. Springer, Cham (2019). https://doi.org/10.1007/978-3-030-21074-8_14
19. Simard, P., Steinkraus, D., Platt, J.: Best practices for convolutional neural networks applied to visual document analysis. In: ICDAR (2003). https://doi.org/10.1109/ICDAR.2003.1227801
20. Vats, E., Hast, A., Singh, P.: Automatic document image binarization using Bayesian optimization. In: HIP, p. 89–94 (2017). https://doi.org/10.1145/3151509.3151520
21. Zhu, J.Y., Park, T., Isola, P., Efros, A.A.: Unpaired image-to-image translation using cycle-consistent adversarial networks. In: ICCV, October 2017
22. Öfverstedt, J., Lindblad, J., Sladoje, N.: Fast and robust symmetric image registration based on distances combining intensity and spatial information. EEE Trans. Image Process. **28**(7), 3584–3597 (2019). https://doi.org/10.1109/TIP.2019.2899947

Iterative Weighted Transductive Learning for Handwriting Recognition

George Retsinas[1], Giorgos Sfikas[2(✉)], and Christophoros Nikou[2]

[1] School of Electrical and Computer Engineering,
National Technical University of Athens, Athens, Greece
gretsinas@central.ntua.gr
[2] Department of Computer Science and Engineering, University of Ioannina,
Ioannina, Greece
{sfikas,cnikou}@cse.uoi.gr

Abstract. The established paradigm in handwriting recognition techniques involves supervised learning, where training is performed over fully labelled (transcribed) data. In this paper, we propose a weak supervision technique that involves tuning the trained network to perform better on a specific test set, after having completed its training with the standard training data. The proposed technique is based on the notion of a reference model, to which we refer as the "Oracle", which is used for test data inference and retraining in an iterative manner. As test data that are erroneously labelled will be a hindrance to model retraining, we explore ways to properly weight Oracle labels. The proposed method is shown to improve model performance as much as 2% for Character Error Rate and 5% for Word Error Rate. Combined with a competitive convolutional-recurrent architecture, we achieve state-of-the-art recognition results in the IAM and RIMES datasets.

Keywords: Weak supervision · Handwriting recognition · Transductive learning

1 Introduction

Today, the vast majority of state-of-the-art methods for document image processing tasks are learning-based methods [17]. This rule has very few exceptions [19], while the form of learning employed in most cases is supervised learning. Handwriting Text Recognition (HTR) adheres to this rule, and the most succesful HTR methods can be described as supervised deep learning methods. The immediate implication is that large labelled sets are required for succesful training. Labels in HTR are manually created transcriptions that can be accurate in

This research has been partially co-financed by the EU and Greek national funds through the Operational Program Competitiveness, Entrepreneurship and Innovation, under the call OPEN INNOVATION IN CULTURE, project *Bessarion* (T6YBΠ-00214).

J. Lladós et al. (Eds.): ICDAR 2021, LNCS 12824, pp. 587–601, 2021.
https://doi.org/10.1007/978-3-030-86337-1_39

terms of correspondence to the image data up to the level of either word, line or character, depending on the specifics of the method at hand. Creating large annotated datasets is understably time-consuming, and even more so if one takes also into account the labour required to manually record token pixel boundaries.

Using deep and supervised models for HTR comes thus with the shortcoming that large annotated datasets are required for model training. An issue of no lesser importance than this first issue, is that of model transferability. In particular, we cannot be a priori certain that our model trained on set A will generalize well enough to be practically applicable on set B. For example, set B may be written using a style that is not covered by the writing styles included in set A, or the topic of the text in set B may be such that a language model trained on set A does not apply to B well enough, and so on.

A multitude of techniques have been proposed to create more transferable models, either in or out of the context of document image processing [20, 25]. A strategy that can be helpful is to treat test data as if they were unlabelled training data, and proceed to employ techniques from the arsenal of semi-supervised learning. This strategy is referred to as transductive learning [28], and is closely related to self-training [4] and (standard) semi-supervised learning [21]. The main difference from the latter strategies is that the unlabelled data is not necessarily the test data. In all cases, care must be taken as using unlabelled data may just as well lead to a degradation of performance [2].

Concerning the application of transductive learning, let us note that the character of the task does play a role with respect to whether this type of learning can be effective. In HTR in practice, a test set is available for inference by the learning machine as a whole: for example, the test set can be made up of consecutive scanned lines of a given manuscript. This is not the case in other tasks, where the norm is to have test data appearing sequentially to the inference pipeline. Hence, HTR can be categorized as a task where we realistically have the option to treat our test set as a whole, instead of a disjoint set of objects. In the context of using transductive learning this is important, as it enables the option to adapt the initial model to the domain represented by the test set.

In this work, we propose and explore the applicability of a transductive learning technique, suitable for HTR. The proposed technique is based on multiple iterations of using the test data with updated labels. Also, we experiment on using a weight term on newly labelled text lines, in order to formulate a rule for including in the transduction retraining loss. We explore two different ways to perform weighting, and compare versus not using any weighting and including all new labels for retraining. We show that the proposed iterative, weighted transductive learning method can significantly improve results for a competitive CNN-RNN architecture. Our experiments show that the proposed model leads to state-of-the-art performance for HTR line recognition on the IAM and RIMES datasets.

This paper is structured as follows. In Sect. 2, we review related work for transductive learning and HTR. In Sect. 3, we present the proposed method and the base HTR architecture that we use for our experiments, presented and discussed in Sect. 4. We close the paper with concluding remarks in Sect. 5.

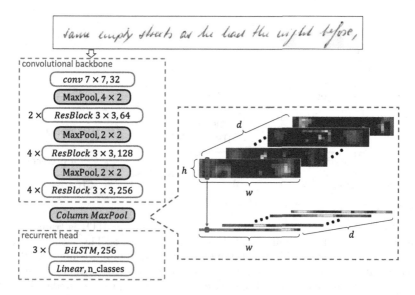

Fig. 1. Employed model architecture. We use a CNN-RNN architecture, with a column-wise max pooling layer forming the link between the convolutional and the recurrent-sequential component.

2   Related Work

Techniques that are closely related to transductive learning come from the domain of semi-supervised learning. As a notable example, self-training has been previously in the context of semi-supervised learning and HTR [4]. The trained model is used to first infer labels on unlabelled data, of which the most confident labels are used for retraining the original model. Different retraining rules are tested, according to which part of the newly labelled words are picked to train the model. These rules use a different confidence threshold, ranging from a more conservative value to a rule that simply uses all words for retraining.

Transductive learning in HTR, in the sense of including a training subprocess that uses labelled *test* data (in contrast to classical semi-supervised learning, that assumes a separate subset that is used for that purpose) has been shown to be especially useful in cases where the size of the dataset considered is very limited [6]. Concerning character-level recognition, transductive learning has been shown to improved performance for the task of digit recognition on the well-known MNIST dataset [12]. A more recent work for Tibetan handwritten character recognition has validated the usefulness of transductive learning [9], by modifying a number of semi-supervised and domain adaptation algorithms in the task context. Along with domain adaptation-based approaches, the authors use self-supervision and a method based on adversarial training.

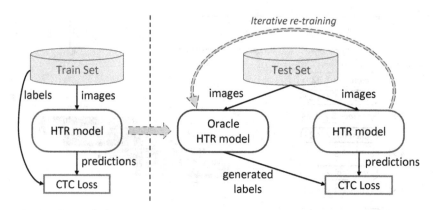

Fig. 2. Proposed learning pipeline. Left: a model suitable for HTR is trained on the training set (see Fig. 1 for details about the model architecture). This model will subsequently be the "Oracle model". Right: the Oracle is used to produce labels for the test set. These labels are used as to train a copy of the Oracle. After training completes, the newly trained model replaces the Oracle, which in turn produces updated test set labels.

3 Proposed HTR Model and Transductive Learning Technique

In this section, we describe a transduction-based technique. We start by the assumption of a specific HTR architecture, the results of which will be shown to be further improved with the proposed technique.

3.1 Proposed Model Architecture

The model that we will use to test the proposed technique can be characterized as a CNN-RNN architecture (overview in Fig. 1). CNN-RNN can be broadly defined as a convolutional backbone being followed by a recurrent head, typically connected to a CTC loss. CNN-RNN variants have given routinely very good results for HTR [3,24].

In our model, the convolutional backbone is made up of standard convolutional layers and ResNet blocks, interspersed with max-pooling and dropout layers. In particular, we have a 7×7 convolution with 32 output channels, a series of 2 3×3 ResNet blocks with 64 output channels, 4 3×3 ResNet blocks with 128 output channels and 4 3×3 ResNet blocks with 256 output channels. The standard convolution and the ResNet blocks are all followed by ReLU activations, Batch Normalization 2×2 max pooling of stride 2 and dropout. Overall, the convolutional backbone accepts a line image and outputs a tensor of size $h \times w \times d$. This output is then squashed by column-wise max-pooling into a $w \times d$-sized tensor. With this operation, the feature map is ready to be processed by the recurrent component, while achieving model translation invariance in the vertical direction [24]. The recurrent component consists of 3 stacked Bidirectional Long Short-Term Memory (BiLSTM) units of hidden size 256. These are followed by a linear projection layer, which converts the sequence to a size equal to the number of possible tokens, $n_{classes}$. Finally, the softmax output which tops the recurrent head converts these feature sequence into a sequence of probability distributions. The softmax output, sized $w \times n_{classes}$, is fed into a Connectionist Temporal Classification (CTC) loss. Note that the CTC logic provides translation invariance in the horizontal sense, complementing the vertical invariance that is related to the pooling operation at the end of the convolutional backbone. Decoding of the output during evaluation is performed by a greedy selection of the characters with the highest probability at each step.

3.2   Transductive Learning Cycle

The motivation behind the proposed methodology is clear: adapt an already trained network to the test set characteristics (e.g. writing style) by iteratively re-training a new network with labels generated by the pre-trained network. Of course, we cannot be guaranteed that the generated labels are correct. Therefore, such an approach has an important prerequisite regarding the performance of the pre-trained network. Specifically, the network should be able to recognize an adequately large percentage of the existing characters. Having a model that provides more character hits than misses, i.e. a per-character percentage of over 50% accuracy, can potentially improve the overall performance. We consider that such a bound can be easily achieved by modern HTR systems.

Considering the task at hand, the proposed transductive learning can be described as follows. First off, we assume that a training set with labelled data (text lines) is available, along with a test set with unknown labels. The CNN-RNN model is first trained on the training data, whereupon after training is complete, we freeze this model's weights and from now this model is referred to as the "Oracle" model. We use the Oracle on the test set on inference mode, in order to produce labels for the test set. The test set adaptation is performed by training a new network with identical architecture to the Oracle. Note that the only prior information about the text manifold, i.e. the character distributions, is induced by the Oracle model. Training a new network with the generated labels may lead to performance degradation if this manifold cannot be preserved to some extend. To this end, the new network is initialized with the Oracle weights. Thus, its initialization is already a local optimum with respect to the training set, which encodes information about that set's character manifold. Learning rate selection is crucial in order to attain a well-performing adapted network. Small learning rates could trap the network to the initial local optimum of the Oracle, while large learning rate could, as usual, lead to convergence issues. In practice, we used 5× larger learning rate compared to the one used for the initial Oracle, proven to be capable of discovering nearby local optima corresponding to adapted manifolds.

Generated Labels and Losses: Considering the unique aspects of the hand-written text, we adopt the CTC loss along with a greedy decoding on the output of the Oracle in order to generate the corresponding labels. Therefore, a text string is generated for each text line, following the typical HTR training logic. An alternative approach to the CTC-based strategy is the usage of the per-column (of the RNN input feature map) character probability scores, as produced by the softmax operation over the output of the network. Since we can generate per-column character predictions using the Oracle network, such formulation can be addressed by a cross-entropy loss, minimizing these predictions.

As misclassified data are harmful for using with the transduction cycle, previous works have opted to use a criterion that selects only part of the data for retraining [4]. In this work, we avoid using a hard decision threshold, and instead we use a weighted version of the aforementioned loss based on the confidence of the Oracle predictions. This way we take into account possible unseen writing styles which score poorly with the Oracle model.

Considering the CTC loss, we cannot straightforwardly define a per-character weight. Nevertheless, we can generate a holistic score over the entire text line using the CTC loss and the produced decoding/label. Consequently, the new model will tend to be optimized with respect to the more confident decodings, while less confident decodings will contribute less to the total. In order to incorporate these scores into a weighted average formulation (considering batches of line images, as feed to the optimized) we apply a softmax operation to them. Given a set (batch) of images $\{I_i\}$ and their corresponding decodings $\{s_i\}$ ($i = 1, \ldots, N$) and the networks $h(\cdot)$ and $h_o(\cdot)$, which correspond to the training network and the Oracle network respectively, the weighted CTC formulation can be expressed as follows:

$$a_i = L_{CTC}(h(I_i), s_i) \tag{1}$$

$$w_i = \frac{e^{-c \cdot a_i}}{\sum_i e^{-c \cdot a_i}} \tag{2}$$

$$\text{loss} = \sum_i w_i \cdot L_{CTC}(h_o(I_i), s_i) \tag{3}$$

The c parameter of the softmax operation defines the divergence of the weights. We select a value of 0.1 in order to avoid extreme values which resemble a hard thresholding operation.

The weighted version of cross-entropy variant is rather intuitive, since each per-column prediction already has a clear-defined Oracle output probability score, which is used as a weight. Scaling is therefore performed on a per-column instead of a per-datum basis.

Iterative Re-Training: If training with generated labels produced by the Oracle can successfully improve the network's performance, it is natural to extend the rationale by simply iteratively performing the same scheme in order to further increase performance. Specifically, the proposed method is proposed in training cycles. During each cycle the Oracle network is fixed and used only to produce the labels. At the end of the cycle the new model is fully trained and, ideally, has better performance than the Oracle. Therefore the Oracle is replaced by the trained model and a new model is trained during the next cycle using the enhanced Oracle network. This iterative procedure is executed until a predefined number of cycles or until no further loss improvement is observed. The entire proposed procedure is depicted in Fig. 2.

Additionally, the initialization of the new network to be trained at the start of each cycle may affect this iterative approach. At the start of the whole procedure, the network to-be-trained is initialized with the same weights as the Oracle. Similarly, one may choose to have both Oracle and the new network initialized with the same parameters at each cycle. Typically, according to this strategy, the new network retains the weights of the previous cycle and thus no initialization is applied. However, we proposed a different approach; at the start of each cycle the network to-be-trained is initialized according to the initial Oracle, trained on the train set. The reasoning behind this choice is related to the manifold properties that we already mentioned. Specifically, if we continuously shift the initial underlying manifold by using as starting point the newly trained weights, we may considerably diverge from the initial distribution of the characters and lose important information from the train set. Therefore, if we consider the initial pre-trained network as the starting point of our method, we can converge to nearby solutions with similar properties, while using the updated labels of the iterative process. In other words, the proposed initialization acts as an implicit constraint on the structure of the final solution with respect to the character manifold on the initial train set.

Connection to Knowledge Distillation: The main idea, i.e. (partially) retaining the initial behavior of the network while training another network, closely resembles knowledge distillation approaches [8]. In fact, identical to our approach, distillation includes a pair of models: the teacher (corresponds to the Oracle) and the student model. Distillation consists of two losses, the task related one and the distillation loss, typically defined as the Kullback-Leibler (KL) divergence:

$$KL(p|q) = \sum_i p_i \log \frac{p_i}{q_i} \tag{4}$$

where p_i and q_i are the softmax activations at the output of the models. Typical distillation methods scale the outputs, before the softmax operation, by a temperature factor $1/T$ that "softens", when $T > 1$, the comparison of the two distributions. The distillation loss aims to bring the distributions of the models' outputs as close as possible.

The per-column weighting variant of the proposed method can be viewed as a distillation loss, since cross entropy is closely related to KL divergence: $H(p, q) = \sum_i p_i \log q_i$. Contrary to distillation logic, we want to mine existing, yet hidden, knowledge and not retain the Oracle's output distribution. Therefore the cross entropy approach is a hard-assignment variant of distillation, or equivalently, very small temperature factor values ($T \to 0$) are considered. As an extension, the CTC version of the proposed algorithm can be viewed as an sequence-wise alternative of a distillation loss[1].

4   Experimental Evaluation

Evaluation of our methods is performed on two widely used datasets, IAM [15] and RIMES [7]. The ablation study, considering different settings of the proposed methodology, is performed on the challenging IAM dataset, consisted of handwritten text from 657 different writers and partitioned into writer-independent train/test sets.

All experiments follow the same setting: line-level recognition using a lexicon-free greedy CTC decoding scheme. Character Error Rate (CER) and Word Error Rate (WER) metrics are reported in all cases (lower values are better).

The pre-processing steps, applied to every line image, are: 1. All images are resized to a resolution of 128×1024 pixels. Initial images are padded (using the image median value, usually zero) in order to attain the aforementioned fixed size. If the initial image is greater than the predefined size, the image is rescaled. The padding option aims to preserve the existing aspect ratio of the text images. 2. A simple global affine augmentation is applied at every image, considering only rotation and skew of small magnitude in order to generate valid images. Additionally, elastic net deformations are also applied, in order to simulate local spatial variation. 3. Each line transcription has spaces added at the beginning and at the end, i.e. "He rose from" is changed to " He rose from ". This operation aims to assist the system to predict the marginal spaces that exist in the majority of the images.[2]

Training is performed in two stages. *1) Training the initial Oracle, $t = 0$:* training for 240 epochs with cosine annealing and restarts every 40 epochs [13], with learning rate 0.001. *2) Transductive learning cycle:* each cycle consists of 20 epochs, while the learning rate is initialized at 0.005 and gradually reduced according to cosine annealing. We set an increased learning rate compared to the initial network in order to discover neighboring local optima and not be constrained to the one found by the initial training.

In the following, we explore different settings of the proposed self-training system, considering a variety of generalization schemes, losses and optimization strategies. To further understand the capabilities of the proposed method, we also consider sub-optimal pre-trained networks, trained only on a subset of the initial train dataset. Finally, we compare our approach to other methods that attain state-of-the-art accuracy on either the IAM and RIMES datasets.

4.1   Ablation Study

The impact of three different optimization aspects is explored for the case of a single-cycle self-training step.

---

[1] A detailed analysis on sequence-level knowledge distillation has been presented in [10].

[2] Note that during evaluation we do not take into account these added spaces, as this would bias the result in favour of our method.

Table 1. Impact of (a) generalization strategies and (b) part-optimization of the proposed transductive method.

Strategy	CER %	WER%
Initial	4.55	15.71
None	4.51	15.47
Dropout	4.41	15.24
Augmentation	4.29	14.99
Dropout + Augmentation	4.24	14.75

(a)

	CER %	WER%
CNN Backbone	4.37	15.04
Recurrent Head	4.48	15.54
Whole Network	4.24	14.75

(b)

Impact of Generalization Strategies: We have first experimented on how the proposed transductive method combines with two different strategies to boost the model ability to generalize to new data. The strategies considered are Dropout and Augmentation, and the corresponding models are compared on Table 1(a). Conceptually, the proposed algorithm may benefit from an overfitting to the test data in order to fully utilize the existing data and thus generalization scheme may lead to worse performance. Nonetheless, from the results we can conclude that these strategies together with the proposed method combine well. Especially augmentation provides a considerable boost in performance, hinting that we may discover correct labels when slightly transforming the image, while we can better formulate the underlying character manifold/distribution of the test data by adding more (synthetically generated) instances.

Impact of Optimizing Parameters: Re-training the whole network may in principle lead to worse-performing local optima, especially in the context of transductive learning. Keeping some layers fixed may act as a constraint in order to retain desirable properties from the initial network. In order to check whether one network component is more robust than the rest in this respect, we consider re-training either the CNN-backbone part or the recurrent head part, while keeping the other part fixed. Error rate results are presented in Table 1(b). Results suggest that the best option is to leave no part of the model frozen.

Impact of Type and Weighting of the Loss Function: As described in Sect. 3.2, we use a weighting scheme when using Oracle labels. We have experimented in using each of the described alternatives: using CTC-based scores versus CE-based scores. Also, we compare against using equal weights to all Oracle labels. The results can be examined in Table 2. Weighting variants appear marginally better than their unweighted counterparts. More importantly, CTC-based approaches significantly outperform CE-based ones. This can be attributed to the fact that the CE method assigns a label to each output column, including transitional columns between two characters that may not correspond to a specific character. Weighting does not appear to solve this problem, neither preliminary experiments with predefined thresholds. On the other hand the CTC-based approach is better suited for the problem at hand and it can be effective even when the unweighted version is considered.

Table 2. Comparison of weighting variants. CTC: Results using a CTC-based loss. CE: Results using a CE-based loss. wCTC: Results using a CTC-based loss with each datum loss weighted according to the Oracle CTC decoding score. wCE: Results using a CE-based loss, with each feature map column loss weighted according to the Oracle CE output.

Loss	CER %	WER %
Initial	4.55	15.71
CTC	4.24	14.75
CE	4.37	15.16
wCTC	4.23	14.74
wCE	4.35	15.09

4.2   Effect of Using Multiple Iterations

Theoretically, after performing a self-training cycle, we acquire a better-performing network, ideal for acting as the new Oracle for an upcoming cycle, as we have already described in Sect. 3. The ability of multiple iterations of the proposed self-training approach (wCTC variant) to further increase the network's performance is observed in Table 3.

Table 3. HTR performance using different numbers of transductive cycle iterations.

Iteration	CER %	WER %
Initial	4.55	15.71
1	4.24	14.75
2	4.07	14.26
3	4.01	14.09
4	3.99	14.03
5	3.95	13.92

We further investigate the impact of the initial learning rate at the start of each cycle. The results are presented in Fig. 3. As suggested, selecting small learning rates provides a minor boost in performance. Note that both values of $5 \cdot 10^{-3}$ and 10^{-2}, provide a similar boost after first cycle. However, the latter seems to diverge from meaningful solutions in the upcoming cycles.

4.3   Using a Suboptimal Oracle

Transductive learning has been proved to be a good option before, especially when data is very limited [6,9]. In order to test how the proposed method fares under a limited-resource paradigm, we have tested the method with Oracles that were trained with only a small percentage of the originally available labelled training data. We have trained our Oracle using a number of IAM training lines that correspond to percentages equal to 10%, 25%, 50%. Results can be

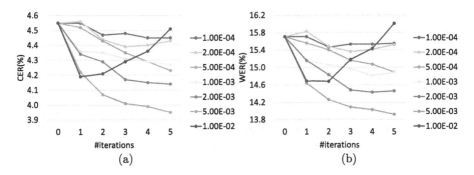

Fig. 3. Impact of learning rate for the iterative procedure. Different metrics are depicted in (a) and (b).

examined in Table 4 and Fig. 4. In all cases, the proposed method is shown to improve results, steadily up to a margin of 1–2% for CER and 3–5% for WER.

Table 4. Tests under conditions where the Oracle has trained on limited data. Percentages on the top row correspond to percentages of total training lines of the IAM dataset that where used to train the Oracle. Per-row iterations and scores correspond to different iterations of the transduction learning cycle.

Percentage	10%		25%		50%	
Iteration	CER %	WER %	CER %	WER %	CER %	WER %
Initial	10.85	33.24	7.77	25.57	5.72	19.40
1	10.02	30.95	7.22	24.22	5.26	17.99
2	9.65	29.96	6.69	22.55	5.07	17.42
3	9.36	29.40	6.63	22.45	4.96	17.22
4	9.18	28.96	6.55	22.37	4.91	17.05
5	9.04	28.72	6.52	22.24	4.87	16.88

Figure 4 provides additional insights on the iterative approach and specifically its behavior for a large number of transductive cycles. The performance gain is quickly saturated, while small performance degradation can be observed for over 10 iterations. Such degradation is expected, since the Oracle may produce many missclassified predictions for a specific character, but the degraded networks still retain significantly better results than the initial network trained on the train set. The case of 10% appears to be continuously improving up to 10 iterations, unlike the other cases, due to the larger capability of improvement.

Furthermore, we evaluate the proposed initialization scheme, described in Sect. 3, at the start of each transductive cycle. Initializing with the pre-trained network or retaining the weights (*no-init*), appear to have similar performance for most of the cases. Their difference can be detected at the more demanding 10% case as shown in Fig. 5, where the proposed method saturates slower than the typical no initialization scheme.

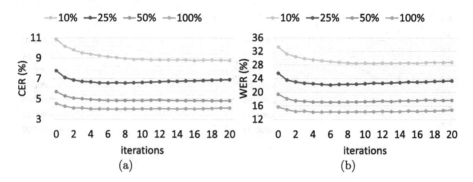

Fig. 4. Behavior of HTR performance when considering a large number of iterations. Two different metrics are reported: (a) CER and (b) WER.

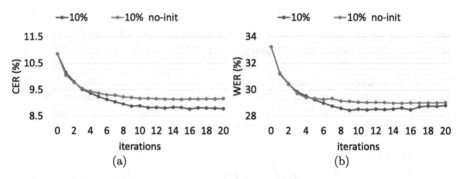

Fig. 5. Impact of initialization strategy at each cycle. Both metrics are considered: (a) CER and (b) WER.

Finally, the per-character improvement is shown in Fig. 6 for two cases: 10% and 100% (whole train set). Specifically we report the number of corrected characters after 5 cycles of the proposed methodology. Negative values correspond to worse performance compared to the initial pre-trained network. We can observe that, as expected, large number of corrected characters are reported for the 10% case. Nevertheless, there are characters that may further deteriorate their performance. A characteristic case is the character i, where a notable decrease in performance is observed. The sub-optimal initial network may often confuse i with characters of similar appearance, such as j and l. Such behavior can be intensified by the proposed iterative procedure.

4.4   Comparison with the State of the Art

Finally, we have compared the proposed HTR model with the proposed learning scheme against state-of-the-art methods in the literature. The variant used for the proposed method uses weighted CTC loss and 5 retraining iterations. Comparative results can be examined in Table 5. We can observe that the proposed

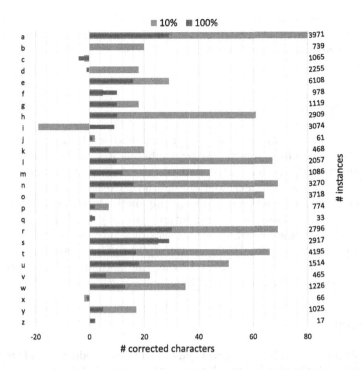

Fig. 6. Per-character analysis of improvements using the proposed method. Results for two settings are compared: using all of the IAM training set or a part of it (10%), in order to simulate the case where training data are limited. The horizontal axis denotes number of extra corrected characters using the proposed method versus not using it. The vertical axis denotes each different character and number of total test instances.

Table 5. Performance comparison for IAM/RIMES datasets.

Architecture	IAM		RIMES	
	CER %	WER %	CER %	WER %
Puigcerver [18]	6.2	20.2	2.6	10.7
Krishnan et al. [11]	9.78	32.89	–	–
Dutta et al. [3]	5.8	17.8	5.07	14.7
Chowdhury and Vig [1]	8.10	16.70	3.59	9.60
Michael et al. [16]	4.87	–	–	–
Tassopoulou et al. [24]	5.18	17.68	–	–
Yousef et al. [27]	4.9	–	–	–
Markou et al. [14]	6.14	20.04	3.34	11.23
Xiao et al. [26]	3.97	12.90	1.56	12.61
Proposed w/out transduction	4.55	15.71	3.16	11.09
Proposed w/ transduction, @5 iter	3.95	13.92	2.90	10.17

CNN-RNN model is already very competitive, with or without the transduction scheme being applied. The proposed HTR architecture, combined with the transduction scheme leads to results that are on par with the state of the art in both IAM and RIMES.

5 Conclusion

We have proposed a transductive learning method for Handwriting Recognition models, and we have shown that by combining it with a competitive CNN-RNN architecture the model reaches state-of-the-art recognition results. By using repeated transduction iterations we have shown that results can further improve. Furthermore, we have compared two weighting methods for more precise tuning of the reference model loss. Regarding future work, a possible research perspective could be exploring transduction in terms of a projection between training and test set geometry, or considering restating the method within a probabilistic framework [21,23]. Also, we envisage testing the method on scripts with more limited data available than latin [5,22].

References

1. Chowdhury, A., Vig, L.: An efficient end-to-end neural model for handwritten text recognition. In: British Machine Vision Conference (BMVC) (2018)
2. Cozman, F.G., Cohen, I., Cirelo, M.: Unlabeled data can degrade classification performance of generative classifiers. In: Flairs Conference, pp. 327–331 (2002)
3. Dutta, K., Krishnan, P., Mathew, M., Jawahar, C.: Improving CNN-RNN hybrid networks for handwriting recognition. In: 2018 16th International Conference on Frontiers in Handwriting Recognition (ICFHR), pp. 80–85. IEEE (2018)
4. Frinken, V., Bunke, H.: Self-training strategies for handwriting word recognition. In: Perner, P. (ed.) ICDM 2009. LNCS (LNAI), vol. 5633, pp. 291–300. Springer, Heidelberg (2009). https://doi.org/10.1007/978-3-642-03067-3_23
5. Giotis, A.P., Sfikas, G., Nikou, C., Gatos, B.: Shape-based word spotting in handwritten document images. In: 13th International Conference on Document Analysis and Recognition (ICDAR), pp. 561–565. IEEE (2015)
6. Granet, A., Morin, E., Mouchère, H., Quiniou, S., Viard-Gaudin, C.: Transfer learning for a letter-ngrams to word decoder in the context of historical handwriting recognition with scarce resources. In: 27th International Conference on Computational Linguistics (COLING), pp. 1474–1484 (2018)
7. Grosicki, E., Carre, M., Brodin, J.M., Geoffrois, E.: Rimes evaluation campaign for handwritten mail processing (2008)
8. Hinton, G., Vinyals, O., Dean, J.: Distilling the knowledge in a neural network. arXiv preprint arXiv:1503.02531 (2015)
9. Keret, S., Wolf, L., Dershowitz, N., Werner, E., Almogi, O., Wangchuk, D.: Transductive learning for reading handwritten tibetan manuscripts. In: 2019 International Conference on Document Analysis and Recognition (ICDAR), pp. 214–221. IEEE (2019)
10. Kim, Y., Rush, A.M.: Sequence-level knowledge distillation. arXiv preprint arXiv:1606.07947 (2016)

11. Krishnan, P., Dutta, K., Jawahar, C.: Word spotting and recognition using deep embedding. In: 2018 13th IAPR International Workshop on Document Analysis Systems (DAS), pp. 1–6. IEEE (2018)
12. Lefakis, L.: Transductive learning for document classification and handwritten character recognition. Ph.D. thesis, Utrecht university (2008)
13. Loshchilov, I., Hutter, F.: SGDR: stochastic gradient descent with warm restarts. arXiv preprint arXiv:1608.03983 (2016)
14. Markou, K., et al.: A convolutional recurrent neural network for the handwritten text recognition of historical Greek manuscripts. In: International Workshop on Pattern Recognition for Cultural Heritage (PATRECH) (2020)
15. Marti, U.V., Bunke, H.: The IAM-database: an English sentence database for offline handwriting recognition. Int. J. Doc. Anal. Recogn. 5(1), 39–46 (2002)
16. Michael, J., Labahn, R., Grüning, T., Zöllner, J.: Evaluating sequence-to-sequence models for handwritten text recognition. In: 2019 International Conference on Document Analysis and Recognition (ICDAR), pp. 1286–1293. IEEE (2019)
17. Mühlberger, G., et al.: Transforming scholarship in the archives through handwritten text recognition. J. Doc. 75(5), 954–976 (2019)
18. Puigcerver, J.: Are multidimensional recurrent layers really necessary for handwritten text recognition? In: 2017 14th IAPR International Conference on Document Analysis and Recognition (ICDAR), vol. 1, pp. 67–72. IEEE (2017)
19. Retsinas, G., Louloudis, G., Stamatopoulos, N., Gatos, B.: Efficient learning-free keyword spotting. IEEE Trans. Pattern Anal. Mach. Intell. 41(7), 1587–1600 (2018)
20. Retsinas, G., Sfikas, G., Gatos, B.: Transferable deep features for keyword spotting. In: Multidisciplinary Digital Publishing Institute Proceedings, vol. 2, p. 89 (2018)
21. Sfikas, G., Gatos, B., Nikou, C.: Semicca: a new semi-supervised probabilistic CCA model for keyword spotting. In: 2017 IEEE International Conference on Image Processing (ICIP), pp. 1107–1111. IEEE (2017)
22. Sfikas, G., Giotis, A.P., Louloudis, G., Gatos, B.: Using attributes for word spotting and recognition in polytonic Greek documents. In: 2015 13th International Conference on Document Analysis and Recognition (ICDAR), pp. 686–690. IEEE (2015)
23. Sfikas, G., Nikou, C., Galatsanos, N., Heinrich, C.: Majorization-minimization mixture model determination in image segmentation. In: CVPR 2011, pp. 2169–2176. IEEE (2011)
24. Tassopoulou, V., Retsinas, G., Maragos, P.: Enhancing handwritten text recognition with n-gram sequence decomposition and multitask learning. arXiv preprint arXiv:2012.14459 (2020)
25. Tensmeyer, C., Wigington, C., Davis, B., Stewart, S., Martinez, T., Barrett, W.: Language model supervision for handwriting recognition model adaptation. In: 2018 16th International Conference on Frontiers in Handwriting Recognition (ICFHR), pp. 133–138. IEEE (2018)
26. Xiao, S., Peng, L., Yan, R., Wang, S.: Deep network with pixel-level rectification and robust training for handwriting recognition. SN Comput. Sci. 1(3), 1–13 (2020)
27. Yousef, M., Hussain, K.F., Mohammed, U.S.: Accurate, data-efficient, unconstrained text recognition with convolutional neural networks. Pattern Recogn. 108, 107482 (2020)
28. Zhou, Z.H.: A brief introduction to weakly supervised learning. Natl. Sci. Rev. 5(1), 44–53 (2018)

Competition Reports

ICDAR 2021 Competition on Scientific Literature Parsing

Antonio Jimeno Yepes[1,2]([✉]), Peter Zhong[1,3], and Douglas Burdick[4]

[1] IBM Research, Melbourne, VIC, Australia
[2] University of Melbourne, Melbourne, VIC, Australia
antonio.jimeno@gmail.com
[3] Oracle, Melbourne, VIC, Australia
peter.zhong@oracle.com
[4] IBM Research Almaden, San Jose, CA, USA
drburdic@us.ibm.com

Abstract. Scientific literature contain important information related to cutting-edge innovations in diverse domains. Advances in natural language processing have been driving the fast development in automated information extraction from scientific literature. However, scientific literature is often available in unstructured PDF format. While PDF is great for preserving basic visual elements, such as characters, lines, shapes, etc., on a canvas for presentation to humans, automatic processing of the PDF format by machines presents many challenges. With over 2.5 trillion PDF documents in existence, these issues are prevalent in many other important application domains as well.

A critical challenge for automated information extraction from scientific literature is that documents often contain content that is not in natural language, such as figures and tables. Nevertheless, such content usually illustrates key results, messages, or summarizations of the research. To obtain a comprehensive understanding of scientific literature, the automated system must be able to recognize the layout of the documents and parse the non-natural-language content into a machine readable format.

Our ICDAR 2021 Scientific Literature Parsing Competition (ICDAR2021-SLP) aims to drive the advances specifically in document understanding. ICDAR2021-SLP leverages the PubLayNet and PubTabNet datasets, which provide hundreds of thousands of training and evaluation examples. In Task A, Document Layout Recognition, submissions with the highest performance combine object detection and specialised solutions for the different categories. In Task B, Table Recognition, top submissions rely on methods to identify table components and post-processing methods to generate the table structure and content. Results from both tasks show an impressive performance and opens the possibility for high performance practical applications.

Keywords: Document layout understanding · Table recognition · ICDAR competition

© Springer Nature Switzerland AG 2021
J. Lladós et al. (Eds.): ICDAR 2021, LNCS 12824, pp. 605–617, 2021.
https://doi.org/10.1007/978-3-030-86337-1_40

1   Introduction

Documents in Portable Document Format (PDF) are ubiquitous with over 2.5 trillion documents [12] available from several industries, including insurance documents to medical files to peer-review scientific articles. PDF represents one of the main sources of knowledge both online and offline. While PDF is great for preserving the basic elements (characters, lines, shapes, images, etc.) on a canvas for different operating systems or devices for humans to consume, it's not a format that machines can understand.

Most of the current methods for document understanding rely on deep learning, which requires a large number of training examples. We have generated large data sets automatically using PubMed Central[1] that have been used in this competition. PubMed Central is a large collection of full text articles in the biomedical domain provided by the US NIH/National Library of Medicine.

As of today, PubMed Central has almost 7 million full text articles from 2,476 journals, which offers the possibility to study the problem of document understanding over a large set of different article styles. Our data set has been generated using a subset of PubMed Central that is distributed under a Creative Commons license available for commercial use.

The competition is split in two tasks that address the understanding of document layouts by asking participants to identify several categories of information in document pages (Task A) and the understanding of tables by asking participants to produce an HTML version of table images (Task B). The IBM Research AI Leaderboard system was used to collect and evaluate the submissions of the participants. This system is based on EvalAI[2]. In task A, participants had access to all the data except for the ground truth of the final evaluation test set, the test set was released when PubLayNet was made available. In task B, we released the final evaluation test set three days before submitting the final result by the participants.

We had a large number of participant submissions with 281 for the Evaluation Phase of Task A from 78 different teams. Results from both tasks show an impressive performance by current state-of-the-art algorithms, improving significantly over previously reported results, which opens the possibility for high performance practical applications.

2   Task A - Document Layout Recognition

This task aims to advance the research in recognizing the layout of unstructured documents. Participants of this competition need to develop a model that can identify the common layout elements in document images, including text, titles, tables, figures, and lists, with confidence score for each detection. The competition site is available from[3].

---

[1] https://www.ncbi.nlm.nih.gov/pmc.

[2] https://eval.ai/.

[3] Task A website: https://aieval.draco.res.ibm.com/challenge/41/overview.

2.1 Related Work

There has been several competitions for document layout understanding, with many organised as ICDAR competitions. Examples of these competitions include [1], which cover as well complex layouts [2,3], which are limited in size. There are as well data sets for document layout understanding outside competitions, for example the US NIH/National Library of Medicine Medical Article Records Groundtruth (MARG) that was obtained from scanned article pages.

Overall, the previous data sets available for document layout understanding are of limited size, typically several hundred pages. The main reason for the limited size is that ground-truth data is annotated manually, which is a slow, costly, tedious process. In our Task A competition, we provide a significantly larger data set, several orders of magnitude larger, that has been generated automatically in which the validation and test sets have been manually verified.

2.2 Data

This task used the PubLayNet dataset[4] [17]. The annotations in PubLayNet are automatically generated by matching the PDF format and the XML format of the articles in the PubMed Central Open Access Subset as described in [17].

The competition had two phases. The Format Verification Phase spanned the entire competition, for participants to verify their results file met our submission requirements with the provided mini development set. The Evaluation Phase also spanned the whole competition. In this phase, participants could submit results on the test samples for evaluation. Final ranking and winning teams were decided by the performance in the Evaluation Phase. Table 1 shows the statistics of the data sets used in the different phases of the Task A competition.

Table 1. Task A data set statistics

Split	Size	Phase
Training	335,703	N/A
Development	11,245	N/A
Mini development	20	Format Verification Phase
Test	11,405	Evaluation

The results submitted by the participants have been objectively and quantitatively evaluated using the mean average precision (MAP) @ intersection over union (IoU) [0.50:0.95] metric on bounding boxes, which is used in the COCO object detection competition[5]. We calculated the average precision for a sequence of IoU thresholds ranging from 0.50 to 0.95 with a step size of 0.05. Then, the mean of the average precision on all element categories was computed as the final score.

---

[4] https://github.com/ibm-aur-nlp/PubLayNet.
[5] http://cocodataset.org/#detection-eval.

2.3 Results

In the Evaluation Phase, we had more than 200 submissions from over 80 teams. Table 2 shows the top 9 results for the Evaluation Phase of the competition. Overall results and individual results are significantly higher compared to previously reported results [17]. The three top systems manage to have an overall performance above 0.97.

The top performing systems, as described in the next section, relied on object detection approaches, which is similar to previous work on this data set. In addition, the predictions from object detection were compared to information extracted from the PDF version of the page or from specialized classifiers. This seems to be applied in most cases to the title and text categories, which significantly improve the performance of previously reported results.

Table 2. Task A results

Team Name	Text	Title	List	Table	Figure	Overall
Davar-Lab-OCR	**0.9838**	**0.9607**	0.9680	0.9735	0.9804	**0.9733**
TAL	0.9823	0.9420	**0.9700**	**0.9775**	**0.9833**	0.9710
Simo	0.9810	0.9536	0.9636	0.9738	0.9796	0.9703
BIT-VR Lab	0.9778	0.9270	0.9645	0.9762	0.9816	0.9654
IOD	0.9774	0.9251	0.9620	0.9773	0.9814	0.9647
小牛刀	0.9797	0.9515	0.9575	0.9635	0.9709	0.9646
JHL	0.9774	0.9245	0.9620	0.9754	0.9814	0.9642
刷不了	0.9778	0.9248	0.9634	0.9734	0.9803	0.9639
SRK	0.9767	0.9200	0.9599	0.9737	0.9800	0.9621

2.4 Systems Description

These are the descriptions of the top systems provided by the participants for Task A[6].

Team: Davar-Lab-OCR, Hikvision Research Institute. The system is built based on a multi-modal Mask-RCNN-based object detection framework. For a document, we make full use of the advantages from vision and semantics, where the vision is introduced in the form of document image, while semantics (texts and positions) is directly parsed from PDF. We adopt a two-stream network to extract modality-specific visual and semantic features. The visual branch processes document image and semantic branch extracts features from text embedding maps (text regions are filled with the corresponding embedding vectors, which are learned from scratch). The features are fused adaptively as the complete representation of document, and then are fed into a standard object detection procedure.

---

[6] Not all descriptions for the top systems were provided.

To further improve accuracy, model ensemble technique is applied. Specifically, we train two large multimodal layout analysis models (a. ResNeXt-101-Cascade DCN Mask RCNN; b. ResNeSt-101-Cascade Mask RCNN), and inference the models in several different scales. The final results are generated by a weighted bounding-boxes fusion strategy. The code and related paper will be published in https://davar-lab.github.io/news.html.

Team: Tomorrow Advancing Life (TAL). TAL[7] used HTC (Hybrid Task Cascade for Instance Segmentation) as the baseline, which is an improved version of cascade mask rcnn. We first used some general optimization:

(1) carefully designed the ratio of anchor;
(2) add deformable convolution module and global context block to the backbone;
(3) replace FPN with PAFPN;
(4) extract multi-level features instead of one-level features; (5) adopt IOU-balanced sampling to make the training samples more representative.

To tackle the difficulty of precise localization, we use two methods:

(1) we implement the algorithm SABL (Side-Aware Boundary Localization), where each side of the bounding box is respectively localized with a dedicated network branch;
(2) we train an expert model for the 'title' category to further improve the localization precision

In the post-processing stage, a classification model and self-developed text line detection model are used to solve the problem of missing detection in specific layout. In order to solve the problem of false detection of non target text, LayoutLM[8] is used to classify each line of text and remove the non target class.

At last, we ensemble multiple backbone models such as resnest200, resnext101, etc., and set different nms threshold for different categories. References[9][10].

Team: Simo, Shanghai Jiao Tong University. We treat the document layout analysis as an object detection task, and achieve it based on the framework of mmdetection. We first train a baseline model (Mask-RCNN). Afterwards, we improved our model from the following aspects:

1. *Annotations*: We find that for the "text" category, some samples in the train dataset are unannotated, which leads to low recall of this category. Thus we design heuristic strategies to replenish the annotations in the training dataset, which can increase the overall AP on category of "text".

---

[7] http://www.100tal.com/about.html.
[8] https://github.com/microsoft/unilm/tree/master/layoutlm.
[9] LayoutLM: https://github.com/microsoft/unilm/tree/master/layoutlm.
[10] mmdetection:https://github.com/open-mmlab.

2. *Large models*: To improve performance, the network is trained based on a large backbone (ResNet-152), together with GCB and DCN blocks, which can improve our performance largely.
3. *Results refinement*: For categories of "text" and "title", we use the coordinates extracted from the PDF to refine the final results. Specifically, we parse the text line coordinates through PDFMiner, and refine the layout prediction (large box) using the above line coordinates.
4. *Model ensemble*: Finally, we use model ensemble techniques to ensemble the above results as our final result.

Team: SRK. Our final solution is based on the ensemble of Mask Cascade R-CNN with ResNeSt-50 FPN backbone. First model was used for "Title" detection (small objects) and the second one for detection entities of other classes: "Text", "List", "Figure" and "Table". There were no any image augmentation techniques during models train and inference. Inference optimization was done by choosing NMS threshold parameter. The best result was obtained with a value of 0.9. We've used Detectron2 library for implementation and checking of our models. This solution is a continuation of our previous research on Document Layout Analysis problem, published in [6].

Team: BIT-VR Lab. In this work, our base detection method follows the two-stage framework of DetectoRS that employs HTC branch to make full use of instance and segmentation annotation to enhance the feature flow in the feature extraction stage. We train a series of CNN models based on this method with different backbones, larger input image scales, customized anchor size, various loss functions, rich data augmentation and soft-NMS method. More specifically, we use NAS technique to obtain optimal network architecture and optimal parameter configuration. Another technique is that we use OHEM to make training more effective and efficient and improve the detection accuracy of difficult samples like the "Title" category.

Besides, we trained Yolo-v5x model as our one-stage objection detection method, and CenteNet2 to take advantage of different characteristics in both one-stage and two-stage methods. To obtain the final ensemble detection results, we combine three different network frameworks as above and different multi-scale testing approaches with specific ensemble strategy.

3   Task B - Table Recognition

Information in tabular format is prevalent in all sorts of documents. Compared to natural language, tables provide a way to summarize large quantities of data in a more compact and structured format. Tables provide as well a format to assist readers with finding and comparing information. This competition aims to advance the research in automated recognition of tables in unstructured formats.

Participants of this task need to develop a model that can convert images of tabular data into the corresponding HTML code, which follows the HTML table representation from PubMed Central. The HTML code generated by the task participants should correctly represent the structure of the table and the content of each cell. HTML tags that define the text style including bold, italic, strike through, superscript, and subscript should be included in the cell content. The HTML code does NOT need to reconstruct the appearance of tables such as border lines, background color or font, font size, or font color. The competition site is available from[11].

3.1   Related Work

There are other table recognition challenges, which are mainly organized at the International Conference on Document Analysis and Recognition (ICDAR). ICDAR 2013 Table Competition is the first competition on table detection and recognition [5]. A total of 156 tables are included in ICDAR 2013 Table Competition for evaluation of table detection and table recognition methods; however, no training data is provided. ICDAR 2019 Competition on Table Detection and Recognition provides training, validation, and test samples (3,600 in total) for table detection and recognition [4]. Two types of documents, historical hand-written and model programmatic, are offered in image format. The ICDAR 2019 competition includes three tasks: 1) identifying table regions; 2) recognizing table structure with given table regions; 3) recognizing table structure without given table regions. The ground truth only includes the bounding box of table cell, without the cell content.

Our Task B competition proposed a more challenging task: the model needs to recognize both the table structure and the cell content of a table solely relying on the table image. In another word, the model needs to infer the tree-structure of the table and the properties (content, row-span, column-span) of each leaf node (table header/body cells). In addition, we do not provide intermediate annotations of cell position, adjacency relations, or row/column segmentation, which are needed to train most of the existing table recognition models. We only provide the final results of the tree representation for supervision. We believe this will motivate participants to develop novel models for image-to-structure mapping.

3.2   Data

This task used the PubTabNet dataset (v2.0.0)[12] [16]. PubTabNet contains over 500k training samples and 9k validation samples, of which the ground truth HTML code, and the position of non-empty table cells are provided. Participants can use the training data to train their model and the validation data for model selection and hyper-parameter tuning. The 9k+ final evaluation set (image only, no annotation) was released 3 d before the competition ended for

---

[11] Task B website: https://aieval.draco.res.ibm.com/challenge/40/overview.
[12] https://github.com/ibm-aur-nlp/PubTabNet.

the Final Evaluation Phase. Participants submitted their results on this set in the final phase.

Submissions were evaluated using the TEDS (Tree-Edit-Distance-based Similarity) metric[13] [16]. TEDS measures the similarity between two tables using the tree-edit distance proposed in [11]. The cost of insertion and deletion operations is 1. When the edit is substituting a node n_o with n_s, the cost is 1 if either n_o or n_s is not td. When both no and n_s are td, the substitution cost is 1 if the column span or the row span of n_o and n_s is different. Otherwise, the substitution cost is the normalized Levenshtein similarity [9] (in $[0, 1]$) between the content of n_o and n_s. Finally, TEDS between two trees is computed as

$$TEDS(Ta, Tb) = 1 - \frac{EditDist(Ta, Tb)}{max(|Ta|, |Tb|)} \tag{1}$$

where EditDist denotes tree-edit distance, and $|T|$ is the number of nodes in T. The table recognition performance of a method on a set of test samples is defined as the mean of the TEDS score between the recognition result and ground truth of each sample.

The competition had three phases. The Format Verification Phase spanned the whole competition, for participants to verify if their results file met our submission requirements with the mini development set that we provided. The Development Phase spanned from the beginning of the competition to 3 d before the competition ended. In this phase, participants could submit results on the test samples to verify their model. The Final Evaluation Phase run in the final 3 d of this competition. Participants could submit the inference results on the final evaluation set in this phase. Final ranking and winning teams were decided by the performance in the Final Evaluation Phase. Table 3 shows the size of the different data sets used in the different Task B phases.

Table 3. Task B data set statistics

Split	Size	Phase
Training	500,777	N/A
Development	9,115	N/A
Mini development	20	Format Verification Phase
Test	9,138	Development
Final evaluation	9,064	Final evaluation

3.3 Results

For Task B, we had 30 submissions from 30 teams for the Final Evaluation Phase. Top 10 systems ranked using their TEDS performance on the final evaluation

---

[13] https://github.com/ibm-aur-nlp/PubTabNet/tree/master/src.

set are shown in Table 4. Due to a problem with the final evaluation data set, bold tags <b> where not considered in the evaluation.

The first four systems have similar performance, while we see a more significant different thereafter. As it is shown in the description of the systems, they rely on the combination of several components that identify relevant components from table images and then compose them. The performance is better than compared to previously reported result of 91 in the TEDS metric using an image to sequence approach [17]. In [17], the data set is comparable to the test set of this competition and was derived as well from PubMed Central.

Table 4. Task B top TEDS results. The overall result (TEDS all) is decompose into simple and complex tables [16]

Team Name	TEDS Simple	TEDS Complex	TEDS all
Davar-Lab-OCR	97.88	94.78	**96.36**
VCGroup	**97.90**	94.68	96.32
XM	97.60	**94.89**	96.27
YG	97.38	94.79	96.11
DBJ	97.39	93.87	95.66
TAL	97.30	93.93	95.65
PaodingAI	97.35	93.79	95.61
anyone	96.95	93.43	95.23
LTIAYN	97.18	92.40	94.84

3.4   Systems Description

These are the descriptions of the top systems provided by the participants for Task B[14].

Team: Davar-Lab-OCR, Hikvision Research Institute. The table recognition framework contains two main processes: table cells generation and structure inference[15].

(1) Table cells generation is built based on the Mask-RCNN detection model. Specifically, the model is trained to learn the row/column aligned cell-level bounding boxes with corresponding mask of text content region. We introduce the pyramid mask supervision and adopt a large backbone of HRNet-W48 Cascade Mask RCNN to obtain the reliable aligned bounding boxes. In addition, we train a single-line text detection model with an attention-based text recognition model to provide the OCR information. This is simply achieved by selecting the instances that only contain single-line text. We

---

[14] Not all descriptions for the top systems were provided.
[15] Davar-Lab-OCR paper and source code: https://davar-lab.github.io.

also adopt multi-scale ensemble strategy on both the cell and single-line text detection models to further improve performance.

(2) In the structure inference stage, the bounding boxes for cells can be horizontally/vertically connected according to their alignment overlaps. The row/column information is then generated via a Maximum Clique Search process, during which empty cells can be easily located.

To handle some special cases, we train another table detection model to filter out text not belonging to the table.

Team: VCGroup. In our method [7,10,14][16], we divide the table content recognition task into four sub-tasks: table structure recognition, text line detection, text line recognition and box assignment. Our table structure recognition algorithm is customized based on MASTER, a robust image text recognition algorithm. PSENet is used to detect each text line in the table image. For text line recognition, our model is also built on MASTER. Finally, in the box assignment phase, we associated the text boxes detected by PSENet with the structure item reconstructed by table structure prediction, and fill the recognized content of the text line into the corresponding item. Our proposed method achieves a 96.84% TEDS score on 9,115 validation samples in the development phase, and a 96.32% TEDS score on 9,064 samples in the Final Evaluation Phase.

Team: Tomorrow Advancing Life (TAL). The TAL system consists of two schemes:

1. Rebuild table structure through 5 detection models, which are table head-body detection, row detection, column detection, cell detection and text-row detection. Mask R-CNN is selected as the baseline for these 5 detection models, with targeted optimization for different detection tasks. In the recognition part, the results of cell detection and text-row detection are inputted into the CRNN model to get the recognition result corresponding to each cell.

2. The restoration of table structure is treated as an img2seq problem. To shorten the decoding length, we replace every cell content with different numbers. The numbers are obtained from text-row detection results. Then we use CNN to encode the image and use a transformer model to decode the structure of the table. The corresponding text-line content can then be obtained by using the CRNN model.

The above two schemes can be used to get the complete table structure and content recognition results. We have a set of selection rules, which combine the advantages of both schemes, to output the one best final result.

Team: PaodingAI, Beijing Paoding Technology Co., Ltd. PaodingAI's system is divided into three main parts: text block detection, text block recognition and table structure recognition. The text block detector is trained by the

---

[16] VCGroup Github repo: https://github.com/wenwenyu/MASTER-pytorch.

Detectors_cascade_rcnn_r50_2x model provided by MMDetection. The text block recognizer is trained by the SAR_TF[17] model. Table structure recognizer is our own implementation of the model proposed in [13]. In addition to the above model, we also use rules and a simple classification model to process <thead>, <b>, and blank characters. Our system is not an end-to-end model and does not use an integrated approach.

Team: Kaen Context, Kakao Enterprise[18]

To resolve the problem of table recognition in an efficient way, we use the 12-layer decoder-only linear transformer architecture [8].

Data preparation: We use RGB images without rescaling as input conditions and the merged HTML code is used as target text sequences. We reshape a table image into a sequence of flattened patches with shape (N, 8*8*3), where 8 is the width and height of each image patch, and N is the number of patches. Then, we map the image sequence to 512 dimensions with a linear projection layer. The target text sequence is converted into a 512-dimensional embedding through an embedding layer and appended at the end of the projected image sequence. Finally, we add different positional encodings to the text and image sequences to allow our model to distinguish them.

Training: The concatenated image-text sequence is used as the input of our model and the model is trained by the cross-entropy loss under the teacher forcing algorithm.

Inference: The outputs of our model are sampled with beam search (beam = 32).

4   Conclusions

We have proposed two tasks for document understanding using large data sets derived from PubMed Central for the training and evaluation of participant systems. These tasks address two important problems, understanding document layouts and table identification, including both table border and cell structure.

We had a large participation for both tasks, which was quite significant for Task A with 281 submissions from 78 teams. Results from top participant submissions significantly improve the performance of previously reported results.

Results from both tasks show an impressive performance and opens the possibility for high performance practical applications. There are still some aspects to improve from Task A, such as a better identification of titles, and better processing of complex tables in Task B. Both tasks have used a data set derived from scientific literature. The generated data sets are quite large and diverse in the formats in which the information is represented. This diversity should help using the trained models in other domains, which could be evaluated using new data sets generated for other domains such as FinTabNet [15] for the financial domain.

---

[17] https://github.com/Pay20Y/SAR_TF.

[18] Company located in Seongnam-si, Gyeonggi-do, South Korea.

Acknowledgements. We would like to thank Sundar Saranathan for his help with the competition system.

We would like to thank the US NIH/National Library of Medicine for making available the data sets used in this competition.

References

1. Antonacopoulos, A., Bridson, D., Papadopoulos, C., Pletschacher, S.: A realistic dataset for performance evaluation of document layout analysis. In: 2009 10th International Conference on Document Analysis and Recognition, pp. 296–300. IEEE (2009)
2. Clausner, C., Antonacopoulos, A., Pletschacher, S.: ICDAR 2017 competition on recognition of documents with complex layouts-rdcl2017. In: 2017 14th IAPR International Conference on Document Analysis and Recognition (ICDAR), vol. 1, pp. 1404–1410. IEEE (2017)
3. Clausner, C., Papadopoulos, C., Pletschacher, S., Antonacopoulos, A.: The enp image and ground truth dataset of historical newspapers. In: 2015 13th International Conference on Document Analysis and Recognition (ICDAR), pp. 931–935. IEEE (2015)
4. Gao, L., Huang, Y., Li, Y., Yan, Q., Fang, Y., Dejean, H., Kleber, F., Lang, E.M.: ICDAR 2019 competition on table detection and recognition. In: 2019 International Conference on Document Analysis and Recognition (ICDAR), pp. 1510–1515. IEEE (September 2019). https://doi.org/10.1109/ICDAR.2019.00166
5. Göbel, M., Hassan, T., Oro, E., Orsi, G.: ICDAR 2013 table competition. In: 2013 12th International Conference on Document Analysis and Recognition, pp. 1449–1453. IEEE (2013)
6. Grygoriev, A., et al.: HCRNN: a novel architecture for fast online handwritten stroke classification. In: Proceedings of the International Conference on Document Analysis and Recognition (2021)
7. He, Y., et al.: Pingan-vcgroup's solution for ICDAR 2021 competition on scientific table image recognition to latex. arXiv (2021)
8. Katharopoulos, A., Vyas, A., Pappas, N., Fleuret, F.: Transformers are RNNs: fast autoregressive transformers with linear attention. In: International Conference on Machine Learning, pp. 5156–5165. PMLR (2020)
9. Levenshtein, V.I.: Binary codes capable of correcting deletions, insertions, and reversals. In: Soviet physics doklady, vol. 10, pp. 707–710. Soviet Union (1966)
10. Lu, N., et al.: Master: Multi-aspect non-local network for scene text recognition. Pattern Recognit. **117**, 107980 (2021)
11. Pawlik, M., Augsten, N.: Tree edit distance: robust and memory-efficient. Inf. Syst. **56**, 157–173 (2016)
12. Staar, P.W., Dolfi, M., Auer, C., Bekas, C.: Corpus conversion service: a machine learning platform to ingest documents at scale. In: Proceedings of the 24th ACM SIGKDD International Conference on Knowledge Discovery & Data Mining, pp. 774–782 (2018)
13. Tensmeyer, C., Morariu, V.I., Price, B., Cohen, S., Martinez, T.: Deep splitting and merging for table structure decomposition. In: 2019 International Conference on Document Analysis and Recognition (ICDAR). pp. 114–121. IEEE (2019)
14. Ye, J., et al.: Pingan-vcgroup's solution for icdar 2021 competition on scientific literature parsing task b: Table recognition to html. arXiv (2021)

15. Zheng, X., Burdick, D., Popa, L., Zhong, X., Wang, N.X.R.: Global table extractor (gte): a framework for joint table identification and cell structure recognition using visual context. In: Proceedings of the IEEE/CVF Winter Conference on Applications of Computer Vision, pp. 697–706 (2021)
16. Zhong, X., ShafieiBavani, E., Jimeno Yepes, A.: Image-based table recognition: data, model, and evaluation. arXiv preprint arXiv:1911.10683 (2019)
17. Zhong, X., Tang, J., Jimeno Yepes, A.: Publaynet: largest dataset ever for document layout analysis. In: 2019 International Conference on Document Analysis and Recognition (ICDAR), pp. 1015–1022. IEEE (2019)

ICDAR 2021 Competition on Historical Document Classification

Mathias Seuret[1]([envelope]) [ID], Anguelos Nicolaou[1] [ID], Dalia Rodríguez-Salas[1] [ID],
Nikolaus Weichselbaumer[2] [ID], Dominique Stutzmann[3] [ID], Martin Mayr[1] [ID],
Andreas Maier[1] [ID], and Vincent Christlein[1] [ID]

[1] Pattern Recognition Lab, Friedrich-Alexander-Universität Erlangen-Nürnberg,
Erlangen, Germany
{mathias.seuret,anguelos.nicolaou,dalia.rodriguez,
martin.mayr,andreas.maier,vincent.christlein}@fau.de
[2] Gutenberg-Institut für Weltliteratur und schriftorientierte Medien,
Johannes Gutenberg-Universität Mainz, Mainz, Germany
weichsel@uni-mainz.de
[3] Institut de Recherche et d'Histoire des Textes, Paris, France
dominique.stutzmann@irht.cnrs.fr

Abstract. This competition investigated the performance of historical document classification. The analysis of historical documents is a difficult challenge commonly solved by trained humanists. We provided three different classification tasks, which can be solved individually or jointly: font group/script type, location, date. The document images are provided by several institutions and are taken from handwritten and printed books as well as from charters. In contrast to previous competitions, all participants relied upon Deep Learning based approaches. Nevertheless, we saw a great performance variety of the different submitted systems. The easiest task seemed to be font group recognition while the script type classification and location classification were more challenging. In the dating task, the best system achieved a mean absolute error of about 22 years.

Keywords: Document classification · Document analysis · Dating · Historical document images

1 Introduction

Several aspects of historical documents are of interest for scholars in the humanities. The most important one is the textual content of these documents, however no text can be used to understand the past without knowing at least some of its context, such as when it was written, by whom, or where.

The competition shared similarities with previous competitions. In particular, with previous ICFHR and ICDAR competitions on Competition on the Classification of Medieval Handwritings in Latin Script (CLaMM)'16 [3] and

© Springer Nature Switzerland AG 2021
J. Lladós et al. (Eds.): ICDAR 2021, LNCS 12824, pp. 618–634, 2021.
https://doi.org/10.1007/978-3-030-86337-1_41

CLaMM'17 [2] focusing on Latin medieval manuscripts. While the first competition [3] focused purely on script type classification, the second one [2] had two tasks: (a) script type classification and (b) dating.

In this competition, we changed the setting. First, we used different material (handwritten and printed; charters and manuscripts) and second extend to a third task. The idea behind the use of different tasks was to enable multi-task or self-supervised learning in the field of document analysis.

In particular, we provided the following subtasks: (1) Font group recognition in case of printed documents, and script identification in case of handwritten documents. Knowing this information is useful for selecting adequate Optical Character Recognition (OCR) or Handwritten Text Recognition (HTR) models for extracting the text content of the document. (2) Dating the documents, which allows to link the document to its historical context. (3) And finally, determining the place of origin of a handwritten document – for this last task, we focused on documents produced in several French monasteries. To the best of our knowledge, this is the first dataset on locations.

The paper is organized as follows. In Sect. 2, we give details about the datasets used for the different tasks. The submitted methods are explained in Sect. 3. Section 4 explains the error metrics used and shows the results of the competition accompanied with some additional performance analysis. The paper is concluded in Sect. 5.

2 Data

2.1 Font Group

As training data, we suggested the participants to use a dataset that we previously published [14], which contains over 35k document images, with one to five labels corresponding to ten font group classes, and two noise classes. As test data[1], we obtained document images from the Bayerische Staatsbibliothek München, which gives access to scans of several hundred thousands early-modern printed books. In order to balance the test data, we used a classifier for discarding pages with extremely frequent font groups, such as Fraktur, and thus increase the odds to find pages with rarer fonts, such as Textura. Note that while it helped to increase significantly the proportion of rare fonts, a perfectly balanced set was not obtained. The training data contains pages from all books known to have been printed in Gotico-Antiqua, hence we could not include document-independent pages for this font group in the test data.

Additionally, in order to investigate the robustness of the submissions, we augmented a copy of the data with augmentations, such as rescaling, without informing the participants about this.

Our test data is composed of 2753 original images, plus one modified version of each of these, for a total of 5506 images. Out of these, we have the class distribution given in Table 1a.

---

[1] Available on Zenodo: https://doi.org/10.5281/zenodo.4836551

Table 1. Class distribution for the (a) font group classification task and (b) the script classification task. Both original and augmented versions have been counted.

Font	Page count		Script	Page count
Antiqua	788		Caroline	120
Bastarda	376		Cursiva	116
Fraktur	704		Half-Uncial	108
Greek	602		Humanistic	108
Hebrew	976		Hybrida	92
Italic	704		Praegothica	86
Rotunda	792		Semihybrida	100
Schwabacher	440		Semitextualis	118
Textura	124		Southern Textualis	142
			Textualis	130
			Uncial	136
	5506			1256
(a)			(b)	

2.2 Script Type

For the script type classification, we suggested the participants to use the CLaMM 2017 dataset [2] for training. The test data[2] was obtained from the same manuscripts following a two-step approach. First, for each manuscript, many random subsets of pages have been generated, and the one with the largest spread over the manuscript and maximizing the distance to training pages was selected. Second, a manual inspection of the pages made sure that pages have the expected label, and that no other script, such as glosses, is present. Due to the source of the images, we decided to augment them slightly in order to avoid possible bias due to factors such as luminosity or average paper color. The class distribution for the script classification task is given in Table 1b.

2.3 Dating

The scripts used for the task of date identification were acquired from e-codices.[3] From the original meta data, we used the *not before* and *not after* dates as labels. From all available documents, only those of which their date of origin were available and/or not ambiguous were considered. A date was said to be ambiguous if two or more scholars differed more than 50 years on either *not before* or *not after* dates. Additionally, we restricted our selection to manuscripts whose dating intervals begin no later than the XVII century. From over 2500 manuscripts

---

[2] Available on Zenodo: https://doi.org/10.5281/zenodo.4836659
[3] Virtual Manuscript Library of Switzerland (http://e-codices.unifr.ch/)

Table 2. Distribution per century of the manuscripts for the date identification task.

Century	IX	X	XI	XII	XIII	XIV	XV	XVI	XVII	Total
Documents	259	73	105	158	128	206	548	153	69	1698

made available on e-codices, 1698 matched our requirements. We downloaded random pages from these manuscripts, and automatically cropped the textual content in the following way. (1) The Laplacian of the image was computed. (2) The areas corresponding to the black borders around the image were set to 0. (3) A convolution of a maximum filter (kernel size: image width/25, image height/50) was applied. (4) Then, horizontal and vertical projection profiles (PP) were computed using the output of the previous step as input. (5) Otsu thresholding was applied to both PPs. (6) A second convolution of a maximum filter (kernel size: PP size/50) was applied to each of the the PPs to fill small gaps. (7) The largest positive segment of the PPs was selected as respective horizontal and vertical limits for cropping the image. Afterwards, all crops were reviewed manually to discard the ones containing no or little text.

The documents were clustered per century of the beginning of their date interval in order to split them in train and test subsets with enough samples distributed over time. This distribution of the documents over time from both training and test sets is given in Table 2. For all centuries, 41 manuscripts are assigned to the test set, except for the two centuries with the fewest number of documents; for centuries X and XVII only 29 and 28 documents, respectively, are assigned for testing. Thus, no pages from one manuscript got scattered in both subsets. The other manuscripts are provided for training.

In total, the training and test sets[4] contain respectively 11294 and 2516 pages.

2.4 Location

The location modality is modeled as a classification task. The location classes were selected by experts so that they will represent corpora that are highly localized; whether they represent cities, regions, or monasteries. The selection combines several sources:

– the series of catalogues of dated manuscripts for France, published since 1959 and converted into XML files,[5]
– specific palaeographical studies on book production in scriptoria, as Corbie [4] or Fontenay [16], or workshops for book trade [13],
– a study on the royal chancery in Paris and its notaries [5],
– an ongoing project on Saint-Bertin.[6]

---

[4] Available on Zenodo: https://doi.org/10.5281/zenodo.4836687
[5] https://github.com/oriflamms/CMDF
[6] https://saint-bertin.irht.cnrs.fr/

Table 3. Location dataset classes

Location	Train Samples	Val. Samples	Test Samples
Cluny	180	4	25
Corbie	795	5	25
Cîteaux	140	5	25
Florence	45	5	25
Fontenay	250	5	25
Himanis (Paris chancery)	230	5	25
Milan	55	5	25
Mont-Saint-Michel	182	5	25
Paris (book trade)	467	5	25
Saint-Bertin	2988	5	25
Saint-Germain-des-Prés	55	5	25
Saint-Martial-de-Limoges	75	5	25
Signy	55	5	25
Total:	5517	60	300

For every corpus, we searched for digitizations of the manuscripts listed with their context of production in the scholarly studies. For every corpus, 5 documents were randomly selected for the test-set, one for the validation-set, and the remaining ones for the train-set[7] – with an exception for Cluny, which has one less test manuscript due to an issue with its images detected at a late stage. From each document, up to five pages were extracted as samples. In Table 3, specific information about the corpora and in Fig. 2 location approximations of the 13 corpora employed can be seen.

3   Methods

3.1   Baseline

As baselines, we use a system inspired by our previous work on font group identification [14]. Images wider than 1000 pixels are resized to this width. We assume that this size normalization makes the text of the documents have similar sizes – maybe different by an order of magnitude, but not more. To train the network, we extract 15 000 random 320×320 pixels patches from each class. During training, these patches are augmented with (1) affine transforms, (2) random resized crops (resizement in $[0.25, 1[$) of 224×224 pixels, (3) random modifications of the brightness, contrast and hue, (4) JPEG compression artifacts, and (5) binarization of 15 % of the crops (10 % with Otsu, 5 % with Sauvola). We train a ResNet50 for 20 epochs with stochastic gradient descent, excepted for the script classification, task for which we doubled this amount.

---

[7] Available on Zenodo: https://doi.org/10.5281/zenodo.4836707

3.2   NAVER Papago

Seungjae Kim from the company Naver Papago submitted the following methods for the font group classification task.

Method 1. The basic strategy was to fix the network to ResNet50, and try different ideas to generalize and fit the network to images having different sizes, quality, and class imbalance, which are challenging problems of the data. For data preprocessing, the images were rescaled along the largest side and padding was added to make all images having the same size of 768 × 576. Images that did not have a main font were filtered out. For training with cross-entropy, a batch size of 256 with the RAdam optimizer [11] was used. The first 15 epochs were run with an initial learning rate of 0.001, afterwards it is divided by 10 at epochs 5 and 10. Random brightness, contrast, shifts and scales for data augmentation using the Albumentations[8] library were applied (Details: RandomBrightnessContrast(p=0.3, contrast_limit=0.15, brightness_limit=0.15), ShiftScaleRotate(p=0.3, shift_limit=0.2, scale_limit=0.2, rotate_limit=0, border_mode=cv2.BORDER_CONSTANT)).

Method 2. For this method, the team tried to get the highest score on a custom validation set (10 %) with hyperparamter tuning. The batch size was decreased to 112 using synchronized batch normalization [12]. Additionally, a stratified sampling was used. In contrast to the first method, the images were also resized using the longest side and padded such that all images had the same size of 1024 × 768.

Method 3. Having reached the highest private validation score with the ResNet-50 model, the team trained another model. For inference, a simple average ensemble was conducted. For the second model, a SE-ResNext-50-32x4d network [7] was chosen. It was trained with a batch size of 80. The remaining hyper-parameters were set equal to the ResNet-50 model (see method 1/2).

Method 4. The major changes for this method were the use of NFNet [1] models. In particular, an NFNet-F0 model was combined with an NFNet-F1 formning an ensemble approach. The NFNet-F0 model was trained with image input sizes of 1024 × 768 and a batch size of 112. In comparison, the image sizes were reduced to 768 × 576 and the batch size reduced to 96 for the NFNet-F1 model. In both cases, the models were trained for 12 epochs with an initial learning rate of 0.001, which is divided by 10 at epochs 2 and 7. In addition to training augmentations, test time augmentations with scale and brightness variations were applied and the results averaged.

---

[8] https://github.com/albumentations-team/albumentations

Method 5. For the last submission, the team decided to ensemble an NFNet-F0 model with a SE-ResNext [7] model. Both models used an input size of 1024×768. The NFNet-F0 model was trained similarly as in method 4 while the SE-ResNext model was trained as described in method 3. Additionally, the team applied weakly supervised learning with pseudo labels. After inference on the test set, predictions that had a confidence higher than 0.99 were labeled with those predictions and were used to finetune the model. Afterwards, predictions were made once again with the finetuned model.

3.3   PERO

Martin Kišš, Jan Kohút, Karel Beneš, Michal Hradiš from the Faculty of Information Technology of the Brno University of Technology submitted the following methods.

They divided the datasets of the font and the script tasks in such a way that the distribution of all classes of the validation set was uniform. For the date task, they randomly selected 1000 pages as the validation set.

Depending on the tasks and methods, they used different loss functions:

Classification Losses. For classification tasks, the output y of the system is a vector of probabilities, which is compared with the set $T = \{t_1, t_2, \ldots, t_N\}$ of labels associated with the input:

$$L_{\min} = \min_t \mathrm{CE}(\mathbf{y}, t) \tag{1}$$

$$L_{\mathrm{avg}} = \sum_t y_t \mathrm{CE}(\mathbf{y}, t) \tag{2}$$

where y_t is the probability attributed to label t and $\mathrm{CE}(y, t)$ is the cross-entropy between the categorical distribution y and the (one-hot represented) target t.

Interval Regression Loss. While they used the cross-entropy loss for classification, they used an interval regression loss for the dating task. That means, the output y is a scalar and measures the loss w. r. t. the interval $[a, b]$:

$$L_{\mathrm{date}}(y, a, b) = \begin{cases} a - y - r & \text{if } y \leq a \\ a - y - r & \text{if } y \geq a \\ \left(\frac{y-m}{r}\right)^2 r & \text{otherwise} \end{cases}, \tag{3}$$

where $m = (a+b)/2$ is the midpoint and $r = (a-b)/2$ the radius of the interval. This is technically just a Huber loss rescaled to the target interval.

Method 1 and Method 3. Both methods use a ResNeXt-50 model operating on 224×224 patches. Method 1 is the best model as per validation data while method 3 is the average of ±3 checkpoints 1000 updates apart. For the classification tasks, the L_{avg} (Eq. (2)) loss function was used. Every page was evaluated in four different scales, each scale was divided into a grid of non-overlapping patches. For classification tasks, the output in each scale is calculated as the mean of outputs on the 10 most confident patches. For the script and font tasks, the overall page-level output is the average of the scale-level outputs. For the location task, the most confident scale-level output is the overall page-level output. For the date task, the scale-level output is calculated as the median of all scaled outputs and the overall page-level output is then the mean of these scale-level outputs.

Method 2. System 2 is a CNN operating on text lines obtained by an in-house segmentation method [10]. Global average pooling is applied to deal with variable text line length. For classification tasks, the $L_{\min}$ (Eq. (1)) loss function was used. The final model was obtained by merging the last N checkpoints, which were 1000 iterations away from each other; $N < 10$ was tuned to optimize the validation accuracy. To obtain a page-level output for the classification tasks, the team calculated the mean vector of output probabilities of text lines longer than 64 pixels. When dealing with the date task, the team calculated the page-level output as the median of text line-level outputs of text lines longer than 128 pixels.

Method 4. System 4 is a fusion of systems 2 and 3 by linear interpolation, tuned on the validation set. With the meaning of 0 being System 2 only and 1 being System 3 only, the team set the interpolation for font, script, location, and date to 0.25, 0.25, 0.33, and 0.1, respectively.

Method 5. System 5 is a fusion of systems 2 and 3 implemented by a trained (multi-class) logistic regression. The team trained it until convergence with ℓ^2-regularization towards a point of averaging outputs. The optimal strength of the regularization was obtained by a k-fold cross-validation. This system is not applied to the date task.

3.4   The North LTU

Konstantina Nikolaidou from the Luleå University of Technology submitted the following method. For all tasks, before the training process, the images were cropped around their borders and then downscaled by a factor of 2. For the script, date, and location classification tasks, all provided datasets were merged (for each task separately) and a stratified split of 90 %/10 % was made to create the final training and validation sets. For the font classification, the team used the existing training set. For all training sets, crops of 224 × 224 with a stride of 224 were used for every image and all grayscale images were converted to RGB. For the script classification, an 18-layer pre-trained ResNet [6] was used while for the rest of the tasks a 50-layer ResNet.

The following data augmentation was used during training: random horizontal flipping, random rotation with a maximum of 10 degrees, random perspective distortion with a maximum of 0.2 degree, as well as jitter of brightness, contrast, and saturation each by a maximum factor of 0.5. For validation and testing, 224 × 224 crops were used. The team used the Adam optimizer [9] with a weight decay regularization of 0.0004 and early stopping when the validation accuracy did not improve for 10 epochs. The date classification was treated as a multi-label classification task with labels not before and not after. In that case, early stopping was applied when both validation accuracies were not improving for 10 consecutive epochs. For the script task, a scheduler reduced the learning rate by a factor of 0.1 every 2 epochs while for the rest of the tasks it was reduced every single epoch. The initial learning rate in all cases was 0.001. Furthermore, in the tasks of font and location classification, a balanced batch sampler that over-samples the least frequent classes was used.

3.5 CLUZH

Phillip Ströbel of the Department of Computational Linguistics, University of Zurich, submitted the following methods.

Method 1 and 2. The approach makes use of the fast.ai[9] library, which facilitates building, applying, and fine tuning models for a variety of tasks. For the font group classification they used 21,178 scans for training and 9,675 scans for validation, while for classifying script types we applied a 80/20 split (3030 for training and 758 for validation). The skew in label distribution is more heavily pronounced in the font group data set than it is in the script type data set, which is a fact we ignored for the time being. Fast.ai provides numerous methods for data augmentation. Each classification task was trained with and without data augmentation. Such data augmentation included rotation, warping, flipping, zooming, adjusting contrast, etc. The team hypothesised that especially in the case of script type classification, where there was substantially less data, data augmentation of any kind would benefit the task.

All images of the font group data set were resized to 512 pixels, and the images of the script type data to 256 in batch sizes of 64. A pre-trained ResNet-34 model was used as base model. FastAI's learning rate finder [15] was used and the learning rate for both subtasks was set to 0.02. The ResNet-34 models were finetuned for one epoch each and trained 15 epochs on the font group data set, while 50 training epochs were needed for the script type data set (in any case, the model for the script type data started overfitting early on in the training procedure). In the finetuning approach (method 2), the lower layers in the network were frozen and only the weights in the last layer (head) were adjusted. After the finetuning phase, the training procedure unfroze all layers and adjusted the weights of all layers accordingly. We found no difference in performance by adding more finetuning epochs before training the models.

---

[9] https://www.fast.ai/

Method 3. In a last approach, the team trained a model on all data (combination of the two data sets) and found that while the performance on font group classification did not suffer, there could be benefits for script type classification. The identification of font groups seems much easier than the classification of script types. The hypothesis that data augmentation helps in the case of script type classification was not clearly confirmed in the internal experiments of the team, although a little improvement for models trained with augmented images was observed. The team believes, however, that those tasks are very similar, and thus could (should) be combined. Moreover, more data for script classification would be beneficial.

4 Evaluation

4.1 Evaluation Metrics

For the classification tasks, we report two error metrics: (1) Accuracy (in our case equal to recall) and (2) Unweighted Average Recall (UAR). Accuracy denotes the overall number of true positives P per number of images in the dataset N. UAR is also known as average accuracy or recall using macro averaging, where the accuracy is computed for each of the K classes consisting of possibly different number of samples N_k and then averaged:

$$\text{Acc} = \frac{P}{N} \qquad (4) \qquad\qquad \text{UAR} = \frac{1}{K} \sum_{k=1}^{K} \frac{P_k}{N_k} \, . \qquad (5)$$

For the dating task, we report the Mean Absolute Error (MAE), i.e., the average distance for the estimated year $\hat{y}_i$ in years to the ground truth year interval $I_i = (a_i, b_i)$ for the sample i. Given the indicator function that denotes if an estimated year falls into the interval:

$$\text{ind}(\hat{y}_i, I_i) = \begin{cases} 1 & \text{if } a_i \leq \hat{y}_i \leq b_i \\ 0 & \text{else} \end{cases}, \qquad (6)$$

it follows:

$$\text{MAE} = \frac{1}{N} \sum_{i}^{N} (1 - \text{ind}(\hat{y}_i, I_i)) \, \min(|\hat{y}_i - a_i|, |\hat{y}_i - b_i|) \, . \qquad (7)$$

Additionally, we report the accuracy that the reported number is in the ground truth range, denoted as *In Interval* (InI):

$$\text{InI} = \frac{1}{N} \sum_{i}^{N} \text{ind}(\hat{y}_i, I_i) \, . \qquad (8)$$

The participants are ranked according to the UAR and MAE.

The evaluation code is publicly available.[10]

---

[10] https://github.com/seuretm/icdar21-evaluation

Table 4. Classification results for the (a) font group classification and (b) the script type classification tasks (in percentage).

Method	Acc.	UAR
PERO 5	99.04	98.48
PERO 2	99.09	98.42
PERO 4	98.98	98.27
NAVER Papago 3	98.33	97.17
Baseline	97.89	96.77
NAVER Papago 2	97.86	96.47
PERO 1	97.93	96.36
NAVER Papago 4	97.69	96.24
NAVER Papago 5	97.64	96.01
NAVER Papago 1	97.64	95.86
PERO 3	97.44	95.68
CLUZH 3	97.13	95.66
CLUZH 1	96.84	95.34
CLUZH 2	96.80	95.16
The North LTU	87.00	82.80

(a)

Method	Acc.	UAR
PERO 4	88.77	88.84
PERO 5	88.46	88.60
PERO 2	88.46	88.54
PERO 1	83.04	83.11
PERO 3	80.33	80.26
The North LTU	73.96	74.12
Baseline	55.81	55.22
CLUZH combo	36.86	35.25
CLUZH v1	26.11	25.39
CLUZH v2	22.29	21.51

(b)

4.2   Results

Task 1: Font and Script Classification. The most systems were submitted for the font group classification task. Table 4a shows that three systems of team PERO ranked first, and Table 5 shows a confusion matrix of the winning method. Since all these three methods contain a method based on line segmentation, it seems an explicit line detection/segmentation is beneficial for this task. In contrast, the global methods PERO 1 and PERO 3, which are operating on patches ranked behind NAVER Papago 3, which is a method based on ResNeXt-50. Interestingly, the very deep and efficient models used for NAVER Papago 4 and 5 ranked in the middle field. Our baseline model, which is based on patches seemed to be competitive ranking and ended up in the upper third. The CLUZH methods ranked just behind PERO 3 and in front of The North LTU. Nevertheless, CLUZH 3 achieved better values than the other two methods, which shows that the combination of datasets, which was only done by the team CLUZH, seems to be beneficial.

This phenomenon can also be recognized in Table 4b dealing with script type classification. However, the methods fell behind the other competitors. Once more, the systems PERO 2,4,5 performed by far the strongest, and a confusion matrix for the winning method is given in Table 6.

Table 5. Confusion matrix for the font group classification of the method PERO 5.

Predicted →	An.	Ba.	Fr.	GA.	Gr.	He.	It.	Ro.	Sc.	Te.
Antiqua	788									
Bastarda		327		2					47	
Fraktur			704							
Gotico-Antiqua										
Greek	4				598					
Hebrew						976				
Italic							704			
Rotunda								792		
Schwabacher									440	
Textura										124

Table 6. Confusion matrix for the script classification of the method PERO 4.

Predicted →	Un.	Ha.	Ca.	Hu.	Pr.	St.	Se.	Te.	Hy.	Se.	Cu.
Uncial	96		40								
Half-Uncial	2	102	4								
Caroline		2	118								
Humanistic				108							
Praegothica					86						
South. Textualis						142					
Semitextualis							92	26			
Textualis					2	4	7	117			
Hybrida							14		78		
Semihybrida									24	68	8
Cursiva					1		1	3		3	108

An interesting result in the script classification task is that while the training data was in grayscale, most methods appear to have had better results on the color test images than on the grayscale ones. We first assumed that this could be due to image quality – grayscale being, e. g., of lower resolution. However, the grayscale images have a median dimension of 5388 pixels, against 3918 for the color ones, so the accuracy differences are most likely due difficulty differences. Thus, training on grayscale seems not to hinder good results on color data.

The two notable confusions in Table 5 that are caused by the relative stylistic proximity of the two confused classes. Schwabacher is a very formalized form of Bastarda. There are some versions of Bastarda that are very similar to it, which leads to some misclassifications. The few confusions of Greek as Antiqua can simply be explained by the fact that our data contains mainly cursive Greek, but also some capital Greek, which shares a number of glyphs with Antiqua.

Table 7. Results for the dating task.

Method	MAE	InI
PERO 2	21.91	48.25
PERO 4	21.99	47.97
PERO 3	32.45	39.75
PERO 1	32.82	40.58
The North LTU	79.43	28.74

Table 8. Results for the place of origin classification task (in percentage).

Method	Acc.	UAR
PERO 5	79.38	78.69
PERO 3	74.69	74.15
PERO 1	74.38	73.77
PERO 4	70.31	69.23
PERO 2	69.38	68.31
Baseline	63.12	62.46
The North LTU	42.81	42.85

Table 6 shows confusions that are similar to the ones seen in previous studies [2,8]. Two domains appear intertwined: scripts before the year 1000 (Uncial, Half-Uncial, Caroline) and the scripts of the cursive tradition (Cursiva, Semihybrida, Hybrida). The diversity of the large Textualis family is evidenced in some misclassifications onto related scripts. Not symmetrical confusions from Hybrida to Semitextualis and from Semitextualis to Textualis correspond to formal similarities (Hybrida and Semitextualis differ only by their 'f' and 's' letter forms, Semitextualis and Textualis by their 'a' letter forms). Yet, the asymmetry is revealing, as each script is confused with the one that was its historical model and inspiration.

Task 2: Dating. Table 7 shows that about half of the estimates landed in the ground truth interval (InI). On average, the PERO 5 system made an error of only about 22 years. Also Fig. 1a shows that the distribution of the MAEs (number of bins: 50) for all pages whose estimated date was not in the ground truth interval is long-tailed, with very few results with an error above a century.

Moreover, Fig. 1 shows a deeper investigation of the errors based on two points of view: the century to which the document is attributed to (Fig. 1b), and how many pages belong to the page's century (Fig. 1c). While it is clear that the method performs better for some centuries, the second illustration leaves no doubt that this is due to the amount of training manuscripts belonging to each century. Indeed, there is a well visible negative correlation between the amount of data and the MAE.

Task 3: Location Classification. In the location task, cf. Table 8, PERO 5 achieved the highest results; the corresponding confusion matrix is given in Table 9 and we can see the confusion graph localised on a map on Fig. 2. In contrast to all other results, PERO's systems 2 and 4 ranked behind 1 and 3, i. e., the line segmentation methods were outperformed by the global patch-based systems. This could hint upon more differences coming from the background of the documents than the writing style. In this task the tuned logistic-regression-based system PERO 5 obtained a quite large performance gain in comparison

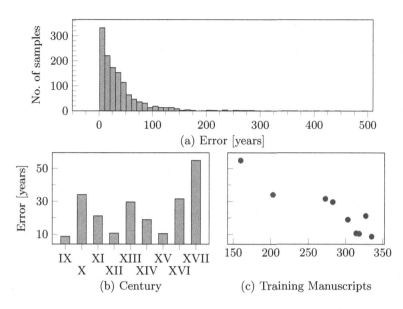

Fig. 1. PERO 2 dating submission statistics: absolute error histogram per (a) years (bin size: 10 years), (b) century and (c) training manuscripts. Note that '0'-entries, i. e., points that were correctly identified, were removed.

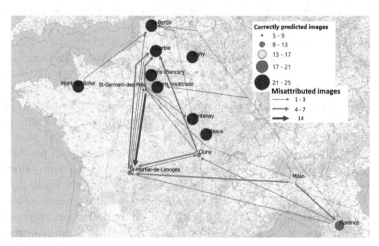

Fig. 2. Locations and confusions of method PERO 5 marked as a connectivity graph over them. The size of dots correspond to the correctly attributed images, size of lines illustrates the number of misattributed images in each direction. Background map: © OpenStreetMap contributors.

Table 9. Confusion matrix for the location classification of the method PERO 5.

Predicted →	Ci.	Cl.	Co.	Fl.	Fo.	Hi.	Mi.	MS.	SM.	Pa.	SB.	SG.	Si
Citeaux	25												
Cluny		7	5							7		1	
Corbie			24							1			
Florence		1		18					3	1	2		
Fontenay					24					1			
Himanis (Paris chancery)						25							
Milan				4			16			4		1	
Mont-St.-Michel								23			2		
Paris (book trade)									25				
St. Martial de Limoges		2	5							12	3	3	
St. Bertin											25		
St. Germain des Pres			5	1						14		5	
Signy													25

to PERO 4, which was just linearly averaging methods 2 and 3. This stands in contrast to the font/script type classification tasks where both ensembling methods performed similarly well.

Table 9 shows that three coeval productions from the same cultural environment as is the Cistercian order, are clearly recognized (Cîteaux, Fontenay, Signy), a clue to their inner coherence, as are the additional monastic productions of Corbie, Mont-St.-Michel and St.-Bertin in the 12th century. The inner homogeneity of two separate, although contemporaneous productions, appear in Paris, in the royal chancery and in the book trade. Less coherent productions seem to be uncovered in Cluny, St.-Martial de Limoges and St. Germain des Prés, the latter being even mostly confused with the general production in Paris. Errors are concentrated on images from single manuscripts which stem from a very diverse production, such as the one of Cluny, esp. Paris BnF NAL 1435, an outlier of the 15th century, or on outliers within a homogeneous production (ms. Paris BnF lat. 7433 for Florence in an Italian semitextualis and not a humanistic script). Some MSS will deserve a further examination as they create errors while not being obvious outliers (e. g., two MSS for Milan, Paris BnF lat. 8381-8382).

5   Conclusion

Overall, the team PERO won all four subtasks. Most of the time, improved results were achieved by PERO's systems working on the line level except the location task. Furthermore, rescaling the images might have affected the other systems negatively. The easiest task seemed to be the font group classification. A reason could be the availability of a large training dataset in contrast to the other tasks. The dating task was also solved surprisingly well with mean absolute errors of about 22 years. For this task, we also found a very strong correlation between the amount of training document and the accuracy of the winning method – this hints that although we used every suitable manuscripts of e-codices, more data would lead to even better performance, especially for the centuries which are the least represented in our dataset.

References

1. Brock, A., De, S., Smith, S.L., Simonyan, K.: High-performance large-scale image recognition without normalization (2021)
2. Cloppet, F., Eglin, V., Helias-Baron, M., Kieu, C., Vincent, N., Stutzmann, D.: ICDAR2017 competition on the classification of medieval handwritings in Latin script. In: ICDAR, pp. 1371–1376. Kyoto (November 2017)
3. Cloppet, F., Églin, V., Kieu, V.C., Stutzmann, D., Vincent, N.: Icfhr 2016 competition on the classification of medieval handwritings in Latin script. In: ICFHR, pp. 590–595 (October 2016)
4. Ganz, D.: Corbie in the Carolingian Renaissance : Untersuchung zur monastischen Kultur der Karolingerzeit am Beispiel der Abtei Corbie. No. 20 in Beihefte der Francia, J. Thorbecke, Sigmaringen (1990). http://www.perspectivia.net/content/publikationen/bdf/ganz_corbie
5. Glénisson, J., Guérout, J.: Registres du trésor des chartes: inventaire analytique, 1: Règne de Philippe le Bel. Inventaires et documents, Impr. Nationale, Paris (1958)
6. He, K., Zhang, X., Ren, S., Sun, J.: Deep residual learning for image recognition. In: CVPR, pp. 770–778. Las Vegas (June 2016)
7. Hu, J., Shen, L., Albanie, S., Sun, G., Wu, E.: Squeeze-and-excitation networks. IEEE Trans. Pattern Anal. Mach. Intell. **42**(8), 2011–2023 (2020)
8. Kestemont, M., Christlein, V., Stutzmann, D.: Artificial paleography: computational approaches to identifying script types in medieval manuscripts. Speculum **92**(S1), S86–S109 (2017)
9. Kingma, D.P., Ba, J.: Adam: a method for stochastic optimization. In: 3rd ICLR, pp. 1–15. San Diego (May 2015)
10. Kodym, O., Hradiš, M.: Page layout analysis system for unconstrained historic documents (2021)
11. Liu, L., et al.: On the variance of the adaptive learning rate and beyond. In: ICLR (2020). https://openreview.net/forum?id=rkgz2aEKDr
12. Peng, C., et al.: Megdet: a large mini-batch object detector. In: CVPR, pp. 6181–6189 (June 2018)
13. Rouse, R.H., Rouse, M.A.: Manuscripts and their makers : commercial book producers in medieval Paris 1200–1500. No. 25 in Studies in Medieval and Early Renaissance art history, H. Miller, London (2000)
14. Seuret, M., Limbach, S., Weichselbaumer, N., Maier, A., Christlein, V.: Dataset of pages from early printed books with multiple font groups. In: Proceedings of the 5th International Workshop on Historical Document Imaging and Processing, pp. 1–6. HIP 2019, Association for Computing Machinery, New York, USA (2019)

15. Smith, L.N.: Cyclical learning rates for training neural networks. In: WACV, pp. 464–472 (March 2017)
16. Stutzmann, D.: Écrire à Fontenay. Esprit cistercien et pratiques de l'écrit en Bourgogne (XIIe-XIIIe siècles). Ph.D. thesis, Université Paris 1 Panthéon-Sorbonne, Histoire, Paris (2009)

ICDAR 2021 Competition on Document Visual Question Answering

Rubèn Tito[1(✉)], Minesh Mathew[2], C. V. Jawahar[2], Ernest Valveny[1], and Dimosthenis Karatzas[1]

[1] Computer Vision Center, UAB, Barcelona, Spain
{rperez,ernest,dimos}@cvc.uab.cat
[2] CVIT, IIIT Hyderabad, Hyderabad, India

Abstract. In this report we present results of the ICDAR 2021 edition of the Document Visual Question Challenges. This edition complements the previous tasks on Single Document VQA and Document Collection VQA with a newly introduced on Infographics VQA. Infographics VQA is based on a new dataset of more than 5,000 infographics images and 30,000 question-answer pairs. The winner methods have scored 0.6120 ANLS in Infographics VQA task, 0.7743 ANLSL in Document Collection VQA task and 0.8705 ANLS in Single Document VQA. We present a summary of the datasets used for each task, description of each of the submitted methods and the results and analysis of their performance. A summary of the progress made on Single Document VQA since the first edition of the DocVQA 2020 challenge is also presented.

Keywords: Infographics · Document understanding · Visual Question Answering

1 Introduction

Visual Question Answering (VQA) seeks to answer natural language questions asked on images. The initial works on VQA focused primarily on images in the wild or natural images [1,10]. Most models developed to perform VQA on natural images, make use of (i) deep features from whole images or objects (regions of interest within the image), (ii) a Question Embedding module which make use of word embeddings and Recurrent Neural Networks (RNN) or Transformers [29], and (iii) a fusion module which fuses the image and text modalities using attention [2,27].

Images in the wild often contain text and reading this text is critical to semantic understanding of natural images. For example, Veit et al. observe that nearly 50% of images in MS-COCO dataset have text present in it [30]. TextVQA [26] and ST-VQA [4] datasets evolved out of this actuality and both the datasets feature questions where reading text present on the images is important to arrive

---

R. Tito and M. Mathew—Equal contribution.

J. Lladós et al. (Eds.): ICDAR 2021, LNCS 12824, pp. 635–649, 2021.
https://doi.org/10.1007/978-3-030-86337-1_42

at the answer. Another track of VQA which require reading text on the images is VQA on charts. DVQA [13], FigureQA [14] and LEAF-QA [5] comprise a few types of standard charts such as bar charts or line plots. Images in these datasets are rendered using chart plotting libraries using either dummy data or real data and the datasets contain millions of questions created using question templates.

On the other hand, Reading Comprehension [7,12,25] and Open-domain Question Answering [16,22] on machine readable text is a popular research topic in Natural Language Processing (NLP) and Information Retrieval (IR). Unlike VQA where the context on which the question is asked is an image, here questions are asked on a given text passage or a text corpus, say the entire Wikipedia. Recent developments in large-scale pretraining of language models using huge corpora of unlabeled text and later transferring pretrained representations to downstream tasks enabled considerable progress in this space [6,35].

Similar to QA involving either images or text, there are multimodal QA tasks where questions are asked on a context of both images and text. Textbook QA [15] is a QA task where questions are based on a context of text and diagrams and images. RecipeQA [34] questions require models to read text associated with a recipe and understand related culinary pictures. Both datasets have multiple choice style questions and the text modality is presented as machine readable text passages, unlike unstructured text in VQA involving scene text or chart VQA, which need to be first recognized from the images using an OCR.

Document analysis and recognition aims to automatically extract information from document images. DAR research tends to be bottom up and focus on generic information extraction tasks (character recognition, layout analysis, table extraction, etc.), disconnected from the final purpose the extracted information is used for. The DocVQA series of challenges aims to bridge this gap, by introducing Document Visual Question Answering (DocVQA) as a high-level semantic task dynamically driving DAR algorithms to conditionally interpret document images. Automatic document understanding is a complicated endeavour that implies much more than just reading the text. In designing a document the authors order information in specific layouts so that certain aspects "stand out", organise numerical information into tables or diagrams, request information by designing forms, and validate information by adding signatures. All these aspects and more would need to be properly understood in order to devise a generic document VQA system.

The DocVQA challenge comprises three tasks. The Single Document VQA and Document Collection VQA tasks were introduced in the 2020 edition of the challenge as part of the 'Text and Documents in the Deep Learning Era' workshop in CVPR 2020 [21], and have received continuous interest since then. We discuss here important changes and advancements in these tasks since the 2020 edition. The Infographics VQA task is a newly introduced task for the 2021 edition.

The Single Document VQA task requires answering questions asked on a single-page document image. The dataset for Single Document VQA comprises business documents, which include letters, fax communications reports with

Question: What is the extension number as per the voucher? **Answers:** (910) 741-0673	**Question:** In which years did Anna M. Rivers run for the State senator office? **Answers:** 2016, 2020 **Doc. Evidences:** 454, 10901	**Question:** What percentage of Americans are online? **Answers:** 90%, 90
(a) Single Document VQA	(b) Document Collection VQA	(c) Infographics VQA

Fig. 1. Examples for the three different tasks in DocVQA. Single Document VQA task (a) is a standard VQA on business/industry documents. Instead, in Document Collection VQA task (b), questions are asked on a document collection comprising documents of same template. Finally, Infographics VQA task (c) is similar to Single Document VQA task, but the images in the dataset are infographics.

tables and charts and others. These documents mostly contain text in the form of sentences and passages [20]. We discuss here new submissions for this task since the 2020 edition.

In the Document Collection VQA task, questions are posed over a whole collection of documents which share the same template but different content. More than responding with the right answer, in this scenario it is important to also provide the user with the right evidence in collection to support the response. For the 2021 edition, we have revisited the evaluation protocols to better deal with unordered list answers and to provide more insight into the submitted methods.

In the newly introduced Infographics VQA task, questions are asked on a single image, but unlike Single Document VQA where textual content is dominating the document images, here we focus on infographics, where the structure, graphical and numerical information are also important to convey the message. For this task we introduced a new dataset of infographics and associated questions and answers annotations [19].

More details about the datasets, the future challenges and updates can be found in https://docvqa.org.

2 Competition Protocol

The Challenge ran from November 2020 to April 2021. The setup of the Single Document VQA and Document Collection VQA tasks was not modified with respect to the 2020 edition, while for Infographics VQA, which is a completely new task, we released the training and validation sets between November 2020 and January 2021, and the test set on February 11, 2021, giving participants two months to submit results until April 10th. We relied at all times on the scientific integrity of the authors to follow the established rules of the challenge that they had to agree with upon registering at the Robust Reading Competition portal.

The Challenge is hosted at the Robust Reading Competition (RRC) portal[1]. All submitted results are evaluated automatically, and per-task ranking tables and visualization options to explore results are offered through the portal. The results presented in this report reflect the state of submissions at the closure of the 2021 edition of the challenge, but the challenge will remain open for new, out-of-competition, submissions. The RRC portal should be considered as the archival version of results, where any new results, submitted after this report was compiled will also appear.

3 Infographics VQA

The design of the new task on Infographics VQA was informed by our analysis of results from the 2020 edition of the Single Document VQA task. In particular, we noticed that most state of the art methods for Single Document VQA followed a simple pipeline consisting on applying OCR (that was also provided) followed by purely text-based QA approaches. Although such approaches give good overall results as running text dominates the kind of documents of the Single Document VQA task, a closer look at the different question categories reveals that those methods performed quite worse on questions that rely on interpreting graphical information, handwritten text or layout information. In addition, these methods work because the Single Document VQA questions were designed so that they can be answered in an extractive manner, which means that the answer text appears in the document image and usually can be inferred from the surrounding context.

For the new Infographics VQA task, we focused on a domain where running text is not dominating the document, while we downplayed the amount of questions based on purely textual evidence and defined more questions which require the models to interpret the layout and other visual aspects. We also made sure that a good number of questions require logical reasoning or elementary arithmetical operations to arrive at the final answer.

We have received and evaluated a total of 337 different submissions on this task from 18 different users. Those submissions include different versions from the same methods. In Sect. 3.4 can be found the ranking of the 6 methods that finally participated in the competition.

---

[1] https://rrc.cvc.uab.es/?ch=17.

3.1   Evaluation Metric

To evaluate and rank the submitted methods on this task we use the standard ANLS [3] metric for reading-based VQA tasks. There exist a particular case in Infographics VQA task where the full answer is actually a set of items for which the order is not important, like a list of ingredients in a recipe. In the dataset description (Sect. 3.2) we name this as 'Multi-Span' answer. In this case we create all possible permutations of the different items and accept it as correct answers.

3.2   The InfographicsVQA Dataset

Infographic images were downloaded from various online sources using the Google and Bing image search engines. We removed duplicate images using a Perceptual Hashing technique implemented in Imagededup library [11]. To gather the question-answer pairs, instead of using online crowdsourcing platforms we used an internal web based annotation tool, hired annotators and worked closely with them through continuous interaction to ensure the annotations quality and balance the amount of questions for each type defined. In cases where there are multiple correct answers due to language variability, we collected more than one answer per question. For example in the case of the example shown in Fig. 1c the question has two valid ground-truth answers.

In total 30,035 question-answer pairs were annotated on 5,485 infographics. We split the data into train, validation and test splits in an 80-10-10 ratio, with no overlap of images between different splits. We also collected additional information regarding questions and answers in the validation and test splits which help us better analyze VQA performance. In particular, we collected the type of answer, which indicates whether the answer can be found as (i) a single text-span in the image ('image-span'), (ii) a concatenation of different text-spans in the image ('multi-span'), (iii) a span of the question text ('question span') or (iv) a 'non-span' answer. The evidence type for each answer is also annotated, and there could be multiple evidences for some answers. The different evidence types are Text, Table/list, Figure, Map (a geographical map) and Visual/layout. Additionally, for questions where a counting, arithmetic or sorting operation is required we collect the operation type as well. More details of the annotation process, statistics of the dataset, analysis of images, questions and answers and examples for each type of evidence, answer, and operation are presented in [19].

Along with the InfographicsVQA dataset, we also provided OCR transcriptions of each of the images in the dataset obtained using Amazon Textract OCR.

3.3   Baselines

In order to establish a baseline performance, we used two models built on state of the art methods. The first model is based on a layout aware language modelling approach—LayoutLM [33]. We take a pretrained LayoutLM model and continue pretraining it on the train split of the InfographicsVQA dataset using a Masked

Language Modelling task. Later we finetune this model for an extractive question answering using a span prediction head at the output. We name this baseline as "(Baseline) LayoutLM".

The second baseline is the state of the art Scene Text VQA method M4C [9]. This method embeds all different modalities – question, image text and image, into a common space, where a stack of transformer layers is applied allowing each entity to attend to inter- and intra- modality features providing them with context. Then, the answer is produced by an iterative decoder with a dynamic pointer network that can provide an answer either from a fixed vocabulary or from the OCR tokens spotted on the image. This baseline is named as "(Baseline) M4C". In [19] we give more details of these models, various ablations we try out and detailed results and analysis.

We also provide a human performance evaluation carried out by two volunteers "(Baseline) Human".

3.4   Submitted Methods

We received 6 submissions in total. All of them are based on pretrained language representations. However, we can appreciate that the top ranked methods use multi-modal pretrained architectures combining visual and textual information extracted from the image, while the lower ranked methods use representations based only on natural language.

#1 - Applica.ai TILT [24]: The winning method learns simultaneously layout information, visual features and textual semantics with a multi-modal transformer based method, and rely on an encoder-decoder architecture that can generate values that are not included in the input text.

#2 - IG-BERT (single model): A visual and language pretrained model on infographic text pairs. The model was initialized from BERT-large and trained on InfographicsVQA training and validation data. The visual features are extracted using a Faster-RCNN trained on Visually29K [18]. Also, they used OCR tokens extracted by the Google Vision API instead of the ones provided in the competition.

#3 - NAVER CLOVA: This method uses and extractive QA method based on the HyperDQA [21] approach. First they pre-train BROS [8] model on the IIT-CDIP [17] dataset but sharing the parameters between projection matrices during self-attention. Then, they perform additional pretraining on SQuAD [25] and WikitableQA [23] datasets. Finally, they also fine-tune on the DocVQA [21] dataset.

#4 - Ensemble LM and VLM: An ensemble between two different methods from extractive NLP QA method (Method 1) and scene-text VQA method (Method 2). The first one is based on BERT-large pretrained on SQuAD + DocVQA and fine-tuned on InfographicsVQA. The final output is vote-based by three trained models with different hyper-parameters. The second model is the SS-Baseline [36] trained on TextVQA [26], ST-VQA [4] and InfographicsVQA. At the end a rule-based post-processing is performed for three types of questions, i.e., selection, inverse percentage, and summation to get the final result by filling the empty results of Model 1 with answers predicted by Model 2.

#5 - bert baseline: BERT-large model pretrained on SQuAD [25]. It also uses a fuzzy search algorithm to better find the start and end index of the answer span.

#6 - BERT (CPDP): BERT-large model pretrained on SQuAD [25].

Table 1. Infographics VQA task results table. Top section show methods withing the ICDAR 2021 competition. Bottom section shows baseline methods proposed by the competition organizers.

Method	ANLS
(Baseline) Human	0.9800
Applica.ai TILT	**0.6120**
IG-BERT (single model)	0.3854
NAVER CLOVA	0.3219
Ensemble LM and VLM	0.2853
(Baseline) LayoutLM	0.2720
bert baseline	0.2078
BERT (CPDP)	0.1678
(Baseline) M4C	0.1470

3.5 Performance Analysis

In Table 1 the final competition ranking of the Infographics VQA task is shown. In addition, in Fig. 2 we show the methods' Average Normalized Levenshtein Similarity (ANLS) score breakdown by type of answer (top), type of evidence (middle) and type of operations and reasoning required to answer the posed questions (bottom). As it can be seen, the winning method Applica.ai TILT outperforms the rest in all the categories. In addition, the plots show a drop in performance of all methods when the text of the answer is a set of different text spans in the image (Multi-Span) or it does not appear in the image (Non-Span). The easiest questions are the ones that can be directly answered with running text from the documents (Fig. 2-middle: Text), this should be expected given the extractive QA nature of most of the methods. Finally, the hardest ones are all the questions that require to perform an operation to come up with the answer as the last plot illustrates comparing the Counting, Sorting and Arithmetic against the performance on None.

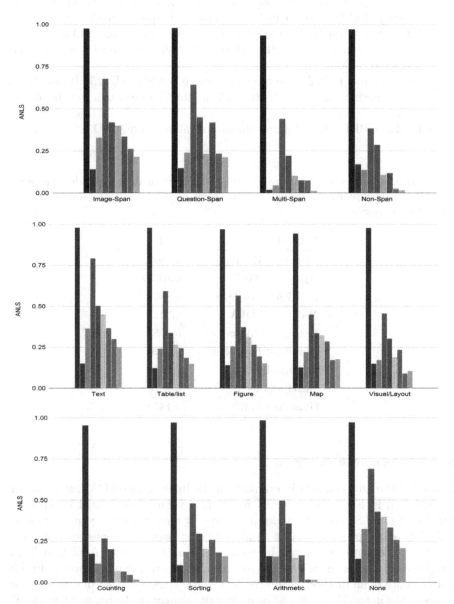

• (Baseline) Human ▪ (Baseline) M4C ▪ (Baseline) LayoutLM ▪ Applica.ai TILT ▪ IG BERT (single model)
▪ NAVER CLOVA ▪ Ensemble LM and VLM ▪ bert base ▪ BERT

Fig. 2. ANLS score of Infographics VQA task methods breakdown by answer types (top), evidence type (middle) and operations required to answer the posed questions (bottom).

4   Document Collection VQA

In Document Collection VQA task, the questions are posed over a whole collection of 14K document images. The objective in this task is to come up with the answers to the given questions, but also with the positive evidences. We consider positive evidence the document IDs from which the answer can be inferred from.

This task was first introduced in the scope of the CVPR 2020 DocVQA Challenge edition, and we have received and evaluated 60 submissions from 7 different users, out of which 5 have been made public by their authors and feature in the ranking table. 3 of these submissions are new, tackling the problem of answering the questions and not only providing the evidences.

4.1   Evaluation Metric

In this task, the whole answer to a given question is usually a set of sub-answers extracted from different documents, and consequently the order in which those itemized sub-answers are provided is not relevant. Thus, it presents a problem similar to the Multi-Span answers in Infographics VQA. However, the amount of items in this task varies from 1 to >200 depending on the question, and therefore is not smart to just create all possible permutations. For this reason the metric used to assess the answering performance in this task is the Average Normalized Levenshtein Similarity for Lists (ANLSL) presented in [28], an adaptation of the ANLS [3], to work with a set of answers for which the order is not important. Thus, captures smoothly OCR recognition errors and evaluates reasoning capability at the same time while handles unordered set of answers by performing the Hungarian Matching algorithm between the method's provided answer and the ground truth.

In addition, even though the retrieval of the positive evidences is not used to rank the methods, we evaluate them according to the Mean Average Precision (MAP) which allows to better analyze the method's performance.

4.2   Baselines

We defined two baseline methods to establish the base performance for this task. Both baselines work in two steps. First they rank the documents in the collection to find which ones are relevant to answer the given question and then, they extract the answer from those documents marked as relevant. The first method "(Baseline) TS - BERT" ranks the documents by performing text spotting of specific words in the question over the documents in the collection. Then using an extractive QA pretrained BERT model extracts the answers from the top ranked documents. The second method named "(Baseline) Database", makes use of the commercial OCR Amazon Textract to extract the key-value pairs from the form's fields for all the documents in the collection. With this information a database-like structure is built. Then, the questions are manually parsed into Structured Query Language (SQL) from which the relevant documents and answers are retrieved. Note that while the first baseline is generic in nature and could potentially be applied on different collections, the second one relies on the nature of the documents (forms). Detailed information for both methods can be found in [28].

4.3   Submitted Methods

In this task, only one new method has submitted. The winning method Infrrd-RADAR (Retrieval of Answers by Document Analysis and Re-ranking) first applies OCR to extract textual information from all document images and the combines results with image information to extract key information. The question is parsed into SQL queries by using the spaCy library to split and categorize the question chunks into predefined categories. The SQL queries are then used to retrieve a set of relevant documents and BERT-Large is used to re-rank them again. Afterwards, the re-ranked document IDs are used to filter the extracted information and finally, based on the parsed questions a particular field is collected and posted as an answer.

4.4   Performance Analysis

Table 2 shows the competition result ranking comparing the submitted method in this challenge with the baselines and the methods from the CVPR 2020 challenge version. The Infrrd-RADAR method outperforms all baselines in the ranking metric (ANLSL). However, methods from the 2020 edition show better performance on the retrieval of positive evidence. This indicates a potential for improvement since given the collection nature of this dataset, a better performance in finding the relevant documents is expected to lead to better answering performance. In Fig. 3 we show a breakdown of the evidence and answer scores by query. On one hand there is a set of questions for which most of the methods performs well (Q10, Q15, Q16) while Infrrd-RADAR is the only one to come up with the answers for the questions Q11 and Q13 and providing their evidences. However, it performs worse than Database in questions Q8, Q9 and Q18, which is probably a consequence of the performance drop when finding the relevant documents as it can be seen in the top plot. Notice also that some questions refer to only one single document (Q13, Q15, Q16, Q19), which facilitates methods to score either 0 or 1.

Table 2. Document Collection VQA task results table. New submissions after CVPR challenge edition [21] are indicated by [†] icon.

Method	Answering ANLSL	Retrieval MAP
[†]Infrrd-RADAR	**0.7743**	74.66%
(Baseline) Database [28]	0.7068	71.06%
(Baseline) TS-BERT [28]	0.4513	72.84%
DQA [21]	0.0000	**80.90%**
DOCR [21]	0.0000	79.15%

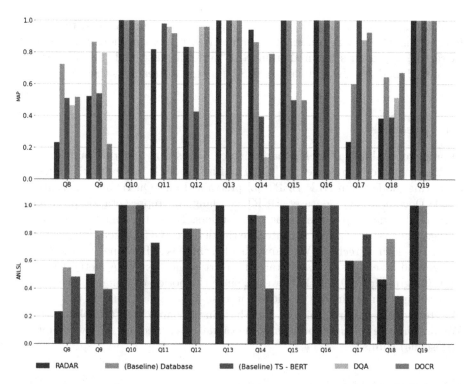

Fig. 3. Document Collection VQA methods performance by questions. Top figure shows the retrieval MAP score. Bottom figure shows ANLSL answering score. In bottom figure methods which didn't provide answers are omitted for clarity.

5 Single Document VQA

The Single Document VQA task was first introduced in the scope of the CVPR 2020 DocVQA Challenge edition. Since then, we have received and evaluated >750 submissions from 44 different users, out of which 28 have been made public by their authors and feature in the ranking table. 21 of these submissions are new, and have pushed overall performance up by ~2%. In this task questions were posed over single page document images and the answer to the questions was in most of the cases text that appears within the image. The documents of this dataset were sourced from the industry documents library[2] and the dataset finally comprised 30, 035 question-answer pairs and 5, 485 images.

---

[2] https://www.industrydocuments.ucsf.edu/.

Although the documents feature complex layouts and include tables and figures, running text is dominant. Combined with way questions were defined, extractive approaches inspired from NLP perform quite well. To evaluate the submitted methods the standard ANLS [3] metric for reading-based VQA tasks was used.

5.1   Baselines

In order to establish a baseline performance, we used two models built on state of the art methods presented in [20]. The first model is the extractive question answering model from NLP BERT [6] pretrained on SQuAD and finetuned on Single Document VQA dataset. BERT is a language representation pre-trained from unlabeled text passages using transformers. To get the context necessary to extract the answer span we concatenate the detected OCR tokens following left-right, top-down direction. We name this baseline as "(Baseline) BERT Large" The second baseline is the same state of the art Scene Text VQA method M4C [9] as in Infographics VQA task. However, since the notion of visual objects in real word images is not directly applicable in case of document images, the features of detected objects are omitted. This baseline is named as "(Baseline) M4C". We also provide a human performance evaluation carried out by few volunteers "(Baseline) Human".

5.2   Submitted Methods

Here we briefly describe the current top 3 ranked methods (new submissions) while the previous edition's methods description can be found in [21]. The complete ranking can be found in Table 3.

#1 - Applica.ai TILT [24]: The same team and method as the winners of the Infographics VQA task.

#2 - LayoutLM 2.0 [32]: Pretrained representation that also learns the text, layout and image in a multi-modal framework using new pretraining tasks where cross-modality interaction are better learned than in previous LayoutLM [33]. It also integrates a spatial-aware mechanism that allows the model to better exploit the relative positional relationship among different text blocks.

#3 - Alibaba DAMO NLP: An ensemble of 30 models and a pretrained architecture based on StructBERT [31].

Table 3. Single Document VQA task results table. New submissions after CVPR challenge edition [21] are indicated by [†] icon.

Method	ANLS
(Baseline) Human	0.9811
[†]Applica.ai TILT	**0.8705**
[†]LayoutLM 2.0 (single model)	0.8672
[†]Alibaba DAMO NLP	0.8506
PingAn-OneConnect-Gammalab-DQA	0.8484
Structural LM-v2	0.7674
[†]QA_Base_MRC_2	0.7415
QA_Base_MRC_1	0.7407
HyperDQA_V4	0.6893
(Baseline) Bert Large	0.6650
Bert fulldata fintuned	0.5900
[†]UGLIFT v0.1 (Clova OCR)	0.4417
(Baseline) M4C	0.3910
Plain BERT QA	0.3524
HDNet	0.3401
CLOVA OCR	0.3296
DocVQAQV_V0.1	0.3016

6   Conclusions and Future Work

The ICDAR 2021 edition of the DocVQA challenge is the continuation of a long-term effort towards Document Visual Question Answering. One year after we first introduced it, the DocVQA Challenge has received significant interest by the community. The challenge has evolved, by improving evaluation and analysis methods and by introducing a new complex task on Infographics VQA. In this report, we have summarised these changes, and presented new submissions to the different tasks. Importantly, we note that while the standard approach one year ago was a pipeline combining commercial OCR with NLP-inspired QA methods, state of the art approaches in 2021 are multi-modal. This confirms that the visual aspect of documents is indeed important, even in more mundane layouts such as the documents of Single Document VQA. This is especially true when analysing specific types of questions. The DocVQA challenge will remain open for new submissions in the future. We plan to extend this further with new tasks and insights in the results. We also hope that the VQA paradigm will give rise to more top-down, semantically-driven approaches to document image analysis.

Acknowledgments. This work was supported by an AWS Machine Learning Research Award, the CERCA Programme/Generalitat de Catalunya, and UAB PhD scholarship No B18P0070. We thank especially Dr. R. Manmatha for many useful inputs and discussions.

References

1. Agrawal, A., et al.: VQA: Visual Question Answering (2016)
2. Anderson, P., et al.: Bottom-up and top-down attention for image captioning and visual question answering (2017)
3. Biten, A.F., et al.: ICDAR 2019 competition on scene text visual question answering. In: 2019 International Conference on Document Analysis and Recognition (ICDAR), pp. 1563–1570. IEEE (2019)
4. Biten, A.F., et al.: Scene text visual question answering. In: Proceedings of the IEEE/CVF International Conference on Computer Vision, pp. 4291–4301 (2019)
5. Chaudhry, R., Shekhar, S., Gupta, U., Maneriker, P., Bansal, P., Joshi, A.: Leaf-QA: locate, encode attend for figure question answering. In: WACV (2020)
6. Devlin, J., Chang, M.W., Lee, K., Toutanova, K.: BERT: pre-training of deep bidirectional transformers for language understanding. In: ACL (2019)
7. Dua, D., Wang, Y., Dasigi, P., Stanovsky, G., Singh, S., Gardner, M.: DROP: a reading comprehension benchmark requiring discrete reasoning over paragraphs. In: NAACL-HLT (2019)
8. Hong, T., Kim, D., Ji, M., Hwang, W., Nam, D., Park, S.: Bros: a pre-trained language model for understanding texts in document (2021)
9. Hu, R., Singh, A., Darrell, T., Rohrbach, M.: Iterative answer prediction with pointer-augmented multimodal transformers for TextVQA. In: Proceedings of the IEEE/CVF Conference on Computer Vision and Pattern Recognition (2020)
10. Hudson, D.A., Manning, C.D.: GQA: a new dataset for compositional question answering over real-world images. CoRR abs/1902.09506 (2019). http://arxiv.org/abs/1902.09506
11. Jain, T., Lennan, C., John, Z., Tran, D.: Imagededup (2019). https://github.com/idealo/imagededup
12. Joshi, M., Choi, E., Weld, D., Zettlemoyer, L.: TriviaQA: a large scale distantly supervised challenge dataset for reading comprehension. In: ACL (2017)
13. Kafle, K., Price, B., Cohen, S., Kanan, C.: DVQA: understanding data visualizations via question answering. In: CVPR (2018)
14. Kahou, S.E., Michalski, V., Atkinson, A., Kádár, Á., Trischler, A., Bengio, Y.: FigureQA: an annotated figure dataset for visual reasoning. arXiv preprint arXiv:1710.07300 (2017)
15. Kembhavi, A., Seo, M., Schwenk, D., Choi, J., Farhadi, A., Hajishirzi, H.: Are you smarter than a sixth grader? Textbook question answering for multimodal machine comprehension. In: CVPR (2017)
16. Kwiatkowski, T., et al.: Natural questions: a benchmark for question answering research. Transactions of the Association of Computational Linguistics (2019)
17. Lewis, D., Agam, G., Argamon, S., Frieder, O., Grossman, D., Heard, J.: Building a test collection for complex document information processing. In: Proceedings of the 29th Annual International ACM SIGIR Conference on Research and Development in Information Retrieval, pp. 665–666 (2006)
18. Madan, S., et al.: Synthetically trained icon proposals for parsing and summarizing infographics. arXiv preprint arXiv:1807.10441 (2018)

19. Mathew, M., Bagal, V., Tito, R.P., Karatzas, D., Valveny, E., Jawahar, C.: InfographicVQA. arXiv preprint arXiv:2104.12756 (2021)
20. Mathew, M., Karatzas, D., Jawahar, C.V.: DocVQA: a dataset for VQA on document images. In: WACV (2020)
21. Mathew, M., Tito, R., Karatzas, D., Manmatha, R., Jawahar, C.: Document visual question answering challenge 2020. arXiv preprint arXiv:2008.08899 (2020)
22. Nguyen, T., et al.: MS MARCO: a human generated machine reading comprehension dataset. CoRR abs/1611.09268 (2016)
23. Pasupat, P., Liang, P.: Compositional semantic parsing on semi-structured tables. In: Proceedings of the 53rd Annual Meeting of the Association for Computational Linguistics and the 7th International Joint Conference on Natural Language Processing (Volume 1: Long Papers), pp. 1470–1480 (2015)
24. Powalski, R., Borchmann, Ł., Jurkiewicz, D., Dwojak, T., Pietruszka, M., Pałka, G.: Going full-tilt boogie on document understanding with text-image-layout transformer. arXiv preprint arXiv:2102.09550 (2021)
25. Rajpurkar, P., Zhang, J., Lopyrev, K., Liang, P.: Squad: 100,000+ questions for machine comprehension of text. In: Proceedings of the 2016 Conference on Empirical Methods in Natural Language Processing, pp. 2383–2392 (2016)
26. Singh, A., et al.: Towards VQA models that can read. In: Proceedings of the IEEE/CVF CVPR, pp. 8317–8326 (2019)
27. Teney, D., Anderson, P., He, X., van den Hengel, A.: Tips and tricks for visual question answering: learnings from the 2017 challenge (2017)
28. Tito, R., Karatzas, D., Valveny, E.: Document collection visual question answering. arXiv preprint arXiv:2104.14336 (2021)
29. Vaswani, A., et al.: Attention is all you need. In: Proceedings of the 31st International Conference on NeurIPSal Information Processing Systems, pp. 6000–6010 (2017)
30. Veit, A., Matera, T., Neumann, L., Matas, J., Belongie, S.: Coco-text: dataset and benchmark for text detection and recognition in natural images (2016)
31. Wang, W., et al.: StructBERT: incorporating language structures into pre-training for deep language understanding. arXiv preprint arXiv:1908.04577 (2019)
32. Xu, Y., et al.: LayoutLMv2: multi-modal pre-training for visually-rich document understanding. arXiv preprint arXiv:2012.14740 (2020)
33. Xu, Y., Li, M., Cui, L., Huang, S., Wei, F., Zhou, M.: LayoutLM: pre-training of text and layout for document image understanding. In: Proceedings of the 26th ACM SIGKDD International Conference on Knowledge Discovery & Data Mining, pp. 1192–1200 (2020)
34. Yagcioglu, S., Erdem, A., Erdem, E., Ikizler-Cinbis, N.: RecipeQA: a challenge dataset for multimodal comprehension of cooking recipes. In: EMNLP (2018)
35. Yang, Z., Dai, Z., Yang, Y., Carbonell, J., Salakhutdinov, R.R., Le, Q.V.: XLNet: generalized autoregressive pretraining for language understanding. In: NeurIPS (2019)
36. Zhu, Q., Gao, C., Wang, P., Wu, Q.: Simple is not easy: a simple strong baseline for TextVQA and TextCaps. arXiv preprint arXiv:2012.05153 (2020)

ICDAR 2021 Competition on Scene Video Text Spotting

Zhanzhan Cheng[1,2(✉)], Jing Lu[2], Baorui Zou[3], Shuigeng Zhou[3], and Fei Wu[1]

[1] Zhejiang University, Hangzhou, China
{11821104,wufei}@zju.edu.cn
[2] Hikvision Research Institute, Hangzhou, China
lujing6@hikvision.com
[3] Fudan University, Shanghai, China
{18210240270,sgzhou}@fudan.edu.cn

Abstract. Scene video text spotting (SVTS) is a very important research topic because of many real-life applications. However, only a little effort has been put to spotting scene video text, in contrast to massive studies of scene text spotting in static images. Due to various environmental interferences like motion blur, spotting scene video text becomes very challenging. To promote this research area, this competition introduces a new challenge dataset containing 129 video clips from 21 natural scenarios in full annotations. The competition contains three tasks, that is, video text detection (Task 1), video text tracking (Task 2) and end-to-end video text spotting (Task3). During the competition period (opened on 1st March, 2021 and closed on 11th April, 2021), a total of 24 teams participated in the three proposed tasks with 46 valid submissions, respectively. This paper includes dataset descriptions, task definitions, evaluation protocols and results summaries of the ICDAR 2021 on SVTS competition. Thanks to the healthy number of teams as well as submissions, we consider that the SVTS competition has been successfully held, drawing much attention from the community and promoting the field research and its development.

Keywords: Scene video text spotting · Video text detection · Video text tracking · End-to-End

1 Introduction

Scene video text spotting (SVTS) is a text spotting system for localizing and recognizing text from video flowing, which usually contains multiple modules: video text detection, text tracking and the final recognition. SVTS has become an important research topic due to many real-world applications, including license plate recognition in intelligent transportation system, road sign recognition in

---

Z. Cheng, J. Lu and B. Zou—Contributed equally to this competition.

J. Lladós et al. (Eds.): ICDAR 2021, LNCS 12824, pp. 650–662, 2021.
https://doi.org/10.1007/978-3-030-86337-1_43

advanced driver assistance system or even online handwritten character recognition, to name a few. With the rapid development of deep learning techniques, great progress has been made in scene text spotting from static images. However, spotting text from video streams faces more serious challenges than the static OCR tasks in applications. Concretely, SVTS has to cope with various environmental interferences (e.g., camera shaking, motion blur and immediate illumination changing etc.) and meet the real-time response requirement. Therefore, it is necessary to develop efficient and robust SVTS systems for practical applications.

In recent years, only a little effort has been put to spotting scene video text, in contrast to massive studies of text reading in static images. And the studies of SVTS obviously fall behind its increasing applications. This is mainly due to: (1) Though 'Text in Videos' challenges have been recognized since 2013, the dataset is too small (containing only 49 videos from 7 different scenarios), which constrains the research on SVTS. (2) The lack of uniform evaluation metrics and benchmarks, as described in the literature [3,4]. For example, many methods only evaluate their localization performance on YVT and 'Text in Videos' [13,14], but few methods pay attention to the end-to-end evaluation.

Considering the importance of SVTS and the challenges it faces, we propose the ICDAR 2021 competition on SVTS, aiming to draw attention on this problem from the community and promote its research and development. The proposed competition could be of interest to the ICDAR community from two main aspects:

- Inherited from LSVTD [3,4], the video text dataset is further extended, containing 129 video clips from 21 real-life scenarios. Compared to the existing ICDAR video text reading datasets, the extended dataset has some special features and challenges. (1) More accurate annotations compared to existing video text datasets. (2) A general dataset with a large range of scenarios, which is collected with different kinds of video cameras: mobile phone cameras in various indoor scenarios (*e.g.*, bookstore and office building) and outdoor street views; HD cameras in traffic and harbor surveillance; and Car-DVR cameras in fast-moving outdoor scenarios (*e.g.*, city road, highway). (3) Some video clips are overwhelming of low-quality images caused by blurring, perspective distortion, rotation, poor illumination or motion inferences (*e.g.* object/camera moving or shaking). To address the potential privacy issue, some sensitive fields (such as person face and vehicle plate license etc.) of the video frames are blurred. The datasets can be an effective complement to the existing ICDAR datasets.
- Three specific tasks are proposed: video text detection, tracking and the end-to-end recognition. Comprehensive evaluation metrics are used for the three competition tasks, i.e., $Recall_d$, $Precision_d$ and $F\text{-}score_d$ [3] used for detection, ATA_t, $MOTA_t$, $MOTP_t$ used for tracking, both sequence-level metrics [3] like $Recall_e$, $Precision_e$, $F\text{-}score_e$ and the traditional metrics like ATA_t, $MOTA_t$, $MOTP_t$ used for the end-to-end evaluation. In combination with the extended dataset, it enables wide development, evaluation and enhancement of video

text detection, tracking and end-to-end recognition technologies for SVTS. It will help attract wide interests (expected to exceed 50 submits) on SVTS, inspire new insights, ideas and approaches.

The competition opened on 1st March, 2021 and closed on 11th April, 2021. There are a total of 24 teams participated in the three proposed tasks with 22, 13, 11 valid submissions, respectively. This competition report provides the motivation, dataset description, task definition, evaluation metrics, results of submitted methods and their discussion. Considering a large number of teams and submissions, we think that the ICDAR 2021 competition on SVTS is successfully held. We hope that the competition draws more attention from the community and further promote the field research and its development.

2   Competition Organization

ICDAR 2021 competition on SVTS is organized by a joint team of Zhejiang University, Hikvision Research Institute and Fudan University. The competition make use of Codalab web[1] portal to maintain information of the competition, download links for the datasets, and user interfaces for participants to register and submit their results. The schedule of the SVTS competition is as follows:

- 5 January 2021: Registration is started for competition participants. Training and validation datasets are available for downloads.
- 1 March 2021: Submissions of all the tasks are open for participants. Test data is released (without ground-truth).
- 31 March 2021: Registration deadline for competition participants.
- 11 April 2021: Submissions deadline of all the tasks for participants.

Overall, we received 46 valid submissions from 24 teams from both research communities and industries for the three tasks. Note that duplicate submissions are removed.

3   Dataset

The dataset has 129 video clips (ranging from several seconds to over 1 min long) from 21 real-life scenes. It was extended on the basis of LSVTD dataset [2] by adding 15 videos for 'harbor surveillance' scenario and 14 videos for 'train watch' scenario, for the purpose of addressing video text spotting problem in industrial applications (Fig. 1).

Characteristics of the dataset are as follows.

- Large scale and diversified scenes. Videos are collected from 21 different scenes, larger than most existing scene video text datasets. It contains 13 indoor scenes (reading books, digital screen, indoor shopping mall, inside shops, supermarket, metro station, restaurant, office building, hotel

---

[1] https://competitions.codalab.org/competitions/27667.

Fig. 1. Examples of SVTS for the competition tasks.

bus/railway station, bookstore, inside train and shopping bags) and 8 outdoor scenes (outdoor shopping mall, pedestrian, fingerposts, street view, train watch, city road, harbor surveillance and highway).

- Videos are collected with different kinds of video cameras: mobile phone cameras in various indoor scenarios (e.g. bookstore and office building) and outdoor street views, HD cameras in traffic and harbor surveillance, and Car-DVR cameras in fast-moving outdoor scenarios (e.g. city road, highway).
- Different difficulty levels. Hard: videos are overwhelmed by low-quality texts(e.g., blurring, perspective distortion, rotation, poor illumination or even with motion inferences like object/camera moving or shaking). Medium: some of the text regions are of low-quality while others are not interfered by artifacts. Easy: only a few text regions are polluted in these videos.
- Multilingual instances: alphanumeric and non-alphanumeric.

Dataset Split. The dataset[2] is divided into training set, validation set and testing set, in which separately contains 71, 18 and 40 videos. The train set contains at least one video from each scenario.

Annotations. The annotation strategy is same to LSVTD [4]. For each text region, the annotation items is as follows: (1) Polygon coordinate represents text location. (2) ID means the unique identification for each text among consecutive frames, i.e., the same text in consecutive frames shares the same ID. (3) Language is categorized as Latin and Non-Latin. (4) Quality coarsely indicates the quality level of each text region, which can be qualitatively labeled as three quality levels: 'high' (recognizable, clear and without interferences), 'moderate' (recognizable but polluted with one or several interferences) or 'low' (one or more characters are unrecognizable). (5) Transcripts mean text string for each text region. We parsed videos (ranging from 5 s to 1 min) to frames and then instructed 6 experienced annotation workers to label them, and conducted cross-checking on each text region.

4   Tasks

The competition has three tasks: the video text detection, the video text tracking and the end-to-end video text spotting, in which only 'alphanumeric text instances' are considered to be evaluated by tools[3]. In the future, we intend to release the more challenging multilingual SVTS competition.

4.1   Task 1-Video Text Detection

The task is to obtain the locations of text instances in each frame in terms of their affine bounding boxes. Results are evaluated based on the Intersection-over-Union (IoU) with a threshold of 0.5, which is similar to the standard metrics in general object detection like the Pascal VOC challenge. Here, $Recall_d$, $Precision_d$ and $F\text{-score}_d$ are used as the evaluation metrics. The participants are required to prepare a JSON file (named as 'detection_predict.json') containing detection results of all test videos, and then compress and name it as 'answer.zip' to upload it. The JSON file is illustrated as Fig. 2(a).

---

[2] https://competitions.codalab.org/competitions/27667#learn_the_details-datasets.
[3] https://competitions.codalab.org/competitions/27667#learn_the_details.

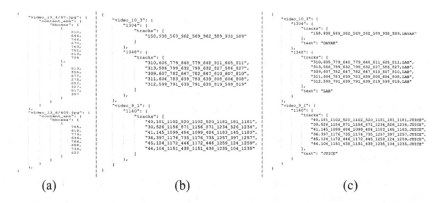

(a)   (b)   (c)

Fig. 2. Illustration of the submitted JSON files in the three tasks. For each task, all results are saved in a dict. (a) JSON format of video text detection. The field 'bboxes' correspond to all the detected bounding boxes in each frame. Each bounding box is represented by 4 points in the clock-wise order. (b) JSON format of video text tracking, which contains all tracked text sequences in the video. That is, each sequence is represented with a unique ID like '1304' in 'Video_10_3', and the field 'tracks' indicates the tracked bounding boxes in the text sequence. (c) JSON format of end-to-end video text spotting, in which a 'recognition' word is appended to its bounding box.

4.2   Task 2-Video Text Tracking

This task intends to track all text streams from testing videos. Following [14], ATA_t, $MOTA_t$ and $MOTP_t$ are used as the evaluation metrics. Note that ATA is selected the main metric because ATA measures the tracking performance over all the text instances. Similar to Task 1, the participants are required to prepare a JSON file (named as 'track_predict.json') containing predictions of all test videos, and then compress and name it as 'answer.zip' to upload it. The JSON file is illustrated as Fig. 2 (b).

4.3   Task 3-End-to-End Video Text Spotting

This task aims to evaluate the performance of end-to-end video text spotting. It requires that words should be correctly localized, tracked and recognized simultaneously. Concretely, a predicted word is considered as a true positive if and only if its IoU with a ground-truth word is larger than 0.5, and its recognition result is correct at the same time.

In general, ATA_e, $MOTA_e$ and $MOTP_e$ can be used as the traditional evaluation metrics according to the recognition results. However, in many real-world applications, the sequence-level spotting results are the most urgently needed for users, while it's not what the user cares about for the framewise recognition results. Therefore, we propose the sequence-level evaluation protocols to evaluate the end-to-end performance, i.e., $Recall_s$, $Precision_s$, $F\text{-}score_s$ as used in [4].

Here, a predicted text sequence is regarded as a true positive if and only if it satisfies two constraints:

– The spatial-temporal localisation constraint. The temporal locations of text regions should fall into the interval between the annotated starting and ending frame. In addition, the given candidate should have a spatial overlap ratio (over 0.5) with its annotated bounding box.
– The recognition constraint: for text sequences satisfying the first constraint, their predicted results should match the corresponding ground truth text transcription.

In order to perform the evaluation, the participants are required to submit a JSON file containing all the predictions for all test videos. The JSON file should be named as 'e2e_predict.json', and then is submitted by compressing it as 'answer.zip'. The JSON file is illustrated as Fig. 2 (c).

Some other details that the participants need to pay attention: (1) Text areas annotated as 'LOW' or '###' will not be taken into account for evaluation. (2) Words with less than 3 characters are not taken into account for evaluation. (3) Word recognition evaluation is case-insensitive. (4) All the symbols contained in recognition results should be removed before submitted. (5) The sequence level recognition results should be COMPLETE words.

5   Submissions

The submission results are as shown in Table 1, 2 and 3.

5.1   Top 3 Submissions in Task 1

Tencent-OCR team propose the multi-stage text detector by following Cascade Mask R-CNN [2] equipped with multiple backbones like HRNet-W48 [17], Res2Net101 [8], ResNet101 [11], and SENet101 [12]. Combined with polygon-NMS [16], the segmentation branch is used to obtain multi-oriented text instances. Besides, a CTC [9]-based recognition branch is incorporated into the RCNN stage. The whole model is trained in an end-to-end learning pipeline. To better cope with the challenge of diverse scenes, they employ two approaches in the training phase: (1) Some strong data augmentation strategies are adopted like photometric distortions, random motion blur, random rotation, random crop, and random horizontal flip. (2) Various open-source dataset, e.g., IC13, IC15, IC15 Video, the Latin part of MLT19, COCO-Text, and Synth800k, are involved in the training phase. In the inference phase, they make inference by considering to multiple resolutions of 600, 800, 1000, 1333, 1666 and 2000. Considering that the image/video quality significantly affects the performance, they design a multi-quality TTA (test time augmentation) approach. With the aid of recognition and tracking results, some detected boxes are removed to achieve higher precision. Finally, four models with distinct backbones are leveraged by ensembling.

Table 1. Video text detection results and rankings after removing the duplicate submissions in task 1. Note that * denotes missing descriptions in affiliations.

User ID	Rank	F-score$_d$	Precision$_d$	Recall$_d$	Affiliations
tianqihenhao	1	0.8502	0.8561	0.8444	TEG, Tencent
wfeng	2	0.8159	0.8787	0.7615	IA, CAS
DXM-DI-AI-CV-TEAM	3	0.7665	0.8253	0.7155	DuXiaoman Financial
tangyejun	4	0.7582	0.8088	0.7136	*
wangsibo	5	0.7522	0.8377	0.6825	*
weijiawu	6	0.7298	0.7508	0.7098	Zhejiang University
yeah0110	7	0.7276	0.7314	0.7238	*
BOE_AIoT_CTO	8	0.7181	0.7133	0.7229	BOE
colorr	9	0.7172	0.7101	0.7245	*
qqqyd	10	0.7140	0.7045	0.7238	*
yucheng3	11	0.6749	0.8622	0.5544	University of Chinese Academy of Sciences
superboy	12	0.6704	0.8336	0.5607	*
seunghyun	13	0.6219	0.6897	0.5663	NAVER corp
hanquan	14	0.5881	0.6252	0.5552	*
sabrina_lx	15	0.5691	0.5463	0.5939	*
gywy	16	0.5640	0.5432	0.5865	*
gogogo_first	17	0.5390	0.6521	0.4594	*
SkyRanch	18	0.4766	0.4432	0.5153	NUCTech
Steven2045	19	0.4648	0.3614	0.6514	*
enderloong	20	0.3968	0.3034	0.5732	*
wuql	21	0.3941	0.3039	0.5607	*
AlphaNext	22	0.3373	0.2323	0.6158	*

Table 2. Video text tracking results and rankings after removing the duplicate submissions in task 2. Note that * denotes missing descriptions in affiliations.

User ID	Rank	ATA$_t$	MOTA$_t$	MOTP$_t$	Affiliations
tianqihenhao	1	0.5372	0.7642	0.8286	TEG, Tencent
DXM-DI-AI-CV-TEAM	2	0.4810	0.6021	0.8017	DuXiaoman Financial
panda12	3	0.4636	0.7009	0.8277	IA, CAS
lzneu	4	0.3812	0.5647	0.8198	*
wangsibo	5	0.3778	0.5657	0.8200	*
yucheng3	6	0.3116	0.5605	0.8203	University of Chinese Academy of Sciences
tangyejun	7	0.2998	0.5027	0.8196	*
yeah0110	8	0.2915	0.4811	0.8218	*
sabrina_lx	9	0.2436	0.3757	0.7667	*
seunghyun	10	0.1415	0.2183	0.6949	NAVER corp
enderloong	11	0.0918	0.1820	0.7520	*
tiendv	12	0.0676	0.2155	0.7439	University of Information Technology
weijiawu	13	0.0186	0.1530	0.7454	Zhejiang University

Table 3. End-to-end video text spotting results and rankings after removing the duplicate submissions in task 3. Note that * denotes missing descriptions in affiliations.

User ID	Rank	F-score$_s$	Precision$_s$	Recall$_s$	ATA$_s$	MOTA$_s$	MOTP$_s$	Affiliations
tianqihenhao	1	0.5308	0.6655	0.4414	0.4549	0.5913	0.8421	TEG, Tencent
DXM-DI-AI-CV-TEAM	2	0.4755	0.6435	0.3770	0.4188	0.4960	0.8142	DuXiaoman Financial
panda12	3	0.4183	0.5243	0.3479	0.3579	0.5179	0.8427	IA, CAS
lzneu09	4	0.3007	0.3611	0.2576	0.2737	0.4255	0.8330	Northeastern University
yucheng3	5	0.2964	0.3506	0.2567	0.2711	0.4246	0.8332	University of Chinese Academy of Sciences
tangyejun	6	0.2284	0.2527	0.2084	0.2121	0.3676	0.8337	*
tiendv	7	0.0813	0.1402	0.0572	0.0802	0.0887	0.7976	University of Information Technology
enderloong	8	0.0307	0.0239	0.0429	0.0357	0.0159	0.7813	*
colorr	9	0.0158	0.0085	0.1225	0.0146	0.0765	0.8498	*
weijiawu3	10	0.0077	0.0041	0.0550	0.0088	-0.1530	0.7670	Zhejiang University
BOE_AIoT_CTO	11	0.0000	0.0000	0.0000	0.0000	-0.0003	0.0000	BOE

CASIA_NLPR_PAL Team propose the semantic-aware video text detector, which is an end-to-end trainable video text detector(SAVTD [7]) performing detection and tracking at the same time. In the video text detection task, they adopt Mask R-CNN [10] to predict axis-aligned rectangular bounding boxes and the corresponding instance segmentation masks, and then fit a minimum enclosing rotated rectangle to each mask for oriented texts.

DXM-DI-AI-CV-TEAM use ResNet50 as the backbone. The input scale is 640*640(random crop) and the decoder is Upsample+Conv. Regarding the strategy, the detection results are thoroughly examined by the recognition module in the first place. Secondly, a text classifier is trained by utilizing the data from ICDAR2019-LSVT in order to detect and identify the non-text results in the text boxes. The final detection results are then obtained.

5.2   Top 3 Submissions in Task 2

Tencent-OCR team propose the multi-metric text tracking method, which uses 4 different metrics to compare the matching similarity between the current frame detection boxes and the existing text trajectories, i.e., box IoU, text content similarity, box size similarity, and text-geometry neighbor relationship metric. The weighted sum of these matching confidence scores are employed as a matching cost between the currently detected box and a tracklet. Starting from the first frame, they construct a cost matrix for detected boxes in each frame and existing tracklets. They utilize the Kuhn-Munkres algorithm to obtain matching pairs, and then a grid search is executed to find better parameters. Each box that is not linked with existing tracklets is regarded as a new trajectory. They also design a post-processing strategy to reduce ID switches by considering both text regions and recognition results. Finally, low-quality trajectories with low text confidence are removed.

DXM-DI-AI-CV-TEAM tackle this task by using a two-fold matching strategy. The ECC [6] algorithm is also utilized to estimate the motion between two video frames and to calculate the estimated bbox for each trajectory. On the first stage of matching, ResNet34 is employed to extract appearance feature from detection results and estimated bboxes. Next, according to the cosine distance between each feature, a cost matrix is calculated. The matching is then completed in a cascade manner similar to deepsort. On the second stage, the overlap ratio between estimated bbox and unmatched bounding boxes are used for matching. Strategically, the trajectories with length less than 4 are removed in order to improve the precision on the sequence level.

CASIA_NLPR_PAL Team handles this task via their semantic-aware video text detector. Instead of performing text tracking with appearance features extracted from text RoIs directly, they use two fully connected layers to project the roi-features into new ones to get descriptor for each instance, and then matching current frame instance descriptors and previous frame instance descriptors to get current frame tracking identities. In addition to using a new end-to-end trainable method, they also use following strategies to improve the performance. First, in order to train a powerful model, they combine the train set videos and validation set videos for training, in which the base model is pre-training on scene text datasets, such as ArT, MLT. Second, since videos in the train set containing different scenes and cover several size ranges, in order to improve the robustness, they also adopt deformable convolution [5]. In the end, they use ResNet-DCN 50 with multi-scale train/test as the final model.

5.3   Top 3 Submissions in Task 3

Tencent-OCR team develop a method named as Convolutional Transformer for Text Recognition and Correction. Two types of networks are leveraged in the recognition stage, including the CTC [9]-based model and the 2D attention sequence-to-sequence model. The backbone networks consist of convolutional networks and context extractors. They train multiple CNNs including VGGNet, ResNet50, ResNet101, and SEResNeXt50 [12]. Then they extract contextual information using BiLSTM, BiGRU, and transformer models. For the CTC-based method, they integrate an end-to-end trainable ALBERT [15] as a language model. The models are pre-trained on 60 million synthetic data samples, and are further fine-tuned on open-source datasets including SVTS, ICDAR-2013, ICDAR-2015, CUTE, IIIT5k, RCTW-2017, LSVT, ReCTS, COCO-Text, RCTW, MLT-2021, and ICPR-2018-MTWI. Data augmentation tricks are also employed, such as Gaussian blur, Gaussian noise, and brightness adjustment. In the end-to-end text spotting stage, they predict all detected boxes of a trajectory using different recognition methods. The final text result corresponding to the trajectory is then selected among all recognition results, considering both confidence and character length. Finally, low-quality trajectories whose text result scores are low or whose results contain Chinese characters are removed.

DXM-DI-AI-CV-TEAM achieves the recognition model by using ResNet feature extractor with TPS(Thin Plate Spines) [1]. A relation attention module is employed to capture the dependencies of feature maps and a parallel attention module is used for decoding all characters in parallel. The data for training base recognition model contains public datasets such as SynthText, Syn90k, CurvedSynth and SynthText_Add. To finetune the model, the data with MODERATE and HIGH quality label from the competition datasets are of good use. Compared with solutions from other paper to increase the quality scores, this recognition model used the combination of voting and confidence of the results to obtain the final text of the text stream.

CASIA_NLPR_PAL Team handles the scene text recognition task with sliding convolutional character models. For each detected text line, they firstly use a classifier to determine the text direction. Then a sliding-window-based method simultaneously detects and recognizes characters by sliding the text line image with character models, which are learned end-to-end on text line images. The character classifier outputs on the sliding windows are normalized and decoded with CTC-based algorithm. The output classes of the recognizer include all the ASCII and the commonly used Chinese characters. The final adopted model is trained on all competition training and validation data, some publicly released data sets, and a large number of synthetic samples.

6   Discussion

In the video text detection task, most participants employ the semantic-based Mask R-CNN framework to capture regular and irregular text instance. Tencent-OCR team achieves the best score in F-score$_d$, Recall$_d$ and Precision$_d$ with multiple backbones (e.g., HRNet, Res2Net and SENet, and so on) and model ensembles. Data augmentation strategies, rich opensource data and multi-scale train/test strategies are useful and important for getting better results. Besides, many participants employ the end-to-end trainable learning frameworks, which is an obvious research tendency in video text detection.

For video text tracking task, most methods focus on the trajectory estimation by calculating a cost matrix referring to the extracted appearance features. After that, the matching algorithms are employed to generate the final text streams. Tencent-OCR team achieves the best score in ATA$_t$, MOTA$_t$ and MOTP$_t$ with multiple metrics (e.g., box IoU, text content similarity, box size similarity and text-geometry neighbor relationship) which are further integrated as the final matching score. Besides, some post-processing strategies like removing low-quality text instance are also important for achieving better results.

For the final text recognition in Task 3, many methods first attempt to train a general model by using various public datasets (e.g., SynthText, Syn90k), and then further finetuned on the released training dataset with MODERATE and HIGH quality. Different teams employ different recognition decoders including the CTC-based, 2D attention-based or even the transformer based decoder. Tencent-OCR team achieving the 1-st rank in F-score$_e$, Recall$_e$ and Precision$_e$ with various data augmentation, opensource datasets, network backbones and model ensembles.

From the performance of the three tasks, we find that the first two submitted results on detection achieving F-score$_d$ of more than 0.8. It indicates that the existing approaches for general text detection are performing well for single video frames. While in Task 2 and 3, most submitted methods obtain relatively low scores (less than 0.55) because the video text spotting is very challenging due to various environment interferences. Therefore, there are still large space for improvement for the important research topic. We also note that, many top ranking methods use ensemble of multiple backbones or metrics to improve performance, and some methods use a wide range of opensource datasets for training a good pre-train model. Besides, the end-to-end trainable framework for video text spotting becomes an obvious research tendency. Most of submitted methods use different ideas and strategies, and we expect more innovative approaches will be proposed after this competition.

7   Conclusion

This paper summarizes the organization and results of ICDAR 2021 competition on SVTS, detailed on the Codalab website. Large scale video text dataset was collected and annotated in full annotations, respectively, containing 21 different scenes. There have been a number of 24 teams participating in the three tasks and 46 valid submissions in total, which have shown great interest from both research communities and industries. Submitted results has shown the abilities of state-of-the-art video text spotting systems. On one hand, we intend to keep on maintaining the ICDAR 2021 SVTS competition leaderboard to encourage more participants to submit and improve their results. On the other hand, we will extend this competition to multilingual competition for further promoting the research community.

References

1. Bookstein, F.L.: Thin-plate splines and the decomposition of deformations. IEEE TPAMI **16**(6), 567–585 (1989)
2. Cai, Z., Vasconcelos, N.: Cascade R-CNN: delving into high quality object detection. In: CVPR, pp. 6154–6162 (2018)
3. Cheng, Z., Lu, J., Niu, Y., Pu, S., Wu, F., Zhou, S.: You only recognize once: towards fast video text spotting. In: ACM MM, pp. 855–863 (2019)
4. Cheng, Z., et al.: FREE: a fast and robust end-to-end video text spotter. IEEE Trans. Image Process. **30**, 822–837 (2020)
5. Dai, J., et al.: Deformable convolutional networks. In: Proceedings of the IEEE International Conference on Computer Vision, pp. 764–773 (2017)

6. Maes, F., Collignon, A.: Multimodality image registration by maximization of mutual information. IEEE TMI **16**(2), 187–198 (1997)
7. Feng, W., Yin, F., Zhang, X., Liu, C.: Semantic-aware video text detection. In: CVPR. IEEE (2021)
8. Gao, S., Cheng, M.M., Zhao, K., Zhang, X.Y., Yang, M.H., Torr, P.H.: Res2Net: a new multi-scale backbone architecture. IEEE TPAMI **PP**(99), 1 (2019)
9. Graves, A., Fernández, S., Gomez, F.: Connectionist temporal classification: labelling unsegmented sequence data with recurrent neural networks. In: ICML, pp. 369–376 (2006)
10. He, K., Gkioxari, G., Dollár, P., Girshick, R.: Mask R-CNN. In: ICCV, pp. 2961–2969 (2017)
11. He, K., Zhang, X., Ren, S., Sun, J.: Deep residual learning for image recognition. In: CVPR, pp. 770–778 (2016)
12. Jie, H., Li, S., Gang, S.: Squeeze-and-excitation networks. In: CVPR, pp. 7132–7141 (2018)
13. Karatzas, D., et al.: ICDAR 2015 competition on robust reading. In: ICDAR, pp. 1156–1160 (2015)
14. Karatzas, D., et al.: ICDAR 2013 robust reading competition. In: ICDAR, pp. 1484–1493 (2013)
15. Lan, Z., Chen, M., Goodman, S., Gimpel, K., Sharma, P., Soricut, R.: ALBERT: a lite BERT for self-supervised learning of language representations. arXiv preprint arXiv:1909.11942 (2019)
16. Liu, Y., Jin, L., Zhang, S., Sheng, Z.: Detecting curve text in the wild: new dataset and new solution. arXiv preprint arXiv:1712.02170 (2017)
17. Sun, K., Xiao, B., Liu, D., Wang, J.: Deep high-resolution representation learning for human pose estimation. In: CVPR, pp. 5693–5703 (2019)

ICDAR 2021 Competition on Integrated Circuit Text Spotting and Aesthetic Assessment

Chun Chet Ng[1,2]([✉]), Akmalul Khairi Bin Nazaruddin[2], Yeong Khang Lee[2],
Xinyu Wang[3], Yuliang Liu[4], Chee Seng Chan[1], Lianwen Jin[5], Yipeng Sun[6],
and Lixin Fan[7]

[1] University of Malaya, Kuala Lumpur, Malaysia
[2] ViTrox Technologies Sdn. Bhd., Bandar Cassia, Penang, Malaysia
ictext@vitrox.com
[3] The University of Adelaide, Adelaide, Australia
[4] The Chinese University of Hong Kong, Kowloon, China
[5] South China University of Technology, Guangzhou, China
[6] Baidu, Beijing, China
[7] Webank, Shenzhen, China
https://ictext.v-one.my/

Abstract. With hundreds of thousands of electronic chip components are being manufactured every day, chip manufacturers have seen an increasing demand in seeking a more efficient and effective way of inspecting the quality of printed texts on chip components. The major problem that deters this area of research is the lacking of realistic text on chips datasets to act as a strong foundation. Hence, a text on chips dataset, ICText is used as the main target for the proposed Robust Reading Challenge on Integrated Circuit Text Spotting and Aesthetic Assessment (RRC-ICText) 2021 to encourage the research on this problem. Throughout the entire competition, we have received a total of 233 submissions from 10 unique teams/individuals. Details of the competition and submission results are presented in this report.

Keywords: Character spotting · Aesthetic assessment · Text on integrated circuit dataset

1 Introduction

Recent advances in scene text researches, including related works of text detection and recognition indicate that these research fields have received a lot of attentions, and hence pushing scene text community to progress with a strong momentum [3,17]. However, this momentum is not shared in the research field of text spotting on microchip components, which attracts far less attention than scene text research. Moreover, with the 5G and Internet of Things (IoT) trend flowing into almost all technical aspects of our daily life, the demands of electronic chip components are increasing exponentially [5]. In line of how chip

© Springer Nature Switzerland AG 2021
J. Lladós et al. (Eds.): ICDAR 2021, LNCS 12824, pp. 663–677, 2021.
https://doi.org/10.1007/978-3-030-86337-1_44

manufacturers and automated testing equipment service providers carry out chip quality assurance process, we believe now is the right time to address the text detection and recognition on microchip component problem.

As a starting point, this competition is hosted to focus on text spotting and aesthetic assessment on microchip components or *ICText* in short. It includes the study of methods to spot character instances on electronic components (printed circuit boards, integrated circuits, etc.), as well as to classify the multi-label aesthetic quality of each characters. Apart from the similar end goals of both scene text and *ICText* to achieve a high accuracy score, the proposed methods are also expected to excel in resource constrained environments. Thus, the end solution is required to be light weight and able to carry out inference in high frame per second so that it can lower the cost of chip inspection process. Such research direction is mainly adopted by microchip manufacturers or machine vision solution providers, where one of their primary goal is to ensure that the text is printed correctly and perfectly without any defects in a timely manner.

This competition involves a novel and challenging text on microchip dataset, followed by a series of difficult tasks that are designed to test capability of the character spotting models to spot the characters and classify their aesthetic classes accurately. The submitted models are also evaluated in terms of inference speed, allocated memory size and accuracy score.

1.1   Related Works

Deep learning methods have been widely applied in the microchip component manufacturing processes, to reduce human dependency and focus on automating the entire manufacturing pipeline. Particularly, vision inspection and automatic optical inspection system plays an important role in this pipeline. There are a number of existing works discussing the usage of machine learning based chip component defect inspection and classification methods. For examples, [1,23,24] are making use of traditional image processing methods to detect and classify defects on printed circuit boards. While recent works such as [18,26,29] are using deep learning based object detection models to reach similar objectives.

However, there are a limited number of related works [13–15,19] on text detection and recognition on electronic chip components as majority of the related works focusing more on the defect detection or component recognition. For example, the authors of [14] binarise the printed circuit board images by using a local thresholding method before text recognition and the method was tested on 860 cropped image tiles with text contents from printer circuit board. Their proposed method is separated into two stages, hence the final accuracy are subjected to their respective performance and it fails to generalise to new unseen images. Besides, the scope of [19] is very limited as the authors focused on recognising broken or unreadable digits only on printed circuit boards. The involved dataset consists of only digits, 0 to 9 with a set of 50 images each, resulting in a total of 500 character images. [13] integrated OCR library into their printed circuit board reverse engineering pipeline to retrieve the printed part number on the housing package of an integrated circuit board. While in

[15], they relied on convolution neural network to classify blurry or fractured characters on 1500 component images with roughly 20 characters and symbols. The aforementioned related works are able to achieve good performance in each of their own scenarios and datasets, but the results are not sufficiently convincing as these models are not tested on benchmark datasets within a controlled test environment settings.

First and foremost, there is not a large scale public dataset with thousands of chip components images available at the time of writing. This leads to the lacking of a fair and transparent platform for all the existing methods to compare and compete with each other, where researchers cannot verify that which models are superior than others. As shown by related works that aesthetic or cosmetic defects can be found commonly on electronic chip components and printed circuit boards, it is obvious that such defects occurs frequently on printed characters. Unfortunately, this problem has not derived interest as much as chip components level counterpart, and we believe that now is the right time to do so.

Additionally, [17] showed that recent scene text researches are paying more attentions on the inference speed of their models because a faster model correlates to lower cost and high efficiency in production environment. Before us, not a single one of the existing Robust Reading Competitions (RRC) [11, 12, 20] have held similar challenges, related to deployability of deep learning models. Based on the submissions to the previous ICDAR RRC 2019 challenges with large scale datasets, especially ArT [4] and LSVT [25], we found out that most models are very deep and complex. They are also exposed to the weaknesses of low FPS and resource intensive models due to the nature of most submissions are coming from ensemble of multiple models. Such models are also likely to require a high amount of GPU memory during inference which make it more difficult to deploy them to end/embedded devices with low computing power.

1.2   Motivation and Relevance to the ICDAR Community

Text instances are usually printed on electronic components as either serial number or batch number to represent significant information. Theoretically, the printed characters should be in a simpler and cleaner format than text instances in the wild, with the assumptions that the printings are done perfectly or the images are captured under well-controlled environment. Unfortunately, this is not always the case. In production line, the quality of the printed characters is subject to various external factors, such as the changes of component material, character engraving methods, etc. Furthermore, the lighting conditions are a non trivial factor when the components are being inspected. If it is too dim then the captured images will have low contrast, while high brightness will lead to a reflection on the electronic components and yield over exposed images. Moreover, the text instances are variegated in terms of orientations, styles, and sizes across the components as different manufacturers have different requirements for character printings. Coupled with the inconsistency of printed characters with the condition of captured images, they are slightly different from what current scene text images can offer and pose problems for current text spotting models.

Convolutional neural network is used by [15] to verify printed characters on the integrated circuit components; however, the proposed method still relies heavily on handcrafted operation and logic, which further strengthens our arguments that the aforementioned problems are valid and worth further exploration. In addition, the lack of a high quality and well annotated dataset to serve as a benchmark for stimulating works is one of the major hurdle faced by researchers in this sub-field of text spotting. Secondly, well designed tasks are also needed to fairly evaluate the proposed models in terms of the applicability and feasibility to deploy the methods for real life application, where we aim to bridge the gap between research and industrial requirements. With the proposed tasks that are focusing on the inference speed (frame per second) and also occupied GPU memory, we hope to inspire researchers to develop a light and memory efficient model. As such model is required to fit in low computing devices that are usually used in the automated optic inspection systems.

We believe that this new research direction is beneficial to the general scene text community. Through the introduction of this novel dataset and also challenges based on an industrial perspective, we aim to drive research works that are feasible to output tangible production values. With the release of this dataset, we hope to drive a new research direction towards text spotting on industrial products in the future. The proposed *ICText* dataset can be served as a benchmark to evaluate existing text spotting methods. Challenges introduced in this proposed competition focus on the study of the effectiveness and efficiency of text spotting methods are also the first among previous RRC [11,12,20], where we believe such practical requirements are crucial to drive the development of lightweight and efficient deep learning text spotting models.

2   Competition Description

2.1   *ICText* Dataset

The novel dataset proposed for this competition, *ICText* is collected by ViTrox Corporation Berhad[1] or ViTrox in short. The company is a machine vision solution provider with the focus on development and manufacturing of automated vision inspection solutions. The images are collected and characters are annotated according to the following standards:

- Images in this dataset contain multi-oriented, slanted, horizontal and vertical text instances. There are also combinations of multiple text orientations.
- Different electronic components are involved during image collection under different lighting conditions so that *ICText* represents the problems faced by the industry realistically.
- All text instances are annotated with tight character level bounding boxes, transcriptions and also aesthetic classes.
- Sensitive information is censored wherever appropriate, such as proprietary chip designs and company information.

---

[1] https://vitrox.com/.

Type/Source of Images. Images in *ICText* dataset were collected by ViTrox's automated optical inspection machine, where the microchip components are coming from different clients. In general, images are captured from top angle with a microchip component with characters. Different automated optical inspection systems have variable configurations, leading to cases where the microchip component images are captured under different lighting conditions. We can also notice that certain images are under-exposed and over-exposed due to these settings. Sample images can be found in Fig. 1a, b and c, where we can observe a low contrast between the printed characters and the image background, making it very difficult to spot the characters on the images. Apart from the usual vision-related challenges (illumination, blurriness, perspective distortion, etc.), *ICText* stands out in challenging scene text understanding models with the combination of different text orientations in one single image and never seen before background (different microchip components). As illustrated in Fig. 1d, e and f, the multi-oriented characters are incomplete/broken with separated body/branches and also inseparable with the background. The variation of chip component backgrounds indirectly affect the quality of printed characters and leading to multiple font styles on the images. These uncontrollable defects are subjected to the character engraving methods, chip component's base materials or printing defects during manufacturing.

***ICText* versus Scene Text Dataset.** *ICText* includes different type of images as compared to existing scene text datasets, where the major difference lies in the image background. Characters in *ICText* are printed on microchip components, whereas text instances in scene text dataset have much more complex and vibrant colour backgrounds. Because the images in *ICText* are captured under different camera settings, there exist cases where some images have extreme illuminations conditions (low-contrast versus over-exposure) and image quality (blurry versus well-focused). We believe that such characteristic are challenging to text spotting models, especially in the cases where the text and background colour has low contrast. Besides, printed characters in our chip component images involve a huge variety of font styles and sizes due to the fact that different manufacturers have different requirements in character printing. Hence the challenging aspect of font variations in scene text images are also well observed in *ICText*. The diversity of images are ensured by collecting images of components with different structures and text orientations. To prevent the same set of components with similar structures or characters exist in between training and testing set, we carry out careful filtering on the collected images.

***ICText* Dataset Splits.** There are a total of 20,000 images in the *ICText* dataset. It is split into a training set (10,000 images), validation set (3000 images) and testing set (10,000 images). Images and annotations for training set will be released by on-demand request basis, and a Non Disclosure Agreement has to be signed prior to the release. However, the annotations for the validation set and the entire test set are kept private due to privacy and copyright protection. A comparison between *ICText* dataset and other related datasets can be found

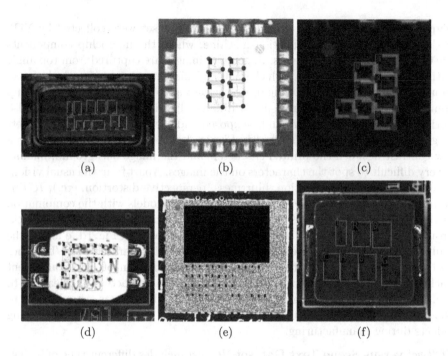

Fig. 1. Images sampled from the entire dataset. We show that they contain characters with combination of multiple font style, sizes orientations and aesthetic classes. Red dot stands for characters with low contrast compared to background; green dot represents blurry characters; blue dot highlights broken characters. (Color figure online)

in Table 1, where we can see that *ICText* is the most complete and largest text on chips dataset at the time of writing.

Ground Truth Annotation Format. All text instances in the images are annotated at the character level with a tight bounding box. *ICText* follows the MS COCO's annotation JSON format for object detection. There is an entry for each image in the dataset and match with their respective annotation entries. Apart from the usual keys in MS COCO's JSON format, we also provide additional information at the character level, such as the illegibility, degree of rotation and aesthetic classes. The "bbox" key represents the 4 pairs of polygon coordinates, and the "rotation_degree" key represents the angle the image can be rotated to get human-readable characters. The counter-clockwise rotation angle is provided to ease the training of models, so that rotation sensitive characters like 6 and 9 will not be confused. Multi labels for aesthetic classes are one-hot encoded, where each column is a class. Similar to COCO-text [7] and ICDAR2015 [11], annotations marked with ignore flag are do not care regions, they will not be taken into consideration during evaluation.

Table 1. A quantitative comparison between *ICText* and other microchip component text datasets. All four datasets report the presence of character defects (low contrast, blurry text, broken text).

Year	Paper	Number of Images
2014	[14]	860 cropped text images
2015	[19]	500 character images
2019	[15]	1,500 component images
2021	*ICText*	20,000 component images

2.2 Competition Tasks

The RRC-*ICText* consists of three main tasks: text spotting, end-to-end text spotting and aesthetic assessment, and lastly inference speed, allocated memory size and accuracy score assessment. This section details them.

Task 1: Text Spotting. The main objective of this task is to detect the location of every character and then recognise it given an input image, which is similar to all the previous RRC scene text detection tasks. The input modal of this task is strictly constrained to image only, no other form of input is allowed to aid the model in the process of detecting and recognising the characters.

Input: Microchip component text image.
Output: Bounding box with its corresponding class (a−z, A−Z, 0−9).

Task 2: End-to-end Text Spotting and Aesthetic Assessment. The main objective of this task is to detect and recognize every character and also their respective multi-label aesthetic classes in the provided image.

Input: Microchip component text image.
Output: Bounding box with the corresponding class/category (a−z, A−Z, 0−9) and multi-label aesthetic classes.

Task 3: Inference Speed, Model Size and Score Assessment. The main objective of this task is to evaluate the submitted models on their feasibility and applicability of deploying the methods for real life applications. In order to evaluate the submitted methods under previous tasks (Task 1 and 2), we decided to separate this task into two sub-tasks:

Task 3.1 - This sub-task takes the score of **Task 1** into consideration, we intend to evaluate the speed, size and text spotting score of the submitted model.
Task 3.2 - This sub-task is meant to evaluate the model on inference speed, occupied GPU memory size and multi-label classification score as stated in **Task 2**.

Submission to these two tasks are evaluated on the private test set through a docker image with inference code, model and trained weights. This is to study the relationship between speed, size, and score, where the end goal is to research and develop a well balanced model based on these three aspects. The submitted model is expected to achieve an optimal performance with minimal inference time and occupied GPU memory, a representation of efficiency and effectiveness.

Input: Docker image with inference code, model and trained weights.
Output: Inference speed, model size and respective Score.

2.3   Evaluation Protocol

Evaluation servers for all the tasks are prepared at Eval.ai with separate pages for Task 1 & 2[2] and Task 3[3]. Do note that due to privacy concerns and prevention of malicious use of the *ICText* dataset, the Task 1 & 2 are evaluated on validation set and only Task 3 is evaluated on the entire private test set. Relevant evaluation code can be found at this GitHub repository[4].

Evaluation Metric for Task 1. We follow the evaluation protocol of the MSCOCO dataset [16]. As we have separated the characters to 62 classes (a−z, A−Z, 0−9), MSCOCO's evaluation protocol can give us a better insight than the commonly used Hmean score in scene text spotting. We rank the submissions based on mean Average Precision or mAP as to MSCOCO challenges.

Evaluation Metric for Task 2. We first evaluate whether the predicted character matches the ground-truth at IoU 0.5 and then only we calculate precision, recall and F-2 Score value for the predicted multi-label aesthetic classes. If the predicted character is wrong, then the respective aesthetic classes will not be evaluated. A default value of [0,0,0] is given to predictions that have no aesthetic labels and we flip the aesthetic labels of both ground truth and prediction when all the values are zero, to prevent zero division error when calculating precision and recall. The submitted results are ranked by the average of multi-label F-2 Score across all images. Besides, F-2 Score is selected because recall is more important than precision when it comes to defect identification during chip inspection, as missed detection would yield high rectification costs.

Evaluation Metric for Task 3. We evaluate the submitted model in terms of inference speed (in frame per second or FPS) and also occupied GPU memory size (in megabytes or MB) together with the respective score of each sub-tasks. The evaluation metric of this task is named as Speed, Size and Score or $3S$ in short. It is defined using the following equation:

$$3S = 0.2 \times normalised\_speed + 0.2 \times (1 - normalised\_size) + 0.6 \times normalised\_score \quad (1)$$

---

[2] https://eval.ai/web/challenges/challenge-page/756/overview.
[3] https://eval.ai/web/challenges/challenge-page/757/overview.
[4] https://github.com/vitrox-technologies/ictext_eval.

We decided to give a higher priority to the respective task's score, because we believe that an optimal model should have a solid performance as a foundation, as well as able to strike a balance between speed and size. Firstly, *normalised_speed* of the submitted model is calculated using the following equation:

$$normalised\_speed = \min(\frac{actual\_FPS}{acceptable\_FPS}, 1) \tag{2}$$

We execute the inference script in the submitted Docker image and record total time taken for the model to carry out inference on the entire test set, which is then used to calculate the *actual_FPS*. For the *acceptable_FPS*, we decided to set it to 30FPS due to the fact that it is an average FPS recorded in modern object detectors on GPU and edge computing devices like Nvidia's Jetson Nano [6] and Intel's Neural Compute Stick 2 [22]. Furthermore, we calculate *normalised_size* of the submitted model using the following equation:

$$normalised\_size = \min(\frac{allocated\_memory\_size}{acceptable\_memory\_size}, 1) \tag{3}$$

The *allocated_memory_size* is defined based on the maximum GPU memory usage recorded during model inference. Then, we set the *acceptable_memory_size* to 4000 MB or 4 GB because it is an average memory size available to mobile and edge computing devices. The *normalised_score* is referred to mean Average Precision for Task 3.1 and multi-label F-2 Score for Task 3.2, both scores have to be normalized to the range of 0 to 1. We then rank the participants based on the final $3S$ score from the highest to lowest. All submitted docker images to Task 3 are evaluated on the Amazon EC2 p2.xlarge instance.

3 Results and Discussion

At the end of the competition, we have received a total of 233 submissions from 10 unique teams/individuals. In Task 1, there are 161 submissions from all 10 teams, whereas we received 50 submissions from 3 teams in Task 2. While in Task 3.1, there are 20 submissions from 2 teams and only 1 submissions for Task 3.2. Due to a high number of repetitive submissions from the participants, we only consider the one with the best result out of all submissions from the same team in this report. The final rankings are detailed in the Table 2.

3.1 Result Analysis

Based on the results, we are able to analyse the average performance of participants on *ICText*, as this would help us to understand the overall upper threshold scores of all tasks. Firstly, we observe that the participants achieve an average mAP score of 0.41 in Task 1. They have an Average Precision (AP) score of 0.551 and 0.505 for Intersection over Union (IoU) at 0.5 and 0.75. With the

Table 2. Final ranking table for all the tasks in RRC-*ICText* 2021.

Rank	Team (Affiliation)	Scores		
Task 1 - Validation Set		mAP	AP IOU@0.5	AP IOU@0.75
1	gocr (Hikvision)	0.60	0.78	0.73
2	SMore-OCR (SmartMore)	0.53	0.75	0.65
3	VCGroup (Ping An)	0.53	0.72	0.65
4	war (SCUT)	0.48	0.60	0.58
5	hello_world (Uni. of Malaya)	0.46	0.59	0.56
6	maodou (HUST)	0.44	0.58	0.54
7	WindBlow (Wuhan Uni.)	0.43	0.57	0.53
8	shushiramen (Intel)	0.31	0.45	0.39
9	XW (-)	0.21	0.31	0.27
10	Solo (-)	0.12	0.16	0.15
Task 2 - Validation Set		Precision	Recall	F-2 Score
1	gocr (Hikvision)	0.75	0.70	0.70
2	shushiramen (Intel)	0.30	0.65	0.51
3	hello_world (Uni. of Malaya)	0.22	0.29	0.26
Task 3.1 - Test Set		mAP	FPS;GPU Mem.(MB)	3S
1	gocr (Hikvision)	0.59	31.04;305.88	0.74
2	SMore-OCR (SmartMore)	0.53	2.69;10970.75	0.33
Task 3.2 - Test Set		F-2 Score	FPS;GPU Mem.(MB)	3S
1	gocr (Hikvision)	0.79	29.68;305.88	0.85

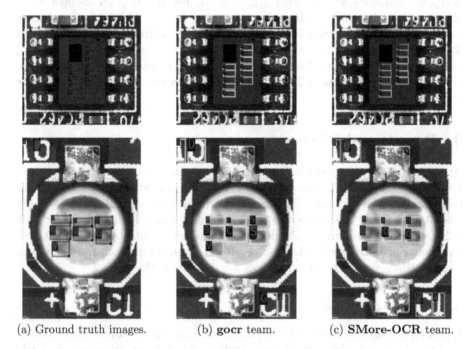

(a) Ground truth images.   (b) **gocr** team.   (c) **SMore-OCR** team.

Fig. 2. A comparison between ground truth and predictions by **gocr** and **SMore-OCR** team for Task 3.1 on the test set. Ground truth bounding boxes are in blue colour, while predicted bounding boxes are plotted in yellow. (Color figure online)

highest mAP score of 0.60, we believe that this shows how challenging *ICText* dataset is given that the participants' results are only slightly higher than the average mAP score. Besides, the average multi-label precision and recall score of participants in Task 2 is 0.42 and 0.55, with the average multi-label F-2 score reaching 0.49. As the number of submissions for Task 3.1 and 3.2 are limited, we report the average of FPS and GPU memory usage (in Megabytes) for both tasks, which are 21.14 and 3860.8 MB.

In Task 2, it is worth noting that the score of **shushiramen** team is higher than **hello_world** team, although the latter team achieves better score in Task 1, with **shushiramen** team scoring 0.31 for mAP and **hello_world** team has the mAP score of 0.46. This shows that the multi-label classification capability of the models are equally important as the detectors, although the predicted multi-label aesthetic classes are evaluated sequentially when the character is localised and recognized accurately. Given marginally higher mAP of **hello_world** team as compared to the **shushiramen** team, the multi-label F-2 Score of **hello_world** team is being dragged down by poor classification module, which will be discussed in Sect. 3.2. Moreover, we also notice how the evaluation metric of Task 3.1 plays its parts on emphasizing the deployability of models on the results of **SMore-OCR** and **gocr**, where their difference in mAP is 0.06, but their 3S score is different by a margin of 50%. This illustrates that the model of **gocr** team is more suitable to be deployed to end/edge devices as it requires less computing resources and has high inference speed. Some qualitative examples are shown in Fig. 2, where we observe that the models of both **gocr** (Fig. 2b) and **SMore-OCR** (Fig. 2c) are unable to spot multi-oriented characters accurately with the majority of predictions having wrong character classes, and their models fail drastically when the printed characters have aesthetic defects.

3.2   Participants' Methods

Gocr (Hikvision). Being the winning team for all three tasks, **gocr** team presents an interesting solution. Extra 84,400 images have been generated synthetically using SynthText [8] and text style transfer method introduced in [28] to tackle the lower case character class imbalance problem in our training set. YoloV5s [10] is used as the main model for character spotting and an additional classification branch of 3 aesthetic classes are added to it, so that the model is catered for all tasks. In short, there are a few tips and tricks leading to the superior performance of this model, including the use of synthetic images for pretraining and fine tuning, generation of pseudo labels based on validation images for semi supervised learning and the employment of teacher-student learning strategy for knowledge distillation. The team has also taken additional measures to deploy their model using TensorRT so that it can achieve faster speed and utilize less GPU resources, which is well aligned with our objectives in Task 3. We recommend readers to check out their main paper [27] for more information.

SMore-OCR (SmartMore). Mask-RCNN [9] based character spotting model has been proposed by this team. Before the model's training commences, the team generated 30,000 synthetic images based on our training set. They first mask out the training set images by using the provided bounding boxes, and then the EdgeConnect method [21] is used to fill up the blanks. After that, they follow SynthText's methodology [8] to write characters back to the synthetic images. Once the process is completed, they pre-trained a Mask R-CNN model on the synthetically generated data, before fine tuning on our training set. The Mask R-CNN is initialized with default hyper-parameters and trained for 6 epochs with the batch size of 32. In order to recognize multi-oriented characters based on the output of Mask-RCNN, two ImageNet pre-trained ResNet-50 are fine-tuned on the synthetic data and our training set as the character orientation and classification models. Both models use mini-batch stochastic gradient decent (SGD) with momentum of 0.9 and weight decay of 5×10^{-4}. The learning rate is dynamically adjusted during the training process with the following equation of $lr = lr_0 \div (1 + \alpha p)^\tau$, where $lr_0 = 10^{-3}$, $\alpha = 10$ and $\tau = 0.75$, with p being linearly adjusted from 0 to 1. However, the team claims that the addition of these two models do not improve the end results and subsequently drops it out of the entire pipeline, hence the final results are solely based on Mask-RCNN.

Hello_world (Uni. of Malaya). The MS-COCO pre-trained YoloV5m [10] is selected as the main model, which is fine-tuned on our dataset for 120 epochs with the settings of 576 as the image size and using the batch size 40. Besides, the default learning rate is used and the training data is augmented using mosaic image augmentation. However, the character rotation angle information is not used during training. In order to improve the character spotting performance, the team generates synthetic binary images of random characters as extra training data. Furthermore, a multi-label aesthetic model is separately trained using ImageNet pre-trained mobilenet_v2 on the cropped 64×64 character bounding box images for 4 epochs with binary cross entropy loss. During inference, the fine-tuned YoloV5m first detects, recognizes the characters, crops them out and passes to the aesthetic model for multi-label classifications.

3.3   Error Analysis

Following the method introduced in [2], TIDE is selected as the main error analysis tool because it provides insightful explanations on the weaknesses of participants' models and able to pinpoint the specific errors that are commonly made on our *ICText* dataset. Hence, TIDE is used to analyse all submissions to Task 1 and Task 3.1. Results are shown in Table 3, where the values in the table are mAP scores in the range of 0 to 100 if that particular error is fixed. We determine that the rate of occurrence for a specific error is high when the mAP increment is high. In this work, we focus on the main error types, as the special error types (FalsePos and FalseNeg) might have overlapping cases with them.

We observe that the most common error made by all models is the Classification error (Cls). We find it interesting that the mAP increment of lower ranking models are a few times higher than the top 3 models, this shows that they are performing badly to classify the detected characters. On the other hand, out of the 484285 predictions by the **gocr** team's Yolov5s model, 41% of them or 200694 predictions are categorised under Classification error. With this high number of predictions, we can also see that the model contributes the most to False Positive error (FalsePos). This scenario happens when the model might confuse between similar characters such as 6 or 9, and even in between case-insensitive characters like v and V. While the Mask R-CNN used by **SMore-OCR** team produce 62054 bounding boxes, which is 7.8 times less than what **gocr** team outputted. And only 18148 of them are categorised under Classification error, constituting 29% of the total predictions. This huge difference on number of predictions can be explained by the difference in adopted model architecture by these two teams, where Yolov5 is a one-stage object detection model that produces anchor boxes at more locations in the image compared to the two stage Mask R-CNN model, where the number of anchor boxes is limited. Furthermore, the second most frequent error made by the participants' models is Missing error, where the models fail to detect certain ground truth characters. The **Solo** team's model suffers the most under this error, and it can only matched 8059 out of 30400 ground-truth bounding boxes, which is only 26.5%. Coupling with the high False Negative error, this helps to explain why the model performs badly on our dataset. Based on the results reported on both Task 1 and Task 3.1 in Table 3, we can conclude that all models are facing difficulties to accurately recognize multi-oriented and defective characters, which are the two main aspects of *ICText* dataset.

Table 3. TIDE results of all submissions in Task 1 and Task 3.1.

Team	Cls	Loc	Both	Dupe	Bkg	Miss	FalsePos	FalseNeg
Task 1 - Validation Set								
gocr	7.27	0.2	0.02	0	5.44	1.06	14.86	3.65
SMore-OCR	11.14	0.67	0.14	0	6.12	1.94	10.37	9.05
VCGroup	6.01	0.38	0.09	0	5.49	4.97	12.99	8.77
war	28.07	0.14	0.03	0	2.09	2.56	6.99	17.72
hello_world	32.08	0.18	0.02	0	2.52	0.85	4.81	14.32
maodou	33.44	0.09	0.02	0	1.72	1.85	3.36	29.26
WindBlow	31.77	0.07	0.01	0	1.7	3.1	3.39	28.97
shushiramen	42.55	0.25	0.23	0.01	1.41	2.44	10.02	30
XW	40.53	0.22	0.37	0	0.93	6.58	4.87	29.02
Solo	11.97	0.03	0.01	0	0.28	38.74	0.94	77.94
Task 3.1 - Test Set								
gocr	10.85	0.21	0.03	0	0.7	5.67	7.58	9.87
SMore-OCR	11.14	0.17	0.12	0	1.8	5.92	6.56	13.45

4 Conclusion

In summary, the RRC-ICText has introduced a new perspective to the scene text community with the focus on industrial OCR problems and a never seen before large scale text on integrated circuits dataset. The competition includes challenges that represent problems faced by component chip manufacturers, involving the need to recognize characters and classify multi-label aesthetic defects accurately in a timely manner under limited computing resources. The proposed *ICText* dataset, tasks, and evaluation metrics present an end-to-end platform for text spotting researchers to evaluate their methods for performance and production deployability. Results by the participants show that there is still a huge room for improvement on the dataset, albeit creative ideas have been applied to tackle the posed problems. Lastly, the RRC-ICText is a good starting point to inspire researches on scene text models that are not only emphasizing high accuracy but at the same time have low computing requirements.

References

1. Anitha, D.B., Rao, M.: A survey on defect detection in bare PCB and assembled PCB using image processing techniques. In: WiSPNET 2017, pp. 39–43 (2017)
2. Bolya, D., Foley, S., Hays, J., Hoffman, J.: Tide: a general toolbox for identifying object detection errors. arXiv preprint arXiv:2008.08115 (2020)
3. Chen, X., Jin, L., Zhu, Y., Luo, C., Wang, T.: Text recognition in the wild: a survey. ACM Comput. Surv. **54**(2) (2021). https://doi.org/10.1145/3440756
4. Chng, C.K., et al.: ICDAR 2019 robust reading challenge on arbitrary-shaped text - RRC-ArT. In: ICDAR 2019, pp. 1571–1576 (2019)
5. Deloitte: Semiconductors - the next wave opportunities and winning strategies for semiconductor companies (2020). https://www2.deloitte.com/cn/en/pages/technology-media-and-telecommunications/articles/semiconductors-the-next-wave-2019.html
6. NVIDIA Developer: Jetson nano: deep learning inference benchmarks (2019). https://developer.nvidia.com/embedded/jetson-nano-dl-inference-benchmarks
7. Gomez, R., et al.: ICDAR 2017 robust reading challenge on coco-text. In: ICDAR 2017, pp. 1435–1443. IEEE (2017)
8. Gupta, A., Vedaldi, A., Zisserman, A.: Synthetic data for text localisation in natural images. In: CVPR 2016 (2016)
9. He, K., Gkioxari, G., Dollár, P., Girshick, R.: Mask R-CNN. In: ICCV 2017, pp. 2980–2988 (2017). https://doi.org/10.1109/ICCV.2017.322
10. Jocher, G.: ultralytics/yolov5 (2021). https://doi.org/10.5281/zenodo.3908559
11. Karatzas, D., et al.: ICDAR 2015 competition on robust reading. In: ICDAR (2015)
12. Karatzas, D., et al.: ICDAR 2013 robust reading competition. In: ICDAR (2013)
13. Kleber, S., Nölscher, H.F., Kargl, F.: Automated PCB reverse engineering. In: 11th USENIX Workshop on Offensive Technologies (WOOT 17) (2017)
14. Li, W., Neullens, S., Breier, M., Bosling, M., Pretz, T., Merhof, D.: Text recognition for information retrieval in images of printed circuit boards. In: IECON, pp. 3487–3493 (2014)
15. Lin, C.H., Wang, S.H., Lin, C.J.: Using convolutional neural networks for character verification on integrated circuit components of printed circuit boards. Appl. Intell. **49**(11), 4022–4032 (2019). https://doi.org/10.1007/s10489-019-01486-5

16. Lin, T., et al.: Microsoft COCO: common objects in context. CoRR abs/1405.0312 (2014). http://arxiv.org/abs/1405.0312
17. Long, S., He, X., Ya, C.: Scene text detection and recognition: the deep learning era. arXiv preprint arXiv:1811.04256 (2018)
18. Munisankar, S.N.N., Rao, B.N.K.: Defect detection in printed board circuit using image processing. IJITEE **9**(2), 3005–3010 (2019)
19. Nava-Dueñas, C.F., González-Navarro, F.F.: OCR for unreadable damaged characters on PCBs using principal component analysis and Bayesian discriminant functions. In: CSCI 2015, pp. 535–538 (2015)
20. Nayef, N., et al.: ICDAR 2017 robust reading challenge on multi-lingual scene text detection and script identification-RRC-MLT. In: ICDAR 2017, vol. 1, pp. 1454–1459. IEEE (2017)
21. Nazeri, K., Ng, E., Joseph, T., Qureshi, F., Ebrahimi, M.: EdgeConnect: structure guided image inpainting using edge prediction. In: ICCV 2019 (2019)
22. OpenVINO: Get a deep learning model performance boost with intel® platforms - openvino™ toolkit (2021). https://docs.openvinotoolkit.org/latest/openvino_docs_performance_benchmarks.html#intel-ncs2
23. Pramerdorfer, C., Kampel, M.: A dataset for computer-vision-based PCB analysis. In: MVA 2015, pp. 378–381 (2015)
24. Ray, S., Mukherjee, J.: A hybrid approach for detection and classification of the defects on printed circuit board. IJCA **121**(12), 42–48 (2015)
25. Sun, Y., et al.: ICDAR 2019 competition on large-scale street view text with partial labeling - RRC-LSVT. In: ICDAR 2019, pp. 1557–1562 (2019)
26. Tang, S., He, F., Huang, X., Yang, J.: Online PCB defect detector on a new PCB defect dataset. CoRR abs/1902.06197 (2019). http://arxiv.org/abs/1902.06197
27. Wang, Q., Li, P., Zhu, L., Niu, Y.: 1st place solution to ICDAR 2021 RRC-ICTEXT end-to-end text spotting and aesthetic assessment on integrated circuit. arXiv preprint arXiv:2104.03544 (2021)
28. Wu, L., et al.: Editing text in the wild. In: ACMMM 2019, pp. 1500–1508 (2019)
29. Zhang, Q., et al.: Deep learning based defect detection for solder joints on industrial x-ray circuit board images. arXiv preprint arXiv:2008.02604 (2020)

ICDAR 2021 Competition on Components Segmentation Task of Document Photos

Celso A. M. Lopes Junior[1] , Ricardo B. das Neves Junior[1] ,
Byron L. D. Bezerra[1(✉)] , Alejandro H. Toselli[2] , and Donato Impedovo[3]

[1] Polytechnic School of Pernambuco - University of Pernambuco, Pernambuco, Brazil
{camlj,rbnj}@ecomp.poli.br, byron.leite@upe.br
[2] PRHLT Research Center, Universitat Politécnica de Valéncia, Camí de Vera, s/n,
46022 Valéncia, Spain
a.toselli@northeastern.edu
[3] Department of Computer Science, University of Bari, Bari, Italy
donato.impedovo@uniba.ite

Abstract. This paper describes the short-term competition on "Components Segmentation Task of Document Photos" that was prepared in the context of the "16th International Conference on Document Analysis and Recognition" (ICDAR 2021). This competition aims to bring together researchers working on the filed of identification document image processing and provides them a suitable benchmark to compare their techniques on the component segmentation task of document images. Three challenge tasks were proposed entailing different segmentation assignments to be performed on a provided dataset. The collected data are from several types of Brazilian ID documents, whose personal information was conveniently replaced. There were 16 participants whose results obtained for some or all the three tasks show different rates for the adopted metrics, like "Dice Similarity Coefficient" ranging from 0.06 to 0.99. Different Deep Learning models were applied by the entrants with diverse strategies to achieve the best results in each of the tasks. Obtained results show that the current applied methods for solving one of the proposed tasks (document boundary detection) are already well stablished. However, for the other two challenge tasks (text zone and handwritten sign detection) research and development of more robust approaches are still required to achieve acceptable results.

Keywords: ID document images · Visual object detection and segmentation · Processing of identification document images

1 Introduction

This paper describes the short-term competition on "Components Segmentation Task of Document Photos" organized in the context of the "16th International Conference on Document Analysis and Recognition" (ICDAR 2021).

Supported by University of Pernambuco, CAPES and CNPq.

J. Lladós et al. (Eds.): ICDAR 2021, LNCS 12824, pp. 678–692, 2021.
https://doi.org/10.1007/978-3-030-86337-1_45

The traffic of identification document images (ID document) through digital media is already a common practice in several countries. A large amount of data can be extracted from these images through computer vision and image processing techniques, as well machine learning approaches. The extracted data can serve for different purposes, such as names and dates retrieved from text fields to be processed by OCR systems or extracted handwritten signatures to be checked by biometric systems, in addition to other characteristics and patterns present in images of identification documents. Actually there are few developed image processing applications that focus on the treatment of ID document images, specially related with image segmentation of ID documents.

The main goal of this contest was to stimulate researchers and scientists in the search for new techniques of image segmentation for the treatment of these ID document images. The availability of an adequate experimental dataset of ID documents is another factor of great importance, given that there are few of them free-available to the community [26]. Due to privacy constraints, for the provide dataset all text with personal information was synthesized with fake data. Furthermore, the original signatures were substituted by new ones collected randomly from different sources. All these changes were performed by well-designed algorithms and post-processed by humans whenever needed to keep the real-world conditions as much as possible.

2   Competition Challenges and Dataset Description

In this contest, entrants have to employ their segmentation techniques on given ID document images according to tasks defined for different levels of segmentation. Specifically, the goal here is to evaluate the quality of the applied image segmentation algorithms to ID document images acquired by mobile cameras, where several issues may affect the segmentation task performance, such as: location, texture and background of the document, camera distance to the document, perspective distortions, exposure and focus issues, reflections or shadows due to ambient light conditions, among others.

To evaluate tasks entailing different segmentation levels, the following three challenges have to be addressed by the participants as described below.

2.1   Challenge Tasks

1st Challenge - Document Boundary Segmentation: The objective of this challenge is to develop boundary detection algorithms for different kinds of documents [21]. The entrants should develop an algorithm that takes as input an image containing a document, and return a new image of the same size with the background in black pixels and the region occupied by the document in white pixels. Figure 1-left shows an example of this detection process.

2nd Challenge - Zone Text Segmentation: This challenge encourages the development of algorithms for automatic text detection in ID documents [21]. The entrants have to develop an algorithm capable of detecting text patterns in

Fig. 1. Left: example of images containing a document and their boundary detection output images. Right: an example of a document image and its output image with detected text regions masked.

the provided set of images; that is, to process an image of a document (without background), and return a new image of the same size with non-interest regions in black pixels and regions of interest (text regions) in white pixels. This detection process is illustrated in Fig. 1-right.

3rd Challenge - Signature Segmentation: This challenge aims at developing algorithms to detect and segment handwritten signatures on ID documents [12, 28]. Given an image of a document, the model or technique applied should return an image with the same size of the input image with the handwritten signature strokes as foreground in white pixels, and everything else in black pixels. Figure 2 displays an example of this detection.

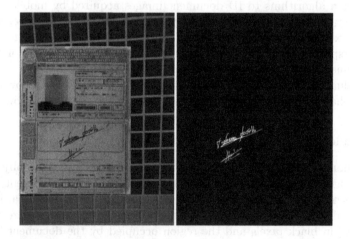

Fig. 2. Example of a document image and its output image with a detected handwritten signature masked.

2.2   ID Document Image Dataset

For this competition, the provided dataset is composed of thousands of images of ID documents from the following Brazilian document types: National Driver's License (CNH), Natural Persons Register (CPF) and General Registration (RG). The document images were captured through different cell phone cameras at different resolutions. Documents that appear in the images are characterized by having different textures, colors and lighting, on different real-world backgrounds with non-uniform patterns.

The images are all in RGB color mode and PNG format. The ground truth of this dataset is different for each challenge. For the first challenge, the interest regions are the document areas themselves, while the printed text areas are the ones for the second challenge. For the third challenge, we are only interested in the pixels of detected handwritten signature strokes. As commented before, for all the challenges, the interest regions/pixels are represented by white pixel areas, while non-interest ones in black pixel areas.

As personal information contained in the documents cannot be made public, the original data was replaced with synthetic data. We generate fake information for all fields present in the document (such as name, date of birth, affiliation, number of document, among others), as described in [26].

On the other hand, handwritten signatures were acquired from the MCYT [22] and GPDS [9] datasets, which were synthetically incorporated into the images of ID documents.

Both the ID document images and the corresponding ground truth of the dataset were manually verified, and they are intended to be used for model training and for computing evaluation metrics. Table 1 summarizes the basic statistics of the dataset. The whole dataset is available for the research community under request (visit: https://icdar2021.poli.br/).

Table 1. Statistics of the dataset and experimental partition used in the competition, where C1, C2, and C3 corresponds to 1st, 2nd and 3rd challenge, respectively.

Number of:	Train-C1	Test-C1	Train-C2	Test-C2	Train-C3	Test-C3
Images	15,000	5,000	15,000	5,000	15,000	5,000

3   Competition Protocol

As introduced in the previous section, this competition presents three challenges:

- 1st Challenge: Document Boundary Segmentation
- 2nd Challenge: Zone Text Segmentation
- 3rd Challenge: Signature Segmentation

Competitors can decide how many tasks they want to compete for, choosing between one or more challenges. In addition to the competition data, it was allowed to use additional datasets for model training. In this sense, some participants used the BID Dataset [26], and Imagenet [7] for improving the prediction of their approaches.

Participants could also use solutions from a challenge task as support to develop solutions for other challenges. Some participants used image segmentation predictions of the 1st challenge to remove background from images before they were used for the 2nd and 3rd challenge tasks.

Participants could also use pre or post-image processing techniques to develop their solutions.

3.1 Assessment Methods

The Structural Similarity Index (SSIM) metric was employed in the evaluation process. However, the results obtained using SSIM were very similar in all systems and techniques. Therefore we did not report their results in this work.

The following similarity metrics were used for evaluating the approaches applied by the participants:

Dice Similarity Coefficient (DSC). The Dice Similarity Coefficient (DSC), presented by Eq. 1 is a statistical metric proposed by Dice, Lee R [8], which can be used to measure the similarity between two images.

$$DSC = \frac{2|X \cap Y|}{|X| + |Y|} \tag{1}$$

where $|X|$ and $|Y|$ are elements of the ground truth and segmented image, respectively.

Scale Invariant Feature Transform (SIFT). The scale-invariant feature transform (SIFT) is a feature extraction algorithm, proposed by David Lowe [20], that describe local features in images.

The feature extraction performed by SIFT consists in the detection of Key-Points at the image. These key-points denote information related to the regions of interest (represented by the white pixels) of the ground truth and the segmented image. Once mapped, the key-points of the ground truth images are stored and compared with the key-points identified in the predicted image. From this comparison, it is possible to define a similarity rate.

4   Briefly Description of the Entrant's Approaches

In total there were 16 research groups from different countries that have participated in the contest. The entrants submitted their segmentation hypotheses computed on the corresponding test sets for some or all of the three challenge tasks. The main characteristics of the approaches used by each entrant in each challenge task are described below.

1. **58 CV Group (58 Group, China) - Task 1:** The approach is based on "Efficient and Accurate Arbitrary-Shaped Text Detection with Pixel Aggregation Network" (PANNet) [34] with improvements for edge segmentation. For model training, the segmentation effect has been improved by employing two different loss functions: the edge segmentation loss and the focal loss [17].

2. **Ambilight Lab Group (NetEase Inc., China) - Task 1:** Document boundary detection has been carried out by using the composed model HRnet+OCR+DCNv2 [30, 36, 42]. **Task 2:** On the segmentation results in Task 1, a text angle classify model based on ResNet-50 [11] and a text segmentation model HRnet+OCR+DCNv2 were applied. **Task 3:** On the segmentation results in Task 1, an approach based on YOLO-v5[1] for signature detection and HRnet+OCR+DCNv2 for signature segmentation were applied.

3. **Arcanite Group (Arcanite Solutions LLC, Switzerland) - Tasks 1, 2 & 3:** U-Net [25] is employed as model architecture, where the encoder is based on ResNet-18 [11] and ImageNet [7], while the decoder is composed basically by transpose convolutions for up-sampling the outputs. Cross-entropy [39] and Dice [29] was used as loss functions in the training process.

4. **Cinnamonq AI Group (Hochiminh City Univ. of Tech., Vietnam) - Tasks 1, 2 & 3:** Methods in these Tasks employ U-Net [25] as model architecture with Resnet-34 [11] backbone. This model has been trained to resolve the assignment proposed in each Task, where corresponding data augmentation has been applied for improving generalization and robustness.

5. **CUDOS Group (Anonymous) - Tasks 1, 2 & 3:** The approaches are based on encoder-decoder SegNet [1] architecture and "Feature Pyramid Network" (FPN) [16] for object detection with MobileNetV2 [27] backbone of the network. Data augmentation and cross-entropy [39] loss were employed in the training of the network.

6. **Dao Xianghu light of TianQuan Group (CCB Financial Technology Co. Ltd, China) -** The approaches performed on the three tasks employ encoder-decoder model architectures. **Task 1:** The approach uses two DL models (both based on ImageNet pretrained weights): DeepLabV3 [5] (with ResNet-50 backbone) and Segfix [37] for refinement. **Task 2:** The applied approach is based on a model named DeepLabV3+ [6]. **Task 3:** In this case, the used approach is the Unet++ [41]. In general, training of the models for the three tasks were carried out with data augmentation, using pretrained weights initialized on ImageNet, and SOTA (or FocalDice) [17] loss function.

7. **DeepVision Group (German Research Center for AI, Germany) - Tasks 1 & 2:** The assignments in these Tasks were formulated as an instance segmentation problem. Basically, the approach uses Mask R-CNN [10] for

---

[1] https://zenodo.org/record/4679653#.YKkfQSUpBFQ.

instance segmentation, with backbone of Resnet-50 [11] pretrained on ImageNet Dataset [7].

8. **dotLAB Group (Universidade Federal de Pernambuco, Brazil) - Tasks 1, 2 & 3:** The applied approaches are based on a network architecture called "Self-Calibrated U-Net" (SC-U-Net) [18] which uses self-calibrated convolutions to learn better discriminative representations.

9. **NAVER Papago Group (NAVER Papago, South Korea) - Tasks 1, 2 & 3:** The approaches are based on Unet++ [41] with encoder based on ResNet [11]. Training of models were conducted with data augmentation, using cross-entropy [39] and Dice [29] as loss functions. The models were also finetuned using the Lovász loss function [2].

10. **NUCTech Robot Group (NUCTech Limited Co., China) - Tasks 1 & 2** and **1 & 3** are resolved simultaneously through two unified models named "Double Segmentation Network" (DSegNet). Feature extraction of DSegNet is carried out by the ResNet-50 [11] backbone. These networks have two output channels: one outputs the document boundary and text segmentation zones and the other outputs document boundary and signature segmentation. Data augmentation has also been used in the training of the models with Dice [29] as loss function.

11. **Ocean Group (Lenovo Research and Xi'an Jiaotong University, China) - Task 1:** Ensemble framework consisting of two segmentation models, Mask R-CNN [10] with PointRend module [13] and DeepLab [4] with decouple SegNet [14], whose outputs are merged to get the final result. The models were also trained with data augmentation. **Task 3:** A one-stage detection model is used to detect handwritten signature bounding box. Then a full convolutional encoder-decoder segmentation network combined with a refinement block was used to predict accurately the handwritten signature location.

12. **PA Group (AI Research Institute, OneConnect, China) - Tasks 1 & 3:** The approaches are based on ResNet-34 followed by a "Global Convolution Network" (ResNet34+GCN) [11,40]. **Task 2:** The approach is based on ResNet-18 followed by a "Pyramid Scene Parsing Network" (ResNet18+PSPNet) [11,40]. Dice loss function [29] was used in the training of all task's models.

13. **SPDB Lab Group (Shanghai Pudong Development Bank, China) - Task 1:** The approach based on three deep networks to predict the mask and vote on the prediction results of each pixel: 1) Mask R-CNN [10] which uses HRNet [31] as backbone, HR feature pyramid networks (HRFPN) [35] and PointRend [13]; 2) Multi-scale DeepLabV3 [5]; and 3) U2Net [23]. **Task 2:** The approach has three different DL models participate in the final pixel-wise voting prediction: 1) Multi-scale OCR-Net [33,36]; 2) HRNet+OCRNet [36]; and 3) DeepLabV3+ [6], which uses ResNeSt101 [38] as backbone. **Task 3:** The approach employs object detector YOLO [24], whose cropped image output is sent to several segmentation models to get the final pixel-wise voting prediction: 1) U2Net [23];

2) HRNet+OCRNet [36]; 3) ResNeSt+DeepLabV3 +[6,38]; and 4) Swin-Transformer [19].

14. **SunshineinWinter Group (Xidian University, China) - Tasks 1 & 2:** Addressed approaches use the so-called EfficientNet [32] as the backbone of the segmentation method and the "Feature Pyramid Networks" (FPN) [16] for object detection. Balanced cross-entropy [39] was used as a loss function. **Task 3**: The approach is based on the classical two-stage detection model Mask R-CNN [10], while the segmentation model is the same as the ones in Tasks 1 & 2.

15. **USTC-IMCCLab Group (University of Science and Technology of China, Hefei Zhongke Brain Intelligence Technology Co. Ltd., China) - Tasks 1 & 2:** The approached methods are based on a combination of a "Cascade R-CNN" [3] and DeepLabV3+ [6] for pixel prediction of document boundary and text zones.

16. **Wuhan Tianyu Document Algorithm Group (Wuhan Tianyu Information Industry Co. Ltd., China) - Task 1:** DeepLabV3 [5] model was used for document boundary detection. **Task 2:** A two-stage approach which uses a DeepLabV3 [5] for roughly text field detection and OCRNet [15] for locating accurately text zones. **Task 3:** OCRNet [15] was employed for handwritten signature location.

5 Results and Discussion

The set of tests made available to the participants covers the three tasks of the competition. The teams were able to choose between solving one, two, or three tasks. For each Task, we had the following distribution: Task 1 with 16 teams, Task 2 with 12 teams, and Task 3 with 11 teams. Thus, 39 submissions were evaluated for each metric.

We present the results for each task of the challenge in the following tables: Table 2 shows the results for Task 1; in Table 3 are the results of Task 2; and Table 4 shows Task 3 results. The results of the teams were evaluated using the metrics described in Sect. 3. We removed the SSIM metric from our analysis, since it did not prove to represent (all participants achieved results above 0.99999). A possible reason for this behaviour is that the images resulting from the segmentation have predominantly black pixels for all Tasks. This shows that none of the teams had problems with the structural information of the image. Therefore, the metrics used to evaluate all systems were DSC and SIFT.

For ranking the submissions, we used at first the DSC metric and as a tiebreaker criteria the SIFT one. The reason for this decision is that SIFT is invariant to rotation and scale and less sensitive to small divergences between the output images and the Ground Truth images. The results of each task will be described further and commented on the best results.

5.1 Results of the Task 1: Document Boundary Segmentation

In Task 1, all competitors achieved excellent results for the two metrics, DSC and SIFT. Table 2 shows the results and the classification of the teams. The three teams with the best classification were Ocean (1st), SPDB Lab (2nd), and PA (3rd), in this order. The first two teams achieved a result above 0.99, and both approach a similar architecture, the Mask-RCNN model. More details on the approach of the teams can be seen in Sect. 4.

Table 2. Result: 1st Challenge - Document Boundary Segmentation

Teams results and methods—Mean;($\pm std$)		
Team Name	DSC($\pm std$)	SIFT($\pm std$)
58 CV	0.969594;($\pm$0.062)	0.998698;($\pm$0.020)
Ambilight	0.983825;($\pm$0.034)	0.999223;($\pm$0.021)
Arcanite	0.986441;($\pm$0.018)	0.994290;($\pm$0.037)
Cinnamon AI	0.983389;($\pm$0.006)	0.999656;($\pm$0.014)
CUDOS	0.989061;($\pm$0.002)	0.999450;($\pm$0.007)
Dao Xianghu light of TianQuan	0.987495;($\pm$0.002)	0.998966;($\pm$0.032)
DeepVision	0.976849;($\pm$0.008)	0.999599;($\pm$0.020)
dotLAB Brazil	0.988847;($\pm$0.027)	0.999234;($\pm$0.009)
NAVER Papago	0.989194;($\pm$0.002)	0.999565;($\pm$0.008)
Nuctech robot	0.986910;($\pm$0.007)	0.999891;($\pm$0.040)
Ocean	**0.990971**;($\pm$0.001)	0.999176;($\pm$0.021)
PA	**0.989830**;($\pm$0.003)	0.999624;($\pm$0.007)
SPDB Lab	**0.990570**;($\pm$0.001)	0.999891;($\pm$0.003)
SunshineinWinter	0.970344;($\pm$0.014)	0.998358;($\pm$0.017)
USTC-IMCCLab	0.985350;($\pm$0.006)	0.996469;($\pm$0.039)
Wuhan Tianyu Document Algorithm Group	0.980272;($\pm$0.013)	0.999370;($\pm$0.010)

5.2   Results of the Task 2: Zone Text Segmentation

Table 3 shows the result of Task 2. This task was slightly more complex than Task 1, as it is possible to observe slightly greater distances between the results achieved by the teams. Here the three teams with the best scores were Ambilight (1st), Dao Xianghu light of TianQuan (2nd), and USTC-IMCCLab (3rd). Here, the winner uses a model composition, HRnet+ORC+DCNv2. Based on Task 1 segmentation, the component area is cut out and then rotated to one side. Finally, the High-Resolution Net model is trained to identify the text area.

Table 3. Result: 2nd Challenge - Zone Text Segmentation

Teams results and methods—Mean;($\pm std$)

Team Name	DSC($\pm std$)	SIFT($\pm std$)
Ambilight	**0.926624**;(± 0.051)	0.998829;(± 0.014)
Arcanite	0.837609;(± 0.069)	0.997631;(± 0.006)
Cinnamon AI	0.866914;(± 0.097)	0.995777;(± 0.009)
Dao Xianghu light of TianQuan	**0.909789**;(± 0.040)	0.998308;(± 0.004)
DeepVision	0.861304;(± 0.146)	0.994713;(± 0.015)
dotLAB Brazil	0.887051;(± 0.106)	0.997471;(± 0.006)
Nuctech robot	0.883612;(± 0.051)	0.997728;(± 0.006)
PA	0.880801;(± 0.063)	0.997607;(± 0.006)
SPDB Lab	0.881975;(± 0.061)	0.998220;(± 0.005)
SunshineinWinter	0.390526;(± 0.100)	0.981881;(± 0.025)
USTC-IMCCLab	**0.890434**;(± 0.077)	0.997933;(± 0.005)
Wuhan Tianyu Document Algorithm Group	0.814985;(± 0.097)	0.993798;(± 0.011)

5.3   Results of the Task 3: Signature Segmentation

For Task 3, as reported in the Table 4, results show very marked divergences. The distances between the results are much greater than those observed in the two previous tasks. The three teams that achieved the best results were SPDB Lab (1st), Cinnamon IA (2nd), and dotLAB Brasil (3rd). This task includes the lowest number of participating teams and lower rates results than tasks 1 and 2. These numbers raise the perception that segmenting handwritten signatures presents greater complexity among the 3 tasks addressed in the competition.

The winning group, SPDB Lab, used a strategy that involves several deep models. The Yolo model was used by this group to detect the bounding box (bbox) of the handwritten signature area. Then, the bbox is filled with 5 pixels so that the overlapping bounding box is later merged. These images are sent to various segmentation models for prediction, which are U2Net, HRNet+OCRNet, ResNeSt+Deeplabv3plus, Swin-Transformer. The results of the segmentation of various models are merged by the weighted sum, and then the results are pasted back to the original position according to the recorded location information.

5.4   Discussion

All tasks proposed in this competition received interesting strategies and some of them achieved promising results for the problems posed by the taks.

Task 1 was completed with excellent results, all above 0.96 for DSC metric. With this result of the competition, it is possible to affirm that Document Boundary Segmentation is no longer a challenge for computer vision due to the power of deep learning architectures.

The results of Task 2 show that the challenge for segmenting text zones needs an additional effort of the research community to get a definitive solution, since

Table 4. Result: 3rd Challenge - Signature Segmentation

Teams results and methods—Mean;(±*std*)		
Team Name	DSC(±*std*)	SIFT(±*std*)
Ambilight	0.812197;(±0.240)	0.839524;(±0.360)
Arcanite	0.627745;(±0.291)	0.835515;(±0.359)
Cinnamon AI	**0.841275**;(±0.226)	0.848724;(±0.353)
Dao Xianghu light of TianQuan	0.795448;(±0.262)	0.839367;(±0.361)
dotLAB Brazil	**0.837492**;(±0.412)	0.852816;(±0.324)
Nuctech robot	0.538511;(±0.316)	0.842416;(±0.351)
Ocean	0.768071;(±0.286)	0.838772;(±0.362)
PA	0.407228;(±0.314)	0.836094;(±0.356)
SPDB Lab	**0.863063**;(±0.245)	0.839501;(±0.361)
SunshineinWinter	0.465849;(±0.283)	0.835916;(±0.357)
Wuhan Tianyu Document Algorithm Group	0.061841;(±0.765)	0.589850;(±0.458)

Fig. 3. Sample of the Ocean team's result - Task 1. On the left the input image, in the centre the ground-truth image and on the right the resulting output image.

a minor error in this task may produce substantial errors when recognizing the texts in those documents.

Task 3, Handwritten signature segmentation, showed a more complex task than the other two presented in this competition. Despite the good results presented specially by Top 3 systems, they are far from the rates of page detection algorithms. Therefore, if we need the signature pixels to further verify its similarity against a reference signature, there is still room for new strategies and improvements in this task.

All groups used deep learning models, most of which made use of encoder-decoder architectures like the U-net. Another characteristic worth noting is the strategy of composing several deep models working together or sequentially to detect and segment regions of interest.

In the following, we can see some selected results of the 3 winners teams of each task.

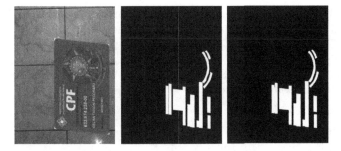

Fig. 4. Sample of the Ambilight team's result - Task 2. On the left the input image, in the centre the ground-truth image and on the right the resulting output image.

Fig. 5. Sample of the SPDB Lab team's result - Task 3. On the left the input image, in the centre the ground-truth image and on the right the resulting output image. Just below the original images are the regions of enlarged signatures for better viewing.

6   Conclusion

The models proposed by the competitors were based on consolidated state-of-the-art architectures. This increased the level and complexity of the competition. Strategies such as using one task as the basis for another task were also successfully applied by some groups.

Nevertheless, we are worried about the response time of such approaches which is important in the case of real-time applications. In addition, one must also consider the application for embedded systems and mobile devices with limited computational resources. A possible future competition may include in the scope these aspects.

To conclude, we see that there is still room for other strategies and solutions for the tasks of text and signature segmentation proposed in this competition.

We expect the datasets produced in this contest (https://icdar2021.poli.br/) can be used by the research community to improve their proposals for such tasks.

Acknowledgment. This study was financed in part by: Coordenação de Aperfeiçoamento de Pessoal de Nível Superior - Brasil (CAPES) - Finance Code 001, and CNPq - Brazilian research agencies.

References

1. Badrinarayanan, V., Kendall, A., Cipolla, R.: Segnet: a deep convolutional encoder-decoder architecture for image segmentation. IEEE Trans. Pattern Anal. Mach. Intell. **39**(12), 2481–2495 (2017)
2. Berman, M., Triki, A.R., Blaschko, M.B.: The lovász-softmax loss: a tractable surrogate for the optimization of the intersection-over-union measure in neural networks. In: Proceedings of the IEEE Conference on Computer Vision and Pattern Recognition, pp. 4413–4421 (2018)
3. Cai, Z., Vasconcelos, N.: Cascade R-CNN: high quality object detection and instance segmentation. IEEE Trans. Pattern Anal. Mach. Intell. **43**(5), 1483–1498 (2021). https://doi.org/10.1109/TPAMI.2019.2956516
4. Chen, L.C., Papandreou, G., Kokkinos, I., Murphy, K., Yuille, A.L.: Deeplab: semantic image segmentation with deep convolutional nets, atrous convolution, and fully connected crfs. IEEE Trans. Pattern Anal. Mach. Intell. **40**(4), 834–848 (2017)
5. Chen, L.C., Papandreou, G., Schroff, F., Adam, H.: Rethinking atrous convolution for semantic image segmentation. arXiv preprint arXiv:1706.05587 (2017)
6. Chen, L.C., Zhu, Y., Papandreou, G., Schroff, F., Adam, H.: Encoder-decoder with atrous separable convolution for semantic image segmentation. In: Proceedings of the European conference on computer vision (ECCV), pp. 801–818 (2018)
7. Deng, J., Dong, W., Socher, R., Li, L.J., Li, K., Fei-Fei, L.: Imagenet: a large-scale hierarchical image database. In: 2009 IEEE conference on Computer Vision and Pattern Recognition, pp. 248–255. IEEE (2009)
8. Dice, L.R.: Measures of the amount of ecologic association between species. Ecology **26**(3), 297–302 (1945)
9. Ferrer, M.A., Diaz-Cabrera, M., Morales, A.: Static signature synthesis: A neuromotor inspired approach for biometrics. IEEE Trans. Pattern Anal. Mach. Intell. **37**(3), 667–680 (2014)
10. He, K., Gkioxari, G., Dollár, P., Girshick, R.: Mask R-CNN. In: Proceedings of the IEEE International Conference on Computer Vision, pp. 2961–2969 (2017)
11. He, K., Zhang, X., Ren, S., Sun, J.: Deep residual learning for image recognition. In: Proceedings of the IEEE Conference on Computer Vision and Pattern Recognition, pp. 770–778 (2016)
12. Junior, C.A., da Silva, M.H.M., Bezerra, B.L.D., Fernandes, B.J.T., Impedovo, D.: FCN+ RL: a fully convolutional network followed by refinement layers to offline handwritten signature segmentation. In: 2020 International Joint Conference on Neural Networks (IJCNN), pp. 1–7. IEEE (2020)
13. Kirillov, A., Wu, Y., He, K., Girshick, R.: Pointrend: image segmentation as rendering. In: Proceedings of the IEEE/CVF Conference on Computer Vision and Pattern Recognition, pp. 9799–9808 (2020)

14. Li, X., et al.: Improving semantic segmentation via decoupled body and edge supervision. arXiv preprint arXiv:2007.10035 (2020)
15. Liao, M., Wan, Z., Yao, C., Chen, K., Bai, X.: Real-time scene text detection with differentiable binarization. In: Proceedings of the AAAI Conference on Artificial Intelligence, vol. 34, pp. 11474–11481 (2020)
16. Lin, T.Y., Dollár, P., Girshick, R., He, K., Hariharan, B., Belongie, S.: Feature pyramid networks for object detection. In: Proceedings of the IEEE Conference on Computer Vision and Pattern Recognition, pp. 2117–2125 (2017)
17. Lin, T.Y., Goyal, P., Girshick, R., He, K., Dollár, P.: Focal loss for dense object detection. In: Proceedings of the IEEE International Conference on Computer Vision, pp. 2980–2988 (2017)
18. Liu, J.J., Hou, Q., Cheng, M.M., Wang, C., Feng, J.: Improving convolutional networks with self-calibrated convolutions. In: Proceedings of the IEEE/CVF Conference on Computer Vision and Pattern Recognition, pp. 10096–10105 (2020)
19. Liu, Z., et al.: Swin transformer: Hierarchical vision transformer using shifted windows. arXiv preprint arXiv:2103.14030 (2021)
20. Lowe, D.G.: Object recognition from local scale-invariant features. In: Proceedings of the Seventh IEEE International Conference on Computer Vision, vol. 2, pp. 1150–1157. IEEE (1999)
21. das Neves Junior, R.B., Verçosa, L.F., Macêdo, D., Bezerra, B.L.D., Zanchettin, C.: A fast fully octave convolutional neural network for document image segmentation. In: 2020 International Joint Conference on Neural Networks (IJCNN), pp. 1–8. IEEE (2020)
22. Ortega-Garcia, J., et al.: MCYT baseline corpus: a bimodal biometric database. IEE Proc.-Vis. Image Signal Process. **150**(6), 395–401 (2003)
23. Qin, X., Zhang, Z., Huang, C., Dehghan, M., Zaiane, O.R., Jagersand, M.: U2-Net: going deeper with nested u-structure for salient object detection. Pattern Recognit. **106**, 107404 (2020)
24. Redmon, J., Divvala, S., Girshick, R., Farhadi, A.: You only look once: unified, real-time object detection. In: Proceedings of the IEEE Conference on Computer Vision and Pattern Recognition, pp. 779–788 (2016)
25. Ronneberger, O., Fischer, P., Brox, T.: U-Net: convolutional networks for biomedical image segmentation. In: Navab, N., Hornegger, J., Wells, W.M., Frangi, A.F. (eds.) MICCAI 2015. LNCS, vol. 9351, pp. 234–241. Springer, Cham (2015). https://doi.org/10.1007/978-3-319-24574-4_28
26. de Sá Soares, A., das Neves Junior, R.B., Bezerra, B.L.D.: Bid dataset: a challenge dataset for document processing tasks. In: Anais Estendidos do XXXIII Conference on Graphics, Patterns and Images, pp. 143–146. SBC (2020)
27. Sandler, M., Howard, A., Zhu, M., Zhmoginov, A., Chen, L.C.: Mobilenetv 2: inverted residuals and linear bottlenecks. In: Proceedings of the IEEE Conference on Computer Vision and Pattern Recognition, pp. 4510–4520 (2018)
28. Silva, P.G., Junior, C.A., Lima, E.B., Bezerra, B.L., Zanchettin, C.: Speeding-up the handwritten signature segmentation process through an optimized fully convolutional neural network. In: 2019 International Conference on Document Analysis and Recognition (ICDAR), pp. 1417–1423. IEEE (2019)
29. Sudre, C.H., Li, W., Vercauteren, T., Ourselin, S., Jorge Cardoso, M.: Generalised dice overlap as a deep learning loss function for highly unbalanced segmentations. In: Cardoso, M.J., et al. (eds.) DLMIA/ML-CDS -2017. LNCS, vol. 10553, pp. 240–248. Springer, Cham (2017). https://doi.org/10.1007/978-3-319-67558-9_28
30. Sun, K., et al.: High-resolution representations for labeling pixels and regions. arxiv 2019. arXiv preprint arXiv:1904.04514 (2019)

31. Sun, K., Xiao, B., Liu, D., Wang, J.: Deep high-resolution representation learning for human pose estimation. In: Proceedings of the IEEE/CVF Conference on Computer Vision and Pattern Recognition, pp. 5693–5703 (2019)
32. Tan, M., Le, Q.: Efficientnet: rethinking model scaling for convolutional neural networks. In: International Conference on Machine Learning, pp. 6105–6114. PMLR (2019)
33. Tao, A., Sapra, K., Catanzaro, B.: Hierarchical multi-scale attention for semantic segmentation. arXiv preprint arXiv:2005.10821 (2020)
34. Wang, W., et al.: Efficient and accurate arbitrary-shaped text detection with pixel aggregation network. In: Proceedings of the IEEE/CVF International Conference on Computer Vision, pp. 8440–8449 (2019)
35. Wei, S., et al.: Precise and robust ship detection for high-resolution sar imagery based on hr-sdnet. Remote Sens. $12(1)$, 167 (2020)
36. Yuan, Y., Chen, X., Wang, J.: Object-contextual representations for semantic segmentation. arXiv preprint arXiv:1909.11065 (2019)
37. Yuan, Y., Xie, J., Chen, X., Wang, J.: SegFix: model-agnostic boundary refinement for segmentation. In: Vedaldi, A., Bischof, H., Brox, T., Frahm, J.-M. (eds.) ECCV 2020. LNCS, vol. 12357, pp. 489–506. Springer, Cham (2020). https://doi.org/10.1007/978-3-030-58610-2_29
38. Zhang, H., et al.: Resnest: Split-attention networks. arXiv preprint arXiv:2004.08955 (2020)
39. Zhang, Z., Sabuncu, M.R.: Generalized cross entropy loss for training deep neural networks with noisy labels. arXiv preprint arXiv:1805.07836 (2018)
40. Zhao, H., Shi, J., Qi, X., Wang, X., Jia, J.: Pyramid scene parsing network. In: Proceedings of the IEEE Conference on Computer Vision and Pattern Recognition, pp. 2881–2890 (2017)
41. Zhou, Z., Rahman Siddiquee, M.M., Tajbakhsh, N., Liang, J.: UNet++: a nested u-net architecture for medical image segmentation. In: Stoyanov, D., et al. (eds.) DLMIA/ML-CDS -2018. LNCS, vol. 11045, pp. 3–11. Springer, Cham (2018). https://doi.org/10.1007/978-3-030-00889-5_1
42. Zhu, X., Hu, H., Lin, S., Dai, J.: Deformable convnets v2: more deformable, better results. In: Proceedings of the IEEE/CVF Conference on Computer Vision and Pattern Recognition, pp. 9308–9316 (2019)

ICDAR 2021 Competition on Historical Map Segmentation

Joseph Chazalon[1]([✉]), Edwin Carlinet[1], Yizi Chen[1,2], Julien Perret[2,3],
Bertrand Duménieu[3], Clément Mallet[2], Thierry Géraud[1],
Vincent Nguyen[4,5], Nam Nguyen[4], Josef Baloun[6,7], Ladislav Lenc[6,7],
and Pavel Král[6,7]

[1] EPITA R&D Lab. (LRDE), EPITA, Le Kremlin-Bicêtre, France
`joseph.chazalon@lrde.epita.fr`
[2] Univ. Gustave Eiffel, IGN-ENSG, LaSTIG, Saint-Mande, France
[3] LaDéHiS, CRH, EHESS, Paris, France
[4] L3i, University of La Rochelle, La Rochelle, France
[5] Liris, INSA-Lyon, Lyon, France
[6] Department of Computer Science and Engineering, University of West Bohemia,
Univerzitní, Pilsen, Czech Republic
[7] NTIS - New Technologies for the Information Society, University of West Bohemia,
Univerzitní, Pilsen, Czech Republic

Abstract. This paper presents the final results of the ICDAR 2021 Competition on Historical Map Segmentation (MapSeg), encouraging research on a series of historical atlases of Paris, France, drawn at 1/5000 scale between 1894 and 1937. The competition featured three tasks, awarded separately. Task 1 consists in detecting building blocks and was won by the L3IRIS team using a DenseNet-121 network trained in a weakly supervised fashion. This task is evaluated on 3 large images containing hundreds of shapes to detect. Task 2 consists in segmenting map content from the larger map sheet, and was won by the UWB team using a U-Net-like FCN combined with a binarization method to increase detection edge accuracy. Task 3 consists in locating intersection points of geo-referencing lines, and was also won by the UWB team who used a dedicated pipeline combining binarization, line detection with Hough transform, candidate filtering, and template matching for intersection refinement. Tasks 2 and 3 are evaluated on 95 map sheets with complex content. Dataset, evaluation tools and results are available under permissive licensing at https://icdar21-mapseg.github.io/.

Keywords: Historical maps · Map vectorization · Competition

1 Introduction

Motivation. This competition consists in solving several challenges which arise during the processing of images of historical maps. In the Western world, the

This work was partially funded by the French National Research Agency (ANR): Project SoDuCo, grant ANR-18-CE38-0013. We thank the City of Paris for granting us with the permission to use and reproduce the atlases used in this work.

J. Lladós et al. (Eds.): ICDAR 2021, LNCS 12824, pp. 693–707, 2021.
https://doi.org/10.1007/978-3-030-86337-1_46

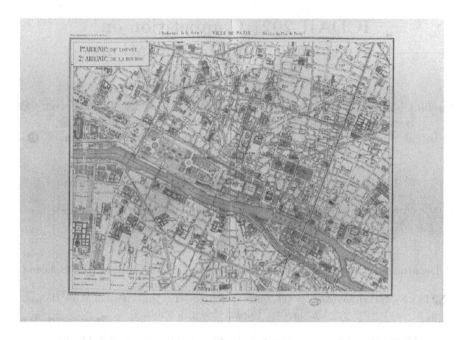

Fig. 1. Sample map sheet. Original size: 11136 × 7711 pixels.

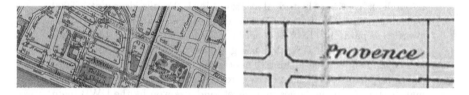

Fig. 2. Some map-related challenges (*left*): visual polysemy, planimetric overlap, text overlap. . . and some document-related challenges (*right*): damaged paper, non-straight lines, image compression, handwritten text. . .

rapid development of geodesy and cartography from the 18th century resulted in massive production of topographic maps at various scales. City maps are of utter interest. They contain rich, detailed, and often geometrically accurate representations of numerous geographical entities. Recovering spatial and semantic information represented in old maps requires a so-called *vectorization* process. Vectorizing maps consists in transforming rasterized graphical representations of geographic entities (often maps) into instanced geographic data (or vector data), that can be subsequently manipulated (using Geographic Information Systems). This is a key challenge today to better preserve, analyze and disseminate content for numerous spatial and spatio-temporal analysis purposes.

Tasks. From a document analysis and recognition (DAR) perspective, full map vectorization covers a wide range of challenges, from content separation to text recognition. *MapSeg* focuses on 3 critical steps of this process. (i) Hierarchical information extraction, and more specifically the *detection of building blocks* which form a core layer of map content. This is addressed in *Task 1.* (ii) *Map content segmentation*, which needs to be performed very early in the processing pipeline to separate meta-data like titles, legends and other elements from core map content. This is addressed in *Task 2.* (iii) Geo-referencing, i.e. mapping historical coordinate system into a recent reference coordinate system. Automating this process requires the detection of well-referenced points, and we are particularly interested in detecting *intersection points between graticule lines* for this purpose. This is addressed in *Task 3.*

Dataset. *MapSeg* dataset is extracted from a series of 9 atlases of the City of Paris[1] produced between 1894 and 1937 by the Map Service (*"Service du plan"*) of the City of Paris, France, for the purpose of urban management and planning. For each year, a set of approximately 20 sheets forms a tiled view of the city, drawn at 1/5000 scale using trigonometric triangulation. Figure 1 is an example of such sheet. Such maps are highly detailed and very accurate even by modern standards. This material provides a very valuable resource for historians and a rich body of scientific challenges for the document analysis and recognition (DAR) community: map-related challenges (Fig. 2, *left*) and document-related ones (Fig. 2, *right*). The actual dataset is built from a selection of 132 map sheets extracted from these atlases. Annotation was performed manually by trained annotators and required 400 h of manual work to annotate the 5 sheets of the dataset for Task 1 and 100 h to annotate the 127 sheets for Tasks 2 and 3. A special care was observed to minimize the spatial and, to some extent, temporal overlap between train, validation and test sets: except for Task 1 which contains districts 1 and 2 both in train and test sets (from different atlases to assess the potential for the exploitation of information redundancy over time), each district is either in the training, the validation or the test set. This is particularly important for Tasks 2 and 3 as the general organization of the map sheets are very similar for sheets representing the same area.

Competition Protocol. The MapSeg challenge ran from November 2020 to April 2021. Participants were provided with training and validation sets for each 3 tasks by November 2020. The test phase for Task 1 started with the release of test data on April 5th, 2021 and ended in April 9th. The test phase for Tasks 2 and 3 started with the release of test data on April 12th, 2021 and ended in April 16th. Participants were requested to submit the results produced by their methods (at most 2) over the test images, computed by themselves.

Open Data, Tools and Results. To improve the reliability of the evaluation and competition transparency, evaluation tools were released (executable and open source code) early in the competition, so participants were able to

---

[1] *Atlas municipal des vingt arrondissements de Paris.* 1894, 1895, 1898, 1905, 1909, 1912, 1925, 1929, and 1937. *Bibliothèque de l'Hôtel de Ville.* City of Paris. France. Online resources for the 1925 atlas: https://bibliotheques-specialisees.paris.fr/ark:/73873/pf0000935524.

evaluate their methods on the validation set using exactly the same tools as the organizers on the test set. More generally, all competition material is made available under very permissive licenses at https://icdar21-mapseg.github.io/: dataset with ground truth, evaluation tools and participants results can be downloaded, inspected, copied and modified.

Participants. The following teams took part in the competition.

CMM Team: *Team:* Mateus Sangalli, Beatriz Marcotegui, José Marcio Martins Da Cruz, Santiago Velasco-Forero, Samy Blusseau; *Institutions:* Center for Mathematical Morphology, Mines ParisTech, PSL Research University, France; *Extra material:* morphological library Smil: http://smil.cmm. minesparis.psl.eu; code: https://github.com/MinesParis-MorphoMath.

IRISA Team: *Competitor:* Aurélie Lemaitre; *Institutions:* IRISA/Université Rennes 2, Rennes, France; *Extra material:* http://www.irisa.fr/intuidoc/.

L3IRIS Team: *Team:* Vincent Nguyen and Nam Nguyen; *Institutions:* L3i, University of La Rochelle, France; Liris, INSA-Lyon, France; *Extra material:* https://gitlab.univ-lr.fr/nnguye02/weakbiseg.

UWB Team: *Team:* Josef Baloun, Ladislav Lenc and Pavel Král; *Institutions:* Department of Computer Science and Engineering, University of West Bohemia, Univerzitní, Pilsen, Czech Republic; NTIS - New Technologies for the Information Society, University of West Bohemia, Univerzitní, Pilsen, Czech Republic; *Extra material:* https://gitlab.kiv.zcu.cz/balounj/21_icdar_ mapseg_competition.

WWU Team: *Team:* Sufian Zaabalawi, Benjamin Risse; *Institution:* Münster University, Germany; *Extra material:* Project on automatic vectorization of historical cadastral maps: https://dhistory.hypotheses.org/346.

2   Task 1: Detect Building Blocks

This task consists in detecting a set of closed shapes (building blocks) in the map image. Building blocks are coarser map objects which can regroup several elements. Detecting these objects is a critical step in the digitization of historical maps because it provides essential components of a city. Each building block is symbolized by a closed shape which can enclose other objects and lines. Building blocks are surrounded by streets, rivers fortification wall or others, and are never directly connected. Building blocks can sometimes be reduced to a single spacial building, symbolized by diagonal hatched areas. Given the image of a complete map sheet and a mask of the map area, participants had to detect each building block as illustrated in Fig. 3.

2.1   Dataset

Inputs form a set of JPEG RGB images which are quite large ($\approx$8000 $\times$ 8000 pixels). They are complete map sheet images as illustrated in Fig. 1, cropped to the relevant area (using the ground truth from Task 2). Remaining non-relevant

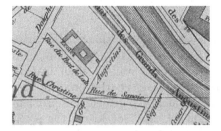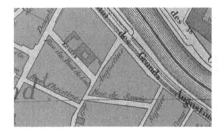

Fig. 3. Input (*left*) and output (*right*, orange overlay) image excerpts for task 1. (Color figure online)

pixels were replaced by black pixels. Expected output for this task is a binary mask of the building blocks: pixels falling inside a building block must be marked with value 255, and other pixels must be marked with value 0. The dataset is separated as follows: the train set contains 1 image (sheet 1 of 1925 atlas—903 building blocks), the validation set contains 1 image (sheet 14 of 1925 atlas—659 building blocks), and the test set contains 3 images (sheet 1 of 1898 atlas, sheet 3 of 1898 and 1925 atlases—827, 787 and 828 building blocks).

2.2 Evaluation Protocol

Map vectorization requires an accurate detection of shape boundaries. In order to assess both the detection and segmentation quality of the target shapes, and therefore to avoid relying only on pixel-wise accuracy, we use the COCO Panoptic Quality (PQ) score [8] which is based on an instance segmentation metric. This indicator is computed as follows:

$$PQ = \underbrace{\frac{\sum_{(p,g)\in TP} \text{IoU}(p,g)}{|TP|}}_{\text{segmentation quality (SQ)}} \times \underbrace{\frac{|TP|}{|TP| + \frac{1}{2}|FP| + \frac{1}{2}|FN|}}_{\text{recognition quality (RQ)}}$$

where TP is the set of matching pairs $(p, g) \in (P \times G)$ between predictions (P) and reference (G), FP is the set of unmatched predicted shapes, and FN is the set of unmatched reference shapes. Shapes are considered as matching when:

$$\text{IoU}(p,g) = \frac{p \cap g}{p \cup g} > 0.5.$$

COCO SQ (segmentation quality) is the mean IoU between matching shapes: matching shapes in reference and prediction have an IoU > 0.5. **COCO RQ** (detection/recognition quality) is detection F-score for shape: a predicted shape is a true positive if it as an IoU > 0.5 with a reference shape. The resulting **COCO PQ** indicator ranges from 0 (worst) to 1 (perfect). An alternate formulation of this metric is to consider it as the integral of the detection F-score over all possible IoU thresholds between 0 and 1 [4]. We also report the "F-Score vs IoU threshold" curves for each method to better assess their behavior.

2.3   Method Descriptions

L3IRIS—*Winning Method*. From previous experiments, the L3IRIS team reports that both semantic segmentation and instance segmentation approaches led to moderate results, mainly because of the amount of available data. Instead of trying to detect building blocks directly, they reversed the problem and tried to detect non-building-blocks as a binary semantic segmentation problem which facilitated the training process. Additionally, they considered the problem in a semi/weakly supervised setting where training data includes 2 images (val+train): the label is available for one image and is missing for the other image. Their binary segmentation model relies on a DenseNet-121 [7] backbone and is trained using the weakly supervised learning method from [12]. Input is processed as 480×480 tiles. During training, tiles are generated by random crops and augmented with standard perturbations: noise, rotation, gray, scale, etc. Some post-processing techniques like morphological operators are used to refine the final mask (open the lightly connected blocks and fill holes). The parameter for the post-processing step is selected based on the results on the validation image. Training output is a single models; not ensemble technique is used. The weakly supervised training enables the training of a segmentation network in a setting where only a subset of the training objects are labelled. The overcome the problem of missing labels, the L3IRIS team uses two energy maps $w+$ et $w-$ which represent the certainty that a pixel being positive (object's pixel) or negative (background pixel). These two energy maps are evolved during the training at each epoch based on prediction scores in multiple past epochs. This technique implicitly minimizes the entropy of the predictions on unlabeled data which can help to boost the performance of the model [5]. For Task 1, they considered labeled images (from training set) and unlabeled ones (from validation set) for which the label was retained for validation. The two energy maps are applied only for images from the validation set.

CMM Method 1. This method combines two parallel sub-methods, named **1A** and **1B**. In both sub-methods, the RGB images are first processed by a U-Net [14] trained to find edges of the blocks. Then, a morphological processing is applied to the image of edges in order to segment the building blocks. Finally, the results of both sub-methods are combined to produce a unique output. In method **1A**, the U-Net is trained with data augmentation that uses random rotations, reflections, occlusion by vertical and horizontal lines and color changes. The loss function penalizes possible false positive pixels (chosen based on the distance from edge pixels in the color space) with a higher cost. To segment the blocks, first a top-hat operator is applied to remove large vertical and horizontal lines and an anisotropic closing [3] is used to remove holes in the edges. Edges are binarized by a hysteresis threshold and building blocks are obtained by a fill-holes operator followed by some post-processing. The result is a clean segmentation, but with some relatively large empty regions. In method **1B**, the U-Net is trained using the same augmentation process, but the loss function weights errors to compensate class imbalance. Method 2 (described hereafter) is applied to the inverse of the edges obtained by this second U-Net. To combine both methods, a closing by a large structuring element is used to find the dense regions in the output of **1A**, the complement of which are the empty regions of **1A**. The final output is equal to **1A** in the dense region and to **1B** in the empty regions.

Method 2. This method is fully morphological. It works on a gray-level image, obtained by computing the luminance from the RGB color image. The main steps are the following: (i) *Area closing* with 1000 pixels as area threshold; this step aims at *removing small dark components* such as letters. (ii) *Inversion of contrast* of the resulting image followed by a *fill-holes* operator; The result of this step is an image where most blocks appear as flat zones, and so do large portions of streets. (iii) The *morphological gradient* of the previous image is computed and then *reconstructed by erosion* starting from markers chosen as the minima with dynamic not smaller than $h = 2$. This produces a new, simplified, gradient image. (iv) The *watershed transform* is applied to the latter image, yielding a labelled image. (v) A *filtering of labelled components* removes (that is, sets to zero) components smaller than 2000 pixels and larger than 1M pixels; and also removes the components with area smaller than 0.3 times the area of their bounding box (this removes large star-shaped street components). All kept components are set to 255 to produce a binary image. (vi) A *fill-holes* operator is applied to the latter binary image. (vii) *Removal of river components*: A mask of the river is computed by a morphological algorithm based on a large opening followed by a much larger closing, a threshold and further refinements. Components with an intersection of at least 60% of their area with the river mask are removed.

WWU Method 1. This method relies on a binary semantic segmentation U-Net [14] trained with Dice loss on a manual selection of 2000 image patches of size $256 \times 256 \times 3$ covering representative map areas. The binary segmentation ground truth is directly used as target. This U-Net architecture uses 5 blocks and each convolutional layer has a 3×3 kernel filter with no padding. The expending path uses transpose convolutions with $2\times$ stride. Intermediate activations use RELU and the final activation is a sigmoid.

Method 2. This method relies on a Deep Distance transform. The binary label image is first transformed into a Euclidean *distance transform map* (DTM), which highlights the inner region of the building blocks more than the outer edges. Like for method 1, 2000 training patch pairs of size 256×256 are generated. The U-Net architecture is similar to the previous method, except that the loss function is the mean square error loss (MSE) and the prediction is a DTM. This DTM is then thresholded and a watershed transform is used to fill regions of interest and extract building block shapes.

2.4   Results and Discussion

Results and ranking of each submitted method is given in Fig. 4: table on the left summarizes the COCO PQ indicators we computed and the plot on the right

Rank	Team (method)	COCO PQ (%) ↑
1	L3IRIS	74.1
2	CMM (1)	62.6
3	CMM (2)	44.0
4	WWU (1)	06.4
5	WWU (2)	04.2

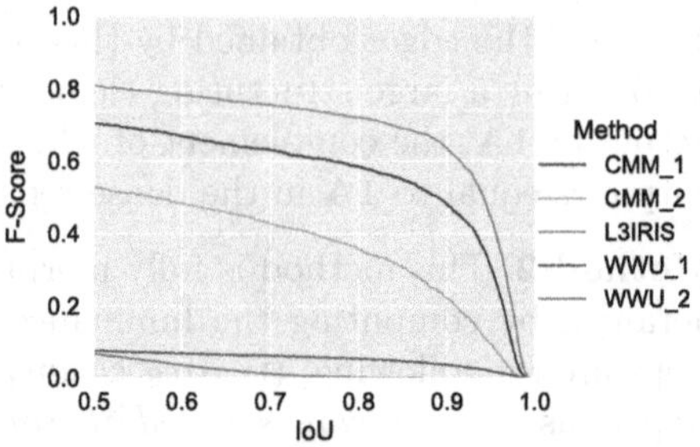

Fig. 4. Final COCO PQ scores for Task 1 (*left*)—score ranges from 0 (worst) to 100% (best), and plot of the F-Score for each threshold between 0.5 and 1 (*right*). COCO PQ is the area under this curve plus $\frac{1}{2}$ times the intersect at 0.5.

illustrates the evolution of the retrieval F-Score for each possible IoU threshold between 0.5 and 1. While the WWU 1 and 2 approaches get low scores, it must be noted that this is due to local connections between shapes which led to a strong penalty with the COCO PQ metric. As they implement very natural ways to tackle such problem, they provide a very valuable performance baseline. CMM method 2 is fully morphological and suffers from an insufficient filtering of noise content, while CMM method 1 gets much better results by combining edge filtering using some fully convolutional network with morphological filtering. L3IRIS winning approach improves over this workflow by leveraging weakly-supervised training followed by some morphological post-processing.

3   Task 2: Segment Map Content Area

This task consists in segmenting the map content from the rest of the sheet. This is a rather classical document analysis task as it consists in focusing on the relevant area in order to perform a dedicated analysis. In our case, Task 1 would be the following stage in the pipeline. Given the image of a complete map sheet (illustrated in Fig. 1), participants had to locate the boundary of the map content, as illustrated in Fig. 5.

3.1   Dataset

The inputs form a set of JPEG RGB images which are quite large ($\approx$10000 × 10000 pixels). They are complete map sheet images. Expected output for this task is a binary mask of the map content area: pixels belonging to map contents are marked with value 255 and other pixels are marked with value 0. The dataset is separated as follows: the train set contains 26 images, the validation set contains 6 images, and the test set contains 95 images. As in the atlas series we have multiple occurrences of similar sheets representing the same area, we made sure that each sheet appeared in only one set.

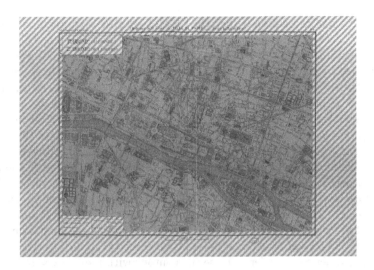

Fig. 5. Illustration of expected outputs for Task 2: green area is the map content and red hatched area is the background. (Color figure online)

3.2 Evaluation Protocol

We evaluated the quality of the segmentation using the Hausdorff distance between the ground-truth shape and the predicted one. This measure has the advantage over the IoU, Jaccard index and other area measures that it keeps a good "contrast" between results in the case of large objects (because there is no normalization by the area). More specifically, we use the "Hausdorff 95" variant which discards the 5 percentiles of higher values (assumed to be outliers) to produce a more stable measure. Finally, we averaged indicators for all individual map images to produce a global indicator. The resulting measure is an error measure which ranges from 0 (best) to a large value depending on image size.

3.3 Method Descriptions

UWB—*Winning Method*. This method is based on the U-Net-like fully convolutional networked used in [1]. This network takes a whole downsampled page as an input and predicts the border of the expected area. Border training samples were generated from the original ground truth (GT) files. The values are computed using a Gaussian function, where $\sigma = 50$ and the distance from border is used. Training is performed using the training set and augmentation (mirroring, rotation and random distortion) [2] to increase the amount of the training samples. To improve the location of the detected border, a binarized image is generated using a recursive Otsu filter [13] followed by the removal of small components. Network prediction and the binarized image are post-processed and combined with the use of connected components and morphological operations, which parameters are calibrated on the validation set. The result of the process is the predicted mask of a map content area.

CMM. This method assumes that the area of interest is characterized by the presence of black lines connected to each other all over the map. These connections being due either to the external frame, graticule lines, streets, or text superimposed to the map. Thus, the main idea is detecting black lines and reconstructing them from a predefined rectangle marker in the center of the image. The method is implemented with the following steps. (i) Eliminate map margin (M) detected by the quasi-flat zone algorithm and removed from the whole image (I_0): $I = I_0 - M$. (ii) Then, black lines (B), are extracted by a thresholded black-top-hat ($B = Otsu(I - \varphi(I))$) (iii) A white rectangle (R) is drawn centered in the middle of the image and of dimensions $\frac{W}{2} \times \frac{H}{2}$, with W and H the image width and height respectively. (iv) Black lines B are reconstructed by dilation from the centered rectangle: $B_s = Build(R, B)$. Several dark frames surround the map. Only the inner one is connected to the drawn rectangle R and delimits the area of interest. (v) The frame (black line surrounding the map) may not be complete due to lack of contrast or noise. A watershed is applied to the inverse of the distance function in order to close the contour with $markers = R \cup border$. The region corresponding to marker R becomes the area of interest. (vi) Finally, legends are removed from the area of interest. Legends are the regions from the inverse of B_s that are rectangular and close to the border of the area of interest.

IRISA. This approach is entirely based on a grammatical rule-based system [9] which combines visual clues of line segments extracted in the document at various levels of a spatial pyramid. The line segment extractor, based on a multi-scale Kalman filtering [10], is fast and robust to noise, and can deal with slope and curvature. Border detection is performed in two steps: (i) Line segments are first extracted at low resolution level (scale 1:16) to provides visual clues on the presence of the double rulings outside the map region. Then, the coarse enclosing rectangle is detected using a set of grammar rules. (ii) Line segments are extracted at medium resolution level (scale 1:2) to detect parts of contours of the maps, and of title regions. Another set of grammar rules are then used to describe and detect the rectangular contour with smaller rectangle (title and legend) in the corners. The performance is this approach is limited by the grammar rules which do not consider, though it would be possible, the fact that the map content may lay outside the rectangular region, nor it considers that some legend components may not be located at the corners of the map.

L3IRIS. This approach leverages the cutting edge few-shot learning technique HSNet [11] for image segmentation, in order to cope with the limited amount of training data. With this architecture, the prediction of a new image (query image) will be based on the trained model and a training image (support image). In practice, the L3IRIS team trained the HSNet [11] from scratch with a Resnet 50 backbone from available data to predict the content area in the input map with 512×512 image input. Because post-processing techniques to smooth and fill edges did not improve the evaluation error, the authors kept the results from the single trained model as the final predicted maps.

Rank	Team	Hausdorff 95 (pix.) ↓
1	UWB	19
2	CMM	85
3	IRISA	112
4	L3IRIS	126

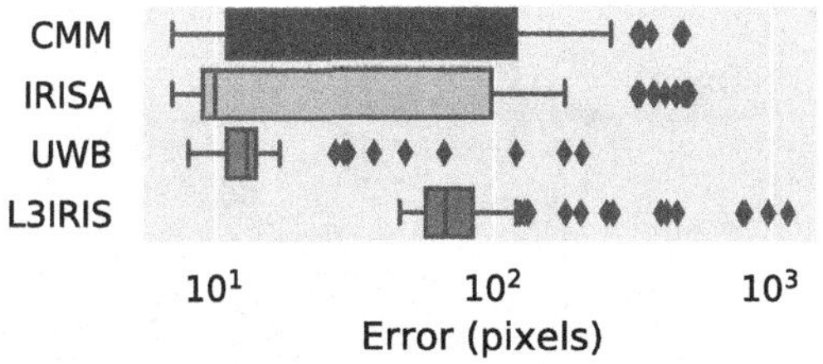

Fig. 6. Final Hausdorff 95 errors for Task 2 (*left*)—values range from 0 pixels (best) to arbitrarily large values, and error distribution for all test images (*right*).

3.4 Results and Discussion

Results and ranking of each submitted method is given in Fig. 6: table on the left summarizes the Hausdorff 95 indicators we computed and the plot on the right illustrates the error distribution (in log scale) for all test images. The L3IRIS method, based on a single FCN leveraging recent few-shot learning technique produces overly smoothed masks. The IRISA method, based on a robust line segment detector embedded in a rule-based system, was penalized by unseen frame configurations. However, it produced very accurate results for know configurations. The CMM approach, based on morphological processing, also produced very accurate frame detection and generalized better. The remaining errors are due to missed regions considered as background close to the boundary. The UWB winning method, finally, efficiently combines a coarse map content detection thanks to a deep network, with a re-adjustment of detected boundaries using a recursive Otsu binarization and some morphological post-processing.

4 Task 3: Locate Graticule Lines Intersections

This task consists in locating the intersection points of graticule lines. Graticule lines are lines indicating the North/South/East/West major coordinates in the map. They are drawn every 1000 m in each direction and overlap with the rest of the map content. Their intersections are very useful to geo-reference the map image, i.e. for projecting map content in a modern geographical coordinate reference system. Given the image of a complete map sheet, participants had to locate the intersection points of such lines, as illustrated in Fig. 7.

4.1 Dataset

The inputs for this task are exactly the same as Task 2. Expected output for this task is, for each input image, the list intersection coordinates. The dataset is separated exactly as for Task 2 in terms of train, validation and test sets.

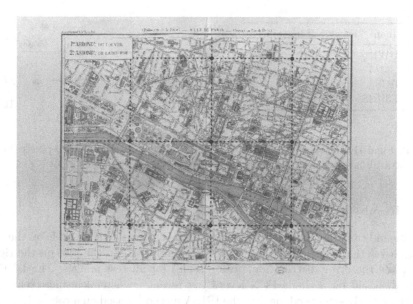

Fig. 7. Illustration of expected outputs for Task 3: dashed green lines are the graticule lines and red dots are the intersections points to locate. (Color figure online)

4.2 Evaluation Protocol

For each map sheet, we compared the predicted coordinates with the expected ones. We used an indicator which considers detection and localization accuracy simultaneously, like what we did for Task 1. A predicted point was considered as a correct detection if it was the closest predicted point of a ground truth (expected) point and if the distance between the expected point and the predicted one is smaller than a given threshold. We considered all possible thresholds between 0 and 50 pixels, which roughly corresponds to 20 m for these map images, and is an upper limit over which the registration would be seriously disrupted. In practice, we computed for each map sheet and for each possible threshold the number of correct predictions, the number of incorrect ones and the number of expected elements. This allowed us to plot the F_β score vs threshold curve for a range of thresholds. We set $\beta = 0.5$ to weights recall lower than precision because for this task it takes several good detections to correct a wrong one in the final registration. The area under this "$F_{0.5}$ score vs threshold" curve was used as performance indicator; such indicator blends two indicators: point detection and spatial accuracy. Finally, we computed the average of the measures for all individual map images to produce a global indicator. The resulting measure is a value between 0 (worst) and 1 (best).

4.3 Method Descriptions

UWB—*Winning Method*. This method is based on three main steps. (i) First, a binary image is generated using the recursive Otsu approach described in Sect. 3.3. This image is then masked using the map content area predicted for Task 2. (ii) Graticule lines are then detected a Hough Line Transform. While the accumulator bins contain a lot of noise, the following heuristics were used to filter the candidates: graticule lines are assumed to be straight, to be equally spaced and either parallel or perpendicular, and there should be at least four lines in each image. To enable finding, filtering, correcting, fixing and rating peak groups that represents each graticule line in Hough accumulator, each candidate contains information about its rating, angle and distance between lines. Rating and distance information is used to select the best configuration. (iii) Intersections are finally coarsely estimated from the intersections between Hough lines, then corrected and filtered using the predicted mask from the Task 2, using some template matching with a cross rotated by the corresponding angle. The approach does not require any learning and the parameters are calibrated using both train and validation subsets.

IRISA. This method is based on two main steps. (i) The same line segment detector as the one used by the IRISA team for Task 2 (Sect. 3.3) is used to detect candidates. This results in a large amount of false positives with segments detected in many map objects. (ii) The DMOS rule grammar system [9] is used to efficiently filter candidates using a dedicated set of rules. Such rules enable the integration of constraints like perpendicularity or regular spacing between the lines, and the exploration of the hypothesis space thanks to the efficient back-tracking of the underlying logical programming framework.

CMM. This method combines morphological processing and the Radon transform. Morphological operators can be disturbed by noise disconnecting long lines while the Radon transform can integrate the contribution of short segments that may correspond to a fine texture. The process consists in the following six steps. (i) First, the image is sub-sampled by a factor 10 in order to speed up the process and to make it more robust to potential line discontinuities. An erosion of size 10 is applied before sub-sampling to preserve black lines. (ii) Then the frame is detected. Oriented gradients combined with morphological filters are used for this purpose, inspired by the method proposed in [6] for building façade analysis. (iii) Line directions are then found. Directional closings from 0 to 30 degrees with a step of 1 degree, followed by black top-hat and Otsu threshold are applied. The angle leading to the longest detected line is selected. (iv) The Radon transform of the previous black top-hat at the selected direction and its orthogonal are computed. Line locations correspond to the maxima of the Radon transform. (v) Lines are equally spaced in the map. The period of this grid is obtained as the maximum of the autocorrelation of the Radon transform. Equidistant lines are added on the whole map, whether they have been detected by the Radon transform or not. Applied to both directions, this generates the graticule lines. (vi) Finally a refinement is applied to localize precisely the line intersections at the full scale. In the neighborhood of each detected intersection, a closing is applied with a cross structuring element (with lines of length 20)

Rank	Team	Detection score (%) ↑
1	UWB	92.5
2	IRISA	89.2
3	CMM	86.6
4	L3IRIS	73.6

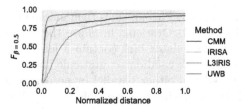

Fig. 8. Final detection score for Task 3 (*left*)—score ranges from 0 (worst) to 100 (best), and plot of the $F_{\beta=0.5}$ value for each distance threshold (*right*)—normalized by the maximum distance of 50 pixels.

at the previously detected orientations. The minimum of the resulting image provides the precise location of the intersection. If the contrast of the resulting closing is too low (<10), the point is inferred from the other points that were successfully found in the refinement step on the corresponding intersecting lines.

L3IRIS. This method is based on a deep intersection detector followed by a filtering technique based on a Hough transform. The process consists in the following five steps. (i) First, cross signs (all the points that are similar to the target points) are detected using a U-Net model [14] for point segmentation, trained with the 26 training images. The ground truth was generated by drawing a circle at each point's location over a zero background. The resulting model detects many points, among which the target points. (ii) From these candidates, a Hough transform is used (with additional heuristics) to detect graticule lines. The parameters of the Hough transform are selected automatically based on the number of final points in the last step. (iii) A clustering algorithm is then used to select line candidates forming a regular grid (parallel and orthogonal lines). (iv) Finally, intersections between these lines are filtered using the predicted mask from Task 2 (Sect. 3.3) to get the final set of points.

4.4 Results and Discussion

Results and ranking of each submitted method is given in Fig. 8: table on the left summarizes the detection score we computed and the plot on the right illustrates the evolution of the retrieval F_β score for each possible normalized distance threshold. The L3IRIS approach is penalized by a lack of local confirmation of intersection hypothesis, leading to false positives, misses, and sub-optimal location. The CMM approach produced very accurate detections thanks to morphological filters, but the Radon transform seems to generate extra hypothesis or discard groups of them and therefore has some stability issues. The IRISA approach performed globally very well thanks to is robust line segment detector and its rule-based system, despite some false positives. The UWB winning approach, finally, efficiently combined a binary preprocessing with a coarse Hough transform and a final refinement leading to superior accuracy.

5   Conclusion

This competition succeeded in advancing the state of the art on historical atlas vectorization, and **we thank all participants for their great submissions**. Shape extraction (Task 1) still require some progress to automate the process completely. Map content detection (Task 2) and graticule line detection (Task 3) are almost solved by proposed approaches. Future work will need to improve on shape detection, and start to tackle shape classification and text recognition.

References

1. Baloun, J., Král, P., Lenc, L.: ChronSeg: novel dataset for segmentation of hand-written historical chronicles. In: Proceedings of the 13th International Conference on Agents and Artificial Intelligence (ICAART), pp. 314–322 (2021)
2. Bloice, M.D., Roth, P.M., Holzinger, A.: Biomedical image augmentation using Augmentor. Bioinformatics **35**(21), 4522–4524 (2019)
3. Blusseau, S., Velasco-Forero, S., Angulo, J., Bloch, I.: Tropical and morphological operators for signal processing on graphs. In: Proceedings of the 25th IEEE International Conference on Image Processing (ICIP), pp. 1198–1202 (2018)
4. Chazalon, J., Carlinet, E.: Revisiting the coco panoptic metric to enable visual and qualitative analysis of historical map instance segmentation. In: 16th International Conference on Document Analysis and Recognition (ICDAR) (2021, to appear)
5. Grandvalet, Y., Bengio, Y.: Semi-supervised learning by entropy minimization. In: Proceedings of the 17th International Conference on Neural Information Processing Systems (NIPS), pp. 529–536 (2004)
6. Hernández, J., Marcotegui, B.: Morphological segmentation of building façade images. In: Proceedings of the 16th International Conference on Image Processing (ICIP), pp. 4029–4032. IEEE (2009)
7. Huang, G., Liu, Z., Van Der Maaten, L., Weinberger, K.Q.: Densely connected convolutional networks. In: Proceedings of the IEEE Conference on Computer Vision and Pattern Recognition (CVPR), pp. 4700–4708 (2017)
8. Kirillov, A., He, K., Girshick, R., Rother, C., Dollár, P.: Panoptic segmentation. In: Proceedings of the IEEE Conference on Computer Vision and Pattern Recognition (CVPR), pp. 9404–9413 (2019)
9. Lemaitre, A., Camillerapp, J., Coüasnon, B.: Multiresolution cooperation makes easier document structure recognition. Int. J. Doc. Anal. Recognit. (IJDAR) **11**(2), 97–109 (2008)
10. Leplumey, I., Camillerapp, J., Queguiner, C.: Kalman filter contributions towards document segmentation. In: International Conference on Document Analysis and Recognition (ICDAR), pp. 765–769 (1995)
11. Min, J., Kang, D., Cho, M.: Hypercorrelation squeeze for few-shot segmentation. arXiv preprint arXiv:2104.01538 (2021)
12. Nguyen, N., Rigaud, C., Revel, A., Burie, J.: A learning approach with incomplete pixel-level labels for deep neural networks. Neural Netw. **130**, 111–125 (2020)
13. Nina, O., Morse, B., Barrett, W.: A recursive OTSU thresholding method for scanned document binarization. In: 2011 IEEE Workshop on Applications of Computer Vision (WACV), pp. 307–314. IEEE (2011)
14. Ronneberger, O., Fischer, P., Brox, T.: U-Net: convolutional networks for biomedical image segmentation. In: Navab, N., Hornegger, J., Wells, W.M., Frangi, A.F. (eds.) MICCAI 2015. LNCS, vol. 9351, pp. 234–241. Springer, Cham (2015). https://doi.org/10.1007/978-3-319-24574-4_28

ICDAR 2021 Competition on Time-Quality Document Image Binarization

Rafael Dueire Lins[1,2(✉)], Rodrigo Barros Bernardino[1], Elisa Barney Smith[3], and Ergina Kavallieratou[4]

[1] Universidade Federal de Pernambuco, Recife, PE 50730-120, Brazil
{rdl,rbb4}@cin.ufpe.br
[2] Universidade Federal Rural de Pernambuco, Recife, PE 52.171-900, Brazil
[3] Boise State University, Boise, ID, USA
ebarneysmith@boisestate.edu
[4] University of the Eagian, Mytilene, Greece
kavallieratou@aegean.gr

Abstract. The ICDAR 2021 Time-Quality Binarization Competition assessed the performance of 12 new and 49 other previously published binarization algorithms for scanned document images. Four test sets of "real-world" documents with different features were used. For each test set, the top twenty algorithms in the quality of the resulting two-tone images had their average processing time presented, yielding an account of their time complexity.

Keywords: Document binarization · DIB-dataset · Binarization competition

1 Introduction

Binarization algorithms take as input a color or grayscale image and produce a bi-tonal image as result. Binarization is a key preprocessing step in document analysis solutions such as automatic character recognition and layout analysis, as the quality of the monocromatic document has a strong effect on the performance for such document processing tasks. Besides that, document binarization can increase its readability and work as a file compression strategy [58], as the size of binary images is often orders of magnitudes smaller than the original gray or color images, requiring much less storage space or computer bandwidth for network transmission. Thus, assessing the quality of document binarization algorithms is of importance, a concern witnessed by contests that started more than a decade ago [14,44].

Any algorithm makes design assumptions about the nature of the data being processed. Document binarization algorithms are no exception. The DIB Team (https://dib.cin.ufpe.br/) presented evidence that "no binarization algorithm is good for all kinds of documents". Thus, depending on the intrinsic features of each document, a binarization algorithm may perform better than another.

© Springer Nature Switzerland AG 2021
J. Lladós et al. (Eds.): ICDAR 2021, LNCS 12824, pp. 708–722, 2021.
https://doi.org/10.1007/978-3-030-86337-1_47

The second point made by the DIB team is that "time matters". As binarization algorithms are part of several document processing pipelines, it is also important to balance the quality-time binomial. Several algorithms may produce binary images of similar quality with huge differences in time complexity.

Recently, a new series of document binarization contests were created not only comparing the enrolled participants among themselves, but also comparing the quality-time performance of the new with "classical" algorithms. Besides that, much larger datasets of "real-world" and synthetic documents were tested in different feature clusters. Scanned documents either historic (machine typed or handwritten, 200 or 300 dpi, with different hues and textures of paper, and with different degrees of back-to front interference) and bureaucratic (200 dpi, offset printed, and with light back-to-front interference) have their binarization quality-time assessed in contests [32, 33].

This ICDAR 2021 competition assessed the quality-time performance of 12 new algorithms from seven teams from all over the world. The competitors' algorithms were also compared with the quality-time performance of 49 other binarization algorithms that may be considered "classic". Four test sets of "real-world" 200 and 300 dpi scanned documents were used in this assessment.

2 Participants

Seven research groups from six different countries, from the Americas, Asia, and Europe enrolled in this 2021 competition. In what follows, the teams are presented together with a brief outline of their approaches.

2.1 AutoHome Corp, Haidian Strict, Beijing, China
Team: Huang Xiao, Liu Rong, Xu Chengshen, Li Lin, Ye Mingdeng

A combination of binary cross entropy and dice loss is chosen as the loss function of a deep-learning algorithm. Data augmentation is performed in the training process so as to improve the scores. The original colored or gray images are divided into patches with the same dimension (e.g. 128*128). For each colored patch, a trained Unet model is utilized to obtain a binarized patch. A binarized large image with the same size as the original image can be obtained with the combination of those binarized patches. In this method, the model stacking technique is performed via two Unet models with patch dimensions of 128*128 and 256*256. Further, a global view with a patch dimension of 512*512 is also combined to obtain the final results. The model with a global view is trained aiming to capture the global context and the character locations.

This team presented two different methods to this competition:

Method (1) - HuangBCD: The segmentation model is BCD-Unet based [2].

Method (2) - HuangUnet: The segmentation model is Unet based.

2.2   Universitas Syiah Kuala, Indonesia
Team: Khairun Saddami

Method (1) - iNICK: An extension of the NICK binarization method [52]. The image standard deviation is used to determine the k value which is calculated as $k = -\sigma/(255 - 1.5\sigma)$, where σ is the image standard deviation that represents the image contrast.

Method (2) - CNW: Combination of Niblack and Wolf [49]. The threshold $T = (2m + mk((\sigma/m) - (\sigma/R) - 1))/2$, where σ is the image standard deviation, m is the mean of local window, R is the maximum standard deviation, $k = 0.35$.

Method (3) - CLD: Combined the local adaptive and global thresholding formulas, as described in [50].

2.3   West Pomeranian University of Technology, Poland
Team: Hubert Michalak and Krzysztof Okarma

Method (1) - Michalak21_1: The first step of the algorithm is related to image downsampling where one of well-known interpolation methods may be applied. For this purpose the MATLAB function imresize was used with bilinear and the simple nearest neighbour method. Application of a relatively large kernel during downsizing of the image results in the loss of details related to shapes of individual characters. Therefore only the low frequency image data is preserved representing the overall distribution of the image brightness, being in fact mainly the downsampled background information. After resizing back the downsampled image to the original size using the same kernel, the image containing only the low frequency information is obtained, representing the approximated high resolution background. In the next step of the proposed method the subtraction of this image from the original is made to enhance the text data, followed by simple contrast increase and logical negation. The image obtained is subjected to the fast global thresholding using Otsu method.

Method (2) - Michalak21_2: The proposed method is based on the equalization of the illumination of an image, increasing also its contrast, making it easier to conduct the proper binarization. It is based on the analysis of the local entropy, assuming its noticeably higher values in the neighbourhood of the characters. Hence, only the relatively high entropy regions should be further analysed as potentially containing some characters, whereas low entropy regions may be considered as the background. The additional steps of the morphological dilation, increase of contrast and final binarization using Bradley's method are made during the final stage.

Method (3) - Michalak21_3: The initial idea of the application of the region based binarization for text recognition was presented assuming the application of document images containing predefined text. The proposed improved method assumes the division of the image into regions of $N\hat{I}N$ pixels. For each of the regions the local threshold can be determined as $T = a * mean(X) - b$, where $mean(X)$ is the average brightness of the image region and the parameters a and b are subjected to optimization. The algorithm is based on the same idea of calculation of the local thresholds as the average brightness corrected by two parameters, however the number of regions is higher than would result from the resolution of the image and therefore they partially overlap each other. In this case for each sub-region several threshold values are calculated depending on the number of overlapping blocks covering the sub-region. The resulting local threshold is determined as the average of the threshold values calculated for the number of regions dependent on the assumed number of layers and the overlapping factor. The rationale for such an approach is a better tolerance of rapid illumination changes with the ability of a correct image binarization.

2.4 University of São Paulo (USP), Brazil
Team: Diego Pavan Soler

The "DiegoPavan" binarization method chooses to downscale the input image, rather than using patching, and then rescaling the network output to the input original size. The network architecture used is based on DE-GAN [60], where we changed the input image to HSV, adjusted the hyperparameters and the training process, including image augmentation.

2.5 University of Fribourg, Sweden
Team: Jean-Luc Bloechle

The "YinYang21" binarization algorithm detects the background of the original image using small overlapping windows. First, each window calculates its median color using a quantized color palette. Then, the estimated background image is generated by interpolating the computed median pixels of the overlapping windows. Next, the background image is subtracted from the original image, and the resulting difference image is transformed into grayscale, keeping only the lowest RGB component. A binarization is done by Otsu's algorithm. Detection and removal of small isolated connected components is made. The algorithm submitted in this competition is a faster and more accurate version of the previously submitted one in DocEng 2020 Binarization Competition [33].

2.6 Berlin State Library - Prussian
Cultural Heritage Foundation, Germany
Team: Vahid Rezanezhd, Clemens Neudecker
and Konstantin Baierer

The "Vahid" algorithm is based on machine learning and it is in fact a pixel-wise segmentation model. The dataset used for training is a combination of training sets for binarization competitions in different years with pseudo-labeled images from our dataset in the Berlin State Library. A specific dataset has been produced for very dark or bright images. The model is based on a Resnet50-Unet [42].

2.7 Hubei University of Technology, China
      Team: Xinrui Wang, Wei Xiong, Min Li, Chuansheng Wang and Laifu Guan

The DocUNet method comprises three main steps. Firstly, a morphological bottom-hat transform is carried out to enhance the document image contrast, and the size of a disk-shaped structural element is determined by the stroke width transform (SWT). Secondly, a hybrid pyramid U-Net convolutional network [27] is performed on the enhanced document images for accurate pixel classification. Finally, the Otsu algorithm is applied as an image post-processing step to yield the final image.

3   Quality Evaluation Methods

To evaluate the binarization algorithms relative to image quality, the scanned documents were clustered according to their features (print type and paper texture luminosity). This produced five document sets. The quality of the binary images was compared using the PSNR, DRDM, F-Measure (FM) and pseudo-FMeasure (Fps) [41], and Cohen's Kappa [6,12]. The final ranking is defined by sorting the ranking summation in ascending order, following the methodology introduced by [46]. The consistency of the global ranking with a carefully made visual inspection was also checked.

   Cohen's Kappa [12]

$$k = \frac{P_O - P_C}{1 - P_C}, \tag{1}$$

compares the observed accuracy with an expected accuracy, indicating how well a given classifier performs. P_O is the number of correctly mapped pixels (accuracy) and P_C is calculated by using:

$$P_C = \frac{n_{bf} \times n_{gf} + n_{bb} \times n_{gb}}{N^2}, \tag{2}$$

where n_{bf} and n_{bb} are the number of pixels mapped as foreground and background on the binary image, respectively, while n_{gf} and n_{gb} are the number of foreground and background pixels on the GT image and N is the total number of pixels. The Kappa coefficient has an excellent correspondence with the image-quality perception by human visual inspection of the resulting images. As indicated in [45], κ may be a good and easy to interpret image-quality evaluation measure for binary classifiers [6]. Thus, the top-twenty algorithms in image quality, ranked following [46], will have their κ coefficient and standard deviation (shown in parenthesis), together with the mean processing time and its standard deviation (also shown in parenthesis) presented in the tables of the results.

4 Processing-Time Evaluation

The 12 new algorithms assessed here were implemented by their authors. The purpose of the processing time evaluation here is to provide an order of magnitude of time elapsed for binarizing the whole datasets. The training-times for the AI-based algorithms were not considered. The competing 12 algorithms, printed in blue, were implemented using different programming languages, operating systems, and even for specific hardware platforms such as GPUs. They are compared against the other 49 algorithms in the literature, most of which were implemented by their authors or are available in image processing environments such as MatLab or ImageJ.

In the benchmaking, the following hardware was used:

- CPU: Intel(R) Core(TM) i7-10750H CPU @ 2.6 GHz RAM: 32 GB
- GPU: GeForce GTE 1650 4 GB

The algorithms were executed in the following platforms:

- **Windows 10 (version 1909):**
  - **Matlab:** Akbari_1 [1], Akbari_2 [1], Akbari_3 [1], ElisaTV [3], Ergina-Global [23], Ergina-Local [24], Gosh [5], Howe [18], Ghosh [5], Jia-Shi [20], Gattal [15], Lu-Su [36], Michalak [33], Yasin [32], iNICK [52], CNW [49], CLD [51], Michalak21_1, Michalak21_2, Michalak21_3.
- **Linux Pop!_OS 20.04**
  - **Python with GPU:** DilatedUNet [33]
  - **Python Machine Learning with CPU:** Calvo-Zaragoza [10], DiegoPavan [60], Doc-DLinkNet [68], DocUNet [32], HuangBCD, HuangUnet, Vahid, Yuleny [32].
  - **C++:** Bataineh [4], Bernsen [7], ISauvola [17], Niblack [40], Nick [25], Otsu [43], Sauvola [54], Singh [59], Su-Lu [61], WAN [39], Wolf [65].
  - **Java:** Bradley [9], dSLR [57], Huang [35], Intermodes [47], IsoData [63], Johannsen [21], KSW [22], Li-Tam [28], Mean [16], Mello-Lins [38], MinError [26], Minimum[47], Moments [62], Percentile [13], Pun [48], RenyEntropy [53], Shanbhag [56], Triangle [67], Wu-Lu [37], Yean-CC [66], YinYang [33], YinYang21.

Some experiments have been done with the algorithms that can be executed on both OS. No significant processing time difference was noticed. That is likely due to the fact that the exact same hardware and up to date compilers have been used in all cases. Although the user programming languages were Matlab, Python and Java, it is known that they are often used as an API to lower level implementations, leading to smaller differences in time purely due to the programming language used. Also, modern compilers are all nearly equally efficient. That can be easily verified by the results, as some algorithms using GPU performed fast, while others were slow. As for the Matlab implemented ones, some were among the fastest, while others among the slowest methods. Thus, if some optimization is made, using modern versions of the compilers, most algorithms should have similar performance in different languages.

The mean processing time was used in the analysis. It is most important to remark that the primary purpose here is to provide the order of magnitude of the processing time elapsed by each of the algorithms. The mean processing time figures are presented only to the 20 top **quality** algorithms, for each of the datasets tested.

5   Test Sets and Results

If one does not take into account some external physical noises such as stains, fungi, dirt that may affect document images [29], two aspects play fundamental role in complicating the binarization of scanned documents: the luminosity and texture of the paper background, that suffers the natural aging process, and the back-to-front interference. The back-to-front interference [31], later called bleeding or show-through, happens when a document is typed or written on both sides of a sheet of paper and the opacity of the paper is such as to allow the back printing or writing to be visualized on the front side. Given the importance of such a noise, most of the selected images from the Nabuco and LiveMemory datasets have such types of noise, with different degrees of interference α [30]. As already mentioned, four test sets of "real-world" 200 and 300 dpi scanned documents were used in this assessment.

The Nabuco and LiveMemory datasets used in the experiments here are part of the DIB - Document Image Binarization data set (https://dib.cin.ufpe.br/), which is part of the IAPR-TC10/TC11 open repository of document images [30].

From the Nabuco bequest of historical documents from the late XIX century, 20 images have been selected, which were subdivided into three clusters according to the average luminosity level of the background texture. Dark textures have an average luminosity of 147, mid texture of 193 and light texture of 220. A total of seven dark, seven light texture handwritten and six mid-dark texture typewritten documents were selected. From the LiveMemory project, five images with various configurations were selected. From the PRImA project, four images that belong to the Europeana Newspapers Project Dataset were used. The images have been selected in order to provide some variability between the datasets, but similar images within the datasets. The chosen datasets are representative of a large number of "real-world" documents of interest.

The ground-truth images used here were obtained by binarizing the original images with the ten best quality algorithms from the previous competitions [32,33] in images similar to the ones chosen to this competition. Such images underwent a careful visual inspection. The three best binary images were merged by applying the AND logical operator. The resulting image underwent salt and pepper filtering. The resulting image was visually re-inspected and underwent an eventual manual cleaning.

5.1   Nabuco Dataset

The letters of Joaquim Nabuco (b. 1849/d. 1910), a Brazilian stateman who was the first Brazilian ambassador to the USA, and one of the most expressive figures in freeing black slaves in the Americas, are of great historical importance, and some of them are available in the DIB dataset. Three sets of images of the Nabuco bequest were used in the assessment here.

Table 1. Quality-time results for Nabuco, Light Texture, Handwritten Documents

# Team	Kappa (SD)	Time (SD)	Example Image
1 Vahid	0.89 (0.06)	10.18 (4.49)	
2 HuangUnet	0.87 (0.13)	24.91 (7.91)	
3 Akbari_1 [1]	0.84 (0.21)	4.91 (1.98)	
4 HuangBCD	0.87 (0.10)	113.29 (35.16)	
5 Akbari_2 [1]	0.84 (0.21)	4.95 (2.12)	
6 Akbari_3 [1]	0.84 (0.21)	4.89 (1.99)	
7 Jia-Shi [20]	0.84 (0.21)	4.87 (1.99)	
8 Wolf [64]	0.86 (0.05)	0.06 (0.03)	
9 Sauvola [55]	0.86 (0.06)	0.04 (0.02)	
10 DocDLink [32]	0.81 (0.18)	55.60 (26.86)	
11 Yasin	0.83 (0.10)	1.18 (0.99)	
12 Gosh [5]	0.81 (0.15)	31.84 (16.58)	
13 Su-Lu [8]	0.85 (0.06)	0.41 (0.18)	
14 Lu-Su [36]	0.81 (0.12)	16.15 (7.06)	
15 Minimum [47]	0.84 (0.10)	0.01 (0.01)	
16 iNICK [52]	0.81 (0.11)	5.32 (4.09)	
17 DilatedUNet [33]	0.80 (0.12)	44.43 (15.47)	
18 Intermodes [47]	0.80 (0.11)	0.01 (0.00)	
19 Mello-Lins [38]	0.79 (0.21)	0.01 (0.00)	
20 ElisaTV [3]	0.76 (0.20)	2.41 (1.06)	

5.2   The LiveMemory Dataset

The LiveMemory Project [34] was a pioneering initiative to build a digital library of the entire collection of proceedings of the technical events of the Brazilian Telecommunications Society (SBrT) led by the first author of this paper, back in 2007. The real challenge was to scan all the printed-only volumes, semi-automatically index all the papers, enhance image quality, and to binarize the images in way such as to allow all the volumes to be stored in one single DVD, which was handed to all members of the SBrT. The documents were scanned in 200 dpi, true-color and stored using the jpeg file-format with standard (1% loss). The LiveMemory dataset is clearly the one with smaller variation among images, as they are all "modern" documents, offset printed and have an uniform background with some back-to-front interference.

Table 2. Quality-time results for Nabuco, Dark Texture, Handwritten Documents

# Team	Kappa (SD)	Time (SD)	Example Image
1 Sauvola [55]	0.91 (0.04)	0.03 (0.00)	
2 Gosh [5]	0.89 (0.03)	20.97 (2.09)	
3 Wolf [64]	0.89 (0.03)	0.04 (0.00)	
4 DocDLink [32]	0.88 (0.05)	42.12 (2.31)	
5 HuangBCD	0.89 (0.02)	89.30 (7.21)	
6 Su-Lu [8]	0.90 (0.06)	0.32 (0.04)	
7 HuangUnet	0.89 (0.03)	19.81 (1.54)	
8 Yasin	0.89 (0.04)	0.82 (0.24)	
9 iNICK [52]	0.89 (0.03)	3.19 (0.51)	
10 Nick [25]	0.89 (0.03)	0.03 (0.00)	
11 Singh [59]	0.89 (0.03)	0.03 (0.00)	
12 YinYang21	0.86 (0.07)	0.51 (0.07)	
13 DocUNet [32]	0.85 (0.07)	37.33 (4.53)	
14 Li-Tam [28]	0.86 (0.05)	0.01 (0.00)	
15 Vahid	0.86 (0.06)	7.39 (0.49)	
16 Shanbhag [56]	0.85 (0.10)	0.01 (0.00)	
17 Howe [18]	0.85 (0.07)	15.59 (7.72)	
18 DilatedUNet [33]	0.85 (0.07)	31.96 (3.14)	
19 Ergina_L [24]	0.86 (0.07)	0.12 (0.02)	
20 Ergina_G [23]	0.85 (0.08)	0.08 (0.01)	

Table 3. Quality-time results for Nabuco, Mid Texture, Typewritten Documents

# Team	Kappa (SD)	Time (SD)	Example Image
1 Gosh [5]	0.92 (0.07)	51.82 (6.28)	
2 HuangUnet	0.91 (0.05)	37.67 (1.81)	
3 Yasin	0.90 (0.06)	1.03 (0.14)	
4 HuangBCD	0.91 (0.04)	167.59 (7.49)	
5 iNICK [52]	0.89 (0.07)	3.70 (0.52)	
6 Wolf [64]	0.92 (0.03)	0.10 (0.01)	
7 Singh [59]	0.92 (0.04)	0.13 (0.01)	
8 Michalak21a	0.87 (0.10)	0.02 (0.00)	
9 Li-Tam [28]	0.88 (0.08)	0.02 (0.00)	
10 Minimum [47]	0.90 (0.03)	0.02 (0.00)	
11 Nick [25]	0.91 (0.05)	0.08 (0.00)	
12 Su-Lu [8]	0.91 (0.02)	0.71 (0.07)	
13 Intermodes [47]	0.87 (0.06)	0.02 (0.00)	
14 Michalak21c	0.85 (0.10)	0.47 (0.04)	
15 ElisaTV [3]	0.86 (0.08)	4.27 (0.20)	
16 Akbari_1 [1]	0.86 (0.06)	8.45 (0.85)	
17 Akbari_2 [1]	0.86 (0.06)	8.45 (0.87)	
18 Bradley [9]	0.84 (0.09)	0.14 (0.01)	
19 Akbari_3 [1]	0.86 (0.06)	8.46 (0.87)	
20 Jia-Shi [20]	0.86 (0.06)	8.46 (0.88)	

Table 4. Quality-time results for LiveMemory Test Set

# Team	Kappa (SD)	Time (SD)	Example Image
1 Michalak [33]	0.94 (0.04)	0.08 (0.05)	
2 Bradley [9]	0.94 (0.05)	0.29 (0.01)	
3 Wolf [64]	0.94 (0.05)	0.22 (0.02)	
4 ElisaTV [3]	0.93 (0.06)	9.55 (1.15)	
5 Gosh [5]	0.94 (0.03)	111.80 (21.29)	
6 IsoData [63]	0.90 (0.12)	0.14 (0.02)	
7 Gattal [15]	0.91 (0.11)	54.40 (1.48)	
8 Otsu [43]	0.90 (0.13)	0.02 (0.00)	
9 Li-Tam [28]	0.91 (0.10)	0.14 (0.01)	
10 Yasin	0.93 (0.05)	2.05 (0.99)	
11 iNICK [52]	0.93 (0.03)	3.48 (0.35)	
12 Michalak21_1	0.94 (0.03)	0.08 (0.05)	
13 Intermodes [47]	0.92 (0.07)	0.14 (0.02)	
14 Michalak21_3	0.92 (0.06)	1.32 (0.67)	
15 Johannsen [21]	0.92 (0.05)	0.14 (0.02)	
16 Su-Lu [8]	0.93 (0.02)	1.67 (0.10)	
17 YinYang21	0.91 (0.07)	1.60 (0.13)	
18 HuangBCD	0.92 (0.07)	316.87 (25.66)	
19 HuangUnet	0.92 (0.07)	316.78 (26.17)	
20 WAN [39]	0.92 (0.07)	1.01 (0.09)	

5.3 PRImA

Europeana Newspapers: [11] Its main goal is to provide a representative collection of all the types of newspapers which are and/or might be subject of ongoing or future digitisation activities. As such, it is hosting scanned images, metadata and ground truth (a representation of the ideal result of a processing step like OCR or layout analysis) on the level of individual newspaper pages. Three of the selected images are from the Royal Library, in the Netherlands, and one is from the Austrian National Library, Austria.

6 Result Analysis

This contest is designed to look at the tradeoff between binarization performance and computational time. There is no single winner. The 20 top performing teams are reported by dataset in Tables 1, 2, 3, 4 and 5. Yasin and HuangBCD appeared in the top ranked algorithms for all five datasets and the sister algorithm HuangUnet appeared in the top ranked for four of the datasets. Michalak21's first and third algorithms both appeared three times in the rankings. YinYang21 appeared 3 times and Vahid appeared twice.

The average kappa values for the top 20 reported for each dataset fell in a narrow range from 0.75 to 0.94, The binarized images produced using the best quality algorithm for the test images, as one may expect, had very high

Table 5. Quality-time results for PRImA Data Set

# Team	Kappa (SD)	Time (SD)	Example Image (cropped)
1 Gosh [5]	0.90 (0.09)	159.77 (92.16)	
2 Bradley [9]	0.90 (0.08)	0.43 (0.35)	
3 Michalak21a	0.89 (0.08)	0.10 (0.06)	
4 Intermodes [47]	0.89 (0.14)	0.19 (0.23)	
5 Michalak [33]	0.91 (0.08)	0.10 (0.06)	
6 Li-Tam [28]	0.87 (0.17)	0.19 (0.23)	
7 DocDLink [32]	0.92 (0.06)	292.46 (223.60)	
8 ElisaTV [3]	0.88 (0.04)	13.56 (10.63)	
9 IsoData [63]	0.87 (0.14)	0.19 (0.23)	
10 Su-Lu [8]	0.87 (0.10)	2.93 (1.95)	
11 Moments [62]	0.85 (0.16)	0.19 (0.23)	
12 Michalak21c	0.89 (0.05)	1.83 (1.05)	
13 Yasin	0.87 (0.08)	2.42 (1.24)	
14 Ergina_L [24]	0.87 (0.10)	1.28 (0.68)	
15 Gattal [15]	0.87 (0.13)	56.94 (3.80)	
16 Akbari_1 [1]	0.86 (0.06)	32.40 (20.71)	
17 Ergina_G [23]	0.86 (0.14)	0.85 (0.62)	
18 Huang [19]	0.86 (0.10)	0.19 (0.22)	
19 Akbari_2 [1]	0.86 (0.06)	32.39 (20.68)	
20 HuangBCD	0.86 (0.08)	445.08 (301.37)	

visual quality. The ten top quality images for each of the sample images will be made available at the DIB website (https://dib.cin.ufpe.br/) immediately after ICDAR 2021.

The execution times varied more significantly than the performance as measured by kappa value. The median run time of the top performing algorithms was 1 s with 21% of the algorithms taking less than 0.1 ss. Michalak21_1 was the fastest of the ranked new algorithms competing this year. Nine of the algorithms took more than a minute on average to process the page, which for most applications will not be practical for the small performance benefit the algorithm may offer. HuangBCD that appeared in the top rankings for all datasets was also the algorithm that had the longest run time of all the ranked algorithms. The median run time for the algorithms published before 2010 was 0.11 s, whereas the median of the algorithms published 2010–2019 increased to 4.95 s and the median of those published in 2020 and 2021 is 7.39. The performance does not vary significantly between those groups.

7   Conclusions

This ICDAR 2021 Competition on Time-Quality Document Image Binarization shows that document image binarization is still a challenging task. The number of ways the problem can be made more difficult leads to demand to develop a new algorithm that can handle that one outlier case which others could not properly binarize. Machine-learning binarization algorithms are rising in providing better quality images, but some of the classic algorithms like IsoData [63] and Savoula [55] continued to appear in the top ranked algorithm list and they still provide very good, if not the best quality bitonal-image at a much lower time complexity.

Machine-learning binarization algorithms are rising in providing better quality images, but some of the classic algorithms still provide very good, if not the best quality bitonal-image at a much lower time complexity. It is important to remark that the training-time for the machine-learning based algorithms was not computed. Another point worth remarking is that some of those ML algorithms require computational resources that may be considered prohibitive, as some of the competing algorithms in the ICDAR 2019 Competition on Time-Quality Document Image Binarization [32] were still unable to run to all test images of the test sets used here.

Acknowledgements. The authors grateful to all the participants of this ICDAR competition, and to all researchers who made the code for their binarization algorithms publically available. This research was partly sponsored by CNPq – Brazilian Government.

References

1. Akbari, Y., Britto Jr., A.S., Al-Maadeed, S., Oliveira, L.S.: Binarization of degraded document images using convolutional neural networks based on predicted two-channel images. In: ICDAR (2019)
2. Azad, R., Asadi-Aghbolaghi, M., Fathy, M., Escalera, S.: Bi-directional ConvLSTM U-net with densley connected convolutions. In: ICCVW 2019, pp. 406–415 (2019)
3. Barney Smith, E.H., Likforman-Sulem, L., Darbon, J.: Effect of pre-processing on binarization. In: Document Recognition and Retrieval XVII, p. 75340H (2010)
4. Bataineh, B., Abdullah, S.N.H.S., Omar, K.: An adaptive local binarization method for document images based on a novel thresholding method and dynamic windows. Pattern Recogn. Lett. **32**(14), 1805–1813 (2011)
5. Bera, S.K., Ghosh, S., Bhowmik, S., Sarkar, R., Nasipuri, M.: A non-parametric binarization method based on ensemble of clustering algorithms. Multimed. Tools Appl. **80**(5), 7653–7673 (2020). https://doi.org/10.1007/s11042-020-09836-z
6. Bernardino, R., Lins, R., Jesus, D.M.: A quality and time assessment of binarization algorithms. In: 2019 15th IAPR ICDAR, pp. 1444–1450 (2019)
7. Bernsen, J.: Dynamic thresholding of gray-level images. In: International Conference on Pattern Recognition, pp. 1251–1255 (1986)
8. Su, B., Lu, S., Tan, C.L.: Robust document image binarization technique for degraded document images. Trans. Image Process. **22**(4), 1408–1417 (2013)
9. Bradley, D., Roth, G.: Adaptive thresholding using the integral image. J. Graph. Tools **12**(2), 13–21 (2007)
10. Calvo-Zaragoza, J., Gallego, A.J.: A selectional auto-encoder approach for document image binarization. Pattern Recogn. **86**, 37–47 (2019)
11. Clausner, C., Papadopoulos, C., Pletschacher, S., Antonacopoulos, A.: The ENP image and ground truth dataset of h. newspapers. In: ICDAR'15, pp. 931–935 (2015)

12. Congalton, R.G.: A review of assessing the accuracy of classifications of remotely sensed data. Remote Sens. Environ. **37**(1), 35–46 (1991)

13. Doyle, W.: Operations useful for similarity-invariant pattern recognition. J. ACM **9**(2), 259–267 (1962)

14. Gatos, B., Ntirogiannis, K., Pratikakis, I.: ICDAR 2009 document image binarization contest (DIBCO 2009). In: 2009 10th ICDAR, pp. 1375–1382. IEEE (2009)

15. Gattal, A., Abbas, F., Laouar, M.R.: Automatic parameter tuning of K-means algorithm for document binarization. In: 7th ICSENT, pp. 1–4. ACM Press (2018)

16. Glasbey, C.: An analysis of histogram-based thresholding algorithms. Graph. Model. Image Process. **55**(6), 532–537 (1993)

17. Hadjadj, Z., Meziane, A., Cherfa, Y., Cheriet, M., Setitra, I.: ISauvola: improved Sauvola's algorithm for document image binarization. In: Campilho, A., Karray, F. (eds.) ICIAR 2016. LNCS, vol. 9730, pp. 737–745. Springer, Cham (2016). https://doi.org/10.1007/978-3-319-41501-7_82

18. Howe, N.R.: Document binarization with automatic parameter tuning. IJDAR **16**(3), 247–258 (2013). https://doi.org/10.1007/s10032-017-0293-7

19. Huang, L.K., Wang, M.J.J.: Image thresholding by minimizing the measures of fuzziness. Pattern Recogn. **28**(1), 41–51 (1995)

20. Jia, F., Shi, C., He, K., Wang, C., Xiao, B.: Degraded document image binarization using structural symmetry of strokes. Pattern Recogn. **74**, 225–240 (2018)

21. Johannsen, G., Bille, J.: A threshold selection method using information measures. In: International Conference Pattern Recognition, pp. 140–143 (1982)

22. Kapur, J., Sahoo, P., Wong, A.: A new method for gray-level picture thresholding using the entropy of the histogram. Comput. Vis. Graph. Image Process. **29**(1), 140

23. Kavallieratou, E.: A binarization algorithm specialized on document images and photos. In: ICDAR 2005, no. 1, pp. 463–467 (2005)

24. Kavallieratou, E., Stathis, S.: Adaptive binarization of historical document images. In: Proceedings - International Conference on Pattern Recognition, vol. 3, pp. 742–745 (2006)

25. Khurshid, K., Siddiqi, I., Faure, C., Vincent, N.: Comparison of Niblack inspired binarization methods for ancient documents. In: SPIE, p. 72470U (2009)

26. Kittler, J., Illingworth, J.: Minimum error thresholding. Pattern Recogn. **19**(1), 41–47 (1986)

27. Kong, X., Sun, G., Wu, Q., Liu, J., Lin, F.: Hybrid pyramid U-Net model for brain tumor segmentation. In: Shi, Z., Mercier-Laurent, E., Li, J. (eds.) IIP 2018. IAICT, vol. 538, pp. 346–355. Springer, Cham (2018). https://doi.org/10.1007/978-3-030-00828-4_35

28. Li, C., Tam, P.: An iterative algorithm for minimum cross entropy thresholding. Pattern Recogn. Lett. **19**(8), 771–776 (1998)

29. Lins, R.D.: A taxonomy for noise in images of paper documents - the physical noises. In: Kamel, M., Campilho, A. (eds.) ICIAR 2009. LNCS, vol. 5627, pp. 844–854. Springer, Heidelberg (2009). https://doi.org/10.1007/978-3-642-02611-9_83

30. Lins, R.D., Almeida, M.M.D., Bernardino, R.B., Jesus, D., Oliveira, J.M.: Assessing binarization techniques for document images. In: DocEng'17, pp. 183–192 (2017)

31. Lins, R.D., Guimarães Neto, M., França Neto, L., Galdino Rosa, L.: An environment for processing images of historical documents. Microprocess. Microprogram. **40**(10–12), 939–942 (1994)

32. Lins, R.D., Kavallieratou, E., Barney Smith, E., Bernardino, R.B., de Jesus, D.M.: ICDAR 2019 Time-Quality Binarization Competition. In: ICDAR, pp. 1539–1546 (2019)
33. Lins, R.D., Simske, S.J., Bernardino, R.B.: DocEng'2020 time-quality competition on binarizing photographed documents. In: DocEng'20, pp. 1–4. ACM (2020)
34. Lins, R.D., Torreão, G., Pereira e Silva, G.: Content recognition and indexing in the LiveMemory platform. In: Ogier, J.-M., Liu, W., Lladós, J. (eds.) GREC 2009. LNCS, vol. 6020, pp. 220–230. Springer, Heidelberg (2010). https://doi.org/10.1007/978-3-642-13728-0_20
35. Lu, D., Huang, X., Sui, L.X.: Binarization of degraded document images based on contrast enhancement. IJDAR **21**(1–2), 123–135 (2018)
36. Lu, S., Su, B., Tan, C.L.: Document image binarization using background estimation and stroke edges. IJDAR **13**(4), 303–314 (2010)
37. Lu, W., Songde, M., Lu, H.: An effective entropic thresholding for ultrasonic images. In: 14th ICPR, vol. 2, pp. 1552–1554 (1998)
38. Mello, C.A.B., Lins, R.D.: Image segmentation of historical documents. In: Visual 2000 (2000)
39. Mustafa, W.A., Abdul Kader, M.M.M.: Binarization of document image using optimum threshold modification. J. Phys.: C. Ser. **1019**(1), 012022 (2018)
40. Niblack, W.: An Introduction to Digital Image Processing. Strandberg Publishing Company, Copenhagen (1985)
41. Ntirogiannis, K., Gatos, B., Pratikakis, I.: Performance evaluation methodology for historical document image binarization. IEEE Trans. Image Process. **22**(2), 595–609 (2013)
42. Oliveira, S.A., Seguin, B., Kaplan, F.: dhSegment: a generic deep-learning approach for document segmentation. CoRR abs/1804.1 (2018)
43. Otsu, N.: A threshold selection method from gray-level histograms. IEEE Trans. Syst. Man Cybern. **9**(1), 62–66 (1979)
44. Paredes, R., Kavallieratou, E., Lins, R.D.: ICFHR 2010 contest: quantitative evaluation of binarization algorithms. In: 12th ICFHR, pp. 733–736. IEEE (2010)
45. Powers, D.M.W.: Evaluation: from precision, recall and F-measure to ROC, informedness, markedness & correlation. J. M. L. Technol. **2**(1), 37–63 (2011)
46. Pratikakis, I., et al.: ICDAR 2019 competition on document image binarization. In: ICDAR, No. November 2019 (2019)
47. Prewitt, J.M.S., Mendelsohn, M.L.: The analysis of cell images. Ann. N. Y. Acad. Sci. **128**(3), 1035–1053 (2006)
48. Pun, T.: Entropic thresholding, a new approach. Comput. Graph. Image Process. **16**(3), 210–239 (1981)
49. Saddami, K., Afrah, P., Mutiawani, V., Arnia, F.: A new adaptive thresholding technique for binarizing ancient document. In: INAPR, pp. 57–61. IEEE (2018)
50. Saddami, K., Munadi, K., Away, Y., Arnia, F.: Combination local and global thresholding method for binarizing ancient Jawi document. JTIIK (2019)
51. Saddami, K., Munadi, K., Away, Y., Arnia, F.: Effective and fast binarization method for combined degradation on ancient documents. Heliyon (2019)
52. Saddami, K., Munadi, K., Muchallil, S., Arnia, F.: Improved thresholding method for enhancing Jawi binarization performance. In: ICDAR, vol. 1, pp. 1108–1113. IEEE (2017)
53. Sahoo, P., Wilkins, C., Yeager, J.: Threshold selection using Renyi's entropy. Pattern Recogn. **30**(1), 71–84 (1997)
54. Sauvola, J., Pietikäinen, M., Pietikainem, M.: Adaptive document image binarization. Pattern Recogn. **33**(2), 225–236 (2000)

55. Sauvola, J., Seppanen, T., Haapakoski, S., Pietikainen, M.: Adaptive document binarization. In: ICDAR, vol. 1, pp. 147–152. IEEE Comput. Soc (1997)
56. Shanbhag, A.G.: Utilization of information measure as a means of image thresholding. CVGIP: Graph. Models Image Process. **56**(5), 414–419 (1994)
57. Silva, J.M.M., Lins, R.D., Rocha, V.C.: Binarizing and filtering historical documents with back-to-front interference. ACM SAC **2006**, 853–858 (2006)
58. da Silva, J.M.M., Lins, R.D.: Color document synthesis as a compression strategy. In: ICDAR (ICDAR), pp. 466–470 (2007)
59. Singh, T.R., Roy, S., Singh, O.I., Sinam, T., Singh, K.M.: A new local adaptive thresholding technique in binarization. IJCSI Int. J. Comput. Sci. Issues **08**(6), 271–277 (2011)
60. Souibgui, M.A., Kessentini, Y.: DE-GAN: a conditional generative adversarial network for document enhancement. Trans. Pattern Anal. Mach. Intell. 1 (2021)
61. Su, B., Lu, S., Tan, C.L.: Binarization of historical document images using the local maximum and minimum. In: 8th IAPR DAS, pp. 159–166. ACM Press (2010)
62. Tsai, W.H.: Moment-preserving thresolding: a new approach. Comput. Vis. Graph. Image Process. **29**(3), 377–393 (1985)
63. Velasco, F.R.: Thresholding using the Isodata clustering algorithm. Technical report, OSD or Non-Service DoD Agency (1979)
64. Wolf, C., Doermann, D.: Binarization of low quality text using a Markov random field model. In: Object Recognition Supported by User Interaction for Service Robots, vol. 3, pp. 160–163. IEEE Computer Society (2002)
65. Wolf, C., Jolion, J.M., Chassaing, F.: Text localization, enhancement and binarization in multimedia documents. In: Object Recognition Supported by User Interaction for Service Robots, vol. 2, pp. 1037–1040. IEEE Computer Society (2003)
66. Yen, J.C., Chang, F.J.C.S., Yen, J.C., Chang, F.J., Chang, S.: A new criterion for automatic multilevel thresholding. Trans. Image Process. **4**(3), 370–378 (1995)
67. Zack, G.W., Rogers, W.E., Latt, S.A.: Automatic measurement of sister chromatid exchange frequency. J. Histochem. Cytochem. **25**(7), 741–753 (1977)
68. Zhou, L., Zhang, C., Wu, M.: D-linknet: Linknet with pretrained encoder and dilated convolution for high resolution satellite imagery road extraction. In: Computer Vision and Pattern Recognition (2018)

ICDAR 2021 Competition on On-Line Signature Verification

Ruben Tolosana[1]([⊠]), Ruben Vera-Rodriguez[1], Carlos Gonzalez-Garcia[1],
Julian Fierrez[1], Santiago Rengifo[1], Aythami Morales[1], Javier Ortega-Garcia[1],
Juan Carlos Ruiz-Garcia[1], Sergio Romero-Tapiador[1], Jiajia Jiang[2],
Songxuan Lai[2], Lianwen Jin[2,3], Yecheng Zhu[2], Javier Galbally[4], Moises Diaz[5],
Miguel Angel Ferrer[6], Marta Gomez-Barrero[7], Ilya Hodashinsky[8],
Konstantin Sarin[8], Artem Slezkin[8], Marina Bardamova[8], Mikhail Svetlakov[8],
Mohammad Saleem[9], Cintia Lia Szücs[9], Bence Kovari[9], Falk Pulsmeyer[10],
Mohamad Wehbi[10], Dario Zanca[10], Sumaiya Ahmad[11], Sarthak Mishra[11],
and Suraiya Jabin[11]

[1] Biometrics and Data Pattern Analytics Lab, UAM, Madrid, Spain
{ruben.tolosana,ruben.vera,julian.fierrez,santiago.rengifo,
aythami.morales,javier.ortega,juanc.ruiz}@uam.es,
{carlos.gonzalezgarcia,sergio.romerot}@estudiante.uam.es
[2] South China University of Technology, Guangzhou, China
eejiajia_jiang@mail.scut.edu.cn, eelwjin@scut.edu.cn
[3] Guangdong Artificial Intelligence and Digital Economy Laboratory,
Guangzhou, China
[4] European Commission - Joint Research Centre, Ispra, Italy
javier.galbally@ec.europa.eu
[5] Universidad del Atlantico Medio, Las Palmas, Spain
moises.diaz@atlanticomedio.es
[6] Universidad de las Palmas de Gran Canaria, Las Palmas, Spain
miguelangel.ferrer@ulpgc.es
[7] Hochschule Ansbach, Ansbach, Germany
marta.gomez-barrero@hs-ansbach.de
[8] Tomsk State University of Control Systems and Radioelectronics, Tomsk, Russia
hodashn@rambler.ru, sks@security.tomsk.ru
[9] Budapest University of Technology and Economics, Budapest, Hungary
{Mohammad.Saleem,Szucs.CintiaLia,kovari}@aut.bme.hu
[10] Machine Learning and Data Analytics Lab, FAU, Erlangen, Germany
{falk.pulsmeyer,mohamad.wehbi,dario.zanca}@fau.de
[11] Jamia Millia Islamia, New Delhi, India
sjabin@jmi.ac.in

Abstract. This paper describes the experimental framework and results of the ICDAR 2021 Competition on On-Line Signature Verification (SVC 2021). The goal of SVC 2021 is to evaluate the limits of on-line signature verification systems on popular scenarios (office/mobile) and writing inputs (stylus/finger) through large-scale public databases. Three different tasks are considered in the competition, simulating realistic scenarios as both random and skilled forgeries are simultaneously considered on each task. The results obtained in SVC 2021 prove the high potential of

J. Lladós et al. (Eds.): ICDAR 2021, LNCS 12824, pp. 723–737, 2021.
https://doi.org/10.1007/978-3-030-86337-1_48

deep learning methods. In particular, the best on-line signature verification system of SVC 2021 obtained Equal Error Rate (EER) values of 3.33% (Task 1), 7.41% (Task 2), and 6.04% (Task 3).

SVC 2021 will be established as an on-going competition (https:// sites.google.com/view/SVC2021), where researchers can easily benchmark their systems against the state of the art in an open common platform using large-scale public databases such as DeepSignDB (https:// github.com/BiDAlab/DeepSignDB) and SVC2021_EvalDB (https:// github.com/BiDAlab/SVC2021_EvalDB), and standard experimental protocols.

Keywords: SVC 2021 · Biometrics · Handwriting · On-line signature · Benchmark · DeepSignDB · SVC2021_EvalDB · Deep learning

1   Introduction

On-line handwritten signature verification has always been a very active area of research due to its high popularity for authentication scenarios [9] and the variety of open challenges that are still under research nowadays [14], e.g., one/few-shot learning [10,20,27,44], device interoperability [2,30,35,43], aging [22,40], types of impostors [21,39], signature complexity [24,42,46], template storage [8], etc. Despite all these challenges, the performance of on-line signature verification systems has been improved in the last years due to several factors, especially: *i)* the evolution in the acquisition technology going from devices specifically designed to acquire handwriting and signature in office-like scenarios through a pen stylus (e.g. Wacom devices) to the current touch screens of mobile scenarios in which signatures can be captured anywhere using our own personal smartphone through the finger [3,13,35,37], and *ii)* the extended usage of deep learning technology in many different areas, overcoming traditional handcrafted approaches and even human performance [1,26,38,45,47]. So, with all these aspects in mind, the question is: what are the current performance limits of the on-line signature verification technology under realistic scenarios?

This paper describes the experimental framework and results of the ICDAR 2021 Competition on On-Line Signature Verification (SVC 2021). The goal of SVC 2021 is to evaluate the limits of on-line signature verification systems on popular scenarios (office/mobile) and writing inputs (stylus/finger) through large-scale public databases. Three different tasks are considered in the competition:

- **Task 1**: Analysis of office scenarios using the stylus as input.
- **Task 2**: Analysis of mobile scenarios using the finger as input.
- **Task 3**: Analysis of both office and mobile scenarios simultaneously.

In addition, we simulate in SVC 2021 the following realistic operational conditions, to the best of our knowledge, not considered in previous on-line signature verification competitions [4,25,28,29,48]:

– Over 1,700 subjects and 100 different acquisition devices are considered in the competition, using both Wacom devices (office scenarios) and general purpose devices such as tablets and smartphones (mobile scenarios).
– Random and skilled forgeries are simultaneously considered in each task. In addition, different types of skilled forgeries are considered in the competition such as static (i.e., only the image of the signature to forge is available) and dynamic forgeries (i.e., both image and dynamics are available), in both trained and blueprint cases [39].
– High intra-subject variability (a.k.a. aging) as different acquisition set-ups are considered in the competition ranging from 1 to 5 sessions, and with a time gap between sessions from days to months.

This realistic scenario has been achieved thanks to the public DeepSignDB database [45] and the novel SVC2021_EvalDB database (this later one specifically acquired for SVC 2021). Besides, we have designed realistic and challenging experimental protocols making public the corresponding signature comparisons files and the benchmarking platform.

SVC 2021 will be established as an on-going competition[1], where researchers can easily benchmark their systems against the state of the art using a common experimental protocol and an open computing platform (CodaLab[2]).

The remainder of the paper is organised as follows. Sections 2 and 3 describe the details of the databases and the set-up considered in the competition, respectively. Section 4 provides a description of the submitted on-line signature verification systems. Section 5 describes the results of the competition. Finally, Sect. 6 draws the final conclusions.

2   SVC 2021: Databases

Two databases are considered in SVC 2021: DeepSignDB and SVC2021_EvalDB. These databases are publicly available for the research community and can be downloaded following the instructions included in[3,4]. We provide next a description of them.

2.1   DeepSignDB

The DeepSignDB database [45] comprises on-line signatures from a total of 1,526 subjects from four different well-known databases: MCYT (330 subjects) [33], BiosecurID (400 subjects) [16], Biosecure DS2 (650 subjects) [25], eBioSign (65 subjects) [37], and a novel signature database composed of 81 subjects. Deep-SignDB comprises more than 70K signatures acquired using both stylus and finger writing inputs in both office and mobile scenarios. A total of 8 different

---

[1] https://sites.google.com/view/SVC2021.
[2] https://competitions.codalab.org/competitions/27295.
[3] https://github.com/BiDAlab/DeepSignDB.
[4] https://github.com/BiDAlab/SVC2021_EvalDB.

devices are considered in the acquisition (i.e., 5 Wacom devices and 3 Samsung general purpose devices). In addition, different types of impostors and number of acquisition sessions are considered along the database. For more details about DeepSignDB, we refer the reader to the published article [45].

2.2   SVC2021_EvalDB

The SVC2021_EvalDB database is a novel database specifically acquired for SVC 2021. Two acquisition scenarios are considered: office and mobile scenarios.

- **Office scenario:** on-line signatures from 75 total subjects were acquired using a Wacom STU-530 device with the stylus as writing input. Regarding the acquisition protocol, the device was placed on a desktop and subjects were able to rotate it in order to feel comfortable with the writing position. It is important to highlight that the subjects considered in the acquisition of SVC2021_EvalDB are different compared to the ones considered in the Deep-SignDB database.

  Signatures were collected in two separated sessions with a time gap between them of at least 1 week. For each subject, there are 8 total genuine signatures (4 signatures/session) and 16 skilled forgeries (8 signatures/type) performed by four different subjects in two different sessions. Regarding the skilled forgeries, both static and dynamic forgeries were considered in the first and second acquisition sessions, respectively. Information related to X and Y spatial coordinates, pressure, and timestamp is recorded for the Wacom device. In addition, pen-up trajectories are also available.

- **Mobile scenario:** on-line signatures from 119 total subjects were acquired using the same acquisition framework considered in MobileTouchDB [41]. Regarding the acquisition protocol, we implemented an Android App and uploaded it to the Play Store in order to study an unsupervised mobile scenario. This way all subjects could download the App and use it on their own devices without any kind of supervision, simulating a practical scenario in which subjects can generate touchscreen on-line signatures in any possible scenario, e.g., standing, sitting, walking, indoors, outdoors, etc. As a result, 94 different smartphone models from 16 different brands were collected during the acquisition.

  Regarding the acquisition protocol, between four and six separated sessions in different days were considered with a total time gap between the first and last session of at least 3 weeks. For each subject, there are at least 8 total genuine signatures (2 signatures/session) and 16 skilled forgeries (8 signatures/type) performed by four different subjects. Regarding the skilled forgeries, both static and dynamic forgeries were considered, similar to the office scenario. Information related to X and Y spatial coordinates, and timestamp is recorded for all devices. Pen-up information is not available in this case.

3   SVC 2021: Competition Set-Up

3.1   Tasks

The goal of SVC 2021 is to evaluate the limits of on-line signature verification systems on popular scenarios (office/mobile) and writing inputs (stylus/finger) through large-scale public databases. As a result, the following three tasks are considered in the competition:

- **Task 1**: Analysis of office scenarios using the stylus as input.
- **Task 2**: Analysis of mobile scenarios using the finger as input.
- **Task 3**: Analysis of both office and mobile scenarios simultaneously.

In addition, SVC 2021 simulates realistic operational conditions considering random and skilled forgeries simultaneously in each task.

3.2   Evaluation Criteria

The SVC 2021 competition follows a ranking based on points. Each task is evaluated separately, having three winners with their corresponding points (gold medal: 3, silver medal: 2, and bronze medal: 1). The participant/team that gets more points in total (Task 1, 2, and 3) in the final evaluation stage of the competition is the winner of SVC 2021.

The evaluation metric considered is the popular Equal Error Rate (%) similar to most on-line signature verification studies in the literature.

3.3   Experimental Protocol

The two following stages are considered in SVC 2021:

- **Development:** the goal of this stage is to provide the participants with the data needed to train the on-line signature verification systems. Only the DeepSignDB database is provided to the participants in this stage of the competition. In addition, participants can freely use other databases to train their systems.

  In order to allow the participants to test their trained systems under similar conditions considered in the final evaluation stage of the competition, we divide the DeepSignDB database into training and evaluation datasets. The training dataset is based on 1,084 subjects whereas the evaluation dataset comprises the remaining 442 subjects of the database. For the training of the systems (1,084 subjects), no instructions are given to the participants. They can use the data as they like. Nevertheless, for the evaluation of the systems (442 subjects), we provide the participants with the signature comparisons to run. Participants can run their on-line signature verification systems using the signature comparisons files provided to obtain the scores and test the EER performance on the public web platform (CodaLab) created for the competition[5]. This way participants can obtain a quantitative measure of the performance of the developed systems for the final evaluation stage of the competition.

In this development stage of the competition, participants can submit up to 300 system evaluation trials in total for all three tasks together. Results are updated in CodaLab in real time and they are visible to everyone in a ranking dashboard.

– **Final Evaluation:** the final evaluation of SVC 2021 is carried out using only the novel SVC2021_EvalDB database. The database together with the corresponding signature comparisons files (one file per task) are sent to the participants after signing the corresponding license agreement. It is important to highlight that all signatures are included in a single folder, and both the nomenclature of the signatures and the signature comparisons files are randomized to avoid cheating. Ground-truth labels are not provided to the participants. In addition, and in order to consider a very challenging impostor scenario, the skilled forgery comparisons included in the corresponding files are optimised using machine learning methods, selecting only the best high-quality forgeries.

In this final evaluation of SVC 2021, participants are allowed to submit the scores achieved by up to 3 different signature verification systems for each of the tasks considered in the competition.

4   SVC 2021: Description of Evaluated Systems

A total of 54 participants/teams initially registered in SVC 2021. However, only 6 teams finally submitted their scores with a total of 12 different on-line signature verification systems. Next, we describe briefly the systems provided by each of the teams of the competition.

4.1   DLVC-Lab Team

The DLVC-Lab team is composed of members of the South China University of Technology, and the Guangdong Artificial Intelligence and Digital Economy Laboratory.

The DLVC-Lab team proposed an end-to-end trainable deep soft-DTW (DSDTW) model, which greatly enhances the classical Dynamic Time Warping (DTW) method with the capability of deep representation learning. In particular, they use neural networks to learn deep time functions as inputs for DTW. As DTW is not fully differentiable with regards to its inputs, they introduce its smoothed formulation, soft-DTW [7], and incorporate the soft-DTW distances of signature pairs into a triplet loss function for optimization. As soft-DTW is differentiable, the entire system is end-to-end trainable and achieves a perfect integration of neural networks and DTW.

---

[5] https://competitions.codalab.org/competitions/27295.

Three different approaches were submitted to SVC 2021. System 1 is based on Convolutional Recurrent Neural Networks (CRNN) whereas System 2 and 3 are based on fully Convolutional Neural Networks (CNN). Systems 2 and 3 only differ in the training data. Concretely, Systems 1 and 2 use the development set of the DeepSignDB database for training (1,084 subjects), including both stylus-written and finger-written signatures. System 3 uses only finger-written signatures for training.

Regarding the feature extraction, 12 total time functions are extracted for each signature, considering information such as velocity and acceleration. These time functions are fed to the DSDTW.

4.2 SIG Team

The Spanish-Italian-German (SIG) team is composed of members of the European Commission (Italy), Universidad del Atlantico Medio (Spain), Universidad de las Palmas de Gran Canaria (Spain), and Hochschule Ansbach (Germany).

The signature verification system presented is based on the main principle laid out in [19]: the generation of synthetic off-line signatures from the real on-line samples and the fusion of both types of data can lead to the overall improvement of the on-line verification performance. Following that rationale, the system submitted is based on the combination of on- and off-line signature information.

The **on-line signature approach** is based on local features and the well-known DTW algorithm. In particular, the system is based on a subset of the initial 27 time functions introduced in [32] and selected using the Sequential Floating Forward Selection (SFFS) algorithm. The specific implementation of the DTW algorithm uses the Euclidean Distance to compute the optimal path in between signatures and outputs as score s_{on} the last value of the optimal path, normalised by the path length. Please be aware that, for the cases where pressure p is not available (i.e., mobile scenario in Task 2 of the competition), the time signal is simply discarded, together with any other time function derived from it.

Regarding the **off-line signature approach**, the first step performed is the generation of the synthetic off-line data starting from the real on-line signatures. Two different methods are used for this purpose: *i)* continuous trace [12,19], and *ii)* dotted trace [11]. Once the two synthetic off-line signatures are created (for each dynamic signature given as input), three different handcrafted features are extracted: *i)* run-length distribution [5], *ii)* geometrical features [15], and *iii)* quad-tree implementation of histogram of templates [36].

The score for each of the three feature sets is obtained by comparing the reference and probe vectors using the DTW algorithm followed by the cityblock distance. This process leads to six off-line intermediate scores ($s1_{off}$, $s2_{off}$,..., $s6_{off}$) for each on-line comparison defined in the competition (recall that each individual on-line signature is converted to two off-line synthetic signatures, defined by three different feature sets).

The six intermediate scores obtained by the off-line approach are finally fused into one unique off-line score s_{off} using a weighted sum. The weights for the fusion are empirically calculated on the training databases of the competition optimising the EER for each of the tasks considered in the assessment.

Finally, the on- and off-line scores (s_{on} and s_{off}) are normalised to the [0,1] range using the tanh-estimators and fused into the final score s given as output by the system based on the weighted sum.

Only the DeepSignDB database provided in the development stage of SVC 2021 was considered for training and evaluating the system.

4.3   TUSUR KIBEVS Team

The TUSUR KIBEVS team is composed of members of the Tomsk State University of Control Systems and Radioelectronics.

The on-line signature verification system presented is based on the use of global features and a gradient boosting classifier. First, a set of 100 global features is extracted for each enrolled and test signatures ($F_{enrolled}$ and F_{test}) based on previous approaches in the literature [17]. Then, a new feature vector F is obtained based on the subtraction of the previous enrolled and test feature vectors: $F = |F_{enrolled} - F_{test}|$. The resulting feature vector F is introduced to CatBoost [34], a fast, scalable, and high performance Gradient Boosting on Decision Trees (GBDT) that is available as an open source library[6].

Regarding the training procedure, only the DeepSignDB database provided in the development stage of SVC 2021 is considered. A total of 10K signature comparisons are randomly selected (5K genuine and 5K forgeries), considering both office (stylus) and mobile (finger) scenarios simultaneously. Forgery comparisons included 2.5K skilled forgeries and 2.5K random forgeries.

4.4   SigStat Team

The SigStat team is composed of members of the Budapest University of Technology and Economics.

Three different on-line signature verification systems were presented. All of them are implemented using the SigStat framework[7]. First, all signatures go through a preprocessing stage. Time samples with zero pressure are removed from the stylus-based signatures to reduce noise and remove some artifacts. Finally, X, Y, and pressure information are scaled to the [0,1] range and shifted by the average of their values. After this preprocessing stage, the biometric information is used to calculate different distance scores between signature pairs, considering three different approaches.

The first system considers local thresholds to detect whether the query signature is genuine of forgery. In particular, it uses DTW to calculate signature distances and the k-Nearest Neighbours (k-NN) approach to set a lower and an

---

[6] https://catboost.ai/.
[7] http://www.sigstat.org.

upper threshold for each reference signature. During the development stage, the system is tested on the evaluation subset of the DeepSignDB (442 users). The distances and comparisons between the signatures are used to calculate and tune several parameters, selecting the optimal values of the genuine G_{th} and forgery F_{th} thresholds and a scaling parameter s for the classification purpose.

For testing, the distance d between the questioned signature (S_q) and the reference signature (S_r) is obtained using DTW. The final score P_q is calculated as follows:

$$P_q = \frac{s \cdot F_{th} - d}{s \cdot F_{th} - G_{th}} \tag{1}$$

The second system considers global thresholds and is based on 4 classifiers and a linear fusion of them. The first three classifiers take advantage of global features such as the standard deviation of X and Y spatial coordinates, and the signing time duration. The last classifier is based on the DTW distance of signature pairs.

In the development stage, the evaluation subset of DeepSignDB is used to make genuine-genuine and genuine-forgery comparisons. For each comparison, the calculated DTW distance, the device input, and the expected prediction are stored. Next, the comparisons and their results are sorted into four different groups based on expected prediction and input device (genuine finger, genuine stylus, forgery finger, and forgery stylus). For each group some statistical parameters such as the minimum and median values are calculated and used to set the global thresholds for the system.

For testing, the score of the questioned signature P_q is calculated based on the DTW distance of the reference-questioned pair d, the minimum distance of genuine comparisons $d_{g_{min}}$ and the median distance of forgery comparisons $d_{f_{med}}$:

$$P_q = 1 - \frac{d_{f_{med}} - d}{d_{f_{med}} - d_{g_{min}}} \tag{2}$$

In case of $d < d_{g_{min}}$, the score P_q is automatically 0 and when $d > d_{f_{med}}$ is 1. A similar approach is considered for the remaining three classifiers based on global features.

Finally, the third system extends the set of global features considered in the second system, for example including the DTW distance as feature. Contrary to previous systems, a gradient boosting classifier (XGBoost) is considered for the final prediction.

4.5   MaD-Lab Team

The MaD-Lab team is composed of members of the Machine Learning and Data Analytics Lab (FAU).

The proposed system consists of a 1D CNN trained to classify pairs of signatures as matching or not matching. Features are extracted using a mathematical concept called *path signature* together with statistical features. These features are then used to train an adapted version of ResNet-18 [23].

Regarding the preprocessing stage, the X and Y spatial coordinates are normalised to a $[-1, 1]$ range whereas the pressure information to $[0, 1]$. In case that no pressure information is available (Task 2, mobile scenario), a vector with all one values is considered.

For the feature extraction, a set of global features related to statistical information is extracted for each signature. Besides, additional features are extracted using the signature path method [6]. This is a mathematical tool that extracts features from paths. It is able to encode linear and non-linear features from the signature path. The path signature method is applied over the raw X and Y spatial coordinates, their first-order derivatives, the perpendicular vector to the segment, and the pressure.

Finally, for classification, a 1D adapted version of the ResNet-18 CNN is considered. To adapt the ResNet-18 image version, every 2D operation is exchanged with a 1D one. Also, a sigmoid activation function is added in the last layer to output values between 0 and 1. Pairs of signatures are presented to the network as two different channels.

Regarding the training parameters of the network, binary cross-entropy is used as the loss function. The network is optimised using stochastic gradient descent (SGD) with a momentum of 0.9 and a learning rate of 0.001. The learning rate is decreased by a factor 0.1 if the accumulated loss in the last epoch is larger than the epoch before. In case the learning rate drops to 10^{-6}, the training process is stopped. Also, if the learning rate does not decrease below 10^{-6}, the training process is stopped after 50 epochs.

4.6   JAIRG Team

The JAIRG team is composed of members of the Jamia Millia Islamia.

Three different systems were presented, all of them focused on Task 2 (mobile scenarios). The on-line signature verification systems considered are based on an ensemble of different deep learning models training with different sets of features. The ensemble is formed using a weighted average of the scores provided by five individual systems. The specific weights to fuse the scores in the ensemble approach are obtained using a Genetic Algorithm (GA) [18].

For the feature extraction, three different approaches are considered: *i)* a set of 18 time functions related to X and Y spatial coordinates [43], *ii)* a subset of 40 global features [32], and *iii)* a set of global features extracted after applying 2D Discrete Wavelet Transform (2D-DWT) over the image of the signatures.

For classification, Bidirectional Gated Recurrent Unit (BGRU) models with a Siamese architecture are considered [38]. Different models are studied varying the number of hidden layers, input features, and training parameters. Finally, an ensemble of the best BGRU models in the evaluation of DeepSignDB is considered, selecting the fusing weight parameters through a GA.

Table 1. Final evaluation results of SVC 2021 using the novel SVC2021_EvalDB database acquired for the competition. For each specific task, we include the points achieved by each team depending on the ranking position (gold medal: 3, silver medal: 2, and bronze medal: 1).

Task 1: Office Scenario			Task 2: Mobile Scenario			Task 3: Office/Mobile Scenario		
Points	Team	EER(%)	Points	Team	EER(%)	Points	Team	EER(%)
3	DLVC-Lab	3.33%	3	DLVC-Lab	7.41%	3	DLVC-Lab	6.04%
2	TUSUR KIBEVS	6.44%	2	SIG	10.14%	2	SIG	9.96%
1	SIG	7.50%	1	SigStat	13.29%	1	TUSUR KIBEVS	11.42%
0	MaD	9.83%	0	TUSUR KIBEVS	13.39%	0	MaD	14.21%
0	SigStat	11.75%	0	Baseline DTW	14.92%	0	SigStat	14.48%
0	Baseline DTW	13.08%	0	MaD	17.23%	0	Baseline DTW	14.67%
			0	JAIRG	18.43			

Table 2. Global ranking of SVC 2021.

Position	Team	Total Points
1	DLVC-Lab	9
2	SIG	5
3	TUSUR KIBEVS	3
4	SigStat	1
5	MaD	0
6	JAIRG	0

5 SVC 2021: Experimental Results

This section describes the final evaluation results of the competition using the novel SVC2021_EvalDB database acquired for SVC 2021. It is important to highlight that the winner of SVC 2021 is based only on the results achieved in this stage of the competition as described in Sect. 3.3. Tables 1 and 2 show the results achieved by the participants in each of the three tasks, and the final ranking of SVC 2021 based on the total points, respectively. For completeness, we include in Table 1 a Baseline DTW system (similar to the one described in [31]) based on X, Y spatial coordinates, and their first- and second-order derivatives for a better comparison of the results.

As can be seen in Tables 1 and 2, DLVC-Lab is the winner of SVC 2021 (9 points), followed by SIG (5 points) and TUSUR KIBEVS (3 points). It is important to highlight that the on-line signature verification systems proposed by DLVC-Lab achieve the best results in all three tasks. In particular, an EER absolute improvement of 3.11%, 2.73%, and 3.92% is achieved in each of the tasks compared to the results obtained by the second-position team. Also, it is interesting to compare the best results achieved in each task with the results obtained using traditional approaches in the field (Baseline DTW). Concretely, for each of the tasks, DLVC-Lab achieves relative improvements of 74.54%, 50.34%, and 58.3% EER compared to the Baseline DTW. These results prove the high potential of deep learning approaches for the on-line signature verification field, as commented in previous studies [44,45].

Other approaches like the ones presented by the SIG team based on the use of on- and off-line signature information have provided very good results, achieving points in all three tasks (5 total points). The same happens with the system proposed by the TUSUR KIBEVS team based on global features and a gradient boosting classifier (CatBoost [34]), achieving 3 points in total. In particular, the approach presented by TUSUR KIBEVS has outperformed the approach proposed by the SIG team for the office scenario (6.44% vs. 7.50% EER). Nevertheless, much better results are obtained by the SIG team for the mobile and office/mobile scenarios (10.14% and 9.96% EERs) compared to the TUSUR KIBEVS results (13.39% and 11.42% EERs).

6   Conclusions

This paper has described the experimental framework and results of the ICDAR 2021 Competition on On-Line Signature Verification (SVC 2021). The goal of SVC 2021 is to evaluate the limits of on-line signature verification systems on popular scenarios (office/mobile) and writing inputs (stylus/finger) through large-scale public databases. The following tasks are considered in the competition: *i)* Task 1, analysis of office scenarios using the stylus as input; *ii)* Task 2, analysis of mobile scenarios using the finger as input; and *iii)* Task 3, analysis of both office and mobile scenarios simultaneously. In addition, both random and skilled forgeries are simultaneously considered in each task in order to simulate realistic scenarios.

The results achieved in the final evaluation stage of SVC 2021 have proved the high potential of deep learning methods compared to traditional approaches such as Dynamic Time Warping (DTW). In particular, the winner of SVC 2021 has been the DLVC-Lab team that proposed an end-to-end trainable deep soft-DTW (DSDTW). The results achieved in terms of Equal Error Rates (EER) are 3.33% (Task 1), 7.41% (Task 2), and 6.04% (Task 3). These results prove the challenging conditions of SVC 2021 compared to previous international competitions [4, 25, 28, 29, 48], specially for the mobile scenario (Task 2).

SVC 2021 will be established as an on-going competition[8], where researchers can play fair by benchmarking easily their systems against the state of the art in an open common platform using large-scale public databases such as Deep-SignDB[9] and SVC2021_EvalDB[10], and standard experimental protocols.

Acknowledgements. This work has been supported by projects: PRIMA (H2020-MS CA-ITN-2019-860315), TRESPASS-ETN (H2020-MSCA-ITN-2019-860813), BIBECA (RTI2018-101248-B-I00 MINECO/FEDER), Orange Labs, and by UAM-Cecabank.

---

[8] https://sites.google.com/view/SVC2021.

[9] https://github.com/BiDAlab/DeepSignDB.

[10] https://github.com/BiDAlab/SVC2021_EvalDB.

References

1. Ahrabian, K., Babaali, B.: Usage of autoencoders and Siamese networks for online handwritten signature verification. Neural Comput. Appli. **31**, 9321–9334 (2018). https://doi.org/10.1007/s00521-018-3844-z
2. Alonso-Fernandez, F., Fierrez-Aguilar, J., Ortega-Garcia, J.: Sensor interoperability and fusion in signature verification: a case study using tablet PC. In: Li, S.Z., Sun, Z., Tan, T., Pankanti, S., Chollet, G., Zhang, D. (eds.) IWBRS 2005. LNCS, vol. 3781, pp. 180–187. Springer, Heidelberg (2005). https://doi.org/10.1007/11569947_23
3. Antal, M., Szabó, L.Z., Tordai, T.: Online signature verification on MOBISIG finger-drawn signature corpus. Mobile Inf. Syst. (2018)
4. Blankers, V.L., van den Heuvel, C.E., Franke, K., Vuurpijl, L.: ICDAR 2009 Signature Verification Competition. In: Proceedings of International Conference on Document Analysis and Recognition (2009)
5. Bouamra, W., Djeddi, C., Nini, B., Diaz, M., Siddiqi, I.: Towards the design of an offline signature verifier based on a small number of genuine samples for training. Expert Syst. Appl. **107**, 182–195 (2018)
6. Chevyrev, I., Kormilitzin, A.: A primer on the signature method in machine learning. arXiv preprint arXiv:1603.03788 (2016)
7. Cuturi, M., Blondel, M.: Soft-DTW: a differentiable loss function for time-series. In: Proceedings of International Conference on Machine Learning (2017)
8. Delgado-Mohatar, O., Fierrez, J., Tolosana, R., Vera-Rodriguez, R.: Biometric template storage with blockchain: a first look into cost and performance tradeoffs. In: Proceedings of IEEE/CVF Conference on Computer Vision and Pattern Recognition Workshops (CVPRw) (2019)
9. Diaz, M., Ferrer, M.A., Impedovo, D., Malik, M.I., Pirlo, G., Plamondon, R.: A perspective analysis of handwritten signature technology. ACM Comput. Surv. **51**, 1–39 (2019)
10. Diaz, M., Fischer, A., Ferrer, M.A., Plamondon, R.: Dynamic signature verification system based on one real signature. IEEE Trans. Cybern. **48**(1), 228–239 (2016)
11. Diaz, M., Ferrer, M.A., Impedovo, D., Pirlo, G., Vessio, G.: Dynamically enhanced static handwriting representation for Parkinson's disease detection. Pattern Recogn. Lett. **128**, 204–210 (2019)
12. Diaz-Cabrera, M., Gomez-Barrero, M., Morales, A., Ferrer, M.A., Galbally, J.: Generation of enhanced synthetic off-line signatures based on real on-line data. In: Proceedings of International Conference on Frontiers in Handwriting Recognition (ICFHR), pp. 482–487 (2014)
13. Ellavarason, E., Guest, R., Deravi, F., Sanchez-Riello, R., Corsetti, B.: Touch-dynamics based behavioural biometrics on mobile devices-a review from a usability and performance perspective. ACM Comput. Surv. (CSUR) **53**(6), 1–36 (2020)
14. Faundez-Zanuy, M., Fierrez, J., Ferrer, M.A., Diaz, M., Tolosana, R., Plamondon, R.: Handwriting biometrics: applications and future trends in e-security and e-health. Cogn. Comput. **12**(5), 940–953 (2020). https://doi.org/10.1007/s12559-020-09755-z
15. Ferrer, M.A., Alonso, J.B., Travieso, C.M.: Offline geometric parameters for automatic signature verification using fixed-point arithmetic. IEEE Trans. Pattern Anal. Mach. Intell. **27**(6), 993–997 (2005)
16. Fierrez, J., et al.: BiosecurID: a multimodal biometric database. Pattern Anal. Appl. **13**(2), 235–246 (2010). https://doi.org/10.1007/s10044-009-0151-4

17. Fierrez-Aguilar, J., Nanni, L., Lopez-Peñalba, J., Ortega-Garcia, J., Maltoni, D.: An on-line signature verification system based on fusion of local and global information. In: Kanade, T., Jain, A., Ratha, N.K. (eds.) AVBPA 2005. LNCS, vol. 3546, pp. 523–532. Springer, Heidelberg (2005). https://doi.org/10.1007/11527923_54
18. Galbally, J., Fierrez, J., Freire, M.R., Ortega-Garcia, J.: Feature selection based on genetic algorithms for on-line signature verification. In: Proceedings of IEEE Workshop on Automatic Identification Advanced Technologies (2007)
19. Galbally, J., Diaz-Cabrera, M., Ferrer, M.A., Gomez-Barrero, M., Morales, A., Fierrez, J.: On-line signature recognition through the combination of real dynamic data and synthetically generated static data. Pattern Recogn. **48**(9), 2921–2934 (2015)
20. Galbally, J., Fierrez, J., Martinez-Diaz, M., Ortega-Garcia, J.: Improving the enrollment in dynamic signature verification with synthetic samples. In: Proceedings of International Conference on Document Analysis and Recognition (2009)
21. Galbally, J., Gomez-Barrero, M., Ross, A.: Accuracy evaluation of handwritten signature verification: rethinking the random-skilled forgeries dichotomy. In: Proceedings of IEEE International Joint Conference on Biometrics (IJCB) (2017)
22. Galbally, J., Martinez-Diaz, M., Fierrez, J.: Aging in biometrics: an experimental analysis on on-line signature. PLOS ONE **8**(7), e69897 (2013)
23. He, K., Zhang, X., Ren, S., Sun, J.: Deep residual learning for image recognition. In: Proceedings of IEEE/CVF Conference on Computer Vision and Pattern Recognition (2016)
24. Houmani, N., Garcia-Salicetti, S., Dorizzi, B.: A novel personal entropy measure confronted to online signature verification systems performance. In Proceedings of International Conference on Biometrics: Theory, Applications and System, BTAS, pp. 1–6 (2008)
25. Houmani, N., et al.: BioSecure signature evaluation campaign (BSEC'2009): evaluating on-line signature algorithms depending on the quality of signatures. Pattern Recogn. **45**(3), 993–1003 (2012)
26. Lai, S., Jin, L.: Recurrent adaptation networks for online signature verification. IEEE Trans. Inf. Foren. Secur. **14**(6), 1624–1637 (2018)
27. Lai, S., Jin, L., Lin, L., Zhu, Y., Mao, H.: SynSig2Vec: learning representations from synthetic dynamic signatures for real-world verification. In: Proceedings of AAAI Conference on Artificial Intelligence (2020)
28. Malik, M.I., et al.: ICDAR2015 competition on signature verification and writer identification for on- and off-line skilled forgeries (SigWIcomp2015). In: Proceedings of International Conference on Document Analysis and Recognition (ICDAR) (2015)
29. Malik, M.I., Liwicki, M., Alewijnse, L., Ohyama, W., Blumenstein, M., Found, B.: ICDAR 2013 competitions on signature verification and writer identification for on- and offline skilled forgeries (SigWiComp 2013). In: Proceedings of International Conference on Document Analysis and Recognition (2013)
30. Martinez-Diaz, M., Fierrez, J., Galbally, J., Ortega-Garcia, J.: Towards mobile authentication using dynamic signature verification: useful features and performance evaluation. In: Proceedings of International Conference on Pattern Recognition (2008)
31. Martinez-Diaz, M., Fierrez, J., Hangai, S.: Signature matching. In: Li, S.Z., Jain, A. (eds.) Encyclopedia of Biometrics, pp. 1382–1387. Springer, Heidelberg (2015). https://doi.org/10.1007/978-1-4899-7488-4
32. Martinez-Diaz, M., Fierrez, J., Krish, R.P., Galbally, J.: Mobile signature verification: feature robustness and performance comparison. IET Biom. **3**(4), 267–277 (2014)

33. Ortega-Garcia, J., Fierrez-Aguilar, J., et al.: MCYT baseline corpus: a bimodal biometric database. In: Proceedings of IEEE Vision, Image and Signal Processing, Special Issue on Biometrics on the Internet (2003)
34. Prokhorenkova, L., Gusev, G., Vorobev, A., Dorogush, A., Gulin, A.: CatBoost: unbiased boosting with categorical features. In: Proceedings of Advances in Neural Information Processing Systems (2018)
35. Sae-Bae, N., Memon, N.: Online signature verification on mobile devices. IEEE Trans. Inf. Forensics Secur. **9**(6), 933–947 (2014)
36. Serdouk, Y., Nemmour, H., Chibani, Y.: Handwritten signature verification using the quad-tree histogram of templates and a support vector-based artificial immune classification. Image Vis. Comput. **66**, 26–35 (2017)
37. Tolosana, R., Vera-Rodriguez, R., Fierrez, J., Morales, A., Ortega-Garcia, J.: Benchmarking desktop and mobile handwriting across COTS devices: the e-BioSign biometric database. PLoS ONE **12**(5), 1–17 (2017)
38. Tolosana, R., Vera-Rodriguez, R., Fierrez, J., Ortega-Garcia, J.: Exploring recurrent neural networks for on-line handwritten signature biometrics. IEEE Access **6**, 5128–5138 (2018)
39. Tolosana, R., Vera-Rodriguez, R., Fierrez, J., Ortega-Garcia, J.: Presentation attacks in signature biometrics: types and introduction to attack detection. In: Marcel, S., Nixon, M.S., Fierrez, J., Evans, N. (eds.) Handbook of Biometric Anti-Spoofing. ACVPR, pp. 439–453. Springer, Cham (2019). https://doi.org/10.1007/978-3-319-92627-8_19
40. Tolosana, R., Vera-Rodriguez, R., Fierrez, J., Ortega-Garcia, J.: Reducing the template ageing effect in on-line signature biometrics. IET Biom. **8**(6), 422–430 (2019)
41. Tolosana, R., Vera-Rodriguez, R., Fierrez, J., Ortega-Garcia, J.: BioTouchPass2: touchscreen password biometrics using time-aligned recurrent neural networks. IEEE Trans. Inf. Forensics Secur. **15**, 2616–2628 (2020)
42. Tolosana, R., Vera-Rodriguez, R., Guest, R., Fierrez, J., Ortega-Garcia, J.: Exploiting complexity in pen- and touch-based signature biometrics. Int. J. Doc. Anal. Recognit. **23**, 129–141 (2020). https://doi.org/10.1007/s10032-020-00351-3
43. Tolosana, R., Vera-Rodriguez, R., Ortega-Garcia, J., Fierrez, J.: Preprocessing and feature selection for improved sensor interoperability in online biometric signature verification. IEEE Access **3**, 478–489 (2015)
44. Tolosana, R., Delgado-Santos, P., Perez-Uribe, A., Vera-Rodriguez, R., Fierrez, J., Morales, A.: DeepWriteSYN: on-line handwriting synthesis via deep short-term representations. In: Proceedings of AAAI Conference on Artificial Intelligence (2021)
45. Tolosana, R., Vera-Rodriguez, R., Fierrez, J., Ortega-Garcia, J.: DeepSign: deep on-line signature verification. IEEE Trans. Biom. Behav. Identity Sci. **3**(2), 229–239 (2021)
46. Vera-Rodriguez, R., et al.: DeepSignCX: signature complexity detection using recurrent neural networks. In: Proceedings of International Conference on Document Analysis and Recognition (ICDAR) (2019)
47. Wu, X., Kimura, A., Iwana, B.K., Uchida, S., Kashino, K.: Deep dynamic time warping: end-to-end local representation learning for online signature verification. In: Proceedings of International Conference on Document Analysis and Recognition (ICDAR) (2019)
48. Yeung, D.-Y., et al.: SVC2004: first international signature verification competition. In: Zhang, D., Jain, A.K. (eds.) ICBA 2004. LNCS, vol. 3072, pp. 16–22. Springer, Heidelberg (2004). https://doi.org/10.1007/978-3-540-25948-0_3

ICDAR 2021 Competition on Script Identification in the Wild

Abhijit Das[1,2]([⊠]), Miguel A. Ferrer[3], Aythami Morales[4], Moises Diaz[5],
Umapada Pal[1], Donato Impedovo[6], Hongliang Li[7], Wentao Yang[7],
Kensho Ota[8], Tadahito Yao[8], Le Quang Hung[9], Nguyen Quoc Cuong[9],
Seungjae Kim[10], and Abdeljalil Gattal[11]

[1] Indian Statistical Institute, Kolkata, India
[2] Thapar University, Patiala, India
abhijit.das@thapar.edu
[3] Univ. de Las Palmas de Gran Canaria, Las Palmas, Spain
[4] Universidad Autonoma de Madrid, Madrid, Spain
aythami.morales@uam.es
[5] Universidad del Atlántico Medio, Las Palmas, Spain
moises.diaz@atlanticomedio.es
[6] Università degli Studi di Bari Aldo Moro, Bari, Italy
donato.impedovo@uniba.it
[7] South China University of Technology, Guangzhou, China
[8] Canon IT Solutions Inc., Tokyo, Japan
[9] University of Information Technology, Ho Chi Minh City, Vietnam
[10] NAVER Papago, Seoul, Korea
[11] Larbi Tebessi University, Tebessa, Algeria

Abstract. The paper presents a summary of the 1st Competition on
Script Identification in the Wild (SIW 2021) organised in conjunction
with 16th International Conference on Document Analysis and Recognition (ICDAR 2021). The goal of SIW is to evaluate the limits of script
identification approaches through a large scale in the wild database
including 13 scripts (MDIW-13 dataset) and two different scenarios
(handwritten and printed). The competition includes the evaluation over
three different tasks depending of the nature of the data used for training and testing. Nineteen research groups registered for SIW 2021, out
of which 6 teams from both academia and industry took part in the
final round and submitted a total of 166 algorithms for scoring. Submissions included a wide variety of deep-learning solutions as well as
approaches based on standard image processing techniques. The performance achieved by the participants prove the elevate accuracy of deep
learning methods in comparison with traditional statistical approaches.
The best approach obtained classification accuracies of 99% in all three
tasks with experiments over more than 50K test samples. The results
suggest that there is still room for improvements, specially over handwritten samples and specific scripts.

Keywords: Handwritten and printed script identification · Wild ·
Deep learning · Multi-script

© Springer Nature Switzerland AG 2021
J. Lladós et al. (Eds.): ICDAR 2021, LNCS 12824, pp. 738–753, 2021.
https://doi.org/10.1007/978-3-030-86337-1_49

1 Introduction

Due to the ever-increasing demand for the creation of a digital world, many Optical Character Recognition (OCR) algorithms have been developed over the years [1]. Incidentally, script identification plays a vital role in OCR pipeline. It use is also been used for several application such as signature verification [2–4], scene text detection [5,6]. A script can be defined as the graphic form of the writing system used to write a statement [7,8].

The availability of large numbers of scripts makes the development of a universal OCR a challenging task. This is because the features needed for character recognition are usually a function of structural script properties and of the number of possible classes or characters. The extremely high number of available scripts makes the task quite daunting and sometimes deterring, and as a result, most OCR systems are script-dependent [9]. The approach for handling documents in a multi-script environment is divided into two steps: first, the script of the document, block, line or word is estimated, and secondly, the appropriate OCR is used. This approach requires a script identifier and a bank of OCRs, at a rate of one OCR per possible script. Many script identification algorithms have been proposed in the literature. Script identification can be conducted either offline, from scanned documents, or online, if the writing sequence is available [10].

Identification can also be classified either as printed or handwritten, with the latter being the more challenging. Script identification can be performed at different levels: page or document, paragraph, block, line, word, and character. As it is similar to any classical classification problem, the script identification problem is a function of the number of possible classes or scripts to be detected. Furthermore, any similarity in the structure of scripts represents an added challenge [7].

Hence, to elevate state-of-the-art several benchmarking effort by publishing publicly available datasets [11,12] and competition has been organised [13–17]. Consequently, the benchmarking works on script identification in the literature uses different datasets with different script combinations. Therefore, it is difficult to carry out a fair comparison of these different approaches. Moreover, the databases employed in related studies usually include two to four scripts. A few actually include an even higher number of scripts but with not exhaustive combination with both handwritten and printed samples with different level of annotation (word, line and document).

Hence to alleviate this drawback, in this competition we aim to offer a database for script identification, which consists of a wide variety of some the most commonly used scripts, collected from real-life printed and handwritten documents. The competition is also aim to document the recent development in this area of research and attract the attention of the researchers. Specifically, we aim to answer the following questions:

- How do contemporary script identification techniques perform with large scale challenging document images captured in the wild?
- What impact do changes in type of data (handwritten and printed) have on identification performance?

The following contributions that are documented in this report:

- A rigorous evaluation of several contemporary script identification approaches.
- A comprehensive analysis of the identification approaches.
- A public benchmark with more 80K images from 13 scripts obtained from real handwritten and printed documents.

2   Benchmarking Dataset

We developed a large dataset for script identification tasks, consisting of printed and handwriting documents of the following 13 scripts: Arabic, Bengali, Gujarati, Gurmukhi, Devanagari, Japanese, Kannada, Malayalam, Oriya, Roman, Tamil, Telugu, and Thai [18]. Figure 1 shows an example of words in each script.

The printed documents were collected from local newspapers and magazines, whereas mostly native volunteers provided the handwritten. All documents were scanned at 300 dpi of resolution. However, various conditions are included in the databases, like different inks, sheets, font sizes, and styles. As a consequence, controlled background removal and ink equalization was applied to ensure a cleaner database. Furthermore, the word segmentation from the documents was carried out by an automatic system, which was manually fine-tuned later by checking all individual words.

This novel dataset has not shared at the time of the competition. In addition, we only consider words extracted from the texts for this competition. Specifically, the word-based data was divided into *Training* and *Testing* sets for this competition, as summarize Table 1.

- *Training.* All registered participants in the competition had access to the training data. The more than 30K training images were divided into two main subsets with printed (21974) and handwritten (8887) images.
- *Testing.* Once the deadline was reached, we shared the testing data with the participants, which consisted of 55814 unlabelled images to identify in the 13 scripts. The test set includes more than 50K handwritten and printed images. The type of data (handwritten or printed) as well as the script label, were not provided to the participant until the end of the competition.

Script	Printed	Handwritten
Arabic	فيلساعلىمايجرينكلاالأحداث	آتستان سر ین بزر گل
Bangla	এসএসকেএমহাসপা	আগুলের স্কে সম্মান
Gujrati	ગુકશાનીનોરાવૈકરવાકામે	2લીગ૨રાગ અરલી
Gurjmukhi	कारनमॆतदाताहॆरान	ਸਿਤੱੜਟੳ ਡ
Hindi	नयीदिल्लीमेंआयोजितरि	जनिंठल्स हटल्ल लाईका
Japanese	玄関で靴を脱いで、素足	采びしんド阝のる :19
Kannada	ಕಾಗಾಗೆಈಓಔಜಾಂಳ	ಆಂಡ್ಕ್ ೯ೈಡೃ್
Malayalam	യാക്കപ്പെടുന്നത്യൻ	ഇൾൾ ൂൽൻഗ
Oriya	ନିର୍ମାଣକର୍ମିଗାରାମାନେ	ଓଗଣ ଶ୍ରୀ
Roman	Borgesdecíaquecuan	AFTER few days of ink
Tamil	நாட்டிலசம்பகாலமாகஉ	அனுஞுலிஜெஜ்ஞ
Telugu	నందిజెఎన్టీయుఐపువా	అస్టెబరఖరస్ఫకిరల్ర
Thai	ออกมานานกว่าแบบ	การพักบ้านในมากลุ่ม

Fig. 1. Example of image-based words used in the competition.

As can be seen, the number of words is different in each script. It makes the script identification an unbalance challenge with this benchmarking corpus. Finally, the images within *Training* and *Testing* sets were not the same. They were randomly extracted from the words available in the MDIW-13 multiscript document database [19]. In other words, the data in this competition is a subset of the large database. The database is publicly available[1].

---

[1] https://gpds.ulpgc.es/.

Table 1. Summary of the word images included in each script for the *Training* and *Testing* sets.

Script	Training		Testing	
	Printed	Handwritten	Printed	Handwritten
Arabic	1996	570	4206	3370
Bengali	1608	401	949	8919
Gujarati	1229	144	982	37
Gurmukhi	3629	538	5475	135
Devanagari	1706	1203	1076	301
Japanese	1451	352	363	89
Kannada	1183	872	974	1123
Malayalam	2370	575	1950	144
Oriya	1660	333	649	7514
Roman	1574	558	6053	3750
Tamil	451	873	1667	557
Telugu	1261	640	865	161
Thai	1856	1828	1861	2644
Total	21974	8887	27070	28744

3   Evaluation Protocol

SIW 2021 was executed in two stages. During the first stage participants were given the training split of the MDIW-13 datset, including the ground truth and were asked to develop their algorithm. In the second stage, the test split of the MDIW-13 (without the annotation) was provided to the participants to infer the script label on the test images.

The detailed tasks for the competition are as follows: 1) **Task 1:** Script identification in handwritten document; 2) **Task 2:** Script identification in printed document; and 3) **Task 3:** Mixed script identification: Train and tested with handwritten and printed.

The evaluation measure used during the competition was the **Correct Classification Accuracy**. This performance measure was calculated as the percentage of samples correctly classified respect the total number of samples available for each of the tasks. Note that training and test sets present certain class imbalance and the methods need to deal with this challenge.

Participants performed word level script recognition. The submission which achieved the best Correct Classification Accuracy for Task 3 was considered as the winner. The ground truth was manually annotated and segmented according to a semiautomatic process described in [19].

Table 2. Summary of participants and submitted approaches to SIW 2021. The table lists the abbreviations of the models, as used in the experimental section. PR = Pretrained models, EX = External data, HC = Hand-crafted features, AL = Detection and alignment, EM = Ensemble models, DM = Differentiate models, Pre = Pre-processing, post = Post-processing, ✓ = Yes, x = No.

No	Group	PR	EX	HF	AL	EM	DM	Pre	Post
1	Ambilight	✓	✓	x	✓	x	x	✓	✓
2	DLVC-Lab	✓	✓	x	x	✓	x	x	x
3	NAVER Papago	✓	x	x	x	x	x	✓	x
4	UIT MMlab	✓	x	x	x	✓	✓	x	✓
5	CITS	✓	✓	x	x	x	x	✓	✓
6	Larbi Tebessi	x	x	✓	x	x	x	x	x

4   Details of Submitted Approaches

Six different groups submitted their approaches for the final evaluation. The participants include teams from academia and industry. Table 2 presents a summary of the participating groups and their approaches. As can be seen, the proposed approaches show heterogeneous characteristics: with and without pre-processing or post-processing techniques, ensemble or unique models, use of augmented data. We proceed to present a summary of the best systems submitted by each of the participants.

4.1   The Lab of Ambilight, NetEase Inc. (Ambilight)

This team used semantic segmentation method as our baseline model instead of classification method. The semantic segmentation model is more focused on the details of every character and can reduce the disturbances of background, so the semantic segmentation model is better in this task. To fully utilize the classification label, a multi-task training design is introduced to further improve the performance of the segmentation task. Therefore, a classification branch is added in our proposed framework. Also, to fit text geometric features better, attention module and deformable convolution are added into the backbone. Another highlight of our approach lies in that we use lots of synthetic data and grid distortion technique to simulate the handwriting style of different people, which are finally proved valid tricks in this task. During testing phase, we apply semi-supervised learning technique to fit the test data better. All these strategies stated above make us achieve the top performance in the competition.

Introduction and Motivation: As stated above the challenges such as variation in length of texts, ever-changing division, and similar letters even characters existing in different scripts, hence an explicit solution is to design a framework based on the fine-grained classification work. Currently, a popular model of fine-grained classification is mainly based on the attention mechanism, such as the

Table 3. Some insight result on different architecture by team Ambilight.

Model	Printed (mIoU)	Handwritten (mIoU)
VGG19	0.9012 (Acc)	–
ResNet50-FCN	0.9590	0.8327
ResNest50-DeepLab3+	0.9704	0.864
HRnet48-OCRnet	0.9732	0.8732
HRnet48-OCRnet+DCN	**0.9795**	**0.8958**
Swin-Transfromer	0.9654	0.8768

WS-DAN [20] model (proposed by MARA in 2019), where it uses the attention mechanism to crop the partial details of the image to assist the classification. However, its attention mechanism is actually a weak supervision mechanism. For detection-based models, they generally have to use stronger supervision information. For example, for the Part-RCNN [21], the foreground is first detected, and the detected foreground is scored with a discrimination degree. More detailed supervision information is based on segmentation methods. Fine-grained backgrounds are often complex and different, but the foregrounds are very similar. If the foreground can be segmented for classification, better performance will be achieved. Similarly, Mask-CNN [22] is a fine-grained classification model based on strongly supervised segmentation information, but due to the high cost of labelling, it is rarely applied in the industry (Table 3).

Although only the classification information is given in this competition, the image only filled with black characters on a white background allows us to calculate the mask required for segmentation directly through the pixel information. Later, the participants compared different encoder and decoder combinations, tried the classic VGG [23], Resnet [24] as the encoder, FCN [25], Deeplab3+ [26] as the decoder. Also tried the newly proposed Resnest [27], Swin-Transfromer [28], etc. In the end, we combine the current SOTA backbone HRnet [29] and segmentation decoder OCRnet [30] to solve this task. In addition, they also replace part of the normal convolution in the encoder for DCN [31] convolution to play a role of weak attention supervision. The following table demonstrates their comparative study on different methods and backbones.

Detailed Method Description: (a) Since the training data is black on a white background, the mask required for segmentation can be obtained according to the pixel values. After attained the image mask, each pixel has supervision information, and more local features can be extracted for very similar languages. Here the participant use HRnet as the encoder to extract image features, and then use OCRnet as the decoder for the output the segmentation results, and determine the final output category according to the segmentation vote of each pixel. In addition, they additionally designed an auxiliary classifier for training to make the model also pay attention to the global features, and used DCN v2 (Improved Deep & Cross Network) convolution instead of 2D convolution to strengthen the modeling ability of text shapes.

(b) **Pseudo-label Fine-tuning** [32]: The team use the trained model to do inference on test set. If more than 70% of the pixels (foreground pixels) are classified to be the same language, they assume that the predication is correct. Samples that meet this condition will be allocated to the training set for fine-tuning. In this way, iteratively fine-tune the model in multiple rounds.

(c) **Synthetic Data and Data Augmentation:** For the printed scripts, they generate millions of synthetic data in different scripts and fonts using text renderer, which can greatly expand the training data. Firstly, they use the synthetic data to pre-train the model, and then use the given training set for fine-tuning. For the handwritten scripts, they use grid distortion to simulate handwriting changes. This enhancement method can appropriately simulate non-rigid deformations such as changes in the thickness and length of human strokes to enhance the robustness of model for different handwriting.

(d) **Loss Function:** The team did a lot of experiments on the loss function, and finally used focal loss [33] to increase the learning weight of the text part.

(e) **Augmentation during Testing:** They use random-crop-resize during training, which plays a small-scale multi-scale role. Therefore, during the inference on the testing set, multi-scale resizing is performed on the intput image. Results of different sizes are blended together to the final result.

4.2 South China University of Technology (DLCV-Lab)

The three tasks were treated as classification problems and were solved by adopting deep learning methods. In order to improve the diversity of training data, the team utilized data augmentation technique [34] and synthesized a dataset using fonts in different scripts. Finally, they ensemble three CNN-based models, namely, ResNet-101 [24], ResNeSt-200 [27] and DenseNet-121 [35] with CBAM [36] for better classification accuracy.

The first question they considered is whether they should adopt machine learning methods or deep learning methods. In addition to more than 30,000 samples in the MDIW-13 database, they also use the collected fonts to generate a synthetic dataset. With such a relatively large amount of data for training, they thought that deep learning methods may outperform machine learning methods. Among image processing methods in deep learning, Transformer-based models has received lots of recent attention, while CNN-based models are still the mainstream. In the absence of the massive training data required by transformer, they believe CNN-based models may be more suitable. After determining the main technical route, on one hand, we try various CNN-based models and ensemble the best three models for better classification accuracy. On the other hand, they collect some fonts for synthesizing data, and utilize the data augmentation technique for text images to improve the diversity of the training data.

Synthesize Data: To synthesize data, fonts in 13 scripts are collected from the Internet, and corpora are translated from 58,000 English words using the Google translation API. Then they randomly select fonts and the translated word corpora to generate 5,000 images for each script, which are added to the training set.

Data Augmentation: The data augmentation technique they used [34] embodies different transformations for text images, including distortion, stretch and perspective. In the training phase, there is 50% probability for every sample to execute each transformation.

Details of Model: After trying a variety of models, the team choose three separately trained models, Resnet-101, Resnest-200 and Densenet-121 with CBAM to ensemble. We use the Adam optimizer with a weight decay of 1e-4 and a learning rate of 1e-4. The learning rate is set manually to 1/10 of its current value when the loss value no longer drops. The cross-enropy loss function is adopted and image resolution of input is set to 300 × 700. All three models are pretrained on ImageNet. It is worth noting that we did not use the synthetic data for Resnet-101, as it will cause a worse result. Finally, the prediction confidences of the three models are averaged as the output.

4.3 NAVER Papago (NAVER Papago)

Given the time constraint of the competition, the main strategy and motivation was to fix the network and conduct quick experiments to conquer the problems of the data. As many other real world problems do, the data had class imbalance problem. Also, the images didn't have rich pixel level information such as color, contrast, but had much more spatial level information such as shape, font style. To address these issues, first of all, the participant used stratfied data sampling to overcome class imbalance. This helps the model fit to less-frequent-class images and boosts overall score. Secondly, many spatial level augmentations were applied to make the model better recognize the newly seen text shapes in test data. Augmentations such as random shift, scale, stretching, grid distortion were applied, and in accordance, same augmentations were applied at test time which also improved score. The network architecture was fixed to ResNet50 for all the experiments above, and was changed to NFnet-f3 near the end of competition. In conclusion, the overall approach was to leave most of the settings fixed and concentrate on one or two most important issues of the competition.

Preprocessing: All images were rescaled, maintaing the width/height ratio, and padded to have the same size 160 × 320.

Data Split: 10% of each handwritten and printed data were used as validation set. They were sampled by stratified sampling, where the ratio of each class in the sample remains the same as the ratio of each class in the whole data. Then training data and validation data of both handwritten and printed data were mixed, for training all at once.

Inference: 1) pseudo-labeling: test predictions that had confidence over 0.99 were pseudo-labeled and used to finetune model before inference; 2) test time augmentation: test time augmentation with random scale, random horizontal, vertical stretching was applied.

4.4   University of Information Technology (UIT MMLab)

Our approach is building a two-stage deep learning system for script identification. In the first stage, we applied a residual neural network (ResNet) [37] to classify the script as handwriting or print. In the second stage, for each type of these we use our corresponding EfficientNet [38] model to identify the script. For the best result on private test set, we used EfficientNet-B7 for handwritten script and EfficientNet-B4 for printed script.

Handwritten/Printed Type Classification: A Resnet-50 architecture pretrained on ImageNet was used as backbone network. We stacked 1 fully connected layer with 1024 units in front of the output layer. A sigmoid loss function was used as binary classification. An Adam optimizer with learning rate of 0.0001 was used for the training process. The dataset was splitted with the ratio of 9/1 and then trained in 20 epochs with a batch size of 16. We saw that 20 epochs are enough for the model to converge.

We trained separately 2 models with regard to 2 different random seeds in train-validation step. It comes to my attention that one of these 2 models helps to get higher score on printed task. The another model leads to higher score on handwritten task. So, in the inference phase, the predicting result of 2 models are compared with each other. The images that make their result different will be considered manually by a visualizing tool. Following these steps, we can create a quite significant handwritten-printed classification result.

Script Identification: For script identification phase, we decided to use two separate EfficientNet B4 and B7 models for handwritten and printed scripts. The backbone network we used is a pretrained network previously trained on a large ImageNet dataset contains 1000 classes labels, we can take advantage of pretrained weights to extract useful image features without retraining from scratch. With transfer learning approach, we exclude the final fully connected (FC) layer of pretrained model, then we replaced the top layer with custom layers containing FC layers for identify 13 languages of script allows using EfficientNet as a feature extractor in a transfer learning workflow. The features is fed into global average pooling (GAP) to generate a vector whose dimension is the depth of the feature, this vector is the input of the next FC Layer. The GAP layer outputs the mean of each feature map, this drops any remaining spatial information, which is fine because there was not much spatial information left at that point. The final FC layer 1×13 using softmax activation produces the probability of each class ranged from 0 to 1.

Because freezing EfficientNet and training only custom top layers tends to underfit the training data, training both EfficientNet and custom top layers tends to overfit the training data, so the approach is freezing some first layers of EfficientNet to make use of the low level features extracted by pretrained network on ImageNet datasets, then training the remaining layers and top layers. The team use Adam optimizer with a learning rate of 0.0001 to minimize the categorical cross entropy loss function. For each type of script (handwritting and print), They splitted the data into training set and validation set with the ratio of 8/2. Then, they trained two base models i.e. B4, B7, 250 epochs for each model with a batch size of 16 and only save the best weight with the highest validation accuracy.

4.5   Canon IT Solutions Inc. (CITS)

Firstly, the participants from this groups made patches by sliding window. Stride was 56pixels[2]. Secondly, they classified each patch using a Efficient Net [39] as a classifier.

Lastly the participants calculated sum of confidence of each class of all patches, and adopted the class corresponding to maximum argument as the inference result.

For prepossessing, we used shave 20 pixels(up, bottom, left, right), resize height to 224 keeping aspect ratio, normalization. The participants used Tesseract in post-processing for printed Hindi's and Gurmukhi's results.

There were 3 postprocessing steps: 1) trained a CNN model that classifies images are printed or handwritten; 2) collected images which were classified as printed Hindi or printed Gurmukhi; 3) OCR'ed each image in step 2 with Tesseract of Hindi model and Gurmukhi model. Tesseract outputs confidence score with OCR result, and we adopted script with higher confidence score.

The participants generated pseudo handwritten images with CycleGAN [40] and used as a training dataset. This dataset contained 13,000 images (1,000 images for each class).

4.6   Larbi Tebessi University (Larbi Tebessi)

Among the different methods for research purposes, the texture-based descriptor is preferred by many researchers due to its various advantages, strengths and benefits. Proposed research method allows to extract highly informative elements of the printed and handwritten text. Otherwise, they do not involve comparing pieces of text like-for-like. In this way, the computational cost provides good insight into the complexity of the system compared to other systems. However, the proposed research method also has their weak point that must be considered which is sensitive to noise that makes the extracted features sensitive to small changes in the handwriting. The textural information is captured using an oriented Basic Image Feature (oBIF) columns.

In order to increase the performance of the oBIFs descriptor, the participants combine oBIFs at two different scales to produce the oBIF column features by ignoring the symmetry type flat. The oBIFs column features are generated using different values of the scale parameter σ while the parameter ϵ is fixed to 0.001. The generated feature vector is finally normalized.

The oBIFs column histograms are extracted, the both oBIFs column histograms for the scale parameter combination (2,4) and (2,8) are concatenated together to form the feature vector representing each printed and handwritten image. Once the features are extracted, classification is carried out using Support Vector Machine (SVM) classifier. We have employed the Radial Basis Function

---

[2] 56 is 224/4, and 224 is CNN's default input width.

Table 4. Summary of final results for each of the three tasks. Correct classification accuracy (final rank in brackets). The table presents two baseline methods: 1) the Dense Multi-Block Linear Binary Pattern [19] and 2) the Random Chance. T.# Subm = Total submissions

Group	# Subm	Task 1 (handwritten)	Task 2 (printed)	Task 3 (mixed)
Ambilight	16	99.69% (1)	99.99% (1)	99.84% (1)
DLVC-Lab	43	97.80% (3)	99.80% (2)	98.87% (2)
NAVER Papago	26	99.14% (2)	95.06% (5)	97.17% (3)
UIT MMlab	46	95.85% (4)	98.63% (4)	97.09% (4)
CITS	34	90.59% (5)	99.24% (3)	94.79% (5)
Baseline [19]	–	89.78% (–)	95.51% (–)	94.45% (–)
Larbi Tebessi	1	81.21% (6)	86.62% (6)	83.83% (6)
Random	–	7.14% (–)	7.43% (–)	7.22% (–)

(RBF) kernel with the kernel parameter selected to 52 while the soft margin parameter C is fixed to 10. The participants evaluated the oBIF column features to identify scripts from printed and handwritten images on the dataset of the ICDAR 2021 competition on Script Identification in the Wild (SIW 2021). The experiment is carried out by using both printed and handwritten samples in the training and test sets.

5 Benchmarking Results with Analysis and Discussion

In this section we proceed to report and analysis the results obtained from the submission. The Table 4 shows the final rank of the competition for the three different tasks. The results are reported in terms of correct classification accuracy. In order to compare the improvement provided by the competition we released two benchmarking. Firstly, the random chance, which is around 7% for all three tasks. Further, we provided a second benchmark, i.e. the results obtained using the Dense Multi-Block Linear Binary Pattern as per the reported in [19].

We can observe from the Table 4 that the results of the participants ranged from 83.83% to 99.84%. Most of the submission were based on CNN, and they outperformed the baseline [19]. As it is expected, the handwritten task represent a bigger challenge for the participants with accuracy's lower than in the printed scenario for most of the scenarios. Further, for the mixed task the accuracy's are someway between their printed and handwritten results. Incidentally, the submission from Ambilight i.e. the winner team had a marginal difference of .2% between printed and handwritten task. Also the gap is more less i.e. 0.06% while considering mixed and printed. The reason behind this expected to be the approach they chooses while solving this problem. They considered this problem as a semantic segmenting task and which helps to give this leverage while finding the details of the character.

	Ar	Ba	Gu	Gur	Hi	Jap	Kan	Ma	Or	Ro	Ta	Te	Th
Ar	99.97	0	0	0	0	0	0	0	0	0	0	0	0
Ba	0	99.74	0	0.72	0	0	0	0	0.04	0.18	0	0	0
Gu	0	0	100	0	0	0	0	0	0	0	0	0	0.11
Gur	0	0	0	97.83	0	0	0	0	0	0	0	0	0
Hi	0	0	0	0	100	0	0	0	0	0	0	0	0
Ja	0	0	0	0	0	100	0	0	0	0	0	0	0.04
Ka	0	0	0	0	0	0	100	0.70	0	0	0.17	8.87	0.26
Ma	0	0	0	0	0	0	0	98.59	0	0	0.53	0	0.04
Or	0	0.19	0	1.45	0	0	0	0	99.96	0.24	0.53	0	0
Ro	0.03	0.03	0	0	0	0	0	0	0	99.55	0.17	0	0.07
Ta	0	0	0	0	0	0	0	0	0	0	98.24	0	0
Te	0	0.03	0	0	0	0	0	0.70	0	0.03	0.17	91.12	0.04
Th	0	0	0	0	0	0	0	0	0	0	0.17	0	99.43

Fig. 2. Confusion matrix (Task 1) obtained by the best method (Ambilight).

We proceed to further analyse the system submitted by the winning team. The Fig. 2 shows the confusion matrix obtained by the winning approach (Task 1: Handwritten script identification i.e. the most challenging task). The results show that there is room for improvement in some of the scripts, specially Gujarati, Telugu, Tamil, and Malayalam. Incidentally, Gujarati script were mostly mis-classified as Bengali and Oriya, we assume that it is mostly due to the writes writing style. As a fact Bengali and Oriya script do not have much similar outlook to Gujarati. While considering Telugu the biggest mis-classification was with Kannada script nearly 8.8%, it can be considered due to the visual similarity of the two script. This was the highest mis-classification and as a reason Telugu script classification attend the lowest performance. Although it is interesting to find that Kannada script had perfect classification. Similar to Kannada and Telugu, Malayalam and Tamil script have high visual similarity, considering Tamil the highest mis-classification was with Malayalam script, but which is not the case while considering Malayalam highest mis-classification was found between Kannada and Telugu. Hence, we cannot conclude that visual pattern and structure only responsible while considering script identification multi-script scenario.

Summarising the competition, from Fig. 3 we can conclude that the maximum correct classification accuracy obtained by all the participants during the 14 days is nearly 100%. The results obtained during the first week of the competition represent a period where participants adapted their systems to the test set. We can see a performance improvement from 91% to 97.4% during this first week. The second week shows a constant improvement with a final correct classification accuracy of 99.84%. In a competition with two weeks available for submissions, we cannot discard certain overfitting to the test set. However, the large number of samples (more than 50,000) and the different characteristics of the tasks (handwritten vs printed) is a added challenge.

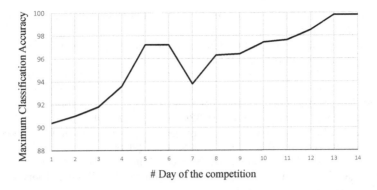

Fig. 3. Maximum correct classification accuracy (Task 3) obtained by all the participants in each day of the competition.

6 Conclusions

The 1st edition of the Script Identification in the Wild Competition was organised to evaluate and benchmark the performance of contemporary script identification techniques captured in the wild and explore the robustness of existing models *w.r.t.* to changes in the font, size, ink and printing quality used for document development and image acquisition as well as changes in the external acquisition conditions. A total of 6 groups participated in the competition and contributed 6 algorithm for the group evaluation. The identification algorithms were compared in terms of printed, handwritten and mixed documents. Most of the submitted models ensured solid identification results in most experimental scenarios. It is worth mentioning that for some the combination of script the performance was slightly lower which requires further attention.

References

1. Eikvil, L.: Optical character recognition, p. 26 (1993). citeseer.ist.psu.edu/14 2042.html
2. Das, A., et al.: Multi-script versus single-script scenarios in automatic off-line signature verification. IET Biometr. **5**(4), 305–313 (2016)
3. Das, A., et al.: Thai automatic signature verification system employing textural features. IET Biometr. **7**(6), 615–627 (2018)
4. Ferrer, M.A., et al.: Multiple training-one test methodology for handwritten word-script identification. In: 2014 14th International Conference on Frontiers in Handwriting Recognition, pp. 754–759. IEEE (2014)
5. Suwanwiwat, H., et al.: Benchmarked multi-script Thai scene text dataset and its multi-class detection solution. Multimed. Tools Appl. **80**(8), 11843–11863 (2021)
6. Keserwani, P., et al.: Zero shot learning based script identification in the wild. In: 2019 International Conference on Document Analysis and Recognition (ICDAR), pp. 987–992. IEEE (2019)

7. Ghosh, D., et al.: Script recognition–a review. IEEE Trans. Pattern Anal. Mach. Intell. **32**(12), 2142–2161 (2010)

8. Bhunia, A., et al.: Indic handwritten script identification using offline-online multimodal deep network. Inform. Fusion **57**, 1–14 (2020)

9. Ubul, K., et al.: Script identification of multi-script documents: a survey. IEEE Access **5**, 6546–6559 (2017)

10. Obaidullah, S.M., et al.: Handwritten Indic script identification in multi-script document images: a survey. IJPRAI **32**(10), 1856012 (2018)

11. Brunessaux, S., et al.: The maurdor project: improving automatic processing of digital documents. In: DAS, pp. 349–354. IEEE (2014)

12. Singh, P.K., et al.: Benchmark databases of handwritten Bangla-Roman and Devanagari-Roman mixed-script document images. Multimed. Tools ad Appl. **77**(7), 8441–8473 (2018)

13. Sharma, N., et al.: ICDAR 2015 competition on video script identification (CVSI 2015). In: ICDAR, pp. 1196–1200. IEEE (2015)

14. Nayef, N., et al.: ICDAR 2019 robust reading challenge on multi-lingual scene text detection and recognition—RRC-MLT-2019. In: ICDAR, pp. 1582–1587. IEEE (2019)

15. Nayef, N., et al.: ICDAR 2017 robust reading challenge on multi-lingual scene text detection and script identification-RRC-MLT. In: ICDAR, vol. 1, pp. 1454–1459. IEEE (2017)

16. Kumar, D., et al.: Multi-script robust reading competition in ICDAR 2013. In: 4th International Workshop on Multilingual OCR, pp. 1–5 (2013)

17. Djeddi, C., et al.: ICDAR 2015 competition on multi-script writer identification and gender classification using 'QUWI' database. In: ICDAR, pp. 1191–1195. IEEE (2015)

18. Ferrer, M.A., et al.: MDIW-13 multiscript document database (2019)

19. Ferrer, M.A. et al.: MDIW-13: New database and benchmark for script identification

20. Hu, T., et al.: See better before looking closer: weakly supervised data augmentation network for fine-grained visual classification. arXiv preprint arXiv:1901.09891 (2019)

21. Zhang, N., Donahue, J., Girshick, R., Darrell, T.: Part-based R-CNNs for fine-grained category detection. In: Fleet, D., Pajdla, T., Schiele, B., Tuytelaars, T. (eds.) ECCV 2014. LNCS, vol. 8689, pp. 834–849. Springer, Cham (2014). https://doi.org/10.1007/978-3-319-10590-1_54

22. Wei, X., et al.: Mask-CNN: localizing parts and selecting descriptors for fine-grained image recognition. arXiv preprint arXiv:1605.06878 (2016)

23. Simonyan, K., Zisserman, A.: Very deep convolutional networks for large-scale image recognition. arXiv preprint arXiv:1409.1556 (2014)

24. He, K., et al.: Deep residual learning for image recognition. In: CVPR, pp. 770–778 (2016)

25. Long, J., et al.: Fully convolutional networks for semantic segmentation. In: CVPR, pp. 3431–3440 (2015)

26. Chen, L., et al.: Encoder-decoder with atrous separable convolution for semantic image segmentation. In: ECCV, pp. 801–818 (2018)

27. Zhang, H., et al.: ResNeSt: split-attention networks. arXiv preprint arXiv:2004.08955 (2020)

28. Liu, Z., et al.: Swin transformer: hierarchical vision transformer using shifted windows. arXiv preprint arXiv:2103.14030 (2021)

29. Sun, K., et al.: Deep high-resolution representation learning for human pose estimation. In: CVPR, pp. 5693–5703 (2019)
30. Yuan, Y., et al.: Object-contextual representations for semantic segmentation. arXiv preprint arXiv:1909.11065 (2019)
31. Dai, J., et al.: Deformable convolutional networks. In: CVPR, pp. 764–773 (2017)
32. Lee, D., et al.: Pseudo-label: the simple and efficient semi-supervised learning method for deep neural networks. In: ICMLW, vol. 3 (2013)
33. Lin, T., et al.: Focal loss for dense object detection. In: CVPR, pp. 2980–2988 (2017)
34. Luo, C., et al.: Learn to augment: joint data augmentation and network optimization for text recognition. In: CVPR, pp. 13746–13755 (2020)
35. Huang, G., et al.: Densely connected convolutional networks. In: CVPR, pp. 2261–2269 (2017)
36. Woo, S., et al.: CBAM: convolutional block attention module. In: ECCV, pp. 3–19 (2018)
37. He, K., et al.: Deep residual learning for image recognition, pp. 2980–2988 (2017)
38. Le, Q., Tan, M.: EfficientNet: rethinking model scaling for convolutional neural networks, pp. 6105–6114 (2019)
39. Tan, M., Le, Q.V.: EfficientNet: rethinking model scaling for convolutional neural networks. CoRR, abs/1905.11946 (2019)
40. Isola, P., et al.: Image-to-image translation with conditional adversarial networks. In: CVPR (2017)

ICDAR 2021 Competition on Scientific Table Image Recognition to LaTeX

Pratik Kayal$^{(\boxtimes)}$, Mrinal Anand, Harsh Desai, and Mayank Singh

Indian Institute of Technology, Gandhinagar, India
{pratik.kayal,mrinal.anand,singh.mayank}@iitgn.ac.in

Abstract. Tables present important information concisely in many scientific documents. Visual features like mathematical symbols, equations, and spanning cells make structure and content extraction from tables embedded in research documents difficult. This paper discusses the dataset, tasks, participants' methods, and results of the ICDAR 2021 Competition on Scientific Table Image Recognition to LaTeX. Specifically, the task of the competition is to convert a tabula r image to its corresponding LaTeX source code. We proposed two subtasks. In Subtask 1, we ask the participants to reconstruct the LaTeX structure code from an image. In Subtask 2, we ask the participants to reconstruct the LaTeX content code from an image. This report describes the datasets and ground truth specification, details the performance evaluation metrics used, presents the final results, and summarizes the participating methods. Submission by team VCGroup got the highest Exact Match accuracy score of 74% for Subtask 1 and 55% for Subtask 2, beating previous baselines by 5% and 12%, respectively. Although improvements can still be made to the recognition capabilities of models, this competition contributes to the development of fully automated table recognition systems by challenging practitioners to solve problems under specific constraints and sharing their approaches; the platform will remain available for post-challenge submissions at https://competitions.codalab.org/competitions/26979.

Keywords: Table recognition · LaTeX · OCR

1 Introduction

Scientific documents contain tables that hold meaningful information such as final results of an experiment or comparisons with earlier baselines in a concise manner, but accurately extracting the information proves to be a challenging task [16, 18]. In recent years, we witness the widespread usage of several typesetting tools for document generation. In particular, tools like LaTeX help in typesetting complex scientific document styles and subsequent electronic exchange documents. It has been shown in a study [1] that 26% of submissions to 54 randomly selected scholarly journals from 15 different scientific disciplines use LaTeX typesetter, and there is a significant difference between LaTeX-using and non-LaTeX-using disciplines. Tables can be presented in a variety of ways visually, such as spanning multiple columns or rows, with or without horizontal and vertical lines, non-standard spacing and alignment, and text formatting [5]. The

© Springer Nature Switzerland AG 2021
J. Lladós et al. (Eds.): ICDAR 2021, LNCS 12824, pp. 754–766, 2021.
https://doi.org/10.1007/978-3-030-86337-1_50

	$\epsilon_{LJ}\left[\frac{Kcal}{mol}\right]$	σ_{LJ} [Å]	r_{cut} [Å]
$C - C$	0.0951	3.473	15.0
$C - H$	0.0380	3.159	15.0
$H - H$	0.0152	2.846	15.0

(a) Scientific table in [20].

	Search Strategies	
	(RMSE, σ)	(PAcc, σ)
SkILL	(0.616, 0.063)	(0.661, 0.045)
SkILL+pruning	(0.581, 0.099)	**(0.663, 0.045)**
Aleph		(0.656, 0.047)

(b) Scientific table in [2].

Fig. 1. Examples of naturally occurring scientific tables.

table detection task involves predicting the bounding boxes around tables embedded within a PDF document, while the table recognition task involves the table structure and content extraction. We find significantly large number of datasets for table detection and table structure recognition [6, 8, 12, 17, 26]. Comparatively, table content recognition is a least explored task with a very few datasets [3, 9, 25]. Interestingly, the majority of these datasets are insufficient to perform end-to-end neural training. For instance, ICDAR 2013 table competition task [9] includes only 150 table instances. Besides, several datasets comprise simpler tables that might not generalize in real-world extraction scenarios. Tables written in LaTeX utilizes libraries like *booktabs, array, multirow, longtable*, and *graphicx*, for table creation and libraries like *amsmath, amssymb, amsbsy*, and *amsthm*, for various scientific and mathematical writing. Such libraries help in creating complex scientific tables that are difficult to recognize by the current systems accurately [18]. Deng et al. [3] proposed a table recognition method for LaTeX generated documents. However, due to minimal data processing like word-level tokenization and no fine-grained task definitions, they predict the entire LaTeX code leading to poor recognition capabilities (Fig. 1).

To understand and tackle this challenging problem of table recognition, we organized a competition with the objective to convert a given image of the table to its corresponding (i) structure LaTeX code and (ii) content LaTeX code. This report presents the final results after analyzing the submissions received.

The Paper Outline: The entire paper is organized as follows. Section 2 describes the organization of the competition. Section 3 describes the tasks descriptions. Section 4 describes the dataset used in the competition. Section 5 details about the performance evaluation metrics. Section 6 presents the description of baselines and methods submitted by the participants. Section 7 presents the final results and discussions.

2 Competition Organization

The Competition on Scientific Table Image Recognition to LaTeX was organized with the objective to evaluate structure recognition capabilities and content recognition capabilities of the current state-of-the-art systems. We use the Codalab[1] platform for submission and automatic evaluation of methods. Multiple submissions were allowed for each

---

[1] https://competitions.codalab.org/.

team. After each submission, evaluation metrics were computed, and the leaderboard was updated accordingly based on the highest score. The competition ran between October 2020 and March 2021 and was divided into three phases, validation, testing, and post-evaluation. In the validation phase, participants were given training and validation dataset and were allowed to test their system on the validation set. In the testing phase held in March 2021, participants submitted their solutions on the test dataset on which final results are evaluated. For the testing phase, the leaderboard was hidden from the participants. In the post-evaluation phase, participants are allowed to submit their solutions and improve their rankings on the leaderboard.

3 Task Description

The task of generating LaTeX code from table images is non-trivial. Similar features of the table can be encoded in different ways in the LaTeX code. While the content in a tabular image is in sequential form, structural information is encoded in the form of structural tokens like \multicolumn and column alignment tokens like c, l, and r in LaTeX. The proposed recognition models generate a sequence of tokens given the input table image. We examine the task of generating LaTeX code from table image by dividing it into two sub-tasks—**Table Structure Recognition** and LaTeX **Optical Character Recognition**. The extracted structure information aids in reconstructing the table structure. In comparison, the extracted content information can help fill the textual placeholders of the table.

3.1 Task I: Table Structure Reconstruction (TSR)

This subtask recognizes the structural information embedded inside the table images. Figure 2b shows a TSR output from a sample tabular image (see Fig. 2a). Structural information such as mutliple columns (defined using the command \multicolumn {cols}{pos}{text}), multiple rows (defined using the command \multirow {rows}{width}{text}), the column alignment specifiers (c, l, or r), horizontal lines (\hline, \toprule, \midrule, or \bottomrule), etc., are recognized in this task. Note that, content inside the third argument ({text}) of \multicolumn or \multirow commands are recognised in L-OCR task (defined in the next section). A special placeholder token *'CELL'* represents a content inside a specific cell of the table. Specifically, the vocabulary of TSR task comprises &, 0, 1, 2, 3, 4, 5, 6, 7, 8, 9, CELL, \\, \hline, \hspace, \multirow, \multicolumn, \toprule, \midrule, \bottomrule, c, l, r, |, \{, and \}.

3.2 Task II: Table Content Reconstruction (TCR)

This subtask extracts content information from the table images. Content information includes alphanumeric characters, symbols, and LaTeX tokens. While the table's basic structure is recognized in the TSR task, the more nuanced structural information is

	Search Strategies	
	(RMSE, σ)	(PAcc, σ)
SkILL	(0.616, 0.063)	(0.661, 0.045)
SkILL+pruning	(0.581, 0.099)	**(0.663, 0.045)**
Aleph	(0.656, 0.047)	

```
{ c | c c c } &
\multicolumn { 3 } { c }
\\ & & & \\ \hline \hline
& & \\ & & & \\ \hline
\multicolumn { 3 } { c }
```

```
& { Search Strategies } \\ & (
RMSE , $ \sigma $ ) & & ( PAcc, $
\sigma $ ) \\ SkILL & ( 0.616,
0.063 ) & & (0.661, 0.045) \\
SkILL + pruning & ( 0.581, 0.099 )
& & \textbf { ( 0.663 , 0.045 ) }
\\ Aleph & { ( 0.656 , 0.047 ) }
```

(a) Table Image   (b) Table structure   (c) Table content

Fig. 2. Sample of the dataset. The words in table content (except LaTeX tokens) are recognized as individual characters and then connected using a separate post-processing method. (a) Input to the models, (b) Sample with structure tokens, (c) Sample with content tokens.

captured by the L-OCR task. Instead of predicting all possible tokens leading to vocabulary size in millions, we demonstrate a character-based recognition scheme. A delimiter token ($|$) is used to identify words among the characters. Figure 2c shows the TCR output from a sample tabular image (see Fig. 2a). The terms (Accuracy, (%)) together produces the first cell in the first row of the table. Accuracy is recognized as A <dim> c <dim> c <dim> u <dim> r <dim> a <dim> c <dim> y. The keyword <dim> (delimiter token) is removed in the post-processing step. Overall, 235 unique tokens comprises the vocabulary of L-OCR sub-task. This includes all the alphabets and their case variants (a–z and A–Z), digits (0–9), LaTeX environment tokens (\vspace, \hspace, etc.), brackets ({, }, etc.), modifier character (\), accents (, , ^, etc.) and symbols (%, #, &, etc.).

4 Dataset

The dataset contains table images and its corresponding LaTeX code representing structure and content information. We only consider papers corresponding to the topics in Computer Science, from the preprint repository *ArXiv*[2]. We preprocess LaTeX code before compiling them to images and post-process the LaTeX code to make them more suitable for the text generation task. The LaTeX source code is processed to generate table images and post-processed to get corresponding ground truth representation. We extract the table snippet during preprocessing, which begins with \begin{tabularx} command and ends with \end{tabularx} command. We remove commands like \cite{}, \ref{} along with the commented text that cannot be predicted from the tabular images. The filtered tabular LaTeX code is compiled as PDF and then converted into JPG images with 300dpi resolution using Wand library[3]. To keep a smaller vocabulary for faster training, we mask LaTeX environment (beginning with \ character) tokens with corpus frequency less than 1000 as a special token \LATEX_TOKEN. We separate the structure and content-related token of the table from the LaTeX source code during post-processing to create the dataset for TSR and TCR tasks, respectively. The structural tokens contain tabular environment parameters, spanning rows and columns tokens, horizontal and line tokens. The content tokens

---

[2] http://arxiv.org/.
[3] https://github.com/emcconville/wand.

Table 1. Summary of datasets. ML denotes the maximum sequence length of target sequence. Samples denote the total number of image-text pairs. T/S is the average number of tokens per sample.

Dataset	ML	Samples	Train	Val	Test	T/S
TSRD	250	46,141	43,138	800	2,203	76.09
TCRD	500	37,917	35,500	500	1,917	213.95

include alphanumeric characters, mathematical symbols, and other LATEX environment tokens. Dataset for the TSR task and TCR task is created by considering ground truth LATEX code with a maximum length of 250 and 500, respectively (Table 1).

5   Evaluation Metrics

We use exact match accuracy and exact match accuracy @ 95% similarity as our core evaluation metrics for evaluating both tasks. Consider that if the model generates a hypothesis y^* consisting of tokens $y_1^* y_2^* \cdots y_{n_1}^*$, during the test phase. We evaluate by comparing it with the ground truth token sequence code y consisting of tokens $y_1 y_2 \cdots y_{n_2}$, where n_1 and n_2 are the number of tokens in model-generated code and ground truth code, respectively.

1. **Exact Match Accuracy (EM):** For a correctly identified sample, $y_1^* = y_1, y_2^* = y_2, \cdots, y_{n_1}^* = y_{n_2}$.
2. **Exact Match Accuracy @95% (EM @95%):** For a correctly identified sample, $y_i^* = y_j, y_{i+1}^* = y_{j+1}, \cdots, y_{i+l}^* = y_{j+l}$ and $l >= 0.95 \times n_2$. In this metric there are no insertions/deletions allowed in between and move is towards $(i + 1, j + 1)$ from (i, j).

In addition, we also employ task-specific metrics that provide intuitive criteria to compare different models trained for the specific task. For the TSR task, we use row accuracy and column accuracy as evaluation metrics.

1. **Row Accuracy (RA):** For a correctly identified sample, the number of rows in y^* is equal to the number of rows in y.
2. **Column Accuracy (CA):** For a correctly identified sample, the number of columns in y^* is equal to the number of columns in y.

We use Alpha-Numeric characters prediction accuracy, LATEX Token accuracy, LATEX symbol accuracy, and Non-LATEX symbols prediction accuracy for the TCR task. Each token in y^* and y is classified into four exhaustive categories: (i) Alpha-Numeric Tokens (Alphabets and Numbers), (ii) LATEX Tokens (tokens that are defined in and used by LATEX markup language like \cdots, \times, \textbf, etc.), (iii) LATEX Symbols (symbols with escape character ('\') like \%, \$, \{, \}, etc.), and (iv) Non-LATEX Symbols (symbols like =, $, {, }, etc.).

1. **Alpha-Numeric Tokens Accuracy (AN):** We form strings y^*_{AN} and y_{AN} from y^* and y, respectively, by keeping the alpha-numeric tokens and discarding the rest and preserving the order. Then, for a correctly identified sample, $y^*_{AN} = y_{AN}$.

2. **LaTeX Tokens Accuracy (LT):** We form strings y^*_{LT} and y_{LT} from y^* and y, respectively, by keeping the LaTeX tokens and discarding the rest and preserving the order. Then, for a correctly identified sample, $y^*_{LT} = y_{LT}$.

3. **LaTeX Symbols Accuracy (LS):** We form strings y^*_{LS} and y_{LS} from y^* and y, respectively, by keeping the LaTeX symbol tokens and discarding the rest and preserving the order. Then, for a correctly identified sample, $y^*_{LS} = y_{LS}$.

4. **Non-LaTeX Symbols Accuracy (NLS):** We form strings y^*_{NLS} and y_{NLS} from y^* and y, respectively, by keeping the non-LaTeX symbol tokens and discarding the rest and preserving the order. Then, for a correctly identified sample, $y^*_{NLS} = y_{NLS}$.

6 Baseline and Participating Methods

6.1 CNN-Baseline

The image-to-markup model proposed by Deng et al. [4] extracts image features using a convolutional neural network (CNN) and arranges the features in a grid. CNN consists of eight convolutional layers with five interleaved max-pooling layers. Each row is then encoded using a bidirectional LSTM with the initial hidden state (called positional embeddings) kept as trainable to capture semantic information in the vertical direction. An LSTM decoder then uses these encoded features with a visual attention mechanism to predict the tokens given the previously generated token history and the features as input.

6.2 Transformer-Baseline

Recently, Transformer [20] based architectures have shown state-of-the-art performance for several Image-to-Text tasks [15,21]. We utilize a ResNet-101 [10] layer that encodes the table image and a Transformer model for text generation. This model is based on architecture proposed in [7].

6.3 VCGroup

Their model is based on MASTER [14] originally developed for scene recognition task. MASTER uses a novel multi-aspect non-local block and fuses it into the conventional CNN backbone, enabling the feature extractor to model a global context. The proposed multi-aspect non-local block can learn different aspects of spatial 2D attention, which can be viewed as a multi-head self-attention module. The proposed solution is a highly optimized version of the original MASTER model. The task wise optimizations and techniques used by them are explained below:

Task I: Table Structure Reconstruction. In the TSR task, they use the following strategies to improve performance

Table 2. Table Structure Reconstruction Results. *Method description unavailable.

Method	EM	EM @95%	RA	CA
VCGroup	0.74	0.88	0.95	0.89
Transformer-Baseline	0.69	0.85	0.93	0.86
CNN-Baseline	0.66	0.79	0.92	0.86
Format*	0.57	0.80	0.91	0.87
asda*	0.50	0.75	0.90	0.86

Table 3. Table Content Reconstruction Results. *Method description unavailable.

Method	EM	EM @95%	AN	LT	LS	NLS
VCGroup	0.55	0.74	0.85	0.75	0.96	0.62
Transformer-Baseline	0.43	0.64	0.74	0.67	0.94	0.54
CNN-Baseline	0.41	0.67	0.76	0.67	0.94	0.48
Format*	0.0	0.52	0.67	0.54	0.92	0.35

- **Ranger Optimizer.** Ranger integrates Rectified Adam [13], LookAhead [24], and GC (gradient centralization) [22] into one optimizer. LookAhead can be considered as an extension of Stochastic Weight Averaging (SWA) in the training stage.
- **Data Augmentation.** Data augmentation methods include shear, affine transformation, perspective transformation, contrast, brightness, and saturation augment is used in our tasks.
- **Synchronized Batch Normalization.** Synchronized BatchNorm (SyncBN) [23] is an effective type of batch normalization used for multi-GPU training. Standard batch normalization only normalizes the data within each device (GPU). SyncBN normalizes the input within the whole mini-batch.
- **Feature Concatenation of Layers in Transformer Decoder.** Different from the original MASTER model, they concatenate the outputs of the last two transformer layers and apply a linear projection on the concatenated feature to obtain the last-layer feature. The last-layer feature is used for the final classification.
- **Model Ensemble.** Model ensemble (by voting and bagging) is a very widely used technique to improve the performance on many deep learning tasks and is used by the participant to improve performance.

Task II: Table Content Reconstruction. For the TCR task, ranger optimizer, SyncBN, data augmentation, feature concatenation, and model ensemble are used along with the following additional strategies.

- **Multiple Resolutions.** They experimented with various input image sizes apart from the 400×400 size images provided, e.g., 400×400, 440×400, 480×400, 512×400, 544×400, 600×400.
- **Pre-train Model.** They use a large-scale data set [3] sharing similar objective (table recognition to Latex) with this competition to pre-train their model. To match the

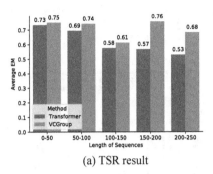

(a) TSR result

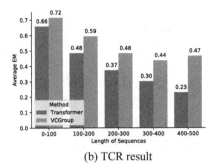

(b) TCR result

Fig. 3. Average EM score of VCGroup and Transformer-Baseline for both (a) TSR task and (b) TCR task.

(a) Example 1: Table Image

(b) Example 1: Outputs

```
Ground Truth:
{ c | c | c } CELL & CELL & CELL \\ CELL & CELL \\ CELL & CELL &
CELL \\ CELL & CELL & CELL \\ CELL & CELL & CELL \\ CELL & CELL &
CELL \\ CELL & CELL & CELL \\ CELL & CELL & CELL \\ CELL & CELL &
CELL \\ CELL & CELL & CELL \\ CELL & CELL & CELL \\ CELL & CELL &
CELL \\ CELL & CELL & CELL \\ CELL & CELL & CELL \\ CELL & CELL &
CELL \\ CELL & CELL & CELL \\ CELL & CELL & CELL
```

```
Transformer:
{ c | c | c } CELL & CELL & CELL \\ CELL & CELL \\ CELL & CELL &
CELL \\ CELL & CELL & CELL \\ CELL & CELL & CELL \\ CELL & CELL &
CELL \\ CELL & CELL & CELL \\ CELL & CELL & CELL \\ CELL & CELL &
CELL \\ CELL & CELL & CELL \\ CELL & CELL & CELL \\ CELL & CELL &
CELL \\ CELL & CELL & CELL \\ CELL & CELL & CELL \\ CELL & CELL &
CELL \\ CELL & CELL & CELL
```

```
VCGroup:
{ c | c | c } CELL & CELL & CELL \\ CELL & CELL \\ CELL & CELL &
CELL \\ CELL & CELL & CELL \\ CELL & CELL & CELL \\ CELL & CELL &
CELL \\ CELL & CELL & CELL \\ CELL & CELL & CELL \\ CELL & CELL &
CELL \\ CELL & CELL & CELL \\ CELL & CELL & CELL \\ CELL & CELL &
CELL \\ CELL & CELL & CELL \\ CELL & CELL & CELL
```

(c) Example 2: Table Image

```
1    If
2        CARPAPolicy(carpa₃) ∧
3        Role(r) ∧ hasRole(carpa₁, r) ∧
4        Permission(p) ∧
5        hasPermission(carpa₁, p) ∧
6        ContextualCondition(exp) ∧
7        hasCondition(carpa₁, exp) ∧
8        has(r, roleIdentity) ∧
9        equal(roleIdentity, "RN00X") ∧
10       Resource(res) ∧ hasResource(p, res) ∧
11       Operation(op) ∧ hasOperation(p, op) ∧
12       has(res, resourceIdentity) ∧
13       equal(resourceIdentity, 3) ∧
14       has(op, action) ∧ equal(action, "read") ∧
15       equal(exp, "c_a")
16   Then
17       carpa(r, p)
```

(d) Example 2: Outputs

```
Ground Truth:
{ | c | c | } \hline CELL & CELL \\ CELL & CELL \\ CELL & CELL
\\ CELL & CELL \\ CELL & CELL \\ CELL & CELL \\ CELL & CELL \\
CELL & CELL \\ CELL & CELL \\ CELL & CELL \\ CELL & CELL \\ CELL
& CELL \\ CELL & CELL \\ CELL & CELL \\ CELL & CELL \\ CELL &
CELL \\ CELL & CELL \\ \hline
```

```
Transformer:
{ | c | c | } \hline CELL & CELL \\ CELL & CELL \\ CELL & CELL
\\ CELL & CELL \\ CELL & CELL \\ CELL & CELL \\ CELL & CELL \\
CELL & CELL \\ CELL & CELL \\ CELL & CELL \\ CELL & CELL \\ CELL
& CELL \\ CELL & CELL \\ CELL & CELL \\ CELL & CELL \\ CELL &
CELL \\ \hline
```

```
VCGroup:
{ | c | c | } \hline CELL & CELL \\ CELL & CELL \\ CELL & CELL
\\ CELL & CELL \\ CELL & CELL \\ CELL & CELL \\ CELL & CELL \\
CELL & CELL \\ CELL & CELL \\ CELL & CELL \\ CELL & CELL \\ CELL
& CELL \\ CELL & CELL \\ CELL & CELL \\ CELL & CELL \\ CELL &
CELL \\ CELL & CELL \\ \hline
```

Fig. 4. Examples of correct cases by VCGroup method in the TSR task.

Methods	Pedestrian
	Moderate
SubCNN	73.70%
MS-CNN	71.33%
SDP+RPN	70.16%
RRC (ours)	75.33%

(a) Example 1: Table Image

Ground Truth:
```
{ * } { M e t h o d s ! } & { P e d e s t r i a n
! } \\ & M o d e r a t e ! \\ S u b C N N ! & $ 7
3 . 7 0 ! \% $ \\ M S ! - C N N ! & $ 7 1 . 3 3 !
\% $ \\ S D P ! + R P N ! & $ 7 0 . 1 6 ! \% $ \\
R R C ! ( o u r s ! ) & $ 7 5 . 3 3 ! \% $
```

Transformer:
```
{ * } { M e t h o d s ! } & P e d e s t r i a n !
\\ & M o d e r a t e ! \\ S u b C N N ! & $ 7 3 . 7
0 ! \% \\ M S ! - C N N ! & $ 7 1 . 3 3 ! \% $ \\ S D
P ! + R P N ! & $ 7 0 . 1 6 ! \% \\ R R C ! ( o u r
s ! ) & $ 7 5 . 3 3 ! \%
```

VCGroup:
```
{ * } { M e t h o d s ! } & { P e d e s t r i a n
! } \\ & M o d e r a t e ! \\ S u b C N N ! & $ 7
3 . 7 0 ! \% $ \\ M S ! - C N N ! & $ 7 1 . 3 3 !
\% $ \\ S D P ! + R P N ! & $ 7 0 . 1 6 ! \% $ \\
R R C ! ( o u r s ! ) & $ 7 5 . 3 3 . 3 3 ! \% $
```

(b) Example 1: Outputs

Solver	Avg	Par10	%Solved
BSS	315	1321	93.8
ISAC	302	1107	95.0
ISAC.filt	289	892	96.3
DASH	251	956	95.7
DASH+	255	858	96.3
DASH+filt	241	643	98.1
VBS	225	326	99.4
VBS_DASH	185	286	99.4

(c) Example 2: Table Image

Ground Truth:
```
\textbf{Solver!}& \textbf{Avg!}& \textbf{Par10!}& \textbf{\%Solved!}\\
\textbf{BSS!}&315!&1321!&93.8!\\ \textbf{ISAC!}&302!&1107!&9
5.0!\\ \textbf{ISAC!\_filt!}&289!&892!&96.3!\\ \textbf{DASH!}&25
1!&956!&95.7!\\ \textbf{DASH!+}&255!&858!&96.3!\\ \textbf{DAS
H!+filt!}& \textbf{241!}& \textbf{643!}& \textbf{98.1!}\\ \textbf{VBS!}
&225!&326!&99.4!\\ \textbf{VBS!\_DASH!}&185!&286!&99.4!
```

Transformer:
```
\bfSolver!& \bfAvg!& \bfPar10!& \bf\%Solved!\\ \bfBSS!&315!&13
21!&93.8!\\ \bfISAC!&302!&1107!&95.0!\\ \bfISAC!\_filt!&289
!&892!&96.3!\\ \bfDASH!&251!&956!&95.7!\\ \bfDASH!+&255!
&858!&96.3!\\ \bfDASH!+filt!& \bf241!& \bf643!& \bf98.1!\\ \bfVB
S!&225!&326!&99.4!\\ \bfVBS!\_DASH!&185!&286!&99.4!
```

VCGroup:
```
\textbf{Solver!}& \textbf{Avg!}& \textbf{Par10!}& \textbf{\%Solved!}\\
\textbf{BSS!}&315!&1321!&93.8!\\ \textbf{ISAC!}&302!&1107!&9
5.0!\\ \textbf{ISAC!\_filt!}&289!&892!&96.3!\\ \textbf{DASH!}&25
1!&956!&95.7!\\ \textbf{DASH!+}&255!&858!&96.3!\\ \textbf{DAS
H!+filt!}& \textbf{241!}& \textbf{643!}& \textbf{98.1!}\\ \textbf{VBS!}
&225!&326!&99.4!\\ \textbf{VBS!\_DASH!}&185!&286!&99.4!
```

(d) Example 2: Outputs

Fig. 5. Examples of correct cases by VCGroup method in the TCR task.

target vocabulary, they trim down this dataset to 58,000 samples from the original 450,000 samples.

More details about their method can be found in [11].

7 Results and Discussion

Table 2 and 3 shows the results of the TSR and TCR tasks, respectively. VCGroup achieved the highest EM score of 74% and 55% for the TSR and TCR task, respectively. This shows the method's superior capability in producing better LATEX code sequences. The difference in RA and CA metrics for all the methods is less compared to EM metric, which shows that while VCGroup has superior overall detection, the other methods have similar row and column detection capabilities. VCGroup showcases significant improvement in recognition capability of Alpha-Numeric characters, LATEX tokens, and non-LATEX symbols, thus justifying their higher EM score for the TCR task. Figure 3 presents the average EM scores of the top two performing methods for different lengths of sequences. The plot shows that VCGroup performs significantly

Fixed Attribute Skyline			Subspace Skyline
Sorting-based	Partition-based		Skycube, Skyey, Subsky, CSC, FMC
	Index	No Index	
BNL, SFS, LESS, SaLSa	NN, BBS, ZSearch	D&C, OSPS, BSkyTree	

(a) Table Image

```
Ground Truth:
{ c | c | c | c } \hline \multicolumn { 3 } { c | } CELL &
CELL \\ \hline CELL & \multicolumn { 2 } { c | } CELL &
\multirow { 3 } CELL \\ \multirow { 2 } CELL & CELL & CELL &
\\ CELL & CELL & CELL & \\ \hline

Transformer:
{ c | c | c | c } \hline \multicolumn { 3 } { c | } CELL &
CELL \\ \multicolumn { 3 } { c | } CELL & CELL \\ \hline CELL
& \multicolumn { 2 } { c | } CELL & CELL \\ CELL &
\multicolumn { 2 } { c | } CELL & CELL \\ CELL & CELL & CELL
& CELL \\ CELL & CELL & CELL & CELL \\ CELL & & CELL & \\
CELL & & CELL & \\ \hline

VCGroup:
{ 1 | 1 | 1 | 1 } \hline \multicolumn { 2 } { c | } CELL &
CELL & CELL \\ & & & CELL \\ \hline CELL & \multicolumn { 2 }
{ c | } CELL & CELL \\ CELL & \multicolumn { 2 } { c | } CELL
& CELL \\ \hline CELL & CELL & CELL & CELL \\ CELL & CELL &
CELL & CELL \\ CELL & CELL & CELL & \\ CELL & & CELL & \\
\hline
```

(b) Outputs

Fig. 6. Example failure case of VCGroup method and Transformer-Baseline in the TSR task.

1	2	3	4	5	6
0	2	2	2	0	2
0	2	2	2	1	2
1	2	2	2	0	2
1	2	2	1	1	2

(a) Example 1: Table Image

```
Ground Truth:
1 & 2 & 3 & 4 & 5 & 6 \\ 0 & 2 & 2 & 2 & 0 & 2 \\ 0 & 2 &
2 & 2 & 1 & 2 \\ 1 & 2 & 2 & 2 & 0 & 2 \\ 1 & 2 & 2 & 1 &
1 & 2

Transformer:
1 & 2 & 3 & 4 & 5 & 6 \\ 0 & 2 & 2 & 2 & 0 & 2 \\ 0 & 2 &
2 & 2 & 1 & 2 \\ 1 & 2 & 2 & 2 & 0 & 2 \\ 1 & 2 & 2 & 1 &
1 & 2

VCGroup:
$ 1 $ & $ 2 $ & $ 3 $ & $ 4 $ & $ 5 $ & $ 6 $ \\ $ 0 $ &
$ 2 $ & $ 2 $ & $ 2 $ & $ 0 $ & $ 2 $ \\ $ 0 $ & $ 2 $ &
$ 2 $ & $ 2 $ & $ 1 $ & $ 2 $ \\ $ 1 $ & $ 2 $ & $ 2 $ &
$ 2 $ & $ 0 $ & $ 2 $ \\ $ 1 $ & $ 2 $ & $ 2 $ & $ 1 $ &
$ 1 $ & $ 2 $
```

(b) Example 1: Outputs

Corpus	Docs	Sentences	Vocab	Tokens
Chemical	965	9,488	4,981	123,119
Elements	118	1,848	5,301	67,484

(c) Example 2: Table Image

```
Ground Truth:
\bf C o r p u s ¦ & \bf D o c s ¦ & \bf S e n t e n c e s
¦ & \bf V o c a b ¦ & \bf T o k e n s ¦ \\ C h e m i c a l
¦ & $ 9 6 5 ¦ $ & $ 9 , 4 8 8 ¦ $ & $ 4 , 9 8 1 ¦ $ & $ 1
2 3 , 1 1 9 ¦ $ \\ E l e m e n t s ¦ & $ 1 1 8 ¦ $ & $ 1 ,
8 4 8 ¦ $ & $ 5 , 3 0 1 ¦ $ & $ 6 7 , 4 8 4 ¦ $

Transformer:
\textbf { C o r p u s ¦ } & \textbf { D o c s ¦ } &
\textbf { S e n t e n c e s ¦ } & \textbf { V o c a b ¦ }
& \textbf { T o k e n s ¦ } \\ C h e m i c a l ¦ & 9 6 5 ¦
& 9 , 4 8 8 ¦ & 4 , 9 8 1 ¦ & 1 2 3 , 1 1 9 ¦ \\ E l e m e
n t s ¦ & 1 1 8 ¦ & 1 , 8 4 8 ¦ & 5 , 3 0 1 ¦ & 6 7 , 4 8
4 ¦

VCGroup:
\textbf { C o r p u s ¦ } & \textbf { D o c s ¦ } &
\textbf { S e n t e n c e s ¦ } & \textbf { V o c a b ¦ }
& \textbf { T o k e n s ¦ } \\ C h e m i c a l ¦ & $ 9 6 5
¦ $ & $ 9 , 4 8 8 ¦ $ & $ 4 , 9 8 1 ¦ $ & $ 1 2 3 , 1 1 9
¦ $ \\ E l e m e n t s ¦ & $ 1 1 8 ¦ $ & $ 1 , 8 4 8 ¦ $ &
$ 5 , 3 0 1 ¦ $ & $ 6 7 , 4 8 4 ¦ $
```

(d) Example 2: Outputs

Fig. 7. Examples of failure cases of VCGroup method and Transformer-Baseline in the TCR task.

better for a longer length of sequences in both tasks. Figure 4 and 5 presents some examples of input tabular image and the corresponding ground truth code, Transformer-Baseline and VCGroup outputs. The examples are chosen to portray the cases where VCGroup outperforms the Transformer-Baseline. For the TSR task, in both examples, the Transformer-Baseline fails to output the last line of the table. For the TCR task, the Transformer-Baseline model fails to predict the dollar ($) token multiple times and, in the other example, outputs the alternative token \bf while the ground truth contains \textbf to bold text. Figure 6 shows an example of a failure case of both VCGroup and Transformer-Baseline in the TSR task. VCGroup incorrectly predicts tabular alignment tokens and the number of columns in the spanning cells. Additionally, it incorrectly predicts \multicolumn token in place of \multirow token. Similarly, the Transformer-Baseline also incorrectly predicts the number of columns in spanning cells and \multicolumn token in place of \multirow token. Figure 7a and 7b shows an example where the VCGroup predicts extra dollar ($) tokens around the numeric tokens as compared to the Transformer-Baseline. Figure 7a and 7b shows an example where both VCGroup and Transformer-Baseline outputs the alternative token \textbf. Additionally, the Transformer-Baseline is not able to correctly predict dollar tokens ($) as present in the ground truth.

8   Conclusion

This paper discusses the dataset, tasks, and participants' methods for the ICDAR 2021 Competition on Scientific Table Image Recognition to LATEX. The participants were asked to convert a given tabular image to its corresponding LATEX source code for this competition. The competition consisted of two tasks. In task I, we ask the participants to reconstruct the LATEX structure code from an image. In task II, we ask the participants to reconstruct the LATEX content code from an image. The competition winner was chosen based on the highest EM score separately for both the tasks. The competition portal received a total of 101 registrations, and finally, three teams participated in the test phase of the TSR task, and two teams participated in the test phase of the TCR task. We attribute this low participation to the difficulty of the proposed tasks. In future, we would increase the size of the dataset and include tables from more sources. We would also run the competition for a longer duration to provide the participants ample time to come up with competing solutions.

Acknowledgments. This work was supported by The Science and Engineering Research Board (SERB), under sanction number ECR/2018/000087.

References

1. Brischoux, F., Legagneux, P.: Don't format manuscripts. Sci. **23**(7), 24 (2009)
2. Côrte-Real, J., Mantadelis, T., Dutra, I., Roha, R., Burnside, E.: Skill-a stochastic inductive logic learner. In: 2015 IEEE 14th International Conference on Machine Learning and Applications (ICMLA), pp. 555–558. IEEE (2015)

3. Deng, Y., Rosenberg, D.S., Mann, G.: Challenges in end-to-end neural scientific table recognition. In: 2019 International Conference on Document Analysis and Recognition (ICDAR), pp. 894–901 (2019)
4. Deng, Y., Kanervisto, A., Ling, J., Rush, A.M.: Image-to-markup generation with coarse-to-fine attention. In: Proceedings of the 34th International Conference on Machine Learning-Volume 70, pp. 980–989. JMLR. org (2017)
5. Embley, D.W., Hurst, M., Lopresti, D., Nagy, G.: Table-processing paradigms: a research survey. IJDAR **8**(2–3), 66–86 (2006). https://doi.org/10.1007/s10032-006-0017-x
6. Fang, J., Tao, X., Tang, Z., Qiu, R., Liu, Y.: Dataset, ground-truth and performance metrics for table detection evaluation. In: 2012 10th IAPR International Workshop on Document Analysis Systems, pp. 445–449. IEEE (2012)
7. Feng, X., Yao, H., Yi, Y., Zhang, J., Zhang, S.: Scene text recognition via transformer. arXiv preprint arXiv:2003.08077 (2020)
8. Gao, L., et al.: ICDAR 2019 competition on table detection and recognition (CTDAR). In: 2019 International Conference on Document Analysis and Recognition (ICDAR), pp. 1510–1515. IEEE (2019)
9. Göbel, M., Hassan, T., Oro, E., Orsi, G.: ICDAR 2013 table competition. In: 2013 12th International Conference on Document Analysis and Recognition, pp. 1449–1453. IEEE (2013)
10. He, K., Zhang, X., Ren, S., Sun, J.: Deep residual learning for image recognition (2015)
11. He, Y., et al.: PingAn-VCGroup's solution for ICDAR 2021 competition on scientific table image recognition to latex (2021)
12. Li, M., Cui, L., Huang, S., Wei, F., Zhou, M., Li, Z.: TableBank: table benchmark for image-based table detection and recognition. In: LREC 2020, May 2020. https://www.microsoft.com/en-us/research/publication/tablebank-table-benchmark-for-image-based-table-detection-and-recognition/
13. Liu, L., et al.: On the variance of the adaptive learning rate and beyond. CoRR abs/1908.03265 (2019). http://arxiv.org/abs/1908.03265
14. Lu, N., Yu, W., Qi, X., Chen, Y., Gong, P., Xiao, R.: MASTER: multi-aspect non-local network for scene text recognition. CoRR abs/1910.02562 (2019). http://arxiv.org/abs/1910.02562
15. Lyu, P., Yang, Z., Leng, X., Wu, X., Li, R., Shen, X.: 2D attentional irregular scene text recognizer. arXiv preprint arXiv:1906.05708 (2019)
16. Niklaus, C., Cetto, M., Freitas, A., Handschuh, S.: A survey on open information extraction. In: Proceedings of the 27th International Conference on Computational Linguistics, pp. 3866–3878 (2018)
17. Siegel, N., Lourie, N., Power, R., Ammar, W.: Extracting scientific figures with distantly supervised neural networks. In: Proceedings of the 18th ACM/IEEE on Joint Conference on Digital Libraries, pp. 223–232 (2018)
18. Singh, M., Sarkar, R., Vyas, A., Goyal, P., Mukherjee, A., Chakrabarti, S.: Automated early leaderboard generation from comparative tables. In: Azzopardi, L., Stein, B., Fuhr, N., Mayr, P., Hauff, C., Hiemstra, D. (eds.) ECIR 2019. LNCS, vol. 11437, pp. 244–257. Springer, Cham (2019). https://doi.org/10.1007/978-3-030-15712-8_16
19. Tsourtis, A., Harmandaris, V., Tsagkarogiannis, D.: Parameterization of coarse-grained molecular interactions through potential of mean force calculations and cluster expansion techniques. In: Thermodynamics and Statistical Mechanics of Small Systems, vol. 19, p. 245 (2017)
20. Vaswani, A., et al.: Attention is all you need. In: Advances in Neural Information Processing Systems, pp. 5998–6008 (2017)
21. Yang, L., et al.: A simple and strong convolutional-attention network for irregular text recognition. arXiv preprint arXiv:1904.01375 (2019)

22. Yong, H., Huang, J., Hua, X., Zhang, L.: Gradient centralization: a new optimization technique for deep neural networks. In: Vedaldi, A., Bischof, H., Brox, T., Frahm, J.-M. (eds.) ECCV 2020. LNCS, vol. 12346, pp. 635–652. Springer, Cham (2020). https://doi.org/10.1007/978-3-030-58452-8_37

23. Zhang, H., et al.: Context encoding for semantic segmentation. CoRR abs/1803.08904 (2018). http://arxiv.org/abs/1803.08904

24. Zhang, M.R., Lucas, J., Hinton, G.E., Ba, J.: Lookahead optimizer: k steps forward, 1 step back. CoRR abs/1907.08610 (2019). http://arxiv.org/abs/1907.08610

25. Zhong, X., ShafieiBavani, E., Jimeno Yepes, A.: Image-based table recognition: data, model, and evaluation. In: Vedaldi, A., Bischof, H., Brox, T., Frahm, J.-M. (eds.) ECCV 2020. LNCS, vol. 12366, pp. 564–580. Springer, Cham (2020). https://doi.org/10.1007/978-3-030-58589-1_34

26. Zhong, X., Tang, J., Yepes, A.J.: PubLayNet: largest dataset ever for document layout analysis. In: 2019 International Conference on Document Analysis and Recognition (ICDAR), pp. 1015–1022. IEEE, September 2019. https://doi.org/10.1109/ICDAR.2019.00166

ICDAR 2021 Competition on Multimodal Emotion Recognition on Comics Scenes

Nhu-Van Nguyen[1,3]($\boxtimes$) (ID), Xuan-Son Vu[2] (ID), Christophe Rigaud[1] (ID),
Lili Jiang[2] (ID), and Jean-Christophe Burie[1] (ID)

[1] L3i, La Rochelle University, La Rochelle, France
{nhu-van.nguyen,christophe.rigaud,jcburie}@univ-lr.fr
[2] Department of Computing Science, Umeå University, Umeå, Sweden
{sonvx,lili.jiang}@cs.umu.se
[3] INSA-Lyon, Lyon, France

Abstract. The paper describes the "Multimodal Emotion Recognition on Comics scenes" competition presented at the ICDAR conference 2021. This competition aims to tackle the problem of emotion recognition of comic scenes (panels). Emotions are assigned manually by multiple annotators for each comic scene of a subset of a public large-scale dataset of golden age American comics. As a multi-modal analysis task, the competition proposes to extract the emotions of comic characters in comic scenes based on visual information, text in speech balloons or captions and the onomatopoeia. Participants were competing on CodaLab.org from December 16th 2020 to March 31th 2021. The challenge has attracted 145 registrants, 21 teams have joined the public test phase, and 7 teams have competed in the private test phase. In this paper we present the motivation, dataset preparation, task definition of the competition, the analysis of participant's performance and submitted methods. We believe that the competition have drawn attention from the document analysis community in both fields of computer vision and natural language processing on the task of emotion recognition in documents.

Keywords: Multimodal fusion · Emotion recognition · Multi-label classification

1 Introduction

Comics is a *multi-billion dollar industry* which is very popular in North America, Europe, and Asia. Initially, comics were printed on paper books, but nowadays, digitized and born-digital comic books become more and more popular and spread culture, education and recreation all over the world even faster.

However, they suffer from a limited automatic content understanding tools which restricts online content search and on-screen reading applications. To deliver digital comics content with an accurate and user-friendly experience on all mediums, it is often necessary to slightly or significantly adapt their content [1].

© Springer Nature Switzerland AG 2021
J. Lladós et al. (Eds.): ICDAR 2021, LNCS 12824, pp. 767–782, 2021.
https://doi.org/10.1007/978-3-030-86337-1_51

These adaptations are quite costly if done manually at large scale, so automatic processing are helpful to keep cost acceptable. This is one of the reasons why the comic book image analysis has been studied by the community of document analysis since about a decade. However, there still exist many challenges to be solved in this domain. While the comic elements such as scenes (panels), balloons, narrative and speech text are quite well detected and segmented now, the character (protagonist) detection, text recognition and element relationship analysis are still challenging, and it is important to draw more efforts from the research community to address these challenges. Moreover, complex tasks such as story understanding or scene analysis have not been well studied yet [1].

1.1   Human Emotions

"Master the human condition through word and image in a brilliantly minimalistic way" is one of important requirements in making comics [12], in order to engage readers. Here we look at how to model human emotions to better analyze and understand emotions in comics in a reversed manner.

Researchers on human emotions have approached the classification of emotions from one of two fundamental viewpoints [24]: (1) discrete model where emotions are discrete arise from separate neural system [4,19]; (2) dimensional model where emotions can be characterized on a dimensional basis in groupings [11,17]. Table 1 presents the four popular models for basic emotions [24]. Researchers have debated over whether a discrete or dimensional model to emotion classification was most appropriate, and studies showed that one model may not apply to all people. *Discrete model* uses a limit set of basic emotions, while *dimensional model* emphasizes the co-occurrence of the basic emotions to contribute to *the individual's affective state*. In this competition, we preferred discrete model by considering the diverse background of crowdsourcing annotators, and the purpose of emotion recognition in comic scene. Further motivated by this Kaggle challenge[1], we decided to add '*neutral*' to the label list since we believe there does not always exist emotions in any given context in daily life as well in comic scene. Additionally, by considering the challenges of explicitly recognizing emotions in comics, we added a label '*others*'. From the above investigations, we finally came out an eight-class label list for this competition including *angry, disgust, fear, happy, sad, surprise, neutral,* and *others*.

Table 1. Four popular basic emotion models (Yadollahi et al. [24])

Study	Basic emotions	Model types
Ekman [4]	Anger, disgust, fear, joy, sadness, surprise	Discrete
Plutchik [17]	Anger, anticipation, disgust, fear, joy, sadness, surprise, trust	Dimensional
Shaver [19]	Anger, fear, joy, love, sadness, surprise	Discrete
Lovheim [11]	Anger, disgust, distress, fear, joy, interest, shame, surprise	Dimensional

---

[1] http://bit.ly/kaggle-challenges-in-representation-learning.

1.2 Emotions in EmoRecCom Challenge

In this competition, we propose to tackle one of the challenge of comic scene analysis: the emotion recognition of comic scene. The emotions come from comic characters feelings in the story and are materialized (to be transmitted to the reader) by the visual information, the text in speech balloons or captions and the onomatopoeia (comic drawings of words that phonetically imitates, resembles, or suggests the sound that it describes), see Fig. 1. While emotion recognition are widely studied in other domains and data such as natural images and multi-modal data issues from social networks, it is not yet exploited in comics images which contain both image and text. Motivated by the value of multi-modality based approaches, the competition task encourages the participants to take advantage of multiple representation of resources to infer the emotions. The task hence is a multi-modal analysis task which can take advantages from both fields: computer vision and natural language processing which are one of the main interests of the document analysis community.

Fig. 1. An example of a comic character in a panel with visual emotion and caption text. It is noted that resulted texts from the automatic transcription method may contain errors (e.g., red underlined words in the example). (Color figure online)

For this competition, we crowd-sourced the image annotation step to get eight binary label for eight target emotions associated to each image of the ground-truth (angry, disgust, fear, happy, sad, surprise, neutral, others). The competition participants were asked to propose up to any number of positive labels for each image. The statistics of emotions in our EmoRecCom dataset are shown in Table 2.

Table 2. Statistics of the EmoRecCom dataset with the number of images for each emotion in the ground truth.

	Angry	Disgust	Fear	Happy	Sad	Surprise	Neutral	Others
#	4005	3678	3485	4197	1525	3435	6914	670

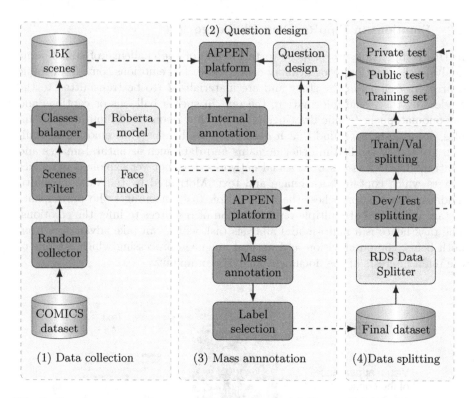

Fig. 2. Our data preparation process has 4 phases: (1) Data collection; (2) Question design (internal annotation); (3) Mass annotation; and (4) Data splitting.

We setup a public website[2] to centralize all related information and participants were competing on the CodaLab platform[3], an open source framework for running competitions, from December 16th 2020 to March 31th 2021. The challenge has attracted 145 registrants, 21 teams have joined the public test phase, and 7 teams have competed in the private test phase.

In the following, we will describe data preparation, proposed challenges, participant's proposed methods and discuss the results.

2   Data Preparation

In the competition, we propose a comic scenes dataset which is composed of comic images from the comic books COMICS public dataset [7]. The COMICS dataset[4] includes over 1.2 million scenes (120 GB) paired with automatic textbox transcriptions (the transcriptions are done by Google Vision OCR, which

[2] https://emoreccom.univ-lr.fr.

[3] https://competitions.codalab.org/competitions/27884.

[4] https://obj.umiacs.umd.edu/comics/index.html.

includes some recognition errors, see Fig. 1). For the overall data preparation process, Fig. 2 shows the workflow of the four phases which are described in next four sections.

2.1 Data Collection

First, we random select 50K images of comic scenes from COMICS dataset. Then we filter out the scenes which do not contain any textbox transcriptions or person faces. To detect faces in scenes, we use the comic face recognition model from the work in [14]. Afterwards, we train a multi-class emotion classification model based on RoBERTA [10], which help to select the potential scenes for the mass annotation phase. The model is trained using EmotionDataset [15] available online[5]. We used this text-based classification model on the transcription to predict the eight-class emotion of 50K comic scenes. We selected up to 2K scenes for each emotion detected by fine-tuning the *RoBERTA_Large* model (in which "disgust" were detected in 5,017 images and other emotions were detected in less than 2K images for each of them). After the previous step, we overall had 8,000 comic scenes. To ensure the balance of scenes containing different emotions, we randomly selected 4,500 scenes where the RoBERTA could not detect any emotion from the remaining scenes of the 50K set. These 12.5K (4.5K+8K) scenes were proposed to the annotators on the crowd-sourcing platform for annotation. To be more precise, at least three annotators were assigned for each scene.

2.2 Question Design and Mass Annotation

We chose to annotate the dataset using crowd-sourcing service platform in order to easily annotate several times each image by different person in order to reduce the subjectivity bias. We experimented different designs and questionnaire to select the most suitable approach for the annotation process. We compared platforms like "Amazon Mechanical Turk" (AMT) - mturk.com, "Appen" - appen.com, and "Toloka" - toloka.ai, which all provide ready-to-use web-based interface for image classification tasks. We selected "Appen", who bought out CrowdFlower few years ago, for its renown quality and to avoid any ethical issues with AMT.

Annotator Selection. Since all comic scene images are originated from American comics, we did not ask to limit the geographical origin of the annotator but instead we required experienced and accurate annotators with good English skills. We remunerate each annotated image 3 and ask annotator to spend at least 10s on each image.

Annotation Tool. We customized the default web tool proposed by *Appen* in order to provide custom guidelines and conditional answers. The basic annotation sequence was as follows:

---

[5] https://github.com/collab-uniba/EmotionDatasetMSR18

– Read ALL TEXTS in the given image.
– Check FACE expressions of all characters.
– Connecting (mentally) TEXTS with the FACE expressions.
– Decide labels: angry, disgust, fear, happy, sad, surprise, neutral, other.

To let new annotators get familiar with the task, we provide six must-read examples of correct/wrong emotion annotations. After reading these instructions, an image from a particular scene is displayed with a questionnaire to fill up and submit. The questionnaire is composed of a main question: "How many actors are in the scene (visible or not)?" with the possibility to answer between 0 and 3 actors (comic character). Then, word sentiment and face emotions are asked for each actor as shown in Fig. 3. The final question is: "Based on the above answers, which emotions are in the scene?". All question is requiring an answer.

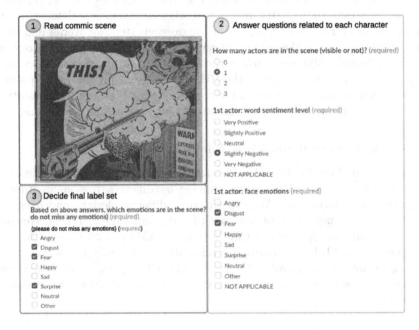

Fig. 3. Interactive annotation tool for the emotion annotation process on *Appen* with three main parts: (1) the scene, (2) questions, and (3) final label set. The number of questions changes dynamically based on the number of actors. For each actor, there will be two corresponding questions: (a) emotion based on textual information, and (b) emotion based on visual information. After answering emotions for all actors, annotators decide final label set of the given scene.

Annotation Quality. We create the final labels by majority voting method, which is straightforward and meaningful. We take common emotions which are chosen by at least 2 annotators. Images that do not have emotion in common (that the three annotators does not have the consent on any labels) are ignored. In total of 2,031 images were ignored, so the EmoRecCom dataset contains 10,199 comic scenes at the end. The final label is binary, an emotion is present in the scene (value 1) or not (value 0).

In addition, we create the probability label (or label certainty) based on the frequency an emotion is chosen by annotators for each scene. This probability are provided to participants but we not used for evaluation.

2.3 Data Splitting

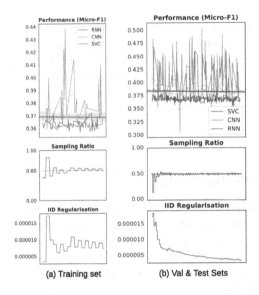

(a) Training set   (b) Val & Test Sets

Fig. 4. Learning Dynamics for splitting the data of EmoRecCom into 3 sets (public training, public testing, and private testing) using RDS^{STO} [13].

Data splitting process is a non-trivial task for linear regression tasks [22]. The same challenge appears in multi-label classification tasks as using random selection method does not guarantee a fair number of samples across multiple labels. Therefore, we apply the reinforced data splitting method (RDS) [13] to result three sets for the data challenge. Moreover, using the fair splitting approach of RDS, we reduce the race for a small gain in performance, but poor in generalization - i.e., try to improve performance on majority labels to gain better performance overall, but not consider minority labels.

Baseline Models. To apply RDS [13], it is required to implement baseline learning models for the given task. Here we choose three different baseline models to act as base-learners on textual data of the task as follows.

1. **RNN.** Recurrent neural network (RNN) is a strong baseline for sequence classification tasks. Here the network is designed with 1 embedding layer, 1 LSTM layer of 64 units, and an output layer of 8 neuron units.
2. **CNN-Text.** The CNN network consists of 1 embedding layer, 4 pairs of Conv2D + MaxPool2D layers, and an output layer of 8 neuron units.
3. **SVC.** This model is built based on *OneVsRestClassifier*[6] of Scikit-Learn [16]. For each of 8 classifiers, data of a label is fitted against all the other labels.

---

[6] http://bit.ly/scikit-learn-multilabel-clf.

Fig. 4 displays the learning patterns of the splitting process. Similar to [9, 22], we applied RDS two times to split the data into three sets including *public train*, *public test*, and *private test* sets.

Table 3. Simple statistics of the EmoRecCom dataset with the number of comic scenes for each competition phase.

	Warm-Up	Public training	Public testing	Private testing
#	100	6,112	2,046	2,041

The final dataset composes of training set, public test set and private test set (see Table 3). There are 6,112 training examples with the respective annotated labels, 2,046 examples (transcriptions + images) of public test set without labels. The private test set contains 2,041 examples without labels. Participants can evaluate the results on the public test set and the private test set by uploading it to the competition website hosted by https://codalab.org.

3   EmoRecCom Challenge

3.1   Multi-label Emotion Classification Task

In this competition, participants designed systems to learn two modalities of data: images and text (automatic transcriptions). The objective is to assign multiple emotions for each data sample. At test phase, their system is presented with a set of comic scenes (each scene is a pair of image and text), and must determine the probability of the 8 emotions to appear in each scene. This is a classic multi-label classification problem.

For participants, we give the access to the private test set only one week to upload the results, before the close of the competition. To be fair between participants, all participants have to register any pre-trained models or external datasets that they have used in their system and they must demonstrate that they did not manually label the private test set. We reserve the right to disqualify entries that use any unregistered models/datatsets or that may involve any manual labeling of the test set.

3.2   Competition Platform

Participants competed on CodaLab.org[7] from 16th December 2020 to 31th March 2021. There are three phases:

1. **Warm Up:** 16th December 2020 to 10th January 2021, participants are given a warm-up dataset of 100 samples to get used with the dataset.

---

[7] https://competitions.codalab.org/competitions/27884.

2. **Public data:** From 10th January 2021 to 24th March 2021, participants are provided 6,112 training examples with the respective annotated labels and a testing set consists of 2,046 examples without labels. Participants can submit their prediction results to the platform and see the results as well as their ranking in the public leader board.

3. **Private Test:** From 24th March 2021 to 31th March 2021, participants are provided 2,041 examples without labels. They must submit the prediction results for this private dataset to the platform before the deadline.

3.3   Evaluation Protocol

Evaluation scripts are made and setup in the platform Codalab.org so participants can upload the predictions and get the evaluation automatically. As mentioned earlier, there are 8 emotion classes including: 0 = Angry, 1 = Disgust, 2 = Fear, 3 = Happy, 4 = Sad, 5 = Surprise, 6 = Neutral, 7 = Others. Participants must submit the result in the same order as the testing set, with the score (probability) indicating the presence of each emotion as the following format:

image_id	Angry	Disgust	Fear	Happy	Sad	Surprise	Neutral	Others
0_27_5	0.5351	0.0860	0.0693	0.1317	0.0443	0.00883	0.2858	0.1947

3.4   Evaluation Metric

The submissions have been evaluated based on the Area Under the Receiver Operating Characteristic Curve (ROC-AUC) score. The ROC curve, is a graphical plot which illustrates the performance of a binary classifier system as its discrimination threshold is varied. The Area Under the ROC Curve (AUC) summarizes the curve information in one number.

The ROC-AUC was calculated between the list of predicted emotions for each image given by the participants and its corresponding target in the ground truth (as described in Sect. 2.2). This is a multi-class classification where the chosen averaging strategy was one-vs-one algorithm that computes the pairwise ROC scores and then the average of the 8 AUCs, for each image [5]. In other words, the score is the average of the individual AUCs of each predicted emotion. To compute this score, we use the Scikit-learn implementation[8].

4   Baselines and Error Analysis

4.1   Baselines

We implemented a text-only baseline model with two variants to act as baselines for the task. The baseline and its variants are implemented with two intuitions

---

[8] http://bit.ly/scikit-learn-auc.

in mind: (1) they should not be the same as those models which were previously used as base-learners in the Data Splitting Process to avoid any biases, (2) they should leverage powerful pre-trained language or language-vision models such as DistilBert [18] or Roberta [10]. Thus, a custom multilabel model was built on top of following language models including Roberta$_{Base}$ and Roberta$_{Large}$. The baseline and its variants were trained using BCEWithLogitsLoss.

4.2 Error Analysis

(a) One modality is not enough. In this example, the text-based baseline model cannot detect the emotion "fear", which is present mostly in the visual modality.

(b) Bad accuracy for "sad, other" : due to the imbalance issue, the accuracy of the two emotions "sad, other" are the worst. Here "sad" was not detected.

(c) Different perceptions about emotions presented in the scenes: this scene can be perceived as either "disgust" (ground truth) or "angry" (prediction).

(d) Complicated scene due to multi-modality: human annotators proposed "angry, disgust, sad, surprise", whereas the baseline model predicted "angry".

Fig. 5. The four most common errors of the baseline.

Our baselines reached the AUC score of 0.5710 for Roberta$_{Base}$ and 0.5812 for Roberta$_{Large}$ for the public test. Therefore, we used the Roberta$_{Large}$ for the private test which reached the AUC score of 0.5867. We showed in Fig. 5 the four most common errors of the baseline model.

5 Submissions and Results

5.1 Participation

Within three and a half months, the challenge has attracted 145 registered participants. During the competition, 21 teams have submitted their results and recorded nearly 600 submission entries. In final, there are 5 top teams that submitted their methods to the competition organizers. Table 4 summarizes the approach and the results obtained from the top 5 teams who submitted documents describing their approach for the final evaluation.

Table 4. Top 5 teams on private test data with submitted papers describing their final approaches.

No	Team name	Team members	Affiliation	AUC
1	S-NLP	Quang Huu Pham, Viet-Hoang Phan, Viet-Hoang Trinh, Viet-Anh Nguyen, and Viet-Hoai Nguyen	RnD Unit, Sun Asterisk Inc	0.6849
2	DeepblueAI	Chunguang Pan, Manqing Dong, Zhipeng Luo	DeepblueAI Technology	0.6800
3	NELSLIP	Xinzhe Jiang*, Chen Yang*, Yunqing Li*, Jun Du (* equal contribution)	NET-SLIP lab, University of Science and Technology of China	0.6692
4	DETA	Quang-Vinh Dang, Guee-Sang Lee	Artificial Intelligence Convergence, Chonnam National University, Gwangju, Korea	0.6324
5	Gululu	Xiaoran Hu, Masayuki Yamamura	Department of Computer Science Tokyo Institute of Technology Tokyo, Japan	0.5716
-	Baseline	EmoRecCom organizers	-	0.5867

1st *position: S-NLP - team from Sun Asterisk Inc, Japan.* The team experiments different approaches to fuse both image and text modalities. Using specialized multi-modal framework such as MMBT (MultiModal BiTransformer) framework [8] does not give the best results, this team uses the conventional multimodal fusion architectures for this task. Three fusion levels are performed: the feature level or early fusion, the decision level or late fusion, and the mid-fusion. This method used EfficientNetB3 [20], Resnet [6] as the backbone for visual feature extraction, and RoBERTa [10] for textual features extraction. At the end, the average of prediction scores from 5 different models (image only, text only, early fusion, mid-fusion, late fusion) is used as the final score. Different experiments of this team are shown in Table 5.

2nd *position: DeepblueAI - team from DeepblueAI Technology, China.* This method uses the average prediction score from two models as the final prediction. The first model leverages the BERT-base [3] for the textual information only. The second model takes both image and textual information as the inputs. This method integrates the image embedding (Resnet50 [6]) with the textual information as the inputs of the BERT-based model, where the image embedding is considered as a special token of the BERT-based model. The average of the token's embedding is further processed with a multi-sample dropout module for getting the prediction. The early fusion architecture is illustrated in Fig. 6.

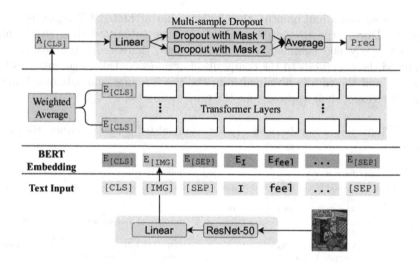

Fig. 6. Early fusion approach by the 2nd team - DeepblueAI.

3rd *position: NELSLIP - team from NEL-SLIP lab, University of Science and Technology of China.* This team uses pretrained models like BERT [3], Distil-BERT [18] and XLNet [25] to extract the feature of text modality, and ResNet50 to extract the feature of visual modality. The concatenation of a visual feature and a text feature were then processed by a 4-layer self-attention module and a Fully Connected (FC) classifier. Finally, the overall performance is boosted by the stacking ensemble of 100 different models (see Table 6).

4th *position: DETA - team from Department of Artificial Intelligence Convergence, Chonnam National University, Korea.* This method fuses the text features and image features from pre-trained models, RoBERTa [10] for text and EfficientNet [20] for images. The concatenation of the text features and the image

Table 5. Different experiments of the winner team - S-NLP.

Method	Backbone	ROC-AUC on test
Text Only	RoBerta	0.6423
Image Only	Efficienet (B3)	0.5412
Early fusion	RoBerta	0.6358
Late fusion	Efficienet (B3) + RoBERTa	0.6288
Mid-fusion	Efficienet (B3) + RoBERTa	0.6654

Table 6. Different experiments by the 3rd team - NETSLIP

Methods	ROC-AUC on val	ROC-AUC on test
Vision-Text model (ResNet-50 & BERT)	0.6320	–
XLNet	0.6302	–
BERT	0.6478	–
DistilBERT	0.6528	–
Averaging ensemble of 100 models	–	0.6674
Stacking ensemble of 100 models	–	0.6692

features goes through a series of 1×1 convolutions to decrease channels of the features. It is then processed by the BI-GRU (Bi-directional Gated Recurrent Unit) module [2] before a FC classifier to produce the final prediction.

5th position: Gululu - team from Department of Computer Science Tokyo Institute of Technology, Tokyo, Japan. This team crop the center of the comic image before extracting the visual feature. The text feature is extracted by pre-trained BERT model. Then the Visual attention network (VAN) [21] is used to learn attention weights for the features. A FC classifier is used at the end to produce the prediction scores.

5.2 Discussions

The emotion recognition in comic scenes is a hard problem. We have experienced the difficulty and ambiguity by doing the internal annotation and by observing the external annotation. This is the reason why we asked at-least three annotators to give their decision for each comic scene.

All the methods leverage both image and text modalities to get the final predictions. However, all teams have consent that the text information is dominant, but the visual information can help improve the performance. The common approach of these methods is to extract text feature; visual feature from pre-trained models such as BERT [3], RoBERTa [10] for text and ResNet [6], EfficientNet [20] for image. And then merge those features. Early fusion and late fusion are experimented by some methods, but only one method (the winner) fuses the two features at mid-level and it performed very well compared to the early and late fusion. Early fusion is better than late fusion. This remark is confirmed further by the fact that both the winner and the second team used the early fusion approach while other three teams only used the late fusion.

While the late fusion and mid-level fusion give good performance, as demonstrated in the methods of the first team, we believe that the early fusion is more compelling and have more room to improve. The model can learn more about the underlying structure of the two modalities if they do not yet undergo significant independent processing. Another common technique is the ensemble (average or stacking) where we combine different models to get the final score. The number of models used among methods varies from 2 to 100 models.

Some flawed approaches have been shared by the teams. First, unsupervised learning on the original COMICS dataset (1.2M comic scenes) based on Masked Language Model (MLM) or comic book classification tasks performs worse than existing pre-trained models. We believe the main reason is that the amount of text in COMICS dataset is still not enough to pre-train a language model. Second, text data augmentation by using transcriptions from nearby comic scenes does not help. Finally, eight independent binary classification models cannot outperform the multilabel classification model.

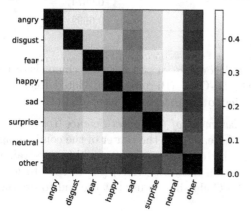

Fig. 7. Cosine similarity correlation of emotions based on multi-hot encoded vectors of 8 emotions in training data.

One important remark is that, we have found none of the teams exploited separately the two emotions relevant information in the image modality: the face of comic characters and the onomatopoeia. All the team considered the visual modality by the whole image. We believe this is one of the main reason that the visual modality give worse performance compared to text modality. The onomatopoeia and the face are the two most important information in the scenes that describe the emotions of characters, if we can explicitly focus on those two objects, it will be easier to learn the relevant information from visual features. Moreover, it is needed to have further investigations on how to explore the correlation of emotions (see Fig. 7) (e.g., using graph-learning [23]) to enhance knowledge for modeling the interplay between visual and textual modalities.

6   Conclusion

We organized the first competition on the multi-modal emotion recognition in images of documents. The problem is challenging and we put a lot of effort into building a high-quality benchmark dataset. The competition has attracted many participants and teams who submitted numerous results of their algorithms to the public leaderboard. Through method descriptions of the top teams on the private leaderboard, we observed many findings which are very important to facilitate future research in this domain. In particular, to combine the image and text modalities, early, mid-level and late fusions have been experimented. Based on the performance of the submissions, we believe that multimodal approaches have great potential to improve the performance of not only emotion recognition but also other tasks in the field of document analysis. The competition dataset and website will be open to the public even after the conference. Also, a legacy version of the competition will remain open without limitation of time to encourage future result comparison[9]. We strongly believe that this competition and the contributed benchmark data will play an important role in the development of future research for the multimodal document analysis community.

References

1. Augereau, O., Iwata, M., Kise, K.: A survey of comics research in computer science. J. Imaging **4**, 87 (2018)
2. Chung, J., Gülçehre, Ç., Cho, K., Bengio, Y.: Empirical evaluation of gated recurrent neural networks on sequence modeling. CoRR abs/1412.3555 (2014)
3. Devlin, J., Chang, M.W., Lee, K., Toutanova, K.: BERT: pre-training of deep bidirectional transformers for language understanding. In: Proceedings of the 2019 NAACL, pp. 4171–4186. Association for Computational Linguistics (2019)
4. Ekman, P.: An argument for basic emotions. Cogn. Emot. **6**(3–4), 169–200 (1992)
5. Hand, D.J., Till, R.J.: A simple generalisation of the area under the ROC curve for multiple class classification problems. Mach. Learn. **45**(2), 171–186 (2001). https://doi.org/10.1023/A:1010920819831
6. He, K., Zhang, X., Ren, S., Sun, J.: Deep residual learning for image recognition. In: 2016 IEEE CVPR 2016, Las Vegas, NV, USA, 27–30 June 2016, pp. 770–778. IEEE Computer Society (2016)
7. Iyyer, M., et al.: The amazing mysteries of the gutter: drawing inferences between panels in comic book narratives. In: 2017 IEEE Conference on Computer Vision and Pattern Recognition (CVPR), pp. 6478–6487 (2017)
8. Kiela, D., Bhooshan, S., Firooz, H., Testuggine, D.: Supervised multimodal bitransformers for classifying images and text. arXiv preprint arXiv:1909.02950 (2019)
9. Le, D.T., et al.: ReINTEL: a multimodal data challenge for responsible information identification on social network sites, pp. 84–91. Association for Computational Lingustics, Hanoi (2020)
10. Liu, Y., et al.: Roberta: a robustly optimized BERT pretraining approach. arXiv preprint arXiv:1907.11692 (2019)
11. Lovheim, H.: A new three-dimensional model for emotions and monoamine neurotransmitters. Med. Hypoth. **78**(2), 341–348 (2012)
12. McCloud, S.: Making Comics: Storytelling Secrets of Comics, Manga and Graphic Novels. Harper, New York (2006)
13. Nguyen, H.D., Vu, X.S., Truong, Q.T., Le, D.T.: Reinforced data sampling for model diversification (2020)
14. Nguyen, N., Rigaud, C., Burie, J.: Digital comics image indexing based on deep learning. J. Imaging **4**(7), 89 (2018)
15. Novielli, N., Calefato, F., Lanubile, F.: A gold standard for emotion annotation in stack overflow. In: Proceedings of the 15th International Conference on Mining Software Repositories, MSR 2018, pp. 14–17. Association for Computing Machinery, New York (2018). https://doi.org/10.1145/3196398.3196453
16. Pedregosa, F., et al.: Scikit-learn: machine learning in Python. J. Mach. Learn. Res. **12**, 2825–2830 (2011)

---

[9] https://competitions.codalab.org/competitions/30954.

17. Plutchik, R., Kellerman, H.: Emotion: Theory, Research and Experience. Academic Press, Cambridge (1986)
18. Sanh, V., Debut, L., Chaumond, J., Wolf, T.: DistilBERT, a distilled version of BERT: smaller, faster, cheaper and lighter. CoRR abs/1910.01108 (2019)
19. Shaver, P., Schwartz, J., Kirson, D., Oćonnor, C.: Emotion knowledge: further exploration of a prototype approach. J. Pers. Soc. Psychol. **52**(6), 1061 (1987)
20. Tan, M., Le, Q.V.: EfficientNet: rethinking model scaling for convolutional neural networks. In: Proceedings of the 36th International Conference on Machine Learning, ICML 2019, 9–15 June 2019, Long Beach, California, USA. Proceedings of Machine Learning Research, vol. 97, pp. 6105–6114 (2019)
21. Truong, Q.T., Lauw, H.W.: VistaNet: visual aspect attention network for multimodal sentiment analysis. In: Proceedings of the AAAI Conference on Artificial Intelligence, vol. 33, no. 01, pp. 305–312 (2019). https://doi.org/10.1609/aaai.v33i01.3301305
22. Vu, X.S., Bui, Q.A., Nguyen, N.V., Nguyen, T.T.H., Vu, T.: MC-OCR challenge: mobile-captured image document recognition for Vietnamese receipts. In: RIVF 2021. IEEE (2021)
23. Vu, X.S., Le, D.T., Edlund, C., Jiang, L., Nguyen D.H.: Privacy-preserving visual content tagging using graph transformer networks. In: ACM International Conference on Multimedia. In: ACM MM 2020. ACM (2020)
24. Yadollahi, A., Shahraki, A.G., Zaiane, O.R.: Current state of text sentiment analysis from opinion to emotion mining. ACM Comput. Surv. **50**(2), Article 25 (2017)
25. Yang, Z., Dai, Z., Yang, Y., Carbonell, J.G., Salakhutdinov, R., Le, Q.V.: XlNet: generalized autoregressive pretraining for language understanding. In: Advances in Neural Information Processing Systems, pp. 5754–5764 (2019)

ICDAR 2021 Competition on Mathematical Formula Detection

Dan Anitei[iD], Joan Andreu Sánchez[(✉)][iD], José Manuel Fuentes[iD],
Roberto Paredes[iD], and José Miguel Benedí[iD]

Pattern Recognition and Human Language Technologies Research Center,
Universitat Politècnica València, 46022 Valencia, Spain
{danitei,jandreu,jofuelo1,rparedes,jmbenedi}@prhlt.upv.es

Abstract. This paper introduces the Competition on Mathematical
Formula Detection that was organized for the ICDAR 2021. The main
goal of this competition was to provide the researchers and practitioners a common framework to research on this topic. A large dataset was
prepared for this contest where the GT was automatically generated and
manually reviewed. Fourteen participants submitted their results for this
competition and these results show that there is still room for improvement especially for the detection of embedded mathematical expressions.

1 Introduction

Currently, huge amounts of documents related to science, technology, engineering, and mathematics (STEM) are being published by online digital libraries
worldwide. Searching in STEM documents is one of the most usual activities for
researchers, scholars, and scientists worldwide. Searching for plain text in large
electronic STEM collections is considered a solved problem when queries are
simple strings or regular expressions. However, searching for complex structures
like chemical formulas, plots, draws, maps, tables, and mathematical expressions, among many others, remains scarcely explored [6,11]. This paper describes
the competition that was performed in the context of the International Conference on Document Analysis and Recognition 2021 for searching mathematical
expressions (MEs) in large collections of STEM documents. This competition
was stated as a visual detection problem and therefore the dataset was just
provided with visual information.

The searching can be based just on visual features or it can also use textual information around the ME. This paper focuses on the former approach,
while the latter approach is left for future research. It is worth mentioning that
searching in large datasets may require performing some pre-processing since
recognition and search in query time can be prohibitive.

Searching for MEs requires as a first step locating them in the document.
MEs can be either embedded along the lines of the text or displayed. We refer to

© Springer Nature Switzerland AG 2021
J. Lladós et al. (Eds.): ICDAR 2021, LNCS 12824, pp. 783–795, 2021.
https://doi.org/10.1007/978-3-030-86337-1_52

the first ones as *embedded* MEs and the second ones as *displayed* MEs. Locating displayed MEs can be easily performed with profile projection methods since these expressions are separated from the text, although they can be confused with other graphic elements (e.g. tables, figures, plots, etc.). However, it is more difficult to locate embedded MEs since they can be easily confused with running text. Given the relevance of this problem for the STEM research community we considered it very challenging and proposed this first competition to know how far the current technology is from providing good results for this problem.

Current technology for ME recognition in documents is based on Machine Intelligent methods that need a large amount of data [1,6] with the necessary ground truth, both for training and testing. However, this ground truth has to be prepared manually. In fact, this is one of the bottlenecks for researching automatic methods. Developing techniques for preparing large datasets with ground truth (GT) is a real need. This paper introduces the IBEM dataset[1] for this competition that is currently made up of 600 research papers, more than 8 000-page images, and more than 160 000 MEs. The dataset has been automatically generated from the LaTeX version of the documents and consequently can be enlarged easily. The ground truth includes the position at the page level for each ME both embedded in the text and displayed.

This paper is organized as follows: first, the literature about the problem of searching MEs is reviewed. Then, we describe the dataset used in this competition and how it is structured. The competition is described in Sect. 4 and then we present the results and the conclusions.

2   Related Work

Searching MEs in documents is not a very researched problem [11] although STEM researchers devote a lot of time to look for information for their daily research. Looking for MEs in the printed text has recently received attention as some competitions have shown [12].[2] One problem that the authors of this paper have identified related to the searching of MEs is that no paper exists about searching in large datasets of STEM documents and this is one of the important targets in this competition.

The prevalent technology for pattern recognition is based on machine learning techniques and these techniques require large amounts of training sets. There are several datasets of typeset MEs that have been used in the past for various research projects. The UW-III dataset [7] is well-known, but the amount of data it contains is limited. It provides GT for symbol classification and the LaTeX version is also available. A second well-known dataset is the InftyCDB-1 dataset [10] that contains 21 056 MEs. This dataset does not include matrices, tables, and figures. The relationship among symbols in a ME was defined manually, and the markup language is not included in the GT. Another important dataset is the

---

[1] http://ibem.prhlt.upv.es/en/.

[2] https://ntcir-math.nii.ac.jp/introduction/.

IM2LATEX-100K [1], that includes 103 556 different LaTeX MEs along with rendered pictures. The MEs were extracted by parsing the LaTeX sources of papers from tasks I and II of the 2003 KDD cup [2]. The problem that we identified in this dataset is that it is useful for researching the recognition of MEs, but not for the detection of MEs in the context of the article they come from. This issue is pointed in [6], where the proposed solution was to build a new dataset that contains 47 articles with 887 pages, but the total number of MEs is not provided. Finally, it is worth mentioning that this last dataset has been used in a competition on MEs detection [5].

The limitations pointed above, lead to the conclusion that a dataset that includes large numbers of images of pages from scientific documents, where MEs are located and annotated, and where the markup language is available, needed to be built.

3 Dataset Description

We chose the KDD Cup dataset [2], which has been used for knowledge extraction and data-mining purposes. More importantly, this collection of documents is publicly available and it allowed us to overcome copyright issues.

The KDD Cup dataset is a large collection of research papers ranging from 1992 to 2003 inclusive, with approximately 29 000 documents in total. The LaTeX sources of all papers are available for download. We chose 600 documents to create the dataset presented in this paper.

The GT prepared for the documents includes the location and dimension of MEs (upper left corner coordinates, width, and height), and their type (embedded or displayed). The bounding boxes of the MEs were highlighted, which allowed to filter out the running text, resulting in images containing only ME definitions. Highlighting the MEs would also provide a way to visually verify and validate the GT. An example of the output described before can be seen in Fig. 1.

To obtain the GT and the output shown in Fig. 1, we divided the extraction process into two parts. The first part consisted of designing LaTeX macros for highlighting and extracting the MEs. The second part consisted of building regular expressions that would detect the delimiters used to define LaTeX mathematical environments and automate the insertion of the LaTeX macros created in the first part. For this reason, we decided to only focus on documents that did not rename these delimiters.

For each type of MEs, embedded or displayed, we used different approaches to compute the location of the bounding boxes. We made use of the `savepos` module provided by the `zref` [3] package to obtain the absolute coordinates of the starting and ending points of the expressions as rendered on the page image. Once these coordinates were calculated, the dimensions of the bounding box were computed.

(a) Original page.

(b) Highlighted bounding boxes.

(c) Color inverted.

Fig. 1. An example of the output obtained from processing a page from the collection. (Color figure online)

In the case of displayed MEs, the first and last symbols of the expression would not always be rendered in the upper left and lower right corner of the bounding box. Figure 2 shows an example of this situation.

$$\sum_{Y_1 \sqcup Y_2 = Y, X_1 \sqcup X_2 = X} (-1)^{(|Y_1 \cap Y_4| + |X_1 \cap X_4|)} [R^*(Y_1 \cap Y_4, X_1 \cap X_4) \times_\hbar$$
$$A \times_\hbar R(Y_2 \cap Y_4, X_2 \cap X_4)] \cdot$$
$$(-1)^{(|Y_1 \cap Y_3| + |X_1 \cap X_3|)} [R^*(Y_1 \cap Y_3, X_1 \cap X_3) \times_\hbar R(Y_2 \cap Y_3, X_2 \cap X_3)] = 0. \ \square$$

Fig. 2. Example of a displayed ME in which the coordinates of the first symbol don't correspond to the upper leftmost pixel of the ME bounding box.

To take into account situations like the one in Fig. 2, macros were inserted to take coordinate measurements before and after each newline symbol and also between the symbols of some mathematical elements such as fractions, sums, products, integrals, etc., which could have superscript or subscript elements making them be rendered in an upper or lower position than the first or last symbol, respectively. As in the case of embedded MEs, we had to take into account that the ME could be split over two pages or columns.

Once the LaTeX macros for highlighting and extracting MEs were created, we proceeded with the second phase, which implied designing regular expressions for automating the insertion of these macros. Since the number of regular expressions would be quite large given the many alternatives of environments that LaTeX provides for defining MEs, in this project we used the **sed** tool [8] as the preferred text stream editor.

Once the MEs were highlighted (green for embedded, yellow for displayed), we compiled the resulting LaTeX file to obtain a pdf format, which we broke down into images of the corresponding pages. Each such image was later processed[3] and passed through a yellow and green color filter to obtain a negative image similar to the one that is shown in Fig. 1.

Bearing in mind that the GT of the dataset presented in this paper was generated automatically from a collection of scientific papers, there was no previous information about the position of MEs in these documents. In light of this fact, data validation had to be done by visually checking that the bounding boxes of the MEs present were in the correct position and of the right dimension.

We manually chose 600 documents that were visually validated and we proceeded with extracting the GT. Table 1 highlights the characteristics of the resulting dataset[4]. It is important to remark that the GT includes some ME that may seem erroneous. For example, sometimes multi-line ME are marked with just one bounding box while on other occasions, there is a separated bounding box for each line. We do not consider these two situations as an error because in the end the GT just represents how the LaTeX users are using this editor for writing MEs. This fact was relevant enough in the competition since distinguishing these situations affected the results obtained in the competition as we illustrate in the Results section.

Table 1. Statistics about the dataset.

Total no. of documents	600
Total no. of pages	8 272
No. of displayed MEs	29 593
No. of embedded MEs	136 635
Average no. of pages per document	13.79
Average no. of displayed MEs per document	49.32
Average no. of embedded MEs per document	227.73

4 Competition Description

4.1 Protocol

The protocol for the competition followed these steps:

1. A web page was prepared for the competition. This web was used to provide information, to register the participants, to deliver the data, and for results submission.

---

[3] https://opencv.org/.

[4] The dataset is free available at http://doi.org/10.5281/zenodo.4757865.

2. The training set was provided to the participants at the beginning of the competition, approximately two months in advance of the deadline of results submission.
3. The evaluation tool was available at the beginning of the competition.
4. One week before the deadline of the competition, the test set was made available to the entrants. This test set did not have the associated ground truth available.

The test set provided to the participants was merged with several thousands of page images for which there was no ground truth. The participants were not able to distinguish the actual test set from the other page images. This was done for two reasons, namely:

- To prevent participants to overfit their system on the actual test set.
- To disseminate the idea that this type of task has to be defined for large datasets.

The participants had to provide the results on the merged data set, but they were ranked according to the actual test set.

The participants had to use only the provided data for training their systems. External data was not allowed for training the systems. This was done to make it easier to discriminate among systems, and therefore the results should not depend on the amount of training data.

The participants did not receive any feedback about their results on the test set. Providing evaluation results while the competition was open may help the participants to fit their systems. Besides, several submissions per participant were allowed, but just the best one was used to rank the participants.

4.2   Evaluation of Systems

The evaluation was performed using Intersection-over-Union (IoU), and systems were ranked based on their F-measure after matching output formula boxes to ground truth formula regions. IoU overlap threshold was 0.7. Any predicted bounding box that surpassed this threshold was considered as a true positive (TP), while those that did not were considered as false positives (FP). For a detection box to be considered as TP, it also had to be of the correct class. The GT bounding boxes that were not detected were considered false negatives (FN). Thus, the precision of the system was calculated by the formula: TP/(TP + FP), while the recall of the system was calculated by the formula: TP/(TP + FN).

Considering that detection models usually output various candidates for a given region, before matching the output formula boxes with the GT boxes, the entrants were recommended to apply a bounding box suppression algorithm. In case that two or more bounding boxes overlapped in the submitted solution, a non-maximum suppression strategy was applied as a sanity check to avoid multiple true positives corresponding to the same GT box. This algorithm only allowed us to keep the bounding boxes with the highest confidence score for each of the classes while removing bounding boxes that had an overlap of more than 0.25 score of IoU with lower confidence. Care had to be given to not providing overlapped bounding boxes if the entrants wanted to avoid a non-maximum suppression strategy.

4.3   Dataset Partition

The IBEM dataset was divided into two sets (training and test sets) and delivered in two phases (training and test phases) to allow for performing different types of experiments. A first training set was provided in the training phase, and a second small training set was provided in the evaluation phase. The test set was released in the test phase.

First, the 600 documents contained in the dataset were shuffled at the document level. The set of documents prepared for the training phase was created by choosing the first 500 documents. The set of 100 documents prepared for the evaluation phase were distributed as follows:

1. the first 50 documents were used for testing (Ts10).

Then, the remaining 50 documents were shuffled at the page level.
2. 50% (329 pages) of these images could be used for training (Tr10);
3. 50% (329 pages) of these images were used for testing (Ts11).

Note that 1 was used for performing a task-independent evaluation and 3 was used for performing a task-dependent evaluation (about 25 documents).

The available data for the training datasets consisted of:

1. The original images of all the training pages.
2. A text file per training page, containing the corresponding ground truth.

The goal of the competition was to obtain the best mathematical expression detection rate on the Ts10 and Ts11 datasets.

5   Participant Systems

More than 47 participants registered in the competition but in the end, only 14 participants submitted some results. The list of participants that submitted some result are listed in chronological order as they registered to the competition:

1. Lenovo Ocean from Lenovo Research, China (Lenovo).
2. HW-L from Huazhong University of Science and Technology, China (HW-L).
3. PKU Founder Mathematical Formula Detection from State Key Laboratory of Digital Publishing Technology, China (PKUF-MFD).
4. PKU Study Group from Peking University, China (PKUSG).

5. Artificial Intelligence Center of Institute of New Technology from Wuhan Tianyu Information Industry Co., Ltd., China (TYAI).
6. TAL Education Group, China(TAL).
7. SCUT-Deep Learning and Vision Computing Lab from Netease Corporation, China (DLVCLab).
8. Visual Computing Group from Ping An Property & Casualty Insurance Company of China Ltd., China (PAPCIC).
9. Autohome Intelligence Group from Autohome Inc., China (AIG).
10. Vast Horizon from Institute of Automation, Chinese Academy of Sciences, China (VH).
11. Shanghai Pudong Development Bank, China (SPDBLab).
12. University of Gunma, Japan (Komachi).
13. Augmented Vision from German Research Center for Artificial Intelligence, Germany (AV-DFKI).
14. University of Information Technology, China (UIT).

All participants used CNN for their solutions by adapting pre-trained models and/or architectures in most cases. The winner system was based on the Generalized Focal Loss (GLF) [4] since the scale variation is huge in this task between the embedded formula and the displayed formula. GFL can well eliminate the imbalance issue of positive/negative sampling on large or small objects. They adopted Ranger as an optimizer. Several GFL models were assembled via Weighted Box Fusion (WBF) [9].

6   Results

Results obtained in the competition are shown in Table 2 sorted according to the F1 score. Note that the best results were obtained by the PAPCIC group that was able to keep good results for embedded formulas. This was a key difference since detecting displayed MEs was easily solved for most of the participants. Note also that the PAPCIC system was able to get better results in the task-dependent scenario for embedded MEs than in the task-independent scenario. It is important to remark that the context can be relevant to detect embedded MEs. Consequently, PAPCIC was nominated as the winner of the competition.

Figure 3 shows the differences obtained by the winner of the competition and the second competitor. The differences are remarked with red circles. We can observe that the significant differences were due to the detection of embedded MEs. Note that this type of expression could be better detected if the context was taken into account.

It is also interesting to remark a key difference between the systems that were ranked in first and second positions. The PAPCIC system was able to better distinguish multi-row expressions than the Lenovo system, as Fig. 4 shows. As

Table 2. Results obtained in the competition. The *E, I, S* characters of the *Type* column stand for *Embedded, Isolated,* and whole *System,* respectively. For each type of evaluation, the table shows the F1 score with the corresponding precision and recall scores enclosed in parentheses.

Group ID	Type	F1 score (Ts10 + Ts11)	F1 score Task dependent (Ts11)	F1 score Task independent (Ts10)
PAPCIC	E	94.79 (94.89, 94.69)	95.11 (95.11, 95.11)	94.64 (94.79, 94.50)
	I	98.76 (98.25, 99.28)	98.70 (98.34, 99.07)	98.79 (98.21, 99.37)
	S	**95.47** (95.47, 95.47)	**95.68** (95.62, 95.73)	**95.37** (95.40, 95.35)
Lenovo	E	94.29 (95.36, 93.25)	93.98 (95.41, 92.60)	94.44 (95.34, 93.56)
	I	98.19 (98.26, 98.12)	97.85 (98.04, 97.67)	98.33 (98.35, 98.31)
	S	**94.96** (95.86, 94.08)	**94.60** (95.84, 93.40)	**95.13** (95.87, 94.39)
DLVCLab	E	93.79 (94.54, 93.05)	93.88 (94.70, 93.07)	93.75 (94.46, 93.04)
	I	98.54 (98.19, 98.89)	98.61 (98.33, 98.88)	98.51 (98.13, 98.90)
	S	**94.60** (95.17, 94.04)	**94.64** (95.29, 93.99)	**94.59** (95.12, 94.07)
TYAI	E	93.39 (94.43, 92.38)	93.94 (95.05, 92.86)	93.13 (94.13, 92.15)
	I	98.55 (98.19, 98.92)	98.42 (98.15, 98.70)	98.61 (98.21, 99.02)
	S	**94.28** (95.08, 93.49)	**94.66** (95.55, 93.79)	**94.1** (94.86, 93.35)
SPDBLab	E	92.80 (93.25, 92.36)	92.14 (92.83, 91.46)	93.12 (93.45, 92.79)
	I	98.06 (98.06, 98.06)	97.76 (97.85, 97.67)	98.19 (98.15, 98.23)
	S	**93.70** (94.08, 93.33)	**93.03** (93.63, 92.44)	**94.01** (94.28, 93.74)
PKUF-MFD	E	91.94 (92.18, 91.70)	92.32 (92.93, 91.72)	91.76 (91.82, 91.69)
	I	96.56 (96.88, 96.24)	96.87 (97.28, 96.46)	96.43 (96.72, 96.15)
	S	**92.72** (92.98, 92.47)	**93.04** (93.62, 92.47)	**92.57** (92.68, 92.47)
HW-L	E	90.53 (91.55, 89.53)	90.57 (91.82, 89.35)	90.51 (91.42, 89.61)
	I	98.94 (98.81, 99.06)	98.61 (98.33, 98.88)	99.08 (99.02, 99.13)
	S	**91.97** (92.81, 91.15)	**91.86** (92.88, 90.86)	**92.02** (92.77, 91.28)
Komachi	E	90.39 (90.92, 89.86)	89.69 (90.43, 88.97)	90.72 (91.16, 90.28)
	I	98.57 (98.27, 98.87)	98.6 (98.69, 98.51)	98.55 (98.09, 99.02)
	S	**91.79** (92.19, 91.39)	**91.11** (91.75, 90.48)	**92.10** (92.39, 91.81)
AIG	E	89.71 (90.18, 89.25)	89.19 (89.96, 88.44)	89.95 (90.28, 89.63)
	I	95.95 (99.00, 93.09)	96.07 (99.01, 93.30)	95.90 (99.00, 93.00)
	S	**90.75** (91.61, 89.90)	**90.26** (91.34, 89.21)	**90.97** (91.74, 90.22)
PKUSG	E	89.10 (90.26, 87.97)	88.59 (90.23, 87.00)	89.34 (90.27, 88.42)
	I	97.96 (97.85, 98.06)	97.94 (98.31, 97.58)	97.96 (97.66, 98.27)
	S	**90.62** (91.58, 89.68)	**90.09** (91.54, 88.68)	**90.87** (91.60, 90.15)
TAL	E	87.87 (88.56, 87.20)	88.51 (89.12, 87.92)	87.57 (88.29, 86.85)
	I	96.85 (96.31, 97.40)	97.09 (96.33, 97.86)	96.75 (96.30, 97.21)
	S	**89.42** (89.91, 88.93)	**89.89** (90.29, 89.49)	**89.19** (89.73, 88.67)
UIT	E	86.04 (85.60, 86.49)	85.64 (85.62, 85.65)	86.23 (85.58, 86.89)
	I	97.05 (95.12, 99.06)	98.11 (97.08, 99.16)	96.60 (94.31, 99.02)
	S	**87.94** (87.26, 88.63)	**87.63** (87.47, 87.80)	**88.08** (87.16, 89.02)
AV-DFKI	E	85.35 (87.37, 83.41)	84.75 (86.96, 82.66)	85.63 (87.57, 83.77)
	I	97.48 (97.12, 97.84)	97.40 (97.30, 97.49)	97.52 (97.04, 97.99)
	S	**87.45** (89.10, 85.87)	**86.80** (88.67, 85.01)	**87.76** (89.30, 86.27)
VH	E	84.25 (83.49, 85.02)	84.39 (83.76, 85.02)	84.18 (83.37, 85.01)
	I	98.59 (98.51, 98.67)	98.51 (98.42, 98.60)	98.62 (98.55, 98.70)
	S	**86.67** (86.01, 87.34)	**86.61** (86.06, 87.18)	**86.70** (85.99, 87.41)

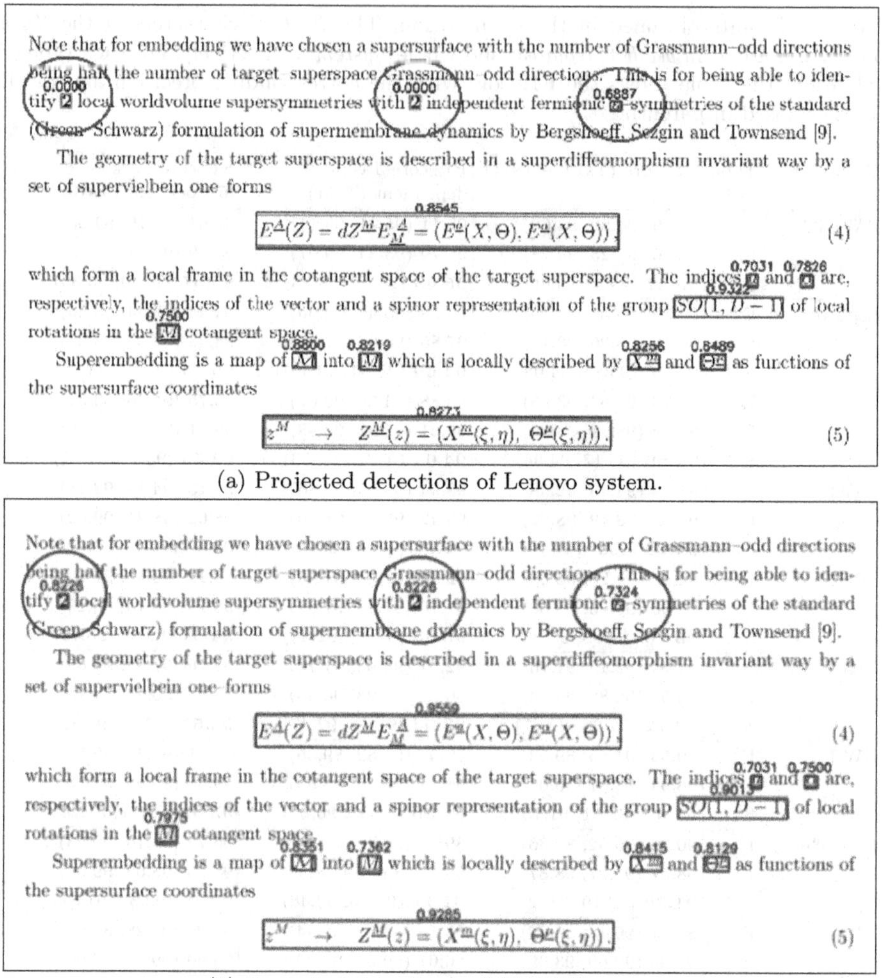

Note that for embedding we have chosen a supersurface with the number of Grassmann–odd directions being half the number of target superspace Grassmann odd directions. This is for being able to identify 2 local worldvolume supersymmetries with 2 independent fermionic symmetries of the standard (Green–Schwarz) formulation of supermembrane dynamics by Bergshoeff, Sezgin and Townsend [9].

The geometry of the target superspace is described in a superdiffeomorphism invariant way by a set of supervielbein one forms

$$E^A(Z) = dZ^M E_M^A = (E^a(X, \Theta), E^{\underline{\alpha}}(X, \Theta)),$$ (4)

which form a local frame in the cotangent space of the target superspace. The indices a and α are, respectively, the indices of the vector and a spinor representation of the group $SO(1, D - 1)$ of local rotations in the M cotangent space.

Superembedding is a map of M into M which is locally described by $X^{\underline{m}}$ and Θ as functions of the supersurface coordinates

$$\xi^M \quad \to \quad Z^{\underline{M}}(z) = (X^{\underline{m}}(\xi, \eta), \; \Theta^{\underline{\mu}}(\xi, \eta)).$$ (5)

(a) Projected detections of Lenovo system.

Note that for embedding we have chosen a supersurface with the number of Grassmann–odd directions being half the number of target-superspace Grassmann–odd directions. This is for being able to identify 2 local worldvolume supersymmetries with 2 independent fermionic symmetries of the standard (Green–Schwarz) formulation of supermembrane dynamics by Bergshoeff, Sezgin and Townsend [9].

The geometry of the target superspace is described in a superdiffeomorphism invariant way by a set of supervielbein one-forms

$$E^A(Z) = dZ^M E_M^A = (E^a(X, \Theta), E^{\underline{\alpha}}(X, \Theta)),$$ (4)

which form a local frame in the cotangent space of the target superspace. The indices a and α are, respectively, the indices of the vector and a spinor representation of the group $SO(1, D - 1)$ of local rotations in the M cotangent space.

Superembedding is a map of M into M which is locally described by $X^{\underline{m}}$ and Θ as functions of the supersurface coordinates

$$\xi^M \quad \to \quad Z^{\underline{M}}(z) = (X^{\underline{m}}(\xi, \eta), \; \Theta^{\underline{\mu}}(\xi, \eta)).$$ (5)

(b) Projected detections of PAPCIC system.

Fig. 3. Part of a page from the Ts11 set, shown as a comparison between the two best systems that have been submitted, in which Lenovo system did not detect the first two embedded MEs and partially detected (IoU < 0.7) the third embedded ME encircled in red. For this example the PAPCIC system outperformed the other systems, having reached a perfect F1 score. (Color figure online)

we mentioned in Sect. 3, we did not include any change in the GT when dealing with multi-row ME, that is, if the authors decided to write the ME in several consecutive math environments rather than in just one math environment in the LaTeX source, we left them in that way. In the example that is shown in Fig. 4, the PAPCIC system got 100% F1 score while the Lenovo system got 83.72%.

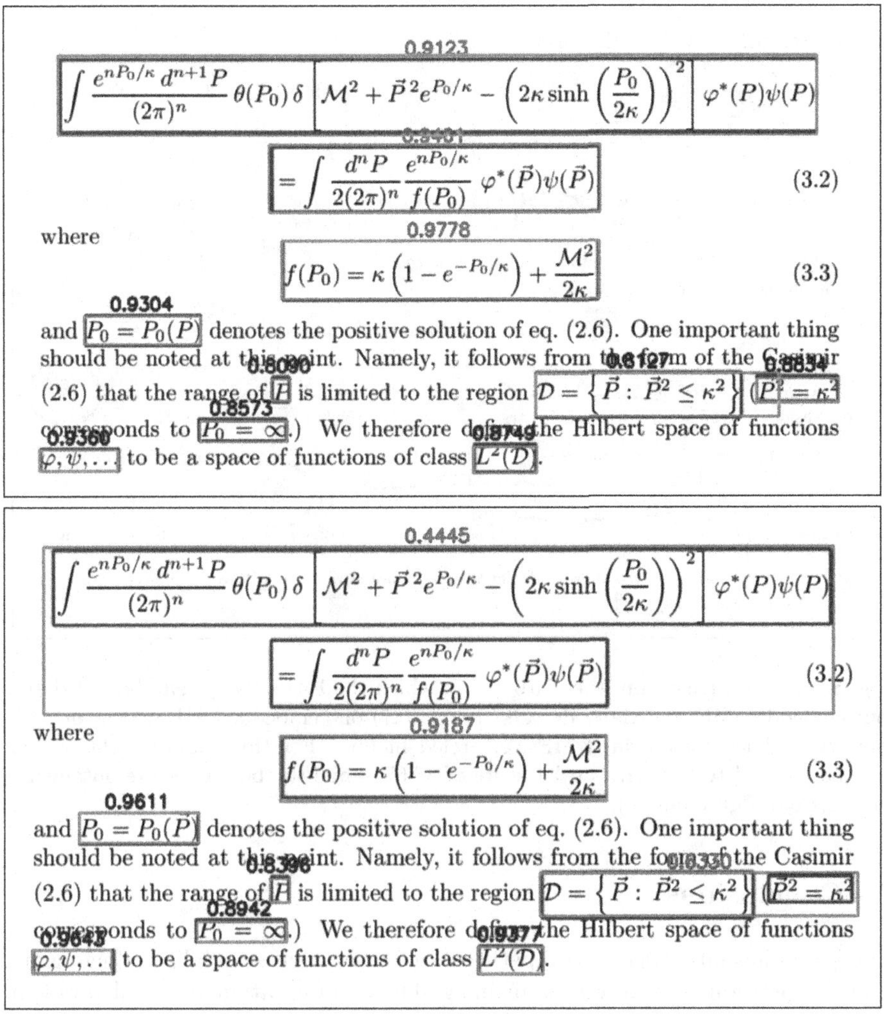

0.9123

$$\int \frac{e^{nP_0/\kappa}\, d^{n+1}P}{(2\pi)^n}\, \theta(P_0)\, \delta\left(\mathcal{M}^2 + \vec{P}^2 e^{P_0/\kappa} - \left(2\kappa \sinh\left(\frac{P_0}{2\kappa}\right)\right)^2\right)\varphi^*(P)\psi(P)$$

0.8401

$$= \int \frac{d^n P}{2(2\pi)^n}\, \frac{e^{nP_0/\kappa}}{f(P_0)}\, \varphi^*(\vec{P})\psi(\vec{P}) \tag{3.2}$$

where

0.9778

$$f(P_0) = \kappa\left(1 - e^{-P_0/\kappa}\right) + \frac{\mathcal{M}^2}{2\kappa} \tag{3.3}$$

0.9304

and $P_0 = P_0(P)$ denotes the positive solution of eq. (2.6). One important thing should be noted at this point. Namely, it follows from the form of the Casimir (2.6) that the range of $\vec{P}$ is limited to the region $\mathcal{D} = \left\{\vec{P} : \vec{P}^2 \le \kappa^2\right\}$ ($\vec{P}^2 = \kappa^2$ corresponds to $P_0 = \infty$.) We therefore define the Hilbert space of functions $\varphi, \psi, \ldots$ to be a space of functions of class $L^2(\mathcal{D})$.

0.4445

$$\int \frac{e^{nP_0/\kappa}\, d^{n+1}P}{(2\pi)^n}\, \theta(P_0)\, \delta\left(\mathcal{M}^2 + \vec{P}^2 e^{P_0/\kappa} - \left(2\kappa \sinh\left(\frac{P_0}{2\kappa}\right)\right)^2\right)\varphi^*(P)\psi(P)$$

$$= \int \frac{d^n P}{2(2\pi)^n}\, \frac{e^{nP_0/\kappa}}{f(P_0)}\, \varphi^*(\vec{P})\psi(\vec{P}) \tag{3.2}$$

where

0.9187

$$f(P_0) = \kappa\left(1 - e^{-P_0/\kappa}\right) + \frac{\mathcal{M}^2}{2\kappa} \tag{3.3}$$

0.9611

and $P_0 = P_0(P)$ denotes the positive solution of eq. (2.6). One important thing should be noted at this point. Namely, it follows from the form of the Casimir (2.6) that the range of $\vec{P}$ is limited to the region $\mathcal{D} = \left\{\vec{P} : \vec{P}^2 \le \kappa^2\right\}$ ($\vec{P}^2 = \kappa^2$ corresponds to $P_0 = \infty$.) We therefore define the Hilbert space of functions $\varphi, \psi, \ldots$ to be a space of functions of class $L^2(\mathcal{D})$.

Fig. 4. Differences between multi-row ME located by the PAPCIC system (top) and the Lenovo system (bottom).

Figure 5 shows an example in the PAPCIC system that did not get good results. We can observe that most of the errors were produced in the embedded MEs for which the layout was complicated and consequently the context was not helpful.

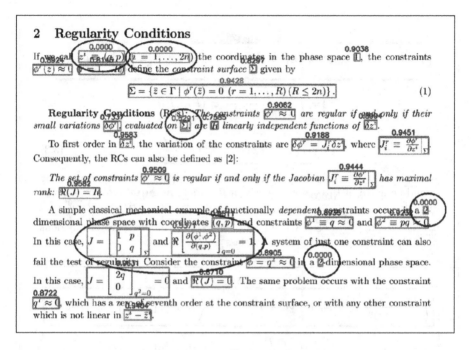

Fig. 5. Part of a page from the Ts10 set, in which the PAPCIC system did not detect four embedded MEs, partially detects (IoU < 0.7) one embedded ME and incorrectly joins two adjacent embedded MEs, encircled in blue. For this example, the system ranks as second to last with an F1 score of 87.67, while the best F1 score obtained is 94.74. (Color figure online)

7   Conclusions

This paper has introduced a competition for MEs detection in STEM documents. The competition was raised as an image-based information retrieval problem. Several participants sent their results that were tested on a dataset that was not seen in the training phase. We observed that excellent results can be achieved in terms of F1 score, but there is still room for improvement especially in embedded formulas.

For future work we plan to extend this competition in several dimensions: i) enlarging the dataset with more STEM documents; ii) making possible the searching by combining visual information and text information around the MEs; and iii) making possible the searching by using semantic information. This last feature means that when looking for MEs, dummy variables should be ignored.

Acknowledgements. This work has been partially supported by the Ministerio de Ciencia y Tecnología under the grant TIN2017-91452-EXP (IBEM) and by the Generalitat Valenciana under the grant PROMETEO/2019/121 (DeepPattern).

References

1. Deng, Y., Kanervisto, A., Rush, A.M.: What you get is what you see: a visual markup decompiler. arXiv abs/1609.04938 (2016)
2. Gehrke, J., Ginsparg, P., Kleinberg, J.: Overview of the 2003 KDD cup. SIGKDD Explor. Newsl. (2), 149–151 (2003)
3. Oberdiek, H.: The zref package. https://osl.ugr.es/CTAN/macros/latex/contrib/zref/zref.pdf
4. Li, X., et al.: Generalized focal loss: Learning qualified and distributed bounding boxes for dense object detection (2020)
5. Mahdavi, M., Zanibbi, R., MouchÃšre, H., Viard-Gaudin, C., Garain, U.: ICDAR 2019 CROHME + TFD: competition on recognition of handwritten mathematical expressions and typeset formula detection. In: International Conference on Document Analysis and Recognition (2019)
6. Ohyama, W., Suzuki, M., Uchida, S.: Detecting mathematical expressions in scientific document images using a U-Net trained on a diverse dataset. IEEE Access **7**, 144030–144042 (2019)
7. Phillips, I.: Methodologies for using UW databases for OCR and image understanding systems. In: Proceedings of the SPIE, Document Recognition V, vol. 3305, pp. 112–127 (1998)
8. Pizzini, K., Bonzini, P., Meyering, J., Gordon, A.: GNUsed, a stream editor. https://www.gnu.org/software/sed/manual/sed.pdf
9. Solovyev, R., Wang, W., Gabruseva, T.: Weighted boxes fusion: ensembling boxes from different object detection models. Image Vis. Comput. **107**, 104117 (2021)
10. Suzuki, M., Uchida, S., Nomura, A.: A ground-truthed mathematical character and symbol image database. In: Proceedings of the 8th International Conference on Document Analysis and Recognition (ICDAR 2005), pp. 675–679 (2005)
11. Zanibbi, R., Blostein, D.: Recognition and retrieval of mathematical expressions. Int. J. Doc. Anal. Recogn. **14**, 331–357 (2011)
12. Zanibbi, R., Oard, D.W., Agarwal, A., Mansouri, B.: Overview of ARQMath 2020: CLEF lab on answer retrieval for questions on math. In: Arampatzis, A., et al. (eds.) CLEF 2020. LNCS, vol. 12260, pp. 169–193. Springer, Cham (2020). https://doi.org/10.1007/978-3-030-58219-7_15

Author Index

Printed in the United States
by Baker & Taylor Publisher Services